First Conference of Computational Methods in Offshore Technology (COTech2017)

IOP Conference Series: Materials Science and Engineering
Volume 276

Stavanger, Norway
30 November – 1 December 2017

ISBN: 978-1-5108-5497-0
ISSN: 1757-8981

Printed from e-media with permission by:

Curran Associates, Inc.
57 Morehouse Lane
Red Hook, NY 12571

CURRAN ASSOCIATES INC. proceedings.com

Some format issues inherent in the e-media version may also appear in this print version.

Printed by Curran Associates, Inc. (2018)

For permission requests, please contact the Institute of Physics
at the address below.

Institute of Physics
Dirac House, Temple Back
Bristol BS1 6BE UK

Phone: 44 1 17 929 7481
Fax: 44 1 17 920 0979

techtracking@iop.org

Additional copies of this publication are available from:

Curran Associates, Inc.
57 Morehouse Lane
Red Hook, NY 12571 USA
Phone: 845-758-0400
Fax: 845-758-2633
Email: curran@proceedings.com
Web: www.proceedings.com

Table of contents

Volume 276

First Conference of Computational Methods in Offshore Technology (COTech2017)

30 November to 1 December 2017, Stavanger, Norway

Accepted papers received: 21 November 2017
Published online: 12 December 2017

Preface

First Conference of Computational Methods in Offshore Technology (COTech2017)

Peer review statement

Papers

Wind Engineering

What scaling means in wind engineering: Complementary role of the reduced scale approach in a BLWT and the full scale testing in a large climatic wind tunnel

Olivier Flamand.....1

Implementation and application of the actuator line model by OpenFOAM for a vertical axis wind turbine

L Riva, K-E Giljarhus, B Hjertager and S M Kalvig.....11

A high-fidelity weather time series generator using the Markov Chain process on a piecewise level

K Hersvik and O-E V Endrerud.....24

3D WindScanner lidar measurements of wind and turbulence around wind turbines, buildings and bridges

T Mikkelsen, M Sjöholm, N Angelou and J Mann.....33

Wind effects on long-span bridges: Probabilistic wind data format for buffeting and VIV load assessments

K Hoffmann, R G Srouji and S O Hansen.....47

Aeroelastic Response from Indicial Functions with a Finite Element Model of a Suspension Bridge

O Mikkelsen and J B Jakobsen.....62

Full-scale monitoring of wind and suspension bridge response

J T Snæbjörnsson, J B Jakobsen, E Cheynet and J Wang.....79

Super-long bridges with floating towers: the role of multi-box decks and Hardware-In-the-Loop technology for wind tunnel tests

A Zasso, T Argentini, I Bayati, M Belloli and D Rocchi.....92

Advanced computational Methods & Applications in Marine Technology

CFD Analysis in Subsea and Marine Technology

Hrvoje Jasak.....103

Multi-fluid CFD analysis in Process Engineering

B H Hjertager.....114

Accurate green water loads calculation using naval hydro pack

H Jasak, I Gatin and V Vukčević.....136

Study of lifting operation of a tripod foundation for offshore wind turbine

H Zhu, L Li and M C Ong.....144

Vortex-induced vibration (VIV) effects of a drilling riser due to vessel motion

R S Joseph, J Wang, M C Ong and J B Jakobsen.....163

Large eddy simulation of the tidal power plant deep green using the actuator line method

S T Fredriksson, G Broström, M Jansson, H Nilsson and B Bergqvist.....173

Fluid flow in steady and oscillatory lid-driven square cavities

J Zhu, L E Holmedal, D Myrhaug and H Wang.....185

Marine Operation for Cold Climate

Limitations related to marine operations in the Barents Sea

O T Gudmestad.....193

Implications caused by SARex on the implementation of the IMO polar code on survival at sea

K E Solberg.....204

On weather limitations for safe marine operations in the Barents Sea

A P Orimolade and O T Gudmestad.....211

A practical guideline for human error assessment: A causal model

Y Z Ayele and A Barabadi.....225

Limit cycle oscillations at resonances

K Hellevik and O T Gudmestad.....233

Finite Element, Boundary Element & Finite Volume Methods and Design Optimization

Boundary element method solution for large scale cathodic protection problems

D C Rodopoulos, T V Gortsas, S V Tsinopoulos and D Polyzos.....242

Finite element simulation of long wave impact on floating breakwaters with variable stiffness

T K Papathanasiou, A Karperaki and K A Belibassakis.....251

An efficient structural finite element for inextensible flexible risers

T K Papathanasiou, S Markolefas, P Khazaeinejad and H Bahai.....263

Influence of extreme low temperature conditions on the dynamic mechanical properties of carbon fiber reinforced polymers

S P Zaoutsos and M C Zilidou.....273

Numerical assessment of wind turbine blade damage due to contact/impact with tower during installation

Amrit Shankar Verma, Nils Petter Vedvik and Zhen Gao.....285

A case study on topology optimized design for additive manufacturing

A W Gebisa and H G Lemu.....302

A review of fatigue crack propagation modelling techniques using FEM and XFEM

K Rege and H G Lemu.....313

Experimental and numerical study of Bondura® 6.6 PIN joints

I Berkani, Ø Karlsen and H G Lemu.....328

Recursive thoughts on the simulation of the flexible multibody dynamics of slender offshore structures

J Schilder, M Ellenbroek and A de Boer.....344

A nonlinear dynamic corotational finite element model for submerged pipes

F H de Vries, H J M Geijselaers, A H van den Boogaard and A Huisman.....354

Application of foam-extend on turbulent fluid-structure interaction

K Rege and B H Hjertager.....369

Evaluating stress analysis and failure criteria for offshore structures for Pechora Sea conditions

S Nesic, Y Donskoy and A Zolotukhin.....387

Structural Integrity and Health Monitoring

Integrity management of offshore structures and its implication on computation of structural action effects and resistance

T Moan.....398

Analytical and numerical investigation of bolted steel ring flange connection for offshore wind monopile foundations

C A Madsen, J-C Kragh-Poulsen, K J Thage and M J Andreassen.....425

Natural frequency and vibration analysis of jacket type foundation for offshore wind power

Y-C Hung, Y-Y Chang and S-Y Chen.....441

Fracture mechanics analysis for a mooring system subjected to tension and out-of-plane bending

X Xue and N-Z Chen.....449

Effect of corrosion on the buckling capacity of tubular members

F H Øyasæter, A Aeran, S C Siriwardane and O Mikkelsen.....457

An accurate fatigue damage model for welded joints subjected to variable amplitude loading

A Aeran, S C Siriwardane, O Mikkelsen and I Langen.....475

Fatigue strength degradation of metals in corrosive environments

N D Adasooriya, T Hemmingsen and D Pavlou.....490

IOP Conf. Series: Materials Science and Engineering **276** (2017) 011001 doi:10.1088/1757-899X/276/1/011001

First Conference of Computational Methods in Offshore Technology (COTech2017)

November 30 – December 1, 2017, University of Stavanger, Norway

Preface

This First Conference of Computational Methods in Offshore Technology (COTech2017) is a result of a Program Area for Research (COTech) that was established in 2016 at Department of Mechanical and Structural Engineering and Materials Science, University of Stavanger, Norway. The motivation to focus on Computational Methods is because of the fact that numerical computation approach is nowadays not only serving as a means to cultivate and realize innovative ideas, but also it is becoming the primary choice to solve complex engineering problems such as offshore. Based on the defined goals of the COTech Program Area for Research, the aim of this COTech2017 conference is to provide a platform for academics and professionals in application of computational methods with focus on, but not limited to, Offshore Technology to come together, present their recent works in the area, exchange ideas, and establish professional networks. It is intended and will serve as a forum for a multidisciplinary research and brings together Norwegian and invited foreign researchers to enable them exchange their research experience and disseminate their results within application of computational methods in design, analysis and simulation of offshore systems and structures.

The scope of this COTech2017 conference is defined under the following five research topics, which will also serve as conference sessions.

1. *Wind Engineering*
2. *Advanced Computational Methods and Applications in Marine Technology*
3. *Marine Operation for Cold Climate*
4. *Finite Element, Boundary Element and Finite Volume Methods and Design Optimization*
5. *Structural Integrity and Health Monitoring*

Among the submitted manuscripts, nearly 40 full papers passed the review process. The review work was conducted according to the review policy of IOP conference series where each paper was reviewed by at least two reviewers involving both national and international reviewers. Eight known experts in the area were invited to present keynote papers from different countries.

The conference committee, and editors of this proceeding would like to thank all the reviewers and authors of the papers in this proceeding for their valuable contributions to this COTech2017 conference.

IOP Conf. Series: Materials Science and Engineering **276** (2017) 011001 doi:10.1088/1757-899X/276/1/011001

Conference Organizing Committee

Prof. Hirpa G. Lemu,
Prof. Muk Chen Ong,
Prof. Emeritus Ivar Langen
Prof. Bjørn H. Hjertager
Assoc. Prof. Ove Mikkelsen

Topic Area coordinators

Prof. Jasna B. Jakobsen
Prof. Muk Chen Ong
Prof. Emeritus Ove T. Gudmestad
Prof. Dimitrios Pavlou
Assoc. Prof. Sudath C. Siriwardane

Invited Keynote Speakers

Prof. Emeritus Ove T. Gudmestad, University of Stavanger, Norway
Prof. Torgeir Moan, Norwegian University of Science and Technology, NTNU, Norway
Prof. Trond Kvamsdal, Norwegian University of Science and Technology, NTNU, Norway
Prof. Johan Õlvander, University of Linköping, Sweden
Prof. dr. sc. Hrvoje Jasak, University of Zagreb, Croatia
Prof. Demosthenes Polyzos, University of Patras, Greece
Olivier Flamand, Research Head, Centre scientifique et technique du bâtiment (CSTB), France.
Prof. Bjørn Hjertager, University of Stavanger, Norway

Technical Committee Members and Reviewers

1. Prof. Henning Agerskov, Technical University of Denmark (DTU), Denmark
2. Ass. Professor Paul Akangah, North Carolina A&T State University, USA
3. Assoc. Prof. Vladimir P. Balitskiy, Gubkin Oil & Gas University, Russia
4. Prof. Abbas Barabady, The Arctic University of Norway (UiT), Norway
5. Assoc. Prof. Kostas Belibassakis, National Technical Univ. of Athens, Greece
6. Dr. Etienne Cheynet, University of Stavanger, Norway
7. Prof. André De Boer, University of Twente, The Netherlands
8. Prof. Robert Brown, Memorial Univ. of Newfoundland, Canada
9. Dr. Nianzhong Chen, University of Newcastle, UK
10. Prof. Sreekanta Das, University of Windsor, Canada
11. Assoc. Prof. David Fuhrman, Technical University of Denmark (DTU), Denmark
12. Dr. Niels G. Jacobsen, Deltares, The Netherlands
13. Prof. Koh Chan Ghee, National Univ. of Singapore (NUS), Singapore
14. Prof. Rui Miranda Guedes, University of Porto, Portugal
15. Prof. Lars Erik Holmedal, Norwegian Univ. of Sci. and Technology (NTNU), Norway
16. Adj. Assoc. Prof. David Häger, University of Stavanger, Norway
17. Prof. Hrvoje Jasak, University of Zagreb, Croatia
18. Assoc. Prof. Tore M. Jonassen, Oslo University College, Norway

19. Prof. Uday Kumar, University of Luleå, Sweden
20. Dr. Inge Lotsberg, DNV GL Oslo, Norway
21. Dr. Pavel Liferov, Statoil AS, Norway
22. Assoc. Prof. Stylianos Markolefas, Technical Univ. of Central Greece, Greece
23. Dr Per Olav Moslett, DNV, Norway
24. Prof. Håkan Nilsson, Chalmers University of Technology, Sweden
25. Gunnar Noer, Senior Forecaster, The Norwegian metreology institute, DNMI, Norway
26. Assoc. Prof. Pantelis Nikolakopoulos, University of Patras, Greece
27. Dr. Piotr Omenzetter, University of Aberdeen, UK
28. Dr. Theodosios Papathanasiou, Brunel University, London, UK
29. Prof. Demosthenes Polyzos, University of Patras, Greece
30. Prof. Raviliy Safieva from Gubkin Oil & Gas University, Russia
31. Professor Jonas T. Snærbjørnsson, Rykjavik University, Iceland
32. Dr. Srinivas Sriramula, University of Aberdeen, UK
33. Ass. Prof. Hodjat Shiri, Memorial Univ. of Newfoundland, Canada
34. Dr. Jungao Wang, University of Stavanger, Norway
35. Prof. Qian Xudong, National University of Singapore (NUS), Singapore
36. Prof. Stefanos Zaoutsos, University of Applied Sciences of Thessaly, Greece
37. Prof. Anatoly B. Zolotukhin, Gubkin University, Russia
38. Prof. Sigmund Kyrre Ås, Norwegian Univ. of Sci. and Technology (NTNU), Norway

IOP Conf. Series: Materials Science and Engineering **276** (2017) 011002 doi:10.1088/1757-899X/276/1/011002

Peer review statement

All papers published in this volume of *IOP Conference Series: Materials Science and Engineering* have been peer reviewed through processes administered by the proceedings Editors. Reviews were conducted by expert referees to the professional and scientific standards expected of a proceedings journal published by IOP Publishing.

IOP Conf. Series: Materials Science and Engineering **276** (2017) 012001 doi:10.1088/1757-899X/276/1/012001

What scaling means in wind engineering: Complementary role of the reduced scale approach in a BLWT and the full scale testing in a large climatic wind tunnel

Olivier Flamand

CSTB, Centre Scientifique et Technique du Batiment, 11 rue Henri Picherit, 44300 Nantes, France.

E-mail: olivier.flamand@cstb.fr

Abstract. Wind engineering problems are commonly studied by wind tunnel experiments at a reduced scale. This introduces several limitations and calls for a careful planning of the tests and the interpretation of the experimental results. The talk first revisits the similitude laws and discusses how they are actually applied in wind engineering. It will also remind readers why different scaling laws govern in different wind engineering problems. Secondly, the paper focuses on the ways to simplify a detailed structure (bridge, building, platform) when fabricating the downscaled models for the tests. This will be illustrated by several examples from recent engineering projects. Finally, under the most severe weather conditions, manmade structures and equipment should remain operational. What "recreating the climate" means and aims to achieve will be illustrated through common practice in climatic wind tunnel modelling.

1. Introduction

Structures studied in wind engineering are usually of large dimensions. Bridges, towers, slender industrial buildings, cutting edge artistic works etc. all require a careful study of the design wind loads because their unusual dimensions and cost open the opportunity for structural optimization.

Boundary layer wind tunnels (BLWTs) are more often than not the right tools for this problem. They have been used since the middle of 20th century to reproduce the natural turbulent wind at small scale, with size ratios ranging from 1/200 to 1/3000. Turbulence generators, often of rectangular or triangular shape, are placed upstream of the test section in order to generate the reduced-scale boundary layer, characterized by mean wind speed increasing with height and fluctuation of wind velocity in three dimensions, called turbulence. Turbulence scales and spectra are tuned to best fit the target values, using one time scale and one length scale. Because the behavior of air is very similar at such a small scale as in full- scale, modeling the spatial correlation of wind turbulence is often sufficient to reproduce the wind conditions experienced by large structures during a storm.

Looking carefully at this point it was made clear that only part of the natural turbulence can be reproduced in the BLWTs, the one corresponding to the input from kinetic energy, while the thermal part of the turbulence is neglected. BLWT is efficient in reproducing a neutral atmosphere, corresponding to most storm wind characteristics but is not convenient for reproducing everyday types of wind, such as relevant when studying wind power harvesting and pollutant dispersion [3]. Nevertheless, if the experiment is used to reproduce the local dispersion of plumes or the wake of one wind turbine impacting a neighboring one, all considered at short distance, a boundary layer simulation can be very effective. There are of course some other limits in the use of BLWT for reproducing wind effects on buildings and offshore structures, due to downscaling effects, which lead to the use of large wind tunnels in a complementarily full scale approach.

Figure 1. Typical BLWT study of high rise buildings in a 4m x 2.5m section at CSTB (left) and the case of great arch over the Tchernobyl reactor, with Reynolds number effect on a round shape building and on truss (right)

2. Scaling in fluids mechanics

Pioneers in fluid engineering have long time ago faced this difficulty: how to reproduce natural phenomena at smaller scale? In 1780, Euler was probably the first to work seriously on scaling in engineering mechanics, followed by Fourier in the 1800s. In 1878, French mathematician Joseph Bertrand was the first to demonstrate the theorem that was reused by Aimé Vaschy [1] and latter popularized by Edgar Buckingham [2]. It is now known as the "Vaschy-Buckingham" or "Buckingham-Pi" theorem, giving a number of non-dimensional parameters necessary to describe a physical phenomenon. Because they are non-dimensional, these numbers opened the gate to the Dimensional Analysis which gives the rules (Rayleigh's method) to be respected when linking a full-scale prototype and a reduced scale model analysis.

In the 18th century, seven basic physical quantities were chosen as elementary dimensions, the combination of which can represent any physical quantity: Length L, Mass M, Time or duration T, electric intensity I, thermodynamic temperature θ, number of molecules N and light intensity J.

Reynolds, Froude, Prandtl, Scruton, Strouhal, Grashof and Mach have all been immortalized by giving their name to a dimensionless quantity. Does this mean everything has already been said concerning scale effects in wind tunnel simulation?

It is well known that the Froude number must be respected in the reduced scale modeling when gravity is the main force governing the stability of the structure, as for suspension bridges or oscillating pendulum. Main numbers are (Equation (1)):

$$Fr=\frac{u}{\sqrt{g\cdot L}}; \qquad St=\frac{f\cdot L}{u}; \qquad \mathrm{Re}=\frac{u\cdot L}{\nu} \tag{1}$$

When the vortex shedding or the wind turbulence applies an alternated force on an elastic structure, with a given stiffness, Strouhal number is the key parameter to respect. Grashof number is the most important to take into account when buoyancy of hot fumes must be represented experimentally. Let us try to consider the practical rules commonly applied in scaling for wind tunnel studies.

3. Downscaling limits in boundary layer wind tunnels

The first limit for downscaling encountered in boundary layer wind tunnels is related to the Reynolds number. This non dimensional term expresses the balance of viscous forces to inertia forces acting

IOP Conf. Series: Materials Science and Engineering **276** (2017) 012001 doi:10.1088/1757-899X/276/1/012001

on flow particles. It is, in practice, the main parameter governing the reproduction of actual loads on a model. For a reduced geometric scale of 1/200 in wind tunnel, the wind speed should be 200 times the real one to maintain the same value of the Reynolds number, which cannot be achieved. Consequently, wind engineers developed the know-how to compensate for the effects of a Reynolds number 100 to 1000 times lower in the BLWT simulation than the full scale one.

A very common example of the Reynolds Number scaling effects is found in the study of lattice structures. For elements of the lattice having sharp edged profiles it is commonly accepted that the drag force does not change at low Reynolds number. This assumption is based on drag coefficients measured on isolated squares, I, H and U-shaped beams, but is not fully verified when the lattice is so dense that it begins to behave as a mesh structure. Nevertheless, it is characteristic for rounded elements of lattice structure that the drag force is very dependent on the flow regime, i.e. the Reynolds number. In such a case there are two ways to change the lattice shape with the aim of better representing the wind loads on the model than via simply homothetic scaling.

The first possibility lies in decreasing the diameter of the round bars of the lattice by the ratio of drag coefficient at full scale (say 0.6 for a given wind speed and a given roughness) to the drag coefficient at reduced scale in wind tunnel, (say 1.2 because of the subcritical Reynolds regime). In practice, reducing the diameter of round bars by a factor of two is very common, and relies on the assumption of high wind speeds producing a drop in the drag force coefficient when exceeding the critical Reynolds number, usually called drag crisis, at full scale [4]. The second possibility consists of replacing round bars by square bars, with a thickness giving the same drag force. Actually these improvements in the representation of wind loads on reduced scale models cannot replace the accuracy and advantages of the measurements at full scale, in a large wind tunnel reproducing the correct Reynolds number.

Figure 2. On this powder sintered model of a round shape building at a scale 1/250 in BLWT, details of the surface are well reproduced, but it was necessary to add extra roughness trips to achieve the correct "apparent Reynolds number"

The second limitation when studying buildings in small size wind tunnels is the blockage effect, inherent with the presence of the model in the test section, especially in closed loop circuits with solid walls. When the wind collides with a real building, streamlines are deflected around the bluff body with no constraint in the vertical direction which is infinite, and sometimes very few in the lateral direction. In a wind tunnel, ceiling and lateral walls act out as a limit to the expansion of deflected streamlines. Even if correction methods can be applied, as commonly practiced in aeronautics on 2D streamlined bodies for measuring mean loads, these are not effective in boundary layer wind tunnels where wind speed and direction change all the time. Not only the mean flow is deflected here by the model, but additionally the large eddies composing the lowest frequency part of the atmospheric turbulence cannot "extend" vertically and laterally. It is challenging to look for a universal limit of the blockage ratio, as the effect of walls depends on the size of eddies and the shape of the model, but a 10% blockage ratio is often considered the largest tolerable value.

A third boundary in downscaling lies in surface state and precision of current machining techniques. At a geometrical scale of 1/500, a 0.1 mm error on the model of a building corresponds to 0.5 m at full scale, which can be larger than acceptable when leading to major changes in the aerodynamic behavior. This means models should be ideally machined with a precision of 0.01 mm, which is very expensive and often impossible to attain on models with complex shapes, made of materials suitable for easy manufacturing. Today's standard in 3D printing and classical machining gives precision close to 100 microns, what accordingly limits the downscaling to 1 to 10 for a very good geometrical precision and 1 to 100 for a rather good precision. There the machining precision meets the ability of classical boundary layer wind tunnels to recreate turbulence features at a scale not larger than 1/200. Making detailed models of buildings at a geometrical scale 1 to 1000 is often considered as "coarse" and maybe insufficient for delivering accurate results. Another difficulty when fabricating the models to represent the downscaled manmade structures is the surface roughness.

A typical painted steel structure roughness lies in the range 10^{-6} to 10^{-7} compared to its overall size. Reproducing the same roughness ratio on a model downscaled 1 to 100 for wind tunnel purpose would mean making surface with asperities less than 1 µm, which is referenced as a mirror surface state. In real life and with today's techniques, it is at the same time very difficult (if not totally impossible) and costly to achieve such precision. But, is it a real issue for wind engineers? Actually they may find an interest in reproducing rougher surfaces on the small scale model to compensate for a too low Reynolds number at model scale. Similarly to the drag crisis phenomenon on circular cylinder that can be triggered at lower wind speeds, i.e. lower Reynolds numbers, the critical state at low wind speed can be provoked by increasing the roughness of the surface, making a building model rougher than its homothetic downsizing [5]. Various kinds of calibrated roughening have been developed for wind tunnel models, including glued sand, added strips or ligneous material as wood or dots. In all cases, the aim is to counterbalance the lack of inertia forces compared to predominant viscous forces, by forcing the flow separation to occur at a given point of the boundary layer, at the skin of the model. This strategy called "increase of the apparent Reynolds number" must be calibrated by making models of the same structure at different scales and comparing the surface pressure patterns and aerodynamic loads measured in the wind tunnel. This approach requires the use of a large wind tunnel, with the capability of testing models of the same shape with a scale ratio close to one to ten without blockage effects, providing the reference measurements at high Reynolds number for comparison with those at small scale with added roughness. In practice, this strategy works only for subcritical Reynolds numbers, larger than 200 in wind tunnel, and aiming to reproduce a full scale roughness larger than 2.10^{-3} [6].

In summary, challenges related to downscaling effects in wind tunnel modelling cannot be handled by simple recipes that would work in all cases. The effect of scaling in wind tunnel tests should be studied first, for instance by repeating the experiment at various Reynolds numbers, looking for a stable result that demonstrates a supercritical state. Attention must be paid to each aerodynamic phenomenon involved in a complex study: two aerodynamic phenomena acting in a single study in wind tunnel may not have the same downscaling requirements.

4. Practical examples in scaling the drag of cylinder shape buildings

Such a reference study was achieved in the 1990's for a high rise tower in Paris, the shape of which was close to a circular cylinder. This cylindrical tower, designed by the famous French architect Jean Nouvel, was named "la Tour Sans Fins", which means "the endless tower", because its summit was vanishing in the sky, its outer skin becoming more and more porous with altitude. This kind of variable porosity is now frequently seen in recent towers design. At the time it was decided to build 1/300 scaled model for BLWT testing. Following the necessity to achieve concept of previous high Reynolds number referencing, a first model 10 times the diameter of the tower model was built (thus, corresponding to a 1/30 scale of the real tower), but not with the same diameter to height ratio, and subjected to high Reynolds flow in the large high-speed Jules Verne wind tunnel [7] with careful measurement of the pressure pattern around it. With a diameter of 0.8 m and a flow speed of 80 m/s a Reynolds number value of $4.0\cdot 10^6$ was reached in this "high Reynolds number" wind tunnel experiment. This was considered high enough upon the limit of critical state to be representative of the full scale value of $7.0\cdot 10^7$, corresponding to the same supercritical state of the flow. Keeping this pressure pattern as reference, the 1 to 300 scaled model was tested with slower flow, in the wind speed range corresponding to the BLWT, looking for the best arrangement of added roughness on the model surface to give a similar pressure field. It was found that meridional strips with a thickness of 0.6mm and a step of 15° provided the best artificial surface roughening in this case, leading to a mean pressure pattern similar to the high Reynolds number one.

The same question was addressed some years later, when studying wind loads on the European Launcher Ariane 4 on its launching pad, with models at a scale of 1 to 100. Similar tests were conducted, first in high Reynolds number conditions in the large wind tunnel, then at lower wind speed in the boundary layer wind tunnel with calibration of the relevant roughening. It must be underlined that in both cases longitudinal strips proved to be the only efficient manner to reach the required apparent Reynolds number, while sand roughness was not large enough. This artificial roughening method is commonly used on all kind of buildings with rounded shapes.

Figure 3. Roughening cylindrical models in BLWT by use of meridional strips.

5. Simplifying models at reduced scale

Real built structures are made of a collection of small elements, each potentially having an impact on the wind load applied on the whole. What should be the refinement of a reduced scale model, taking into account the many difficulties with machining of such a delicate object?

Such a question was answered for O&G companies TOTAL and DORIS through wind tunnel tests measuring the loads on a FPSO, first with a very detailed model reproducing all elements of this complex structure, then with a model made of simplified bluff bodies and porous screens and finally with a model composed of a single solid volume.

As expected, the simplified model gives results very similar to those obtained with the very detailed one, when wind loads on the single volume are fundamentally different. The simplification process of a complex, three dimension building, regarding the wind effect, must be achieved with advice from a specialized wind engineer.

Ensuring the same drag force whatever the wind direction is one key of this "cleaning" process, a group of small items being easily reproduced by a porous plate having the same plain/void ratio in a mean. Concerning the reproduction of small moldings and facetted effects on a façade, this has no effect on the wind loads when the air stream is detached.

Figure 4. Three levels of simplification for the study of wind loads on an offshore plateform, in boundary layer wind tunnel (courtesy of TOTAL and DORIS)

Any detail of a surface with a size smaller than the thickness of the boundary layer developed on this surface will not influence the wind loads. Therefore, only streamline bodies require the faithful reproduction of their surface details, for most of buildings which have sharp edges, facades can be considered as flat, for the BLWT modelling.

6. Scaling the porosity of cladding elements

There are many recent examples of high-rise buildings incorporating sun shades for the reduction of solar energy input in summer time. Architects also try to use such external structures to give the building a unique appearance, changing the shape and size of these elements across the façade, requiring the design engineer to carefully design them. Due to their exposure at the façade, they are subject to high wind loads, sometimes in accelerated flow areas.

But at the usual scale of buildings in BLWT, even if the detailed shape and porosity of the shades is carefully reproduced following a homothetic design, the porosity of these screens is especially poorly represented from the aerodynamic point of view because the viscous part in the Reynolds number is considerably increased at such reduced scale. Therefore, a two-step strategy is employed. The classical BLWT model is first used to measure global wind forces and local flow fields around the building in its environment. Then a second model of the façade is built, at a larger scale, say 1/50, for the fine modeling of wind loads on the shading structures. This larger model is built to resist high wind speeds and can be set in a large high-speed wind tunnel outside the ground boundary layer. The kind of wind applying loads on this large model is uniform, contrarily to the BLWT wind which is turbulent.

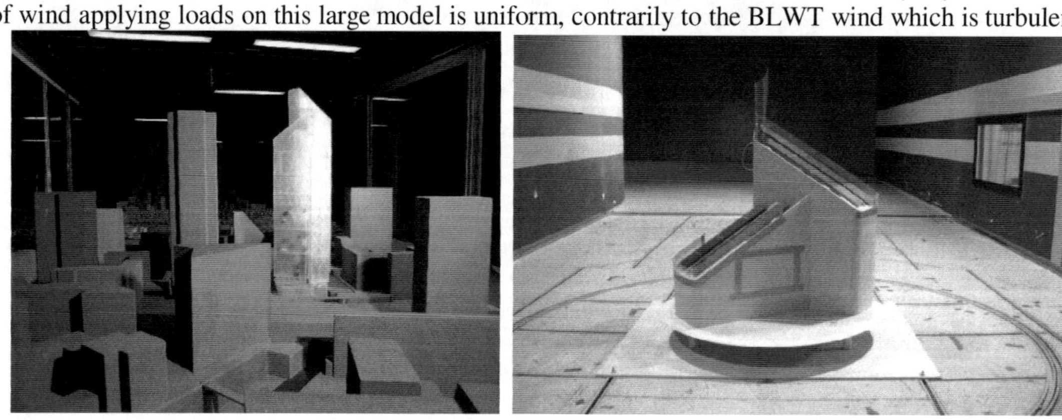

Figure 5. High rise building scaled 1/200 in BLWT (left) and model of the top of the same tower at a scale 1/50 in a high-speed wind tunnel (right) for measurements at high Reynolds number.

The size of the shading elements is assumed to be small compared to the mean dimension of eddies composing the wind turbulence, what justifies the testing in the uniform flow. Loads can be measured directly on the model, as well as pressure drop or local air velocity, with a good accuracy, when high wind speed and a large model size provide sufficiently high Reynolds numbers. This approach is relevant for windbreaks, solar panels assembly, safety barriers, porous canopies and any kind of wall alternating solids and voids.

7. Wind studies of small elements of building

Not only landmark buildings need careful wind resistant design. Looking at post-storm reports by insurance companies, what arises as the main candidate for failure under heavy wind load? Not the buildings themselves, but the small elements constituting our urban environment, such as lighting poles, advertising boards, transportation signs, chimneys, antennas, garden sheds…

These objects are parts of the build environment and, if none of them is as visible as a high rise headquarters of a powerful company, the sum of their individual repair and replacement usually makes up the major part of storm losses. Therefore, there is a real need for an industrial wind proofing service that the large wind tunnels can fulfill with wind load reproduction on full-scale prototypes.

In that case, contrarily to the BLWT approach, there is no need to simulate with care the characteristics of the wind turbulence, its spatial correlation and its length scales. Wind is a local phenomenon, uniformly distributed on an area of several square meters. The main question is not to faithfully reproduce the unique wind field created in a given place by the particular location of surrounding buildings and terrain: the question is "what is the highest speed this type of object will stand in its life" and consequently which way it will be fixed/bound/attached and interact with its environment.

The wind speed representative of such a small element's life is not easy to assess. It mainly depends on the height above ground where it will stand. For instance, a photovoltaic solar panel standing on a roof will not be loaded the same way as the same panel on the ground.

For this reason full-scale testing in a large wind tunnel is often preceded by another study at reduced scale or by CFD modelling, for the assessment of the local wind speed. It is therefore very convenient to associate in the same location a BLWT with a HSWT (High Speed Wind Tunnel), the former being used to determine the wind conditions to be reproduced in the latter. After the maximum local wind speed and its various directions have been determined by small anemometers placed in the vicinity of given building on the reduced scale model coupled with a climate survey, the same level of wind will be applied on a full scale prototype of the item in question in a high speed wind tunnel.

Figure 6. Some examples of full scale testing on prototypes, using an industrial high speed wind tunnel.

Avoiding blockage effects must be a constant concern, whatever the type of the wind tunnel, which frequently fixes the choice of the test section and the power of the facility.

Because the shapes of prototypes to be tested are many and various, from slender vertical light poles to bluff electric boxes, as well as wide, flat solar panels and quasi-cubic individual houses, the shape of the optimal testing section is difficult to define. For this reason, large wind tunnels are usually composed of successive testing chambers of different cross sections, sometimes with moving wall or

IOP Conf. Series: Materials Science and Engineering **276** (2017) 012001 doi:10.1088/1757-899X/276/1/012001

ceiling in order to adapt the test section shape and size to the prototype to be windproof.

8. Scaling snow in climatic wind tunnel

Scaling issues is not reserved to wind modeling; the same questions arise in geotechnical science, in marine studies etc. And when it comes to wind tunnels that not only use air, but a mixture of air and water (rain drops, water proofing, fog...), solid particles (sand storms, dust transportation...), snow, then downscaling the physical properties of phenomena becomes impossible. This is another good reason for full-scale tests in wind tunnels, an example of which will be illustrated hereafter.

There are many questions raised by researchers involved in the investigation of snow loads. Failures induced by snow ingress in systems, snow accretions on structures and vehicles are efficiently studied by full-scale approaches either outdoor or in laboratory environment. On the other hand, questions dealing with large environments, typically interaction between various buildings and their surroundings can only be studied with reduced scale models. Snow accumulation around buildings belongs to this category. A reliable experimental modeling in wind tunnel depends on geometric, dynamic and kinetic criteria: snow particles (shape, density, velocity, drag, lift, liquid water ratio) and wind (temperature, speed, turbulence) must be modeled. Experiments are commonly based on theoretical approaches first introduced by Kind [8] and Iversen [9].

Many attempts have been made to use BLWT with some material replacing snow for the reproduction at reduced scale of snow deposition and snow drift events. Snow particles have been modeled by sand, sawdust, glass balls which do not reproduce all inter-particle forces. Plastics and potato flakes have proved more convenient, showing some similarity with the real snow behavior in the inter-particles affinity. Nevertheless, even if man-made snow does not reach the same tree-like shape as natural one, because snowflakes need a long time to grow and artificial snow consists of small ice oblongs, the most appropriate choice is often to make artificial snow in a cold wind tunnel.

Evaluating snow properties involved at reduced scale by similitude laws yields Table 1. A noticeable disagreement of the full scale Froude number with a small scale model one's, based on the threshold friction velocity weighted by the particle/fluid density ratio, indicates that the trajectory of the model particle will be different from the natural snow particle one. According to Kind [8], if the Froude number is not conserved in the experiment, it is particularly important to verify that saltation length, which represents 10 times the saltation height, is shorter than the building reference length and the typical size of the snow drifts.

Table 1: Similitude parameters for particle transport and accumulation

D_p/L	Geometric ratio particle length/building
$u{*}_t^2/D_p g\left(\rho/(\rho_p-\rho)\right)$	Froude number for particle friction threshold
$u{*}_t^2/Lg\left(\rho/(\rho_p-\rho)\right)$	Froude number for friction on the building
$u{*}/u{*}_t$	Threshold speed/friction speed ratio
$u{*}/u$	Friction /fluid speed ratio
$u^2/D_p g\left(\rho/(\rho_p-\rho)\right)$	Froude number for particle transport
u^2/Lg	Conventional Froude number
ρ_p/ρ	Particle/fluid density ratio

D_p particle diameter, L reference length, g gravity, ρ fluid density, ρ_p particle density, u_t^*threshold friction velocity, u^*friction velocity, u reference wind velocity

In practice, this may be verified when using moderately reduced scale models in a cold wind tunnel with artificial snow guns. Concerning snow transport by wind, forces on particles are better modeled if the Reynolds number based on the saltation height is greater than 30. This also leads to scaling ratio limited to 10 or close to.

Regarding the suspended particles, the particulate Froude number, weighted by the density ratio, is the physical quantity describing the similarity of the transport mechanism at the vicinity of buildings. It also drives the accumulation process, locally. Finally a choice has to be made between the saltation mechanism (close to the ground) or long distance transport processes.

Figure 7. Testing snow accumulation in CSTB climatic wind tunnel [11] at full scale (left) and reduced scale for complex buildings (right).

In practice, the wind/snow engineers set simple rules for common applications, considering essentially the simulation of the drifting volume v0. The basic similitude parameter is v_0/L^3, where L is the reference length of the structure, which leads to the straightforward criterion

$$\left(1-\frac{u^*}{u}\right)\left(\frac{u^2}{Lg}\right)_p = \left(1-\frac{u^*}{u}\right)\left(\frac{u^2}{Lg}\right)_m \qquad (2)$$

where indexes p and m stand respectively for prototype and model. This relationship does not imply any constraint regarding the particle density and size. Hence, the model particle scale does not interfere with the drifting volume simulation.

A major parameter of reduced scale simulation, the experiment duration, can be calculated according to several dimensionless time numbers. Various expressions can be found in the literature. Each of the dimensionless time expressions in Table 2 leads to a different snowstorm duration and there is no real agreement about what criteria should be used.

Table 2: Dimensionless time used to assess the equivalent snow event duration
ut / L
$\rho / \rho_p \cdot ut / L$
$1/2 \cdot \rho/\rho_p \cdot u^2 / gL[1-u_0/u]. \; ut / L$
$tQ\eta / \rho_p L^2$

Therefore, it is advisable to rely on actual measurement of a real outdoor accumulation and tune the snow fall duration in wind tunnel to achieving the same, scaled behavior.

References

[1] Vaschy A 1892 Sur les lois de similitude en physique *Annales Télégraphiques* **19** pp. 25-28
[2] Buckingham E 1914 On physically similar systems. Illustrations of the use of dimensional equations *Phys. Rev.* **4** pp. 345-76
[3] Davenport A G, Vickery B J, Holmes W H 1975 *The structural and environmental effects of wind on buildings and structures* (Post graduate course of the University of Sidney, School of Civil Engineering)
[4] Roulle C 1980 Reproduction du régime hypercritique sur formes rondes en soufflerie à couche limite atmosphérique *CSTB report n°EN-ADIM 80-20L* (Unpublished)
[5] Basu R I 1985 Aerodynamic forces on structures of circular crosssection part 1. model-scale data obtained under twodimensional conditions in low-turbulence streams *J. Wind Eng. Ind. Aerodyn.* **21** pp. 273-94
[6] [6] Szechenyi E 1975 Supercritical Reynolds number simulation for two- dimensional flow over circular cylinders *J. Fluid Mech.* **70**(3) pp. 529-42
[7] Lemoult B et al. 1990 *Caractérisation des efforts aérodynamiques développés sur un cylindre en régime supercritique, simulation en soufflerie (*CSTB report n°EN-SC 90.11-C)
[8] Kind R J 1976 A critical examination of the requirements for model simulation of wind-induced erosion deposition phenomena such as snow drifting *Atmos. Environ.* **10**(3) pp. 219-27
[9] Iversen J D 1979 Drifting snow similitude *Proc.Asce,J.Hydraulics Div.* **105**, pp. 737-53
[10] Delpech P et al. 1998 Snowdrifting simulation around antarctic buildings *J. Wind Eng. Ind. Aerodyn.* **74-76** pp. 567-76
[11] CSTB, *Jules Verne Climatic Wind Tunnel* [Available online] http://recherche.cstb.fr/en/facilities/climatic-wind/

Implementation and application of the actuator line model by OpenFOAM for a vertical axis wind turbine

L Riva[1]*, K-E Giljarhus[2], B Hjertager[2] and S M Kalvig[2]

[1]Department of Engineering Sciences, University of Agder, Norway
[2]Department of Mechanical and Structural Engineering and Material Science, University of Stavanger, Norway

*Contact author: lorenzor@uia.no

Abstract. University of Stavanger has started The Smart Sustainable Campus & Energy Lab project, to gain knowledge and facilitate project based education in the field of renewable and sustainable energy and increase the research effort in the same area. This project includes the future installation of a vertical axis wind turbine on the campus roof. A newly developed Computational Fluid Dynamics (CFD) model by OpenFOAM have been implemented to study the wind behavior over the building and the turbine performance. The online available wind turbine model case from Bachant, Goude and Wosnik from 2016 is used as the starting point. This is a Reynolds-Averaged Navier-Stokes equations (RANS) case set up that uses the Actuator Line Model. The available test case considers a water tank with controlled external parameters. Bachant et al.'s model has been modified to study a VAWT in the atmospheric boundary layer. Various simulations have been performed trying to verify the models use and suitability. Simulation outcomes help to understand the impact of the surroundings on the turbine as well as its reaction to parameters changes. The developed model can be used for wind energy and flow simulations for both onshore and offshore applications.

1. Introduction

Oil export has long been a significant source of income in Norway. The recent drops in the oil price and a considerable export reduction have forced Norway to start questioning about the solidity of its economic structure: it must be updated according to the new political and economic scenario. New-Renewable energies and technology (hydropower not included) have been targeted as one of the possible new economic hubs. In 2014, investment to sustainable energies passed from circa 2-3 billion Euros to 4-5 billion Euros [1].

Aside economic reason, there are also environmental concerns. In December 2015, Norway signed the COP21 agreement. It pledged to reach carbon neutrality by 2030 and to bring greenhouse emissions 40% below the 1990 values [2]. Norway has hence joined European Union (EU) funding projects to foster such transition.

Wind energy is allegedly one of the most promising renewable sources. It has grown dramatically for the last decades. Focusing on EU, if in 2000 electricity production share from wind energy was below 1.5%, in 2014 reached the 14.1% and is forecasted to achieve the 23% by 2030 [3]. An even more dramatic growth has found place in Norway: between 2015 and 2016 the wind power production for year passed from 2.5 TWh to 10 TWh [4]. This transition is related to the will of the principal national companies to open new ways of investment. The intention is to transfer the knowledge of offshore technologies accrued through decades of oil and gas research and development to this sector; trying to keep up with the international market. The Norwegian Wind Energy Association (NORWEA) esteemed investments in Norway on wind energy in 2015 were above 2.4 billion dollars [4]. University of Stavanger has started The Smart Sustainable Campus & Energy Lab project, to gain knowledge and facilitate project based education in the field of renewable and sustainable energy and increase the research effort in the same area. The project envisages the installation of a set of built environment wind turbines (BEWT), instrumented with measurement devices. Ivar Langen's (IL) building, which is indicated in Figure 1, has been identified as possible target point to install one or more turbines. This building represents a good tradeoff between visibility and efficiency.

A flexible CFD set up for future detailed wind flow analysis and wind turbine power performance simulations for the campus was suggested. This need was addressed in Lorenzo Riva's Master Thesis (2016) and this paper is based on Riva's thesis [5]. The Actuator Line Model (ALM) is used for wind turbine representation. This model represents a good trade-off between accuracy and time consumption. Originally developed by Sørensen and Shen [6] for HAWT in 2002, it has been recently updated for VAWT by Shamsoddin and Porté-Angel [7] and by Schito, Bernini and Zasso [8].

Figure 1. UiS Campus and IL building (UiS courtesy)

Bachant et al. [9] released an open-source case based on OpenFOAM [10] with the library necessary to run the case [11]. This model has been selected for this case. Unlike the laboratory experiments with controlled parameters, in the present study the wind flow is simulated according to the atmospheric conditions. No literature has been found documenting ALM outdoor use for VAWT. This is also a big challenge since there are no studies to compare the model implemented with. This model may furnish an example of ambient-condition case, useful for analyses of different contexts and landscapes. Application for onshore as well as offshore surroundings might be considered. In a recent master thesis project, 'Wind Flow Simulation over Fish Farm Feed Barge' by Svein Erik Nuland [12], the same model has been applied for offshore analyses. The work of Nuland demonstrated the feasibility and elasticity of the model.

2. Theory

2.1. Atmospheric boundary layer (ABL)

The behavior of the moving fluid is described by Navier-Stokes equations. In the current study, Reynolds-Averaged Navier-Stokes equations (RANS) with the k- ε model has been adopted. This decision comes from the necessity to find a good compromise between the computational cost and the accuracy of the results. In particular, the RANS constants have been modified to fit better the atmospheric conditions.

There are several challenges related to the CFD modelling of the atmospheric boundary layer, for example, it changes along the flow direction and that makes the flow not homogeneous. As observed by

[13], the k- ε model and the boundary layer need to be modified to account for the wind behavior in atmospheric condition with a greater degree of accuracy. Among several possible solutions, Hargreaves and Wright considerations have become very popular because it is easily applicable [14]. As previously stated, the lowest part of atmosphere is involved. It must be therefore modelled. The horizontally homogeneous turbulent surface layer (HHTSL) suits it well.

This approach relies on the following assumptions:

1 – the vertical velocity is zero;
2 – the pressure is constant in both the vertical and stream-wise directions;
3 – the shear stress is constant throughout the boundary layer;
4 – flow is steady state.

These assumptions lead to:

$$u = \frac{u_\tau}{k}\ln\left(\frac{z+z_0}{z_0}\right) \tag{1}$$

$$k = \frac{u_\tau^2}{\sqrt{C_\tau}} \tag{2}$$

$$\varepsilon = \frac{u_\tau^3}{\kappa(z+z_0)} \tag{3}$$

With u as flow velocity z_0 as roughness length and z as height. C_τ and κ (Von Karman constant) are constrains. Friction velocity u_τ is:

$$u_\tau = \sqrt{\frac{\tau}{\rho}}\ , \tag{4}$$

where ρ is the density. OpenFOAM permits to include directly the ABL conditions. A rough wall function is also used at the ground.

2.2. Actuator line model (ALM)

The ALM is based on the classical blade element theory combined with a Navier- Stokes description of the flow field. It treats turbine blade as actuator line elements, defined by their quarter-chord location, and for which 2-D profile lift and drag coefficients are known [9]. A diagram is shown in Figure 2, where U_{in} is the local inflow wind velocity, U_{rel} is the relative velocity, $-\omega r$ is the apparent wind velocity due to tangential blade speed, M is the pitching moment, θ is the rotational angle of attack (positive counterclockwise), r is the radius, α is the angle of attack, F_l is the lift force, F_d is the drag force.

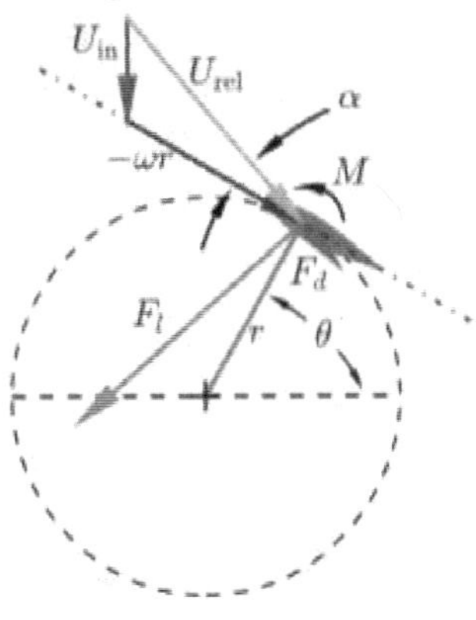

Figure 2. Visualization of the ALM by vector diagram [8]

Once the drag and the lift coefficients are provided by tables, it is possible to compute the lift force, the drag force and the pitching moment. Forces are then projected onto the rotor coordinate system to calculate torque, overall drag and other quantities. The same method is adopted for the turbine shaft and the blade support struts. Finally, the force on the actuator line is added to the Navier-Stokes equations.

For small turbines, the solidity is usually small and the tip speed ratio (TSR), which is the ratio between the relative velocity at the tip of the blade and the free stream velocity, is low. That is the case for instance of the turbine involved here. Unsteadiness will hence be significant and needs to be considered. This will affect mainly angle of attack and relative velocity. In [9], the issue is addressed including dynamic stall implementation and other corrections.

Dynamic stall is computed by a model developed by Sheng et al. [15]. It is derived from the Leishman-Beddoes semi-empirical model. Among the dynamic stall models, this is one of the most popular, probably due to its physical representation of the overall unsteady aerodynamic problem. Basically, it models the dynamic stall introducing a boundary layer delay method and a strategy to take into consideration the dynamic vortex [9]. Sheng at al. have tailored the original model for low Mach numbers. The model is included into the library turbinesFoam [11], which the ALM calls during its application. Inside the ALM, angle of attack is sampled form the flow field. Lift and drag coefficients are then corrected considering the effects due to accelerating the fluid [9].

This model provides also corrections due to the so-called curvature effects, which are caused by the rotation and bring to a variable chord-wise angle of attack distribution; and due to the end effects [9]: vortex lines may not end in a fluid, but must either form closed loops or extend to boundaries. As consequence lift distribution may be affected.

3. Modeling approach

In the following section, a brief description of the model is given. The input values, critical steps and the main assumptions that have been considered and processed during the implementation.

3.1. Turbine description

Since the UiS project has not yet selected any turbine, it has been decided as a preliminary step to choose directly the turbine used into the open-source case [10]. The turbine is named UNH-RVAT (University of the New Hampshire –Reference Vertical Axis Turbine). Design is performed with the aim of furnishing a reference layout for VAWT and can be seen in Figure 3. The UNH-RVAT turbine has three blades made of NACA 0020 profiles with a 0.14 m chord. The blades are mounted at mid-chord with respect to the turbine axis, and the turbine has a height of 1 m and a diameter of 1 m [16].

3.2. Airfoil dataset

Following reference case suggestions, the airfoil profile NACA 0021 has been used, with dataset taken from [17]. This database has some limitations regarding the numerical extrapolations of data from other measurements. Nevertheless, this dataset is likely the most comprehensive available with respect to variety of profiles and ranges of Reynolds numbers [9]. The complete NACA 0020 dataset is not available. However, according to [9], the small difference in foil thickness is assumed to be negligible.

3.3. Campus model

At the beginning of the current study, no existing detailed three-dimensional models of the campus were available. It has been necessary to design a new one. By the utility snappyHexMesh, which is a mesh generation utility, it is possible to generate a three-dimensional mesh from tri-surfaces in STL format. Hence, first a 3D STL file of the campus is necessary. The file has been generated by the collaboration with Ibrahim Mufti Pradityo, an Urban Planning and Design student at UiS. Due to environmental constrains, the expected main wind directions are from South and West. The eastern part of the campus, considered not influential for the observed case, has been neglected. Ivar Langen building, which is the target building, and Kjølv Egeland building, which is the one beside it, have been modelled starting from the basic drawings UiS provided. For the other buildings, no 2D models were available. They were made based of images obtained on Google Maps. Terrain has been reconstructed importing a set of SOSI-points. SOSI (Systematic Organization of Spatial Information) is a geospatial vector data format used in Norway. The overall 3D model was then implemented on Revit 2017, a building information modelling (BIM) software developed by Autodesk. Revit 2017 cannot export STL file. The 3D model has been hence imported into Sketchup, a 3D modelling software. Here it has been located on suitable coordinates in the X-Y-Z system and finally exported as STL file. Coordinates have been selected so that the positive verse of the axis-x coincides with east and the positive verse of the axis-y with north. Figure 4 shows the final design of the campus.

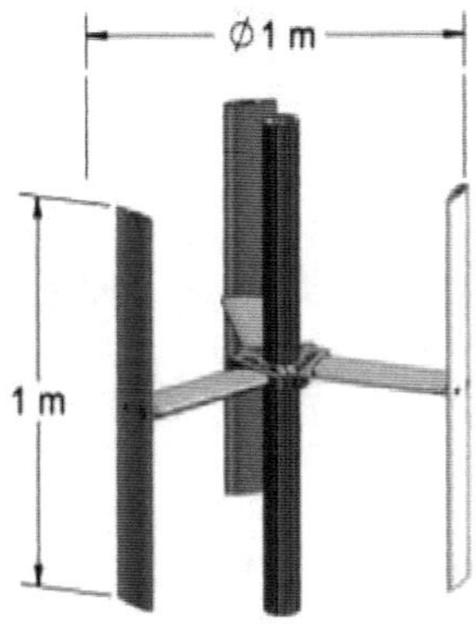

Figure 3. UNH-RVAT turbine [16]

Figure 4. STL 3D Campus Model

3.4. Computational mesh

The computational domain is 300 m x 300 m x 100 m. It is initially divided in 72000 cells of 5 m x 5 m x 5 m. This size has been selected based on tests carried out as preliminary step, where several cells dimension have been proposed, in order to obtain reasonable time of simulation and accuracy [5]. The grid must be refined to resolve the details in the campus models and for the turbine representation. The location of the turbine has been chosen according to preliminary wind simulations over the campus. The center of the turbine is situated in x = 190 m, y = 240 m, z = 25 m (building height is approximately 23.5 m). The turbine is 45 m far from the west-face of the building, 40 m from the east-face, 10 m from the south-face and 15 m from the north-face. Figure 5 displays the selected location, at the final step of simulation, with different streamlines and sights. Streamlines dimension has been accentuated to ease the visibility and color indicates the velocity, which is named UMagnitude and measured in m/s. Campus and ground are colored showing the relative pressure normalized with the density (with zero set at atmospheric pressure). The wind has been simulated in the preliminary study with the inlet value of 15 m/s, in order to provide same Reynolds number of [9]. It has been simulated blowing from west and is strongly affected by the fence located 85 m before the IL building.

Despite the fence is actually not very tall, it is located almost at the same height of the roof. Turbulence is generated, slowing the flow. Wind streamlines look more uniform in the part of the roof where the turbine is located. Wind flow is expected to have therefore less velocity losses and to maintain its velocity profile close to the undisturbed field. The turbine should hence manifest higher performance than if set on other locations over the building. The turbine is inserted into a sequence of four refinement boxes. The outer box is made of 1.25 m cube cells, medium-outer box of 0.3125 m cube cells, medium-inner box of 0.078 m cube cells and inner box of 0.039 m cube cells. The choice of the number of layers and their values comes from the necessity to guarantee a soft transition between the external grid cell dimension and the one characterizing the turbine. A rapid and rough change might bring to errors in the meshing process and in discontinuities in solving, so the collapse of the simulation. A slow and meticulous change instead might slow down the simulation significantly. Turbine is made of 0.039 m cube cells. Overall result after the meshing process is shown in Figure 6. The total number of cells are approximately 1 million.

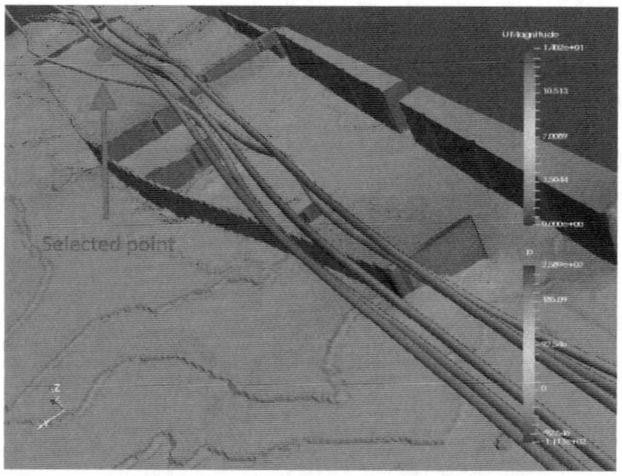

(a)

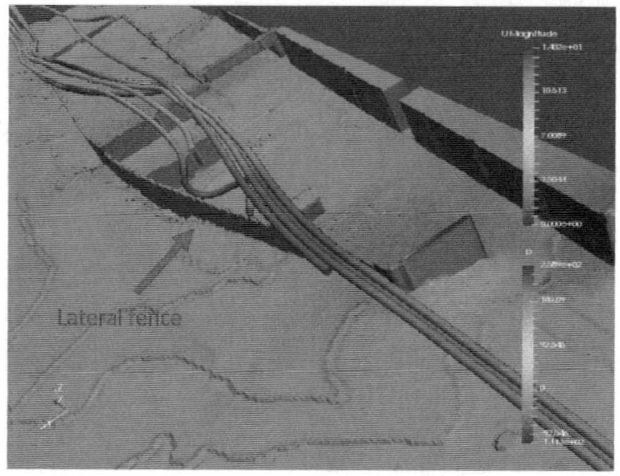

(b)

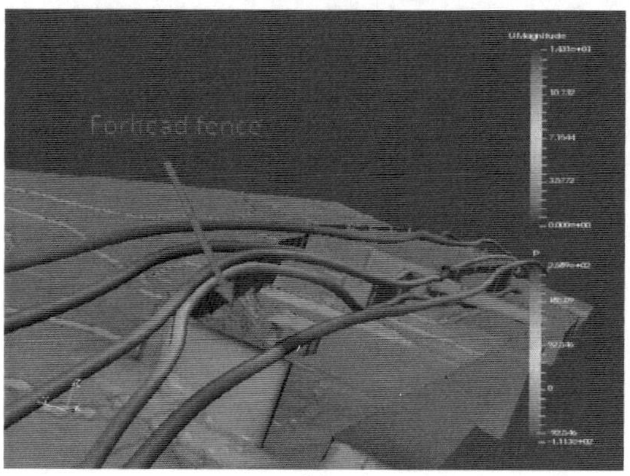

(c)

Figure 5. (a) wind study, streamlines and selected point. Umagnitude range: 0 to 15 m/s. Density normalized pressure range: -1.2e+02 to +2.5e+02 m^2/s^2

(b) Wind study, influence of the lateral fence. Umagnitude range: 0 to 15 m/s. Density normalized pressure range: -1.2e+02 to +2.5e+02 m^2/s^2

(c) Wind study, influence of the forehead fence. Umagnitude range: 0 to 15 m/s. Density normalized pressure range: -1.2e+02 to +2.5e+02 m^2/s^2

3.5. Numerical settings

The case is implemented on OpenFOAM 3. Bachant's case has been modified and adapted. First modification was to adapt the case to air. This can be done through the transportProperties dictionary inside the constant folder. The transportModel selected is Newtonian. and nu is 1.5e-05. Turbulence model is k-ε and coefficients follows [14]. The reference velocity for the ABL is set to 10 m/s, reference height set to 25 m and roughness set to 0.01 m. Following [14], initial values are 1.3 m^2/s^2 for k and 0.01 m^2/s^3 for ε. The time scheme selected is backward, which has second-order accuracy. For the divergence schemes, second-order linear upwind schemes are selected. Pressure solvers are GAMG type.

The principle of this solver is to generate a quick solution on a small number cells mesh, to map then this solution onto a finer mesh and to use it as an initial guess to obtain an accurate solution on the fine mesh [18]. With respect to other common solvers it is faster. In this case it has been also smoothed with a DICGaussSeidel (diagonal incomplete-Cholesky/LU with Gauss-Seidel) smoother. Other parameters are treated with a smoothSolver and as smoother the symGaussSeidel (symmetric Gauss-Seidel) has been used. Besides, nOuterCorrectors enables looping over the entire system of equations within a time step, representing the total number of times the system is solved [18]. If, as in this case, it is 1, solver is replicating PISO algorithm. NCorrectors does the same with only pressure and momentum equations and it is 2.

Time settings require the selection of the overall time of simulation and the duration of each time step. This choice must consider the complex geometry and physics of the case: a turbine is spinning, affected by an evolving flow surrounding. Settings must therefore be chosen in order to ensure that the rotation of the turbine is adequately simulated. The time step is chosen according to the rotational speed of the turbine. Keeping TSR = 1.9 (which is the optimal value in [9]) and v_{wind} = 10 m/s, it will be ω= 38 rad/s. It means turbine takes 0.165 s to complete a rotation. A time step of 0.001 s has been chosen. There are 165 steps per revolution and, according to the reference article [9], this is sufficient to guarantee accuracy.

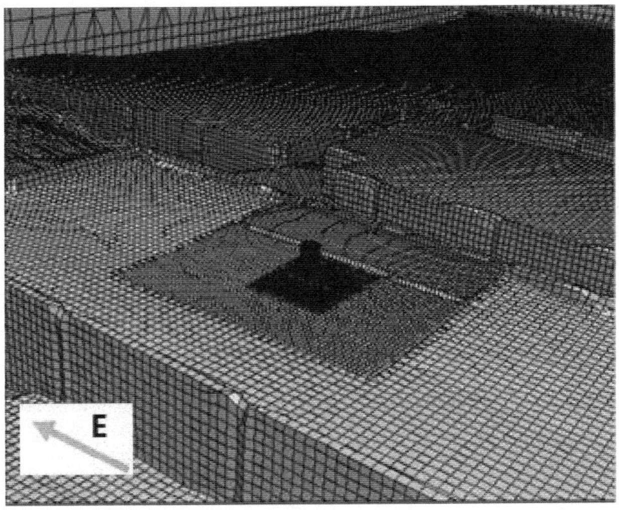

Figure 6. Final mesh and eastward direction.

The total simulation time is chosen based on the time it takes for the overall flow field to reach a stable solution, that is when the wind velocity profile is uniform throughout the mesh. By several simulations runs, it was found that a simulation time of 30 s guarantees good stability and homogeneity throughout the mesh. This value is a compromise between the necessity of guaranteeing acceptable simulation time and a good accuracy. Higher time would be more appropriated to provide precise outputs. Turbine is defined by fvOptions, which calls the turbinesFoam library, selects the model and lets the user activate and change some model parameters. It is important to remind that, according to fvOptions selections, the drag and the lift coefficient will be modified into the turbinesFoam library, before being used. These forces will be used both to solve the discretized Navier-Stokes equations and to compute the torque on the turbine. Torque is then directly used into the library to obtain the torque coefficient by:

$$C_{torque} = \frac{M}{0.5 A_{frontal} v_{freeStream}^2 R} \tag{5}$$

as well as the power coefficient:

$$C_{power} = C_{torque} TSR \tag{6}$$

where $A_{frontal}$ is the frontal area, $v_{freeStream}$ is the free stream velocity, R is the radius and TSR is the Tip Speed Ratio. As evident the torque coefficient uses the free stream velocity. Power coefficient hence refers to this parameter. This decision can fit well a lab surrounding where flow can be easily controlled. In an open air surrounding, velocity changes significantly. Power coefficient might be very different from the value the turbine is actually experiencing. One could state that the surrounding itself is part of the turbine performance and accept to keep this value. Here it has been decided to focus on the performance of the turbine, regardless the surrounding. It has been decided to correct the power coefficient using the average value, of the last 3 s, of the sampled velocities along the z-axis of the turbine (from z = 24.5 to z = 25.5). Turbine obviously affects the surrounding; therefore, this sampling must be done somewhere upstream the turbine location in order to catch the real wind speed that might be exploited. Sampling has been done in x/R = - 10. This distance has been selected since it is enough far not to be influenced by the turbine behavior. The power coefficient then becomes:

$$C_{power,new} = C_{power} \frac{v^2_{freeStream}}{v^2_{sampled}} \tag{7}$$

where $v_{sampled}$ is the sampled average velocity.

3.6. Wind data set collection

During the realization of this study, a set of wind data recorded from 2003 to 2013 has been found. The anemometer is located in the eastern side of the adjacent building. The wind monitoring was developed and measurement data acquired by a former staff at the department at the University of Stavanger. It has been considered useful to analyze the data to possibly enrich the insight of the problem. A wind rose has been therefore drawn from these data and is shown in Figure 7. The dominant wind direction is from south. West and east share a considerable part. North has been neglected because of physical obstacles (part of the building and environment). According to the wind rose, a 10-min mean wind speed range between 4 m/s and 6 m/s is very common. However higher speeds have been recorded, up to 12 m/s.

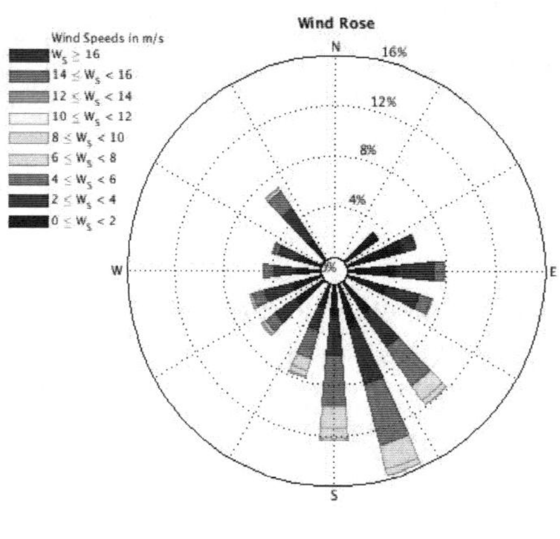

Figure 7. Wind rose

4. Results

This work aims to be the first stone of a path that will be improved and better shaped over the time. Many studies can be still carried out to guarantee a better model. Moreover, as the turbines will be mounted, the computational part will be enriched by the experimental outputs. Therefore, in the meantime, model has been tested by some sensitivity analyses; trying to understand how it would react. Then model has been tested changing the TSR and wind velocity. The performance has been compared installing the turbine in different reasonable locations.

4.1. Model analysis

The case studied has a wind speed at the inlet of 10 m/s at the turbine height (25 m). TSR is 1.9, which was the optimal value in the reference paper. Values are extrapolated as average of the last 3 s of simulation. The velocity field inside the turbine area is characterized by a central low-speed zone, which is typical of VAWT aerodynamics. The velocity profile along the x-axis, U_{wind}, is shown in Figure 8. Compared to the results in e.g. [6] and [7], the profile here does not show an almost symmetric trend. This condition is due to the difference in value of the wind speed along the y-axis. As Figure 8 shows, 1.5 m before hitting the turbine entrance area, wind is less than 10 m/s in y/R = - 3 and below 9 m/s in y/R = 3. Figure 8 shows that the turbine affects wind field even after a relatively long distance: wake is still strong after 1.5 m and evident after 3 m. Figure 9 shows the turbulent kinetic energy along the x-axis. The plot shows that the turbulent kinetic energy follows the expected trend, which for example can be compared to [9] and [19]. As for velocity, also this parameter presents different values on each side of the y-axis. The wake is visible in the plot in terms of an increased kinetic energy. Looking at Figure 9, it seems that the model has failed to catch the usual drop at the center of the turbine. This lack might be related either to an insufficient number of cells along the y-axis. Nevertheless, the curve at x/R = 0 is characterized by two small peaks that might hide the drop in the middle.

Figure 10 shows forces (per unit density, assuming incompressible flow) in central part of the turbine. Figure 10a visualizes the distribution of the forces around the actuator line at half height of the turbine, Figure 10b visualizes it at the center part of the turbine. Forces along the actuator line are computed by the Actuator Line Model and then projected onto the flow field as a source term in the momentum equation. Following [9], the source term is tapered from its maximum value away from the actuator line by means of a spherical Gaussian factor, to avoid instability due to steep gradients. Figure suggests that forces get their maximum value around the windward part of the rotation (rotation in the counter clockwise direction). This phenomenon is a typical aerodynamics feature in VAWT.

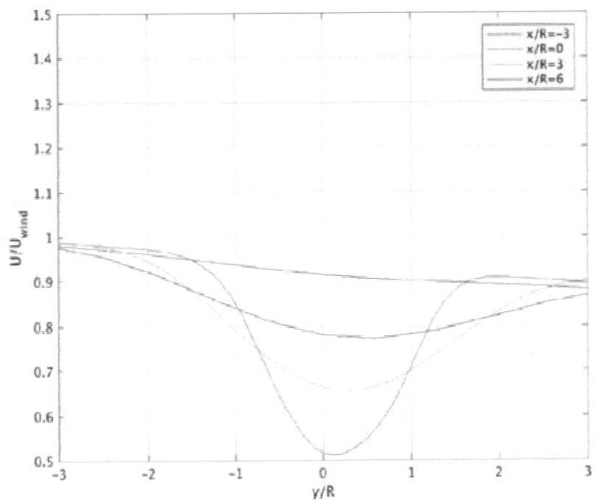

Figure 8. Velocity profile along the x-axis

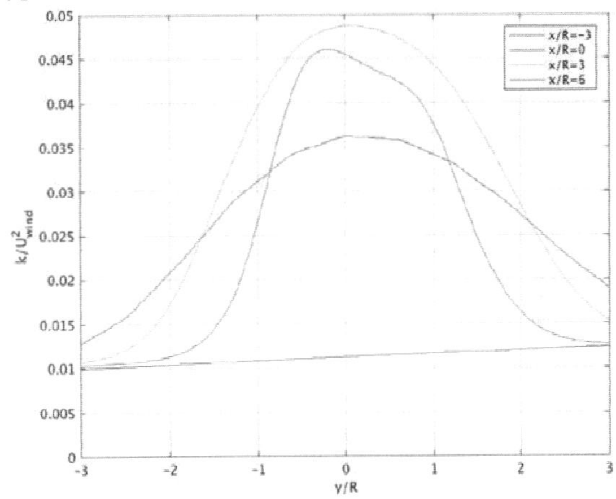

Figure 9. Turbulent kinetic energy profile along the x-axis

4.2. TSR optimization

The first parameter studied is the TSR. Turbine has been tested with six values: 0.9, 1.4, 1.9, 2.4, 2.9, 3.4. All parameters except time step have been kept unmodified. This is a necessary operation in order to have the same number of steps per revolution. The optimization aims to figure out the TSR with the highest production of power. The simulation that has been previously analyzed, has been carried out with the optimized TSR at wind speed of 15 m/s obtained in the reference article [9]. With 10 m/s wind speed, the forces on the turbine are consequentially lower than at 15 m/s wind speed. It is likely the turbine will have to compensate this decrease by increasing TSR during the simulations. Curve might probably move a little rightward. This implies optimized TSR will be slightly higher. The resulting Cp for the considered TSRs are shown in Figure 11.

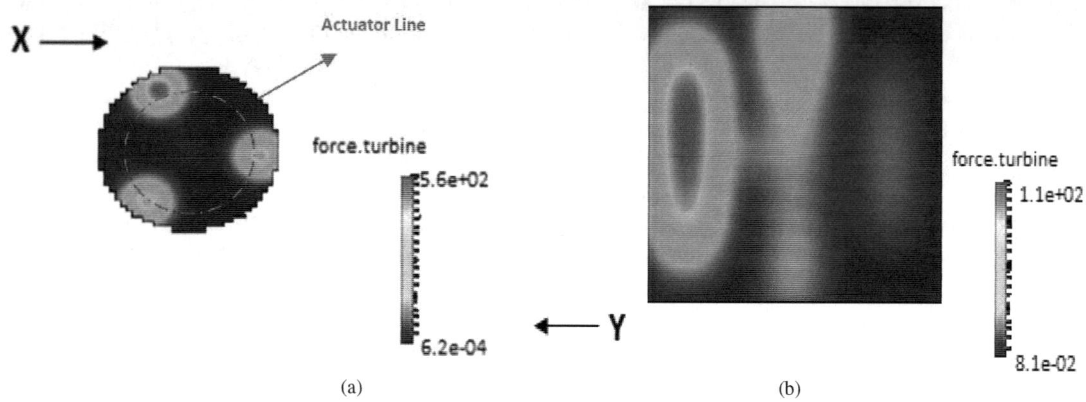

Figure 10. Force distribution (N m^3/kg) (a) around the turbine in the x-y plane and (b) in the x-z plane

4.3. Velocity profile
The model has been tested changing the free wind velocity, with 5 values from 5.5 m/s to 15 m/s. Theoretically, it is easy to predict that the higher the speed the higher the power. Power law should follow a cubic power trend. Since no electric production has been considered, no limit in power production has been taken into account. The resulting power as a function of wind speed is shown in Figure 12. The power follows the theoretical trend. Simulations demonstrated also that Cp increases a little when velocity increases. A higher speed implies a higher Reynolds number. Inertia forces prevail over the viscosity effects shaping a more uniform velocity field. This means higher lift and therefore the turbine works more efficiently.

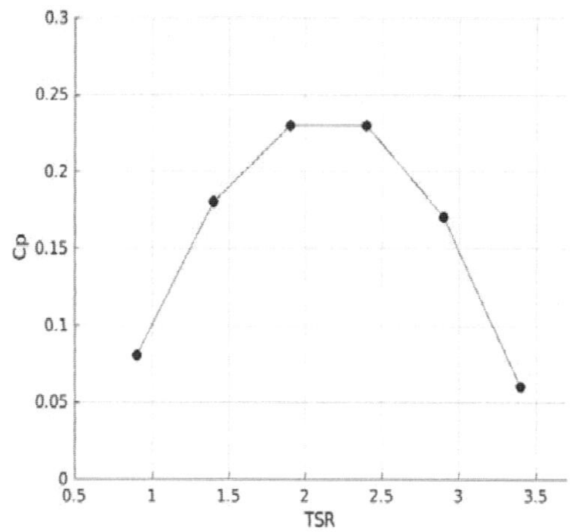

Figure 11. TSR change and Cp

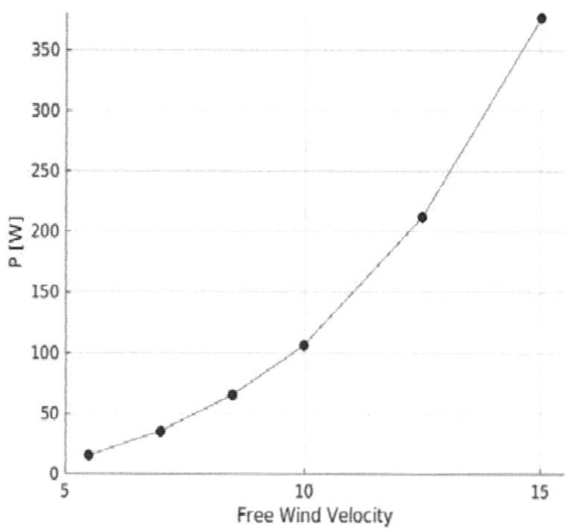

Figure 12. Velocity change and Power

4.4. Study of different locations
For small scale wind turbine, power production may not be the only target. Very often the turbine is associated to the promotion of clean energy and to educational purposes [20]. Energy efficient locations need to be considered in relation to visibility. This is the case of the planned UiS project: research, education and promotion of clean energy are the main objectives and the power production itself has less importance. Two new locations have been tested. They have been named Position 2 and Position 3.

Locations have been chosen in the furthest part of the building with respect to the inlet and the fence, in order to have more uniform wind conditions. In addition, 1.5 m has been selected as distance from the nearest sides, considering this value acceptable to permit the turbine to be seen more easily. As Figure 13 indicates, these locations are closer to the east edge of the roof and, as consequence, easily spotted. Apart from the different position, all the other parameters and setting have been kept unmodified, i.e. the figure presents the flow conditions associated with the inlet wind velocity of 10 m/s from the west.

Compared to the previous position, which was selected according to wind study to maximize efficiency and power observing wind behavior and magnitude over the building, they present a less stable and uniform velocity field and a lower wind velocity. Especially for Position 3, power is expected to reduce dramatically. Wind speed at Position 2 does not change significantly with respect to Position 3, but Position 2 has a significantly lower efficiency, which brings to a reduction in produced power of 22%. Position 3 has a much lower wind speed, mainly due to the fence that the wind hits before flowing on the roof. This fence, creates low instabilities with which the efficiency becomes 0.13 and the power 25 W. Considering that Position 3 would be the overall most visible spot, UiS must take into consideration these results. Position 2 might be considered as a good trade-off between to fulfil the objectives of the project.

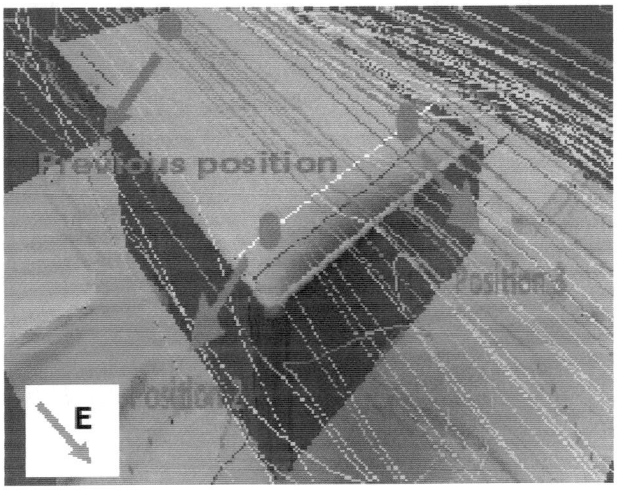

Figure 13. Position 2 and Position 3 locations

5. Conclusions

The aim of the study has been to create a flexible OpenFoam set up for investigation of wind flow over the campus buildings together with vertical axis wind turbine performance modeling. It should be possible to import different terrain model and campus model as well as different wind turbine representations based on the Actuator Line Model.

Built-environment wind turbines are growing in popularity. Nevertheless, examples of the implementation of this model for such type of turbines cannot be found. Therefore, starting from an OpenFOAM case involving laboratory setup, a new version of the model has been implemented. RANS k - ε has been chosen as turbulence model. Parameter studies were performed for TSR, wind velocity, turbine location and wind direction. Results appear to be coherent with the expectations. Power and power coefficient follow the theoretical outputs. As expected the model showed significant differences in power output depending on location and wind direction, which demonstrate the need for a thorough analysis before turbine installation.

Acknowledgement

This research is based on a master thesis project involving the first author [5]. This study has been partially supported by Erasmus program funds.

References

[1] Government.no 2016 Renewable Energy https://www.regjeringen.no/en/topics/energy/renewableenergy/id2000124/
[2] International Energy Agency 2016 Norway statistics https://www.iea.org/statistics/statisticssearch/report/?country=NORWAY=&product=balances
[3] EWEA Business Intelligence 2015 *Aiming high: rewarding ambition in wind energy* (Brussels: The European Wind Energy Association)
[4] Norwegian Wind Energy Association 2016 Wind power news http://www.vindkraftnytt.no/
[5] Riva L 2017 *Implementation and application of the actuator line Model by OpenFOAM for a vertical axis wind turbine set on a University of Stavanger building roof* (Milan: Polytechnic of Milan)
[6] Sørensen J N and Shen W Z 2002 Numerical modelling of wind turbine wakes *J. of Fluids Eng.* **124(2)** 393-399
[7] Shamsoddin S and Porté-Agel F 2014 Large eddy simulation of vertical axis wind turbine wakes *Energies* **7(2)** 890-912
[8] Schito P, Bernini L and Zasso A 2015 OpenFOAM for wind energy: wind turbine as a source term. Workshop HPC: methods for engineering. https://hpc-forge.cineca.it/files/CoursesDev/public/2015/Workshop_HPC_Methods_for_Engineering/OpenFOAM_WindEnergy_schito.pdf
[9] Bachant P, Goude A and Wosnik M 2016 Actuator line modeling of vertical-axis turbines *Wind Energy Rec.* **00** 1–17
[10] Bachant P 2016 UNH-RVAT-3D-OpenFOAM https://github.com/petebachant/UNH-RVAT-3D-OpenFOAM
[11] Bachant P,Goude A and Wosnik M 2016 turbinesFOAM https://github.com/turbinesFoam/turbinesFoam
[12] Nuland S E 2017 *Wind Turbine Installation on offshore fish farm feed barge* (Stavanger: University of Stavanger)
[13] Richards P J and Norris S E 2011 Appropriate boundary conditions for computational wind engineering models revisited *J. of Wind Eng. and Ind. Aerodynamics* **99** 257-266
[14] Hargreaves D M and Wright N G 2007 On the use of the k-epsilon model in commercial CFD software to model the neutral atmospheric boundary layer *J. of Wind Eng. and Ind. Aerodynamics* **95** 355-369

[15] Sheng W, Galbraith R A and Coton F N 2008 A modified dynamic stall model for low Mach numbers *J. of Solar Energy Eng.* **130(3)** 3-13.

[16] Bachant P and Wosnik M 2016 Effects of Reynolds number on the energy conversion and near-wake dynamics of a high solidity vertical-axis cross-flow turbine *Energies* **9(2)** 73

[17] Sheldahl R and Klimas P 1981 *Aerodynamic characteristics of seven symmetrical airfoil sections through 180-degrees angle of attack for use in aerodynamic analysis of vertical axis wind turbines* (Albuquerque: Sandia National Laboratories)

[18] CFD Direct 2016 The Architects of OpenFOAM. OpenFOAM USer Guide. https://cfd.direct/openfoam/user-guide/

[19] Baldassarra G 2015 *Multidimensional CFD Simulation of a H-type Darrieus Vertical-Axis Wind Turbine with OpenFOAM* (Milan: Polytechnic of Milan)

[20] Fields J, Oteri F, Preus R and Baring-Gould I 2016 *Deployment of Wind Turbines in the Built Environment: Risks, Lessons, and Recommended Practices* NREL (Golden: National Renewable Energy Laboratory).

A high-fidelity weather time series generator using the Markov Chain process on a piecewise level

K Hersvik and O-E V Endrerud

Shoreline AS, Stavanger, Norway

E-mail: hersvik@shoreline.no / kjetil@hersvik.net

Abstract. A method is developed for generating a set of unique weather time-series based on an existing weather series. The method allows statistically valid weather variations to take place within repeated simulations of offshore operations. The numerous generated time series need to share the same statistical qualities as the original time series. Statistical qualities here refer mainly to the distribution of weather windows available for work, including durations and frequencies of such weather windows, and seasonal characteristics. The method is based on the Markov chain process. The core new development lies in how the Markov Process is used, specifically by joining small pieces of random length time series together rather than joining individual weather states, each from a single time step, which is a common solution found in the literature. This new Markov model shows favorable characteristics with respect to the requirements set forth and all aspects of the validation performed.

1. Introduction

1.1. Problem description

Weather conditions are a key limiting factor when it comes to offshore operations. For a specific work task to be carried out e.g. in an offshore wind farm, a certain weather window is needed. The weather window is described by a set of operational weather criteria over a certain time span, also known as the window length. The weather criteria in this study include maximum mean wind speed and maximum significant wave height. Detailed analysis of weather window lengths has high economical value. A study of the impact of uncertainty in repair time for offshore wind farm maintenance planning [1] confirms the importance of studying weather uncertainty.

To find suitable weather windows for a certain offshore operation we can use weather forecasts, however forecasts are only applicable when doing short term planning. When doing long term planning, e.g. for wind farm installation or lifespan maintenance operations, weather forecasts are not available. Instead, historical weather data for the relevant site need to be used, and predictions need to be made based on a large set of possible weather scenarios.

1.2. Weather data time series

The historical weather data is stored as a time series. A weather time series consists of measurements at each time step throughout a selected period of time. The measurements at each evenly spaced time step record a predefined set of weather parameters. The measurements can contain observations such as atmospheric pressure, ocean current speeds, visibility, etc., however the type of time series being processed in this study contains observations of the mean wind speed and the significant wave height.

1.3. Aims

As described above, to make the best possible predictions for the future in a statistical way, a somewhat large and appropriate variety of possible weather outcomes is needed. The historical weather data time series is an excellent starting point to use for simulation purposes, however it is desirable to have an endless set of unique, realistic weather data time series for repeated simulations. The generated weather data time series must on average represent the historical weather data well. This has in part been achieved previously with a standard Markov chain model [2]. One major challenge with the standard Markov chain model is related to the discretization of weather parameters into weather state categories, as explained more in section 3.3 "Classification of weather states" below.

The loss of details when categorizing weather parameters into a set of weather states can be made non-relevant if the weather state transition thresholds are aligned with the weather criteria limits. For example, if the significant wave height Hs for a certain operation needs to be below Hs = 2 m, then as long as the weather states distinguish clearly when transitioning across Hs = 2 m, then it becomes non-relevant for that specific operation what the exact Hs value is. However, it is not always easy to tailor the categorization threshold levels. For instance, when analyzing composite offshore operations, many different weather criteria are involved, such as access criteria for different types of vessels, lifting operations, etc. Thus, it can become challenging to tailor the categorization limits according to all the different weather criteria involved in a specific study. Increasing the detail level by using a very fine-grained categorization of weather states is not necessarily a good idea either. It can result in too many weather states, causing too few transition options between each of the weather states.

The generated time series is traditionally limited in detail because of the categorization of weather parameters into weather state categories. The aim of the Markov model presented in this paper is to conserve output details that are normally lost within each weather state category. Although building on the favorable computational strategy from [3], as described in section 3.2 "The proposed Markov model", improvements were also sought with respect to the computational efficiency of the Markov chain implementation.

2. Markov Chain theory

The Markov chain concept is fully relevant for the Markov model in this paper. Markov chains are stochastic chains which fulfill the Markov property. The term stochastic refers to the probabilistic or random nature of a process. The Markov property defines that the probability of a certain state appearing as the next element in the chain is dependent only on the current state in the chain [3]. In other words, when selecting the next state in the chain, only the current state is needed to determine the probability distribution for selecting the next state, and the elements before the most recent element in the chain are ignored. Note that the Markov property does not imply that the coming element in the chain is uncorrelated with respect to the elements that are further away than the nearest neighbor. The probability of a certain state appearing two steps down the road can be calculated using the probability distribution for the first next state combined with the probability distributions for the second next state which are conditional on the first next state.

How does the Markov property fit with the physical real world weather system? The real-world weather system is known as a chaotic system. Chaotic systems are deterministically defined given an exact starting condition for the system, despite their appearance as random. However, the exact condition of the global weather system is far out of reach of modelling, and the available data for a certain analysis will certainly be a far greater limitation. Given the limited data available for each time step in the historical data, it can be tempting to make that next weather state dependent on more than just the current weather state when defining the model to be used. In the end, nonetheless, it is the statistical properties of the output which matter, and using the principles from a Markov chain model proves efficient and suitable.

3. Method

3.1. Input data – Historical weather series

As described in the introduction, the historical weather data serves as input and consists of weather measurements for each time step over several years. The time step Δt is typically 1 or 3 hours, and is the average measured quantity over that time step. The weather measurements always include significant wave height Hs and mean wind speed U, and more weather parameters can be included in the algorithm.

3.2. The proposed Markov model

The Markov model algorithm developed consists of two main parts. The first half parses and analyzes the historical weather data. The last part combines appropriate weather sequences stochastically into a whole new time series, which means using weather transition probabilities determined in the first part of the algorithm.

Again, the first part of the algorithm parses through the historical weather file. It categorizes the weather state at every time step. This is done by classifying the wave height into a category defined by a maximum and minimum value of significant wave height. Equivalently, the mean wind speed is classified into a category of wind speed magnitude. The two categories together define the weather state. More details are described in the "Classification of weather states" section below.

When using the traditional state-by-state Markov process to generate new time series as described in [3], it is common to describe the process by transition probabilities. These transition probabilities can be calculated while parsing the historical time series file, and they can be stored in a large matrix. The matrix will be n by n large, where n is the number of unique weather states available in the mentioned weather state classification. For any reasonable weather state resolution, that matrix will easily be too large to store in computer memory. Adding a few gigabytes of memory will not help much since the memory requirement is proportional to n^2. It is a highly sparse matrix, meaning that the matrix contains for the most part elements with value zero. On the other hand, storing the entire historical time series in memory is far more manageable. It simply means storing a copy of the entire weather data file in memory.

So even though describing the process with transition probabilities is correct and useful for understanding the process, the entire calculation of the transition probabilities can be omitted. Instead, the historical time series can be searched in a strictly random fashion with the right constraints, and the probability of the next weather state will be correctly represented. The process takes place in the last part of the algorithm. This computational strategy was preferred in [3] and is even more suitable with the new piecewise level of using the Markov process which is described in the last part of the algorithm.

The last part of the algorithm starts out making a new time series by selecting a random year from the historical time series and a random start date which is close to the desired start date of the generated time series. A small piece of the historical time series is then copied from the selected starting point and used as the first part of the generated time series. The small piece copied from the historical time series is set to have a random length, uniformly distributed with minimum and maximum length defined to be 6 - 48 time steps, which is conversely 6 - 48 hours in most cases.

The exact choice of 6 – 48 time steps as the building block length is a design choice, which was tried out and found to balance characteristics of the output in a good way. The balance is mainly considering variability in the output vs. keeping imperfections of joining blocks together from affecting the natural weather window distribution. The end validation is needed to confirm if both these considerations are properly achieved. Smaller and larger blocks are assumed to also work well and produce similar quality output, and keeping the block sizes a random length reduces the possibility for systematic impurities. The block sizes do affect the performance of the algorithm slightly since larger block sizes result in fewer search and join operations, however performance does not influence the choice of block sizes for the current application.

Next, a new sample from the historical time series is drawn at random with two restrictions. The first restriction ensures that the new sample starts out with a weather state which is selected from one of the possible state transitions, according to the historical data set. This is ensured by selecting one of the existing weather transitions from the historical file directly. It is a critical criterion for maintaining realistic weather development in the generated time series. It is equivalent to the Markov Chain process, in which the next state is drawn by transition probabilities which depend on the current state. Specifically, the historical weather file is traversed in a random order, looking for an historical point in time where the weather state is identical to the last weather state of the time series being generated. The weather state immediately after the first random matching historical state is selected as the next random weather state. However, for the piecewise process level in this algorithm, the next 6 or up to 48 data points are selected together to become the next piece of the new time series.

As the first search restriction ensured that the new piece for the generated time series matches the previous piece, the second restriction ensures seasonal correctness in the new time series. The second restriction ensures that the new piece originates from the same calendar month as the calendar month it is representing in the generated series, although it can originate from any year.

If no matching weather state is found within the historical data for the specific calendar month, a transition is selected without restriction on calendar month. This can be relevant in transitions from one month to another, in combination with a rare weather state. In that case, there will be at least one matching transition in the historical time development, unless of course the sampling reached the very end of the entire historical file. If no transition is found at all, i.e. at the very end of the imported data, then a completely new state is drawn corresponding to the current season.

A new sample from the historical time series is repeatedly drawn in the same way, based on the preceding weather state, and added to the output time series.

3.3. Classification of weather states

The needed classification of weather states represents a discretization of weather parameters. The representation of each weather parameter becomes constricted to the set of available categories. An important reason for not using a too fine-grained classification of weather states is to avoid significant quantities of deterministic transitions. Deterministic transitions occur when a certain state only has one possible state transition to continue from, namely the next state in the historical data.

For the current Markov model implementation, the weather state categories are defined by the following. The step size for wave height categories ΔHs is $\Delta \mathrm{H_s} = 0.1$ m while the wind speed category step size ΔU is $\Delta \mathrm{U} = 1.0 \frac{\mathrm{m}}{\mathrm{s}}$.

The category steps are widened at the more extreme and rare ends of the spectra. The classification of weather states is both a necessity to solve the task and, at the same time, it cannot be very fine-grained in order to keep the random nature of the process, as already discussed. The main novelty of the developed piecewise approach to using the Markov chain process is to produce time series which display weather details beyond the weather state classification. This can be explained by giving an example from the traditional Markov chain model. In the traditional model, the resolution of the weather measurements is limited by the category step size in the weather state classification. For the current setup, that would result in wind speed values stepping 1 m/s at a time, i.e. no intermediate values.

The core idea in the new development is to process small pieces of the time series as one unit, i.e. batches of a few time steps, with all uncategorized details preserved. This contrasts with traditionally processing individual data points from single time steps which are limited to representing only a general weather state category.

4. Validation analysis

The aim of the validation analysis is to determine both the degree of diversity among the generated weather series and how closely the group averages of key time series properties resemble the historical weather.

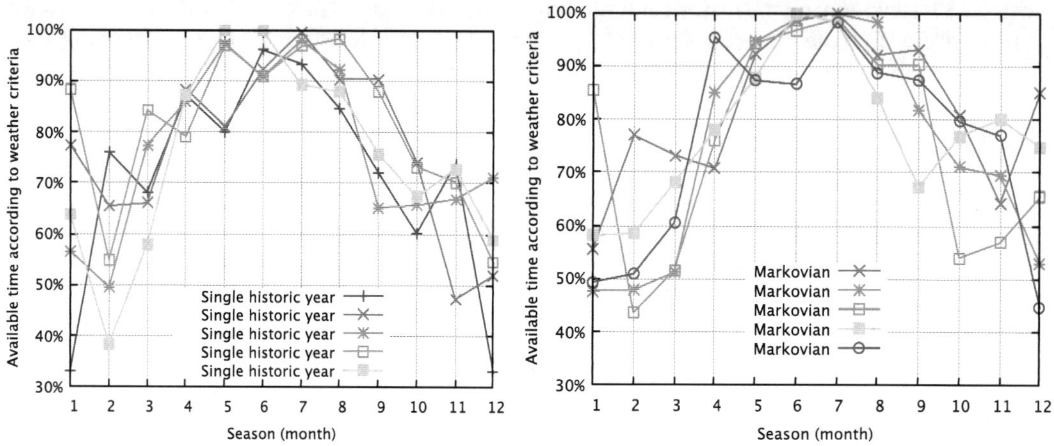

Figure 1. Seasonal available time for work for 5 single-year time series, historic data in left view and Markov model output in right view, given weather criteria: $H_s < 1.5$ (m) ∩ $U < 12$ (m/s).

The key properties relate to weather windows which are time slots available for work. These weather windows are all dependent on the weather criteria, which dictate at what times the certain work tasks can be performed or not, offshore. The criteria consist of maximum allowed significant wave height $H_{s,MAX}$ and maximum allowed mean wind speed U_{MAX} for any specific work task to be possible.

The weather time series used for validation and presentation of results is an anonymous 21-year length weather data time series with time step $\Delta t = 1$ hour, from a relevant site. The characteristics of the data series are illustrated graphically in the figures, where sets of generated time series are compared to the original data series. The algorithm has not in any way been adjusted according to the result of the specific validation performed here. Thus, any weather data time series can be used to reproduce the validation of the algorithm. The key to validating the algorithm is the comparison between generated and original time series, which is the purpose of all the figures.

A typical weather criteria (U_{MAX} and $H_{s,MAX}$) is selected for the validation analysis presented here, and is displayed in the caption of each graph. The most common weather requirements are related to transferring personnel onto the wind turbines using for instance crew transfer vessels. The weather criteria chosen in this validation analysis is a typical criterion, but not related to a specific vessel or task. The significant wave height is typically the limiting factor for traveling offshore to work on wind farms, and most working limits are set between 1 and 2 m significant wave height (H_s).

Regardless of these details, the success criterion of the algorithm is how well all the generated time series compare to the actual historical time series. That is why each of the following analysis graphs compare the generated time series directly to the original historical time series. The comparison is done for a chosen set of key properties.

Key properties of the time series include

- the proportion of time available for work,
- the distribution of weather windows according to window length metric,
- and the averages of these values for each season of the year as measured per calendar month (Figure 3).

Together, these properties give a well-defined picture of the time series on a group level.

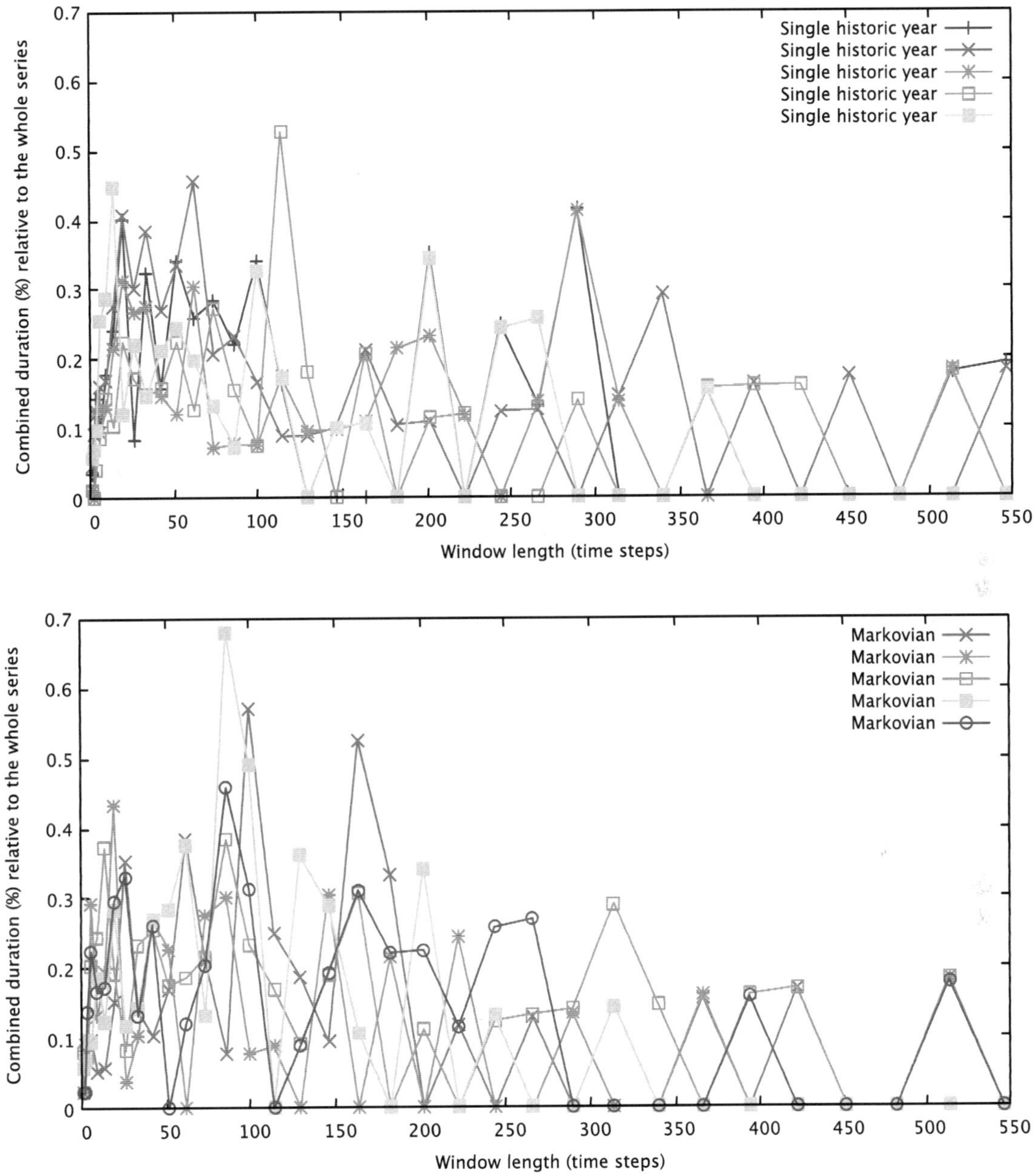

Figure 2. Weather window distribution for individual 1-year series, displaying combined (sum of) weather window durations (as percentage of entire weather series) grouped and arranged as a function of individual weather window lengths, showing historical data in the top view and Markov model data in the bottom view, given weather criteria: $H_s < 1.5$ (m) ∩ $U < 12$ (m/s).

Single-year time series are presented in Figure 1 and Figure 2. Single-year time series offer easy comparison between generated and historical time series because the multi-year historical time series can be broken down into several single-year time series. That way the variation between different historical time series can be compared to the variation within generated time series. Each of Figure 1 and Figure 2 consists of two separate graph displays. They contain results for historical time series and equivalent analysis results for the generated time series, respectively.

Figure 1 displays time available for work in sum for each season of the year. The left display contains results of 5 single-year historical time series, whereas the right display is based on an equivalent set of generated time series.

Figure 2 is based on the same set of single-year time series as in Figure 1. However, Figure 2 shows the combined window duration (sum of window lengths available for work) of all weather windows grouped and arranged by duration of the individual weather windows. Each time series is displayed with a separate graph line and legend indicator. The horizontal axis marks the size of the individual weather windows within the group, at each data point.

From the graphical comparison in Figure 1 and Figure 2, variations within historical and within modelled time series appear similar, which is a highly desired result.

Figure 3 compares average data from 100 generated time series to the historical data. The historical data and each of the generated time series consist of 21 years of weather data. The data has an hourly time resolution, which means a time step value $\Delta t = 1$ h.

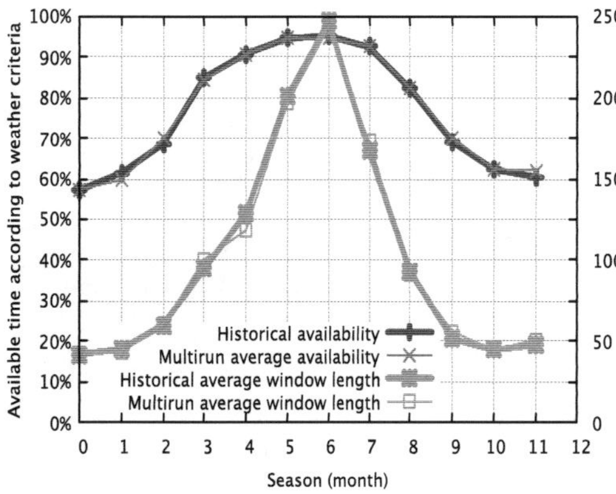

Figure 3. Seasonal average available time for work and average window length – historical time series compared to average of 100 runs, given weather criteria: $H_s < 1.5$ (m) ∩ $U < 12$ (m/s).

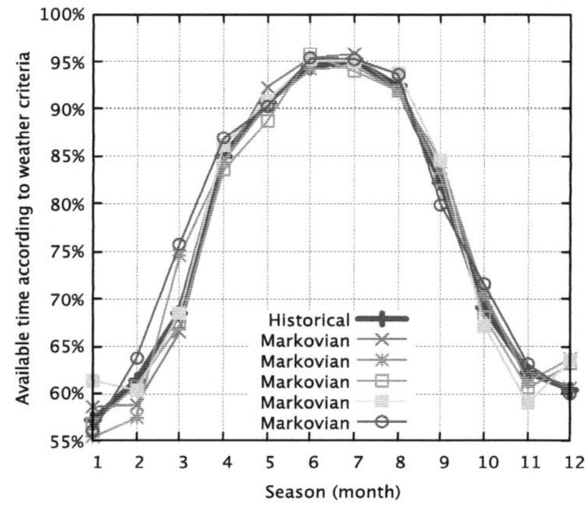

Figure 4. Seasonal average available time for work, comparing five 21 year runs to the equally long historical time series, given weather criteria: $H_s < 1.5$ (m)∩$U < 12$(m/s).

The modelled time series do not distinguish daytime and nighttime. As mentioned in [3], some wind and sea state series have daily components, however these potential aspects are not considered relevant for the current applications of this Markov weather model, namely applications which are simulation of offshore wind farm operations.

Validation analysis of the algorithm shows that in each generated time series, approximately 99 % of the stochastic time series segments (6-48 entries in duration) were joined differently compared to in the historical time series. This applies to the 21-year data and for the granularity of state classification specified above, which is a 0.1 m resolution of wave height and 1 m/s resolution of wind speed. It proves a very good random variability in the output.

Figure 4 shows available time each season according to the weather criteria for a selection of five full 21-year output series. The historical reference values included are identical to the reference data in Figure 3.

Figure 5 shows a distribution of the available time according to individual window length. The durations of similar length weather windows are combined and distributed on the x-axis in the same way as Figure 2.

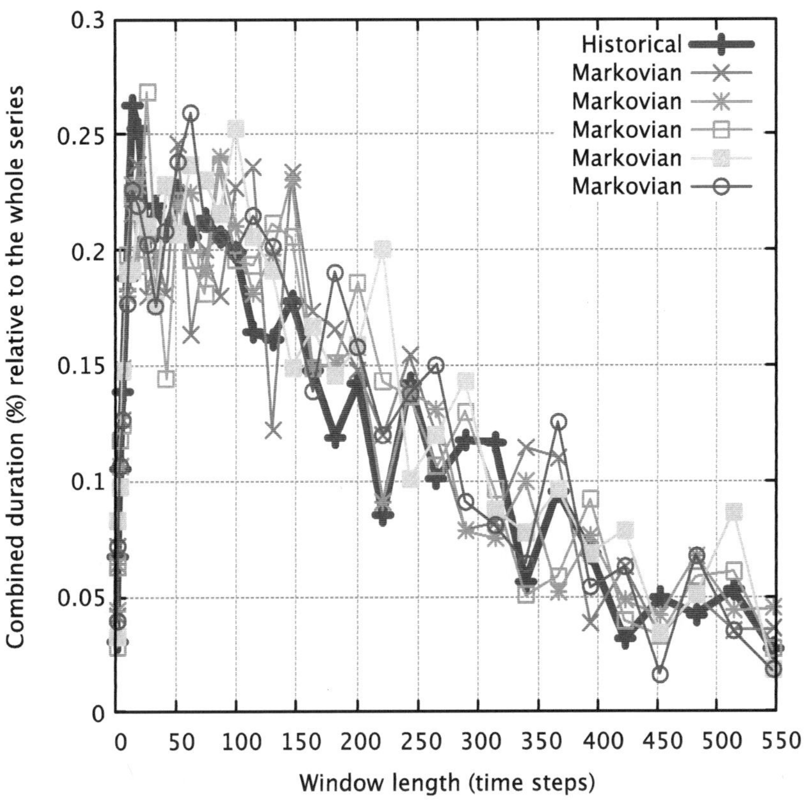

Figure 5. Combined duration (sum) of weather windows from six 21-year series distributed according to individual window length, given weather criteria: $H_s < 1.5$ (m) ∩ $U < 12$ (m/s).

The area under the curve gives a measure of time of the whole series covered by each interval of window lengths. Again, five full 21-year modelled time series are plotted together with the direct representation of the historical data. The combined view of six time 21-year series allows an impression of the variability between runs.

5. Conclusion

The improved Markov Chain model is developed for generating unique weather time-series based on an existing weather series. Despite the undesirable but necessary discretization of weather data into weather states, the method successfully recreates high detail levels in the generated time series. This is achieved by applying the Markov process using very small segments of the time series as building blocks instead of using single data points. That way the natural development within a single weather state category is recreated in the generated time series. The validation performed on the developed Markov model also shows excellent results.

In summary, all aspects of the generated time series are quite indistinguishable from original ones, both in terms of degree of variability between individual time series, as well as group averages. Even when graphing out actual parts of a time series, which is omitted here, it is quite non-trivial to distinguish a modelled time series from an original time series. This is in strong contrast to the traditional Markov chain output, which coarsely outputs average of each weather state. For example, the traditional Markov chain output would in an equivalent setup produce wind speed values with steps of 1 m/s and no intermediate wind speed values.

For the purpose of weather windows available for work, historical and generated time series are seen as interchangeable, based on any of the validation findings known. That satisfies the primary aim of the algorithm.

The developed Markov model has been used with many historical weather files and in countless simulations of offshore wind farm operations during the last year. The computational performance is also improved by more than tenfold, as well as the complexity of the algorithm being simplified to a good extent, especially since no post-processing to smooth the output is needed. From all the use cases known, no issues have been identified so far which can question the validity of the algorithm.

References

[1] Seyr H and Muskulus M 2016 *J. Phys.: Conf. Ser.* 753 09200.
[2] Scheu M, Matha D and Muskulus M 2012 *Proceedings of the 22th International Offshore and Polar Engineering Conference*, pp. 463-8.
[3] Hagen B, Simonsen I, Hofmann M and Muskulus M 2013 *Energy Procedia*, **35**, pp. 137-147.
[4] Monbet V, Alliot P and Prevosto M 2007 *Probab Eng Mech* **22** (2), pp. 113-126.

IOP Conf. Series: Materials Science and Engineering **276** (2017) 012004 doi:10.1088/1757-899X/276/1/012004

3D WindScanner lidar measurements of wind and turbulence around wind turbines, buildings and bridges

T Mikkelsen, M Sjöholm, N Angelou and J Mann

Dep. of Wind Energy; Technical University of Denmark, DTU Campus Risø,Roskilde, Denmark

E-mail: tomi@dtu.dk

Abstract. WindScanner is a distributed research infrastructure developed at DTU with the participation of a number of European countries. The research infrastructure consists of a mobile technically advanced facility for remote measurement of wind and turbulence in 3D. The WindScanners provide coordinated measurements of the entire wind and turbulence fields, of all three wind components scanned in 3D space. Although primarily developed for research related to on- and offshore wind turbines and wind farms, the facility is also well suited for scanning turbulent wind fields around buildings, bridges, aviation structures and of flow in urban environments. The mobile WindScanner facility enables 3D scanning of wind and turbulence fields in full scale within the atmospheric boundary layer at ranges from 10 meters to 5 (10) kilometers. Measurements of turbulent coherent structures are applied for investigation of flow pattern and dynamical loads from turbines, building structures and bridges and in relation to optimization of the location of, for example, wind farms and suspension bridges. This paper presents our achievements to date and reviews briefly the state-of-the-art of the WindScanner measurement technology with examples of uses for wind engineering applications.

1. Introduction

WindScanner is a DTU-lead European Research Infrastructure designed for 3D wind field measurements based on mobile deployable, space-scanning coordinated and time synchronized wind lidars (Figure 1).

The construction of Scanning lidars was motivated by the steadily increasing size of modern wind turbines which are today soaring hundreds of meters into the sky. Their blades sweep through areas bigger than several football fields.

Obviously, the wind field in the rotor plane can no longer be characterized from a single-point measurement but the characterization requires detailed knowledge of the dynamics of the entire 3D wind fields, over the entire rotor plane upwind and in the turbine wakes.

The vision of WindScanner is primarily to develop, establish and operate a mobile joint European distributed Research Infrastructure (RI) for experimental research in wind and turbulence for wind energy. The scope and main focus of the project is primarily the wind energy sector, however there are, as will be described below, several other uses and applications for the WindScanner wind field measurement methodology, including aeronautics, various atmospheric studies, construction and performance of buildings, bridges, street canyon flow etc.

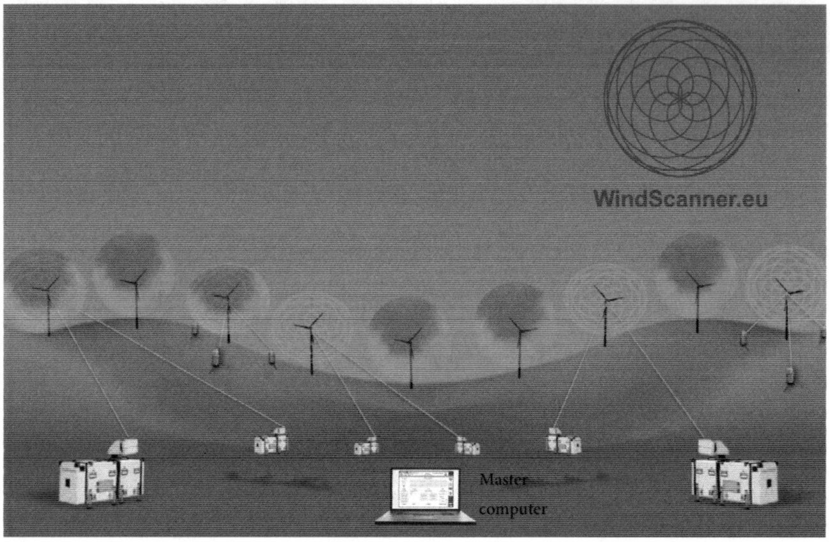

Figure 1. Inflow and wakes around turbines exposed to wind and turbulence in complex terrain are scanned in 3D space and time from WindScanners.

WindScanner.eu was admitted to the EU ESFRI Roadmap in 2010 and since then possible synergies of an enhanced collaboration with similar European RIs in other fields have been promoted.

The mission of fostering a joint European WindScanner project is: (1) Planning and coordination of WindScanner measurement activities, (2) Coordination of purchasing, commissioning, calibration and maintenance of WindScanner equipment, (3) Training and education in WindScanner operation and data analysis, (4) Development of e-science facilities for campaign design, documentation, data management and data analysis, (5) Providing access to the research data, (6) Dissemination of WindScanner opportunities and achievements to foster optimal stakeholder collaboration.

Traditionally, wind speed measurements are made using mechanical devices such as cup or sonic anemometers and wind vanes that need to be mounted at the points of interest, usually using a meteorological mast.

However, WindScanner systems are composed of several ground-based multiple trajectory-scanning and time synchronized Lidars, being remote-sensors using laser light to measure wind speed. The new technology has been successfully tested in Europe and in North America across many different topographies including hot and cold regions, flat and complex terrain sites and on- and offshore.

Today (2017) the original vision, mission, scope and identity of the WindScanner wind and turbulence measurement system based on remote sensing methodologies has been fully defined. During the last decade, the technical and scientific goals for establishing WindScanner as a mobile and joint European research infrastructure project have been achieved.

The WindScanner RI consortium consists of 10 Research Institutions from 7 member states: DTU (Denmark), IPU (Denmark), CENER (Spain), CRES (Greece), Fraunhofer IWES (Germany), ForWind (Germany), ECN (Netherlands), SINTEF (Norway), LNEG (Portugal), University of Porto (Portugal). Today, bound by a common Memorandum of Understanding (MoU) signed in Sep 2015 the aim during the on-going RIs construction phase is now to build-up national WindScanner nodes and to provide infrastructure access to users from industry and the research community.

IOP Conf. Series: Materials Science and Engineering **276** (2017) 012004 doi:10.1088/1757-899X/276/1/012004

The project involves the most significant principal RIs in Europe in the field of wind energy research and WindScanner is therefore considered today by ESFRI to provide a consolidated scientific leadership on the wind field research matter in EU.

Today, the WindScanner research infrastructure offer a new open-access and joint European research infrastructure for promoting research and innovation in atmospheric wind and turbulence via full-scale open air 3-dimensional wind field experimental investigations. Real-time measurements of the 3D wind velocity and turbulence vectors in the atmospheric boundary-layer are obtained using advanced remote sensing-based wind measurement techniques known as wind lidars.

Distributed European WindScanner nodes are being established as national/regional competence centers engaged with a planned central hub located at DTU Wind Energy regarding the continued operation, maintenance and further development, coordination of applications and training of experts to operate the WindScanner wind measurement technology.

WindScanner.eu can lead and disseminate coordinated experimental research for large European-level wind energy measurement campaigns throughout Europe.

At DTU Wind Energy in Denmark, we plan to lead and host the European central hub for coordinating access, planning operation, training and maintenance and also we disseminate detailed WindScanner instrument manufacturing plans, to assist partners to build the mobile WindScanner instruments locally.

The DTU hub will also maintain the scanners and version control software and data processing algorithms to secure data management and manage host servers, and train experts and users from research community and industry.

WindScanner is an open-access experimental facility that also serves atmospheric boundary-layer research on- and offshore, air safety, wind load measurements on buildings and bridges, 3D wind circulation field measurements in street canyons and in the urban environment, etc.

The paper addresses and reviews recent WindScanner based experimental research activities and present several WindScanner experimental setups including 3D wind field measurements from recent WindScanner measurement campaigns.

Additional information can be found on our home pages were also open-access publications and detailed descriptions can be found, cf. `www.WindScanner.dk` and `www.Windscanner.eu`.

2. The 3D scanning windscanner system

The WindScanners were originally designed for 3D wind field and turbulence measurements of the atmospheric airflow around wind turbines. To date, DTU has designed and built two different sets of multiple trajectory-coordinated synchronized WindScanners, i.e a short-range WindScanner system, consisting of three synchronized continuous-wave scanning Doppler lidars, and also a long-range WindScanner system consisting of multiple trajectory-coordinated scanning pulsed Doppler lidars.

The objectives and technical functionality specifications including technical details of the 3D wind field WindScanner facility, both the short-range and the long-range WindScanners are available in [1, 2, 3, 4].

A WindScanner system consists of two or more spatially separated scanners (short- or long-range WindScanners) that are controlled by a central master computer, cf. Figure 1.

The first generations of the short and long-range WindScanners were built at DTU Risø campus in 2007 from modified prototypes of commercially available vertical profiling ZephIR lidars and WindCube 200S lidars, respectively.

During the Danish national infrastructure development project 2009-2013 WindScanner.dk DTU with the support from partners QinetiQ (UK), later ZephIRLidars.com and Leosphere (France) developed steerable scan heads, enabling the lidars to become steerable and synchronized scanning lidars.

In addition, DTU Wind Energy developed specific data acquisition and control software for both systems [5]. Each WindScanner measure the instantaneous line-of-sight wind components along a user-defined steerable scanning trajectory.

The 3D short-range WindScanner systems are built from three continuous-wave lidars and the measurements of the line-of-sight wind components are synchronized via a master computer connected to the WindScanners via an optic fiber network. They generate high-frequency sampling of the line-of-sight wind speed (up to 400 Hz) from spectral Doppler shift measurements scanned in the atmospheric flow with variable probe length, in the range between 0.25 m at the shortest 10 m measurement range and with up to 30 m resolution at their longest measurement range (300 m).

The long-range WindScanners are beam-steered pulsed lidars and have larger but fixed probe length (minimum 30 m) [6]. Long-range WindScanners measurement frequency is typically 1 Hz. However, they can retrieve line-of-sight measurements from a large number of ranges along their line-of-sight path. The long-range WindScanner systems are synchronized with the master computer using 3G network [5].

The maximum range of the biggest, a WindCube 400S based long-range WindScanner is at present about 8 km as observed during the RUNE 2016 offshore experiment [7].

The long-range WindScanner systems are primarily intended for measurements of 2D mean flow fields over a larger area while the short-range WindScanners can measure 3D wind fields with high spatial and temporal resolution in smaller probe volumes, hence enabling scanning of atmospheric small-scale 3D turbulence structures including coherent structures.

3. 2D Windscanner measurements performed on wind turbines, buildings and bridges

Adhering to the primary objectives of the WindScanners, a detailed 3D wind field inflow study on a wind turbine was performed at DTU Risø campus in 2015, focusing on the 3D inflow velocity fields in front of a Vestas V27 turbine [8].

Furthermore, detailed inflow and wake studies were performed on a horizontal axis wind turbine in the induction zone and in the wake of a NEG 550 kW Nordtank wind turbine, also at DTU Risø Campus [9, 10].

Multiple wind lidars also find applications in civil engineering, where they, for instance, hold a potential for studies of two-point statistics of wind turbulence. The coherence is of particular interest since the spatial correlation of wind gusts are of fundamental importance for estimating the total wind load on large structures such as long-span suspensions bridges and wind turbine rotor blades. In the following, we review applications where WindScanners have been used with focus on bridges, wind breaks, urban canopies and offshore applications.

4. Inflow and wake studies around a suspension bridge (Lysefjordbrua) 2014

Two synchronized short-range WindScanner were installed on the bridge in the Lysefjord near Stavanger in Norway during a one-week intensive measurement campaign in May 2014 with the purpose to measure 2D wind field inflow and wake wind fields and coherence of the along- and across-wind velocity components as illustrated in Figures 2-5.

Wind records obtained by five sonic anemometers mounted on the West side of the bridge were used as reference data. Single- and two-point statistics of wind turbulence were studied, with special emphasis on the root-coherence and the co-coherence of turbulence. A good agreement was observed between data obtained by the sonic anemometers and the lidars.

The bridge study has evaluated the potential of the WindScanner wind lidar technology for full-scale, outdoor monitoring in wind engineering. To date, the short-range WindScanner investigation has, together with the in-situ sonic anemometer measurements on the bridge, resulted in the following publications from the Lysefjordbrua measurement campaign in 2014:

Figure 2. Panel (a): the Lysefjord Bridge is a suspension bridge which was equipped with two short-range WindScanners (two red circles) in May 2014. Panel (b): one of two short-range WindScanners installed on the Lysefjord Bridge providing synchronized scanning of the inflow and wake wind field in a two-dimensional plane at the 55 m bridge deck height.

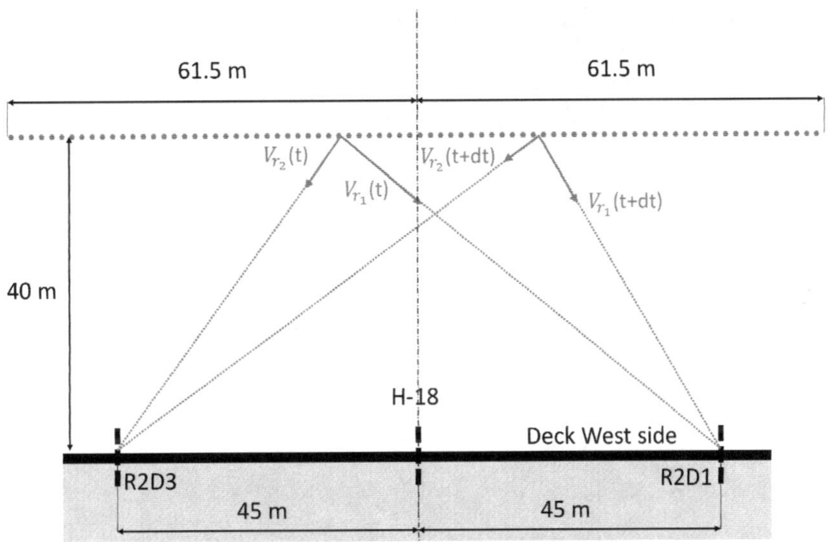

Figure 3. Top view of setup using two Short-range WindScanners at Lysefjordbrua, May 2014.

- Assessing the potential of a commercial pulsed lidar for wind characterisation at a bridge site [11]
- Application of short-range dual-Doppler lidars to evaluate the coherence of turbulence [12].
- Full-scale observation of the wake flow 40 m downstream of a suspension bridge deck [13].

Also the long-range WindScanners provide on- and offshore wind field measurements using synchronized scanning pulsed lidars as seen in Figure 5. Wind fields and wind statistics can be measured in the atmospheric boundary layer wind fields within scanning distances up to 5-8 km, of e.g.: the mean wind velocity, the velocity standard deviation, the turbulence length scales, the wind velocity component spectra and coherence [11, 14].

Recently measurements of mean flow and turbulence were performed over another wide

IOP Conf. Series: Materials Science and Engineering **276** (2017) 012004 doi:10.1088/1757-899X/276/1/012004

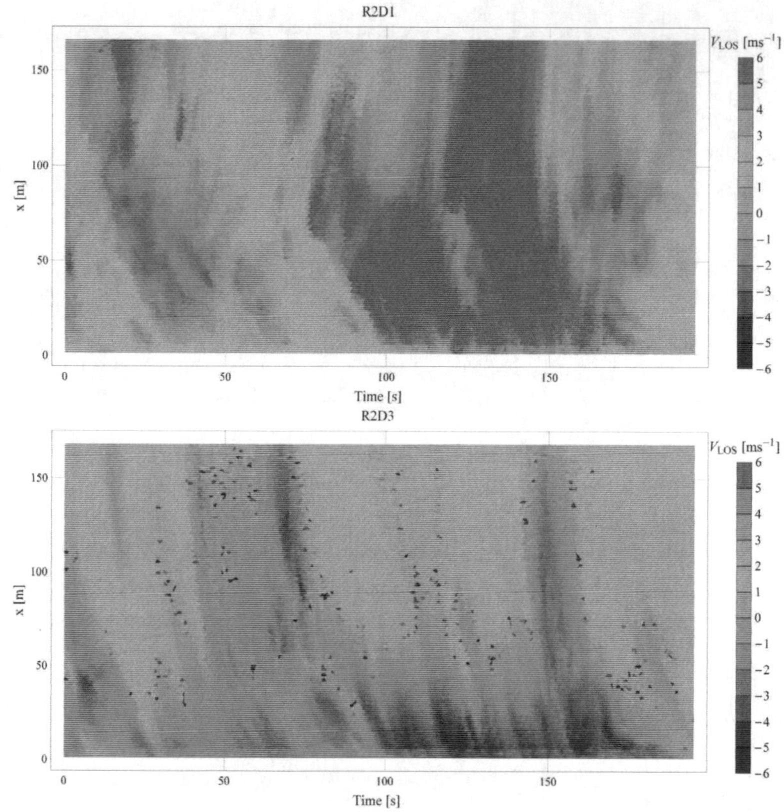

Figure 4. Coherent structures in the line-of-sight wind component measured along an upwind horizontal scan line perpendicular to the bridge by the short-range WindScanner R2D1 (top) and R2D3 (bottom) just after noon on 2014-05-22.

Figure 5. A long-range WindScanner operated by Christian Mikkelsen Research (CMR) during the 2014 Lysefjord bridge measurement campaign.

Norwegian Fjord (Bjørnafjord) using three of DTU Wind Energy developed trajectory-coordinated and synchronized long-range WindScanner lidar systems [14].

5. 3D WindScanner measurements of wakes behind a wind break fence

Short-range WindScanner based 3D wind measurements of the wake region behind a 3 m tall and 30 m long wind break was investigated experimentally in a full-scale experimental study at Risø campus in Denmark in 2015 [15]. The wake measurements were performed with three lidars scanning on a vertical plane downwind of the fence (Figures 6-7). Inflow conditions were monitored by a sonic anemometer installed in a nearby mast. In the free-stream aloft the fence the lidar measurements of the wind speed agreed well with the adjacent sonic anemometer. For near-neutral wind conditions the average inflow conditions were described by a logarithmic vertical mean wind profile.

Figure 6. The fence experimental set-up at DTU Risø campus test station, installed with three short-range WindScanner lidar instruments and a met mast.

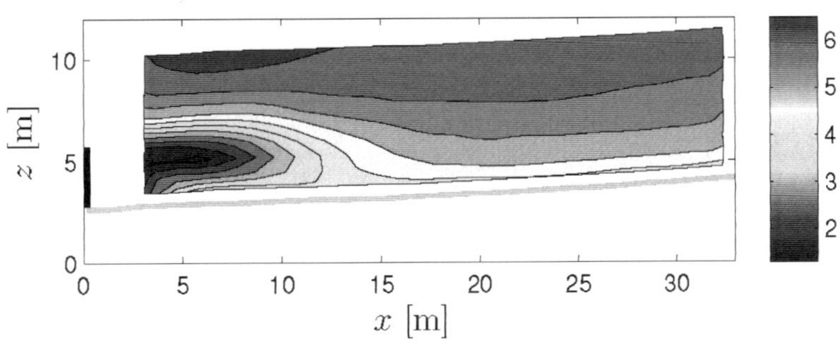

Figure 7. Example of WindScanner mean wind speed measurements of the wake behind the 3 m tall 30 m long solid wind break scanned in a vertical plane at the fence centerline during the period 10 March to 1 April 2015 when the fence was solid. Later, the fence was made porous and scanned again [15]. The grey line indicates the terrain height and the fence are 3 m tall and 30 m long. The measurements were taken in a vertical plane with wind coming almost perpendicular to the fence near the fence centerline.

IOP Conf. Series: Materials Science and Engineering **276** (2017) 012004 doi:10.1088/1757-899X/276/1/012004

Seven measurement cases were defined given the relative wind direction to the fence, the fence porosity, and the inflow conditions. The sheltering effect of the wind break was noticed to be highest below ≈ 1.46 fence heights. The sheltering effect was notably stronger during stable atmospheric conditions. For larger deviation of the mean wind direction from the direction normal to the fence, the effect of the shelter was lower.

For the case with the largest relative directions, no sheltering effect is observed in the far wake (distances $\gtrapprox 6$ fence heights downwind of the fence). Sometimes the wake could be observed at all downwind positions up to 11 fence heights downwind. Below the fence height, the porous fence has a lower impact on the flow close to the fence compared to the solid fence. Velocity profiles in the far wake converge onto each other using the self-preserving forms from two-dimensional wake analysis.

6. TrueWind 3D open air calibration stand

"TrueWind" is a permanent calibration facility using touchless remote sensing methodology to achieve hitherto unprecedented accuracy in wind measurements. To reduce uncertainty of anemometry in the wind energy sector TrueWind is designed to contribute to the advancement of wind metrology engineering, by:

(i) Improved cup anemometry.

(ii) Introduce coherent wind lidar measurement technology to wind tunnel measurement.

(iii) Provide accurate touchless calibrations in wind tunnels.

TrueWind aims to improve significantly the measurement accuracy of commercial cup anemometers used today in wind energy resource assessments, power performance measurements and field calibration of lidars.

Higher accuracy in wind tunnel calibrations has today already resulted from applying the TrueWind touchless remote sensing-based lidar instruments called Lidics (Figure 8). Lidics, similar to the WindScanners but without the scanning parts, measure Doppler shift in the laser radiation backscatter from aerosols via a laser beam stared and focused into a small sounding volume in the wind field.

TrueWind as a remote sensing based wind engineering calibration facility address the objectives delineated in: EU TPWIND - Wind Condition activities addressing advanced measurement

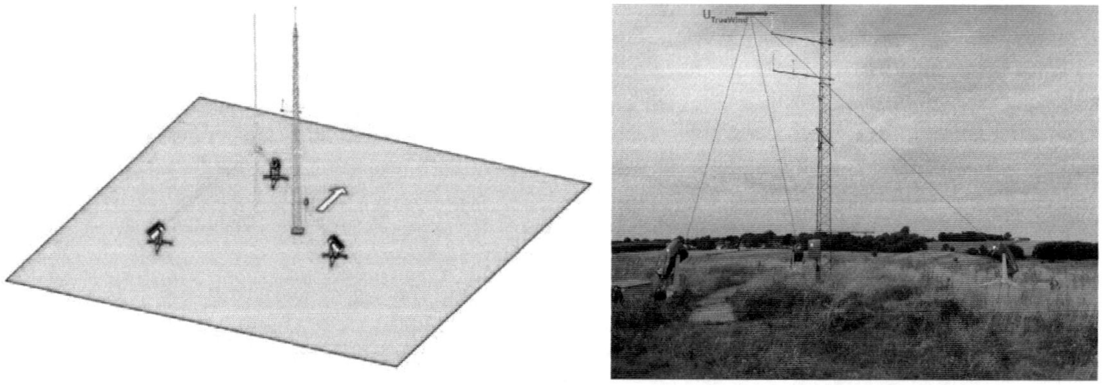

(a) (b)

Figure 8. Panel (a): Permanent TrueWind Calibration Stand at DTU Risø campus Schematic setup. panel (b): TrueWind 3D open air touchless calibrations stand for calibration of cup anemometers and other sensors Oct. 2017.

techniques including remote sensing. (`www.windplatform.eu`).

7. Scanning of wakes from scaled wind turbines in wind tunnel

Two synchronized short-range WindScanners were measuring the wakes behind scaled wind turbines installed in the boundary-layer wind tunnel at Politecnico di Milano (Figure 9) in January 2016 [16].

This research combines wind tunnel experiments with scaled wind turbine models and remote-sensing short-range WindScanner lidar techniques. In this measurement campaign the wind tunnel of the Polytechnic University of Milan was equipped with three wind turbine models and two short-range WindScanner lidars to demonstrate the benefits of lidar in such experimental surroundings.

The WindScanners provided 2D area scans within seconds to minutes, depending on the complexity of the scan pattern, without disturbing the flow. For the initial validation, WindScanner staring mode measurements were compared to hot wire probes commonly used in wind tunnels.

Hub height 2D horizontal area scans as well as wake profiles were measured (Figures 10-11). Compared to hot wire probes the lidars have larger measurement probe volumes and also some loss of measurements due to the moving blades, but the benefits include high flexibility in conducting both point measurements and area scans, in addition to having the benefit of providing touchless undisturbed 2D flow field measurements. The research campaign in 2016 confirmed for the first time a high potential for using short-range WindScanners for mapping of detailed high-resolution flow structures inside a wind tunnel.

The overall objective of applying WindScanners in the wind tunnel was to measure the flow field and, in particular, how the wind turbine wakes interact in case of scaled models in a boundary-layer wind tunnel.

A similar yaw control wake study using a DTU SpinnerLidar has also been investigated in a measurement campaign in 2016- 2017 behind a full-scale V27 wind turbine operated at the Sandia NL test site SWIFT [17].

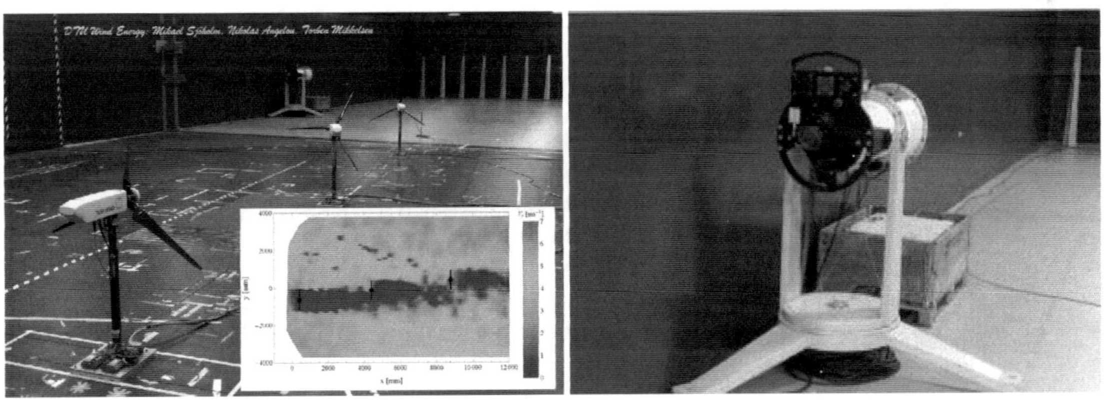

(a) (b)

Figure 9. Panel (a): Wake measurements from three scaled wind turbines by two synchronized WindScanners installed in the wind tunnel at Politecnico di Milano January 2016. The inserted figure shows the mean wind field in a horizontal plane at hub height. The scaled turbines are represented by the black bars. In this figure, the turbines are not yawed. Panel (b): One of the two (R2D3) short-range WindScanners installed in the PoliMi boundary layer wind tunnel.

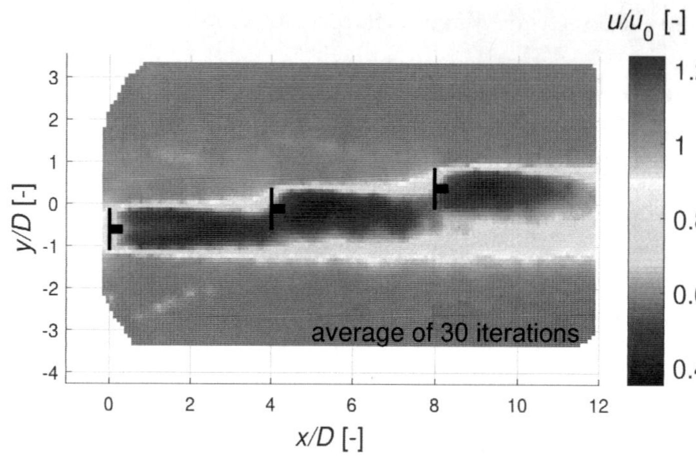

Figure 10. 2D horizontal scan of the mean wind field at the hub height showing the combined wakes from the three scaled turbines aligned without yaw from the mean wind direction.

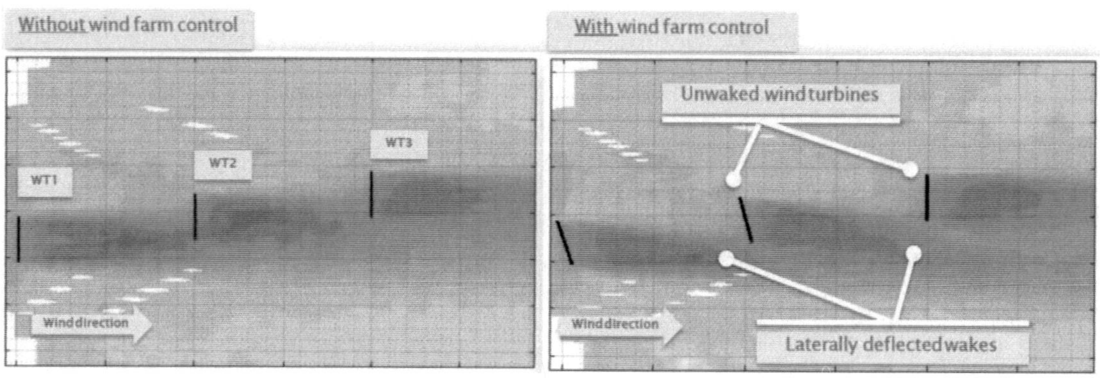

Figure 11. WindScanner measurements of controlled wake steering in the PoliMi Wind tunnel by two WindScanners. The wake field is similar to the setting in Figure 14 above; however the two forefront turbines have on purpose been yawed to steer their wakes around the 2. and 3. turbine in the row [16, 18].

8. Urban street canyon windscanner measurements

For research into detailed wind flow circulation and turbulence coherent structures in urban environment we are also planning to engage the short-range WindScanner measurement methodology with synchronized scanning in an open and full scale urban street canyon study to be conducted in collaboration with the University of Aarhus as part of Danish WindScanner node of the European WindScanner.eu research infrastructure, probably during the year 2019.

From an installation in the ground or aloft in a street canyon the WindScanners can be operated to provide instantaneous velocity measurements of the 3D wind field in a pre-defined volume section of real-life street canyons with high spatial (< 1 m) and temporal (< 1 m) resolution (Figure 12).

Such measurements will enable detailed insight into the microscale flow and transport processes, including turbulence and aerosol concentrations, which then can be visualized, quantified, analyzed and be used for model evaluation as well.

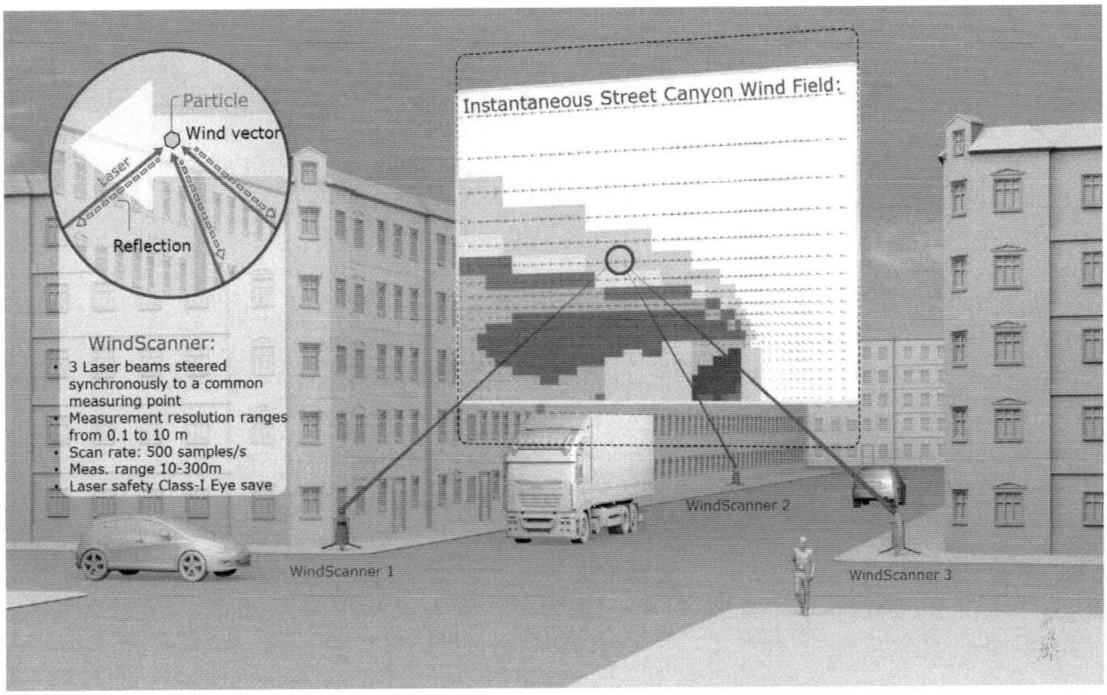

Figure 12. Design of a forthcoming Urban Street Canyon WindScanner experiment in Copenhagen. The figure illustrates how the instantaneous wind field in selected scan planes can be measured, by engaging three synchronized short-range WindScanners.

9. Measurements of inflow and wakes from WindScanners installed on wind turbines.

DTU has in parallel developed a special 2D WindScanner, a so-called SpinnerLidar, which is a single continuous wave lidar (built upon a DM ZephIR lidar) equipped with a DTU designed fixed rosette-pattern scan head, cf. Figure 13.

Installed on the nacelle or directly into the turbine spinner the SpinnerLidar can measure the incoming wind field in about 400 points per second upwind of a wind turbine distributed over a 2D circular scan plane in front of a turbine, and also for measurements of the wake characteristics behind a turbine, on- or offshore [19, 20, 17, 21].

The Danish node of the WindScanner.eu research infrastructure participated centrally in the first IRPWind open call on Joint Experiments 2016 with a joint DTU and ECN project about measurements of high-resolution full-scale wind fields scanned 60 m upwind in front of a 2.5 MW research wind turbine. The measurement campaign involved DTUs 3D WindScanner and a DTU SpinnerLidar.

The aim of the experimental activity was to establish a unique benchmark experiment for comparison of SpinnerLidar inflow measurements by at the same time operating a DTU developed high- resolution nacelle integrated 2D SpinnerLidar installed on the nacelle of the 2.5 MW N9 ECN Nordex 80 research wind turbine in concert with three ground-based profiling short range WindScanners (Figure 14). The benchmark is now available through an open-access e-science platform via the website `www.irpwind-scanflow.eu`.

An intensive measurement campaign was carried out from the 16th of December 2016 until the 20th of February 2017 comprising meteorological mast measurements, ground-based vertical profiling lidar measurements, turbine SCADA data, combined SpinnerLidar rosette scan

Figure 13. A DTU SpinnerLidar installed on the Nacelle of a Nordex 80 test turbine at the ECN test site in Holland during IRPWind ScanFlow 2016-2017 [22].

Figure 14. Experimental Setup at ECN during ScanFlow: One of the Nordex 80 tests Turbine at ECNs proving grounds were equipped with a Nacelle mounted forward-looking DTU SpinnerLidar. At the same time, three short-range WindScanners were operated on the ground for measurement of the vertical wind profile in front of the turbine at the same location where also the DTU SpinnerLidar scanned the inflow [22].

measurements and short-range WindScanner measurements.

The SpinnerLidar operated at ECN from December 16th to 28th, 2016 and from January 16th to February 16th, 2017. All other instruments except the ground-based WindScanners worked continuously. The short-range WindScanners have been measuring from mid-January 2017 to mid-February 2017 when the wind direction was appropriate, i.e. with wind coming in the sector between 185 and 245 degrees. These events occurred on January 29th, February 4th and 5th, 2017.

SpinnerLidars may also be installed on offshore turbines, either on the nacelle or on floating pontoons, cf. Figure 15, for measurement of in-situ power curves and for providing preview inflow for advanced feed-forward turbine control that enables load and tower bending reductions and thereby assist turbine operation offshore and extend the turbines lifetime.

Figure 15. Offshore WindScanner concept: Rotor plane inflow scanned from of an offshore Vertical Axis Wind Turbine equipped with a DTU SpinnerLidar.

10. Conclusions

The paper has briefly reviewed the state-of-the-art of the WindScanner measurement technology with examples of uses for wind engineering applications at short measurement ranges. The technology is suitable for studying turbulent coherent structures of importance for investigation of flow patterns and dynamical loads on wind turbines, building structures and bridges and in relation to optimization of the location of, for example, wind farms and suspension bridges.

Acknowledgment

Etienne Cheynet, UiS, is acknowledged for suggesting many improvements and detailed copy editing of the final manuscript.

References

[1] Mikkelsen T, Mann J, Courtney M and Sjöholm M 2008 *IOP Conf. Ser. Earth Environ. Sci.* **1** U148–U156 ISSN 1755-1307
[2] Vasiljević N, Lea G, Courtney M, Cariou J P, Mann J and Mikkelsen T 2016 *Remote Sens.* **8** 896
[3] Mann J, Cariou J P C, Parmentier R M, Wagner R, Lindelöw P, Sjöholm M and Enevoldsen K 2009 *Meteorol. Z.* **18** 135–140
[4] Mikkelsen T 2014 *J. Phys. Conf. Ser.* vol 524 (IOP Publishing) p 012007
[5] Vasiljević N, Palma J M, Angelou N, Matos J C, Menke R, Lea G, Mann J, Courtney M, Ribeiro L F and Gomes V M 2017 *Atmospheric Meas. Tech.* **10** 3463
[6] Mikkelsen T 2009 *European Wind Energy Conf. and Exhibition 2009*
[7] Floors R, Peña A, Lea G, Vasiljević N, Simon E and Courtney M 2016 *Remote Sens.* **8** 884
[8] Simley E, Angelou N, Mikkelsen T, Sjöholm M, Mann J and Pao L Y 2016 *J. Renew. Sustain. Energy* **8** 013301
[9] Wagner R, Vignaroli A, Angelou N, Sathe A, Meyer Forsting A R, Sjöholm M and Mikkelsen T K 2015 *12th German Wind Energy Conf. DEWEK 2015*
[10] Yazicioglu H, Angelou N, Mikkelsen T and Trujillo J J 2016 *J. Phys. Conf. Ser.* vol 753 (IOP Publishing) p 032032
[11] Cheynet E, Jakobsen J B, Snæbjörnsson J, Reuder J, Kumer V and Svardal B 2017 *J. Wind Eng. Ind. Aerodyn.* **161** 17–26
[12] Cheynet E, Jakobsen J B, Snæbjörnsson J, Mikkelsen T, Sjöholm M, Mann J, Hansen P, Angelou N and Svardal B 2016 *Exp. Fluids* **57** 1–17
[13] Cheynet E, Jakobsen J B, Snæbjörnsson J, Angelou N, Mikkelsen T, Sjöholm M and Svardal B 2017 *J. Wind Eng. Ind. Aerodyn.* **171** 261–272 ISSN 0167-6105
[14] Cheynet E, Jakobsen J B, Snæbjörnsson J, Mann J, Courtney M, Lea G and Svardal B 2017 *Remote Sens.* **9** 977
[15] Peña A, Bechmann A, Conti D and Angelou N 2016 *Wind Energy Sci.* **1** 101–114

[16] van Dooren M F, Campagnolo F, Sjöholm M, Angelou N, Mikkelsen T and Kühn M 2017 *Wind Energy Sci.* **2** 329–341

[17] Herges T, Maniaci D, Naughton B, Mikkelsen T and Sjöholm M 2017 *J. Phys. Conf. Ser.* vol 854 (IOP Publishing) p 012021

[18] Campagnolo F, Schreiber J, Garcia A M, Bottasso C L *et al.* 2017 *The 27th Int. Ocean and Polar Engineering Conf.* (ISOPE)

[19] Mikkelsen T, Angelou N, Hansen K, Sjöholm M, Harris M, Slinger C, Hadley P, Scullion R, Ellis G and Vives G 2013 *Wind Energy* **16** 625–643

[20] Sjöholm M, Pedersen A T, Angelou N, Abari F F, Mikkelsen T, Harris M, Slinger C and Kapp S 2013 *European Wind Energy Association Conference*

[21] Herges T, Maniaci D, Naughton B, Hansen K H, Sjöholm M, Angelou N and Mikkelsen T 2017 *35th Wind Energy Symp.* (AIAA) p 0455

[22] Wagenaar J, Bergman G, Alting I, Hasager C, Mikkelsen T, Angelou T and Sjöholm M 2017 Infrastructure project: IRPWind scanflow

IOP Conf. Series: Materials Science and Engineering **276** (2017) 012005 doi:10.1088/1757-899X/276/1/012005

Wind effects on long-span bridges: Probabilistic wind data format for buffeting and VIV load assessments

K Hoffmann[1,*], R G Srouji[1] and S O Hansen[1,2]

[1]Svend Ole Hansen ApS, Sankt Jørgens Allé 5C, 1615 Copenhagen V, Denmark
[2]SOH Wind Engineering LLC, 141 Leroy Road, Williston, VT 05495, USA

*Corresponding author: kho@sohansen.dk

Abstract. The technology development within the structural design of long-span bridges in Norwegian fjords has created a need for reformulating the calculation format and the physical quantities used to describe the properties of wind and the associated wind-induced effects on bridge decks. Parts of a new probabilistic format describing the incoming, undisturbed wind is presented. It is expected that a fixed probabilistic format will facilitate a more physically consistent and precise description of the wind conditions, which in turn increase the accuracy and considerably reduce uncertainties in wind load assessments. Because the format is probabilistic, a quantification of the level of safety and uncertainty in predicted wind loads is readily accessible. A simple buffeting response calculation demonstrates the use of probabilistic wind data in the assessment of wind loads and responses. Furthermore, vortex-induced fatigue damage is discussed in relation to probabilistic wind turbulence data and response measurements from wind tunnel tests.

1. Introduction

The assessment of expected environmental loads is a key element in the design of engineering structures, influencing the complete chain of load carrying capacity analyses. Most often, the evaluation of governing loads utilizes codified mathematical models and site-specific environmental properties as input parameters. Several models and input parameters are required to cover all structural configurations and the associated risk categories and criteria, extending from the initial construction phase to the complete operational structure. The estimation of probabilities of events that are rare, but not necessarily extreme, is also important for serviceability evaluations.

The large dimensions and distinctive structural dynamics of many modern high-rise and long-span structures often challenge the traditional codified calculation frameworks. This may be manifested by excessive simplifications of the physical and mathematical load/response modelling or inadequate and insufficient descriptive parameters quantifying the environmental input. As a result, the load estimates may be unnecessarily conservative or even fail to capture specific relevant events. In any case, the level of safety in predicted environmental loads and associated structural responses may be very difficult to quantify utilizing traditional approaches.

Detailed information about expected environmental loads is naturally a requirement to obtain an optimal and safe design of any engineering structure. For bridges with span lengths larger than 300 m or natural frequencies below 0.5 Hz wind measurements at bridge site are actually mandatory [1]. The focus of this paper is limited to the assessment of dimensioning buffeting loads and vortex-induced vibrations (VIV), and the associated physical parameters and predictive models.

Within the field of bridge aerodynamics, the fundamental mathematical models were developed more than 50 years ago, where the somewhat limited wind data sources primarily consisted of manually assisted recordings of very few parameters with a rather large spatial and temporal spacing. This is in contrast to present-day automated high-frequency data sampling from globally distributed networks of meteorological weather stations.

The considerable development in advanced wind sensors for measurement campaigns is also important. In the 1980s, cup anemometers were common, and today these are replaced by ultrasonic anemometers. In recent years, the wind characteristics in several Norwegian fjords have been investigated using scanning LIDARs [2]. This has enabled a wind field description with a high spatial and temporal resolution, which was previously impossible to acquire with cup anemometers. In order to fully exploit a probabilistic wind data format, a large amount of wind data measurements are needed on site, i.e. long time series at representative locations. Processing of large amount of data is no longer considered as a limitation as computer software and hardware are continuously being more powerful.

A generally accepted probabilistic wind data format will facilitate that measurements carried out in different Norwegian fjords and at other locations around the world may all contribute to a common database enabling easy comparisons and increased certainty in wind flow descriptions.

2. Wind load assessment

The engineering practice of predicting wind loads on structures is traditionally based on a calculation framework utilizing a combination of site-specific wind characteristics, structural properties, and mathematical models of their interaction. This framework may be visualized by the so-called wind loading chain, introduced by A. G. Davenport in the 1960s; see Figure 1. The concept of the chain illustrates that the different parts of the load assessment are interrelated, and that the reliability of an estimated wind load is at most as large as the least reliable link in the chain.

Figure 1 includes illustrations of fundamental probabilistic distributions representing the uncertainties and stochastic nature of models and parameters associated with each link. The focus of the present paper is probabilistic distributions modelling the wind climate, terrain effects, and the flow to pressure mechanisms, represented by the first three links.

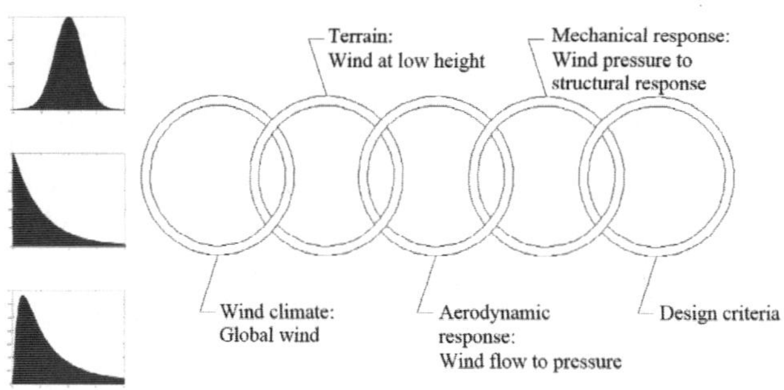

Figure 1. Principal sketch of the wind load design process, as illustrated by the wind loading chain, and different probabilistic distributions related to the uncertainties and stochastic nature of models and parameters associated with each link.

All parts of the wind loading chain consist of stochastic phenomena and simplified mathematical model relations associated with uncertainties. A deterministic approach is, however, often adopted for traditional structures due to its simplicity, since the key features of the wind load generation are often captured when the wind load model framework is calibrated against common load/response characteristics and experience from similar structures. In this framework, the conversion from characteristic values to rare events is attained by partial factors calibrated to the generally accepted level of structural safety. On the other hand, a deterministic approach is often inadequate to assess the extreme wind loads on structures with non-traditional structural dynamics or extreme spatial dimensions. Instead, the wind load calculation format may be extended to cover the actual stochastic phenomena involved, which could not only increase the accuracy in predicted wind loads, but also capture wind effects, which are not covered by the traditional deterministic description. Additionally, the level of safety and uncertainty is implicitly part of the predicted wind loads. The increasing amount of wind data availability for many sites facilitates such a probabilistic description of the governing wind phenomena.

3. Stochastic nature of wind

The description of wind turbulence is of crucial importance for the prediction of fluctuating wind loads on structures. The typical procedure is to model the wind turbulence description by the following site-specific wind data input:

- Turbulence intensities;
- Integral length scales;
- Power-spectral density functions;
- Normalized co-spectra.

The wind data traditionally consist of deterministic parameter values and model functions evaluated at the design mean wind velocity. This is also the underlying wind turbulence description utilized in many wind codes [1], [3] and forms the basis of principles for safety and serviceability of structures [4]. Suggested parameter and model values exist in both wind norms, however, these are not intended to be used for long-span bridge dimensioning without further justification.

Wind data from Horns Rev is utilized in the following section to illustrate the stochastic nature of wind turbulence. The Horns Rev site is located at a reef approx. 14 km off Jutland, west of Denmark, in a very harsh offshore environment. The wind measurements are 12 Hz samples and cover a period of approx. 5 years (2001-2005) and originate from a 3D sonic anemometer positioned 50 m above sea level. Since the site is open waters, a terrain roughness length of approx. 0.003 m may be assumed for high wind velocities. The wind measurements include some time gaps where recordings were not performed. The complete data set consists of 112586 time-series of 10-min duration.

3.1. Turbulence intensity

The relative extent of the along-wind velocity fluctuations at a certain mean wind velocity is represented by the turbulence intensity

$$I_u = \frac{\sigma_u}{U}, \tag{1}$$

where σ_u is the standard deviation of the along-wind turbulence component u and U is the 10-min mean wind velocity. Codified models prescribe a single value of turbulence intensity based on the terrain roughness length and the elevation, implicitly referring to turbulence intensity at high wind velocities [1], [3].

Figure 2 presents wind turbulence intensities measured at Horns Rev. The plot illustrates a few key features related to the stochastic nature of the turbulence intensity. Generally, the scatter is large, especially at low and moderate wind velocities, where the variability of thermal stratification strongly influences the mechanically generated turbulence. Conditions with nominally zero turbulence intensity do only occur at wind velocities, say, below 10-15 m/s, while larger wind velocities seem to ensure a certain amount of wind turbulence. This observation is relevant for the assessment of possible VIV, where wind turbulence components reduce the critical fatigue loads. Similarly, provisions such as Eurocode 1 prescribe a reduction in VIV loads at high wind velocities. At high wind velocities above 20 m/s, where buffeting loads are of importance, the turbulence intensity is in the range 5 - 20 %.

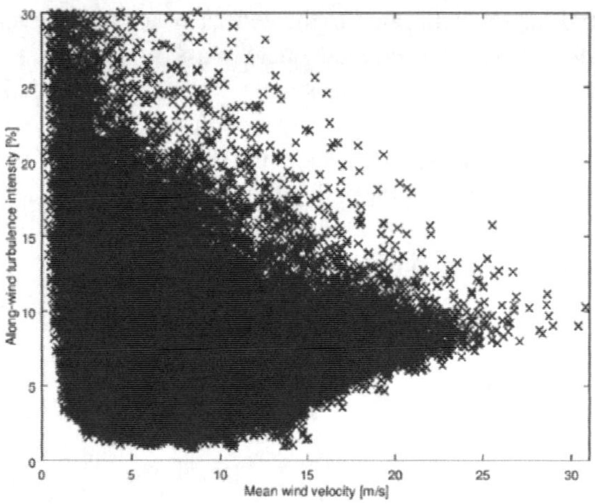

Figure 2. Along-wind turbulence intensity as a function of the 10-min mean wind velocity measured at Horns Rev.

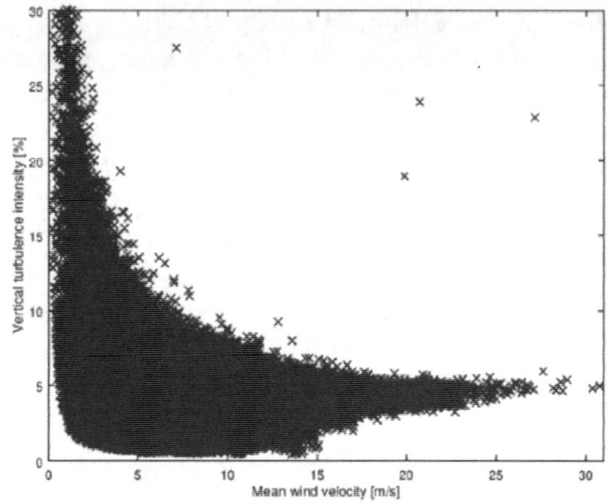

Figure 3. Vertical turbulence intensity as a function of the 10-min mean wind velocity measured at Horns Rev.

The codified value of the along-wind turbulence intensity at the site is approx. 10.3 % [3]. This is an underestimation of the turbulence intensity for some of the extreme events seen in Figure 2. Utilizing the codified prescription without safety factors may in such cases result in underestimated buffeting loads.

The vertical turbulence intensity σ_w/U is presented in Figure 3. Generally, the vertical turbulence intensity seems to produce less scatter than the along-wind turbulence intensity, except for very few outliers. Conditions with nominally zero turbulence intensity seem to occur only at wind velocities, say, below 15 m/s, while larger wind velocities seem to ensure a certain amount of wind turbulence. The vertical turbulence intensity seems to approach approx. 5 % at high wind velocities. This is in agreement with $I_w \approx 0.5I_u$, which is a typical approximation to apply for design purposes [5].

In the field of bridge aerodynamics, low-intensity turbulence conditions at moderate wind velocities is generally governing VIV and possible fatigue damage, while high-intensity turbulence conditions at high wind velocities are dimensioning for buffeting responses. The following subsections will consider these two scenarios and, by examples, illustrate simple probabilistic model relations of the along-wind turbulence intensity.

3.1.1. Low-intensity turbulence at moderate wind velocities. Figure 4 illustrates the distribution of measured along-wind turbulence intensities for mean wind velocities in the interval 5 - 10 m/s. As a first approximation, a gamma distribution is fitted to model the low-turbulence conditions. The gamma probability density function is defined by

$$f_{\text{gam}}(x) := \frac{\beta^{\alpha} x^{\alpha-1} \exp(-\beta x)}{\Gamma(\alpha)}, \tag{2}$$

where α is the shape parameter, β is the rate parameter, and Γ is the gamma function. The fitted distribution corresponds to a shape parameter of 8.0 and rate parameter of 0.8 %.

For a structure prone to VIV, the cumulative fatigue damage may be calculated utilizing the probability distribution of low turbulence intensity events near the critical wind velocity to assess the risk of fatigue failures.

3.1.2. High-intensity turbulence at high wind velocities. Figure 5 shows the distribution of measured along-wind turbulence intensities for mean wind velocities above 15 m/s.

As a first approximation, an exponential distribution is fitted to model high-turbulence conditions. The exponential distribution is a special type of gamma distribution, and the probability density function is

$$f_{\exp}(x) := \beta \exp(-\beta x), \tag{3}$$

where β is the rate parameter. Note that the mean value of the distribution is $1/\beta$ and this will in the following be used to describe the distribution. In the present model framework a translation is also introduced, which corresponds to replacing x by x-x_0 for a translation of x_0.

The fitted distribution of high-turbulence conditions above 8 % corresponds to a mean value of 1.5 % and translation of 6.5 %. This is an example of a probabilistic distribution modelling conditions corresponding to critical buffeting loads.

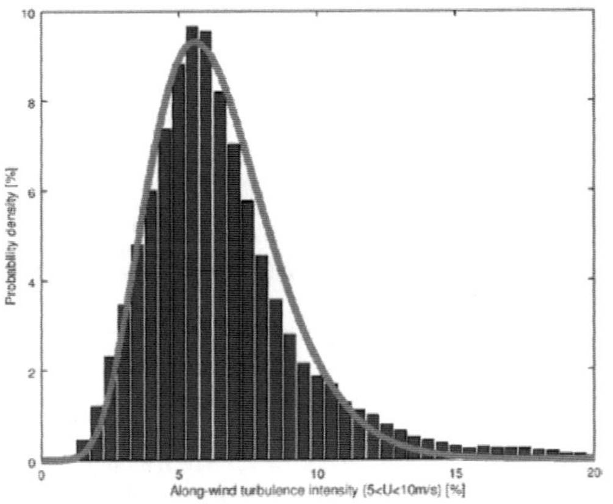

Figure 4. Distribution of measured along-wind turbulence intensities for mean wind velocities in the interval 5 - 10 m/s. Each bar represents the probability for a turbulence intensity interval of 0.5 %. A gamma distribution with a shape parameter of 8.0 and a rate parameter of 0.8 % is fitted to model the low-turbulent conditions below 5 %.

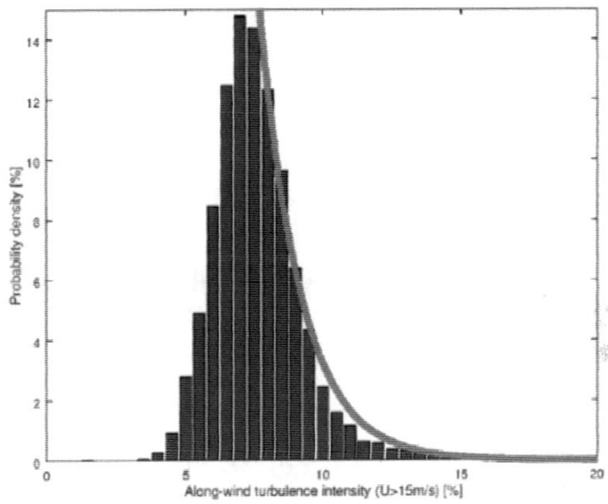

Figure 5. Distribution of measured along-wind turbulence intensities for mean wind velocities above 15 m/s. Each bar represents the probability for a turbulence intensity interval of 0.5 %. An exponential distribution with a mean value parameter of 1.5 % and a translation of 6.5 % is fitted to model the high-turbulent conditions.

3.2. Integral length scale

The integral length scale is a measure of the correlation of along-wind turbulence components, and is used in the traditional modelling framework to prescribe the frequency with the largest turbulent energy at a given mean wind velocity. Integral length scales may be determined by applying Taylor's hypothesis to the time scale found using the auto-correlation function of the along-wind component at a single point [5]. If the process involves time integration of finite measurement periods not much larger than the corresponding time scale, a filtering process, such as the Blackman-Tukey method, should be considered to reduce the error variance, i.e. to avoid an overestimation of the auto-correlation at large time lags [6].

Figure 6 shows the distribution of the integral length scale for along-wind turbulence L_u^x determined by the procedure explained above without the use of a filtering process. As a first approximation, an exponential distribution with a mean value of 125 m has been fitted to model the upper tail of the distribution. By comparison, at the elevation and terrain roughness length of the measurement site, the codified value of the along-wind integral length scale is approx. 177 m [3].

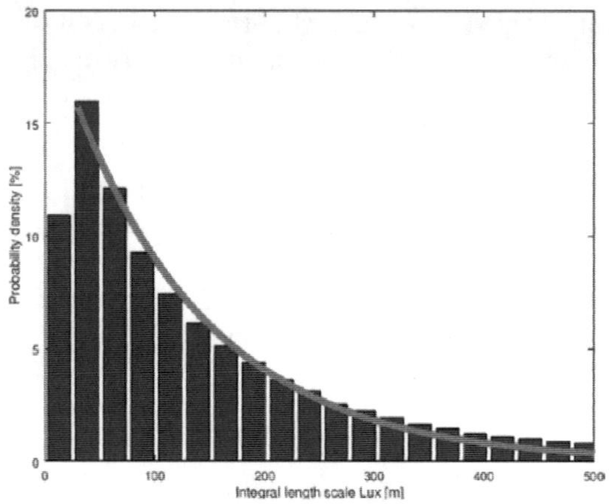

Figure 6. Distribution of measured longitudinal integral length scales for all mean wind velocities. Each bar represents the probability for a length scale interval of 25 m. The fitted exponential distribution with a mean value of 125 m seems to provide a good model of the distribution of integral length scales above approx. 30 m.

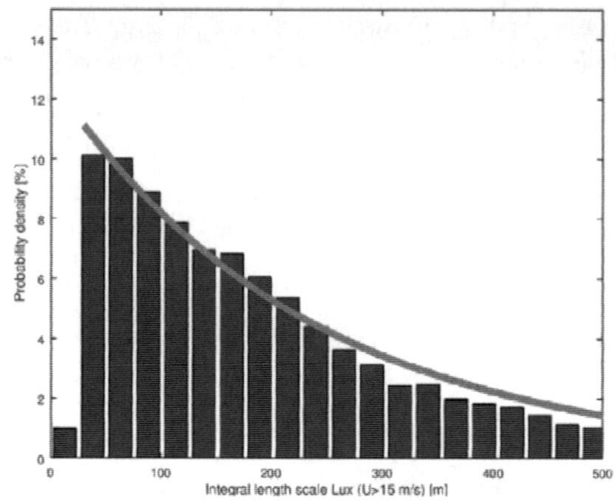

Figure 7. Distribution of measured longitudinal integral length scales for 10-min mean wind velocities above 15 m/s. Each bar represents the probability for a length scale interval of 25 m. The fitted exponential distribution with a mean value of 230 m and a translation of 35 m seems to provide a good model of the distribution of integral length scales above approx. 35 m.

A simple model of the integral length scale at high wind velocities may be estimated by a similar fitting procedure. Figure 7 illustrates such a distribution restricted to 10-min mean wind velocities above 15 m/s. The average integral length scale increase at high wind velocity conditions. The presented exponential fit with a mean value of 230 m and a translation of 35 m approximates the distribution for length scales above approx. 35 m.

3.3. Power-spectral density function

The frequency wise distribution of turbulent energy is modelled by the power-spectral density function. In the Eurocode 1 [3] and Håndbok N400 [1] prescriptions, the normalized single-point power-spectral density is modelled by the expression

$$R_{\mathrm{N,EC1}}(n) := \frac{nS_{u,\mathrm{EC1}}(n)}{\sigma_u^2} = \frac{6.8 f_L}{(1+10.2 f_L)^{5/3}}, \tag{4}$$

where n is the frequency and $f_L := nL_u^x/U$ is the normalized frequency. The presented spectrum is of the Kaimal spectral density form and focuses on a consistent modelling of the high-frequency behavior. The Kaimal spectrum satisfies Kolmogorov's law for the energy distribution in the turbulent fluid. The spectrum shape is fixed, and the integral length scale and the mean wind velocity prescribe the frequency with the largest normalized turbulent energy.

The point-wise power-spectral density at a given location may also be estimated by analyzing wind velocity measurements. This requires a high sampling rate to capture the high-frequency part. Measurements from the first 24 hours of 2002 have been used to produce an example of the spectral density of the turbulence components. This corresponds to approx. 1 million data points given the 12 Hz sampling rate. The power-spectral density of each 10-min block is calculated, and the mean value for each frequency component of these 144 spectra is determined. The power-spectral density above $n = 3$ Hz has been discarded to remove most effects of spectral folding. The normalized frequency is determined using the 10-min mean wind velocity in each 10-min block, varying from 5 to 20 m/s during this specific day. The integral length scale for the along-wind and vertical turbulence model is set to 125 m and 12.5 m, respectively. The normalization of R_{N} for the vertical turbulence is performed using σ_w^2. These two spectra and the Eurocode 1 spectrum are plotted in Figure 8.

The measurements show a good resemblance with the Eurocode 1 prescriptions. The spectrum of the vertical turbulence components seem to match the codified prescriptions by utilizing an integral length scale for the normalized frequency which is 1/10th of the along-wind integral length scale. Note that the actual integral length scale of the vertical turbulence component is likely different than this value. The 1/10 factor is here chosen to produce a model spectrum of the vertical turbulence, expressed by the codified along-wind spectrum.

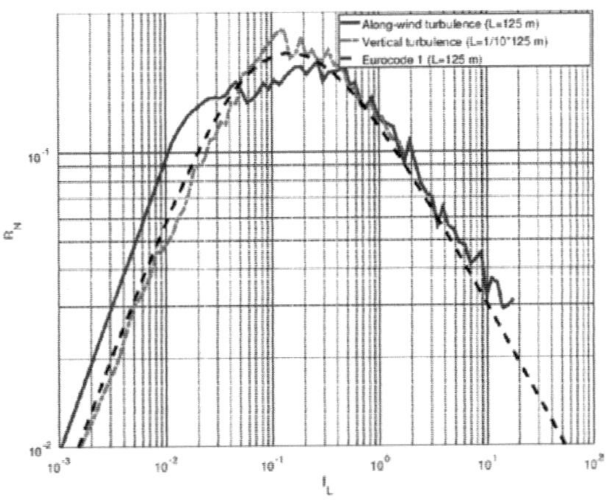

Figure 8. Spectra of measured along-wind and vertical turbulence. The integral length scale used for the normalized frequency is 125 m and 12.5 m, respectively. The black dashed curve is the Eurocode 1 spectrum for the along-wind turbulence with an integral length scale of 125 m.

The integral length scale is not constant during a day. It is therefore expected that a 1-day average spectrum based on a constant integral length scale will have energy redistributed from the peak value to smaller and larger frequencies, i.e. it will be more "flat". Such an effect is slightly visible for the along-wind turbulence spectrum in Figure 8.

The presented codified spectrum gives a frequency-wise representation of the fluctuating energy covering the dynamic effects of relevance for most structures, including modern bridges. In the case of unusual low-frequency dynamics, such as those associated with flexible offshore structures, a spectral representation tuned to match the low-frequency regime is required. For such structures, it is therefore recommended to rely on a different spectral description [7]. The buffeting load calculation presented in Section 5 will utilize a simple probabilistic distribution of the point-wise along-wind power-spectral density based on the codified spectrum.

3.4. Normalized co-spectra

The normalized co-spectrum models the statistical dependence between the turbulence at two points with a certain spatial separation. This statistical dependence is naturally closely related to the spatial dimension of vortices in the wind field. For the along-wind turbulence component, the normalized co-spectrum for two points of lateral separation r_y at a given frequency n_i may be modelled by the Davenport model [5]

$$\psi_u = \exp(-C_{uy} r_y n / U). \tag{5}$$

The non-dimensional decay constant C_{uy} is 10 according to the general prescriptions [1]. The stochastic nature of the normalized co-spectrum or the associated decay constant will not be further addressed in this treatment, but a probabilistic description is introduced as part of the example calculation presented in Section 5.

4. Probabilistic wind data provisions

The stochastic nature of wind turbulence should preferably be accounted for in a physically consistent manner, yet, several rather complicated physical phenomena are responsible for the wind turbulence generation under different climatic conditions. The lack of wind stationarity and the influence of the fetch are additional sources of uncertainty. As a result, the stochastic processes are difficult to model without applying certain simplifications.

The probabilistic models presented in the previous section do not necessarily reflect the physics behind turbulence generation. They are merely introduced as first approximation examples used to describe the governing wind data by standard distributions fitted to a single set of measurement data. These example distributions are also utilized in the example calculation performed in Section 5.

In its simplest form, a physically consistent stochastic wind turbulence model should account for the fact that turbulence at low wind velocities is primarily generated by thermal convection, while turbulence at higher velocities is primarily mechanically generated due to friction forces between the air and any irregular terrain and obstacles. Prescribed stochastic model representations of wind turbulence should therefore represent the distinct physics behind turbulence generation at different climatic conditions, as expressed by the stochastic processes. Such modelling is outside the main field of expertise of the authors, and a consistent treatment of this topic requires further involvement of meteorologist, measurement equipment manufacturers and end-users, and data statisticians.

As mentioned previously, in the field of bridge aerodynamics, wind turbulence at low wind velocities is generally governing VIV, while high-wind turbulence conditions are dimensioning for buffeting responses. A probabilistic assessment of the associated governing structural loads requires probabilistic distributions of the following data:

- Turbulence intensities I_u, I_w as a function of U;
- Integral length scales L_u^x, L_w^y, as a function of U;
- Point-wise power-spectral densities S_u, S_w as a function of U;
- Decay constants C_{uy}, C_{uz}, C_{wx}, C_{wy} as a function of U;
- Mean wind velocity and velocity pressure U, q - Direction-dependent typical and extreme values;
- Air density ρ as a function of U;
- Mean wind inclination from horizontal as a function of U.

This list should be considered an initial provision. It may be advantageous to model some parameters in other formats. For instance, if the non-dimensional decay constants C are used in an exponential format, where they are multiplied by U^{-1}, probabilistic descriptions of C/U are just as useful. A better understanding of the governing wind phenomena, combined with analyses of measurement data, will indicate the type of quantities, which are advantageous to use in the probabilistic modelling. A well-defined probabilistic wind data format will also establish a common basis, promoting a direct comparison of wind data from different measurement sites.

5. Example - Probabilistic wind load assessment

To give an example of a probabilistic wind load assessment, the following section will consider the extreme resonant turbulent response of a simple horizontal line-like structure, positioned in a wind climate similar to the probabilistic descriptions presented in Section 3.

5.1. Buffeting load and response model

The power-spectral density of the resonant turbulent drag load on a line-like structure is given by the relation [5]

$$S_{\mathrm{drag}}(n) = \left(\frac{1}{2}\rho U h l\right)^2 (2C_{\mathrm{D}})^2 \left|J_u(n)\right|^2 S_u(n), \tag{6}$$

where h and l are the height and length of the structure, C_D is the drag coefficient, and S_u is the power-spectral density of the along-wind turbulence. The joint acceptance function is given by

$$|J_u(n)|^2 = \frac{1}{l^2}\int_0^l\int_0^l \xi(y_1)\xi(y_2)\psi_u(r_y,n)\, dy_1 dy_2, \tag{7}$$

where ξ is a mode shape in the direction in the wind. The mode shape is normalized to have a maximum value of 1. The cross-sectional aerodynamic admittance function for drag has been assumed to be 1. Thus, the lack of correlation of pressures on the front and suction on the back has been discarded for simplicity.

By utilizing the white noise approximation, the power-spectral density of the resonant turbulent drag load may be evaluated at a natural frequency n_i. This gives the variance of the corresponding maximum resonant structural acceleration, as expressed by [5]

$$\sigma_{\text{acc}}^2 = \frac{1}{m_i^2}\frac{\pi^2}{2\delta_i} n_i S_{\text{drag}}(n_i). \tag{8}$$

In the expression above, δ_i is the mode-specific sum of the aerodynamic and structural damping given by a logarithmic decrement (LD) and m_i is the generalized mass defined by

$$m_i = \int_0^l \mu(y)\, \xi_i^2(y)\, dy, \tag{9}$$

where μ is the structural mass per unit length.

The standard deviation of the structural acceleration σ_{acc} is the quantity to be estimated in the example calculation. In practical applications, the structural acceleration is usually determined to evaluate corresponding inertial loads, which ultimately are added to the mean and background turbulent loads to model an extreme load scenario.

5.2. Probabilistic wind data

The probabilistic distributions utilized for the wind data in the calculation example are listed in Table 1, including the example distributions for the high-wind turbulence intensity and integral length scales presented in Section 3.

Table 1. Probabilistic distributions for wind data used in the example calculation.

Parameter	I_u **[%]**	L_u^x **[m]**	$S_u(n_i)$ **[m²/s]**	C_{uy} **[-]**	q **[N/m²]**
Distribution type	Exp.	Exp.	Normal	Normal	Gumbel
Mean value	8.0 %[a]	265[a]	$S_{u,\text{EC1}}(n_i)$	10	-
Coeff. of variation	-	-	0.1	0.1	-
Location	-	-	-	-	465
Scale	-	-	-	-	93
Translation	6.5 %	35	-	-	-

[a] After translation.

The mean value of the single-point spectrum S_u at a given frequency is modelled by codified prescriptions [1], [3]. The spectral content at n_i is assumed to be normally distributed, with a coefficient of variation of 0.1. Note that the probabilistic nature of the spectrum is also governed by the probabilistic models of the turbulence intensity and integral length scale. The non-dimensional decay constant C_{uy} for the co-spectrum is assumed to be normally distributed with mean value 10 and coefficient of variation of 0.1. The mean value is in accordance with the prescriptions of Håndbok N400 [1]. The annual extreme 10-min mean wind velocity is implicitly modelled by a Gumbel distribution for the annual extreme mean velocity pressure. The listed Gumbel parameters follow the prescriptions of the Danish national annex of Eurocode 1 and correspond to a 50-year basic wind velocity of 24 m/s, an elevation of 50 m, and a terrain roughness length of 0.003 m. The air density is set to 1.25 kg/m^3 in all the example calculations. It is for simplicity assumed that the direction of the wind is always perpendicular to the structure.

5.3. Structural data

The present probabilistic framework focuses on climatic parameters. All structural parameters are therefore, for simplicity, defined by deterministic values to illustrate the effect of a probabilistic wind data format alone. For the present example, the values listed in Table 2 are applied.

Table 2. Deterministic structural data used in the example calculation.

h	l	C_D	μ	m_i	n_i	ξ_i	δ_i
1 m	100 m	1.0	1,000 kg/m	50,000 kg	1.0 Hz	sin(πy/l)	3 % LD

For most practical applications, the structural parameters have a varying level of uncertainty. For example, often the geometry and mass of the structure are both known quite accurate, while the structural dynamics is governed by a relatively large uncertainty.

5.4. Probabilistic extreme resonant turbulent response

The buffeting load and response model has been evaluated numerically using the probabilistic wind data input to produce 10^8 realizations of the yearly extreme horizontal structural acceleration caused by resonant buffeting loads. The associated distribution is illustrated in Figure 9.

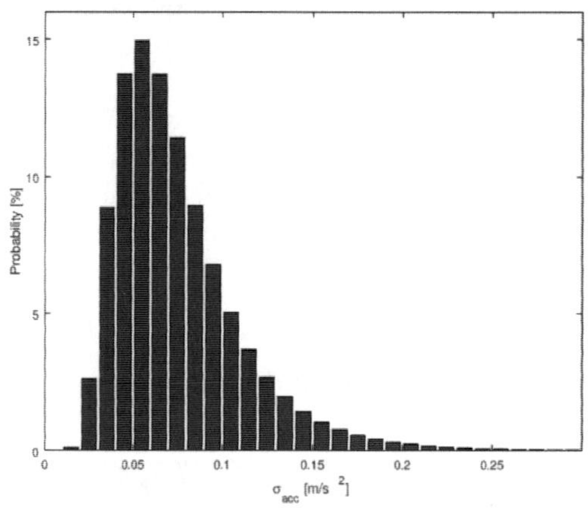

Figure 9. Distribution of yearly extreme standard deviation acceleration due to resonant buffeting loads. Each bar represents the probability for an interval of 0.01 m/s. The plot window covers approx. 99.9 % of the realizations.

The extreme horizontal accelerations associated with different return periods are listed in Table 3, where the yearly probability of a T-year response is defined by 1-exp(-1/T) [5].

Table 3. Estimated extreme horizontal accelerations for different mean return intervals (MRI).

MRI, T [years]	1	50	100	500	1000	5000	10000
$\sigma_{acc,T}$ [m/s^2]	0.058	0.175	0.200	0.263	0.292	0.368	0.403

IOP Conf. Series: Materials Science and Engineering **276** (2017) 012005 doi:10.1088/1757-899X/276/1/012005

The ratio between the 50-year extreme value and larger mean return periods is illustrated in Figure 10. Note that the ratio between the resonant turbulent response and the mean response increases with the return period. While the mean response scales with the square of the mean wind velocity, an increased mean wind velocity also increases load correlation and relative fluctuating energy at the resonant frequency for the present example structure. In the calculation framework, this is manifested by an additional increase in the joint acceptance function and the point-wise power-spectral density of the along-wind turbulence.

The traditional deterministic assessment of a 50-year extreme response is based on codified models of the turbulence intensity and integral length scale. In the present case, this corresponds to $I_u \approx 10.3$ % and $L_u^x \approx 177$ m [3]. Combined with the deterministic Eurocode 1 spectrum and a decay constant $C_{uy} = 10$, this produces a 50-year resonant response of 0.181 m/s^2. This is very similar to the 50-year response of 0.175 m/s^2 estimated in the probabilistic analysis; see Table 3.

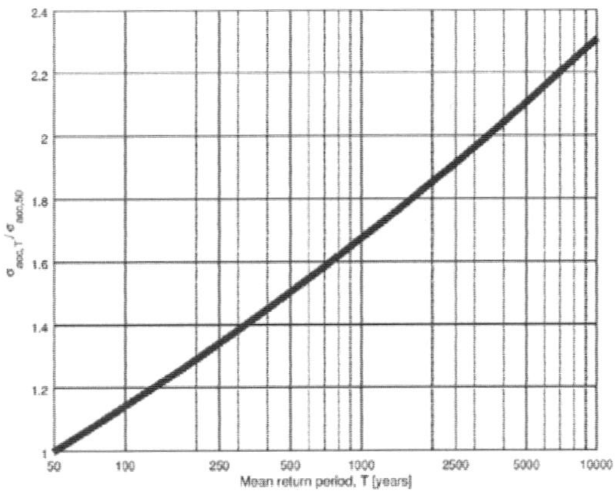

Figure 10. Ratio between 50-year extreme value and larger mean return periods.

Given the estimated extreme response probability distribution, it is straightforward to quantify the level of safety in the predicted responses, due to the correspondence between the return period and the probability of a yearly extreme event. However, it is important to remember that the relative uncertainty in the estimated response probability distribution increases for larger mean return periods. This means that a higher number of realizations is needed to accurately model the risk of extreme events for larger mean return periods.

6. Vertical load correlation for bridge decks

Vertical turbulence components are often the main source of dynamic vertical loads on bridge decks. To describe the interaction between vertical load components along the bridge, the lack of correlation between the generated surface pressures should be taken into account. The spectral response contribution from all pairs of surface points on such a plate-like horizontal structure may be expressed mathematically by an aerodynamic admittance function. In principle, this is nothing but a quadruple integral of all load pairs with a certain spatial separation along the two main axes of the bridge deck, quantifying the average lack of load correlation.

For high frequencies, the simple exponential format of the normalized co-spectrum given by Equation (5) may be adopted. It can then be shown theoretically that in the upper-frequency limit, the plate-like aerodynamic admittance function χ_{xy}^2 may be expressed using two line-like aerodynamic admittance functions χ_x^2, χ_y^2 according to the expression [5]

$$|\chi_{xy}(\phi_{xy})|^2 = \frac{\pi}{2}|\chi_x(\phi_x)|^2|\chi_y(\phi_y)|^2, \tag{10}$$

where $\phi_i = C_i n l_i / U$, utilizing the decay constant C_i, the frequency n, the structural dimension l_i, and the mean wind velocity U. This relationship has also been validated in wind tunnel experiments [8].

Since the line-like aerodynamic admittance function is inversely proportional to the frequency in the upper-frequency limit [5], [9], the expression may be reformulated as

$$\lim_{n\to\infty} \left|\chi_{xy}(\phi_{xy})\right|^2 = \left|\chi_x\left(\frac{2}{\pi}\phi_x\right)\right|^2 \left|\chi_y(\phi_y)\right|^2. \tag{11}$$

The calculation of the two-dimensional aerodynamic admittance function may therefore be significantly simplified in the upper-frequency range by using suitable products of a cross-sectional and a line-like aerodynamic admittance function. The fact that the load pair separations on plate-like structures cover a surface, and not just a line, is therefore included theoretically by a scaling of the argument in one of the aerodynamic admittance functions. The line-like aerodynamic admittance function describes the lack of correlation along the bridge deck, and this is governed by the correlation characteristics of the approaching undisturbed vertical wind velocity component. Thus, the planned full-scale measurements of the wind characteristics in the Norwegian fjords may readily be used in the format given by equation (11).

The simplicity of utilizing two aerodynamic admittance functions in buffeting load calculations for plate-like surfaces in the high-frequency range is a motivation for including such formats in codified prescriptions. It is expected that such a format is to be included in future revisions of Eurocode 1, concerning the buffeting loads on typical structures.

In relation to the aerodynamic modal forces on structures, a similar approach may be applied. In this case, one of the aerodynamic admittance functions quantifies the load correlation in terms of weighted modal loads and is, in principle, expressed by a joint acceptance function.

The simple exponential model in equation (5) is designed to model correlation at typical structural natural frequencies and spatial dimensions. For long-span bridges, the simple exponential modelling of the span-wise correlation of wind turbulence and the associated surface pressure fluctuations is therefore inconsistent given the relatively low natural frequencies and very large dimensions for such structures [10]. For instance, the modelled correlation approaches unity at very small frequencies, even at separations much larger than the average size of wind gusts.

A modified exponential format has been proposed, addressing some of the physical inconsistencies of the simple exponential model [11]. This format seems to provide a better representation, especially at large separations. Modern remote wind measurement systems, such as LIDARS, allow for site-specific large-separation correlation statistics. Further experience from such measurement campaigns will reveal the need for reassessing the modelling of the normalized co-spectrum for long-span bridges. In any case, the format given by equation (11) is valid for any type of large-separation correlation model.

7. Vortex-induced vibrations - Fatigue damage

VIV amplitudes larger than vibrations due to typical buffeting loads are often not accepted in the design criteria for modern long-span bridges. The bridge deck geometry is therefore often thoroughly tested, for instance in wind tunnel section model experiments, to ensure a sufficiently small-amplitude VIV response.

The principal physical phenomenon behind VIV is a motion-induced load in the form of a negative aerodynamic damping [5]. Large-amplitude VIV occur when the corresponding aerodynamic loads exceed the structural damping mechanisms. The aerodynamic load is reduced in the presence of along-wind turbulence components. Conditions with nominally zero wind turbulence cannot generally be discarded at typical mean wind velocities below approx. 15 m/s, similar to the turbulence data presented in Figure 2. The peak VIV amplitude, independent on its magnitude, is therefore generally to be expected on a sub-yearly basis, and experimental procedures should therefore be performed in low-turbulent conditions to outline the expected peak response. The estimated peak VIV amplitude is used in the assessment of serviceability and user comfort.

IOP Conf. Series: Materials Science and Engineering **276** (2017) 012005 doi:10.1088/1757-899X/276/1/012005

Extensive studies have shown that aerodynamic remedies in the form of guide vanes and a centrally located vortex spoiler may be an effective mean for suppressing VIV of bridge decks. For example, wind tunnel experiments previously carried out on a section model of the Hardanger Suspension Bridge have shown that simple aerodynamic remedies reduce the peak vertical vortex-induced response from approx. 15 % to approx. 3 % of the bridge deck height. Similar aerodynamic remedies have been retrofitted on a number of Norwegian and Danish bridges with spans varying from approx. 300 m to 1,600 m, significantly reducing the peak VIV response [12].

The probabilistic aspect of VIV modelling is primarily related to possible fatigue damage by modelling the occurrence of dynamic responses. For long-span bridges, which have relatively low natural frequencies, the bridge interaction with the vortex-shedding frequency may occur a large number of times throughout the bridge's lifetime due to the rather low critical wind velocity.

The VIV load reduction caused by non-zero along-wind turbulence is included in the Eurocode 1, Approach 2 format by a corresponding reduction of the aerodynamic damping parameter K_a [13]. An example of this is presented in Figure 11, where the codified normalized extreme VIV response is plotted as a function of the turbulence intensity for a representative streamlined bridge cross section with and without aerodynamic remedies.

A probabilistic model of low-intensity turbulence near the critical wind velocity, as the one presented in Figure 4, combined with a model of the VIV response as a function of the turbulence intensity, as the one presented in Figure 11, facilitates accurate predictions of load cycles and the corresponding fatigue damage for various structural elements on a bridge. Similarly, the fatigue damage reduction due to aerodynamic remedies is possible to quantify. A more advanced VIV response model could take into account the finite time of the amplitude build-up and decay, which for a long-span bridge may correspond to a time frame much larger than 10 min.

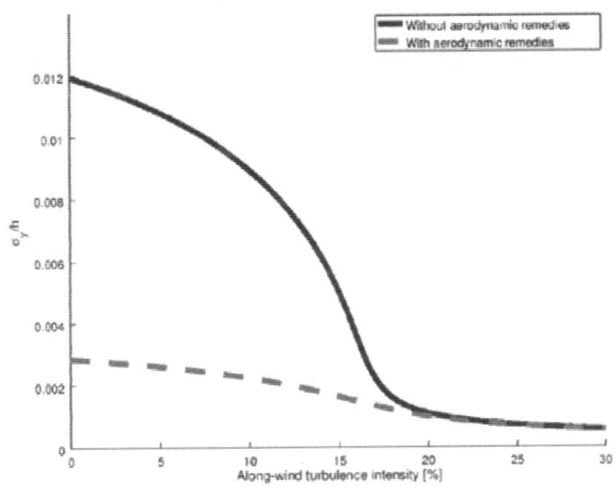

Figure 11. Principal interpretation of wind tunnel tests for a bridge section with and without aerodynamic remedies, illustrating the extreme normalized standard deviation as a function of the along-wind turbulence intensity.

8. Secondary aerodynamic loads

In addition to the mean and buffeting wind loads on a long-span bridge, there exist numerous aerodynamic loads, which are also assessed in the design phase. These loads are here denoted *secondary*, since they are assessed following the VIV, buffeting, and flutter aerodynamics due to their less fundamental influence of the bridge design. Secondary aerodynamic loads include cable loads and loads caused by special combined climatic effects, such as the accretion of ice or snow on bridge deck railings, changing the deck aerodynamics. Joint probability distributions of the wind velocity, direction, temperature, and rain/snow intensity may be utilized to assess such secondary effects. In addition, such distributions would allow for an assessment of the necessity of enforced traffic restrictions and bridge closures due to strong winds. This may be used to evaluate the need of wind shields and their required efficiency.

9. Summary and conclusion

This paper presents initial suggestions for an extended probabilistic wind data format, which may be used to assess VIV and buffeting loads on long-span bridges.

The stochastic nature of wind turbulence is illustrated in a series of plots utilizing wind data from Horns Rev. Initial provision for a probabilistic wind data format applicable to long-span bridges is proposed. A wind load assessment on a simple line-like structure utilizing probabilistic wind data is presented in order to illustrate the main methodology, and how the probabilistic load assessment format establishes a one-to-one correspondence between extreme responses and return periods for a given site and structure. A future development of the extended probabilistic data format and associated wind load calculation frameworks will greatly benefit from continuous dialogue and collaboration between public road authorities, meteorological institutes, and wind load engineers.

Modern long-span bridge concepts will not be realized unless cost-effective design is an important part of the construction process. Wind loading is perhaps the most important environmental load governing the design of long span bridges. Therefore, all efforts invested in establishing safer and more accurate wind load models will be beneficial.

Acknowledgements

The inspiring discussions with Bjørn Isaksen and Kristian Berntsen from the Norwegian Public Roads Administration concerning the extended wind data format, and their experience with the design of long-span bridges and full-scale measurement campaigns are highly appreciated.

The work has benefited from measurements downloaded from the internet database: "Database of Wind Characteristics" located at DTU, Denmark, "http://www.winddata.com/". Wind field time series from the following sites has been applied: Horns Rev, 2001-2005.

References

[1] Håndbok N400 – Bruprosjektering 2015 Norwegian Public Roads Administration
[2] Cheynet E, Jakobsen J B, Snæbjörnsson J, Reuder J, Kumer V and Svardal B 2017 Assessing the potential of a commercial pulsed lidar for wind characterisation at a bridge site *J. of Wind Engineering and Industrial Aerodynamics* **161** pp 17-26
[3] DS/EN 1991-1-4:2007, Eurocode 1: Actions on structures – Part 1-4: General actions – Wind actions 2007 Dansk Standard, Denmark
[4] DS/EN 1990:2007, Eurocode 0: Basis of structural design, 2007, Dansk Standard, Denmark
[5] Dyrbye C and Hansen S O 1997 *Wind loads on structures*, John Wiley & Sons Ltd.
[6] Zaknich A 2005 *Principles of Adaptive Filters and Self-learning Systems. Advanced Textbooks in Control and Signal Processing*, Springer
[7] Højstrup J, Larsen S E and Madsen P H 1990 Power spectra of horizontal wind components in the neutral atmospheric surface boundary layer *AMS 9th Symp. on Turbulence and Diffusion* Roskilde, Denmark pp 305–8
[8] Hoffmann K, Srouji R G and Hansen S O 2016 Buffeting loads on high-rise and long-span structures *8th Int. Colloquium on Bluff Body Aerodynamics and Applications*, Boston, USA
[9] Davenport A G 1977 The prediction of the response of structures to gusty wind *Safety of structures under dynamic loading* ed 1 pp 257–84
[10] Srouji R G, Hoffmann K and Hansen S O 2017 Experimental verification of a generic buffeting load model for high-rise and long-span structures *The 7th European and African Conf. on Wind Engineering* Liege, Belgium
[11] Krenk S 1996 Wind field coherence and dynamic wind forces *Solid mechanics and its applications* **47** pp 269–78

[12] Hansen S O, Srouji R G, Isaksen B and Berntsen K 2015 Vortex-induced vibrations of streamlined single box girder bridge decks *14th International Conference on Wind Engineering* Porto Alegre, Brazil

[13] Hansen S O 2017 Vortex-induced vibrations of structures *Structural Engineers World Congress* Bangalore, India.

Aeroelastic Response from Indicial Functions with a Finite Element Model of a Suspension Bridge

O Mikkelsen[1] and J B Jakobsen[2]

[1]Associate Professor, Department of Mechanical and Structural Engineering and Materials Science, University of Stavanger, Norway

[2] Professor, Department of Mechanical and Structural Engineering and Materials Science, University of Stavanger, Norway

E-mail: ove.mikkelsen@uis.no

Abstract. The present paper describes a comprehensive analysis of the aeroelastic bridge response in time-domain, with a finite element model of the structure. The main focus is on the analysis of flutter instability, accounting for the wind forces generated by the bridge motion, including twisting as well as vertical and horizontal translation, i.e. all three global degrees of freedom. The solution is obtained by direct integration of the equations of motion for the bridge-wind system, with motion-dependent forces approximated from flutter derivatives in terms of rational functions. For the streamlined bridge box-girder investigated, the motion dependent wind forces related to the along-wind response are found to have a limited influence on the flutter velocity. The flutter mode shapes in the time-domain and the frequency domain are consistent, and composed of the three lowest symmetrical vertical modes coupled with the first torsional symmetric mode. The method applied in this study provides detailed response estimates and contributes to an increased understanding of the complex aeroelastic behaviour of long-span bridges.

1. Introduction

The verification of wind-resistant design of a long-span bridge relies on experimentally obtained aerodynamic data from wind tunnel studies, applied in computational models of different levels of complexity. Established wind load models are based on the principle of superposition of wind loads separated into mean wind loads, buffeting loads and motion-induced loads [1]. While the 10 minutes mean wind velocity with an annual probability of exceedance of 0.02 is typically considered when studying the buffeting load effects [2], the aeroelastic stability has to be ensured at more extreme wind speeds.

In the present work, the motion-induced wind loads on the bridge deck are based on the formulation by Scanlan [3]. The load model defines the self-excited forces as linear functions of the displacements and velocities based on the assumption of small oscillations. In the time-domain, the aeroleastic loads can be represented by indicial functions in convolutions with the time-history response of the bridge [4].

Strong winds generate large displacements of the bridge girder and the cables. The time-domain finite element analysis (FEA) performed, include modifications of the structural stiffness due to large displacements, i.e. geometric non-linearity, through iterative solutions of the system equations. While this involves higher computational effort in comparison to modal based analyses in the frequency-domain, time-domain analyses enables a direct inspection of wind-induced stresses in different parts of the structure throughout the simulated time-interval. The analyses can take into account material non-linearity [5] and can be utilized in the predictions of full-scale bridge behaviour at various stages of construction [6]. The assumptions of small oscillations constitute a severe restriction, generally not suitable for buffeting response analysis of long-span bridges. In response to this deficiency, Wu and Kareem [7] proposed an extension of the established linear wind load models in the form of a non-linear convolution scheme to account for the non-linear bridge aerodynamics.

Aerodynamic impulse-response functions, the so called indicial functions, have been previously applied in the time-domain bridge analysis, with reference to modal coordinates or in the full-bridge finite element models by a number of researchers [8-12]. All these studies, only took into account the lift and twisting moment induced from the vertical and twisting motion of the girder, i.e. a set of four indicial functions were implemented. The present work, using a full-bridge FE-model, takes into account the full set of nine indicial functions based on the experimental flutter derivatives. The components of motion-induced drag as well as the lift and torsional moment due to the horizontal motion of the girder are then included in the load description. The objectives of this work are to study the aerodynamic stability of the bridge in smooth flow as well as to identify any influence on the bridge response from motion-induced drag and horizontal motion of the bridge girder. The time-domain responses are compared to results estimated from multimodal analysis in frequency-domain and highlighted in view of previous response analyses of the bridge [12-15] as well as wind-tunnel tests [16].

2. The Hardanger Bridge girder section

The computational method developed is applied in a case study considering the Hardanger Bridge, in service since 2013. The suspension bridge has nearly 200 m tall towers and a main span of 1310 m, connected to the 45 m and 25 m long viaducts. The girder is a wedge-shaped 18.3 m wide steel box with a width to depth ratio of 5.5. The mass per unit length is 12800 kg/m and the mass moment of inertia is 428400 kg·m^2/m. The bridge girder carries two lanes of the road traffic, one in each direction, and a separate lane for bicycles and pedestrians. The bridge girder cross-section is illustrated in figure 1.

A series of wind tunnel tests [17] was dedicated to investigate the box-girder's aerodynamic behaviour, including the vortex-shedding response, the buffeting response and the flutter instability. The aerodynamic characteristics of an updated bridge girder design were investigated by Svend Ole Hansen ApS [16], in collaboration with the Norwegian Public Road Administration (NPRA) [18]. The wind tunnel data was utilized in various studies to estimate the full-scale aerodynamic bridge response [13,14, 15,19]. In parallel with extensive analyses based on modal coordinates, a detailed verification of the bridge design was launched by NPRA, using a finite element model for the response analysis in time-domain. The present work was initiated to implement a comprehensive representation of the motion-dependent loads, through the use of indicial functions, for the time-domain FEA. The original wind-resistant design of the Hardanger Bridge was based on flutter derivatives estimated from ambient vibration data from 2 DOF section model tests [18] in which the lowest vertical and torsional modes were represented. The objective of the present work was to implement all self-excited load components based on experimental data. An additional and complete set of flutter derivatives was provided as part of a cooperation project between Ruhr-University of Bochum and the University of Stavanger (UiS). The forced vibrations tests, performed with a section model in nominally smooth flow, and the identification of a complete set of 18 flutter derivatives were carried out by the UiS partners [20].

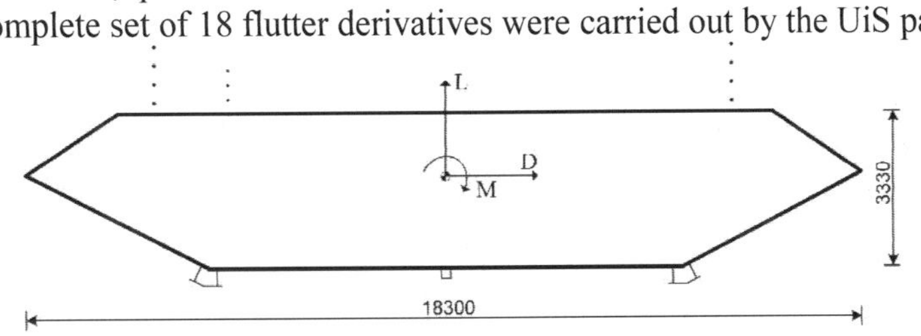

Figure 1. Hardanger Bridge girder section. Positive directions of lift, torsional moment and drag.

IOP Conf. Series: Materials Science and Engineering **276** (2017) 012006 doi:10.1088/1757-899X/276/1/012006

3. Finite element model

The finite element model, illustrated in figure 2, is defined in the general purpose finite element software Abaqus [21]. The model was made available for the present investigation by NPRA. The cables, hangers and the bridge girder are modeled with beam elements, of type B31 in the Abaqus notation. The elements are defined by two nodes, with 6 DOF per node, providing linear interpolation of displacements within the element. The bridge girder and the main cables are discretized in 64 sections corresponding to 20 m long elements, following the spacing between the hangers. The cables of the side spans are discretized with 20 elements. In total, 64 pairs of hangers are suspended from the main cables. Each hanger is modeled by one beam element. The cables and the hangers are assigned a negligible bending stiffness, thus effectively acting as truss elements.

The model devoted to the aeroelastic bridge analyses includes a simplified representation of the bridge towers, by linear springs and point masses. The structural damping of the bridge girder is differentiated for the lateral, the vertical and the twisting response. A damping ratio, relative to critical, was assumed 0.35% for both the fundamental horizontal and the lowest vertical mode and 0.32% for the first torsional mode. The structural damping is implemented with dashpot elements added to the girder and the cable elements in combination with stiffness proportional damping distributed to the beam elements along the girder. Input data i.e. dashpot coefficients and Rayleigh parameters corresponding to the assumed damping ratios is determined by iterative procedures analyzing the forced vibration response of the FE-model.

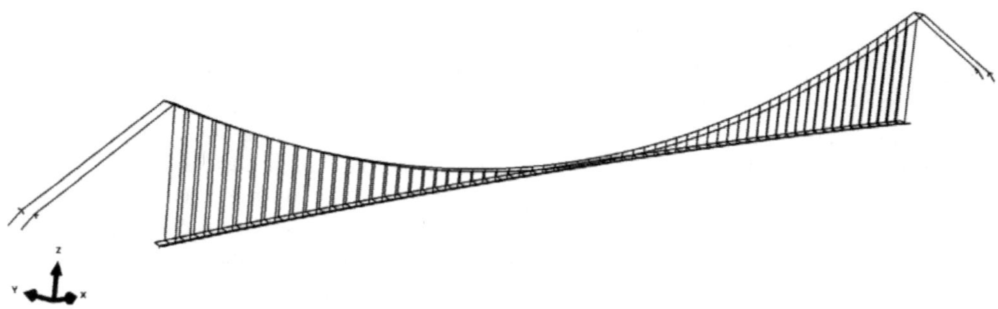

Figure 2. Finite element model.

4. Time-domain solutions

The equation of motion of the bridge-wind system driven by the mean wind loads and the aeroelastic loads, $\boldsymbol{F}_q$ and $\boldsymbol{F}_{ae}$ respectively, is given in equation (1):

$$\boldsymbol{M}\ddot{\boldsymbol{R}}(t)+\boldsymbol{C}\dot{\boldsymbol{R}}(t)+\boldsymbol{K}\boldsymbol{R}(t)=\boldsymbol{F}_q+\boldsymbol{F}_{ae}(t) \quad (1)$$

Here, $\boldsymbol{M}$, $\boldsymbol{C}$, and $\boldsymbol{K}$ are the global mass, damping and stiffness matrices and $\boldsymbol{R}(t)$ is the nodal displacement vector of the structure. The non-linear solver option in Abaqus [21] relevant for direct implicit integration of the equation of motion is an implementation of the Hilber-Hughes-Taylor method [22]. The algorithm has options to use a numerical damping and an adaptive time stepping scheme. In the present analyses, a fixed time step of 0.008 s and zero numerical damping were applied, and the numerical integration carried out using the trapezoidal rule. Wind load on the bridge girder and the cable elements was implemented by developing special purpose elements in Fortran. The *uel* subroutine is called at each time-step during iterations and the update of the self excited loads. The aeroelastic loads, $F_{ae}(t)$, are defined by convolutions implementing indicial functions and the motion of the girder. Self-excited drag loads acting on the cables and the hangers are defined by the quasi-steady theory.

5. Motion-induced forces

The self-excited lift force L_{ae}, per unit length of the bridge girder, is given in equation (2) following the notation by Scanlan [3]:

$$L_{ae} = q_0 B\left[kH_1^*\frac{\dot{r}_z}{U} + kH_2^*\frac{\dot{r}_\theta}{U} + k^2H_3^* r_\theta + k^2H_4^*\frac{r_z}{B} + kH_5^*\frac{\dot{r}_y}{U} + k^2H_6^*\frac{r_y}{B}\right] \tag{2}$$

where dynamic pressure, mean wind velocity and the deck width are denoted q_0, U and B respectively. Flutter derivatives H_i^* are functions of the reduced frequency $k=\omega B/U$.

The motion-induced twisting moment M_{ae} is expressed analogously to equation (2) by replacing H_i^* with A_i^* and the drag D_{ae} by replacing $H_i^*$with P_i^*.

5.1. *Formulation in terms of rational functions*

By defining a non-dimensional displacement amplitude vector $\boldsymbol{r_0}=\{z_0/B\ \theta_0\ x_0/B\}^{\mathrm{T}}$, assuming harmonic oscillations at an arbitrary frequency ω, the motion-induced loads can be expressed with (3) [23]:

$$\boldsymbol{F_{ae}} = \begin{Bmatrix} L_{ae}B \\ M_{ae} \\ D_{ae}B \end{Bmatrix} = q_0 B^2 \begin{Bmatrix} Q_{11} & Q_{12} & Q_{13} \\ Q_{21} & Q_{22} & Q_{23} \\ Q_{31} & Q_{32} & Q_{33} \end{Bmatrix} \boldsymbol{r}_0 e^{i\omega t} = \begin{Bmatrix} k^2(H_4^*+iH_1^*) & k^2(H_3^*+iH_2^*) & k^2(H_6^*+iH_5^*) \\ k^2(A_4^*+iA_1^*) & k^2(A_3^*+iA_2^*) & k^2(A_6^*+iA_5^*) \\ k^2(P_4^*+iP_1^*) & k^2(P_3^*+iP_2^*) & k^2(P_6^*+iP_6^*) \end{Bmatrix} \boldsymbol{r}_0 e^{i\omega t} \tag{3}$$

Equation (3) defines nine complex transfer functions Q_{mn} for the motion-induced loads. Roger [24] proposed a rational function approximation of the aeroelastic loads, given in equation (4) as presented by Abel [25]:

$$\boldsymbol{Q}(p) = \boldsymbol{A}_0 + p\boldsymbol{A}_1 - p^2\boldsymbol{A}_2 + \sum_{i=1}^{np}\frac{p}{p+\gamma_l}\boldsymbol{A}_{l+2} \tag{4}$$

Introducing a complex variable $p=\beta+ik$ and applying the inverse Laplace transform, equation (4) transforms into the expressions for the impulse response functions $\boldsymbol{I}(s)$ in the time-domain. The corresponding step response functions $\boldsymbol{\phi}(s)$ defined in equation (5), called indicial functions, are expressed in terms of a unit step response function $\varepsilon(s)$ and the Dirac delta function $\delta(s)$ [23]:

$$\boldsymbol{\phi}(s) = \boldsymbol{A}_0\varepsilon(s) + \boldsymbol{A}_1\delta(s) + \boldsymbol{A}_2\delta'(s) + \sum_{l=1}^{np}\boldsymbol{A}_{l+2}\exp(-\gamma_l s) \tag{5}$$

The aeroelastic loads determined by convolutions with indicial functions and the structural response are given in equation (6) [26].

$$\boldsymbol{F}_{ae}(s) = \boldsymbol{\phi}(s) * \dot{\boldsymbol{r}}(s) = \boldsymbol{I}(s) * \boldsymbol{r}(s) \tag{6}$$

When transferred to physical time t, the loads take the form of equation (7) where n specifies nodal location along the bridge girder.

$$\boldsymbol{F}^n{}_{ae}(t) = q_0\boldsymbol{B}^2\left\{\boldsymbol{A}_0\,\boldsymbol{r}(t) + \boldsymbol{A}_1\frac{B}{U}\dot{\boldsymbol{r}}(t) + \boldsymbol{A}_2\frac{B^2}{U^2}\ddot{\boldsymbol{r}}(t) + \sum_{l=1}^{np}\boldsymbol{A}_{l+2}\exp(-\gamma_l\frac{U}{B}t) * \dot{\boldsymbol{r}}(t)\right\} \tag{7}$$

The transfer functions in equation (4) can be separated into real and imaginary parts as shown in equation (8) and (9):

$$\Re[\boldsymbol{Q}(ik)] = \boldsymbol{A}_0 - k^2\boldsymbol{A}_2 + \sum_{l=1}^{np}\frac{k^2}{k^2+\gamma_l^2}\boldsymbol{A}_{l+2} \tag{8}$$

IOP Conf. Series: Materials Science and Engineering **276** (2017) 012006 doi:10.1088/1757-899X/276/1/012006

$$\Im[\boldsymbol{Q}(ik)] = k\boldsymbol{A}_1 + \sum_{l=1}^{np} \frac{k\gamma_l}{k^2 + \gamma_l^2} \boldsymbol{A}_{l+2} \tag{9}$$

The transfer functions Q_{mn} are fitted to the experimentally obtained flutter derivatives along the imaginary axis k in the complex plane. The real parts of the functions characterize aeroedynamic stiffness, whereas the imaginary parts characterize aeroedynamic damping. The parameter matrix $\boldsymbol{A}_2$ is interpreted as aerodynamic mass and the corresponding load term is assumed negligible in the context of the bridge response analysis. The coefficient matrices $\boldsymbol{A}_i$ and the parameters γ_l are approximated with the extended minimum state method [23]. The $\boldsymbol{A}_0$ matrix incorporate the steady values of the flutter derivatives. To ensure decaying impulse functions, i. e. stability of the aeroelastic system, γ_l is restricted to have positive values. Mathematically, the γ_l parameters define the pole locations of $Q_{mn}(p)$ along the negative real axis. The transfer functions are constrained through a set of common poles γ_l. The summation terms of the functions, called lag terms, considerably extend the computational effort with respect to the response analyses in time-domain. Furthermore, in a state-space format, the number of lag terms contribute to the dimension of the aeroelastic system. In this context Karpel [27] proposed a minimum-state method to reduce the dimension of the state-space equations, which introduced a coupling between the coefficients in the lag terms. Tiffany and Adams [28] suggested to use the extended minimum state method, splitting up the minimization problem into a linear process to determine the coefficient matrices and a non-linear process for finding the location of the poles [23].

5.2. Previous studies using rational function approximations for aeroelastic response analyses of bridges

Wilde and Fujino [29] and Chen et al. [8] utilized Roger's approximation [24] in the application of wind-loads in the bridge response analyses. Wilde and Fujino based their work on the flutter derivatives corresponding to the theoretical load formulation by Theodorsen [30]. Chen et al. utilized both the experimentally obtained flutter derivatives for a twin-box girder and the Theodorsens functions and presented the selected approximations for H_3^*, A_1^*, A_2^* and A_3^*.

Peil and Kirch [31] investigated the suitability of the rational function approximations to properly describe the motion-induced wind loads on bridge girder sections as well as the flat plate. Based on the use of experimental quantified flutter derivatives, they demonstrated that the approximation error could be reduced by increasing the number of lag terms, but at the same time the approximation itself could be considerably distorted. Distortion, in this context, refers to a notable deviation of the approximations from the experimental data points in wide areas above the pole locations in the complex plane. This behaviour was connected to a strong weighting of the partial fractions in the evaluation process. The effects were found to be pronounced for the Tacoma Bridge girder section but not so significant for the East Great Belt Bridge girder section. In this respect, it should be noted that the The Hardanger Bridge girder section is a streamlined box-girder, similar to the case of the East Great Belt Bridge girder section.

Previous work on multimodal flutter analysis of the Hardanger Bridge [32] addressed the influence of the variable number of pole terms in the rational function approximations on the resulting flutter speed of the bridge. Deviations in flutter speed were found to be within 6% when using the above discussed function approximations, using four, six and nine pole terms.

5.3. Measurements and approximations of self-excited loads

Flutter derivatives representing lift, twisting moment and drag, a full set of 18 derivatives, were determined from tests in nominally smooth flow over a wide range of reduced frequencies (0.3 - 9). Details of the experimental set up for the forced vibration tests and the computational procedures for the identification of flutter derivatives are presented in Neuhaus et al. [20]. In the present investigation, the motion-dependent forces at zero mean wind angle of incidence are applied. Forcing amplitudes were up to 6% of the girder height in the vertical translatory motion and 1° in twisting motion. The tests were performed by forcing each section model degree of freedom separately and quantifying the associated load components in terms of the derivatives H_i^*, A_i^* and P_i^*.

Indicial function approximations with the extended minimum-state method [23], have been used to define the aeroelastic loads in this study. The response analyses with the FE-model in time-domain are performed using the indicial function approximations with six lag terms. With respect to the response analyses, it is to be noted that the reduced torsional frequency of the bridge girder is about 1 at the design wind speed and nearly 0.4 at the estimated flutter speed. It is thus essential that the approximations are most accurate in the corresponding range of reduced frequencies.

The experimental data and the fitted approximations of the aeroelastic transfer functions are presented in figures 3 and 4 in terms of their real and imaginary parts.

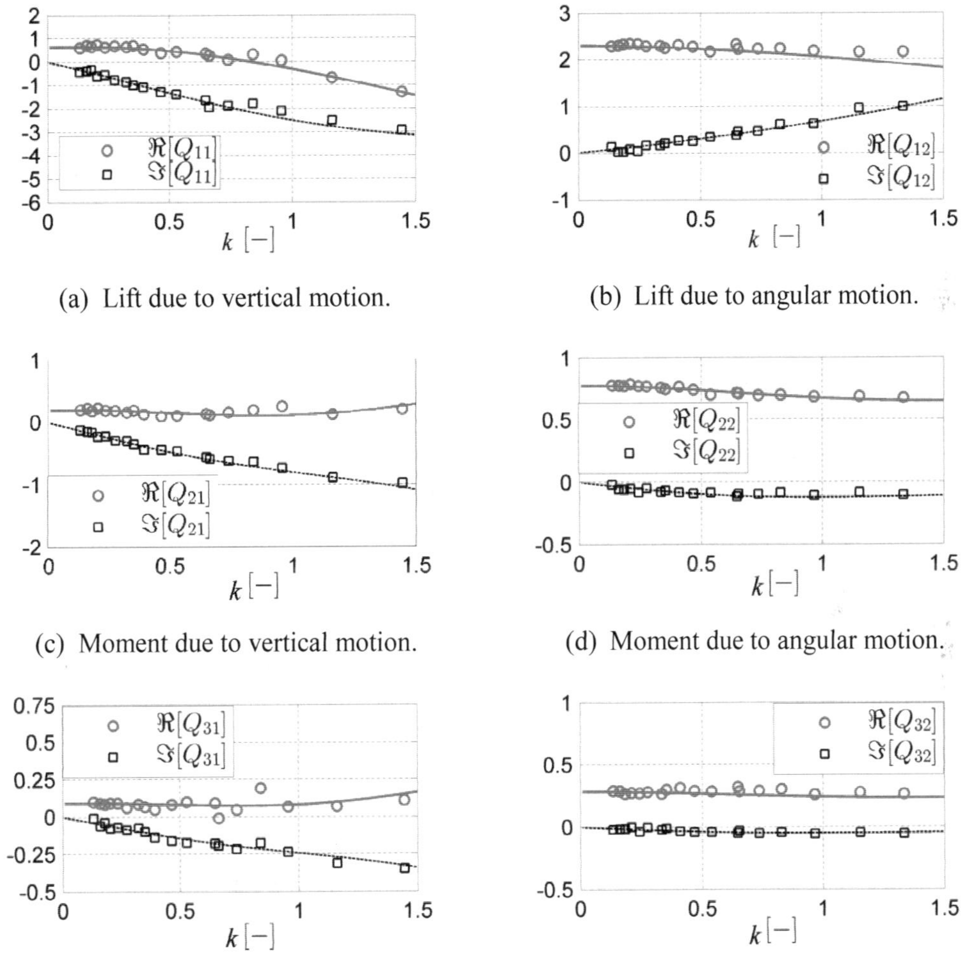

(a) Lift due to vertical motion. (b) Lift due to angular motion.

(c) Moment due to vertical motion. (d) Moment due to angular motion.

(e) Drag due to vertical motion. (f) Drag due to horizontal motion.

Figure 3. Transfer functions, real and imaginary parts due to vertical and angular motion.

The functions Q_{11} and Q_{12} related to the pairs of flutter derivatives H_4^*-H_1^* and H_3^*-H_2^*, exhibit satisfactory fit to the experimental data in the aforementioned essential frequency range. Equally good are the approximations for the functions related to the pairs of flutter derivatives A_4^*-A_1^* and A_3^*-A_2^*, represented by Q_{21} and Q_{22}. It can be noted that the real parts of the functions Q_{12} and Q_{22}, shown in figure 3(b) and 3(d), correspond at zero frequency to the slopes of the static coefficients for lift (2.34) and moment (0.86), respectively.

The experimental data and the rational function approximations associated with the horizontal motion of the girder are presented in figure 4. The associated aeroelastic damping components of lift and drag, represented by the imaginary parts of Q_{13} and Q_{33} in figures 4(a) and 4(c), both have a more or less linear trend and increasing magnitude with increasing reduced frequency. The real parts of the corresponding functions are close to zero in the most relevant reduced frequency range.

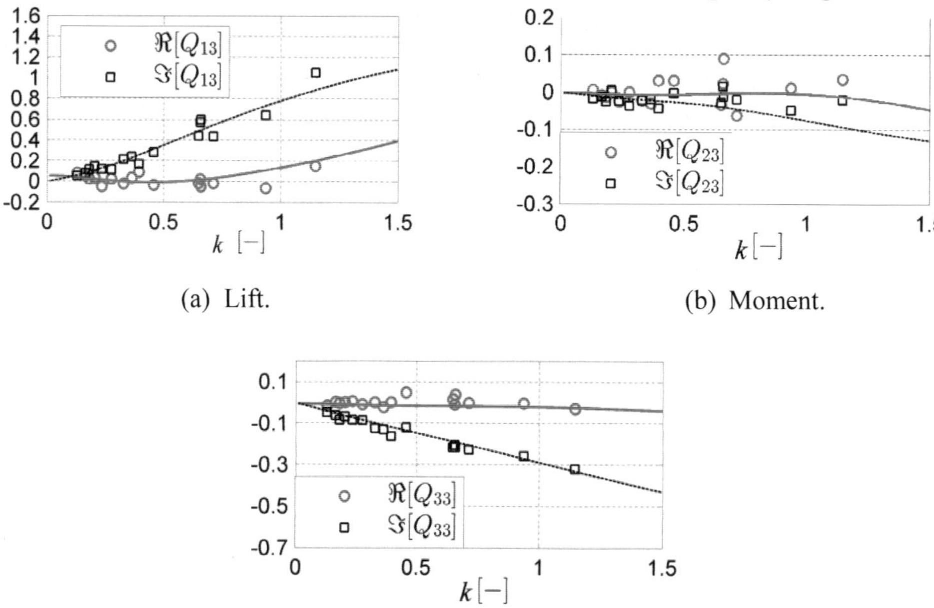

(a) Lift. (b) Moment.

(c) Drag.

Figure 4. Transfer functions, real and imaginary parts due to horizontal motion

The experimental values of the function Q_{23} in figure 4(b) representing the twisting moment due to horizontal motion, fluctuates between ± 0.1 in the reduced frequency range 0.3 -1.5. On average, these values are close to zero at low reduced frequencies. Twisting moment due to horizontal motion has thus negligible contribution to the aeroelastic loads in case of the present girder. The experimental data for the lift and the moment functions related to the horizontal motion, see figures 4(a) and 4(b), have large spreading over the parts of the essential frequency range, and the continuous transfer functions thus do not fit closely to the data points.

Repeated tests and measurements performed at reduced frequency $k \approx 0.7$ can be identified as a cluster of values in the plots of figures 3 and 4. The deviations between experimental values respresenting single test runs can be identified in the aeroelastic stiffness e.g. for $\Re[Q_{13}]$ and $\Re[Q_{33}]$ in figures 4(a) and 4(c) The imaginary parts of the transfer functions, except the aforementioned Q_{23} function, demonstrate satisfactory repeatability between the test runs.

5.4. Indicial functions

The aeroelastic transfer functions approximations presented in figure 4 are associated with nine indicial functions denoted $\phi_{Fi}(s)$, where F indicates the load component and i the contributing displacement. An indicial function expresses the evolution of the respective load component induced by the unit step change in the displacement component i. The non-dimensional time s can be interpreted as the distance, relative to the bridge deck width, travelled by the mean wind during time t. The indicial functions are presented in figure 5.

All functions related to the translatory motion, i.e. ϕ_{Fz} and ϕ_{Fy} presented in figure 5(a) and 5(c) are after sign reversals close to zero for non-dimensional time $s > 1$. The twisting moment and drag induced by the vertical unit step, ϕ_{Mz} and ϕ_{Dz} respectively, as well as the lift due to the horizontal motion, ϕ_{Ly}, are all identified with negative initial values. The functions $\phi_{L\theta}$ and $\phi_{M\theta}$, as shown in figure 4(b) have positive initial values and attain the expected constant values, corresponding to the slopes of the static coefficients, within two units of the nondimensional time.

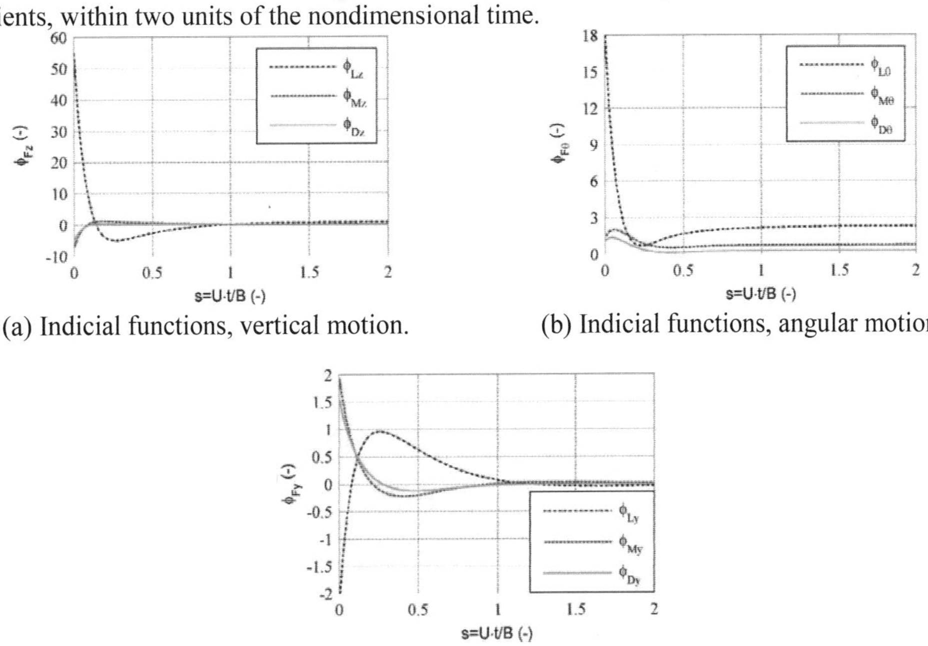

(a) Indicial functions, vertical motion. (b) Indicial functions, angular motion.

(c) Indicial functions, horizontal motion.

Figure 5. Indicial functions $\phi_{Fi}(s)$.

6. Frequency-domain analysis

The results of the time-domain aeroelastic analysis will be compared to those obtained by performing a multi-modal analysis in the frequency.domain.

By denoting the structural modal quantities $\boldsymbol{M}_s$, $\boldsymbol{C}_s$ and $\boldsymbol{K}_s$ as well as the modal displacement vector $\boldsymbol{Z}$ and applying the load model in equation (7), the equation of motion is written:

$$\boldsymbol{M}_s\ddot{\boldsymbol{Z}}+\boldsymbol{C}_s\dot{\boldsymbol{Z}}+\boldsymbol{K}_s\boldsymbol{Z}-q_0B^2\left\{\boldsymbol{A}_0\boldsymbol{Z}+\frac{B}{U}\boldsymbol{A}_1\dot{\boldsymbol{Z}}+\sum_{l=1}^{np}\boldsymbol{A}_{l+2}\boldsymbol{Z}_a\right\}=\boldsymbol{0} \tag{10}$$

$\boldsymbol{Z}_a$ in equation (10) defines the aerodynamic states. With the state vector set equal to $\boldsymbol{Z}_{st}=\{\boldsymbol{Z}\,\dot{\boldsymbol{Z}}\,\boldsymbol{Z}_a\}^{\mathrm{T}}$, the bridge-wind system matrix is established with equation (11):

$$\begin{Bmatrix}\dot{\boldsymbol{Z}}\\ \ddot{\boldsymbol{Z}}\\ \dot{\boldsymbol{Z}}_a\end{Bmatrix}=\begin{bmatrix}\boldsymbol{0} & \boldsymbol{I} & \boldsymbol{0} & \cdots & \boldsymbol{0}\\ -\boldsymbol{M}_s^{-1}\cdot\boldsymbol{K}_q & -\boldsymbol{M}_s^{-1}\cdot\boldsymbol{C}_q & \boldsymbol{M}_s^{-1}\cdot q_0B^2\cdot\boldsymbol{D}\cdot\boldsymbol{A}_3 & \cdots & \boldsymbol{M}_s^{-1}\cdot q_0B^2\cdot\boldsymbol{D}\cdot\boldsymbol{A}_{np+2}\\ \boldsymbol{0} & \boldsymbol{I} & -U/B\cdot\gamma_1\cdot\boldsymbol{I} & \cdots & \boldsymbol{0}\\ \vdots & \vdots & \vdots & \vdots & \vdots\\ \boldsymbol{0} & \boldsymbol{I} & \boldsymbol{0} & \cdots & -U/B\cdot\gamma_{np}\cdot\boldsymbol{I}\end{bmatrix}\begin{Bmatrix}\boldsymbol{Z}\\ \dot{\boldsymbol{Z}}\\ \boldsymbol{Z}_a\end{Bmatrix} \tag{11}$$

where $\boldsymbol{C}_q$ and $\boldsymbol{K}_q$ incorporate both the structural and the aerodynamic modal properties of the system and coefficients D_{ij} in $\boldsymbol{D}$ are equal to $\int\psi_i(x)\cdot\psi_j(x)dx\,/\,\int\psi_i^2(x)dx$ where $\psi_i(x)\cdot\psi_j(x)$ are the combined modes.

IOP Conf. Series: Materials Science and Engineering **276** (2017) 012006 doi:10.1088/1757-899X/276/1/012006

7. Results

7.1. Natural modes

Natural frequencies *f* and periods *T* for the still-air eigen-modes, estimated with the finite element model, are presented in Table 1. *Vs-i*, *Hs-i* and *Ts*-i, with *i*=1-3, denote the vertical, horizontal and the torsional direction. The frequency ratio between the first torsional and the first vertical mode is 2.6. The first three symmetrical modes in each direction are presented in figure 6.

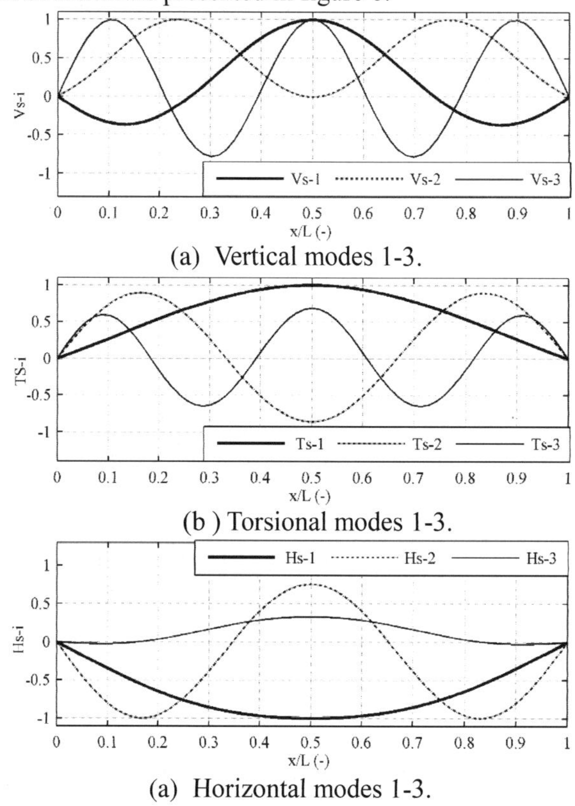

Figure 6. Symmetrical modes. *Vs-i*, *Ts-i* and *Hs-i*, *i*=1-3.

A horizontal component of the first torsional mode, also identified in [18], is included in figure 7. The torsional response corresponding to the vertical motion of cable planes of ±1 m is accompanied by the horizontal displacement of 0.16 m at girder mid-span.

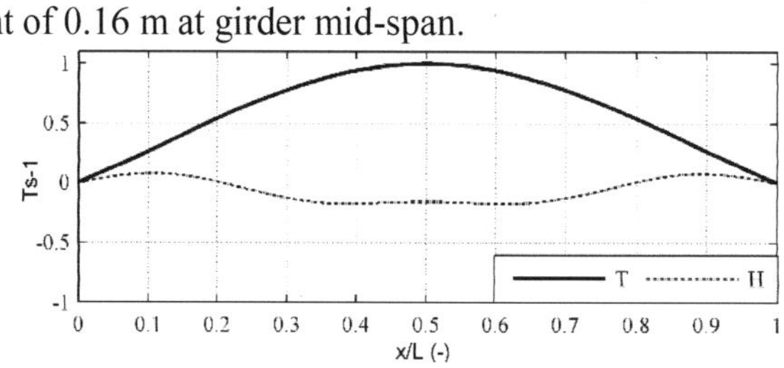

Figure 7. Torsional first symmetric mode *Ts-1*. T: translatory vertical displacement in cable plane at windward side. H: horisontal displacement at girder axis.

Table 1: Natural frequencies and periods.
FE- model results for symmetrical modes

Symmetrical modes *	*f* [Hz]	*T* [s]
Hs-1	0.0503	19.9
Vs-1	0.1408	7.1
Hs-2	0.1736	5.76
Vs-2	0.1982	5.04
Hs-3	0.2449	4.08
Vs-3	0.2728	3.67
Ts-1	0.3602	2.78
Ts-2	0.7829	1.28
Ts-3	1.2395	0.81

* Hs: horizontal, Vs: vertical, Ts: torsional, see Figure 6

7.2. Time-domain analyses

7.2.1. Dynamic response with full set of motion-induced loads

The full set of nine indicial functions in convolutions with the nodal motions of the bridge girder generate the motion-induced loads on the girder and act together with the mean wind loads during the simulations. Depending on the magnitude of the bridge vibrations the simulations are carried out for a time window of t=0-800 s. Initial excitation of the bridge is generated by superposing turbulent loads to the mean wind- and the aeroelastic loads during the first 60 s of the analyses.

The motion-induced drag on the cables and hangers are implemented by quasi-steady assumptions in the form of equation (12):

$$\boldsymbol{F}^{C}{}_{ae}(t) = -\frac{q_0}{U} 2C_C D_C L_C \boldsymbol{R}^{C}_{yd}(t) \tag{12}$$

where C_C is static shape coefficient, D_C is representative cable diameter, L_C the length of cable element and $\boldsymbol{R}^C{}_{yd}(t)$ is horizontal displacements relative to time-averages in the mean wind direction.

The transient response, at design wind speed 38.6 m/s, at the girder mid-span, is shown in figure 8. The directions of structural displacements in the fixed global coordinate system are adopted such that the positive r_{yd} is horizontal displacement in the mean wind direction, positive $r_{\theta d}$ is the rotation of the bridge girder with the windward side moving upwards, and the positive r_{zd} is the upwards vertical displacement.

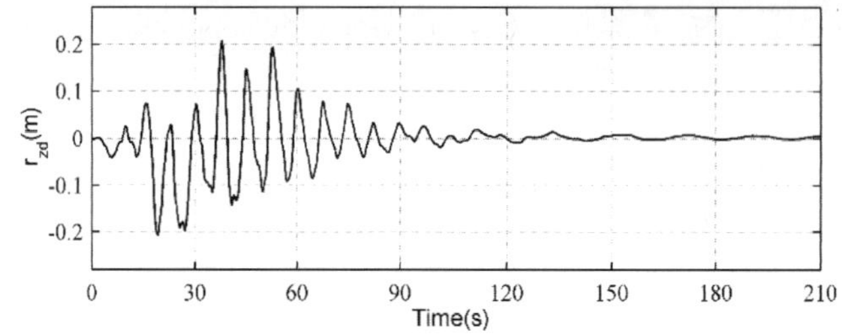

(a) Vertical displacements.

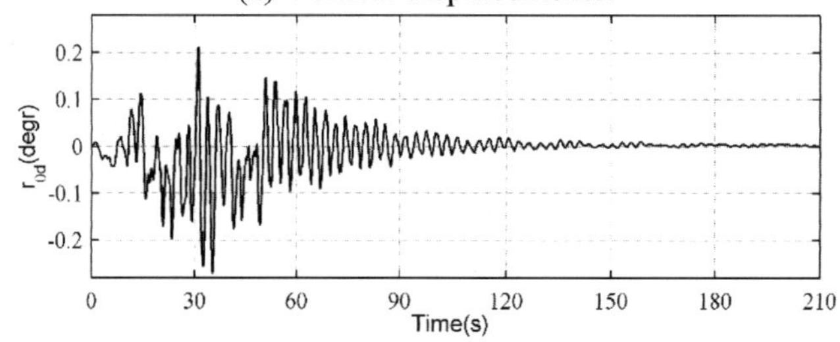

(b) Torsional displacements.

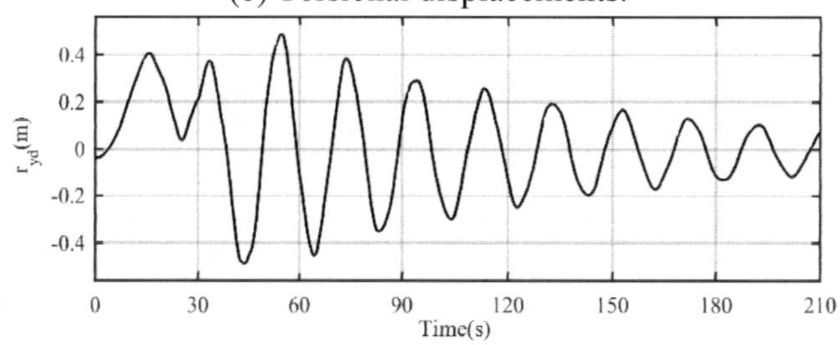

c) Horizontal displacements.

Figure 8. Aeroelastic response at mid-span girder at design wind speed. *U*=38.6 m/s.

The three response components in figure 8 all exhibit significant aerodynamic damping. Total damping, relative to critical, estimated from the time series is 6.0%, 1.5% and 3.9% for the vertical, torsional and the horizontal component, respectively. A minor component at the frequency of the still-air horizontal mode is identified in the vertical response in figure 8(a) in the time interval from 120 s-210 s.

Flutter velocity is estimated to be 75.6 m/s. At this velocity, effectively undamped coupled oscillations develop within 180 s, see figure 9(a) and 9(b), which visualize the in-wind coupling between the torsional and the vertical motion. Figure 9(a) illustrates how the vertical response component at the in-wind torsional frequency fully overshadows the response component at the still-air fundamental frequency, after 60 s. The horizontal motion in figure 9(c) is significantly damped in the flutter oscillations, i.e. the horizontal component is not governing the flutter response. The estimated damping ratio is at 8.1%. Horizontal displacements are reduced from amplitudes of almost 2 m to a negligible motion during 5-6 cycles and a minor undamped component at the torsional frequency is identified at time interval 240-300 s.

Figure 10 presents an expanded view of the flutter response, in the time window 605-615 s. The flutter frequency estimated from the time-series is 0.295 Hz. The twisting component is here presented in terms of the corresponding vertical displacement of one of the hanger planes, $d/2 \cdot r_{\theta d}$, where d is the distance between cable planes. The plot indicates that the vertical motion lags the torsional one by 160°.

A time instance at which the twisting response has a local maximum ($t \approx 614.3$ s) is indicated in figure 10 and the flutter mode at the indicated time instance is presented in figure 11.

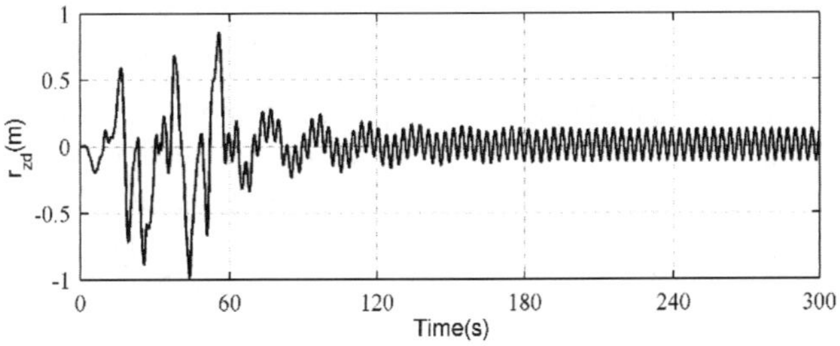

(a) Vertical displacemnents.

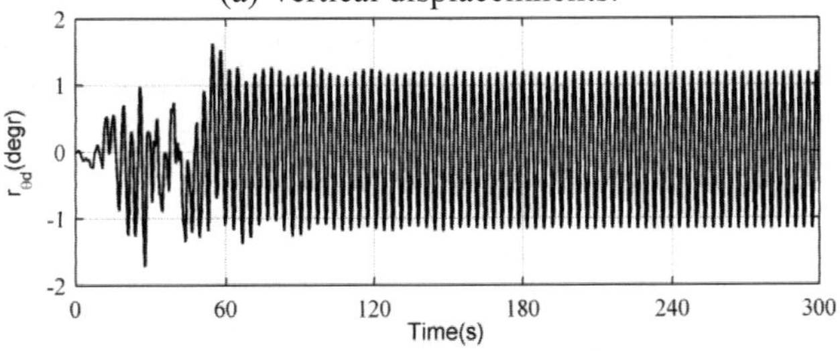

(b) Torsional displacements.

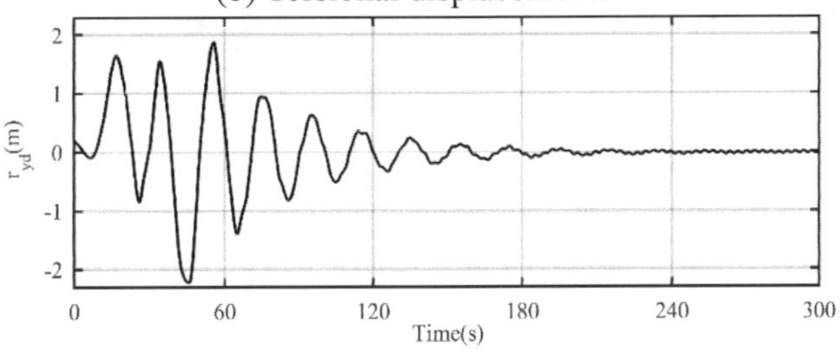

c) Horizontal displacements.

Figure 9. Aeroelastic response at mid-span girder. U=75.6 m/s.

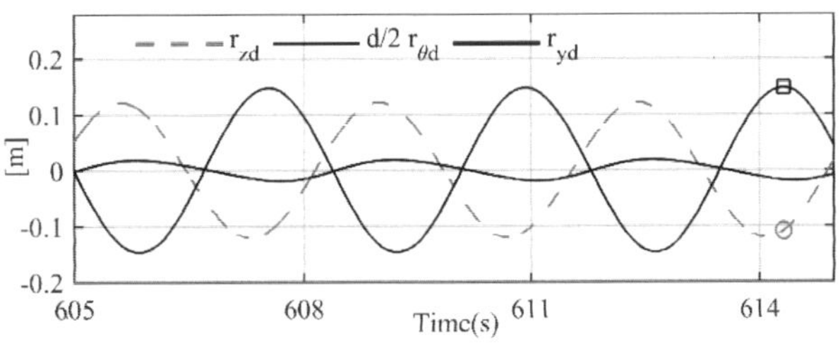

Figure 10. Flutter oscillations at mid span girder. f=0.295 Hz U=75.6 m/s. d=14.5 m.

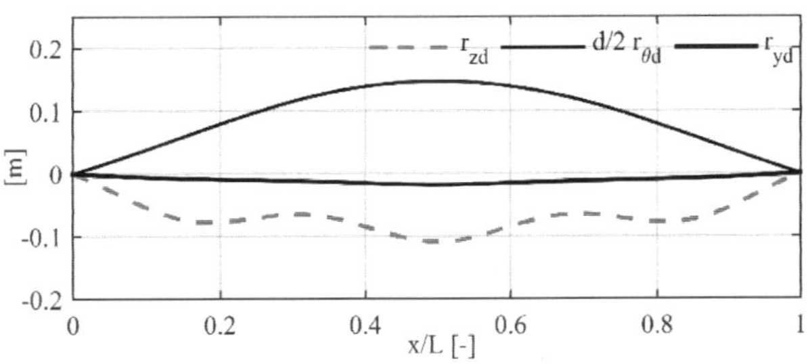

Figure 11. Flutter oscillations. Displacements along bridge girder. U=75.6 m/s. d=14.5 m.

7.2.2. Dynamic response with reduced set of motion-induced loads

In order to separate the effect of the motion-dependent wind load related to the horizontal bridge response, the characteristics of the aeroelastic response based on two different reduced sets of motion-induced loads is presented in the following. The first case concerns the time-domain analyses with 4 indicial functions ϕ_{Lz}, $\phi_{L\theta}$, ϕ_{Mz} and $\phi_{M\theta}$, and the second one involves an additional indicial function for the aeroelastic drag due to the horizontal motion, ϕ_{Dy}. The two reduced sets of motion-dependent loads are denoted "4 i.f." and "5 i.f." The motion-induced loads on cables are included.

Figure 12 presents 300 s long time-histories of responses at mid-span girder, for the two above specified load cases, both at the mean wind velocity of 76.3 m/s. The interaction between the vertical and the twisting motion is recognized in both load cases. In the case, based on the five indicial functions, the flutter oscillations develop within approximately 180 s. In the load case with four indicial functions, the lowest horizontal mode contributes significantly to the vertical response, even after 300 s. The impact of the motion-induced drag damping is demonstrated in figures 12(a) and 12(c). The horizontal motion in figure 12(c) is associated with the damping ratios of 2.5% and 10% for the load cases with 4 i.f. and 5 i.f. respectively.

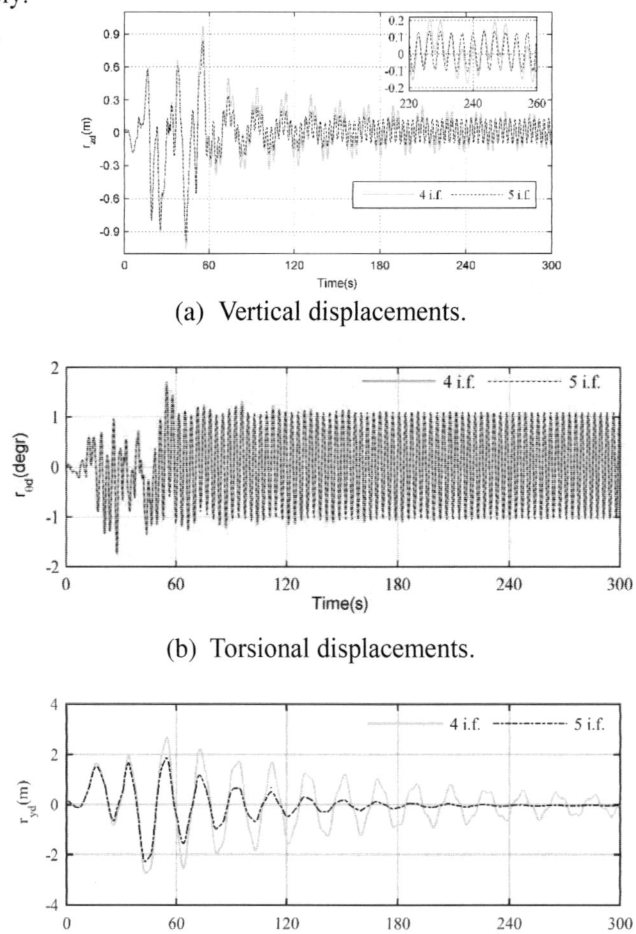

(a) Vertical displacements.

(b) Torsional displacements.

(c) Horizontal displacements.

Figure 12. Flutter oscillations. U_{fl} 76.3 m/s.

IOP Conf. Series: Materials Science and Engineering **276** (2017) 012006 doi:10.1088/1757-899X/276/1/012006

7.3. *Multimode flutter analysis*

The stability of the aeroelastic system is also investigated in the frequency domain, by performing the eigenvalue analysis of the system matrix in equation (11). Eigenvalue solutions are studied by increasing the mean wind speed in steps of 0.1 m/s. Nine lowest symmetric modes are included in the analysis: *Vs-i* (1-4), *Ts-i* (1-2) and *Hs-i* (1-3). The damping ratios of the in-wind bridge modes are presented in figure 13. Critical flutter speed is determined to be 71.9 m/s, the velocity at which damping for one of the modes becomes zero.

The flutter mode is illustrated in the complex plane in figure 14. For clarity, only the most significant contributions are given in the plot. Figure 14 shows that the three lowest vertical modes contribute in the flutter mode. The flutter frequency from the eigenvalue solutions is 0.245 Hz.

Figure 15 shows the time-domain flutter response identified in figure 11 together with the eigenvector from the multimodal response analysis. The two flutter modes have a strong similarity.

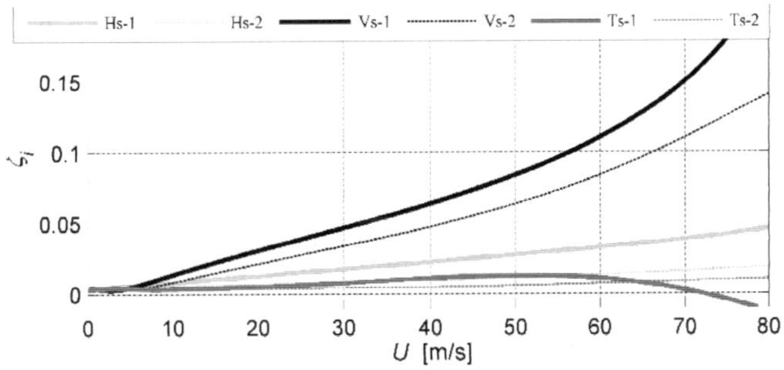

Figure 13. Damping ratios at eigenvalue paths.

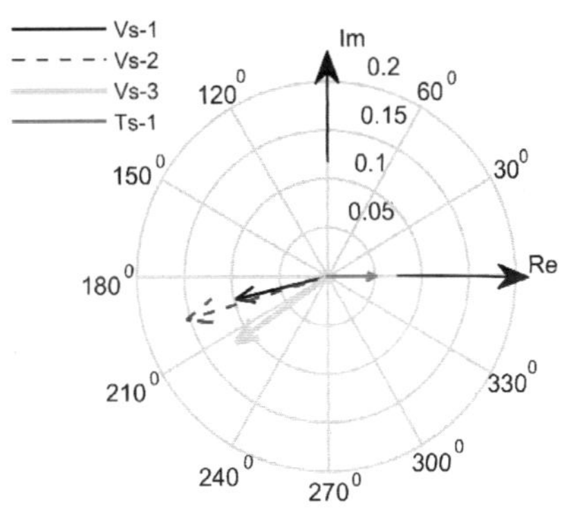

Figure 14. Flutter mode, U_{fl}=71.9 m/s. f=0.245 Hz.

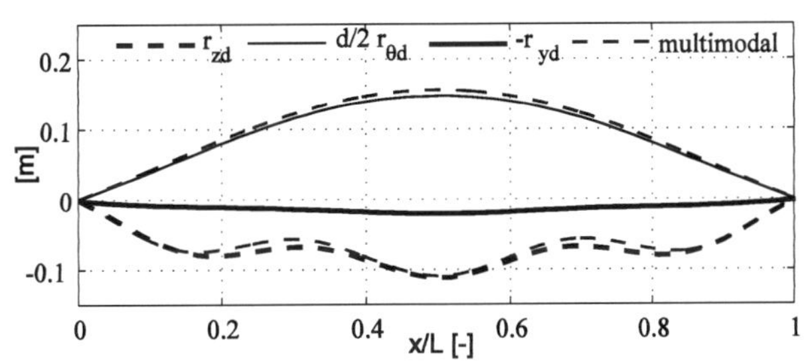

Figure 15. Time-domain response at U_{fl}=75.6. Multimodal response at U_{fl}=71.9 m/s.

8. Discussion

The wind-tunnel tests undertaken for verifications of the aerodynamic design of the Hardanger Bridge girder [16] indicated that the full-scale flutter velocity is 79.5 m/s in smooth flow and 70.3 m/s in turbulent flow, based on the testing of a 2 DOF section model in ambient vibrations. Flutter velocity estimated from a time-domain analysis using 2 DOF model with the set of 4 indicial functions and the methodology presented above was 76.9 m/s [33], or 3% lower than the experimental result reported in smooth flow. The present load model provides thus a critical velocity estimate in a very good agreement with the wind tunnel test results.

In the present study, the coupled flutter oscillations of the prototype bridge involve the three lowest vertical modes of vibration. In previous studies based on multimodal analysis, Jakobsen and Hjorth-Hansen [13] highlighted the contribution of the second vertical mode, but not the third one, to flutter oscillations. Their work, based on the motion-induced forces deduced from the ambient vibration data, also concerned a slightly different, preliminary design of the bridge.

A flutter velocity of 76.3 m/s, as estimated in the present analysis, is also consistent with the values computed by Øiseth et al. [14] for the same structure in the frequency domain, using aerodynamic derivatives (79 m/s) and modified quasi-steady load estimates (76 m/s). A similar value (78.3 m/s) was obtained by Øiseth et al [15] in a time-domain analysis with a finite element model, using the so-called aerodynamic degrees of freedom to model the motion-dependent forces on the bridge represented by a simply supported beam.

9. Conclusions

This paper presents the response analyses of a full-bridge FE-model of the Hardanger Bridge. The objective is to study bridge response in time-domain at high wind speeds considering all components of the motion-dependent loads.

The flutter velocity is found to be more or less unaffected by the motion-induced drag components as well as by the lift and twisting moment induced from the horizontal motion of the bridge girder. On the other hand, the motion-dependent drag loads due to the horizontal motion of the girder contributed to the horizontal and the vertical transient response of the bridge.

The critical flutter velocity was estimated as 75.6 m/s, using the FE-model with a complete set of aeroelastic load components. That is 5% higher than the estimate based on multimodal analysis. The flutter mode shapes in the time-domain and the frequency domain are consistent and involve contributions of the three lowest symmetric vertical modes, together with the lowest torsional mode.

The in-wind torsional frequencies in the time-domain and in the frequency-domain complied within 1% at the design wind-speed. The flutter frequency was 20% higher in time-domain analyses compared to the results from multi-modal analysis. Thus, at the flutter velocity, the apparent reduction in torsional stiffness is distinctly lower using the FE-model compared to the results from the multimodal analysis.

The methodology applied in this work, provides detailed response estimates and can be utilized in general studies of bridge response to increase understanding of the complex aeroelastic behaviour of long-span bridges.

Acknowledgement

The authors express their gratitude to Prof. Höffer of Ruhr-Universität Bochum and his wind engineering team for building the section model and carrying out the forced vibration wind tunnel tests. The bridge structural data provided by the Norwegian Public Road Administration as well as the access to results of separate model tests by Svend Ole Hansen ApS in cooperation with Prof. Hjorth-Hansen of the Norwegian University of Science and Technology are also greatly acknowledged.

References

[1] Scanlan R H 1997 Some observations on the state of bluff-body aeroelasticity, *Journal of Wind Engineering and Industrial Aerodynamics*, **69-71**, 77-90.

[2] Eurocode 1 2005 Actions on structures", Part 1-4: *General Actions-Wind Actions*, European Committee for Standardization, Brussels,.

[3] Dowell E H (Editor), Clark R, Cox D, Curtiss Jr. H C, Edwards J W, Hall K C, Peters D A, Scanlan R H, Simiu E, Sisto F and Strganac T W 2004 *A Modern Course in Aeroelasticity*, 4th Edition, Kluver Academic Publishers, Dordrecht, The Netherlands.

[4] Scanlan R H 1996 Aerodynamics of cable-supported bridges, *Journal of Constructional Steel Research*, **39**(1), 51-68.

[5] Chobsilprakob P, Suthasupradit S and Kim K 2011 Aeroelastic Analysis of Long Span Bridges via Indicial Functions Considering Geometric and Material Nonlinearity, *International Journal of Steel Structures*, **11**(2), 215-226.

[6] Chobsilprakob P, Suthasupradit S, Kim K and Manovachirasan A 2014 Application of Indicial Function for the Flutter Analysis of Long Span Suspension Bridge During Erection, *International Journal of Steel Structures*, **14**(1), 185-194.

[7] Wu T and Kareem A, 2013 A nonlinear convolution scheme to simulate bridge aerodynamics", *Computers and Structures*, **128**, 259-271.

[8] Chen X, Matsumoto M and Kareem A 2000 Time domain flutter and buffeting response analysis of bridges, *Journal of Engineering Mechanics*, **ASCE, 126**(1), 7-15.

[9] Chen X and Kareem A 2002 Advances in modeling of aerodynamic forces on bridge decks, *Journal of Engineering Mechanics*, ASCE, **128**(11), 1193-1205.

[10] Costa C, Borri C, Flamand O and Grillaud G 2007 Time-domain buffeting simulations for wind–bridge interaction", *Journal of Wind Engineering and Industrial Aerodynamics*, **95**, 991–100.

[11] Salvatori L and Borri C 2007 Frequency- and time-domain methods for the numerical modeling of full-bridge aeroelasticity, *Computers and Structures*, **85**, 675-687.

[12] Øiseth O, Rönnquist A and Sigbjörnsson R 2011 Time domain modeling of self-excited aerodynamic forces for cable-supported bridges: A comparative study, *Computers & Structures*, **89**(13-14), 1306–1322.

[13] Jakobsen J B and Hjorth-Hansen E 2007 *Arne Selberg's formula for flutter in light of multi-modal flutter analysis*, Proc. 12th International Conference on Wind Engineering, Cairns, Australia, 143-150.

[14] Øiseth O, Rönnquist A and Sigbjörnsson R 2012 Finite element formulation of the self-excited forces for time-domain assessment of wind-induced dynamic response and flutter stability limit of cable-supported bridges, *Finite elements in analysis and design*, **50**, 173-183.

[15] Øiseth O, Rönnquist A and Sigbjörnsson R 2010 Simplified prediction of wind-induced response and stability limit of slender long-span suspension bridges, based on modified quasi-steady theory: A case study *Journal of Wind Engineering and Industrial Aerodynamics*, **98**(12), 730-741.

[16] Hansen S O Dec. 2006 *The Hardanger Bridge: Static and dynamic wind tunnel tests with a section model*, Prepared for: Norwegian Public Roads Administration, **Rev. 1**.

[17] Hjorth-Hansen E, Strømmen E, Jakobsen J B, Brathaug H P and Solheim E 1993 *Wind tunnel investigations for a proposed suspension bridge across the Hardanger fjord*, Structural Dynamics-EURODYN'93, Moan et al., (eds.), 1011-1018, Balkema.

[18] NPRA, Norwegian Public Road Administration 14.11.2008 *The Hardanger Bridge with approach roads rv.7/rv.13*, Ch. 1-2, Appendices, **Rev. 6.**

[19] Mikkelsen O and Jakobsen J B 2012 *Buffeting resonse analyses of a long-span suspension bridge in time-domain*", Proceedings of the 10th UK Conference on Wind Engineering, Bristol, 185-188.

[20] Neuhaus C H , Roesler S, Höffer R, Hortmanns M and Zahlten W , 2009 *Identification of 18 flutter derivatives by forced vibration tests–A new experimental rig*, Proceedings of the 5th European Conference on Wind Engineering, Firenze Univ. Press, 361-364.

[21] Dassault Systèmes 2009 *Abaqus 6.9 documentation*, Dassault Systèmes Simulia Corp., Providence, RI, USA.

[22] Hilber H M, Hughes T J R and Taylor R L 1977 Improved numerical dissipation for time integration algorithms in structural dynamics, *Earthquake Engineering and Structural Dynamics*, **5,** 283-292.

[23] Neuhaus C H, Mikkelsen O, Jakobsen J B, Höffer R and Zahlten W 2009 *Time domain representations of unsteady aeroelastic wind forces by rational function approximation*, Proceedings of the 5th European Conference on Wind Engineering, Fir. Univ. Press, 339-342.

[24] Roger K 1977 *Airplane math modeling methods for active control design, Structural aspects of active controls*, Agard Conference Proceedings, **4.3.**

[25] Abel I 1979 A*n Analytic Technique for Predicting the Characteristics of a Flexible Wing Equipped With an Active Flutter-suppression System and Comparison with Wind- tunnel Data*, NASA-TP-**1367**(5).

[26] Neuhaus C 2010 *Zur Identifikation selbsterregter aeroelastischer Kräfte", Institut fur Konstruktiven Ingenieurbau*, Thesis, **Heft 15**, Bergische Universität, Wuppertal.

[27] Karpel M 1981 *Design for active and passive flutter suppression and gust alleviation*, NASA Contractor report **3482**

[28] Tiffany S and Adams W 1988 *Nonlinear programming extensions to rational function approximation methods for unsteady aerodynamic forces*, NASA-TP-**2776**.

[29] Wilde K and Fujino Y 1988 Aeodynamic control of bridge deck flutter by active surfaces", *Journal of Engineering Mechanics,* **ASCE 124**(7), 718-727

[30] Theodorsen T 1934 *General theory of aerodynamic instability and the mechanism of flutter*, NACA report **496.**

[31] Peil A and Kirch A 2008 *Approximation of unsteady aerodynamic derivatives by rational functions*, Conference proceeding, 7th European Conference on Structural Dynamics, E181, Southampton

[32] Mikkelsen O and Jakobsen J B 2010 *Flutter Analysis of the Hardanger Bridge*, Proceedings of the 9th UK Conference on Wind Engineering, Southampton, 47-58.

[33] Mikkelsen O 2013 *Modelling of wind loads and wind-induced response of a long-span bridge in time-domain*, Faculty of Science and Technology, PhD Thesis **no. 194**, University of Stavanger, Stavanger.

IOP Conf. Series: Materials Science and Engineering **276** (2017) 012007 doi:10.1088/1757-899X/276/1/012007

Full-scale monitoring of wind and suspension bridge response

J T Snæbjörnsson[1,2], J B Jakobsen[1], E Cheynet[1] and J Wang[1]

[1] Department of Mechanical and Structural Engineering and Materials Science, University of Stavanger, Norway

[2] School of Science and Engineering, Reykjavik University, Iceland

E-mail: `jonasthor@ru.is`

Abstract. Monitoring of real structures is important for many reasons. For structures susceptible to environmental actions, full-scale observations can provide valuable information about the environmental conditions at the site, as well as the characteristics of the excitation acting on the structure. The recorded data, if properly analyzed, can be used to validate and/or update experiments and models used in the design of new structures, such as the load description and modelling of the structural response. Various aspects of full-scale monitoring are discussed in the paper and the full-scale wind engineering laboratory at the Lysefjord suspension bridge introduced. The natural excitation of the bridge comes from wind and traffic. The surrounding terrain is complex and its effect on the wind flow can only be fully studied on site, in full-scale. The monitoring program and associated data analysis are described. These include various studies of the relevant turbulence characteristics, identification of dynamic properties and estimation of wind- and traffic-induced response parameters. The overall monitoring activity also included a novel application of the remote optical sensing in bridge engineering, which is found to have an important potential to complement traditional "single-point" wind observations by sonic anemometers.

1. Introduction

During the last 4 years, an extensive monitoring program has been ongoing on the Lysefjord Bridge in Norway [1, 2, 3, 4, 5]. The aim is, among other things, to develop measurement methods and to increase understanding of the behavior of suspension bridges across deep, and often wide, Norwegian fjords. The project has been cofounded by the University of Stavanger, the Norwegian Center for Offshore Wind Energy (NORCOWE) and the Norwegian Public Road Administration (NPRA) as the main contributor. The project is considered as a part of the multifaceted ongoing preparatory research for the ferry-free Coastal Highway Route E39 project [6, 7].

This paper discusses the role of monitoring loading processes and structural response in modern society. The monitoring program at the Lysefjord suspension bridge is introduced and described in terms of instrumentation, data acquisition, data management and communication systems, data analysis and related research on wind load modelling and structural response analysis.

2. The importance of monitoring

Modern society relies on civil infrastructure for all of its functions. It includes buildings, bridges, tunnels, factories, power plants, offshore installations, heritage structures, port facilities and geotechnical structures, such as dams and foundations.

In-operation structural monitoring is important in assessing the reliability and safety of established infrastructure. The following list introduces a number of cases where structural monitoring may be required [8]:

- Modifications of an existing structure,
- monitoring of structures affected by external works,
- monitoring during demolition,
- structures subject to long-term movement or degradation of materials,
- feedback loop to improve a future design based on experience,
- fatigue assessment,
- novel systems and/or methods of construction,
- assessment of structural integrity after a major event such as an earthquake or storm,
- growth in maintenance needs,
- changes in design philosophy such as the move towards performance-based design.

Bridge monitoring programs have historically been implemented for the purpose of understanding and eventually calibrating models of the load-structure-response chain. Their monitoring, therefore, generally falls into the category of a feedback loop to improve future design, based on experience, or in some cases to verify novel systems of construction. The deployment of sophisticated wind and structural health monitoring (WASHM) systems on cable-suspended bridges has significantly increased since the end of the 90s, in particular in East Asia [8, 9, 10, 11]. In Europe, WASHM systems have been notably deployed in Great Britain [12, 13] and more recently in Norway [1, 14].

Despite great progress in understanding bluff-body aeroelasticity, there continue to be surprises in long-span bridge aerodynamics [15], and the strategic importance and capital investment justify the expense of the most elaborate monitoring systems applied to civil infrastructure [16].

Long-span bridge monitoring systems also provide ideal opportunities to implement and study structural health monitoring (SHM) techniques. In the last decade, permanent bridge monitoring programs implemented in major bridge projects have often evolved into structural health monitoring systems [17]. This allows studies concerning the performance of the bridges themselves, as well as of SHM methodologies.

A direct motivation for monitoring is also found within bridge management programs, bridge upgrading or new bridge projects, where in all cases some form of validation of models and methods is required [18].

Finite-element modelling procedures applied to the Lysefjord Bridge were for instance validated by comparing ambient vibration survey data and finite-element simulations. The same procedures were then applied in a similar mode for simulating the performance of proposed ultra-long multi-span bridges planned for crossings over wide and deep Norwegian fjords [19].

3. Key aspects of monitoring

Structural monitoring has a wide definition and is implemented in different ways for varying motives, as previously mentioned. However, all approaches have the following common components:

(1) sensors,

(2) wired or wireless connections,

(3) data acquisition,
(4) data storage,
(5) data transmission,
(6) database management,
(7) data mining for feature extraction,
(8) load/effect model development from the study of data.

In case of structural health monitoring, additional steps would involve learning from past experience (heuristics), and decision-making based on identified features in combination with identified models.

4. The full-scale wind engineering lab at the Lysefjord suspension bridge

The bridge over the Lysefjord connects the municipality of Forsand, in Rogaland, with the surrounding communities through highway 13 (see Figure 1). The bridge was built in the period of 1995 to 1997 at a cost of about 150 million Norwegian kroner. The bridge is 649 m long in total, but the main span is 446 m long. The deck is a closed steel section, 2.7 m in height and 12.3 m wide. The two concrete towers carrying the bridge deck are 102 m high. The bridge deck stands roughly 55 m above the sea level, allowing ferries and other sailing vessels safe passage under the bridge on their way in and out of the Lysefjord. The bridge deck is oriented from north-west to south-east in a mountainous environment, and is entrenched between two steep hills with slopes ranging from 30° to 45° and a maximum altitude of 350 m to the North and 600 m to the South. The bridge is exposed to winds that may descent from mountains nearby or follow the fjord over a longer path. The West side of the bridge is exposed to a more open and levelled area. Preparations for the monitoring of the bridge began in 2011, but the first instruments were installed in November 2013. Since then, the monitoring system has been expanded and improved several times, last in June 2017.

The main considerations and steps taken in preparing and implementing the monitoring setup, were, and generally are, the following:

- Planning and Preparation (2011-2013, 2014, 2015, 2017):
  - Planning the relevant instrumentation based on what is desired, required and possible.
  - Financing.
  - Selecting suitable equipment depending on availability, robustness, applicability, price.
  - Ordering equipment and accessories, such as cables and connectors.
  - Selecting the final location of each sensor depending on scientific objectives and practicalities.
  - Designing and constructing brackets and fasteners for the instruments.
  - Installation, subject to availability of technicians, lifts, traffic control and other services.
- Monitoring (15.11.2013 onward):
  - Monitoring strategy, for example continuous monitoring vs. triggered monitoring.
  - Data capacity on local hard disks.
  - Data transfer.
  - Data server service at the University.
  - Maintenance, replacement and improvements.
- Data analysis (2014 onward):
  - Strategy.
  - Parameters.
  - Methods.

Figure 2 gives an overview of the instrumentation currently installed on the bridge and the location of individual sensors.

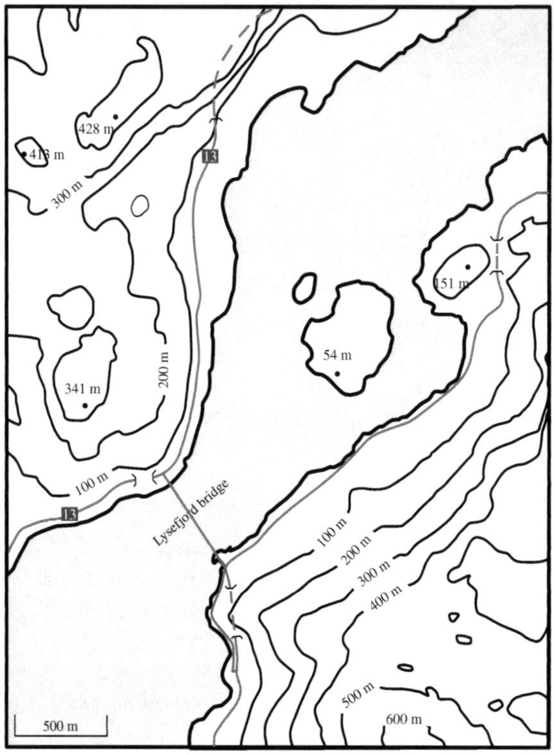

Figure 1. The Lysefjord Bridge and surroundings

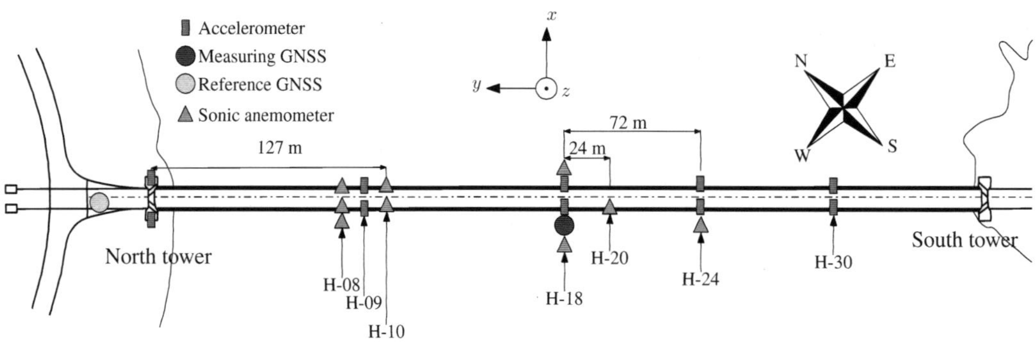

Figure 2. Overview of the current instrumentation of the Lysefjord Bridge.

4.1. Anemometers and weather station

Currently, the instrumentation on the bridge is composed of eight sonic anemometers (3D WindMaster Pro from Gill Instrument Ltd), which can be used with a sampling frequency up to 32 Hz. The anemometers are either mounted directly on brackets that are fixed to the hangers, or placed on the top of a 2.5 m long vertical steel pole placed above the main cables and fixed to the safety railings for the main cables. The sonic anemometers measure the wind along three perpendicular axes, i.e. the along-, across- and vertical wind components, respectively. On the west side of the deck, there is also a weather station on hanger H-10, which measures wind direction, wind speed, temperature, humidity, atmospheric pressure and precipitation at a sampling frequency up to 4 Hz. The anemometers are all located approx. 6 m above the deck,

Figure 3. Anemometers installed on a hanger (left) and on top of a pole above the main cables (right).

except at the hanger 8 (H-08) location, where there are two anemometers located at 6 m and 10 m above the deck, respectively. As the distance between each hanger is 12 m, the distance between the anemometers ranges from 24 m to 168 m.

4.2. Accelerometers and GPS sensor

Inside the bridge deck, four pairs of tri-axial accelerometers have been installed. They are located in four sections, near hangers 9, 18, 24, and 30. Their placement is determined so that both symmetric and asymmetric modes can be captured efficiently. Each of the four sections has two sensors, one on either side of the bridge deck, to monitor the bridge torsional response around its longitudinal axis (y), in addition to the translational responses. The lateral distance between two accelerometers constituting a pair is ca. 7.15 m and the maximal sampling frequency of each accelerometer is 200 Hz.

In addition, there are two accelerometers at the top of the northwest tower, one in each of the two box-shaped columns.

The accelerometers used are triaxial MEMs (microelectromechanical systems) silicon accelerometers from Canterbury Seismic Instruments Ltd. Figure 4 shows an accelerometer within the bridge deck.

In addition, the displacement of the bridge deck is monitored using a Real-Time Kinematic-Global Positioning System (RTK-GPS). More precisely, as the rover is located on top of the main cable, it is the displacement of the main cable in the middle of the bridge span, app. 3 m above the bridge girder that is being observed (see Figure 5b).

The Global Positioning System (GPS) or Global Navigation Satellite System (GNSS) technique has provided new possibilities for continuous monitoring of deflections of structural systems through direct measurements [20, 21, 22]. A differential GPS, which depends on accurate measurement of transit times for radio waves from satellite to receiver, promises direct and absolute measurement of deflection. A base station GPS antenna and logger is set up on a stable (fixed) reference location as close as possible to the monitored structure, with line of sight to at least five orbiting GPS satellites. A rover antenna and logger are installed on the structure,

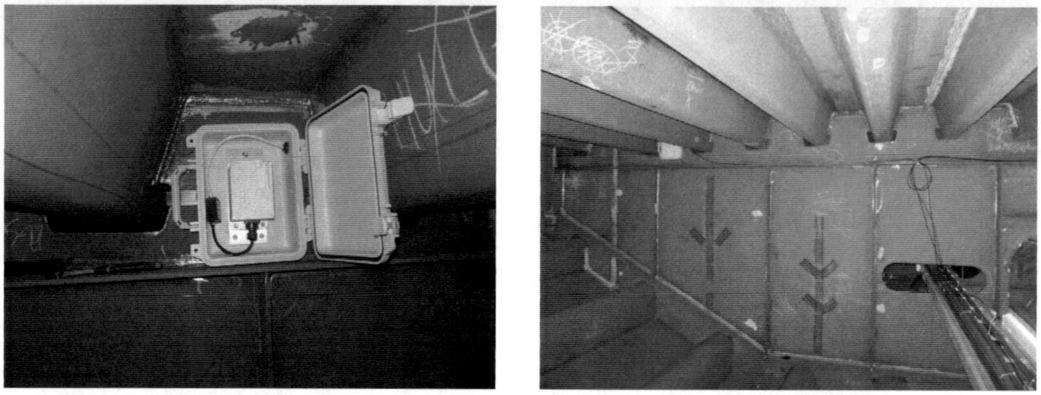

(a) (b)

Figure 4. Accelerometers. (a) Close up of the sensor, (b) Location within the girder.

(a) (b)

Figure 5. GNSS sensors, (a) The base antenna, (b) The rover antenna.

with the same satellite visibility. The locations of rover and base station are obtained in the appropriate geodetic coordinates, but each will have a measure of inaccuracy depending on factors affecting radio transmission times between the satellites and receivers. If the base station and the receiver are close, the error factors should affect the two signals in almost the same way. The RTK-GPS base-rover combination improves the accuracy of measurements by recording the relative displacement between a "fixed" base station on the northwest side (Figure 2, green dot) and a "moving" rover station located on the main cable at mid-span (Figure 2, blue dot). Naturally, the accuracy varies depending on the quality of the instruments used. In the present case, a set of Trimble BD930 GNSS receivers are coupled to Trimble AV33 GNSS antennas. These sensors can handle data sampling at a frequency of 20 Hz, with an accuracy of ± 8 mm $+1$ ppm for the horizontal displacements and ± 15 mm $+1$ ppm for the vertical displacements. The GPS technique is ideal for tracking the performances of long-span suspension bridges due to their low natural frequency and large amplitude displacements.

4.3. Data acquisition, communication and management

The sensors all produce digital signals and are connected to five data acquisition units using Cat5 cables. The DAQ units are three CUSP-3 units, one CUSP-M4 unit and one CUSP-M8 unit from Canterbury Seismic Instruments Ltd. The current system can serve 21 three-channel sensors, i.e. 63 channels in total. Each data acquisition unit is connected to a GPS receiver that provides an accurate timestamp to the data.

The data acquisition units are then connected to a local area network (LAN) that provides communication to the external world through a 3G modem. The local Ethernet network extends from the NW tower to Hanger 24, a distance of ca. 350 m, split up by 4 switches.

The data from each of the five data acquisition stations is then resampled by a separate data acquisition unit (CUSP-M32) which provides a common communications backbone that outputs a single, synchronized, time-aligned data-set for every 10 minutes. The sampling frequency is 50 Hz for all data, which is then reduced to 20 Hz during post-processing by decimation. The data-set is stored locally and simultaneously transferred through a 3G router via the mobile network to a server located at the University of Stavanger. It should be noted that the data sampling is continuous, so the amount of data already stored is enormous. The 3G router and the mobile network enables data transfer and data acquisition control from anywhere in the world.

5. Recorded data and analysis

The studies based on the recorded data, have focused on improving the understanding and modelling of wind loads and wind-induced response of long-span bridges, to facilitate an economic and safe bridge design.

The analysis based on the data recorded at the bridge have evolved around the following subtopics:

- Mapping and interpreting wind conditions and turbulence characteristics using data from Sonic anemometers and long-range lidars [3, 4, 23]. The objective being to improve the understanding of the parameters used in wind to load modelling.
- Monitoring of wake effects using short-range lidars [24].
- Evaluation of the bridge buffeting response based directly on the acceleration data and numerical load-response models [1, 25]. With the purpose of testing and improving available load-response models.
- Evaluation of traffic-induced response and methods to distinguish between traffic-induced response and wind-induced response [26]. For many applications, it is of key importance to be able to isolate the traffic-induced response when evaluating the statistical measure of the wind response.
- System identification of the response is an integral part of all response studies as well as for physical modelling of structures. Natural frequencies and damping ratios have been evaluated and studied for instance with regard to environmental variables such as mean wind velocity [1], temperature [5] or sample duration [27]. Several system identification (SI) methods have been tested and compared in the process.
- Displacement response from GNSS measurements has been systematically compared to displacements evaluated from acceleration data [2].

An important part of that objective comes through measurements and modelling of turbulence and the associated wind-induced bridge vibrations. Figure 6 shows two examples of common wind conditions at the Lysefjord Bridge. In general, the wind direction is strongly affected by the surrounding landscape and the wind direction largely follows the orientation of the fjord, as can be seen by comparison with Figure 1. There is also a pronounced higher turbulence

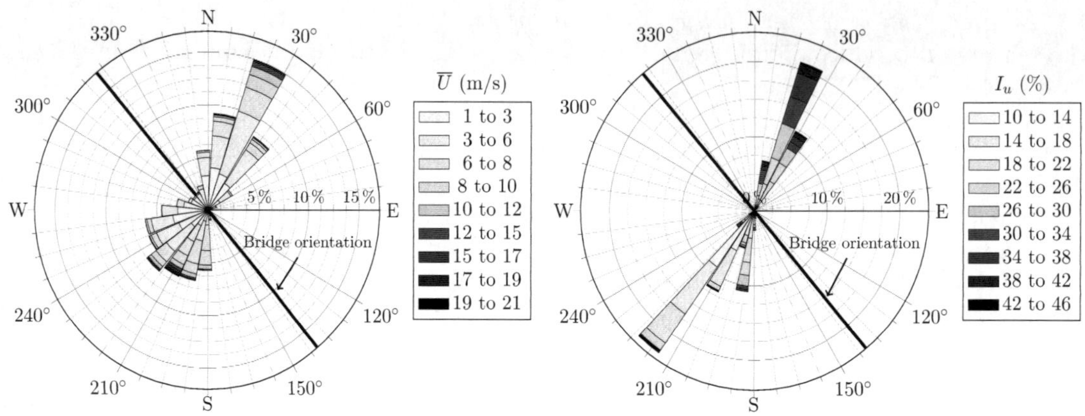

Figure 6. Wind rose of mean wind velocity (top) and turbulence intensity (bottom) recorded in 2015 on the Lysefjord Bridge [23, p. 60-61].

intensity ($I_u = \sigma_u/\overline{U}$) in the flow when the wind comes from the north. It can be explained by the proximity of high mountains that cause significant disturbance of the flow and thereby variability in the velocity, which is reflected by the normalized standard deviation of velocity i.e. the turbulence intensity.

As Figure 6 indicates, the prevailing wind direction is almost never perpendicular to the bridge deck. As a result, the conditions generally modelled in wind tunnels, where the aerodynamic properties of the bridge are primarily explored for wind perpendicular to the bridge deck, rarely or never occur. At the same time, cross-sectional properties are usually always tested with a much smaller turbulence intensity than the measurements portray. This reality is partly what makes full-scale measurements important. That is, to assess the reliability of design and modelling approaches and the validity of the simplifications generally made in simulations of the behavior of real structures, whether they are experimental or numerical.

Figure 7 shows an example of data from the acceleration measurements in the form of the standard deviation of acceleration plotted as a function of the mean wind velocity component normal to the deck, for NNE wind directions and SSV wind directions, respectively. The measured acceleration data are obtained using a sampling frequency of 20 Hz. The computed acceleration of the bridge is obtained using a frequency domain approach, where the turbulence model from the handbook N400 [28] used for bridge design in Norway is applied, with a terrain category (cat.) from 2 to 4. The response shown is the lateral displacement, the vertical displacement and the torsional displacement of the deck, shown across the bridge (x-factor), motion up and down (z-factor) and winding longitudinal axis (θ-factor). As can be seen, the dynamic response of the bridge deck is considerably larger for wind from NNE than wind from SSW, which is mainly associated with higher wind turbulence for wind from NNE. It is also seen that traditional buffeting models tend to underestimate the movements of the bridge when the turbulence intensity is high, i.e. for the wind from NNE.

Figure 8 shows an example of the standard deviation of vertical and horizontal displacement at the center of the deck, on one hand, from accelerometer data and on the other hand, from the GNSS measurements. As can be seen, the comparison is quite good, especially for the lateral displacement, which provides both confirmation of the method used to calculate post-acceleration, as well as the potential of GNSS technology to measure the displacements of structures of this type. However, Figure 8 also shows that the GNSS measurement overestimates the vertical motion compared to the acceleration measurement. Since the accuracy of GNSS measurement technology is generally in the range of several millimeters to centimeters, the movement of the

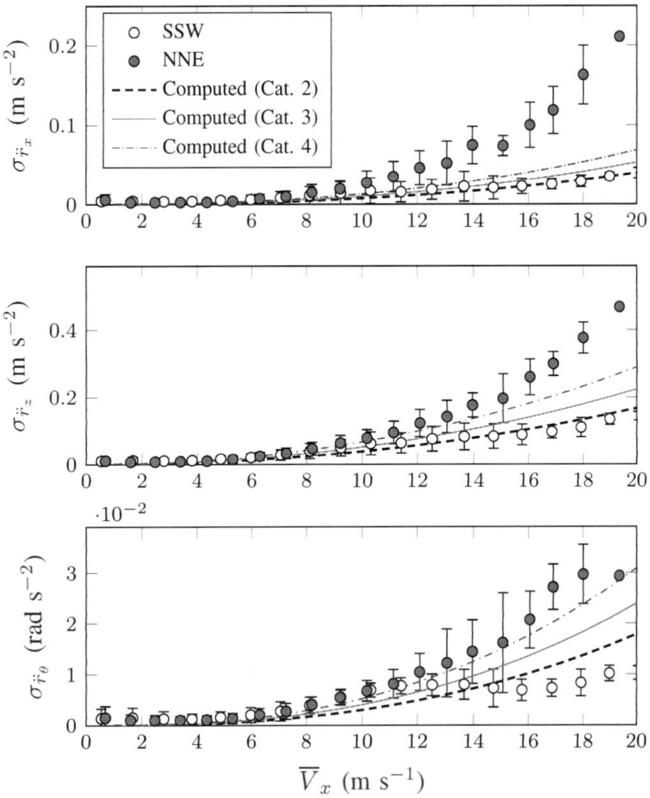

Figure 7. Measured and computed lateral (top panel), vertical (middle panel) and torsional acceleration (bottom panel) response of the Lysefjord Bridge. The data set considered comprises every 10-min bridge acceleration and wind records obtained in 2015.

structure must be several centimeters in order for the technology to provide reliable results. The resolution in the vertical direction is also at only half of that for horizontal motion, which probably explains to some extent why the GNSS measurement gives a slightly higher vertical displacement than the accelerometers.

6. Additional monitoring campaigns at the Lysefjord Bridge

In addition to recording the wind velocity and the bridge response using the permanently installed monitoring system, separate monitoring campaigns have been undertaken to test relatively new measurement techniques based on optical remote sensing. The instruments used were the so-called lidar. Lidar is the acronym for Light Detection And Ranging. It is a remote sensing technology based on the emission of a laser beam and the analysis of the backscattered light by gas and particles in the atmosphere. A lidar is, therefore, made of three main components: a laser, a telescope, and the detection system. These devices are either designed to measure over long distances of several km (long-range lidar) or short distances from 10 m to 200 m (short-range lidar). The Lysefjord Bridge has been used as a test site for temporary experiments using both of the above-described lidar types [3, 4, 24].

A long-range lidar (WindCube 100S by Leosphere) was acquired through NORCOWE and installed approx. 1.8 km from the bridge. Its scanning head was oriented towards the bridge. It was used to observe the wind flow conditions around the bridge, across the entire bridge span, following several different scanning scenarios. The range gate of 25 m and the recording interval

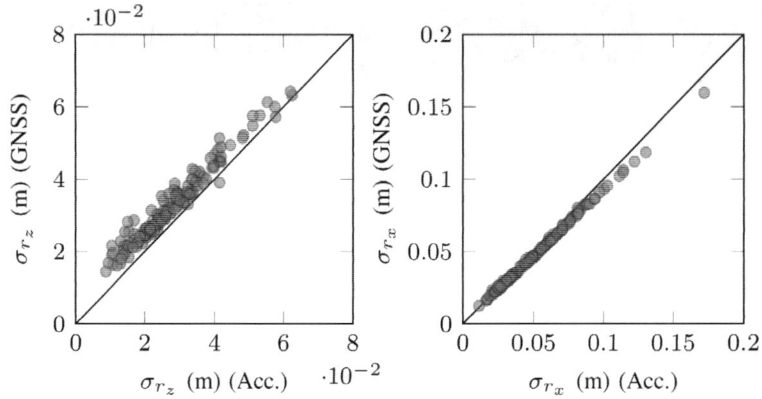

Figure 8. Standard deviation of the GNSS and accelerometric vertical (left) and lateral (right) displacement using a full day of continuous bridge vibration records on 2015-10-07 and a frequency range from 0.1 Hz to 1 Hz. The solid line represents the case of a perfect correlation.

Figure 9. Short-range lidar on the bridge (photo by DTU).

of at least 1 s, enabled characterization of large, low frequency gusts [4].

A short campaign was also undertaken using two synchronized short-range lidars (see Figure 9) to measure two components of the wind velocity along horizontal and vertical lines about 40 m southwest of the deck [3, 24]. The short-range WindScanners are owned and operated by the Technical University of Denmark (see http://www.windscanner.dk/). Figure 10 shows schematically the setup of the instruments, which were located 90 m apart sending out and detecting beams of light that cross at a certain distance away from the bridge. Where the rays cross, two wind speed components can be identified, one in x-direction and the other in y-direction. An example of the measurements obtained can be seen in Figure 11, which shows, among other things, the wind velocity as a function of height. As the figure shows, the bridge deck, which is located about 55 m above sea level, has a significant effect on the flow. i.e. the figure visualizes the wake that forms behind the deck as the flow passes the bridge.

The experience gathered through these two lidar campaigns from an existing bridge has been

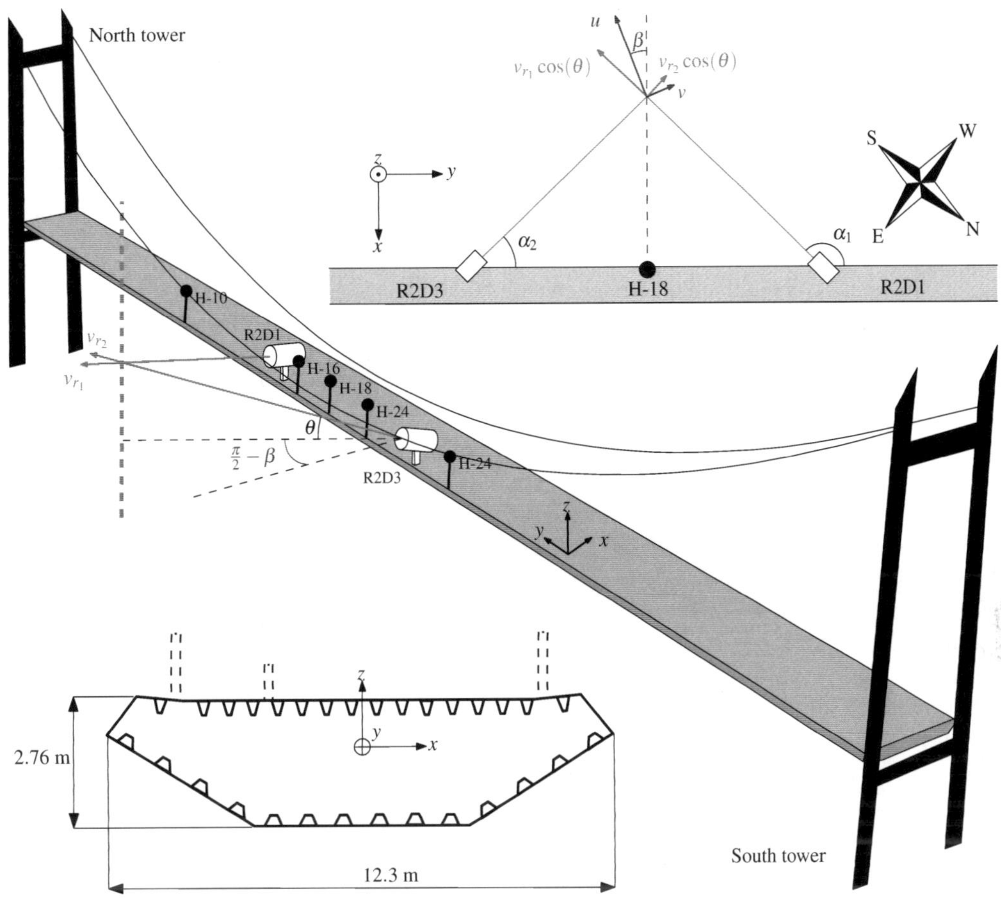

Figure 10. Schematic of the short-range lidar (R2D1 and R2D2) on the bridge.

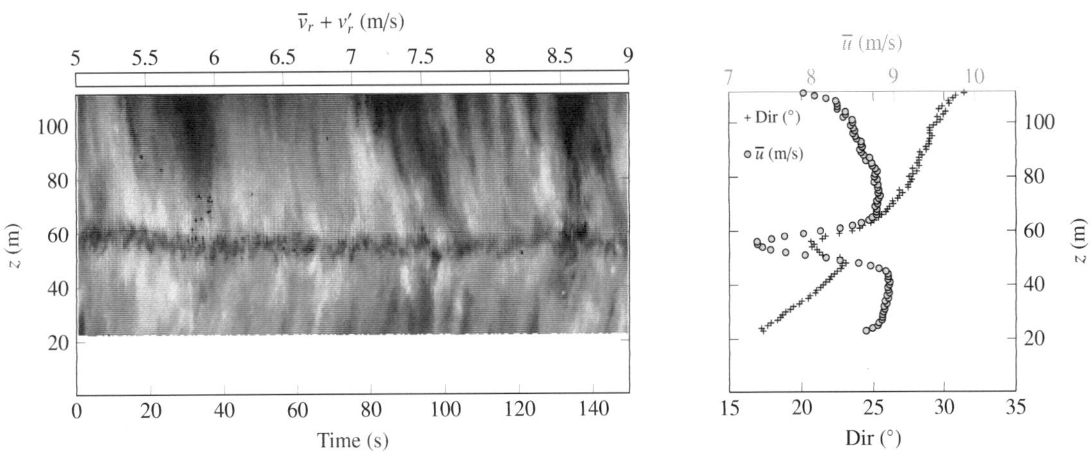

Figure 11. 150 seconds of the along-beam velocity component recorded by one short-range lidar in the wake of the bridge deck, on 23-05-2014 from 05:30, (left) and corresponding 10-min mean wind profile 40 m downwind of the bridge.

applied to develop a novel experiment in Bjørnafjord south of Bergen, where one of the major fjord crossings along the E39 Coastal Highway is being planned. In 2016, three long-range lidars were used to monitor the turbulence conditions in the middle of the 5 km wide fjord, in order to link this information to the turbulence characteristics observed by the anemometers on the shore [29].

7. Final remarks

In this paper, various aspects of full-scale monitoring have been discussed and the full-scale wind engineering laboratory established on the Lysefjord Bridge has been introduced.

Wind and traffic-induced vibrations of a suspension bridge in a complex terrain have been studied in full scale. The focus has been on modal parameter characterization, a central step for proper evaluation of the buffeting response. Environmental effects have been observed, such as the daily and seasonal fluctuations of the natural frequencies. As the monitoring is continuously ongoing it may be possible, with sufficient amount of data, to evaluate probability distributions for the key excitation-, system- and response parameters. These could then subsequently be used to develop performance indicators or criterias for structural health management purposes.

Wind measurement techniques based on optical remote sensing (lidars) have been seen to complement single point wind data from sonic anemometers. Full-scale monitoring of flow around a suspension bridge using a system of small lidars shows promising potential for studies of wind-structure interaction. However, data quality of optical remote sensing instruments is strongly influenced by the day to day environmental conditions, especially for long-range lidars. The data quality will also depend on the type of wind statistic pursued, and the complexity of the scanning pattern.

Monitoring of excitation, created by environmental forces and the related response of real structures is important for many reasons. New information is gathered about the nature and characteristics of the excitation forces that the building is exposed to, as well as the properties and behavior of the structure itself. Both aspects, excitation and response, can be used to validate and/or update experiments and models used in the design of new structures, such as the load description and modelling of the structural response.

Information gathered through monitoring generates increased safety in construction, giving both designers and owners added confidence in the design information and methods applied. In addition, long-term observations of structural behavior can be used for health monitoring and facility management. In fact, structural monitoring should be, and is becoming, an integral part of civil infrastructure management.

It is also important to be able to test and validate the capabilities of new measurement and monitoring technologies through comparison with data from established sensors and monitoring techniques, such as those provided by the full-scale wind engineering lab at the Lysefjord Bridge.

Acknowledgement

The authors are grateful to all those that have contributed to this project. In particular to the NPRA for the support of the measurement campaign at the Lysefjord Bridge, as well as the assistance provided during the installation and maintenance of the monitoring system.

References

[1] Cheynet E, Jakobsen J B and Snæbjörnsson J 2016 *Eng. Struct.* **128** 474–487
[2] Cheynet E, Jakobsen J B and Snæbjörnsson J 2016 *19th Congr. of IABSE, Challenges in Design and Construction of an Innovative and Sustainable Built Environment* (Int. Association for Bridge and Structural Engineering (IABSE).) pp 57–64
[3] Cheynet E, Jakobsen J B, Snæbjörnsson J, Mikkelsen T, Sjöholm M, Mann J, Hansen P, Angelou N and Svardal B 2016 *Exp. Fluids* **57** 1–17

[4] Cheynet E, Jakobsen J B, Snæbjörnsson J, Reuder J, Kumer V and Svardal B 2017 *J. Wind Eng. Ind. Aerodyn.* **161** 17–26
[5] Cheynet E, Snæbjörnsson J and Jakobsen J B 2017 *Dynamics of Civil Structures, Volume 2* (Springer, Cham) pp 87–93
[6] Statens Vegvesen 2012 Ferjefri E39, hovedrapport
[7] Statens Vegvesen 2015 Statusrapport - ferjefri E39
[8] Wong K Y 2004 *Struct. control health monit.* **11** 91–124
[9] Li H, Ou J, Zhao X, Zhou W, Li H, Zhou Z and Yang Y 2006 *Comput. Civ. Infrastructure Eng.* **21** 306–317
[10] Wang H, Hu R, Xie J, Tong T and Li A 2012 *J. Bridge Eng.* **18** 587–600
[11] Li H and Ou J 2016 *J. Civ. Struct. Health Monit.* **6** 43–67
[12] Brownjohn J, Bocciolone M, Curami A, Falco M and Zasso A 1994 *J. Wind Eng. Ind. Aerodyn.* **52** 185–218
[13] Macdonald J H 2003 *J. Wind Eng. Ind. Aerodyn.* **91** 1465–1483
[14] Fenerci A and Øiseth O 2017 *J. Struct. Eng.* **143** 04017131
[15] Andrianne T and De Ville De Goyet V 2016 *8th Int. Colloquium on Bluff Body Aerodynamics and Applications* vol 8 (International Association for Wind Engineering (IAWE))
[16] Moss R and Matthews S 1995 *Struct. Eng.* **73**
[17] Brownjohn J, Moyo P, Omenzetter P and Chakraborty S 2005 *Struct. Control Health Monit.* **12** 227–244
[18] Yanev B 2003 *Proc. of the First Int. Conf. on Structural Health Monitoring and Intelligent Infrastructure, Tokyo, Japan* vol 1 pp 87–95
[19] Wang J, Jakobsen J B, Snæbjörnsson J and Cheynet E 2017 *7th European African Conf. on Wind Engineering (EACWE)* vol 7
[20] Nakamura S 2000 *J. Struct. Eng.* **126** 1413–1419
[21] Tamura Y, Matsui M, Pagnini L C, Ishibashi R and Yoshida A 2002 *J. Wind Eng. Ind. Aerodyn.* **90** 1783–1793
[22] Watson C, Watson T and Coleman R 2007 *J. Surv. Eng.* **133** 23–28
[23] Cheynet E 2016 *Wind-induced vibrations of a suspension bridge: A case study in full-scale* Ph.D. thesis University of Stavanger, Norway
[24] Cheynet E, Jakobsen J B, Snæbjörnsson J, Angelou N, Mikkelsen T, Sjöholm M and Svardal B 2017 *J. Wind Eng. Ind. Aerodyn.* **171** 261–272 ISSN 0167-6105
[25] Wang J, Cheynet E, Jakobsen J B and Snæbjörnsson J 2017 *ASME 2017 36th Int. Conf. on Ocean, Offshore and Arctic Engineering* (American Society of Mechanical Engineers)
[26] Cheynet E, Jakobsen J B and Snæbjörnsson J 2015 *MATEC Web of Conf.s* vol 24 (EDP Sciences) p 04003
[27] Cheynet E, Jakobsen J B and Snæbjörnsson J 2017 *Procedia Eng.* **199** 2047–2053
[28] NPRA 2015 *Bridge Projecting Handbook N400 (in Norwegian)* (The Norwegian Public Roads Administration)
[29] Cheynet E, Jakobsen J B, Snæbjörnsson J, Mann J, Courtney M, Lea G and Svardal B 2017 *Remote Sens.* **9** 977

IOP Conf. Series: Materials Science and Engineering **276** (2017) 012008 doi:10.1088/1757-899X/276/1/012008

Super-long bridges with floating towers: the role of multi-box decks and Hardware-In-the-Loop technology for wind tunnel tests

A Zasso, T Argentini, I Bayati, M Belloli, D Rocchi

Politecnico di Milano: Department of Mechanical Engineering, Milano, Italy

Contact author: tommaso.argentini@polimi.it

Abstract. The super long fjord crossings in E39 Norwegian project pose new challenges to long span bridge design and construction technology. Proposed solutions should consider the adoption of bridge deck with super long spans or floating solutions for at least one of the towers, due to the relevant fjord depth. At the same time, the exposed fjord environment, possibly facing the open ocean, calls for higher aerodynamic stability performances. In relation to this scenario, the present paper addresses two topics: 1) the aerodynamic advantages of multi-box deck sections in terms of aeroelastic stability, and 2) an experimental setup in a wind tunnel able to simulate the aeroelastic bridge response including the wave forcing on the floating.

1. Introduction

The super long fjord crossings in E39 Norwegian project pose new challenges to large span bridge design and construction technology. Proposed solutions should consider the adoption of super long spans for the deck or floating solutions for at least one of the towers, due to the relevant fjord depth. At the same time, the exposed fjord environment, possibly facing the open ocean, calls for higher aerodynamic stability performances, compared to the existing long-span bridges further inland.

The choice of the multi-box deck section concept has reached nowadays a proven technological development and reliability, offering its superior aerodynamic performances as possible key solution for a high aerodynamic stability request [1][2][3]. At the same time the closed box girder streamlined solution with higher aerodynamic performances is also offering the great advantage of limited mass of the deck per unit length, playing therefore a key role in case of floating tower solution. However, multi-box decks may show complex vortex shedding excitation phenomena and they should be thoroughly studied in wind tunnel [4][5][6].

Assessment of a considered suspension bridge solution with one or two towers on a floating support is posing non-trivial challenges in terms of experimental validation of the numerically simulated structure dynamics, based on hybrid codes accounting both for the aero- and the hydro-dynamic loads effect. The new technology of Hardware-In-the-Loop (HIL) testing recently developed at POLIMI in the field of floating offshore wind turbines is proposed as a reliable tool for the experimental validation of the complex numerical approach. A 6 degree of freedom platform driven by linear actuators is available at POLIMI at the purpose of simulating the motion of the floater due to the combined action of the hydrodynamic loads on the floater (numerically simulated in real time) and of the aerodynamic and inertial loads transmitted by the tower (measured by a 6 components dynamometer at the tower-platform interface). The opportunities offered by the mentioned HIL technology are discussed with reference to a possible wind tunnel test experiment on a bridge aeroelastic model focused on the validation of the numerical hybrid tools needed for the floating bridge solution assessment.

2. The aero-hydro-elastic problem

Adopting the modal approach, the dynamics of a bridge can be written as:

$$\left[M_s^*\right]\ddot{\underline{q}}+\left[C_s^*\right]\dot{\underline{q}}+\left[K_s^*\right]\underline{q}=[\Phi]^T\underline{F}_{aero}+[\Phi]^T\underline{F}_{hydro} \tag{1}$$

where $\underline{q}$ is the vector of the modal coordinates; $[\Phi]$ the mode shape matrix, $\left[M_s^*\right]$, $\left[C_s^*\right]$ and $\left[K_s^*\right]$ are the structural inertial, damping, and stiffness modal matrices of the system; $\underline{F}_{aero}$ the external aerodynamic forces that depends on the motion of the deck and on the incoming wind velocity; $\underline{F}_{hydro}$ the external hydrodynamic forces acting on the floating structure, if present. Linearized and non-linear approaches differ in the modelling of both $\underline{F}_{aero}$ and $\underline{F}_{hydro}$.

2.1. The aeroelastic forces

The simulation framework for aeroelastic analyses of long-span bridges is rather well established, and it consists of the following points [7][8][9]:

- The bridge structural dynamics is represented by means of a modal approach, starting from a FEM of the structure.
- The deck is divided into several sections to which aerodynamic forces are applied. Each section, from an aerodynamic point of view, is independent from the other (sectional approach). Cables and towers may be analysed in the same way.
- For each deck section, the turbulent wind field is simulated starting from its stochastic characteristics (e.g. PSD, spatial coherence, turbulence intensity, integral length scale, etc.).
- The dynamic response of the bridge is computed, applying the aerodynamic forces to the deck sections and to other components. The aerodynamic forces depend on the deck state (position and velocity) and on the incoming turbulence.

The standard approach for buffeting response is based on a linearized model of the fluid-structure interaction around a steady configuration of the bridge, which depends on the mean wind speed field. The external aerodynamic forces ($\underline{F}_{aero}$) are computed as the sum of three different effects, namely:

$$\underline{F}_{aero}=\underline{F}_{ST}+\underline{F}_{se}+\underline{F}_{buff} \tag{2}$$

where $\underline{F}_{ST}$ are the stationary aerodynamic forces, $\underline{F}_{se}$ the self-excited, and $\underline{F}_{buff}$ the buffeting forces.

The steady aerodynamic drag, lift and moment forces acting on each deck section are defined through the steady coefficients as a function of the angle of attack:

$$\underline{F}_{ST}=\frac{1}{2}\rho V^2 B\begin{bmatrix} C_D \\ C_L \\ BC_M \end{bmatrix} \tag{3}$$

where B is the deck chord, V the mean wind velocity, and ρ the air density.

The self-excited unsteady aerodynamic terms of drag, lift and moment acting on each deck section (D, L, and M) are modeled using the flutter derivatives coefficients, measured with dedicated forced motion tests on rigid sectional model. The self-excited forces acting on the generic j-th section are defined as:

$$\begin{bmatrix} D_j \\ L_j \\ M_j \end{bmatrix}_{aero} = \left[m^*_{aero,j} \right] \begin{Bmatrix} \ddot{y}_j \\ \ddot{z}_j \\ \ddot{\theta}_j \end{Bmatrix} + \left[c^*_{aero,j} \right] \begin{Bmatrix} \dot{y}_j \\ \dot{z}_j \\ \dot{\theta}_j \end{Bmatrix} + \left[k^*_{aero,j} \right] \begin{Bmatrix} y_j \\ z_j \\ \theta_j \end{Bmatrix} \tag{4}$$

where y_j, z_j and θ_j are respectively the lateral, vertical and torsional displacements of the j-th section, $\left[m^*_{aero,j} \right]$, $\left[c^*_{aero,j} \right]$ and $\left[k^*_{aero,j} \right]$ are the aerodynamic matrices related to the j-th section, expressed as a function of the 18 flutter derivatives coefficients a_i^*, h_i^* and p_i^* with $i = \{1:6\}$.

a_i^*, h_i^* and p_i^* are proportional to the slopes of the static coefficients ($K_L = dC_L / d\alpha; K_M = dC_M / d\alpha$) and they are function both of the reduced velocity ($V^* = V / (fB)$, being f the motion frequency), and of the mean angle of attack.

Buffeting forces are due to the incoming turbulent wind. They are defined through the so-called aerodynamic admittance functions as:

$$\begin{bmatrix} D_j \\ L_j \\ M_j \end{bmatrix}_{buff} = \frac{1}{2}\rho V^2 B \begin{bmatrix} \chi_{D-u}\left(V^*,\alpha\right) & \chi_{D-w}\left(V^*,\alpha\right) \\ \chi_{L-u}\left(V^*,\alpha\right) & \chi_{L-w}\left(V^*,\alpha\right) \\ B\chi_{M-u}\left(V^*,\alpha\right) & B\chi_{M-w}\left(V^*,\alpha\right) \end{bmatrix} \begin{Bmatrix} \frac{u}{V} \\ \frac{w}{V} \end{Bmatrix} \tag{5}$$

where χ are the 6 aerodynamic admittance functions (asymptotic to quasi-steady values at large V*), proportional to the slopes of the static coefficients (dependent upon both of the reduced velocity and of the angle of attack α), while u and w are the turbulent horizontal and vertical components of the turbulent wind field. Admittance functions can be measured with dedicated wind tunnel tests using an active turbulence generator.

2.2. *The hydrodynamic forces*

Collecting the independent coordinates of motion of the floating towers in the x vector, the equations of motion of the global system can be written as

$$\underline{F}_{hydro} = -[A_\infty]\underline{\ddot{x}} - [R_s]\underline{\dot{x}} - [K_s]\underline{x} + \underline{F}_{rad} + \underline{F}_{visc} + \underline{F}_{moor} + \underline{F}^{(1)}_{diff} + \underline{F}^{(2)}_{diff} \tag{6}$$

where $[A_\infty]$ is the wave frequency-independent or "infinite frequency" hydrodynamic added mass matrix due to the radiation of the semisubmersible platform [10], given as output by the panel code WAMIT, $[R_s]$ a diagonal viscous linear damping matrix added to tune the numerical model compared to the results from free decay tests in the ocean basin [10] and $[K_s]$ is the system stiffness matrix containing both the terms related to the hydrostatic as well as gravitation restoring computed with respect to the Sea Water Level (SWL), upon which the x axis lays. The contributions to $\underline{F}_{hydro}$ are summarized in the following paragraphs.

2.2.1. *Radiation forces* $\underline{F}_{rad}$:

The radiation force partially accounted for by the term $[A_\infty]$ in Equation 6 is also due to a convolution term (Equation 7) between the platform velocity and the retardation matrix [K(t)] (impulse response), describing the dissipation of energy of the platform motion due to the radiated waves generated by the platform itself. The matrix [K(t)] can be expressed in the wave frequency domain [K(ω)], as in Equation 10 as a function of the frequency dependent added mass and damping matrices [A(ω)] and [B(ω)] computed with the inviscid and irrotational panel codes such as, WAMIT.

$$\underline{F}_{rad} = \underline{\mu}(t) = \int_0^t K(t-\tau)\underline{\dot{x}}(\tau)d\tau \tag{7}$$

$$\begin{cases} \underline{\dot{q}_r} = [A_r]\underline{q_r} - [B_r]\underline{\dot{x}} \\ \underline{\hat{\mu}} = [C_r]\underline{q_r} \end{cases} \tag{8}$$

$$\underline{F}_{rad} = \underline{\mu}(t) \approx \underline{\hat{\mu}}(t) \tag{9}$$

$$[K(\omega)] = [B(\omega)] + j\omega([A(\omega)] - [A_\infty]) \tag{10}$$

Due the time-consuming computation of a convolution term for control or real-time application like the one presented, the matrix [K(ω)] was identified in the frequency domain [12] by means of another equivalent matrix [K (ω)] leading to the corresponding memory effect term $\hat{\mu}$ m through a state space approximation, Equation 8-10, more effectively implemented in the HIL control, Figure 7.

2.2.2. *Viscous forces F_{visc}*

The viscous forces considered are due to the Morison's contribution derived by the integration along the possible platform's columns of the drag forces depending quadratically on the relative velocities v_{rel} between the platform and the wave particles, either along tangential *t* or axial *ax* direction by the drag coefficients C_d and C_{ax}, respectively

$$F_t(t) = \int_z \frac{1}{2} C_d D_z \,|\, v_{rel,t} \,|\, v_{rel,t} dz \tag{11}$$

$$F_{ax}(t) = \int_z \frac{1}{2} C_{ax} \pi \frac{D_z^2}{4L_z} \,|\, v_{rel,ax} \,|\, v_{rel,ax} dz \tag{12}$$

The forces are collected into the F_{visc} of the Equation 6 with a simple 2D rotation in the z-x plane. The particle velocities (along x and z directions) accounted in V_{rel} , along with the platform's velocities, are computed in the real time model, according the linear Airy theory of wave kinematics. Moreover, the Wheeler wave stretching is implemented along with the linear wave kinematics, to model also the contribution above the free surface

2.2.3. *First order diffraction forces $F^{(1)}{}_{diff}$.*

The first-order wave excitation loads (diffraction) were obtained based on complex frequency dependent wave exciting force vector $X(\omega_k)$ computed by WAMIT (i.e. output file *.3) and for a unitary wave amplitude h, for each i-th degree of freedom, under the assumption of a linear wave surface approximation. Therefore, the linear diffraction force is computed, as in Eq. 13, considering the random phased wave amplitudes spectrum A_k, consistent with a JONSWAP spectrum for a given sea state.

$$F_{diff,i}^{(1)} = \Re\left\{ \sum_{k=1}^{N} A_k X_i(\omega_k) e^{j\omega_k t} \right\} \tag{13}$$

2.2.4. *Second order diffraction forces $F^{(2)}{}_{diff}$*

The second-order difference-frequency potential-flow terms are also taken into account. Similarly to $F_{(1)}$ diff , "sum" and "difference" frequency normalized complex force matrices, $X^{+}(\omega_k\,;\,\omega_l)$ and $X(\omega_k;\omega_l)$, also known as Quadratic Transfer Functions QTF, are taken from WAMIT's output (i.e. .12s/.12d file), which is the state-of-the-art in 3D panel code for offshore hydrodynamics, as function of each combination of incoming wave frequencies ω_k and ω_l . More specifically, full QTF loads taken as reference account for quadratic interaction of first order quantities plus second-order potential derived by the direct method (the formulation of the Boundary Integral Equation - BIE involves integrating the potential itself over the surface). However, the second order diffraction forces, sum (+) or difference (-) frequency forces, Equation 14, are taken into consideration for their importance in the modelling of floating systems, since they excite the rigid natural frequencies typically in the low frequency spectrum range (-) for semisubmersible and spar buoys platforms, as assessed in [13], or in the high frequency range (+) for Tension Leg Platforms (+). Analogously to Equation 8, this contribution can be written as

$$F_{diff,i}^{(2)} = \Re\left\{ \sum_{k=1}^{N} \sum_{l=1}^{N} A_k A_l^* X_i^{+/-}(\omega_k, \omega_l) e^{j(\omega_k - \omega_l)t} \right\} \tag{14}$$

where the complex vector A_l is the same as Ak but transposed and evaluated at the l-th frequency of the JONSWAP spectrum. Also, a second-order correction to the linear sea surface elevation h is applied to take into account low-frequency range non-linearities (difference) of A_i, [14].

2.3. Validation of the hydrodynamic model

The hydrodynamic model, which is implemented on the real-time HIL controller, is designed to be fast but comprehensive of the complex hydrodynamics of a floating system. For this reason, a preliminary code-to-code comparison is carried against a fully non-linear hydro-elastic code (e.g. NREL/HydroDyn). More specifically, free decays along the principal degrees-of-freedom being investigated, the response of the platform due to regular waves (i.e. Response Amplitude Operators, RAO) and irregular random sea, are compared against the reference code. These tests are carried out considering no-wind conditions. On top of this, the corresponding tests might have been carried out also in ocean basin, so that the fine tuning of the model (e.g. linear and quadratic added damping) can be finalized relying also on the experimental data. An example of this procedure can be found in [13] where a two degrees of freedom (2-dof) HIL system (surge and pitch) has been developed for simulating the motion of a 5MW floating wind turbine, showing excellent agreement of the hydrodynamic modelling approach compared either with the reference numerical code and the ocean basin data.

2.4. Aeroelastic stability improvement

The aerodynamic performances of streamlined multi box decks have been extensively studied for many super long span bridges (e.g. [1][2][3]) . Figure 1 shows a comparison of a single-box deck (Storebaelt Bridge), a two-box deck (Stonecutters bridge) and a 3-box deck (Messina bridge).

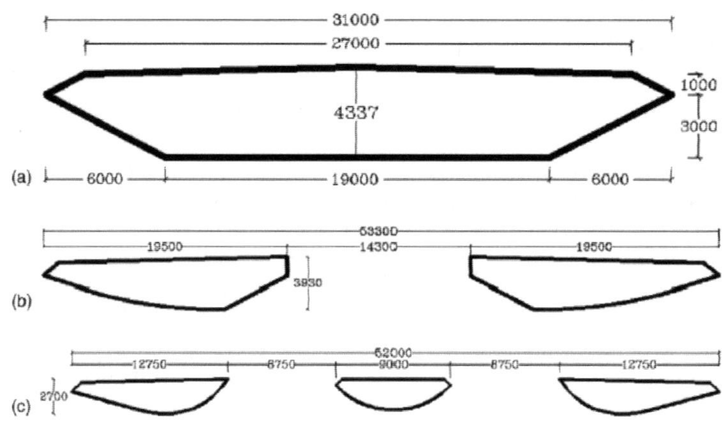

Figure 1. Cross-sections of decks: a) Storebaelt Bridge, b) Stonecutters bridge, c) Messina bridge (from [15])

From an aerodynamic point of view, the multi-box decks have better performances in terms of stability and buffeting responses, because of their smaller aerodynamic coefficients. As an example, Table 1 shows the derivatives of the lift and moment coefficients (K_L and K_M) for the 3 previous sections, having considered for each section its own chord width in the definition of the coefficients. If one compares the effects of these three different aerodynamic behaviors on the same bridge (Storebeaelt) with a very simplified approach, keeping unchanged the structural parameters and changing only the values of K_L and K_M, the results for stability shown in Figure 2 are obtained: predicted flutter speed goes from 70 to 120 to 240 m/s.

2.5. *Issues in multi-box sections*

However, multi-box section must be carefully investigated in wind tunnel for complex vortex shedding phenomena and subsequent vortex-induced vibration, and for the strong non-linear effects that may be relevant.

For vortex shedding, the presence of multiple boxes often introduces multiple lock-in ranges. As an example, the VIVs amplitude for the Messina bridge, which is a 3-box deck, are reported as a function of the reduced wind speed in Figure 3 (see [4]): two lock-in regions are present.

As far as the nonlinear effects are concerned, the most relevant characteristics can be found in the double nonlinear dependence of the aerodynamic forces on the amplitude of the angle of attack and on the reduced velocity (e.g. see Figure 4, where it is shown the nonlinear dependence of the lift coefficient upon the amplitude on the angle of attack and reduced velocity during forced motion tests). This nonlinearity may lead to an underestimation of the buffeting response and should be taken into account during the design of the bridge (e.g. [18])

Table 1. Aerodynamic derivatives KL and KM

Deck	**KM [1/rad]**	**KL [1/rad]**
Storebeaelt (1 box)	1.2	4.4
Stonecuttesrs (2 boxes)	0.55	2.3
Messina (3 Boxes)	0.2	0.7

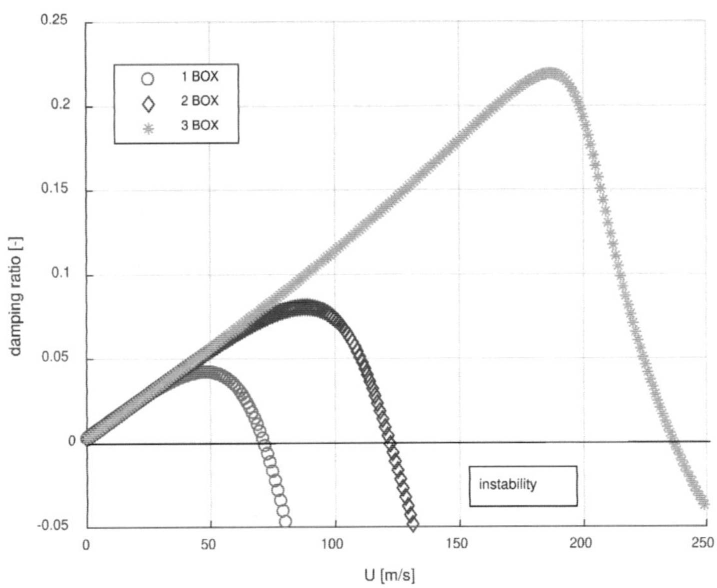

Figure 2. Stability performances of the three different cross-sections: trend of total damping ratio as a function of the mean wind speed

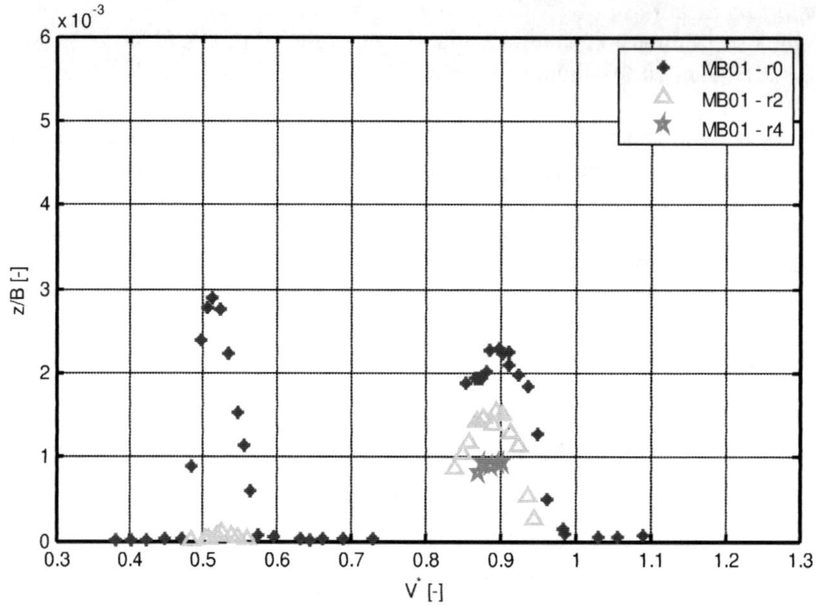

Figure 3. VIV of Messina. Steady state response: non dimensional oscillation amplitude, vertical motion as function of the reduced velocity varying the Scruton number (Sc). r0 corresponds to Sc = 0.1, r2 to Sc = 0.25, r4 to Sc = 0.44 (from [4])

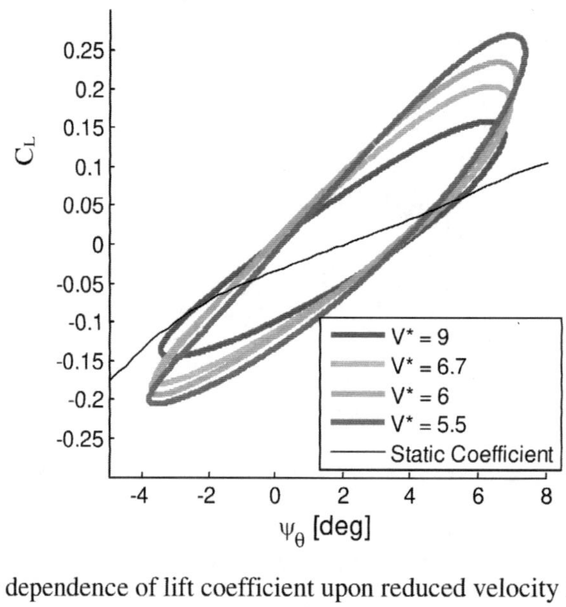

Figure 4. Nonlinear dependence of lift coefficient upon reduced velocity for large variations of the angles of attack (from -4 deg to 6 deg), from [8]

3. The HIL for the simulation of hydrodynamic forces in Wind Tunnel

At present, an experimental laboratory in which both wind and the waves effects on a floating structure can be reproduced is not available. If the numerical design of a floating bridge is required, a feasible option for its validation is to perform wind tunnel tests on a full bridge aeroelastic model, physically modelling only the part of the bridge above the water, and simulating the submerged part with a hardware in the loop (HIL) approach. This approach is currently used for floating wind turbines and could be extended to floating bridges.

3.1. The HIL approach for wind turbines

The Hardware-In-The-Loop (HIL) experimental approach allows to perform hybrid tests in which measurements, computations and actuations are performed in real time. In the specific case of the assessment of the global dynamics of a floating wind turbine through wind tunnel tests, the physical aerodynamic loads on the rotor are measured by means of a 6-components balance, which is mounted on a 2-DoF system (surge and pitch) and bears the whole wind turbine scale model (Rotor-Nacelle-Assembly and tower), see Figure 5.

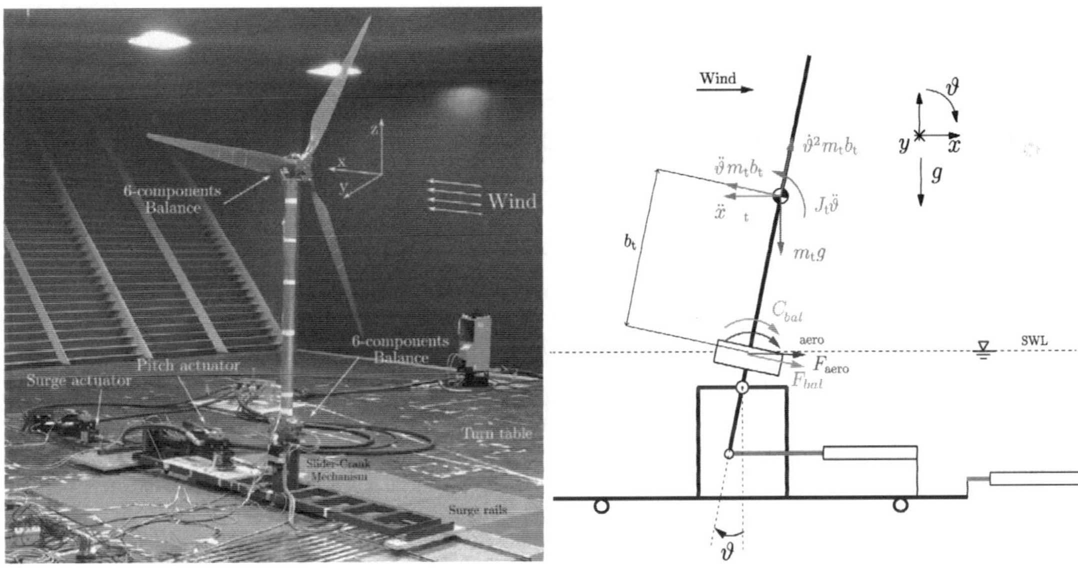

Figure 5. Picture of wind turbine model (top) and sketck of actuation and forces

The measured loads, properly conditioned, contain the aerodynamic loads which are added to computed hydrodynamic loads in order to give the physical displacements at the base of the scale model's tower to be actuated in real time. An example of control scheme is shown in Figure 7.

Both aerodynamic and hydrodynamic forces are clearly dependent on the dynamic state of the system. The combination of such forces need to be performed under specific constraints (real time) and precautions (synchronization), to obtain respectively a stable and consistent motion of the system being analyzed.

Reaching these goals within a target tolerance is possible thanks to a fast but comprehensive numerical model and an efficient real-time signal management.

3.2. Application to aeroelastic model of the bridge

The same actuation system at the base of the turbine tower could be applied at the base of the floating towers, eventually adding additional degrees of freedom. Looking at some preliminary modal simulation of the Bjørnafjorden floating bridge, obtained linearizing the hydrodynamic forces (see Master thesis from NTNU [19]), the floating towers influence mainly the first 20 vibration modes, from 0.0125 Hz to 0.16 Hz.

Considering a Froude scaling for wind tunnel tests, using a geometrical scale of 1:330 for the full bridge, the frequency that should be simulated by the HIL would be between 0.2 and 3, that is in the range of operation of the actuators.

The expected full-scale root mean square lateral amplitude is about 2 m which in model scale would be 6 mm (so about 20 mm peak dynamic displacement), that again is in the range of the actuators. These data are consistent with the capability of the 2DoF [13] and 6DoF [14] HIL systems already available at Politecnico di Milano designed for simulating the motion of a 10MW floating wind turbine [15] with a length scale factor of 75 and a frequency scale factor of 1/25, which makes the motion of the scaled model working the sample displacement and frequency range of a possible floating bridge with the scale factors considered in the paper. For more details about the scaling procedure adopted for floating wind turbines, the reader is referred to [15]. For the sake of completeness in Table 2 surge natural frequencies of Tension Leg and Semisubmersible platform are reported on a model scale.

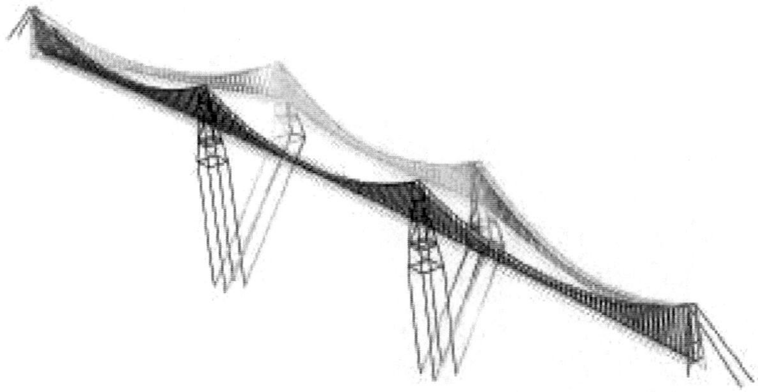

Figure 6. First vibration mode of the Bjørnafjorden floating bridge, natural frequency 0.0125 Hz, from [19]

Table 2. Frequencies of rigid modes for two platforms

Motion	Natural frequency [Hz]
Surge (TLP)	0.75
Surge (Semisub)	0.13

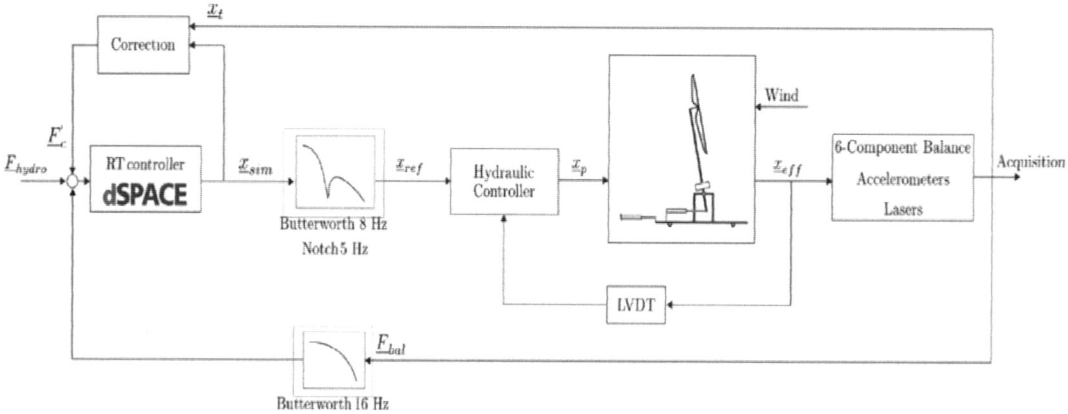

Figure 7. Control scheme for HIL implementation

4. Conclusions

To cross large and deep fjords, very long bridges are being currently studied. Amongst the feasible solutions, very long spans with multi-box sections or shorter spans with floating towers are proposed in several studies.

While the multi-box technology has already been studied and it is rather well established, the floating bridge solution needs more understanding and validation. The HIL approach in wind tunnel could be an efficient tool to validate the combined effects of turbulent winds and ocean waves acting on the structure. So far it has been applied on wind turbines with success and it could be applied analogously to floating bridges.

References

[1] Larose G L and Livesey F M 1997 Performance of streamlined bridge decks in relation to the aerodynamics of a flat plate *J Wind Eng Ind Aerod* **69(Supplement C)** 851-860.

[2] Yang Y, Wu T, Ge Y, and Kareem A 2015 Aerodynamic Stabilization Mechanism of a Twin Box Girder with Various Slot Widths *Journal of Bridge Engineering* **20(3)** 04014067.

[3] Yang Y, Zhou R, Ge Y, Mohotti D and Mendis P 2015 Aerodynamic instability performance of twin box girders for long-span bridges *J Wind Eng Ind Aerod* **145**, 196-208.

[4] Belloli M, Fossati F, Giappino S and Muggiasca S 2014 Vortex induced vibrations of a bridge deck: Dynamic response and surface pressure distribution *J Wind Eng Ind Aerod* **133(Supplement C)** 160-168.

[5] Chen W L, Li H and Hu H 2014 An experimental study on the unsteady vortices and turbulent flow structures around twin-box-girder bridge deck models with different gap ratios *J Wind Eng Ind Aerod* **132** 27-36.

[6] Larsen A, Savage M, Lafrenire A, Hui M C H and Larsen, S 2008 Investigation of vortex response of a twin box bridge section at high and low Reynolds numbers *J Wind Eng Ind Aerod* **96(6-7)** 934-944.

[7] Diana G, Rocchi D and Argentini T 2013 An experimental validation of a band superposition model of the aerodynamic forces acting on multi-box deck sections. *J Wind Eng Ind Aerod* **113** 40-58

[8] Diana G, Rocchi D, Argentini T and Muggiasca S 2010 Aerodynamic instability of a bridge deck section model: Linear and nonlinear approach to force modelling *J Wind Eng Ind Aerod* **98** 363-374 B

[9] Diana G, Resta F and Rocchi D 2008 A new numerical approach to reproduce bridge aerodynamic non-linearities in time domain *J Wind Eng Ind Aerod* **96** 1871-1884

[10] Perez T and Fossen T I 2008 Joint identification ofinfinite-frequency added mass and fluid-memory models of marine structures *Modeling, Identification and Control* **29(3)** 93–102.

[11] Robertson A O 2017 Uncertainty analysis of OC5-DeepCwind floating semisubmersible offshore wind test campaign *The International Society of Offshore and Polar Engineers Conference.*

[12] Perez T and Fossen, T I 2009 A matlab toolbox for parametric identification of radiation-force models of ships and offshore structures *Modeling, Identification and Control* **30(1)** 1–15.

[13] Bayati I, Jonkman J, Robertson A and Platt A 2014 The effects of second-order hydrodynamics on a semisubmersible floating offshore wind turbine *Journal of Physics: Conference Series* **524(1)**.

[14] Bayati I, Gueydon S and Belloli M 2015 Study of the effect of water depth on potential flow solution of the OC4 semisubmersible floating offshore wind turbine *Energy Procedia* **80**.

[15] Ebrahimnejad L, Janoyan K D, Valentine D T and Marzocca P 2014 Investigation of the Aerodynamic Analysis of Super Long Span Bridges by Using ERA-Based Reduced-Order Models *Journal of Bridge Engineering* **19** 04014033.

[16] Argentini T, Diana G, Rocchi D and Somaschini C 2016 A case-study of double multi-modal bridge flutter: Experimental result and numerical analysis *J Wind Eng Ind Aerod* **151** 25-36

[17] Argentini T, Pagani A, Rocchi D and Zasso A 2014 Monte Carlo analysis of total damping and flutter speed of a long span bridge: Effects of structural and aerodynamic uncertainties *J Wind Eng Ind Aerod* **128** 90-104

[18] Diana G, Resta F, Rocchi D and Argentini T 2008 Aerodynamic hysteresis: wind tunnel tests and numerical implementation of a fully non linear model for the bridge aeroelastic forces *The 4th International Conference on Advances in Wind and Structures*

[19] Sondre H, Shun W G 2016 Dynamic response of suspension bridge with floating towers, *Master Thesis* Norwegian University of Science and Technology

[20] Bayati I, Belloli M, Facchinetti A 2017 Wind tunnel 2-DoF hybrid/HIL tests on the OC5 Floating Offshore Wind Turbine *36th International Conference on Ocean, Offshore and Arctic Engineering Trondheim (Norway)* OMAE2017-61763

[21] Bayati I, Belloli M, Facchinetti A Giberti H 2017 A 6DoF/HIL setup for wind tunnel hybrid tests on a 1/75 scale model of a 10 MW floating offshore wind turbine *Aimeta*

[22] Bayati I, Belloli M, Bernini L, Giberti H, Zasso A 2017 Scale model technology for floating offshore wind turbines *IET Renewable Power Generation* **11(9)** DOI:10.1049/iet-rpg.2016.0956.

IOP Conf. Series: Materials Science and Engineering **276** (2017) 012009 doi:10.1088/1757-899X/276/1/012009

CFD Analysis in Subsea and Marine Technology

Hrvoje Jasak[1,2]

[1] Wikki Ltd, 459 Southbank House, Black Prince Road, London SE1 7SJ, United Kingdom
[2] Faculty of Mechanical Engineering and Naval Architecture, University of Zagreb, Ivana Lučića 5, 10 000 Zagreb, Croatia

E-mail: h.jasak@wikki.co.uk, hrvoje.jasak@fsb.hr

Abstract. Computational Fluid Dynamics (CFD) is established in design and analysis for a range of industries, but its use in Marine and Naval Hydrodynamics is behind the trend. This can be attributed to the complexity of modelling needs, including presence of free surface, irregular transient flows, fluid-structure coupling and presence of established modelling tools based on potential theory.

In this paper, state-of-the-art of CFD in Naval Hydrodynamics, wave and offshore applications is given, with an update of recent advances, validation and computing requirements for typical simulation cases.

1. Introduction

Since the beginning of the 21st century, Computational Fluid Dynamics (CFD) has become well established across the range of engineering applications. CFD results are well understood, expectations on accuracy in design studies can be satisfied. For example, automotive industry uses CFD to significantly reduce the use of experimental studies in aerodynamics, internal combustion engine design and exhaust gas after-treatment. Indeed, CFD studies are now regularly performed in applications which were never properly analysed experimentally, such as passenger compartment comfort, vehicle soiling and acoustics-related phenomena, where the design primarily relies on CFD results.

Acceptance of CFD in marine engineering, naval hydrodynamics and evaluation of wave loads appears to be lagging behind the trend of other industries. This is partially due to the complexity of free surface flows, but mainly by the fact that the phenomena under consideration imply significantly higher computational requirements. For example, sea-keeping studies of ships in irregular waves or calculation of transfer functions for wave loading on floating structures still cannot be performed at acceptable speed. “Engineering time-scale” in product design requires turn-over time of 8-12 hours for simulations to be practically useful.

An important factor in the acceptance of CFD is the presence of alternative simulation tools. Numerical models based on potential flow theory are widely accepted, fast and validated and have been in use for more than a decade. The potential flow assumption, whole adequate for a part of industrial needs, brings it own limitations, mainly due to the formulation, or failure to account for coupled/non-linear effects. The best example is the evaluation of added resistance in waves, which may provide adequate loading results on the hull, but cannot account for forward speed of the ship.

The challenges for CFD modelling in naval hydrodynamics can be grouped as follows:

- Handling of discontinuity at the free surface;
- Accurate tracking of free surface position;
- Issues of force evaluation and hydro-mechanical coupling;
- Generation and propagation of surface waves, of various nature: deterministic single- and multi-frequency- wave trains, wave focusing and statistical modelling of realistic irregular sea states;
- Modelling of rigid or flexible floating structures;
- Interaction with global external forces, such as mooring systems and propulsors;
- Modelling of extreme wave events such as freak wave impact or green sea loads;
- Statistics-based models, such as 3-hour-storm or safe return to port simulations and
- Full scale effects.

In what follows, recent advances for each of the listed topics from the author's work shall be presented. The paper will conclude with a readiness assessment of CFD tools for simulation types of varying complexity.

2. CFD Modelling of Free Surface Flows

The two-fluid model for immiscible incompressible free surface flows is well established, consisting of phase continuity, combined momentum equation and global continuity equation.

$$\frac{\partial \alpha}{\partial t} + \nabla\bullet(\mathbf{u}\,\alpha) = 0, \tag{1}$$

$$\frac{\partial(\rho\mathbf{u})}{\partial t} + \nabla\bullet(\rho\mathbf{u}\mathbf{u}) - \nabla\bullet(\mu\nabla\mathbf{u}) = -\nabla p + \rho\mathbf{g}, \tag{2}$$

$$\nabla\bullet\mathbf{u} = 0, \tag{3}$$

where α is the phase fraction, ρ, $\mathbf{u}$ and μ are combined density, velocity and viscosity, by virtue of the immiscibility condition, *e.g.*[1]:

$$\rho = \alpha\rho_1 + (1-\alpha)\rho_2, \tag{4}$$

$$\mathbf{u} = \alpha\mathbf{u}_1 + (1-\alpha)\mathbf{u}_2, \tag{5}$$

$$\mu = \alpha\mu_1 + (1-\alpha)\mu_2. \tag{6}$$

Here, ρ_i, $\mathbf{u}_i$ and μ_i represent phase density, velocity and viscosity, respectively. This model is capable of describing the details of non-linear wave interaction with marine objects.

In practice, the equation set is enhanced by turbulence modelling, dynamic mesh support, coupling with a rigid or elastic motion of solid boundaries, wave generation and absorption *etc.*but sufficiently illustrates the numerical challenges.

3. Discretisation and Numerics for Discontinuous Variables

The Finite Volume Method (FVM) [2] is a method of choice for numerical simulation of fluid flows, due to its combination of strict conservation, simplicity, efficiency and flexibility. Free surface flow modelling brings further challenges, to be addressed below.

3.1. Body Force and Pressure Gradient

The first challenge relates to the handling of body force $\rho\mathbf{g}$ in presence of a step change in ρ. In hydrostatics, the body force and pressure gradient need to balanced and discretisation consistency in the momentum equation requires that *the computational stencil for the balancing terms needs to match.* In (2) this is clearly not the case, resulting in pressure chequerboarding. To remedy the problem, the pressure variable is redefined:

$$p_d = p - \rho\, \mathbf{g}\bullet\mathbf{x}, \tag{7}$$

yielding a reformulated balanced momentum source term

$$-\nabla p + \rho\mathbf{g} = -\nabla p_d - (\mathbf{g}\bullet\mathbf{x})\nabla\rho, \tag{8}$$

with matching computational stencil for ∇p_d and $\nabla\rho$.

3.2. Interface Jump Conditions

While (8) resolves the pressure chequerboarding problem, it still contains the gradient of the (discontinuous) ρ at the interface. A practical consequence of numerical gradient evaluation for a step-function is the parasitic velocity at the interface. Quetey *et al* [3] take into account the discontinuities at the free surface within the FVM, assuming that the free surface always coincides with the mesh faces. Such an approach requires significant mesh refinement at the free surface. The problem is resolved by the Ghost Fluid Method (GFM) [4] developed by Vukčević, accounting for discontinuities at an arbitrary location in the mesh.

The second-order accurate GFM is derived from the assumption of piece-wise constant phase density, allowing (2) to be rewritten in terms of continuous velocity field $\mathbf{u}$ instead of discontinuous linear momentum $\rho\mathbf{u}$:

$$\frac{\partial\mathbf{u}}{\partial t} + \nabla\bullet(\mathbf{u}\mathbf{u}) - \nabla\bullet(\nu\nabla\mathbf{u}) = -\frac{1}{\rho}\nabla p. \tag{9}$$

Presence of the interface is accounted for in *interface jump conditions* with $[\psi] = \psi^- - \psi^+$ denoting the jump of the variable at the interface:

$$[\mathbf{u}] = 0, \tag{10}$$

$$[p] = 0, \tag{11}$$

$$\left[\frac{1}{\rho}\nabla p_d\right] = 0. \tag{12}$$

The jump conditions directly account for discontinuous density and pressure gradient at the interface and are built into the discretisation for pressure gradient and Laplacian. Precise location of the interface is evaluated between the two cell centres straddling the free surface from interface tracking information.

4. Handling of the Free Surface

The primary consideration of modelling of free surface flows is the capturing of position and topology of the free surface. Its position profoundly influences the flow solution to which it is coupled in a non-linear manner. In this section, we shall review some options on free surface handling and implications to the needs of marine CFD.

There are two fundamental approaches. *Surface tracking* deforms the computational mesh at the interface to comply to its current configuration, usually reducing the problem to a moving boundary single-phase flow [5, 6]. *Surface capturing* uses a continuum representation of the free surface and the actual position of the interface is obtained via post-processing. Examples include various forms of marker-in-cell or particle-in-cell simulations, solution of the Volume-of-Fluid (VOF) [1, 7, 8], level-set [9, 10, 11, 12] or similar [13, 14].

4.1. Linearised Free Surface and Surface Tracking

Surface tracking is a straightforward way of dealing with a moving free surface [15], whose position is defined by a double boundary condition:

- Fixed surface pressure, either assuming atmospheric pressure above water level or via pressure continuity in two-phase flows;
- Zero net flux through the surface, defining the surface normal velocity to be equal to the mesh motion flux.

In its simplest form, the free surface boundary condition is linearised in terms of pressure and surface elevation without deforming the mesh, yielding the *linearised free surface model.* Mode complex formulations involving a deforming mesh provide a more accurate description of surface motion with a modest increase in computational cost.

The main advantage of surface tracking is that fact that the free surface problem may be reduced to a single phase, without the need of increased resolution of the computational mesh near the surface. It disadvantages relate to complexity of mesh motion, interaction with solid boundaries of complex shape and most importantly its inability to efficiently handle changes in surface topology such as breaking waves.

4.2. Surface Capturing Models

Surface capturing models use a continuum representation of the free surface, via a volume fraction or a general colouring function. They can be divided into three major groups, illustrated in Fig. 1:

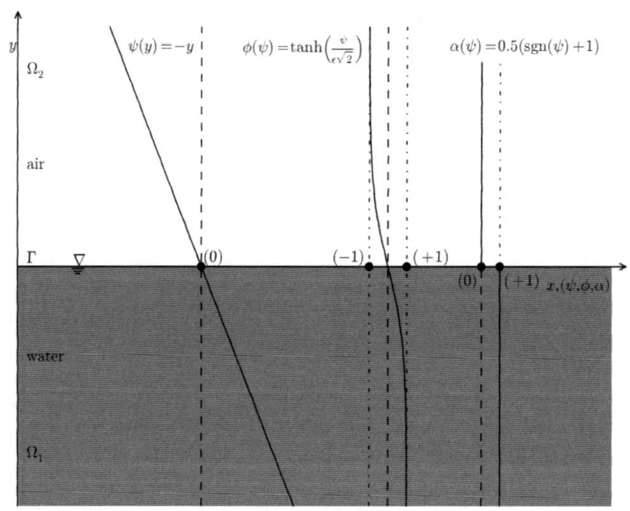

Figure 1. Interface capturing schemes: Level Set, Phase Field and Volume of Fluid.

- Volume of Fluid (VOF) methods (1), [1] is simple to understand and strictly satisfies the phase conservative condition but suffers from the need to advect the step-function in α without unboundedness or smearing;

- Seeking a "smoother" solution variable, Level Set (LS) methods [9] solve for the signed distance to the interface. While the solution no longer contains a step, LS does not guarantee mass conservation either, which is a major drawback. Furthermore, the discretisation error in the transport appears as a loss of unit gradient for the distance function; as a result, the advection step is usually followed by re-initialisation. This can be done by direct calculation of distance [16] in a narrow band, or by solving an additional Eikonal equation to propagate the unit gradient constraint [17]. The best methods use implicit re-initialisation [12] or combine mass conservation properties of VOF and smoothness of LS in some way [11];
- Phase Field (PF) methods are developed on a basis that there exists a transport equation for self-similar advection of a tanh profile [18]. With appropriate mathematical reformulation [12] it is possible to cast the PF equation into the LS form, treating all terms fully implicitly within the FV framework. This effectively combines the advection and re-initialisation step of LS into one.

Based on extensive experience with all of the above, we typically choose VOF for problems where mass conservation is crucial [19]. For solution decomposition methods, it is preferable to choose an unbounded variable and LS methods are preferred [12, 20].

4.2.1. Volume-of-Fluid and Sharp Interfaces While the VOF equation brings advantages such as clear variable bounds and strict phase conservation, its weak point is the shape of the advected profile. For good performance and accurate surface-to-surface interaction it is necessary to preserve the sharpness of the step profile denoting the free surface position as accurately as possible. Various forms of interface compression are used in practice, either based on numerical artefacts of convection schemes [1] or quasi-physical compression [21].

4.2.2. Geometric Reconstruction Methods Alternative formulations of free surface advection exist, where a "perfect" advection of a step profile is achieved with the use of geometric methods, albeit at a considerably higher computational cost. Recently, Ronby *et al* [22] introduced `isoAdvector` method, where accurate VOF advection is achieved through evaluation of VOF face flux with the use of a fluxing iso-surface for each face, Fig. 2. The method is explicit, practically limited by the Courant-Friedrichs-Levy (CFL) condition and significantly more expensive than conventional advection, but it produces remarkable results.

Geometric free surface advection scheme such as `isoAdvector` are used in cases where practical limitations in local mesh size exists and a clean interface is essential. A good example is a green sea simulation or slamming impact.

4.2.3. Level Set Methods As already noted, LS methods suffer from two drawbacks: lack of strong phase conservation and need for re-initialisation as a means to preserve unit gradient of the distance function. Vukčevic *et al* [12] introduce an alternative formulation of the LS, inspired by the properties of the terms in the PF equation which preserve self-similarity. Reformulated PF terms in the LS equation not only remove the need for separate re-initialisation but also improve the conservation properties compared with the original LS formulation written for the signed distance variable ψ:

$$\frac{\partial \psi}{\partial t} + \nabla\cdot(\mathbf{c}\psi) - \psi\nabla\cdot\mathbf{c} - b\nabla\cdot(\nabla\psi) = b\frac{\sqrt{2}}{\epsilon}\tanh\left(\frac{\psi}{\epsilon\sqrt{2}}\right), \tag{13}$$

where ψ is the signed distance function, $\mathbf{c}$ is the combined convective velocity:

$$\mathbf{c} = \mathbf{u} + b\frac{\sqrt{2}}{\epsilon}\tanh\left(\frac{\psi}{\epsilon\sqrt{2}}\right)\nabla\psi + b\kappa\frac{\nabla\psi}{|\nabla\psi|}, \tag{14}$$

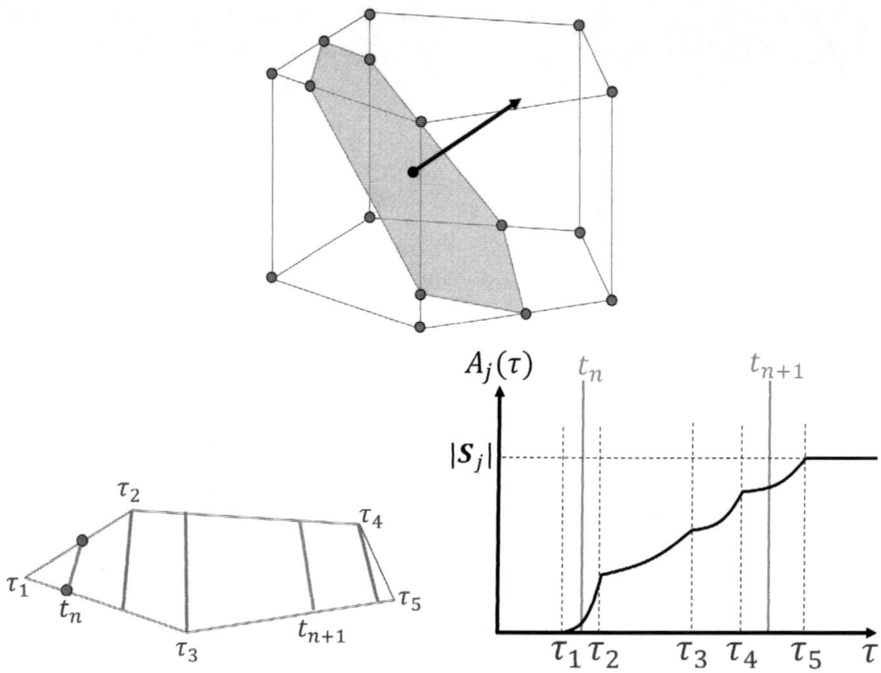

Figure 2. Calculation of the face flux using the `isoAdvector` scheme, [22].

where b is a numerical parameter used to smear out possible local singularities, ϵ is the smearing distance for the PF representation of the LS field and κ is the mean interface curvature. In the absence of curvature driven motion, as is the case in ship hydrodynamics free surface flows, the three terms on the r.h.s. of (13) numerically maintain the signed distance profile implicitly during the transport [18].

Although numerically more complex, this equation incorporates self-similar advection through a balance of additional convection and diffusion terms and is treated in a fully implicit manner. Furthermore, as ψ is not a bounded variable, it is suitable for use in solution decomposition strategies, as described below.

4.3. Hydro-Mechanical Coupling

The case of a floating body supported in free surface flow is often encountered in naval hydrodynamics. Here, the force balance on a floating body is handled via a 6-Degrees-of-Freedom (6-DOF) Ordinary Differential Equation (ODE) with body forces being balanced by pressure and friction acting on the wetted surface. In response to the 6-DOF solution, wall boundary is moved in the CFD domain, changing the hydrostatic and hydrodynamic response in a non-linear manner.

Hydro-mechanical coupling is resolved via outer flow – 6-DOF iterations at considerable cost. A representative number of outer iterations on the system is 6 to 8, with a total of 24 pressure solutions within a single time-step. Gatin *et al* [23, 24] propose an enhanced 6-DOF to pressure coupling, where the boundary motion is included into the fluid pressure solution. The new algorithm resolves the hydro-mechanical coupling in as little as 2 outer and 2 PISO iterations, reducing the CPU time by a factor of 3.

5. Wave Generation and Propagation

With numerical techniques described above it is possible to assemble an efficient CFD tool for free surface flows. In practical applications, a significant part of structural loading results from wave action: it is thus necessary to develop a wave initialisation technique within the CFD framework. The objective is to introduce a realistic wave elevation and velocity either for monochromatic waves or wave spectra into the CFD simulation with minimum distortion.

Wave reflection would naturally occur at the outlet and side boundaries may potentially destroy the solution; it is necessary to devise an appropriate practice to remove undesired wave reflections from domain boundaries.

5.1. Waves and Relaxation Zones

For simple monochromatic waves, waves can be introduced via a conventional wave boundary condition. For more complex waves, *e.g.* in wave focusing or irregular sea states, a simple boundary condition does not provide necessary fidelity. Also, removing reflection at outlet boundaries via a boundary condition works well only when the wave-length and speed of propagation is known. For those reasons, a *relaxation zone* approach is preferred [25], where the CFD solution is blended with the prescribed "far field" wave state within a relaxation zone.

In our simulations we use a modified form of relaxation, where the blending is performed implicitly, at the level of discretisation matrix [12, 20, 26]. The implicit method is not only more consistent and robust, but also works well with relaxation zone length of the order of one significant wave length. Detailed verification and validation data for the use of implicit relaxation zones is reported in [27, 20].

In sea-keeping and global performance simulations, we use relaxation zones with a prescribed "far-field" wave theory across all boundaries. This allows generality in a sense that the waves can propagate in arbitrary directions. With this approach, it is also easier for the analyst to establish how accurately the waves are transported through the CFD region, as any dissipation (loss of amplitude) or dispersion (phase error) becomes immediately visible in the results.

5.2. SWENSE Formulation

In sea-keeping or global performance CFD, far field wave condition carries a bulk of the flow solution. The effect of the floating body on the wave field, via diffracted and radiated waves presents only a fraction of the solution – if the non-linearity may be considered low. It is therefore reasonable to use the SWENSE solution decomposition technique [28, 12], where the CFD solution is decomposed without simplification into two parts:

$$\xi = \xi_I + \xi_P, \tag{15}$$

where

- ξ_I is the incident field, representing the far field wave condition and
- ξ_P is the perturbation field, representing the difference between the complete non-linear flow solution and the chosen incident field.

Under such decomposition, and with a wisely chosen incident field ($|\xi_P| << |\xi|$), it is possible to achieve superior wave propagation and still give a full non-linear CFD solution. In a CFD simulation, the velocity field can be decomposed in a straightforward manner. To describe the free surface position, Vukčević chooses to apply (15) to the LS equation to avoid boundedness problems with other colour functions (*e.g.*VOF, PF). Particular attention is given to the cases where the prescribed incident solution p_I, U_I, ψ_I does not strictly satisfy the discrete form of the Navier-Stokes equations, as it is necessary to compensate for the compatibility error [12].

6. Regular and Irregular Waves

The remaining task in setting up a successful CFD simulation of wave loads consists of prescribing the incoming sea state.

6.1. Regular Waves

In first attempts, analytical forms such as first-, second- and fifth-order Stokes waves were used, with optional superposition of current and/or sea bottom boundary layer profiles. Additional wave forms, such as solitary and c-noidal waves are also implemented.

It was noted that almost all regular waves can be accurately modelled with the fully nonlinear stream–function wave theory [29], at a negligible computational cost compared to CFD. Since the method takes into account the presence of a current implicitly, stream–function wave theory is a preferred choice.

6.2. Wave Focusing and Wave Spectra

While regular wave simulations are already of practical interest, industrial needs regularly require "design waves" or freak waves, modelled as a linear superposition of regular waves optimised to give a desired freak wave height. A freak wave assembled from a wave package is created from a given incoming wave spectrum, by tuning phase shifts over *e.g.* 100 wave components. The resulting wave load is then used in further structural analysis: it is therefore essential to correctly assemble such a wave package and ensure that its shape is preserved at the point of interest/impact.

To model realistic sea states which adhere to the desired spectrum, a selection of spectra is also implemented, This includes the Bretschneider, JONSWAP and Pierson-Moskowitz spectrum, with an additional option of providing an arbitrary list of wave components.

6.3. Higher-Order Spectra

During the propagation of wave spectra using CFD, a distortion of the spectra occurs due to non-linearities arising from wave-to-wave interaction and wave modulation. These non-linearities are present in nature as well and play a major role in evolution of the sea state. In order to resolve these non-linearities outside of the CFD simulation, Gatin *et al* gatinEtAl2016HOS implemented and validated a potential Higher-Order Spectrum (HOS) solver. The solver implements a pseudo-spectral method, solving for the non-linear free surface boundary conditions truncated at arbitrary (user–defined) non-linear mode. Compared to CFD, the HOS solver is extremely fast and is used to propagate a realistic (directional) sea state based on the given spectrum, which can then be screened for realistic freak wave events. The HOS freak wave is superior to wave focusing since it includes non-linear wave-to-wave interaction and single–wave modulation and can be accurately transported within the CFD domain. Furthermore, HOS can be used to calibrate the input spectrum in order to achieved the target spectrum in CFD. The method has been coupled to CFD, with extensive verification and validation [30].

7. Wave-Induced Loads

Having assembled the CFD solver capable of accurately propagating waves provided via the far-field implicit relaxation zones, a detailed study of prediction accuracy for wave-induced loads has been performed. The study starts with the analysis of wave loads on static structures and then proceeds with floating objects in regular and irregular seas.

7.1. Waves on Static Structures

Vukčević [20] provides a detailed verification and validation study for wave loads on a vertical surface-piercing cylinder for various wave steepness, providing guidance on mesh resolution in the vertical and horizontal direction (with respect to the wave height and length), time-step size and number of wave encounters necessary to achieve periodicity. The solution is analysed in terms of wave propagation accuracy without the cylinder and in evaluation of higher-order forces resulting from wave diffraction from the cylinder.

7.2. Floating and Flexible Structures

Having established good force prediction for a static object, a comprehensive sea-keeping study for a ship in head and oblique waves was reported at the Tokyo 2015: A Workshop on CFD in Ship Hydrodynamics [31, 32]. The study starts with the evaluation of steady resistance, sinkage and trim for the KCS hull [33], followed by the verification and validation study for the same hull in head waves of various wave lengths and oblique waves in various directions. For all simulations, mesh resolution and periodic uncertainties are reported, while the time-step size and hydro-mechanical coupling uncertainties are presented for a representative case. As a guidance, a mesh of 1.6 million cells provides sea-keeping results with uncertainties of the order of 4%, with simulation time of approximately 15 minutes per wave encounter period on a small HPC cluster [34].

7.3. Mooring Systems, Propulsion and Sailing

In order to expand the applicability of the method, handling of a mooring systems and propulsion modelling is added to the package. Mooring forces are handled via the interface to the 6-DOF ODE solver, either in terms of linear/torsional spring/damper system, or a direct interface to external mooring model software. Model performance and technical readiness of such a model for practical offshore floater design has been reported by Kim *et al* in [35].

7.4. Green Sea Loads

As the next step, the issue of green water loading is considered. Here, two problems arise. First, the frequency of occurrence of water-on-deck events is relatively rare and a statistical/screening method is required to select specific event of sea-keeping in irregular seas. Secondly, a highly accurate method for the description of free surface on (necessarily) under-resolved meshes is needed. The focus here is in the surface capturing without numerical smearing errors, needed to accurately capture water impact on the structure.

Gatin [19] reports a comprehensive procedure for green sea load evaluation, combining the screening of the HOS spectral solution for irregular seas, sea-keeping simulation for ship motion for the selected events and highly accurate impact simulation.

In green water impact simulation, the `isoAdvector` geometric reconstruction scheme by [22] is used, with remarkable results.

Gatin [36] reports the comparison of numerical and experimental study for the green water impact on a fixed barge, based on experimental results by Lee *et al* [37] for 9 incident waves. A combination of advanced numerical practices produces remarkable good experimental agreement, shown as a comparison of pressure peaks and pressure force integrals for 10 measuring locations in Fig. 3.

Error bars on the CFD results in the figure show the combined periodic and mesh uncertainty, while the experimental error bars show periodic and measuring uncertainty.

8. Full Scale Effects

For a number of years, sea-keeping CFD has been performed at model scale. While this can be somewhat justified by the lack of full-scale experimental data, the need for full-scale data by naval architects can no longer be ignored. For CFD studies, dealing with full-scale ships and floating platforms brings further modelling problems, related with self-propulsion, course-keeping and full-scale turbulence effect.

Vukčević *et al* [38] report the results for a full scale benchmark simulation of a general cargo carrier used in the 2016 Workshop on ship scale hydrodynamic computer simulation. The CFD

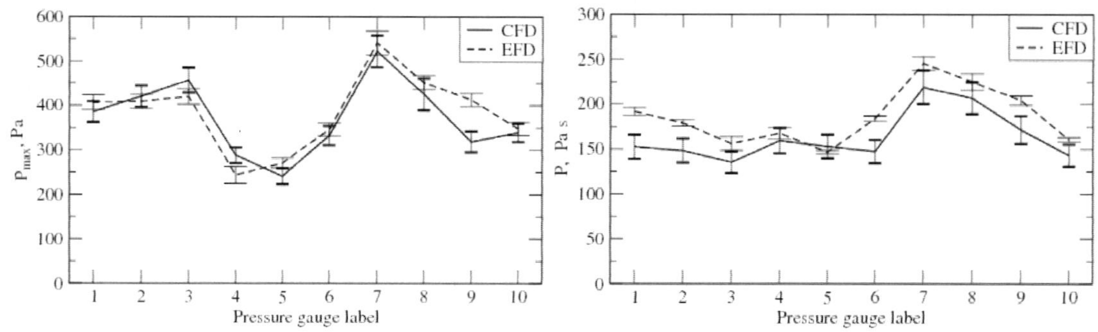

Figure 3. Simulation of Green Sea Impact: pressure peaks (left) and force integrals (right).

model included a full-scale ship hull propelled using the actuator disk propeller model under free sailing condition. The simulation results were compared with sea trial data. The presented results show that the self-propulsion point has been reached, predicting the ship speed to 0.3%, with grid uncertainty of 0.02 knots against the measured mile data, which is remarkable. A test was subsequently repeated on a different geometry in collaboration with Uljanik shipyard, with similar results.

9. Statistically Significant Modelling

At the current point of development, simulation of statistically significant results in a prescribed irregular sea state is coming into focus. Typical simulations will involve global performance CFD simulations of a moored floating platform or FPSO in a 3-hour storm, or a *safe return to port* simulation for a passenger ship.

Safety analysis and novel design criteria, such as designing an optimal ship hull for a given sea state (as opposed to optimisation for steady resistance in calm seas) will pose new challenges on robustness, accuracy and speed of simulation in naval hydrodynamics and off-shore CFD.

10. Summary

This paper presents the state-of-the-art of naval hydrodynamics CFD as seen by the author. While good progress has been made in topics of structural loading by regular waves and steady resistance of ship hulls, significant challenges remain. A contribution to the applicability and practical use of CFD in ship and off-shore design will be made by a combination of improved numerical practices, extensive validation and verification and some improvement in HPC hardware performance.

This paper reports on improvements in numerical discretisation, free surface capturing, coupling of simplified far-field wave models with CFD, generation and propagation of regular and irregular waves and detailed validation and verification of the CFD tools for loads on static structures, sea-keeping in regular waves, simulation of freak wave impact and green sea loads. Further challenges lie in the statistical modelling of irregular seas, where significant further work is needed.

References

[1] Ubbink O and Issa R 1999 *J. Comp. Physics* **153** 26–50
[2] Jasak H 1996 *Error analysis and estimation in the Finite Volume method with applications to fluid flows* Ph.D. thesis Imperial College, University of London
[3] Queutey P and Visonneau M 2007 *Comput. Fluids* **36** 1481–1510
[4] Vukčević V, Jasak H and Gatin I 2017 *Computers & Fluids* **153** 1 – 19 ISSN 0045-7930
[5] Muzaferija S and Perić M 1997 *Numerical Heat Transfer, Part B: Fundamentals* **32** 369–384
[6] Jasak H and Tuković v 2007 *Transactions of FAMENA* **30** 1–18
[7] Aulisa E, Manservisi S, Scardovelli R and Zaleski S 2003 *J. Comput. Phys* **192**
[8] Røenby J, Bredmose H and Jasak H 2016 *Open Science* **3**
[9] Sussman M and Fatemi E 1999 *SIAM J. Sci. Comput.* **20** 1165–1191
[10] Olsson E and Kreiss G 2005 *J. Comput. Phys.* **210** 225–246
[11] Olsson E, Kreiss G and Zahedi S 2007 *J. Comput. Phys.* **225** 785–807
[12] Vukčević V, Jasak H and Malenica v 2016 *Ocean Engineering* **121** 37–46
[13] Sun Y and Beckermann C 2007 *J. Comput. Phys.* **220** 626–653
[14] Sun Y and Beckermann C 2008 *Physica D* **237** 3089–3098
[15] Tuković v and Jasak H 2012 *Computers and Fluids* **55** 70–84
[16] Gómez P, Hernández J and López J 2005 *Int. J. Numer. Methods Eng.* **63** 1478–1512
[17] Hartmann D, Meinke M and Schröder W 2008 *J. Comput. Phys.* **227** 6821–6845
[18] Folch R, Casademunt J and Hernández-Machado A 1999 *Phys. Rev. E* **60** 1724
[19] Gatin I, Vukčević V and Jasak H 2017 *12th OpenFOAM Workshop*
[20] Vukčević V, Jasak H and Malenica S 2016 *Ocean Engineering* **121** 76–88
[21] Rusche H 2003 *Computational fluid dynamics of dispersed two-phase flows at high phase fractions* Ph.D. thesis Imperial College, University of London
[22] Roenby J, Bredmose H and Jasak H 2016 *Open Science* **3**
[23] Gatin I, Vukčević V, Jasak H and Rusche H 2017 *Ocean Engineering* **143** 295–304
[24] Gatin I, Vukčević V and Jasak H 2017 *VII International Conference on Computational Methods in Marine Engineering (MARINE 2017)* pp 663–669
[25] Jacobsen N, Fuhrman D and Fredsoe J 2012 *International Journal for Numerical Methods in Fluids* **70** 1073–1088
[26] Vukčević V, Jasak H and Malenica S 2015 *Proceedings of the 18th Numerical Towing Tank Symposium, NUTTS*
[27] Vukčević V 2016 *Numerical modelling of coupled potential and viscous flow for marine applications - in preparation* Ph.D. thesis Faculty of Mechanical Engineering and Naval Architecture, University of Zagreb PhD Thesis
[28] Ducrozet G, Engsig-Karup A P, Bingham H B and Ferrant P 2014 *J. Comput. Phys.* **257** 863–883
[29] Rienecker M M and Fenton J D 1981 *J. Fluid. Mech.* **104** 119–137
[30] Gatin I, Vukčević V and Jasak H 2016 *Ocean Eng. - under review*
[31] Vukčević V and Jasak H 2015 *Proceedings of the Tokyo 2015: A Workshop on CFD in Ship Hydrodynamics* vol 3 pp 437–442
[32] Vukčević V and Jasak H 2015 *Proceedings of the Tokyo 2015: A Workshop on CFD in Ship Hydrodynamics* vol 3
[33] Gatin I, Jasak H and Vukčević V 2015 *Proceedings of the Tokyo 2015: A Workshop on CFD in Ship Hydrodynamics* vol 3 pp 431–436
[34] Vukčević V, Jasak H, Gatin I and Malenica S 2016 *Proceedings of the 31st Symposium on Naval Hydrodynamics*
[35] Kim J W, Jang H, Bauet A, O'Sullivan J, Lee S, Kim B, Reed A and Jasak H 2016 *Offshore Technology Conference* (Houston, TX, USA) oTC-27294-MS
[36] Gatin I, Vukčević V, Jasak H, Seo J and Rhee S 2017 *Ocean Engineering* **In Press**
[37] Lee H H, Lim H J and Rhee S H 2012 *Ocean Engineering* **42** 47–60 ISSN 0029-8018
[38] Vukčević V, Jasak H, Gatin I and Uroić T 2017 *Proceedings of the VII MARINE 2017 conference* pp 1052–1061

IOP Conf. Series: Materials Science and Engineering **276** (2017) 012010 doi:10.1088/1757-899X/276/1/012010

Multi-fluid CFD analysis in Process Engineering

B H Hjertager
Department of Mechanical and Structural Engineering and Materials Science, University of Stavanger, Stavanger, Norway

E-mail: bjorn.hjertager@uis.no

Abstract. An overview of modelling and simulation of flow processes in gas/particle and gas/liquid systems are presented. Particular emphasis is given to computational fluid dynamics (CFD) models that use the multi-dimensional multi-fluid techniques. Turbulence modelling strategies for gas/particle flows based on the kinetic theory for granular flows are given. Sub models for the interfacial transfer processes and chemical kinetics modelling are presented. Examples are shown for some gas/particle systems including flow and chemical reaction in risers as well as gas/liquid systems including bubble columns and stirred tanks.

1. Introduction

1.1. Basic considerations

Several different approaches are available for setting up the governing equations for multi-phase equipment. One method is the so-called PSIC (Particle-Source-In-Cell) procedure originally presented by Migdal and Agosta [1].This method treats the continuous phase (liquid or gas) in a usual Eulerian description, whereas the dispersed phases (bubbles, droplets or particles) are described in a Lagrangian way. This means that the dispersed phase is tracked through the flow domain from inlet to outlet. In the original method it was assumed no interaction between the various dispersed phases and thus this method was only applicable to low volume fractions of the dispersed phase. In later years this limitation has been removed and the resulting method is called the discrete particle method (DPM).

The second method is the Volume of Fluid (VOF) for gas/liquid systems and was first proposed by Nichols et al. [2]. The method makes it possible to calculate the interphase between the gas and liquid and is thus able to resolve the details of the bubble shapes as they move through the liquid. The method assumes that the flow is fully segregated (resolved), which means that we do not have a mixture of gas and liquid anywhere in the computational cells. We either have pure gas or pure liquid present in the cells. This means that we solve only one set of balance equations: gas or liquid. The inter-phase is tracked by solving a transport equation for the VOF variable. This is a discontinuous variable that is either zero or one. The method is such that in order to resolve sharp interfaces it needs large grid resolution and hence large computer resources.

The third method is the Lattice Boltzman (LB) method that is able to model the flow around solid particles and is therefore able to fully resolve the solids fluid flow. Both the VOF and LB may be characterised by being so-called Direct Numerical Simulation (DNS) method for multi-phase flows.

The fourth method deduces the governing equations based on the Eulerian concept and is normally named the multi-fluid method (Spalding, [3]). This means that the phases are treated as interpenetrating fluids that share the space and interact through the source terms. All equations are such that the volume fraction may take a value between zero and one.

Figure 1 gives an overview of the four various methods for gas-liquid and gas-solids flow. We note that at the top row in the figure gives the most fundamental methods, i.e. methods that need no modelling since all aspects of the flow are calculated. These methods will however need large computational resources. We also see from the figure that DPM/PSIC approach is an intermediate method where some details are lost and needs to be modelled. The third row in the figure gives the multi-fluid method. This is the most demanding method with regard to modelling needs. The important aspect to note is that due to large dimensions and computing demands the multi-fluid method is the optimum choice for industrial scale process equipment. However, the VOF, LB and DPM/PSIC methods are highly needed to feed better sub-models for the multi-fluid technique. We also see that in the last row it is envisaged a combination of multi-fluid and the DPM methods that may be optimum where large bubbles are flowing through a fluidised bed.

MORE PHYSICS / SMALL SCALE (↑)

Gas liquid flows	Gas solids flows
Interphase tracking, VOF	Lattice Boltzman (LB)
Bubble tracking DPM/PSIC	Particle tracking DPM/PSIC
Multifluid	Multifluid
	Multifluid and bubble tracking

MORE MODELLING / INDUSTRIAL SCALE (↓)

Figure 1: Relations between description level for gas liquid and gas solids systems and 1) degree of modelling details needed and 2) scale of reactor (adapted from van der Hoef et al [4]).

1.2. Objective of the Paper

This paper will mainly deal with the multi-fluid technique. Details will be given on the derivation of the multi-fluid equations. Section II will deal with the basics of the multi-fluid approach. Section III will deal with the closures for bubble-liquid flows whereas Section IV will handle the closure framework for solids fluid flows based on the kinetic theory of granular flow (KTGF). Additional sub models for bubble size, interfacial transfer processes, chemical reaction and stirred tank models are dealt with in Section V. Section VI presents some applications of multi-fluid simulations for bubble columns, stirred tanks and fluidised beds. The final section VII gives a summary of the Paper.

2. Basics of the multi-fluid approach

2.1. Introduction

A multiphase flow system consists of a number of single phase regions bounded by moving interfaces. In principle, a multiphase flow model could be formulated in terms of the local instantaneous variables pertaining to each phase and matching boundary conditions at the interfaces. The direct numerical solution of this kind of formulation is not feasible in practice. It would require an extremely fine mesh and a very small time step. The local instantaneous formulation can be used as starting point to derive macroscopic equations by an average procedure, whose numerical solution is feasible. The modelling of the unclosed terms arising from the average procedures is the price to pay for solving averaged multiphase flow systems.

2.2. Averaged Balance Equations

Multi-fluid model: In cases where a detailed description of the interface dynamics is not relevant or is computationally too expensive, the averaging can be performed on a volume larger than the characteristic volume of the secondary phase as illustrated in Figure2. Doing so, all the information lost on the scales smaller than the averaging scale need to be accounted for through sub-grid models, which may be derived empirically, analytically or numerically. The multi-fluid model is derived from the averaged local instantaneous balance equations with an averaging volume larger than the characteristic volume of the secondary phase, and is based on the assumption of inter-penetrating continuum media. That is, different phases can share the same spatial position.

The multi-fluid model balance equations can be written in a compact form:

Mass balance

$$\frac{\partial}{\partial t}(\alpha_k \bar{\rho}_k) + \nabla \cdot (\alpha_k \bar{\rho}_k \hat{\vec{u}}_k) = \Gamma_k \tag{1}$$

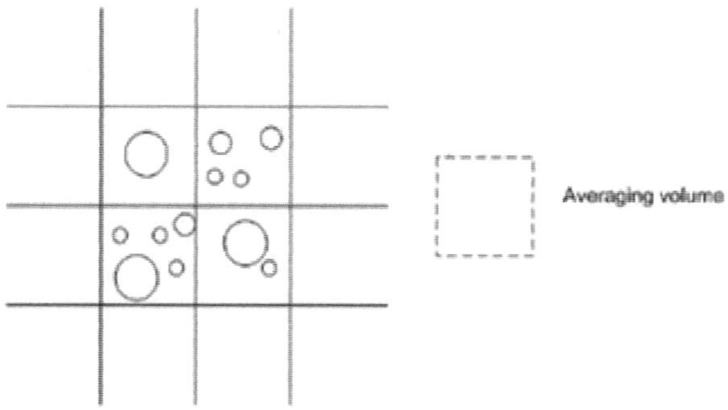

Figure 2: Averaging volume for a multi-fluid flow model framework [13].

Momentum balance

$$\frac{\partial}{\partial t}(\alpha_k \bar{\rho}_k \hat{\vec{u}}_k) + \nabla \cdot (\alpha_k \bar{\rho}_k \hat{\vec{u}}_k \hat{\vec{u}}_k) = \nabla \cdot \left[\alpha_k \left(\hat{\overline{\overline{T}}}_k + \overline{\overline{T}}_k^t \right) \right] + (\alpha_k \bar{\rho}_k \vec{g}) + \Gamma_k \hat{\vec{u}}_k + \vec{M}_{I,k} \tag{2}$$

Here, Γ_k represents the exchange of mass between phases, $\overline{\overline{T}}_k^t$ is the residual stress tensor accounting for the unresolved turbulence scales and $\vec{M}_{I,k}$ accounts for the interfacial forces.

The following constraints hold

$$\sum_{k=1}^{N} \Gamma_k = 0 \quad \text{and} \quad \sum_{k=1}^{N} \alpha_k = 1.0 \tag{3}$$

The average energy equation expressed by enthalpy h as the dependent variable reads (kinetic energy is subtracted):

$$\frac{\partial}{\partial t}(\alpha_k \bar{\rho}_k \hat{h}_k) + \nabla \cdot (\alpha_k \bar{\rho}_k \hat{\vec{u}}_k \hat{h}_k) = -\nabla \cdot \alpha_k (\hat{\vec{q}}_k + \vec{q}_k^t) + \frac{D_k}{Dt}(\alpha_k p_k)$$
$$+ \alpha_k (\hat{\overline{\overline{\tau}}}_k + \overline{\overline{\tau}}_k^t) : \nabla \hat{\vec{u}}_k + \hat{h}_{I,k} \cdot \Gamma_k + \alpha_k \bar{\rho}_k S_k + E_{I,k} \tag{4}$$

Here the terms on the RHS are 1) heat transfer in phase k due to temperature gradients, 2) reversible work in phase k, 3) dissipation due to shear stress in phase k, 4) energy transfer to phase k due to mass transfer between phases, 5) energy generation in phase k due to e.g. chemical reaction and finally 6) heat transfer to phase k from the other phases.

The average balance equation for a chemical specie is expressed as

$$\frac{\partial}{\partial t}(\alpha_k \overline{\rho}_k \tilde{Y}_{j,k}) + \nabla \cdot (\alpha_k \overline{\rho}_k \hat{u}_k \tilde{Y}_{j,k}) = -\nabla \cdot \alpha_k (\tilde{F}_{Y,k} + \tilde{F'}_{Yj,k}) + \tilde{Y}_{I,j,k} \cdot \Gamma_k + \alpha_k \overline{\rho}_k \tilde{\Psi}_{j,k} \tag{5}$$

Here the terms on the RHS are 1) mass transfer in phase k due concentration gradients, 2) mass transfer to phase k due to mass transfer from the other phases, 3) mass generation of specie Y_j in phase k due to e.g. chemical reaction.

3. Closure framework

3.1. General

So far, all the multiphase models described are not yet in a solvable form. They still include unclosed terms that are not explicitly depending on the averaged fluid properties. Unclosed terms, for multiphase flow models, can be grouped in three different classes:

(i) phase interaction terms $\left(\Gamma_k, \overrightarrow{M}_{I,k}\right)$,

(ii) self-interaction terms $\left(\overline{\overline{T_k}}\right)$ and

(iii) turbulence terms $\left(\overline{\overline{T_k'}}\right)$.

Furthermore, the thermodynamic state of the system needs to be specified through state equations, which link the thermodynamic variables. The closures of the terms in (i)-(iii) together with state equations go under the name of closure laws or constitutive laws.

3.2. Bubble Liquid

3.2.1 Self-interaction: The viscous stress tensor of phase k, $\tilde{\overline{\overline{T}}}_k$ can be split up into a pressure term and a shear stress term as

$$\tilde{\overline{\overline{T}}}_k = -p_k \overline{\overline{I}} + \overline{\overline{\tau}}_k \tag{6}$$

The shear stress tensor is often modelled using the Newtonian strain-stress relation

$$\overline{\overline{\tau}}_k = \xi_k \left(\nabla \cdot \hat{u}_k\right) \overline{\overline{I}} + 2\mu_k \left[\overline{\overline{S}}_k - \frac{1}{3} \nabla \cdot \hat{u}_k \overline{\overline{I}}\right] \tag{7}$$

where the strain-rate tensor is defined by

$$\overline{\overline{S}}_k = \frac{1}{2}\left(\nabla \cdot \hat{u}_k + \left(\nabla \cdot \hat{u}_k\right)^T\right) \tag{8}$$

The bulk viscosity, ξ_k, is usually set to zero for all phases, whereas the dynamic viscosity, μ_k, is usually set to a constant value (laminar viscosity) for the primary phase (liquid) and to zero for the secondary phases (classes of bubble). The assumption of constant viscosity for the primary phase may be questionable, since there should be dependence from the concentration of the secondary phases and temperature when non isothermal process is studied. In the present analysis, a constant laminar viscosity is assumed.

3.2.2 Phase interaction: The interactions between the phases are accounted for by the terms $\vec{M}_{I,k}$, Γ_k in the multi-fluid model. The formulation of the exchange mass rate between phases, Γ_k, depends on the particular process under consideration (evaporation, solidification, condensation, etc.) and will not be discussed here.

The interfacial force for phase k is usually decomposed in terms of a generalized drag force and averaged interfacial pressure and shear stress.

$$\vec{M}_{I,k} = \vec{M}_{I,k}^{gd} + p_{I,k}\nabla\alpha_k - \overline{\overline{\tau}}_{I,k}\nabla\alpha_k \tag{9}$$

By this decomposition the RHS of Eq. 2 may be rewritten as follows

$$\begin{aligned}\nabla\cdot\left(\alpha_k \overline{\overline{T}}_k\right)+\left(\alpha_k\overline{\rho}_k\vec{g}\right)+\Gamma_k\hat{u}_k+\vec{M}_{I,k} &= -\alpha_k\nabla p_k+\nabla\cdot\left(\alpha_k\overline{\overline{\tau}}_k\right)+\\ &+\left(p_{I,k}-p_k\right)\nabla\alpha_k-\overline{\overline{\tau}}_{I,k}\nabla\cdot\alpha_k+\left(\alpha_k\overline{\rho}_k\vec{g}\right)+\Gamma_k\hat{u}_k+\vec{M}_{I,k}^{gd}\end{aligned} \tag{10}$$

The generalized drag force accounts for several interaction forces. Here, only the main contributions for bubbly flows, i.e. drag, virtual mass and lift forces, will be considered. In case of bubbly flows, the multiphase system can be modelled with a primary phase (k=1), consisting of the liquid phase, and (N-1) secondary phases each one representing a class of bubbles with volumes ranging within an interval. Assuming that the interfacial forces are only acting between the primary phase and the secondary phases and not among the secondary phases, which is reasonable for dispersed systems, the generalized drag form can be formulated for the primary phase as

$$\vec{M}_{I,1}^{gd} = \sum_{k=2}^{N}\left(\vec{M}_{I,k}^{D}+\vec{M}_{I,k}^{L}+\vec{M}_{I,k}^{VM}\right) \tag{11}$$

and for the secondary phases as

$$\vec{M}_{I,k}^{gd} = -\vec{M}_{I,k}^{D}-\vec{M}_{I,k}^{L}-\vec{M}_{I,k}^{VM} \qquad k=2,\ldots..N \tag{12}$$

This formulation satisfies the averaged jump constraint for the momentum. The drag, lift and virtual mass are modelled as follows:

Drag force

$$\vec{M}_{I,k}^{D} = \frac{3}{4}\overline{\rho}_1\alpha_k\frac{C_{D,k}}{\overline{d}_{b,k}}\left|\hat{u}_k-\hat{u}_1\right|\left(\hat{u}_k-\hat{u}_1\right) \tag{13}$$

Lift force

$$\vec{M}_{I,k}^{L} = -C_{L,k}\overline{\rho}_1\alpha_k\left(\nabla\times\hat{u}_1\right)\times\left(\hat{u}_k-\hat{u}_1\right) \tag{14}$$

Virtual mass force

$$\vec{M}_{I,k}^{VM} = C_{VM,k}\overline{\rho}_1\alpha_k\left(\frac{\mathrm{D}}{\mathrm{D}t}\hat{u}_k-\frac{\mathrm{D}}{\mathrm{D}t}\hat{u}_1\right) \tag{15}$$

3.2.3 Turbulence closure: The averaging process on the convective terms of phase k generates the so called residual stress tensor (or Reynolds stress tensor when a time or ensemble average is performed), $\overline{\overline{T}}_k^t$, which accounts for the residual or unresolved turbulence filtered out by the averaging process. Due to the nature of turbulence, which is still a not well understood problem of physics, the modelling of the residual tensor is not trivial. Kataoka and Serizawa [5] and successively Lopez de Bertodano et al. [6] have derived and extended from single phase to multiphase the k−ε turbulence model. However, in deriving the equations for the turbulent kinetic energy, k, and the turbulent dissipation rate, ε, several cross-correlation terms have been neglected mainly because the lack and the difficulties of understanding them. Hence care must be taken when using this model. The work of Chahed et al. [7] is one of the few if not the only one in which the Reynolds stress model is applied for multiphase flows. In the last years there has been some attempt to use very large eddy simulations (VLES) for computing multiphase flows, in particular bubbly flows, (Milelli [8], Deen et al. [9], Bove et al. [10]. This technique has shown to be able to capture the unsteadiness and the main characteristic of the flow quite successfully. LES and VLES are based on spatial filtered Navier-Stokes equations. Since the volume average of the multiphase model equations is equivalent to filter in space by a particular filter, they can be regarded as LES/VLES. Here VLES will be considered and the residual stress tensor will be modelled consequently. For the continuous phase, k=1, the residual stress tensor can be written as

$$\overline{\overline{T_1'}} = \overline{\rho}_1\left(\widehat{u_1}\widehat{u_1} - \overline{\overline{u_1 u_1}}\right) = -\frac{2}{3}k_1\overline{\overline{I}} + \overline{\overline{\tau_1}} \tag{16}$$

Here, k_1 is the residual turbulent kinetic energy for the continuous phase and is defined as

$$k_1 = \tfrac{1}{2}\overline{\rho}_1\left(\overline{\overline{u_1 u_1}} - \widehat{u_1}\widehat{u_1}\right) \tag{17}$$

τ_1 is the anisotropic residual stress tensor and accounts for two different turbulence contributions, one induced by the shear stress in the liquid phase and the other deriving from the bubble liquid interaction or bubble induced turbulence (Sato and Sekoguchi [11]). The linear eddy-viscosity model is used to relate the anisotropic residual stress to the rate of strain and the superposition of the two turbulence effects is assumed. The anisotropic residual stress tensor can then be modelled as

$$\overline{\overline{\tau_1}} = 2\left(\mu_t + \mu_{t,BIT}\right)\overline{\overline{S_1}} \tag{18}$$

By analogy with the mixing-length hypothesis, the turbulent viscosity, μ_t, is modelled as (Smagorinsky [12])

$$\mu_t = \overline{\rho}_1\left(C_s\Delta\right)^2 S_1 \tag{19}$$

where, C_s is the Smagorinsky coefficient, Δ is the filter length or the volume average characteristic length and S_1 is the characteristic filtered rate of strain. It is defined as

$$S_1 = \left(2\overline{\overline{S_1}} : \overline{\overline{S_1}}\right)^{1/2} \tag{20}$$

The bubble induced viscosity was introduced and modelled by Sato and Sekoguchi [11] as

$$\mu_{t,BIT} = \overline{\rho}_1 C_{\mu,BIT}\sum_{k=2}^{N}\alpha_k d_k\left|\vec{u}_k - \vec{u}_1\right| \tag{21}$$

Where the model constant, $C_{\mu,BIT}$, is set equal to 0.6. For the secondary phases, the residual stress tensor should account for the dispersion of the bubbles (or particles in general) due to the unresolved eddies. Bove [13] shows that the modelling of this term by the linear eddy-viscosity assumption is not able to describe the desired physics. Several works account for the turbulent dispersion in the generalized drag force, among others Lopez de Bertodano [14] and Drew and Passman [15]. This is questionable. In fact, if the turbulent dispersion effect is considered in the generalized drag force, what is the meaning of the residual stress tensor for the secondary phases? When VLES is applied to bubbly flows, eddies down to a length scales comparable with the bubble size (two times the bubble diameter if the grid spacing is set equal to the bubble diameter) are resolved. These eddies are responsible of the turbulent dispersion of the gas phase. With an appropriate choice of the grid spacing the most of the turbulent dispersion effect is then resolved and only a small part needs to be modelled. Therefore, the residual stress tensor for the secondary phases will be neglected. The reader can refer to the work of Moraga et al. [16] for a review of various turbulent dispersion models.

Another popular model for bubble liquid flows is to model the continuous phase using the so-called k-ε turbulence model. The dispersed phase influence is taken account by introducing additional source terms. The turbulence viscosity is modelled as

$$\mu_{t,k} = C_\mu \overline{\rho}_k \frac{k_1^2}{\varepsilon_1} \tag{22}$$

The equation for the turbulent kinetic energy for the continuous phase (k=1) reads

$$\frac{\partial}{\partial t}(\alpha\bar{\rho}k)_1+\nabla\cdot(\alpha\bar{\rho}\hat{\tilde{u}}k)_1=\nabla\cdot\left(\alpha\frac{\mu_t}{\sigma_k}\nabla k\right)_1+\alpha_1(G+P_{b,k}-\bar{\rho}\varepsilon)_1 \tag{23}$$

The equation for dissipation of turbulent kinetic energy for the continuous phase (k=1) reads

$$\begin{aligned}&\frac{\partial}{\partial t}(\alpha\bar{\rho}\varepsilon)_1+\nabla\cdot(\alpha\bar{\rho}\hat{\tilde{u}}\varepsilon)_1=\nabla\cdot\left(\alpha\frac{\mu_t}{\sigma_\varepsilon}\nabla\varepsilon\right)_1+\\&+\alpha_1\left[\left(\frac{\varepsilon}{k}\left(C_1\,G-C_2\bar{\rho}\varepsilon\right)+P_{b,\varepsilon}\right)\right]_1\end{aligned} \tag{24}$$

Production of turbulence due to mean strain gradients is given by:

$$G_1=\bar{\bar{\tau}}_1:\nabla\hat{\tilde{u}}_1 \tag{25}$$

Additional turbulent kinetic energy is produced or dissipated due to the work induced by the bubbles when they move through the liquid phase

$$P_{b,k}=C_b\overrightarrow{M}_1^D\cdot(\hat{\tilde{u}}_b-\hat{\tilde{u}}_1)\qquad P_{b,\varepsilon}=C_{bd}\,\alpha_g\,\bar{\rho}_1\frac{k_1^{3/2}}{d_b} \tag{26}$$

C_b range from 0.02 to 0.75. This means that from 2 to 75 percent of the bubble-induced turbulence goes into the large eddy structure of the continuous phase. C_{bd} range from 0.02 to 0.2. These large variations in the two constants indicate the problems with the k-ε turbulence model for bubble-liquid flows. The values of the other constants applied in the k-ε model are given in the Table 1 below.

Table 1. Values of constants in the k-ε model

C_μ	C_1	C_2	σ_k	σ_ε
0.09	1.44	1.92	1.0	1.3

Lahey et al. [17] propose an extension to the k-ε model given above. Their extension introduces two time scales, namely one which is the normal turbulent turnover time of the liquid eddies $(k/\varepsilon)_l$ and one which is related to the relative velocity and the bubble diameter ($d_b/|u_{rel}|$). The kinetic energy of turbulence is derived from transport equations, one for shear induced turbulence k_{SI}, and one for bubble induced turbulence k_{BI} . These two contributions are added to get the total turbulent kinetic energy in the liquid phase, k_l.

3.2.4 Closed multi-fluid model: Assuming that the surface tension effect is not important, as it is for non-separated flows, the pressure can be considered to have locally the same value for all phases, $p_k = p_{I,k} = p$. The mass conservation and momentum balance yields

Mass conservation

$$\frac{\partial}{\partial t}\left(\alpha_k\bar{\rho}_k\right)+\nabla\cdot\left(\alpha_k\bar{\rho}_k\hat{u}_k\right)=\Gamma_k,\quad k=1,......,N \tag{27}$$

$$\sum_{k=1}^{N}\alpha_k=1;\quad\sum_{k=1}^{N}\Gamma_k=0 \tag{28}$$

Momentum balance for the primary phase

$$
\begin{aligned}
&\frac{\partial}{\partial t}\left(\alpha_1 \overline{\rho}_1 \hat{u}_1\right) + \nabla\cdot\left(\alpha_1 \overline{\rho}_1 \hat{u}_1 \hat{u}_1\right) = -\alpha_1 \nabla p + \alpha_1 \overline{\rho}_1 \vec{g} + \\
&+\nabla\cdot\left(\alpha_1 \mu_{eff,1}\left(\nabla \hat{u}_1 + \left(\nabla \hat{u}_1\right)^T\right)\right) - \nabla\cdot\left(\alpha_1 \mu_1 \left(\frac{1}{3}\nabla \hat{u}_1 \overline{\overline{I}}\right)\right) + \\
&+\Gamma_1 \hat{u}_1 + \sum_{k=2}^{N} \frac{3}{4}\overline{\rho}_1 \alpha_k \frac{C_{D,k}}{\overline{d}_{b,k}} \left|\hat{u}_k - \hat{u}_1\right| \left(\hat{u}_k - \hat{u}_1\right) + \\
&-\sum_{k=2}^{N} C_{L,k} \overline{\rho}_1 \alpha_k \left(\nabla \times \hat{u}_1\right) \times \left(\hat{u}_k - \hat{u}_1\right) + \\
&+\sum_{k=2}^{N} C_{VM,k} \overline{\rho}_1 \alpha_k \left(\frac{\mathrm{D}}{\mathrm{D}t}\hat{u}_k - \frac{\mathrm{D}}{\mathrm{D}t}\hat{u}_1\right)
\end{aligned} \tag{29}
$$

Here, $\mu_{eff,1}$ represents the sum of the dynamic, μ_1, residual, μ_t, and residual bubble induced viscosity, $\mu_{t,BIT}$.

Momentum balance for the secondary phases

$$
\begin{aligned}
&\frac{\partial}{\partial t}\left(\alpha_k \overline{\rho}_k \hat{u}_k\right) + \nabla\cdot\left(\alpha_k \overline{\rho}_k \hat{u}_k \hat{u}_k\right) = -\alpha_k \nabla p + \alpha_k \overline{\rho}_k \vec{g} + \Gamma_1 \hat{u}_1 + \\
&-\frac{3}{4}\overline{\rho}_1 \alpha_k \frac{C_{D,k}}{\overline{d}_{b,k}} \left|\hat{u}_k - \hat{u}_1\right| \left(\hat{u}_k - \hat{u}_1\right) + \\
&+C_{L,k} \overline{\rho}_1 \alpha_k \left(\nabla \times \hat{u}_1\right) \times \left(\hat{u}_k - \hat{u}_1\right) \\
&-C_{VM,k} \overline{\rho}_1 \alpha_k \left(\frac{\mathrm{D}}{\mathrm{D}t}\hat{u}_k - \frac{\mathrm{D}}{\mathrm{D}t}\hat{u}_1\right) \qquad k = 2,\ldots\ldots,N
\end{aligned} \tag{30}
$$

Normally, the virtual mass coefficient, $C_{VM,k}$, is set equal to 0.5. The lift coefficient, $C_{L,k}$, is either set to a constant value or modelled as suggested by Tomiyama [18],

$$
C_{L,k} = \begin{cases} \min\left[0.288\tanh\left(0.121\cdot \mathrm{Re}\right), f\left(Eo\right)\right] & \text{if} \quad \mathrm{Eo} < 4 \\ f(Eo) & \text{if} \quad 4 < Eo \end{cases} \tag{31}
$$

where,

$$
f\left(Eo\right) = 0.00105\cdot Eo^3 - 0.00159\cdot Eo^2 - 0.204\cdot Eo + 0.474 \tag{32}
$$

The drag coefficient, $C_{D,k}$, is modelled either by the distorted model (Ishii and Zuber [19])

$$
C_{D,k} = \frac{2}{3} Eo_k^{1/2} \tag{33}
$$

or by different correlations depending on the mean diameter and the initial deformation of the bubbles as well as the water contamination level, which affects the surface tension (Tomiyama et al. [20], Tomiyama [18]).

For non-contaminated water or high initial shape deformation, the drag coefficient is well correlated by

$$
C_{D,k} = \begin{cases} \frac{16}{\mathrm{Re}_k}\left(1.0 + 0.15\right)\cdot \mathrm{Re}_k^{0.687} & \text{if} \quad 0\,mm < \overline{d}_{b,k} < 0.5\,mm \\ \frac{48}{\mathrm{Re}_k} & \text{if} \quad 0.5\,mm < \overline{d}_{b,k} < 1.3\,mm \\ \frac{8}{3}\frac{\mathrm{Eo}_k}{\mathrm{Eo}_k + 4} & \text{if} \quad 1.30\,mm < \overline{d}_{b,k} \end{cases} \tag{34}
$$

For contaminated water or low initial shape deformation, the following correlation holds

$$
C_{D,k} = \begin{cases} \frac{16}{\mathrm{Re}_k}\left(1.0 + 0.15\right)\cdot \mathrm{Re}_k^{0.687} & \text{if} \quad 0\,mm < \overline{d}_{b,k} < 0.8\,mm \\ \text{Correlation} & \text{if} \quad 0.8\,mm < \overline{d}_{b,k} \end{cases} \tag{35}
$$

Tomiyama [18] expresses the drag coefficient as

$$C_{D,k}=\begin{cases}\frac{8}{3}\frac{\mathrm{Eo}_k}{ER_k^{2/3}\left(1-ER_k^2\right)^{-1}\mathrm{Eo}_k+16ER_k^{4/3}}\mathrm{F}\left(ER_k\right)^{-2} & \text{if}\quad \mathrm{ER}_k<1.0\\ 6 & \text{if}\quad \mathrm{ER}_k=1.0\\ \frac{8}{3}\frac{\mathrm{Eo}_k}{ER_k^{2/3}\left(ER_k^2-1\right)^{-1}\mathrm{Eo}_k-16ER_k^{4/3}}G\left(ER_k\right)^{-2} & \text{if}\quad \mathrm{ER}_k>1.0\end{cases} \tag{36}$$

Where,

$$Eo_k=\frac{\Delta\rho g d_{b,k}^2}{\sigma}\quad(\text{Eotvos number}) \tag{37}$$

$$ER_k=\frac{1}{1+0.163Eo_k^{0.757}}\quad(\text{Mean aspect ratio}) \tag{38}$$

$$\mathrm{Re}_k=\frac{\rho\left|\vec{u}_{s,kl}\right|\bar{d}_{db,k}}{\mu_l}\quad(\text{Bubble Reynolds number}) \tag{39}$$

$$F\left(ER_k\right)=\frac{\cos^{-1}ER_k-ER_k\sqrt{1-ER_k^2}}{1-ER_k^2} \tag{40}$$

$$G\left(ER_k\right)=\frac{ER_k\sqrt{ER_k^2-1}-\tanh^{-1}\left(ER_k^{-1}\sqrt{ER_k^2-1}\right)}{ER_k^2-1} \tag{41}$$

The system of Equations (27)-(29) is determined. In fact it has the same number of equations and unknowns. As stated previously, in order to be solved, initial and boundary conditions need to be specified. In general, they depend on the particular type of process to be solved. Therefore, they will be given when the type of flow and its operative environment is known.

4. Closure framework for particle (solids) gas flows

4.1. Kinetic Theory of Granular flow (KTGF) for mono-sized particles

4.1.1 Introduction: Statistical approach used to formulate the constitutive equations of the solid phase equations come from the interactions of the fluctuating motion of the particles with the mean motion of the particles. Such interactions generate stresses in the solids phase and give rise to an effective viscosity of the solids phase. To be able to calculate these stresses and an effective solids viscosity, a model is proposed by Ding and Gidaspow [21] and Gidaspow [22]. The model is based on the kinetic theory of dense gases, as presented by Chapman and Cowling [23] and the work of Jenkins and Savage [24] and Lun et al. [25].The thermal temperature in the kinetic theory of dense gases is here replaced with a granular temperature for which a transient differential equation is derived. The solid viscosity and solid stresses are a function of this granular temperature that varies with time and space in a fluidized bed. Derivation given is mainly based on the work of Gidaspow [22]. The generalised model for multi-sized particles given by Manger [26] will be summarized below.

In gas/solid systems, particle segregation due to different size and/or density will play a significant role on the flow behaviour. To describe such phenomena, an extension to multiple particle phases is essential. Jenkins and Mancini [27] extended the kinetic theory for granular flow to binary mixtures. The basic assumption was equal turbulent kinetic energy with a small correction for the individual phase temperatures. Mathiesen et al. [28] developed a model based on this work and performed a simulation with one gas and three solid phases. The model predicted segregation effects fairly well, and good agreement with experimental data was obtained. Gidaspow et al. [29] and Manger [26] extended the kinetic theory to binary mixtures of solid with unequal granular temperatures between the phases. Based on their research, a generalized multiphase gas/solid model will be given in the next section.

4.1.2 Summary of Governing Equations for Fluid/Multi-sized particle flows: The three-dimensional, finite-volume, multiphase Eulerian/Eulerian CFD code FLOTRACS-MP- 3D (see also Mathiesen et al., [30] [31], Ibsen et al., [32] and Hansen et al. [33]) uses the generalized fluid/multi-sized particle model mentioned above. The turbulent motion of the particulate phase is modelled using the kinetic theory of granular flow described by Manger [26] and the gas phase turbulence is modelled using a LES/Sub-Grid-Scale model.

4.1.3 Mass balances and momentum balances: The governing equations may be written as:

Continuity equation for phase k without mass transfer:

$$\frac{\partial}{\partial t}(\alpha_k \rho_k) + \frac{\partial}{\partial x_i}(\alpha_k \rho_k U_{i,k}) = 0 \tag{42}$$

Here α, ρ and U are the volume fraction, density and velocity of phase k, respectively. The momentum equation for phase k is written as:

$$\frac{\partial}{\partial t}(\alpha_k \rho_k U_{j,k}) + \frac{\partial}{\partial x_i}(\alpha_k \rho_k U_{i,k} U_{j,k}) = -\alpha_k \frac{\partial p}{\partial x_j} + \frac{\partial}{\partial x_i}\tau_{ij,k} + \alpha_k \rho_k g_j + \sum_{m=1, m\neq k}^{M} \beta_{km}\left(U_{j,m} - U_{j,k}\right) \tag{43}$$

Here p, τ_{ij}, g and β are pressure, stress tensor, gravity and the inter-phase drag coefficient, respectively. There are N solids phases (s) and one gas phase (g) and the total number of phases are therefore M = N+1.

4.1.4 Auxiliary relations: The gas phase stress tensor is given by:

$$\tau_{ij,g} = \mu_{eff,g}\left[\left(\frac{\partial U_j}{\partial x_j} + \frac{\partial U_i}{\partial x_j}\right) - \frac{2}{3}\delta_{ij}\frac{\partial U_k}{\partial x_k}\right]_g \tag{44}$$

where δ_{ij} is the Kroenecker delta. The effective viscosity, $\mu_{eff,g}$, is derived from a sub-grid-scale (SGS) model based on Smagorinsky [12], where the effective viscosity is a sum of a laminar and a turbulent part.

$$\mu_{eff,g} = \alpha_g\left(\mu_{lam,g} + \mu_{turb,g}\right) = \alpha_g \mu_{lam,g} + \alpha_g \rho_g (c_t \Delta)^2 \sqrt{S_{ij,g} : S_{ij,g}} \tag{45}$$

The SGS eddy coefficient, c_t, is set to 0.079 based on Deardoff [34]. The length scale, Δ, and the strain rate tensor of the resolved field, $S_{ij,g}$, are given by:

$$\Delta = (\Delta x \Delta y \Delta z)^{1/3} \text{ and}$$

$$S_{ij,g} = \frac{1}{2}\left[\frac{\partial U_j}{\partial x_i} + \frac{\partial U_i}{\partial x_j}\right]_g \tag{46}$$

The solid phase stress tensor is given by:

$$\tau_{ij,s} = -P_s\delta_{ij} + \xi_s\delta_{ij}\frac{\partial U_{k,s}}{\partial x_k} + \mu_s\left[\left(\frac{\partial U_j}{\partial x_i} + \frac{\partial U_i}{\partial x_j}\right) - \frac{2}{3}\delta_{ij}\frac{\partial U_k}{\partial x_k}\right]_s \tag{47}$$

The solids phase pressure, P_s, the bulk viscosity, ξ_s, and the shear viscosity, μ_s, are derived from the kinetic theory of granular flow. The solids pressure is found from, Ding and Gidaspow [21]:

$$P_s = \sum_{n=1}^{N} P_{C,sn} + \alpha_s\rho_s\Theta_s \tag{48}$$

where $P_{C,n}$ is the pressure caused by collisions between the solids phases s and n and has the expression:

$$P_{C,sn} = \frac{\pi}{3}(1+e_{sn})d_{sn}^3 g_{sn} n_s n_n \left[\frac{m_0\theta_s\theta_n}{(m_s/m_n)\theta_s + (m_n/m_s)\theta_n}\right]\cdot\left[\frac{(m_0/m_s)^2\theta_s\theta_n}{\left(\theta_s + (m_s/m_n)^2\theta_n\right)(\theta_s+\theta_n)}\right]^{3/2}$$

$$e_{sn} = \frac{1}{2}(e_n + e_s),\quad d_{sn} = \frac{1}{2}(d_n + d_s)\quad \text{and}\quad m_0 = m_s + m_n \tag{49}$$

where *e, d, n* and m are coefficient of restitution, diameter of the particle, number of particles and mass of a particle, respectively. The coefficient of restitution is unity for fully elastic, and zero for inelastic collisions. By using the assumption of spherical particles, number of particles and mass of a particle are, respectively:

$$n_s = \frac{6\alpha_s}{\pi d_s^3} \quad \text{and} \quad m_s = \frac{\pi d_s^3\rho_s}{6} \tag{50}$$

where g_{sn} is the radial distribution function, which is close to one for dilute flow and approach infinity for dense flow making motion impossible. Based on the single solid phase model given implicitly by Bagnold [35], a new binary radial distribution function is proposed here:

$$g_0 = \left[1 - \left(\frac{\alpha_s}{\alpha_{s,\max}}\right)^{1/3}\right]^{-1} \qquad g_{sn} = \frac{N}{2}\frac{g_0}{1-\alpha_g}(\alpha_s + \alpha_n) \tag{51}$$

The maximum solids packing, $\alpha_{s,max}$, is often put to 0.65. The solids bulk viscosity is calculated as (Ding and Gidaspow, [21]):

$$\xi_s = \sum_{n=1}^{N} P_{C,sn}\frac{d_{sn}}{3}\left(\theta_s + (m_n/m_s)\theta_n\right)\sqrt{\frac{2}{\pi\theta_s\theta_n\left(\theta_s + (m_n/m_s)^2\theta_n\right)}} \tag{52}$$

The solids phase shear viscosity consists of a kinetic term:

$$\mu_{col,s} = \sum_{n=1}^{N} P_{C,sn}\frac{d_{sn}}{3}\left(\theta_s + (m_n/m_s)\theta_n\right)\sqrt{\frac{2}{\pi\theta_s\theta_n\left(\theta_s + (m_n/m_s)^2\theta_n\right)}} \tag{53}$$

and a collisional part:

$$\mu_{kin,s} = \frac{2\mu_{dil,s}}{\frac{1}{N}\sum_{n=1}^{N}(1+e_{sn})g_{sn}}\left[1 + \frac{4}{5}\sum_{n=1}^{N} g_{sn}\alpha_n(1+e_{sn})\right]^2 \tag{54}$$

where the dilute solids viscosity, $\mu_{dil,s}$, is found from:

$$\mu_{dil,s} = \frac{15}{8d_s^3}\alpha_s l_s\sqrt{\frac{2m_s\theta_{s,av}}{\pi}} \text{ and } l_s = \frac{1}{6\sqrt{2}}\frac{d_s}{\alpha_s} \tag{55}$$

To ensure that the dilute viscosity is finite as the volume fraction of solid approaches zero, the mean free path l_s is limited by a characteristic dimension. The average granular temperature $\theta_{s,av}$ is obtained from:

$$\theta_{s,av} = \frac{2m_s\theta_s}{\left[\sum_{n=1}^{N}\frac{n_n}{n_s}\left(\frac{d_{sn}}{d_s}\right)^2\sqrt{\frac{(m_0/m_s)^2\theta_n}{\left(\theta_s+(m_0/m_s)^2\theta_n\right)}}S^{3/2}\right]^2} \tag{56}$$

$$S = \frac{(m_0/m_s)^2\theta_s\theta_n}{\left(\theta_s+(m_0/m_s)^2\theta_n\right)(\theta_s+\theta_n)}$$

Three solids gas inter-phase drag models will be presented.

Model 1 Gidaspow [22]: For $\alpha_g \leq 0.8$ the inter-phase drag coefficient is found from the Ergun equation (Ergun, [36]):

$$\beta_{sg} = 150\frac{\alpha_s^2\mu_{lam,g}}{\alpha_g(\Phi_s d_s)^2} + 1.75\frac{\alpha_s\rho_g|\vec{u}_g-\vec{u}_s|}{\Phi_s d_s} \tag{57}$$

where Φ_s is the sphericity of the particles (i.e. 1.0 for spheres; 0.81 for cubes; 0.6 - 0.7 for crushed materials; 0.3 for Raschig rings).

For $\alpha_g > 0.8$ the drag formulation of Wen and Yu [37] is used

$$\beta_{sg} = \frac{3}{4}C_D\frac{\alpha_s\alpha_g\rho_g|\vec{u}_g-\vec{u}_s|}{\phi_s d_s}\alpha_g^{-2.65} \tag{58}$$

with a drag coefficient, C_D, from Rowe [38]:

$$C_D = \begin{cases} \frac{24}{\mathrm{Re}_s}\left(1+0.15\mathrm{Re}^{0.687}\right) & \text{for} \quad \mathrm{Re}_s \leq 1000 \\ 0.44 & \text{for} \quad \mathrm{Re}_s > 1000 \end{cases} \tag{59}$$

where the particle Reynolds number is defined as:

$$\mathrm{Re}_s = \frac{d_s\rho_g\alpha_g|\vec{u}_g-\vec{u}_s|}{\mu_{lam,g}} \tag{60}$$

Model 2 Gibilaro et al. [39]:

$$\beta_{sg} = \left(\frac{17.3}{\mathrm{Re}_p}+0.336\right)\frac{\rho_g}{d_p}|\vec{u}_g-\vec{u}_s|\alpha_s\alpha_g^{-1.8} \tag{61}$$

Model 3 Syamlal and O'Brien [40]:

$$\beta_{sg} = \frac{3}{4d_p}C_D\rho_g\frac{1}{R_t^2}\alpha_g(1-\alpha_g)|\vec{u}_g-\vec{u}_s| \tag{62}$$

R_t is the ratio between the falling velocity of a suspension and the terminal velocity of a single sphere. R_t, C_D, A and B are given by the following expressions:

$$2R_t = A - 0.06\,\mathrm{Re}_p + \sqrt{0.0036\,\mathrm{Re}_p^2 + 0.12\,\mathrm{Re}_p\left(2B - A\right) + A^2}$$
$$A = \alpha_g^{4.14} \qquad B = \begin{cases} 0.8\alpha_g^{-1.28} & \text{for } \alpha_g < 0.85 \\ \alpha_g^{2.65} & \text{for } \alpha_g > 0.85 \end{cases}$$
$$C_D = \left(0.63 + 4.8\sqrt{\frac{R_t}{\mathrm{Re}_p}}\right) \qquad \mathrm{Re}_p = \frac{d_s \rho_g \left|\vec{u}_g - \vec{u}_s\right|}{\mu_{lam,g}} \tag{63}$$

The particle/particle drag coefficient may be expressed as (Manger [26]):

$$\beta_{s,g} = P_{C,sn}\left[\begin{array}{l} \dfrac{3}{d_{sn}}\sqrt{\dfrac{2\left(m_s^2\theta_s + m_n^2\theta_n\right)}{\pi m_0^2 \theta_s \theta_n}} + \\ + \dfrac{1}{\left|\vec{u}_n - \vec{u}_s\right|}\left(\nabla\left|\ln\dfrac{\alpha_s}{\alpha_n}\right| + \dfrac{\theta_s\theta_n}{\theta_s + \theta_n}\left|\dfrac{\nabla\theta_n}{\theta_n^2} - \dfrac{\nabla\theta_s}{\theta_s^2}\right| + 3\nabla\left|\dfrac{\ln\left(m_n\theta_n\right)}{\ln\left(m_s\theta_s\right)}\right|\right) \end{array}\right] \tag{64}$$

4.1.5 Granular temperature: A transport equation for the solids phases' turbulent kinetic energy or granular temperature is solved:

$$\frac{3}{2}\left[\frac{\partial}{\partial t}\left(\alpha_s\rho_s\theta_s\right) + \frac{\partial}{\partial x_i}\left(\alpha_s\rho_s U_{i,s}\theta_s\right)\right] = \frac{\partial}{\partial x_i}\left(\kappa_s\frac{\partial\theta_s}{\partial x_i}\right) + $$
$$+\tau_{ij,s} : \frac{\partial U_{j,s}}{\partial x_i} - \gamma_s - 3\beta_{sg}\theta_s \tag{65}$$

Here, the terms on the right side of the equation represent diffusive transport, production due to shear, dissipation due to inelastic collisions and dissipation due to fluid friction. The dissipation due to inelastic collisions is given by Manger [26].The production/dissipation term due to fluctuations in drag has been assumed as negligible. This is a reasonable assumption for the relatively large and heavy particles. Hence, the particle response time is assumed much longer than the characteristic time scale for the turbulent fluid motion.

The conductivity of granular temperature κ_s, and the dissipation due to inelastic collisions γ_s are determined from the kinetic theory for granular flow [26]. The conductivity is given by a dilute and a dense part as:

$$\kappa_s = \frac{2\kappa_{dil,s}}{\frac{1}{N}\sum_{n=1}^{N}\left(1 + e_{sn}\right)g_{sn}}\left(1 + \frac{6}{5}\frac{1}{N}\sum_{n=1}^{N}\left(1 + e_{sn}\right)g_{sn}\alpha_n\right)^2 + $$
$$+2\alpha_s^2\rho_s d_s\sqrt{\frac{\theta_s}{\pi}}\sum_{n=1}^{N}\left(1 + e_{sn}\right)g_{sn}\alpha_n \tag{66}$$
$$\text{where}$$
$$\kappa_{dil,s} = \frac{225}{32}\alpha_s l_s\sqrt{\frac{2m_s\theta_{s,av}}{\pi}}$$

The dissipation due to inelastic collisions is given by (Manger [26]):

$$\gamma_s = \sum_{n=1}^{N}\frac{3}{4}P_{C,sn}\frac{1 - e_{sn}}{d_{sn}}\cdot$$
$$\cdot\left[4\sqrt{\frac{2\theta_s\theta_n}{\pi\left(\left(m_s/m_0\right)^2\theta_s + \left(m_n/m_0\right)^2\theta_n\right)}} - d_{sn}\left(\frac{\left(m_s/m_0\right)\theta_s + \left(m_n/m_0\right)^2\theta_n}{\left(m_s/m_0\right)^2\theta_s + \left(m_n/m_0\right)^2\theta_n}\right)\frac{\partial U_{k,s}}{\partial x_k}\right] \tag{67}$$

The divergence term is often neglected. However, if the term is retained it may cause production instead of dissipation, and must be handled with special care.

5. Additional sub models

5.1. Bubble Size Models

Five different bubble size models are examined. These are denoted models A to E.

Model A: The simplest model assumes that the bubble size is constant and given by overall flow conditions. Johansen and Boyson [41] have used an estimation of the bubble diameter which is dependant on the inlet flow rate:

$$d_b = 0.35\left(\frac{\dot{V}_g}{g}\right)^{0.2} \tag{68}$$

Model B: Jakobsen [42] has used a bubble size model which sets the diameter proportional to the length scale of the turbulent liquid eddies according to:

$$d_b = C_{SMD}\frac{k_l^{\frac{3}{2}}}{\varepsilon_l} \tag{69}$$

The constant C_{SMD} is determined to be 0.04 for flow in bubble columns. The two models given above do not include properties of the interface. This is included in the next model.

Model C: Calderbank [43] has proposed an expression based on dissipation length scale and a critical Weber number as follows:

$$We_{crit} = \frac{\tau\, d_b}{\sigma};\ \tau = \rho_l\, u_{t,l}^2\ ;\ \varepsilon_l = \frac{u_{t,l}^3}{d_b} \tag{70}$$

Here τ is the shear stress, $u_{t,l}$ is the turbulence velocity, σ is the surface tension of the gas liquid interface and ε is the dissipation rate of turbulence. The expression for d_b from the above relations gives:

$$d_b = C_{We}\left(\frac{\sigma}{\rho_l}\right)^{0.6}\frac{1.0}{\varepsilon_l^{0.4}} \tag{71}$$

Experimental verification against data from stirred tanks has given the following correlation based on the above expression:

$$d_b = C_{We}\left(\frac{\sigma}{\rho_l}\right)^{0.6}\frac{1}{\varepsilon_l^{0.4}}\alpha_g^{0.5}\left(\frac{\mu_g}{\mu}\right)^{0.25} + 0.0009 \tag{72}$$

When C_{We} is taken to be 4.15, d_b is given in metres.

Model D: Both models B and C give bubble size as function of local properties. However, in many cases history effects govern the bubble size. Cook and Harlow [44] give a model that takes account of this. This model calculates the transport of the number concentration of bubbles N, according to:

$$\frac{\partial N}{\partial t} + \nabla\cdot(\vec{u}_g\, N) = \omega_b(N_e - N) \tag{73}$$

Here the equilibrium number concentration of bubbles N_e, is calculated based on a critical Weber number as:

$$N_e = \frac{6\,\alpha_g}{\pi\, d_{be}^3} \quad \text{with} \quad We_{crit} = \frac{(\rho_l - \rho_g)|\vec{u}_{rel}|^2 d_{be}}{\sigma} \tag{74}$$

The bubble diameter relates to the void fraction and the bubble number concentration as:

$$d_b = \left(\frac{6\,\alpha_g}{\pi\, N}\right)^{\frac{1}{3}} \tag{75}$$

The two constants in the model, the critical Weber number We_{crit}, and the relaxation parameter ω_b, has the following values: 3.6 and between 4 and 20, respectively. These constants are determined for vertical gas-liquid flow past an obstacle.

Model E: The most general and comprehensive method is to calculate bubble size distribution by solving the Population Balance Equation. Here there are several methods available including: Classes/Sectional Methods (CM/SM) and methods of moments (QMOM, DQMOM, PPDC). Details of these methods will not be discussed here but a full description may be found in Bove [14] and Marchisio and Fox [45].

5.2. Interfacial Heat Transfer

Gas/liquid: Heat transfer between the two phases without mass transfer may be treated as follows:

T_1 and T_2 are the temperatures in the bulk of phase 1 and 2, respectively. T_s is the interface temperature. Energy balance over the phase separation surface: heat in = heat out.

$$Q_{1s} = (\lambda_1 a)(T_1 - T_s); \quad Q_{s2} = (\lambda_2 a)(T_s - T_2)$$

Solving for T_s , $Q_{1s} = Q_{2s}$

$$Q_{1s} = Q_{12} = a\frac{\lambda_1\lambda_2(T_1 - T_2)}{\lambda_1 + \lambda_2}$$

Or as source term in the phase 2 and phase 1 enthalpy equations reads:

$$S_{h_2} = -S_{h_1} = aU(T_1 - T_2) \text{ where } 1/U = 1/\lambda_1 + 1/\lambda_2 \tag{76}$$

Here U is the total heat transfer coefficient between phase 1 and 2; λ_1 and λ_2 are the individual heat transfer coefficients between phase 1 and the interface and phase 2 and the interface, respectively; a is the specific surface area of the interface, i.e. area per unit volume.

Gas/particle: Heat may be generated by the catalytic exothermic solid phase chemical reaction [46]. This heat is transported between the phases (particles, p and gas, g) and may in addition be cooled by e.g. generation of steam in a submerged heat exchanger (wall, w).

Source term for the gas phase

$$S_{h_g} = h_v(T_p - T_g) + h_w(T_w - T_g) \tag{77}$$

Source term for the solids phase

$$S_{h_p} = h_v(T_g - T_p) + h_w(T_w - T_p) + \sum_{k=1}^{k=nreac} \Delta H_{rx,k} \cdot r_k \tag{78}$$

The effective transport coefficients relates to the turbulent viscosities as:

$$\Gamma_{h,g} = \frac{\kappa_{lam,g}}{c_{p,g}} + \frac{\mu_g}{0.7} \quad \text{and} \quad \Gamma_{h,p} = \frac{\mu_p}{0.7} \tag{79}$$

Where $\kappa_{lam,g}$ is the laminar conductivity and $c_{p,g}$ is the specific heat capacity of the gas.

The volumetric heat transfer coefficient h_v may be calculated by using several different correlations for the Nusselt number N_p in different flow regimes, characterized by the Reynolds number Re and the gas phase volume fraction α_g.

For $\alpha_g \leq 0.8$

$$N_p = (2 + 0.106\, Re)\, S_p; \quad Re \leq 200$$

$$N_p = 0.123\left(\frac{4\, Re}{d_p}\right)^{0.83} S_p^{0.17}; \quad 200 < Re \leq 2000 \tag{80}$$

$$N_p = 0.61\, Re^{0.67}\, S_p; \quad Re > 2000$$

For $\alpha_g > 0.8$

$$N_p = (2 + 0.16\,Re^{0.67})\,S_p; \qquad Re \le 200$$
$$N_p = 8.2\,Re^{0.60}\,S_p; \qquad 200 < Re < 1000 \tag{81}$$
$$N_p = 1.06\,Re^{0.457}\,S_p; \qquad Re \ge 1000$$

where

$$Re = \frac{\rho_g d_p |\vec{u}_g - \vec{u}_p|}{\mu_{lam,g}}, \quad S_p = \frac{\alpha_p 6}{d_p}, \quad N_p = \frac{h_v d_p}{\kappa_{lam,g}} \tag{82}$$

The wall to bed heat transfer coefficient, h_w, may for many cases be taken to be a constant, only proportional to the respective volume fractions of the phases. Typical values for the bed to wall heat transfer coefficient may be found in Geldart [47].

5.3. Mass Transfer in Bubble Liquid Flows

The interface mass transfer driven by concentration gradients is taken account of through the source terms for the species equation. For the liquid phase, $Y_{j,l}$, the source term reads:

$$S_{Y_{j,l}} = (k_l a)_l\,(Y_{j,g} - Y^*_{j,l}) \tag{83}$$

The $(k_l\,a)$ is related to the dissipation rate, ε_l gas fraction, α_g and bubble diameter, d_b (Trägårdh, [48]):

$$(k_l a) = C_m \varepsilon_l^{0.2} \frac{6\,\alpha_g}{d_b} \tag{84}$$

Modelling of mass transfer in single component two-phase flow where evaporation/condensation occurs, may be expressed as:

$$\Gamma = \frac{(\lambda_1 a)(T_1 - T_{1s}) + (\lambda_2 a)(T_2 - T_{2s})}{h_{12}} \tag{85}$$

Here T_{1s} and T_{2s} are saturation temperatures, dependant only on pressure; λ_1 is the heat transfer coefficient between phase 1 and the interface surface. λ_2 is the heat transfer coefficient between phase 2 and the interface surface, h_{12} enthalpy of evaporation and a specific surface area of interface.

5.4. Chemical Reaction

Catalytic reaction

The ozone decomposition reaction: The ozone decomposition reaction is a simple irreversible first order catalytic reaction. The reaction can be written as:

$$2\,O_3 \rightarrow 3\,O_2$$

Due to the low concentration of ozone in a riser, the heat produced by the chemical reaction is negligible and the reaction is looked upon as being isothermal.

Species conservation equations: A transport equation for the mass fraction of ozone in the gas phase $Y_{k,g}$ is solved

$$\frac{\partial}{\partial t}(\alpha \rho Y_k)_g + \frac{\partial}{\partial x_i}(\alpha \rho U_i Y_k)_g = \frac{\partial}{\partial x_i}\left[\Gamma_{Y_{k,g}} \frac{\partial Y_k}{\partial x_i}\right] + \alpha_g \cdot r \tag{86}$$

where the transport coefficient Γ_Y is expressed in terms of the diffusion coefficient, D_Y, and the turbulent viscosity of the gas phase:

$$\Gamma_{Y_{k,g}} = D_{Y_{k,g}} + \frac{\mu_{turb,g}}{0.7} \tag{87}$$

The reaction rate constant for the catalytic reaction was measured in a fixed bed reactor. The rate constant is expressed pr. unit volume of catalyst and was measured to be 3.96 s^{-1} (Ouyang et al., [49]). The reaction rate is expressed as:

$$r = -3.96 \cdot \alpha_s \cdot C_{Ozone} \qquad [\text{kg Ozone/m}^3\ \text{s}] \tag{88}$$

6. Applications

The next subsections will give some examples of use of some of the models described above.

6.1. Bubble Columns

In the work of Deen et al. [9] the use of large eddy simulations (LES) in numerical simulations of the gas–liquid flow in bubble columns was studied. The Euler–Euler approach was used to describe the equations of motion of the two-phase flow. It was found that, when the drag, lift and virtual mass forces are used, the transient behaviour that was observed in experiments can be captured. Good quantitative agreement with experimental data is obtained both for the mean velocities and the fluctuating velocities (Figure 3). The LES showed better agreement with the experimental data than simulations using the k–ε model.

Bove [13] extended the above modelling by including solution of the Bubble population equation using three nodes in the size coordinate. This enabled calculation of the bubble size distribution in bubble columns. An example of Bove's results are given in Figure 4.

.

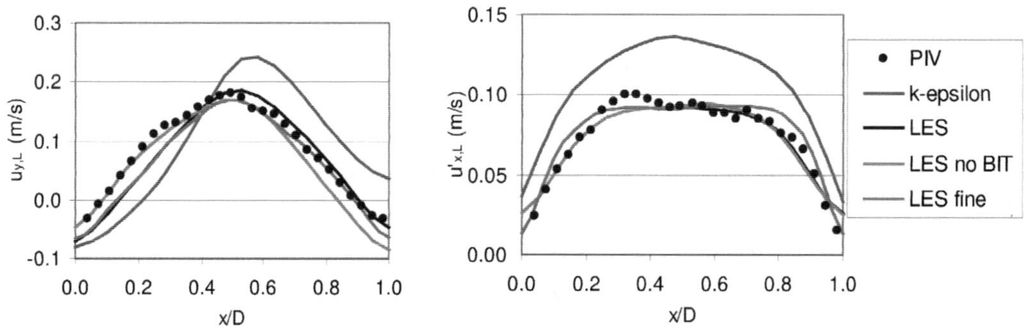

Figure 3: Comparison of simulated and experimental profiles of (left) the axial average liquid velocity and (right) the transverse liquid velocity fluctuations for different conditions.

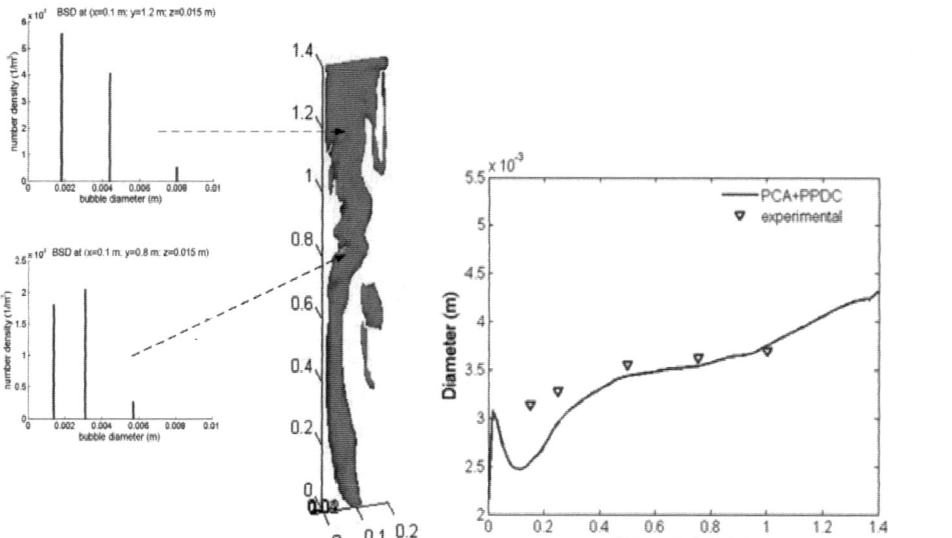

Figure 4: (left) A snap shot of the predicted size distribution at two position in the bubble column of van den Hengel et al [50] and (right) the predicted and measured bubble diameter along the column centreline.

IOP Conf. Series: Materials Science and Engineering **276** (2017) 012010 doi:10.1088/1757-899X/276/1/012010

6.2. Stirred Tanks

Deen et al [51] did a study of flow in a stirred tank. A two-camera PIV technique was used to obtain angle resolved velocity and turbulence data of the flow in a lab-scale stirred tank, equipped with a Rushton turbine

Two cases were investigated: a single-phase flow and a gas-liquid flow. In the former case, the classical radial jet flow pattern accompanied by two trailing vortices was observed. In the latter case, the velocity of the radial jet was reduced, and the vortices were diminished by the presence of the gas. Gas cavities clinging to the back of the impeller blades were observed (Figure 5). Both cases were also investigated with the use of three-dimensional transient CFD simulations. The sliding grid technique was used to describe the movement of the impeller.

For the single-phase flow the simulations in the impeller region corresponded very well with the experimental data. For the gas-liquid flow both the mean and fluctuating liquid velocities in the impeller region were well predicted as seen in Figure 6. This was also the case for the mean radial gas velocities. The largest differences between the simulations and the experiments were found for the mean axial gas velocity.

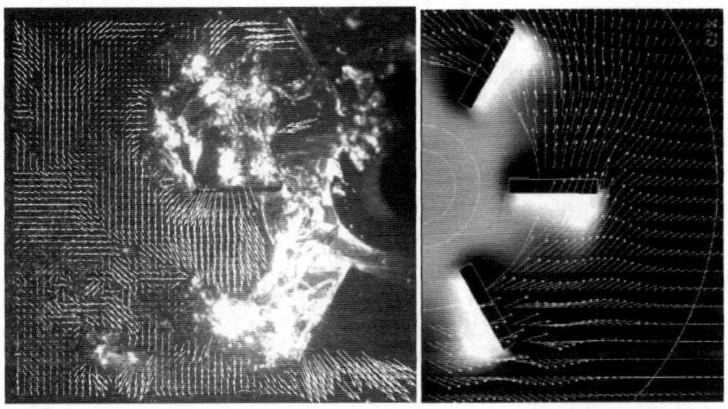

Figure 5: Comparisons between measured (left) and computed (right) distribution of gas around the impeller.

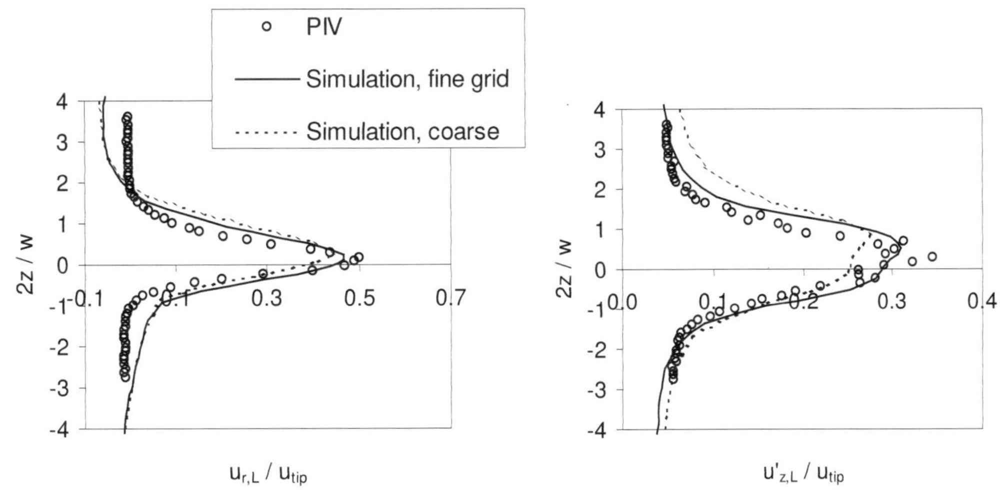

Figure 6: Comparisons of measured and predicted radial liquid velocity (left) and axial liquid turbulent velocity (right) close to the impeller region. The gas liquid case using fine or coarse grids.

IOP Conf. Series: Materials Science and Engineering **276** (2017) 012010 doi:10.1088/1757-899X/276/1/012010

6.3. Circulating fluidized beds

6.3.1 Cold flow: Hansen et al [33] have performed a computational fluid dynamics simulation of a cold flowing riser fluidized with FCC catalysts. The computations were performed using the 3D multiphase computational fluid dynamics code with an Eulerian description of both gas and particle phase, described above. The turbulent motion of the particulate phase was modeled using the kinetic theory for granular flow, and the gas phase turbulence was modeled using a Sub-Grid-Scale model. The complex inlet geometry was approximated using multiple inlet patches. The first results were submitted to a blind-test in connection to the 10th international workshop on two-phase flow prediction held in Merseburg, Germany, 2002.

The results were subsequently validated against experimental findings of particle mass flux across the riser and pressure profile along the riser. The calculations showed good agreement with experimental findings of both mass flux and pressure profile, but further improvements were proposed and investigated. A parameter study showed that mesh refinement, choice of particle diameter and choice of drag model are crucial when simulating FCC riser flow. The result of the submitted blind test and best result after the parameter study is given in Figure 7.

6.3.2 Reactive flow: Hansen et al. [51] have implemented the isothermal decomposition of ozone in the CFD code FLOTRACS-MP-3D described above. The code is a three dimensional (3D) multiphase computational fluid dynamics code with an Eulerian description of both gas and particle phase. The turbulent motion of the particulate phase is modelled using the kinetic theory for granular flow, and the gas phase turbulence is modelled using a sub-grid-scale model.

The decomposition reaction is studied in a 3D representation of a 0.254m I.D. riser, which has been studied experimentally by Ouyang et al. [49]. These authors obtained profiles of ozone concentration in the 10.85-m high riser by the use of a UV detector system. Furthermore, a pressure drop profile was reported. Comparison between measured and simulated time-averaged ozone concentration at different elevations in the riser shows good agreement. The 3D representation of the reactor geometry gives better predictions of the radial variation in concentration than in a similar 2D simulation, Samuelsberg and Hjertager [53]. Typical results are given in Figure 8.

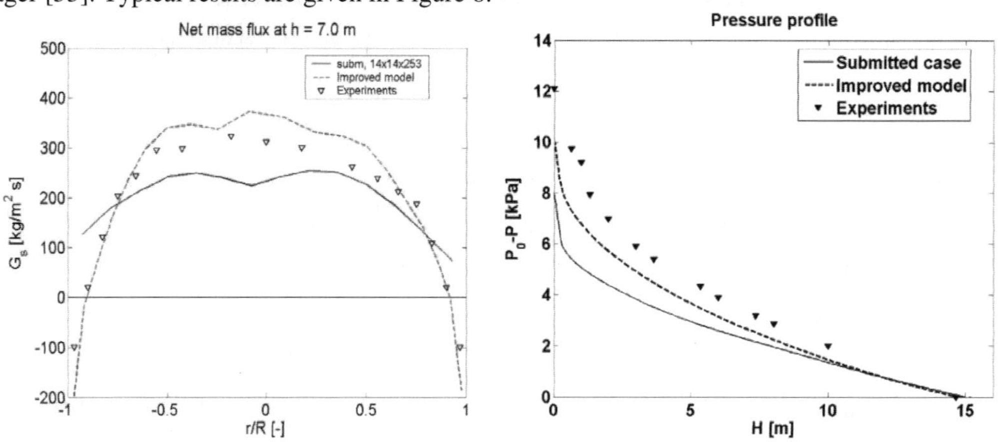

Figure 7: Predicted and measured solids fluxes (left) and pressure profile for blind test and improved cases (right).

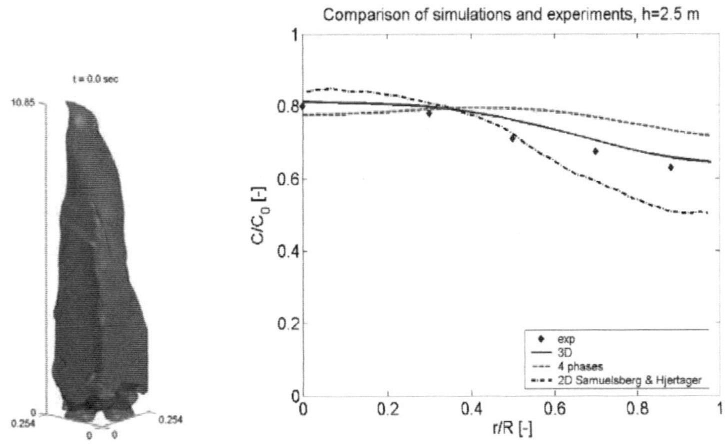

Figure 8: Predicted contours of ozone concentration (iso-surface of $C/C_{inlet} = 0.6$) (left); comparison of measured and predicted radial profile of ozone (right)

7. Summary

The present Paper has presented details of a multi-fluid model for multi-phase flows. Closure laws for both bubble-liquid and solids-gas systems have been exposed. In particular the results using kinetic theory of granular flow (KTGF) have been shown for the solids gas system. Several sub-models are presented including bubble size, chemical reactions as well as heat and mass transfer. Finally, examples of results from bubble-liquid and solids-gas simulations are given.

Acknowledgement

This paper is based on the work that the author and his colleagues have performed in the field of multiphase flow modeling over the last two decades at his two previous affiliations Telemark University College, Porsgrunn, Norway and Aalborg University Esbjerg, Denmark. The author will take this opportunity to thank and acknowledge previous collaborators: Prof., Dr. T. Solberg, Dr. K. Morud, Dr. A. E. Samuelsberg, Dr. E. Manger, Dr. V. Mathiesen, Mr. T. Solbakken, Ms. T. Teppen, Dr. P. Chr. Friberg, Dr. C.H. Ibsen, Dr. N. G. Deen, Dr. K. Granly Hansen, Dr. S. Bove, Dr. J. Madsen, Dr. R. Hansen.

The present paper is based on Hjertager, B.H., 2007. Multi-fluid CFD analysis of chemical reactors, in: Marchisio, D.L., Fox, R.O. (Eds.), *Multiphase Reacting Flows: Modelling and Simulation.* SpringerVerlag, Vienna, Austria. Number 492 in CISM International Centre for Mechanical Sciences, pp. 125–179.

References

[1] Migdal D and Agosta V D 1967 *Appl Mech* **35** pp. 860-5.
[2] Nichols B D, Hirt C W and Hotshkiss R S 1980 *SOLA-VOF: A solution algorithm for transient fluid flow with multiple free boundaries* Report LA-8355 (New Mexico-Los Alamos Sci Lab).
[3] Spalding D B 1985 Computer simulation of two-phase flows with special reference to nuclear reactor systems, in *Computational Techniques in Heat Transfer* ed R W Lewis, K Morgan, J A Johnson,W R Smith (Pineridge Press) pp.1-44.
[4] van der Hoef M A, Van Sint Annaland M and Kuipers J A M 2004 *Chem Eng Sci* **59 (22-23)** pp. 5157–65.
[5] Kataoka I and Serizawa A 1989 *Int. J. Multiphase Flow* **15(5)** pp. 843–55.
[6] Lopez de Bertodano M, Lahey R T and Jones O C 1994 *J. Fluids Eng.* **116** pp. 128-34.
[7] Chahed J, Roig V and Masbernat L 2003 *Int. J. Multiphase Flow* **29** pp. 23-49.

[8] Milelli M 2002 *PhD thesis*: *A numerical analysis of confined turbulent bubble plumes* dissertation eth n. 14799, Swiss Federal Institute of Technology (ETH).
[9] Deen N G, Solberg T and Hjertager B H 2001 *Chem. Eng. Sci.* **56** pp. 6341-49.
[10] Bove S, Solberg T and Hjertager B H 2004 *International J. Chem. Reactor Eng.* **2**(**1**).
[11] Sato Y and Sekoguchi K 1975 *Int. J. Multiphase Flow* **2** pp.79-95.
[12] Smagorinsky J 1963 *Month Weather Rev* **91** pp. 99-164.
[13] Bove S 2005 *PhD thesis-Computational fluid dynamics of gas-liquid flows including bubble population balances* (Aalborg Uni. Esbjerg).
[14] Lopez de Bertodano M 1998 *Nucl. Eng. and Design* **179** pp. 65-74.
[15] Drew D A and Passman S L 1999 *Theory of multicomponent fluids* (New York-Springer-Verlag).
[16] Moraga F J, Larreteguy A E, Drew D A and Lahey R T 2003 *Int. J of Multiphase Flow* **29** pp. 655-73.
[17] Lahey R T, Lopez de Bertodano M and Jones O C 1993 *Nucl Eng Design* **141** pp.177-201.
[18] Tomiyama A 2004 *3rd Int Symp on Two-Phase Flow Modelling and Experimentation* (Pisa).
[19] Ishii M and Zuber N 2002 *AIChE J.* **45** pp 707-21.
[20] Tomiyama A, Celata G P, Hosokawa S and Yoshida S 2002 *Int. J. Multiphase Flow* **28** pp. 1497-519.
[21] Ding J and Gidaspow D 1990 *AIChE J.* **36** pp 523-38.
[22] Gidaspow D 1994 *Multiphase flow and fluidization* (London-Ac. Press).
[23] Chapman S and Cowling G 1970 *The mathematical theory of non-uniform gases*. 3 ed. (Cambridge University Press).
[24] Jenkins J T and Savage S B 1983 *J. Fluid Mech.* **130** pp. 187-202.
[25] Lun C K K, Savage S B, Jeffrey D J and Chepurniy N 1984 *J. of Fluid Mech.* **140** pp. 223-56.
[26] Manger E 1996 *PhD thesis-Modelling and simulation of gas/solids flow in curvilinear coordinates* Telemark Inst. of Techn.
[27] Jenkins J T and Mancini F 1987 *J. Appl. Mech.* **54** pp. 27-34.
[28] Mathiesen V, Solberg T, Manger E and B H Hjertager 1996 *5th Int. conf. on circulating fluidized beds* (Beijing)
[29] Gidaspow D, Huilin L and E. Manger 1996 *Proc. of XIXth Int. Cong. of theoretical and applied mechanics.*
[30] Mathiesen V, Solberg T and Hjertager B H 2000 *Int. J. Multiphase flow* **26** pp. 387-419.
[31] Mathiesen V, Solberg T and Hjertager B H 2000 *Powder Techn.* **112** pp.34-45.
[32] Ibsen C H, Solberg T and Hjertager B H 2001 *Ind. & Eng. Chem. Res.* **40 (23)** pp. 5081-86.
[33] Hansen K G, Ibsen C H, Solberg T and Hjertager B H 2003 *Int. J. of Chem. Reactor Eng.* **1(1)**.
[34] Deardorff J W 1971 *J. of Comp. Phys.* **7** pp.120-133.
[35] Bagnold R A 1954 *Proc. Roy. Soc.* **225** pp. 49-63
[36] Ergun S 1952 *Chem. Eng. Prog.* **48 (2)** pp. 89-98.
[37] Wen CY and Yu Y H 1966 *Chem. Eng. Prog. Symp. Ser.* **62** pp. 100-11.
[38] Rowe P N 1961 *Trans of IChemE* **39** pp.175-80.
[39] Gibilaro L G, di Felice R and Waldram S P 1985 *Chem. Eng. Sci.* **40 (10)** pp.1817-23.
[40] Syamlal M and O'Brien T J 1988 *Int. J. Multiphase Flow,* **14 (4)** pp. 473–81.
[41] Johansen S T and Boysan F 1988 *Met. Trans B*, **19B** pp.755-64.
[42] Jakobsen H A 1993 *Dr. ing. Thesis-On the modelling and simulation of bubble column reactors using a two-fluid Model* (Trondheim-Norwegian Institute of Technology).
[43] Calderbank P H 1958 *Trans of IChemE*, **36** pp.443-63.
[44] Cook T L and Harlow F H 1986 *Int. J. Multiphase Flow*, **12** pp. 35-61.
[45] Marchisio D L 2007 *Multiphase Reacting Flows: Modelling and Simulation* (Vienna SpringerVerlag) Number 492 in CISM Int. Centre for Mechanical Sciences ed R O Fox pp. 125–79.
[46] Samuelsberg A and Hjertager B H 1996 *5th Int. conf. on circulating fluidized beds* (Beijing).
[47] D Geldart ed 1986 *Gas fluidization technology* (Chichester- John Wiley & Sons)

[48] Trägårdh C 1988 *Bioreactor Fluid Dynamics* (Elsevier Appl Sci Publ.) pp. 117-34.
[49] Ouyang S, Lin J and Potter O E 1993 *Powder Techn.* **75** pp.73-78.
[50] van den Hengel E I V, Deen N G and Kuipers J A M 2005 *Ind. Eng. Chem. Res.* 44 (14) pp. 5233-45.
[51] Deen N G, Solberg T and Hjertager B H 2003 *The Can. J. Chem. Eng.* **80(4)** pp. 638-52.
[52] Hansen K G, Solberg T and Hjertager B H 2004 *Chem. Eng. Sci.* **59(22-23)** pp.5217-24.
[53] Samuelsberg A and Hjertager B H 1995 *Adv. in Multiphase Flow* pp. 679-88.

IOP Conf. Series: Materials Science and Engineering **276** (2017) 012011 doi:10.1088/1757-899X/276/1/012011

Accurate green water loads calculation using naval hydro pack

H Jasak*, I Gatin[1], and V Vukčević[2]
Faculty of Mechanical Engineering and Naval Architecture, University of Zagreb, Croatia

*Contact author: hrvoje.jasak@fsb.hr; [1]inno.gatin@fsb.hr; [2]vuko.vukcevic@fsb.hr

Abstract. An extensive verification and validation of Finite Volume based CFD software Naval Hydro based on foam-extend is presented in this paper for green water loads. Two-phase numerical model with advanced methods for treating the free surface is employed. Pressure loads on horizontal deck of Floating Production Storage and Offloading vessel (FPSO) model are compared to experimental results from [1] for three incident regular waves. Pressure peaks and integrals of pressure in time are measured on ten different locations on deck for each case. Pressure peaks and integrals are evaluated as average values among the measured incident wave periods, where periodic uncertainty is assessed for both numerical and experimental results. Spatial and temporal discretization refinement study is performed providing numerical discretization uncertainties.

1. Introduction

Green water loads are of great importance when it comes to designing offshore objects due to their exposure to adverse weather conditions. Predicting green water loads accurately is a challenging task due to the nonlinearity of the flow on deck, causing engineers to turn towards more advance numerical methods. Finite Volume (FV) based Computational Fluid Dynamics (CFD) methods are getting more attention for green water loads prediction, however the confidence among engineers is insufficient at this point due to the lack of verification and validation in the literature.

Temarel et al. [2] concluded that computational and experimental methods lack the maturity for green water load prediction, indicating that further research is needed on the topic. Even though there is a number of publications dealing with numerical methods for green water loads assessment, only a few compare the results with experiments. Greco et al. [3] performed wave load calculations on a patrol ship using CFD, comparing the results to experiments. Joga et al. [4] compared two CFD codes and experimental results for water ingress during green water events. Silva et al. [5] performed simulations of FPSO exposed to beam and quartering waves, where green water loads on deck structure is compared to experimental data. There are more publications dealing with green water loads, however few compare pressure loads with experiments, and even fewer show verification on green water cases.

This paper offers an extensive verification and validation for green water pressure loads of the Naval Hydro software pack, based on foam—extend, the open—source CFD software. FV method with collocated grid arrangement is used, where Volume of Fluid (VOF) method is used for interface

IOP Conf. Series: Materials Science and Engineering **276** (2017) 012011 doi:10.1088/1757-899X/276/1/012011

capturing, with geometric advection method called isoAdvector [6]. Ghost Fluid Method (GFM) is used for treating the kinematic and dynamic free surface boundary conditions at the interface [7]. A fixed FPSO model is used with a breakwater on deck, encountered by regular head waves. Three different waves are simulated, while the results are validated against experimental data published by Lee et al. [1]. Verification is performed using four refinement levels where spatial and temporal resolution is varied simultaneously [8], providing detailed numerical uncertainty assessment.

The paper is organized as follows. First, a brief overview of the numerical model is introduced, followed by the simulation set-up. Next, verification is shown with numerical uncertainties for each wave case, followed by result comparison. Finally, a short conclusion is given.

2. Numerical model

An incompressible, two—phase, viscous and turbulent fluid is modelled using the conservation of mass and momentum equations:

$$\nabla \cdot \boldsymbol{u} = 0, \tag{1}$$

$$\frac{\partial \boldsymbol{u}}{\partial t} + \nabla \cdot (\boldsymbol{u}\boldsymbol{u}) - \nabla \cdot \left(\nu \nabla \boldsymbol{u}\right) = -\frac{1}{\rho}\nabla p_d, \tag{2}$$

where $\boldsymbol{u}$ stands for the velocity field, ν is the effective kinematic viscosity, comprising fluid kinematic viscosity and turbulent eddy viscosity, ρ is fluid density, while p_d stands for dynamic pressure, calculated as $p_d = p - \rho \boldsymbol{g} \cdot \boldsymbol{x}$, where p stands for total pressure, $\boldsymbol{g}$ is gravitational constant, and $\boldsymbol{x}$ is the position vector. Hence, the term denotes the hydrostatic pressure. Here, the free surface boundary conditions are discretized using Ghost Fluid Method [7], which imposes a physical jump of dynamic pressure and density field across the free surface. The method alleviates spurious air velocities from which two—phase FV codes often suffer. Apart from the above equations, a fraction field α transport equation is used to advect the interface:

$$\frac{\partial \alpha}{\partial t} + \nabla \cdot (\alpha \boldsymbol{u}) = 0. \tag{3}$$

Equation (3) is not discretized in conventional FV manner, instead a geometric method called isoAdvector is used to reconstruct the iso—surface representing the interface, which is then used to asses temporal and spatial integrals over cell—faces exactly. The reader is referred to [6] for more information on the isoAdvector method.

Implicit relaxation zones [9] are used to impose regular waves in the computational domain, and to damp diffracted waves towards the end of the domain by forcing them to the incident wave solution.

3. Simulation set up

Simulation set up is performed to comply exactly to the experimental measurements from [1], where three bow shapes are tested with nine regular incident waves each. In this paper, one bow shape is considered (RECT0) with three incident waves, with parameters shown in Table 1. For wave initialisation a nonlinear stream function wave model is used [10]. In the experiments, pressure is measured on ten locations on the horizontal deck. Figure 1 shows the geometry of the FPSO model used in the experiment where the positions of pressure gauges are also indicated.

Table 1. Incident wave parameters.

Wave ID	λ (m)	*a* (cm)	*ka*
1	2.25	6.750	0.188
2	3.00	6.000	0.126
3	3.00	7.500	0.157

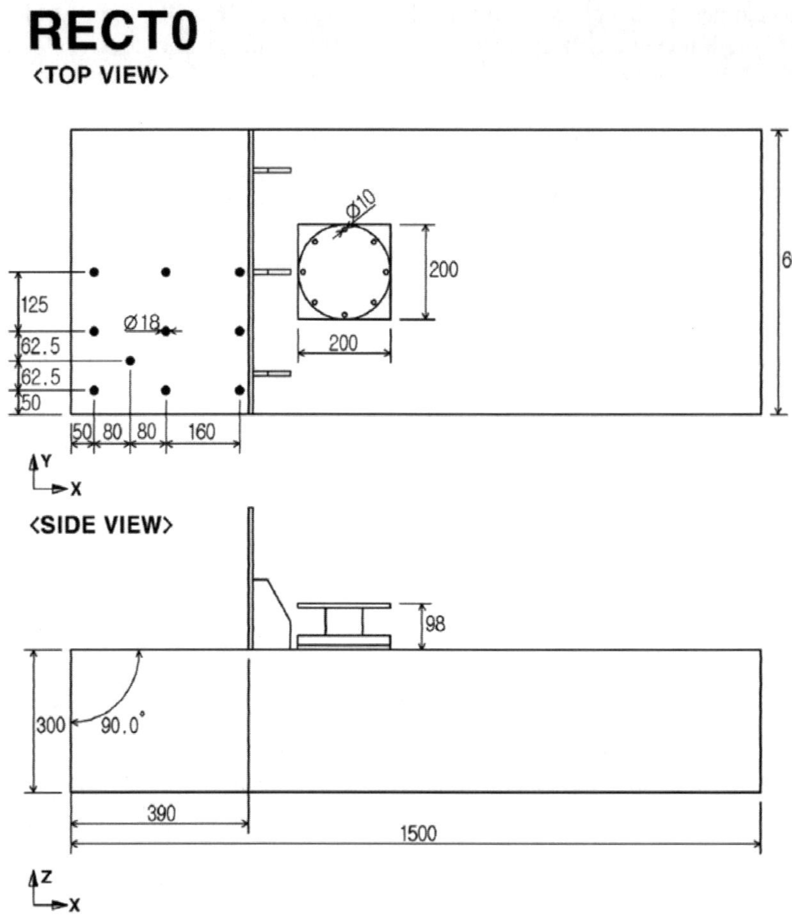

Figure 1. Geometry of the FPSO model from [1].

The CFD simulation domain can be seen in Figure 2, with FPSO hull in the middle of the domain. For each wave, a slightly different computational grid is used depending on the wave height and length, and on the expected run—up against the breakwater. Four grids are generated for each wave case in order to perform a least squares discretization uncertainty analysis by following *Eça* and Hoekstra [8]. Number of cells for each grid is show in Table 2, where a constant refinement ratio $r = \sqrt{2}$ is maintained. Structured, hexahedral cells are used, with refinements close to the free surface, and near the deck. Figure 3 shows the discretized surface of the FSPO model for wave 2 from the coarsest grid. Except the grid resolution, temporal resolution is also varied simultaneously [11]. Consistent variation of the time step is achieved by fixing the Courant—Friedrich—Lewy (C_0) number in all simulations to a same value of 0.75. Since C_0 depends linearly on the time—step and cell size, the refinement ratio is maintained for the time—step size between the simulations.

All simulations were run for 20 periods, while the experimental measuring was performed during 35 wave periods. At time zero the regular wave is initialised, where free surface elevation and velocity profile are imposed as the initial condition. Initial condition for wave 2 simulation is shown on Figure 4.

During the experiment and numerical simulations, absolute pressure is measured at ten pressure gauges. Pressure peak during the wave period is measured and compared, as well as pressure time integral during one wave period. Average values among the measured periods are calculated and compared.

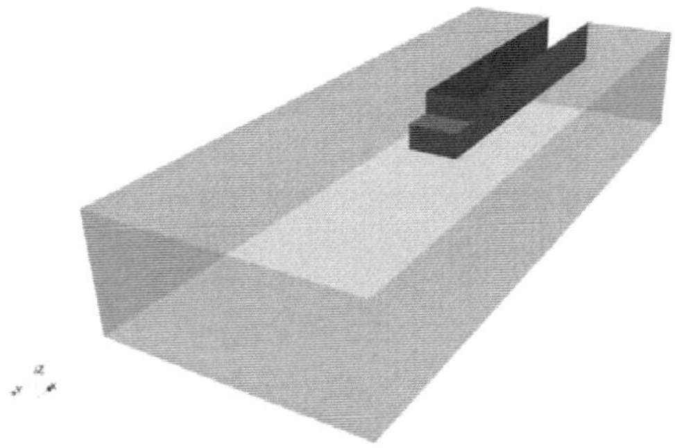

Figure 2. CFD computational domain.

Table 2. Number of cells in individual computational grids.

Wave ID	Grid 1	Grid 2	Grid 3	Grid 4
1	276 699	518 476	1 077 515	2 181 103
2	291 546	546 952	1 140 179	2 299 683
3	319 035	603 876	1 236 052	2 509 667

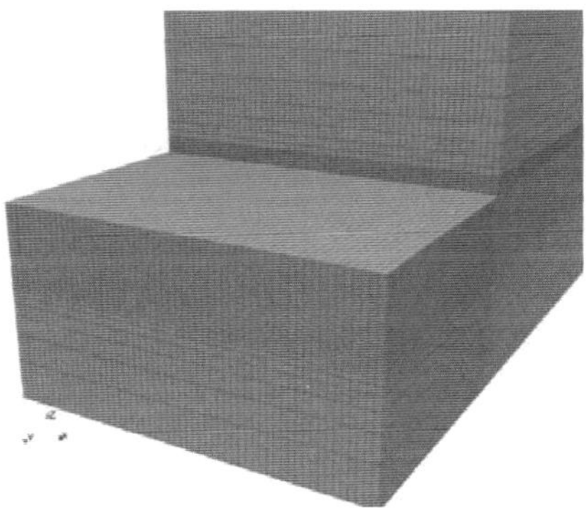

Figure 3. Discretized surface of the FSPO model for wave 2, Grid 1.

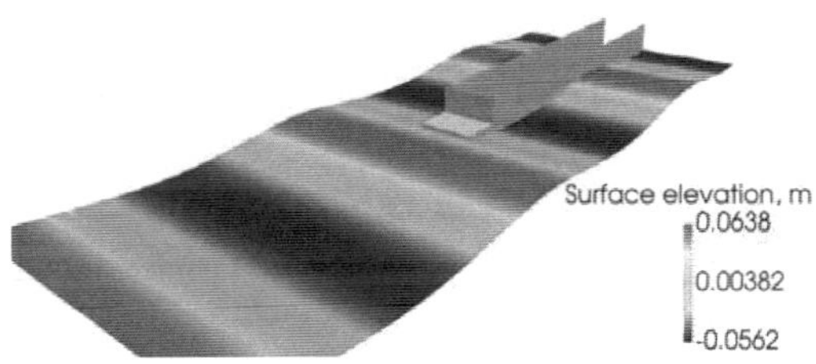

Figure 4. Initial condition for wave 2 simulation.

4. Verification

Numerical discretization uncertainties are calculated using the least squared method proposed by Eça and Hoekstra [8], where four discretization levels are needed. Instead of pressure values, total vertical pressure force on the horizontal deck is used for verification, since unlike pressure, it represents a smooth function in time. On the other hand, it is a spatial integral of pressure, and hence the calculated uncertainties are related to pressure in individual points. Figure 5 shows the time signal of vertical force acting on the deck during the simulation for wave 3. For each refinement level, the average force peak F_{max}, and average force integral in time over one wave period I, are calculated as:

$$F_{max} = \frac{\sum_{i=1}^{N_C} F_{i,max}}{N_C}, \tag{4}$$

$$I = \frac{\sum_{i=1}^{N_C} \int_0^T F_i(t)\, dt}{N_C}, \tag{5}$$

Where N_C represents the number of simulated periods, $F_{i,max}$ is the maximum force in the i-th period, and T is the wave period. These items are used to calculate numerical uncertainties which comprise periodic uncertainty between simulated wave periods, and numerical discretization uncertainty.

Numerical discretization uncertainties for each wave case are shown in Table 3, where $F_{0,max}$ and I_0 stand for the extrapolated theoretical exact solutions of force peak and force integral, respectively. They are obtained using the discretization refinement study [8]. $U_{CD,F}$ and $U_{CD,I}$ are discretization uncertainties of F_{max} and I, respectively. The uncertainties range from 2.2 to 20.7 percent, indicating the unsteady and violent nature of green water impact. It will be shown below that these uncertainties compare to experimental uncertainties.

Figure 6 shows the green sea event from wave 2 case simulation. The wave spills onto deck from the front and the sides, forming a circular water front. The water front converges towards the centreline as it approaches the breakwater, causing water impact and run-up. Next, the water column formed at the breakwater collapses, forming a rebound wave propagating towards the deck edges, and interacting with the subsequent incoming wave.

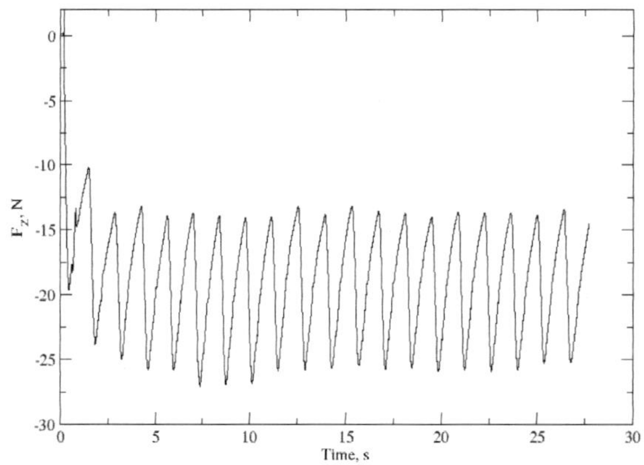

Figure 5. Vertical force measured at the deck for wave 3.

Table 3. Numerical discretization uncertainties.

Wave ID	$F_{0,max}$, N	$U_{CD,F}$, %	I_0, Ns	$U_{CD,I}$, %
1	62.25	5.3	59.45	8.5
2	39.56	12.7	42.21	20.7
3	61.09	2.2	61.03	4.4

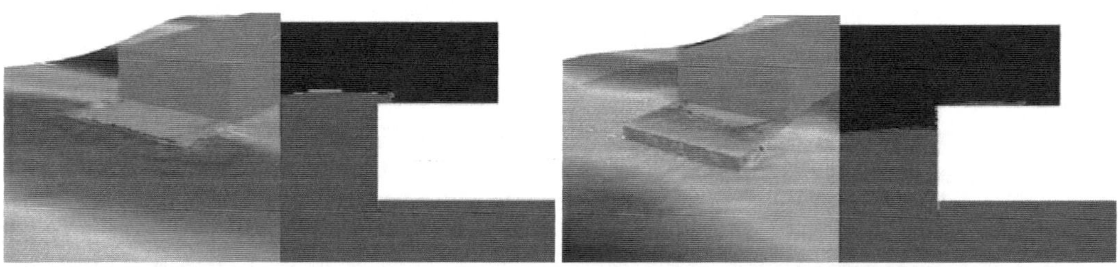

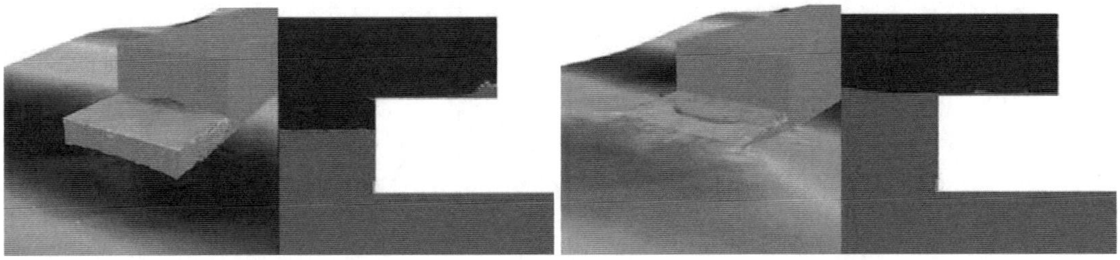

Figure 6. Sequential screen shots from wave 2 simulation on grid 3.

5. Validation

The simulation results are validated against experimental data from [1]. Pressure peaks p_{max}, and integrals in time P are compared to experimental values, calculated as:

$$p_{max} = \frac{\sum_{i=1}^{N_C} p_{i,max}}{N_C}, \tag{6}$$

$$P = \frac{\sum_{i=1}^{N_C} \int_0^T p_i(t)\, dt}{N_C}, \tag{7}$$

where $p_{i,max}$ denotes maximum pressure measured during one wave period.

Result comparison for all wave cases for pressure peaks measured at different pressure gauges is shown in Figure 7, Figure 8 and Figure 9. The error bars represent numerical and experimental uncertainties. Numerical uncertainties comprise periodic and discretization uncertainties, while experimental uncertainties arise from periodic and measuring uncertainties.

For wave 1 (Figure 7), trend and values agree well, where for only two out of ten gauges the uncertainty intervals do not overlap, while the uncertainties are comparable between CFD and the experiment (EFD). For wave 2 (Figure 8), the numerical uncertainties are larger than experimental, mainly due to large discretization uncertainties, Table 3. Nonetheless, values correspond well with the experimental results, where seven out of ten pressure gauges exhibit overlapping uncertainty intervals. Good agreement is observed for wave 3 as well on Figure 9, where the trend is well captured and the values being very close. Numerical uncertainties are generally smaller than experimental, except for pressure gauge 7 and 8, where the largest pressure loads occur. Overall, the agreement between experimental and computational results is acceptable.

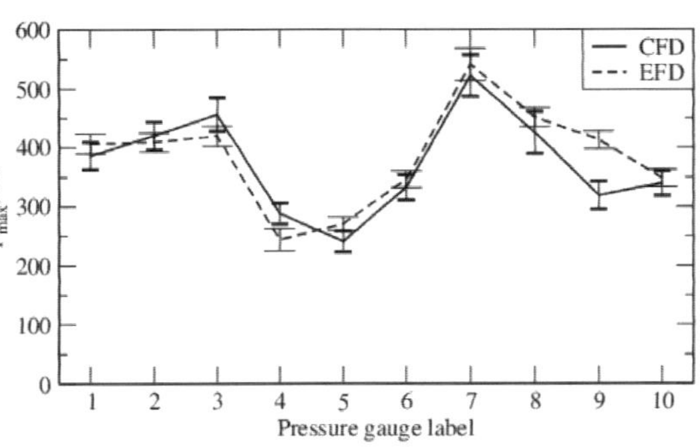

Figure 7. Pressure peak comparison for wave 1.

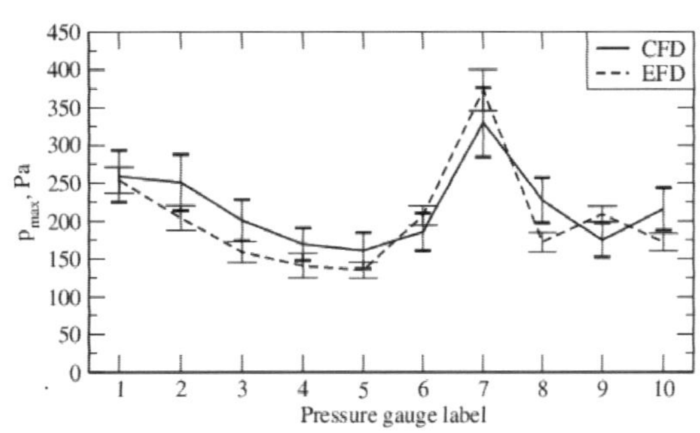

Figure 8. Pressure peak comparison for wave 2.

IOP Conf. Series: Materials Science and Engineering **276** (2017) 012011 doi:10.1088/1757-899X/276/1/012011

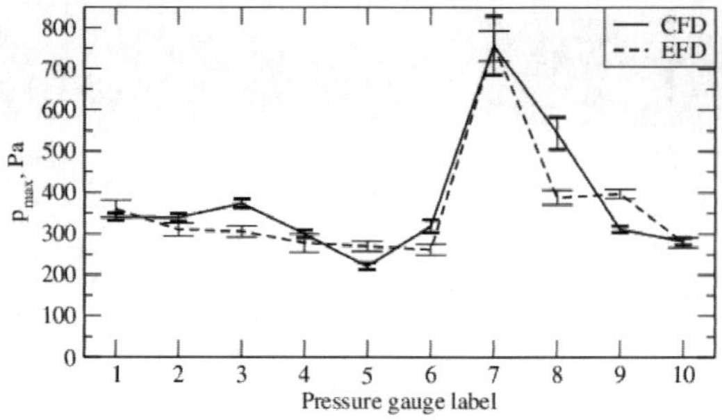

Figure 9. Pressure peak comparison for wave 3.

Comparison of pressure integrals in time is shown on Figure 10, Figure 11 and Figure 12. Numerical uncertainties of pressure integral for wave 1, Figure 10, are larger than experimental, however they remain within acceptable range. Trend and values correspond well, where numerical results mostly underestimate the experimental measurements. Same as for pressure peaks, wave 2, shown on Figure 11, exhibits larger numerical uncertainties with respect to two other wave cases. The values are well predicted for most gauges, with gauge 5 and 8 being an exception. For wave 3, pressure integrals comparison shown in Figure 12 exhibit small numerical uncertainties and good agreement with the experiment with overlap of uncertainty intervals, where pressure gauges 7 and 9 show larger discrepancies comparing to the remaining pressure gauges.

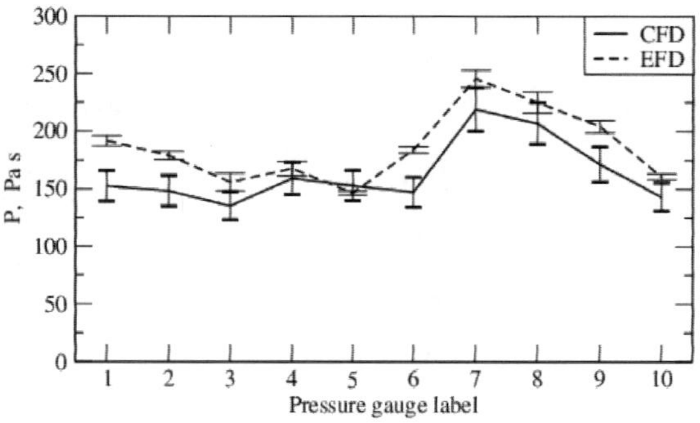

Figure 10. Comparison of pressure integral in time for wave 1.

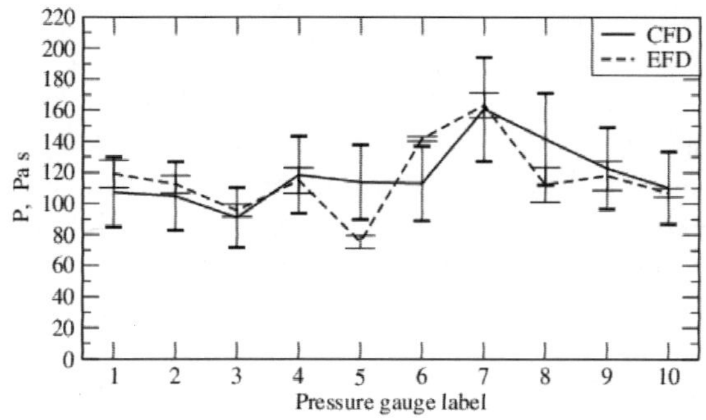

Figure 11. Comparison of pressure integral in time for wave 2.

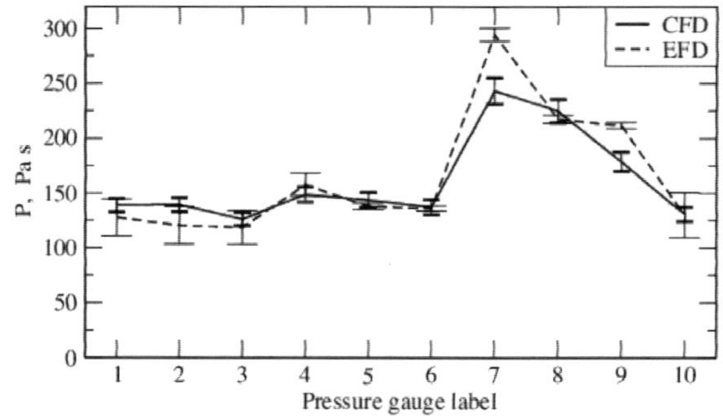

Figure 12. Comparison of pressure integral in time for wave 3.

Comparison between numerical and experimental results show that accurate prediction of green water pressure loads can be obtained at horizontal deck. The differences between the results often fall within the uncertainty range, while the trends over different gauges show good agreement.

6. Conclusion

Verification and validation of Naval Hydro software pack for green water pressure loads are presented in this paper. Fixed model of a simplified FPSO vessel is considered, encountered by three different regular waves causing green water events.

Detailed verification using four spatial and temporal refinement levels is performed in order to determine computational uncertainties. Acceptable uncertainties are obtained ranging from 2 to 20 percent, which are comparable to experimental uncertainties.

The results are validated against experimental data, where pressure peaks and time integrals are compared. Good agreement is achieved for all wave cases for values and trends, where only a few pressure gauges showed larger discrepancies.

This paper presents a step further towards gaining confidence in CFD for prediction of free surface impact loads. In future research, green water loads on vertical breakwater will be considered as well, since the designing of the same poses a challenge from structural load point of view.

Acknowledgement

This research was sponsored by Bureau Veritas under the administration of Dr. Šime Malenica.

References

[1] Lee H H, Lim H J and Rhee S H 2012 Experimental investigation of green water on deck for a CFD validation database *Ocean Eng.*, **42** pp. 47–60.

[2] Temarel P, Bai W, Bruns A, Derbanne Q, Dessi D, Dhavalikar S, Fonseca N, Fukasawa T, Gu X, Nestegård A, Papanikolaou A, Parunov J, Song K H and Wang S 2016 Prediction of wave-induced loads on ships: Progress and challenges *Ocean Eng.* **119** pp. 274–308.

[3] Greco M, Bouscasse B, and Lugni C 2012 3-D seakeeping analysis with water on deck and slamming. Part 2: Experiments and physical investigation *J. Fluids Struct.* **33** pp. 127–47.

[4] Joga R, Saripilli J, Dhavalikar S, and Kar A 2014 Numerical simulations to compute rate of water ingress into open holds due to green waters *In Proc. of the 24th Int. Offshore and Polar Eng. Conf. ISOPE (Busan)*

[5] Silva D F C, Esperança P T T, and Coutinho A L G A 2017 Green water loads on FPSOs exposed to beam and quartering seas, Part II: CFD simulations *Ocean Eng.* **140** pp. 434–52.

[6] Roenby J, Bredmose H, and Jasak H A 2016 Computational Method for Sharp Interface Advection *arXiv* p. 1601.05392.

[7] Vukčević V, Jasak H, and Gatin I 2017 Implementation of the Ghost Fluid Method for free surface flows in polyhedral Finite Volume framework *Comput. Fluids* **153** pp. 1–19.

[8] Eça L and Hoekstra M 2014 A procedure for the estimation of the numerical uncertainty of CFD calculations based on grid refinement studies *J. Comput. Phys.* **262** pp. 104–30

[9] Jasak H, Vukčević V, and Gatin I 2015 Numerical Simulation of Wave Loads on Static Offshore Structures *In CFD for Wind and Tidal Offshore Turbines* pp. 95–105.

[10] Rienecker M M and Fenton J D 1981 A Fourier approximation method for steady water waves *J. Fluid Mech.* **104** pp. 119–37

[11] Eça L, Hoekstra M, Vaz G, Koop A, Pereira F, Abreu H *NPARC Alliance CFD Verification and Validation Web Site*,

[12] Roache P J, Stern F, Wilson R V, Coleman H W, and Paterson E G 2016 Code Verification of Unsteady Flow Solvers with the Method of the Manufactured Solutions," *Int. J. Offshore Polar Eng.* **29(1)** pp. 123–60

IOP Conf. Series: Materials Science and Engineering **276** (2017) 012012 doi:10.1088/1757-899X/276/1/012012

Study of lifting operation of a tripod foundation for offshore wind turbine

H Zhu, L Li* and M C Ong

Department of Mechanical and Structural Engineering and Materials Science
University of Stavanger, Stavanger, Norway

*Corresponding author: lin.li@uis.no

Abstract. This study addresses numerical analysis of the installation of a tripod foundation using a heavy lift vessel (HLV). Limiting sea states are firstly predicted in the frequency domain based on crane tip vertical motions using linear transfer functions. Then, numerical modelling and simulations are carried out in the time domain to analyse the coupled dynamic system taking into consideration of the nonlinearities of the system. In time-domain analysis, two lifting phases are brought into focus, i.e., the lift-off and the lowering phases. For the lift-off phase, two scenarios are considered, i.e., lift-off from the own deck of the HLV and lift-off from a transport barge. Moreover, comparative studies using two types of installation vessels, a floating vessel and a Jack-up, are investigated for the lowering process. Critical responses including the motions of the tripod and the lift wire tensions are presented and compared under various environmental and loading conditions.

1. Introduction

The installation of offshore wind farms continues to increase, which is driven by many factors. The potential energy produced from wind is proportional to the cube of the wind speed. As a result, only a small increment of the wind speed can produce a significantly larger amount of electricity. Compared to the wind plants on land, higher wind speed and low turbulence intensity offshore can result in up to 50% higher energy production [1]. Since more wind energy can be captured offshore, together with other positive factors, such as technology transfer from oil and gas industry, offshore wind industry has boomed around the world, especially in European countries around the North Sea and the Baltic Sea. It is predicted that offshore wind will provide a growing share and reach one third all the wind generated energy by 2050 [2].

Marine lifting operations play a key role in the installation of offshore wind turbines (OWTs) and their bottom-fixed foundations. The installation of OWT monopile support structures has been investigated in some research work through numerical modelling and analysis [3]. The lifting operations are typically carried out by a floating or a jack-up crane vessel. Compared to the jack-up vessel which stands still on the sea bed, the motions of the floating vessel would affect motion responses of the lifted object which is a tripod foundation in the present study. However, floating vessels with onboard heavy cranes are more efficient than the jack-up for mass installations of a wind farm, thanks to their fast transit among the locations of the turbines.

In this paper, lifting operations of a tripod foundation using a heavy lift vessel (HLV), are studied. To author's knowledge, there are limited publications available in open literature, which address the numerical simulation of a tripod foundation. Investigation is first carried out in the frequency domain. Based on the linear wave theory and linear response process [4], wave induced rigid body motions of the vessel alone are calculated. Concerning critical response parameters, limiting sea states are predicted based on probabilistic acceptance criteria [5]. Then, time domain simulations are performed for the coupled dynamic lifting system of the HLV and the tripod. The predicted limiting sea states using the frequency domain method are cross checked in the time domain by running stationary simulations. Attentions are consequently devoted to two time-varying non-stationary processes, i.e., the lowering phase and the lift-off phase. Potential critical events, such as snap load in the lift wire, collision with the lifting vessel or even re-hit during the lift-off phase, are studied. Installation system and theoretical background are introduced first, ahead of the limiting sea states prediction in the frequency domain. Then, clarifications on the numerical modelling are presented prior to the analysis and discussion of the time domain simulation results. In the end, conclusions and future work are listed.

2. Installation system

The weight of a tripod foundation which depends on the capacity of the wind turbine and the water depth, can reach hundreds of tons. In this study, a tripod weighing 920 tons [6] is used and the main parameters are shown in Table 1. Due to the heavy weight and the bulky size of the tripod foundation, a heavy lift floating vessel is chosen for the installation. The main dimensions of the HLV [7] can also be found in Table 1.

Table 1. Main parameters of the HLV and the tripod foundation.

Vessel (HLV)		**Tripod**	
Item	**Value**	**Item**	**Value**
Length overall [m]	183	Total mass [tons]	920
Breadth [m]	47	Total height [m]	63
Moulded depth [m]	18.2	Outer diameter (central column) [m]	5.7
Operational draught [m]	10.2	Outer diameter (braces) [m]	1.2 - 3.15

The onboard crane is capable to perform lifting operation up to 5000 tons with maximum 100 m lifting height from its freeboard deck. The positioning system of the HLV allows for lifting operations in close proximity to other structures, such as lift-off operation of the tripod from a transport barge. For the lifting operation, the position of the crane tip plays a crucial role for the limiting sea states prediction since the crane tip motions are considered as critical responses. Moreover, the motions of the crane tip affect the response of the coupled dynamic lifting system. Hence, the crane tip position should be decided based on the general arrangement of the HLV and the technical specification of the onboard crane.

The global coordinate system is a right-handed coordinate system with the X-axis pointing towards the bow, the Y-axis towards the port side and the Z-axis pointing upwards. The origin of the global coordinate system is located on the ship's centerline, at the cross of the midship section and the still water level during the lifting operation. In addition, the body fixed coordinate system (moving with the body) of the HLV coincides with the global coordinate system when the HLV is at rest. In this study, the coordinates of the crane tip are chosen as $x_p = -81.7\ m,\ y_p = 53.5m, z_p = 88\ m$, relative to the origin of the body fixed coordinate system of the HLV. Moreover, the origin of the body-fixed coordinate of the tripod in the numerical model is set on the axis of its central column, at the point 18m beneath the top of the column (45 m above the lower tip). Figure 1 shows the schematic lifting arrangement of the tripod in the global coordinate system.

In the present study, the offshore site North Sea Centre (Site No. 15) [8] is chosen as a reference installation site. Typical wave conditions from this site would be used as input for the analysis.

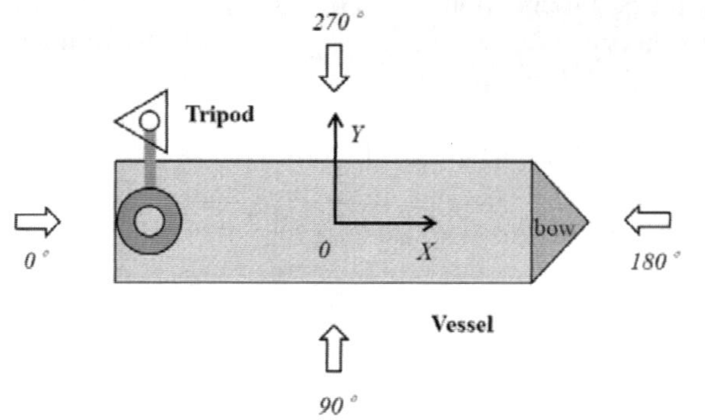

Figure 1. Lifting arrangement of the tripod in the global coordinate system.

3. Theoretical background

3.1. Motions and responses

For the crane tip with the coordinates (x_p, y_p, z_p) in the body fixed coordinate system of the HLV, the motions of the point can be expressed in the following equation [9]:

$$\boldsymbol{s} = (\eta_1 + z_p\eta_5 - y_p\eta_6)\boldsymbol{i} + (\eta_2 - z_p\eta_4 + x_p\eta_6)\boldsymbol{j} + (\eta_3 + y_p\eta_4 - x_p\eta_5)\boldsymbol{k} \tag{1}$$

where, $\boldsymbol{i}$, $\boldsymbol{j}$ and $\boldsymbol{k}$ are the unit vectors and η_1 through η_6 are the rigid body motions of the vessel in surge, sway, heave, roll, pitch and yaw respectively. The crane tip vertical motion is thus formulated as:

$$\boldsymbol{s}_v = \eta_3 + y_p\eta_4 - x_p\eta_5 \tag{2}$$

besides the heave motion η_3, the vertical motion of the crane tip is also affected by the vessel roll η_4 and the vessel pitch η_5. In addition, the positions of the crane tip also affect the vertical motion.

In general, transfer function encompasses both the amplitude scaling and phase shift of the response relative to the wave component. For each degree of freedom, the function can be expressed in complex:

$$H_i(\omega) = A_i(\omega)\exp(i\varphi_i(\omega)) \tag{3}$$

where, $H_i(\omega)$, the transfer function in the i-th degree of freedom; $A_i(\omega)$, amplitude of the transfer function, also denoted as response amplitude operator (RAO) which gives the ratio between the response amplitude to the wave amplitude; $\varphi_i(\omega)$, phase angle of the transfer function; ω, angular frequency of interest.

Given the transfer functions of the vessel, the wave induced rigid body motions at the crane tip can be calculated through motions superposition as in equation (1). For crane tip vertical motion, it can be shown as:

$$\boldsymbol{s}_v = \eta_3 + y_p\eta_4 - x_p\eta_5 = H_3(\omega)\cdot\zeta_a + y_pH_4(\omega)\cdot\zeta_a - x_pH_5(\omega)\cdot\zeta_a$$
$$= [H_3(\omega) + y_pH_4(\omega) - x_pH_5(\omega)]\cdot\zeta_a \tag{4}$$

where, ζ_a is the wave amplitude while the part, $H_3(\omega) + y_pH_4(\omega) - x_pH_5(\omega)$, is combined transfer function for the crane tip vertical motion, denoted as $H_v(\omega)$. By substituting equation (3), it can be further expressed as:

$$H_v(\omega) = H_3(\omega) + y_pH_4(\omega) - x_pH_5(\omega) = A_3(\omega)\exp(i\varphi_3(\omega)) + y_p\cdot A_4(\omega)\exp(i\varphi_4(\omega))$$
$$-x_p\cdot A_5(\omega)\exp(i\varphi_5(\omega)) \tag{5}$$

The calculation of the crane tip vertical motion can be further conducted, ended in the form of response spectrum as in equation (6) and spectral moment as in equation (7).

$$S_R(\omega) = |H_v(\omega)|^2 S(\omega) = [RAO(\omega)]^2 S(\omega) \tag{6}$$

$$m_n = \int_0^\infty \omega^n S_R(\omega)\, d\omega \tag{7}$$

where, RAO is the absolute value (amplitude) of the combined transfer function, S_R is the response spectrum, S is the wave spectrum, ω is the wave angular frequency and m_n is the response spectral moment of general order n.

Response variance σ_R^2, equal to the zeroth response spectral moment, can be expressed as:

$$\sigma_R^2 = m_0 = \int_0^\infty S_R(\omega)\, d\omega \tag{8}$$

Average zero-up-crossing period for the response, T_z, can then be estimated by:

$$T_z = 2\pi \sqrt{\frac{m_0}{m_2}} \tag{9}$$

Consequently, number of response cycles in 1-hour, n_{1h}, can be formulated as:

$$n_{1h} = N = \frac{3600}{T_z} \tag{10}$$

3.2. Limiting sea states based on the critical crane tip motion

Since lifting operations take place in relatively calm weather, the motions of the crane vessel can be computed from linear wave theory and the motions can be assumed to be Gaussian distributed [10] and the global maxima (largest maximum between adjacent zero-up-crossings) follows a Rayleigh distribution [11]. The distribution of 1-hour global response maxima can be expressed as equation (11), by assuming the global maxima in 1-hour as statistically independent and identically distributed.

$$F(X) = \left\{1 - exp\left\{-\frac{1}{2}\left(\frac{X}{\sigma}\right)^2\right\}\right\}^{n_{1h}} \tag{11}$$

where, σ is the standard deviation of the response.

The maximum vertical motion of the crane tip in one hour (Z_{1h}) is chosen as the critical parameter for the lifting operation. Based on limiting criteria with the limiting value Z_{lim}, the probability of exceedance (with $Z_{1h} > Z_{lim}$), can be shown as:

$$P[Z_{1h} > Z_{lim}] = 1 - \left\{1 - exp\left\{-\frac{1}{2}\left(\frac{Z_{\text{lim}}}{\sigma_z}\right)^2\right\}\right\}^N \tag{12}$$

where, σ_z is the standard deviation of crane tip vertical motion. The probability of exceedance is a measurement of risk taken for the marine operation. Low probability of exceedance is preferred for relatively safe operation. In practice, the acceptable probability of exceedance (acceptance criteria) depends on the cost and consequences due to the failure. Given the acceptance criteria (allowable probability of exceeding the limiting value), q_{1h}, the marine operation can only be conducted satisfying the condition as in equation (13).

$$1 - \left\{1 - exp\left\{-\frac{1}{2}\left(\frac{Z_{\text{lim}}}{\sigma_z}\right)^2\right\}\right\}^N \leq q_{1h} \tag{13}$$

Equation (13) can be transformed as:

$$\sigma_z \leq \frac{Z_{lim}}{\sqrt{-2\ln[1-(1-q_{1h})^{1/N}]}} \tag{14}$$

Since the motion standard deviation is a function of the sea state (H_s and T_p), the limiting H_s (significant wave height) and T_p (spectral peak period), can be found as:

$$\sigma_{z,lim}(H_s, T_p) = \frac{z_{lim}}{\sqrt{-2\ln[1-(1-q_{1h})^{1/N}]}} \tag{15}$$

4. Prediction of limiting sea states in the frequency domain

4.1. RAOs of the crane tip vertical motion

Based on the motion transfer functions of the vessel, the RAOs of the crane tip vertical motion can be calculated by referring to equation (5) and they are presented in Figure 2.

It can be found from the figure that the RAO values vary significantly with the wave directions and the lowest occurs in the head sea (180deg). Besides, the curves in various wave directions peak at the frequency around 0.4 rad/s which is close to the natural frequency of the vessel roll.

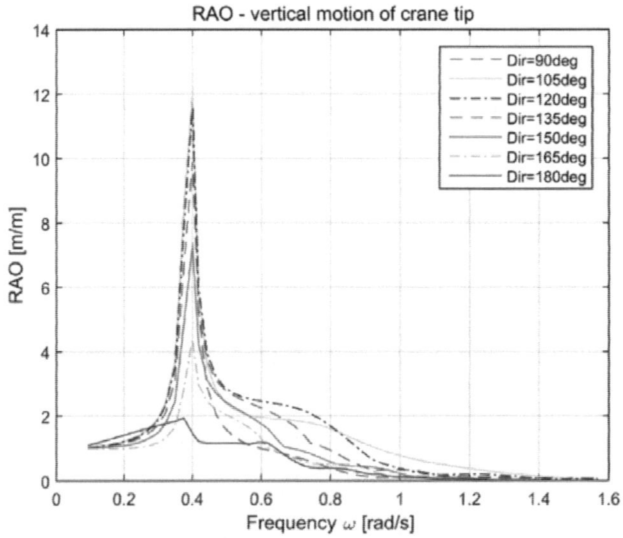

Figure 2. RAOs of the crane tip vertical motion.

4.2. Response spectrum of crane tip vertical motion

The response spectrum of the crane tip vertical motion was calculated from equation (6) using long crested JONSWAP wave spectrum for the reference site. Two wave conditions ($H_s = 1.25m$ and $H_s = 2.5m$ at $T_P = 7s$) with relatively high probability of occurrence according to the 10-year scatter diagram [12], were selected, together with the third one ($H_s = 2.5m$ and $T_P = 16s$) for comparison. In addition to the chosen position for the crane tip, one virtual position (Position 2, closer to the center of gravity of the HLV in X direction) is also used to study the influence from the varied crane tip positions. Coordinates of the crane tip positions are shown in Table 2. Although the assumed Position 2 is practically impossible to reach due to the limited working radius of the onboard crane, the comparative study illustrates the significant influences of crane tip location on the crane tip motions.

Table 2. Coordinates of the crane tip during installation.

Coordinates of crane tip	X	Y	Z
Position 1 (the chosen)	-81.7	53.5	88
Position 2 (the virtual)	-10.5	53.5	88

Figure 3 presents the response spectra of the crane tip vertical motion under the three wave conditions. Huge differences can be observed using different crane tip positions as shown in the Figure 3 (a) and (b). The peak values of the response spectra in the wave direction of 105deg and 120deg using Position 1 is remarkably larger than those using Position 2 which is much closer to the center of gravity (CoG) of the vessel. It demonstrates that the crane tip motions can be greatly mitigated by reasonably locating the crane tip toward mid-ship during installation operation. While, for the wave condition with

$T_p = 16s$ as shown in Figure 3 (c), the differences using the two crane tip positions are minor. The peak values for both cases are extremely high due to resonance. The resonance occurs because the spectral peak period (16s) of the JONSWAP wave spectrum is close to the natural period of the roll motion of the HLV. Moreover, the HLV should better be oriented against the incoming wave (165deg or 180deg) during the operation to avoid large roll induced crane tip motions.

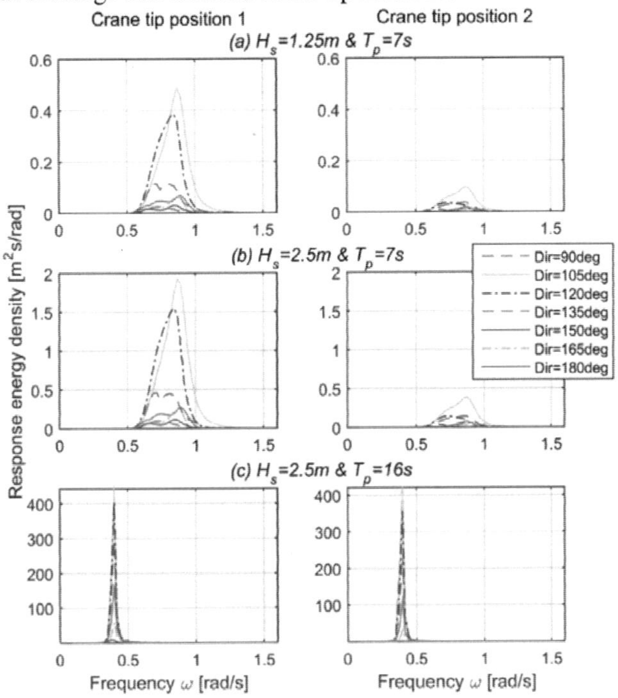

Figure 3. Response spectra of crane tip vertical motion under three wave conditions.

4.3. *Limiting sea states prediction based on the limited vertical motion of crane tip*

The limiting value of the crane tip vertical motion within one hour, Z_{lim} in equation (12), is assumed to be 0.5m. Besides, the acceptable probability of exceedance, q_{1h} in equation (13), is set to be 0.05 with moderate safety allowance. Moreover, it is assumed that the heavy lift operation in this study cannot be carried out under wave conditions with H_s more than 2.5m, which is believed reasonable for such operations.

Figure 4 shows the predicted limiting sea states in different wave heading directions based on the crane tip Position 1. It can be found that the acceptable H_s lowers down dramatically with increasing T_p from 6 s to 15 s. Take the wave heading Dir=165 deg for example, sea states with 2.5m H_s can be tolerated in short waves (e.g., $T_p = 6\ s$), while the acceptable Hs is smaller than 0.5m in the waves with $T_p > 10\ s$. In addition, the limiting sea states depend strongly on the wave headings. It is obvious that the limiting sea state curve becomes lower and lower from the head sea (180deg) to the more oblique sea (from 165deg to 150deg). The lower limiting sea states in the oblique sea result from the larger crane tip responses (spectrum) as shown in Figure 3 due to the influence of vessel roll motions. Specifically, the allowable H_s lowers down from around 1.8m in the wave direction 180deg to about 1.1m in the direction 150deg at $T_p = 7s$. It verifies the importance of the orientation of the vessel relative to the wave direction.

Figure 5 shows the predicted limiting sea states with varying crane tip positions in the wave heading Dir = 165 deg. It is strikingly clear that much severer sea states can be tolerated using crane tip Position 2. It echoes the previous analysis and the judgement that the motion response can be greatly mitigated by locating the crane tip toward mid-ship (closer to the CoG of the vessel).

IOP Conf. Series: Materials Science and Engineering **276** (2017) 012012 doi:10.1088/1757-899X/276/1/012012

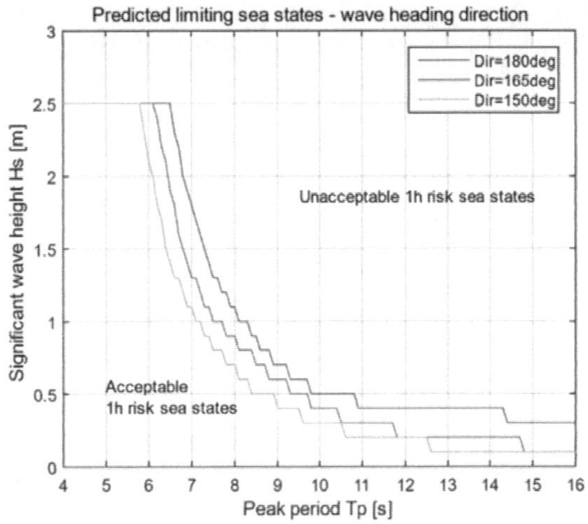

Figure 4. Predicted limiting sea states using three wave headings.

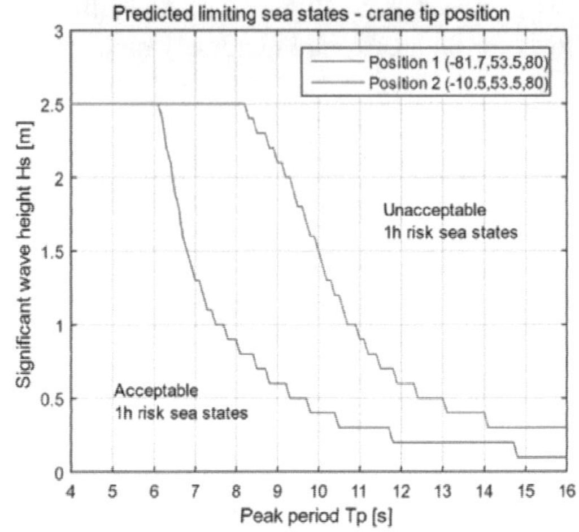

Figure 5. Predicted limiting sea states using two crane tip positions with wave heading Dir = 165°

5. Modelling of the coupled system

Numerical modelling and simulation are carried out using the software SIMA SIMO developed by SINTEF Ocean. SIMO is a non-linear time domain simulation program used for the analysis of rigid body motions and multibody systems [13] [14]. Installation of offshore support structures can be generally divided into the phases, lift-off, lowering (through wave zone) and landing on seabed. This study focuses on two phases, the lowering and the lift-off.

5.1. General setup of the lifting system

The lifting system mainly comprises of two rigid bodies, a floating vessel and a tripod foundation. Onboard the floating vessel, a rigidly connected crane has been modelled. The two rigid bodies are linked through the lift wire which extends from a winch on the crane to the top of the tripod via the crane (boom) tip. For simplicity, hook or sling wires are not modelled. Moreover, a fender coupling is modelled to hold the weight of tripod ahead of the onboard lift-off operation. Figure 6 shows the numerical model for the simulation of lowering process.

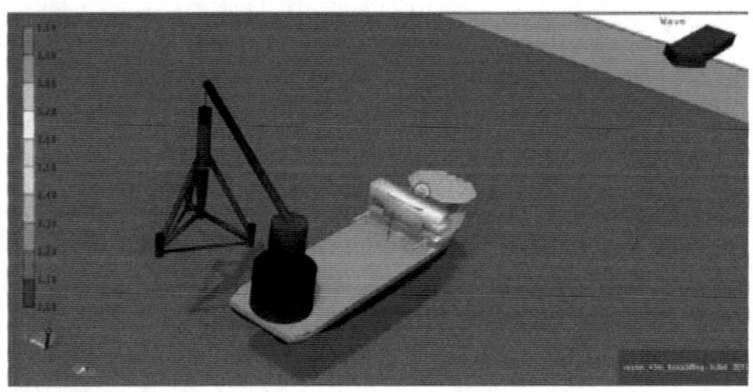

Figure 6. Numerical model for the simulation of lowering process.

Figure 7. Numerical model for the simulation of lift-off from a barge.

For the scenario of lift-off from a transport barge, three rigid bodies are involved. In this case, wire coupling would be applied between the HLV and the tripod while fender coupling would be placed between the tripod and the barge. Figure 7 shows the numerical model for the simulation of the lift-off from a barge.

5.2. Force model on the HLV

The hydrodynamic forces on the vessel include the first order wave excitation forces, the mean wave drift forces and slowly varying forces. Waves are considered as main factor, excluding wind or current forces. The mooring system are simplified by adding linear stiffness terms in surge, sway and yaw by assuming the natural period in the three DOFs equal to 70s which is assumed reasonable for the subject HLV. Then, the stiffness, K, could be obtained following the equation [15]:

$$K_{ii} = \left(\frac{2\pi}{T_{io}}\right)^2 \cdot (m_{ii} + A_{ii}), i = 1,2,6 \tag{16}$$

where, T_{io} is the natural period for the ith DOF and $T_{io} = 70s$, m_{ii} and A_{ii} are the mass and potential added mass, respectively.

Normally, water ballasting is performed in practice to counter the overturning moment from the lifted tripod. For simplicity, constant specified moment is introduced in the simulation of the lowering process. For the case lift-off from a barge, time dependent ballasting is applied, considering the complex relative motion between the two floating vessels.

5.3. Force model on the tripod

For slender body with the ratio between its diameter and the wave length less than 0.2, Morison's formula can be used to calculate the hydrodynamic forces. Member structures of the tripod can be modelled in slender element for the sea states concerned with T_p in the range from 5s to 12s, after checking of the corresponding wave lengths. In the slender body approximation, effects of diffraction and radiation are considered insignificant. The normal force on a moving structure in waves using Morison's formula is as follows,

$$f_N(t) = -\rho C_A A\ddot{r} + \rho C_M A\dot{v} + \frac{1}{2}\rho C_D D(v - \dot{r})|v - \dot{r}| \tag{17}$$

where, v and $\dot{v}$ are fluid particle velocity and acceleration respectively; $\dot{r}$ and $\ddot{r}$ are the sectional velocity and acceleration of the moving structure; D and A are the diameter and cross-section area of structure section; C_M, C_A and C_D are the mass, added mass and drag force coefficients respectively.

C_A and C_D are dependent on many parameters such as the Reynolds number (Re), the Keulegan-Carpenter number (K_C) and the surface roughness ratio (Δ). Outer surface of the newly fabricated foundation can be assumed to be smooth. With Re in the magnitude of 10^6, combined with a small Kc number, the quadratic drag coefficient can be chosen as $C_D = 0.7$ [7].

For cylinders in unbounded fluid with a small K_c, C_A can be assumed to be independent of K_C number and equal to the theoretical value $C_A = 1.0$ for both smooth and rough cylinders [16]. However, the tripod is made of bottomless tube structures with diameters in the order of meters. Following the lowering process, the bottomless hollow structure members will be flushed with sea water following the submergence. The water inside the hollow structure affects the hydrodynamic coefficients and excitation force on the structure members. Referring to the study on monopile lifting operation [7], $C_M = C_A = 1.8$ is applied for all the bottomless tube structures of the tripod. In addition, the slamming forces which can be calculated based on the change of the added mass with time and the relative velocity between the slender element and the wave particle motion, are omitted in the time domain simulations.

5.4. Mechanical couplings

5.4.1. Simple wire coupling. In the numerical model, the lift wire connecting the crane and the tripod is modelled as a simple wire coupling. The simple wire coupling is modelled as a linear spring according to [13]:

$$\mathrm{T} = \mathrm{k} \cdot \Delta \mathrm{l} \tag{18}$$

where T is the wire tension, K is effective axial stiffness and the Δl is the wire elongation. The effective axial stiffness, K, is given by:

$$\frac{1}{\mathrm{k}} = \frac{\mathrm{l}}{\mathrm{EA}} + \frac{1}{k_0} \tag{19}$$

where E is the modulus of elasticity, A is the cross-section area, l is the un-stretched wire length and the $1/k_0$ is crane flexibility. Material damping is set to be 2% of the EA value [13]. Table 3 lists the main parameters of the simple wire coupling.

Table 3. Main parameters of the simple wire coupling.

Lift wire properties		
Flexibility [m/N]	Damping [Ns]	EA [N]
2.00E-09	1.58E+08	7.91E+09

5.4.2. Fender coupling. A fender is defined as a contact element between the tripod and the vessel for the simulation of the lift-off operation. It can give a compressive force normal to the defined sliding plane (freeboard deck of the vessel) and a friction force along the sliding plane. In the model, the fender induced normal force acts upwards to counter the weight of the tripod. As shown in the Table 4, the friction coefficients are chosen based on empirical data between steel and steel. The normal and shear stiffness are decided based on empirical assumption.

Table 4. Main parameters of the fender coupling.

Characteristics of the fender plate				
Friction coefficient		**Shear stiffness [N/m]**	**Normal stiffness [N/m]**	**Internal damping [Ns/m]**
Dynamic	Static			
0.42	0.78	4.68E+07	6.00E+07	1.20E+06

The compressive normal force can then be found by interpolation, from a specified relation between distance and force and from the specified internal damping [13]. Linear interpolation is applied for the relationship between the distance and the normal force and it is assumed reasonable to set the internal damping as 2% of the equivalent stiffness in the normal direction.

6. Time domain simulations

6.1. Simulation of stationary process

The calculated limiting sea states in the frequency domain can only be treated as preliminary assessments since it is carried out based on the vertical crane tip motion from the decoupled HLV. Thus, the calculated limiting sea states in the frequency domain should better be verified by time-domain analysis. Table 5 shows the predicted limiting sea states from the frequency domain analysis, in the wave heading Dir=165deg, which are extracted from the results in Figure 4.

Table 5. Predicted limiting sea states in the wave heading Dir=165deg.

	Predicted limiting sea states with serial number					
	1	2	3	4	5	6
Tp [s]	5	6	7	8	9	10
Hs [m]	2.5	2.5	1.3	0.9	0.6	0.4

Stationary time-domain simulations for the predicted limiting sea states are conducted with the tripod located at two positions, one hanging in air with 13m clearance from the sea level and the other located at 5m above the sea bed (40m in draft). To account for the variability of the stochastic irregular wave conditions, 20 realizations (seeds) of 1-hour simulations for each sea state have been performed. Then, the 1-hour extreme crane tip vertical motions from different seeds are fitted into Gumbel distributions for further assessment.

Figure 8 shows the 95% quantile of the crane tip extreme vertical motion from the fitted Gumbel distribution under the predicted limiting sea states (see Table 5). This quantile corresponds to the acceptable probability of exceedance, $q_{1h} = 0.05$ used in the frequency domain analysis. It can be observed that the simulation results are generally not in line with the predictions by the frequency domain analysis, especially in the case with tripod submerged. For the seas states 1 to 3 with low peak period (from 5s to 7s), the simulated crane tip extreme vertical motions are more violent than the prediction in frequency domain for both the tripod positions, because of the resonance of the coupled system around the T_p range. The resonance mode is dominant by the tripod rotational motions, which cannot be considered from the frequency-domain analysis with HLV alone. It suggests that the predicted limiting sea states in frequency domain have been overestimated. For the sea states 4, 5 and 6 (with T_p equal to 8s, 9s and 10s, respectively), the simulation results quite agree with the frequency domain prediction in the case tripod hanging in air. However, the results are much lower when the tripod is 40m submerged due to the extra damping from the submergence. Moreover, in the submerged case, the simulated most extreme vertical motions are still lower than the pre-set limiting value 0.5m in the sea states 5 and 6, which signifies the underestimate of the limiting sea states in the frequency domain analysis. Hence, the decoupled frequency domain analysis may be used for preliminary assessment of limiting sea state, but it would incorporate some inaccuracy due to the neglecting of the coupling effects of the lifting system.

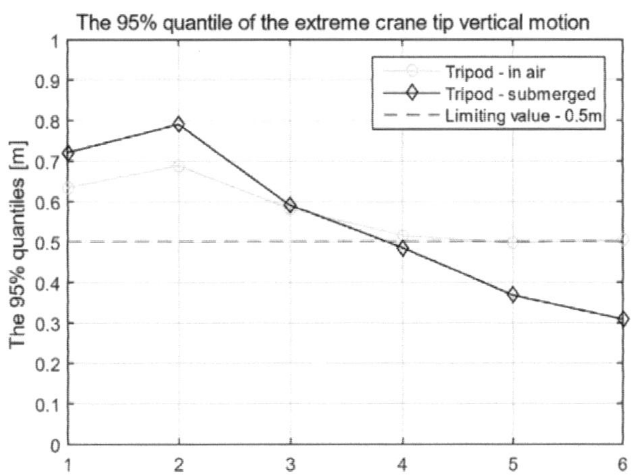

Figure 8. The 95% quantile of the crane tip extreme vertical motion under the predicted limiting sea states.

6.2. Simulation of lowering process

During the nonstationary lowering process, more critical responses (lift wire tension and tripod motions) may occur due to the nonlinear wave loads. Performance of the coupled system is investigated in the wave conditions with H_s of 2.5m, wave direction of 180°, and varying T_p. While, the focus is on the wave condition with $T_p = 7s$, since it is of high probability of occurrence based on the wave data scatter diagram of the reference site. It should be mentioned that the wave conditions here are selected independent of the prediction in frequency domain.

Winch speed of 0.05 m/s is used. During the lowering phase, the winch runs from 300s until 1400s, equivalent to 55m wire releasing, which will lower the tripod from 13m above still water level until 3m above sea bed (42m in draft). Combined with steady state phases with 300s before and after the winch respectively, the length of each simulation is 1700s.

In time domain simulation, the time step shall be set sufficiently small to capture the highest resonant frequency of the system. To capture one cycle of phenomena in the time domain, about 15 time-steps are required [15]. With the natural period of the coupled system in the range of 0.8s to 82s (through eigenvalue analysis), the time step is supposed to be no more than $0.8/15 = 0.06$. It is set to be 0.02s for all the simulations.

6.2.1. Convergence study. Compared to the stationary simulation, the lowering process is subjected to transient effects due to nonlinear wave loads [12]. To verify the convergence of the simulations, 30 different realizations (random seed number) are performed. The extreme motions and responses from each seed are compared with the mean value of all the 30 samples. Besides, a cumulative averaged value for seed number i (the mean value from seed 1 to i), indicating the speed of convergence [15], is calculated.

Figure 9 shows an example of the convergence in the lowering phase from 30 random seeds. The main responses include the lift wire tension and the rotational motions (pitch and roll) of the tripod. It can be seen from the figure that 30 seeds are sufficient to obtain convergent results (with the cumulative average in line with the mean of the sample) for the extreme responses concerned.

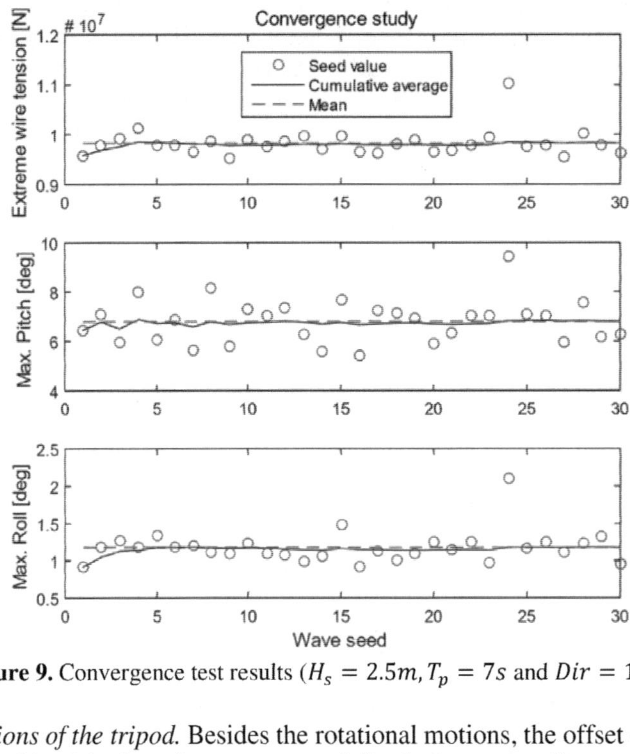

Figure 9. Convergence test results ($H_s = 2.5m, T_p = 7s$ and $Dir = 180°$).

6.2.2. Critical motions of the tripod. Besides the rotational motions, the offset of the tripod lower tips (lower end of the tripod legs) is also of concern, especially the lower tip adjacent to the hull of the vessel (see Figure 1). For the tripod, slight rotation may lead to large horizontal displacement at its lower tip due to the large size of the structure. To guarantee a safe clearance from the vessel during the lowering and ensure precise landing for smooth installation, large tip motions should be avoided.

Figure 10 shows an example (seed 2) of the critical motions of the tripod during the lowering process. Large fluctuations can be observed for the lower tip motions in X direction prior to the submergence (up to around 600s) of the tripod due to the large pendulum motions in air. However, it is damped out following the submergence into water, and the amplitude tends to stabilize until approaching the critical-draft region where the tripod motions are excited [17]. The Y-offset at the lower tip determines the safe clearance from the HLV to avoid potential collisions. Whereas, the results of the tip motion in Y direction are far away from the truth due to the application of the constant specified moment on the HLV. As clarified, specified moment has been applied on the HLV to counter the overturning moment from the hanging tripod. Although equilibrium of moment is guaranteed in air based on the dry weight of the tripod, the balance is broken once the tripod dips into the waves. Following the submergence of the tripod, the decreasing wet weight of the tripod results in the overbalance of the specified moment, which causes continuous positive rolling of the HLV and consequently leads to the abnormal motions of the tripod. In this study, the tripod is originally positioned 8m away from the portside of the vessel, far enough to avoid any collision. But rather than the specified moment, time dependent ballasting is recommended for the future study to avoid the unwanted motions.

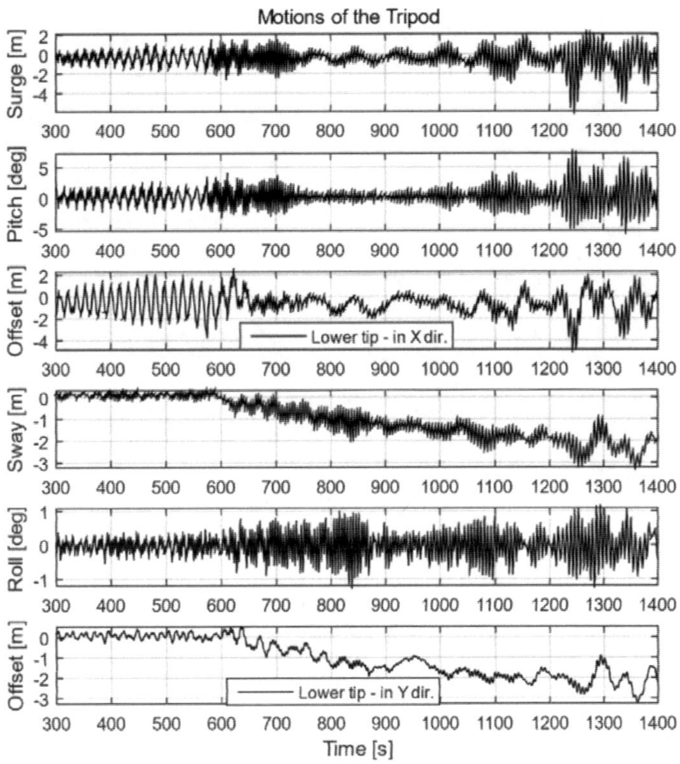

Figure 10. Critical motions of the tripod during the lowering process ($H_s = 2.5m, T_p = 7s$ and $Dir = 180°$).

6.2.3. Influence from the vessel type. Compared to the floating HLV, a jack-up vessel acts as a bottom-fixed structure and provides a stable working platform.

Figure 11 shows an example of the time series of tripod motions and the lift wire tension using the HLV and a jack-up. Compared to the large pitch and tip motion in air using the HLV, no vessel motions are transferred to the tripod from the bottom fixed jack-up vessel. After entering into water, the pitch motions are consistent for the two cases as the lowering continues which suggests that the motions of the tripod are then more affected by the wave forces than the vessel motions. Different from the case using the HLV, there are hardly any roll motion on the tripod using the jack-up, thanks to the long crested, head sea wave condition and the stillness of the jack up vessel. Consequently, the larger roll motion of tripod when using the HLV leads to larger dynamic tension force in the lift wire. Meanwhile, the transversal offset of the tripod lower tip (using the HLV) is also more critical than the case using the jack-up.

Figure 12 shows the statistical maximum lift wire tension and maximum rotational motions of the tripod under the sea states with varying T_p by using two vessels. It can be observed that the rotational motions of the tripod and the tension in the lift wire are smaller by using the jack-up, especially in the long waves. The deviations from the vessel type are enlarged in the long waves with T_p around 11s and 12s, because of increasing motions of the HLV in long waves as the wave peak period approaches to the eigenperiod.

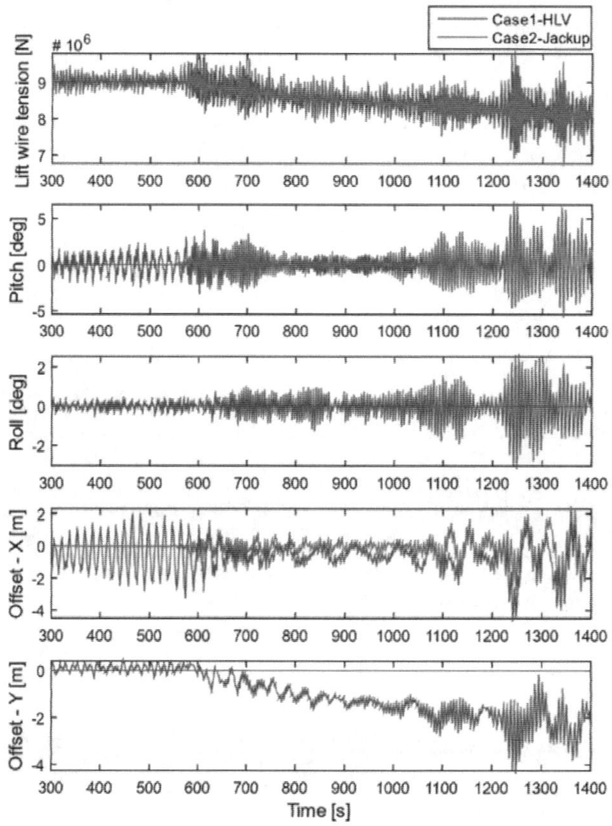

Figure 11. Time series of the lift wire tension and the tripod motions by using different vessels ($H_s = 2.5m$, $T_p = 7s$ and $Dir = 180°$).

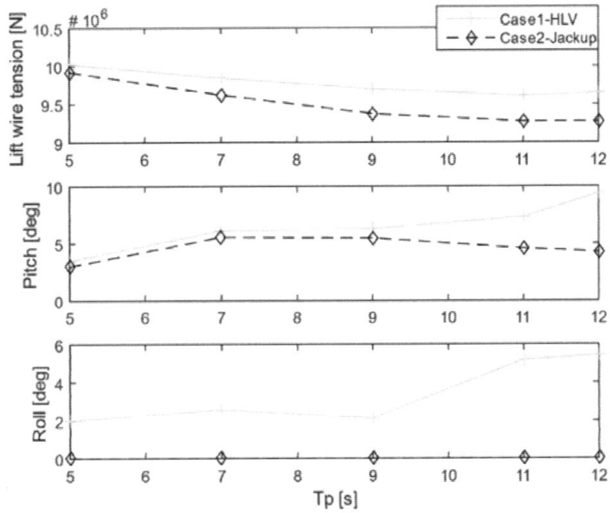

Figure 12. Statistics of extreme responses under various T_p by using different vessels ($H_s = 2.5m$ and $Dir = 180°$).

6.3. Simulation of lift-off operation

Lift-off operation is critical, and many hazards may occur, such as snap loads in lift wire, unacceptable horizontal offset of object following the lift off or even re-hit after lift-off from a barge. To study this critical lifting phase, two cases, lift-off from its own deck and lift-off from a barge, are investigated by performing time-domain simulations. In this study, the barge is modelled with the same hydrodynamic characteristics as the crane vessel (HLV). They are positioned side by side, both heading against the wave. Varying hoisting speeds from 0.1m/s [10] are investigated for the case onboard lift-off.

The winch starts from 200s. It firstly pays in the pre-set loose wire (1.5m) which is long enough to accommodate the relative motion between the crane tip and the seated tripod. The winch is kept engaged until the tripod is lifted 5m upward from its original sitting position. The total length of the dynamic simulation is set to be 450s, so that the system would stay in steady state for certain period after the winch stops.

6.3.1. Lift-off from the deck of the HLV. The simulation of the onboard lift-off is firstly carried out with the winch speed of 0.1m/s, under the sea state $H_s = 2.5m, T_p = 7s$ and $Dir = 180°$. No severe surge or pitch motion are observed, not to mention the negligible sway and roll in the head sea. Thus, attention is brought to the extreme tension force in the lift wire during the operation.

Figure 13 shows the time history and the statistical maximum tension forces in the lift wire during the onboard lift-off with various hoisting speeds. It should be noted that the fluctuation of the wire tension force due to the stop of the winch which is not of concern in this study, has been filtered out. It turns out that the lift-off of the tripod from the deck of the HLV can be smoothly carried out without any slack wire for all the hoisting speeds concerned. Higher peak can be observed with larger hoisting speed, especially when the speed is increased from 0.4m/s to 0.5m/s. Hence, it is critical to control the hoisting speed during the lift-off.

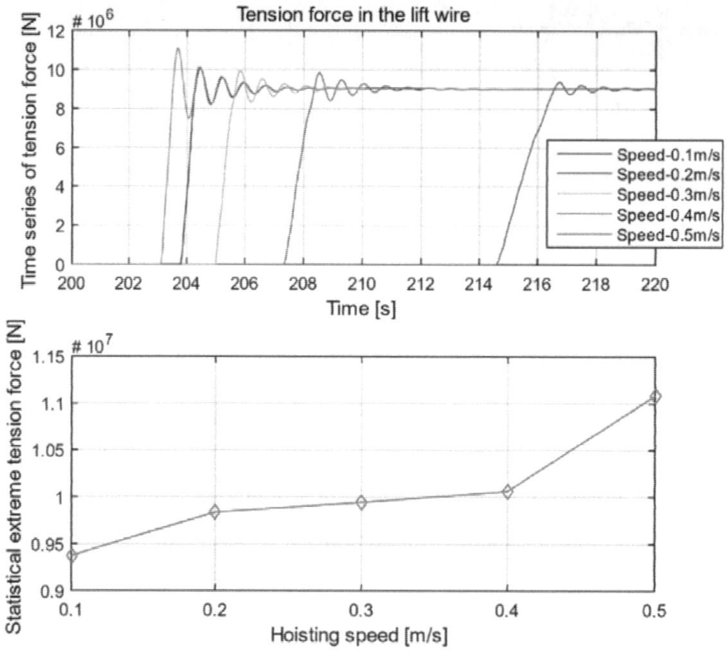

Figure 13. Time series and statistical maximum tension force in the lift wire w.r.t. hoisting speeds ($H_s = 2.5m, T_p = 7s$ and $Dir = 180°$).

6.3.2. Lift-off from a transport barge. Due to the increase of complexity of the coupled system by adding the barge, more wave seeds may be needed for the time-domain simulations to reach a reasonable convergence [18]. After convergence test, it turns out that 45 seeds of time-domain simulations for each sea state are sufficient to achieve convergent results for the extreme lift wire tension.

The process in this case is much more violent compared to lift-off from the own deck of HLV, and high peaks of the coupling forces can be observed as shown in Figure 14. The results are selected from two seeds, seeds 22 and 39 corresponding to the extreme and the moderate situations from the convergence test. The peaks of the lift wire tension are snap loads following the slack in the lift wire. The snap load in the extreme wave seed (seed 22) is almost twice as large as the normal wire tension during hanging in the steady state after successful lift-off. It is the reason that slack slings or wires are recommended to be avoided [10] [19]. Although same settings as the onboard lift-off are applied in this simulation, such as the loose wire length (1.5m) and winch speed (0.1m/s), the lift wire gets stretched as early as 210s with few seconds earlier than the onboard lift-off. It is the result of the relative motions between the HLV and the barge. During the first load transfer, amplitude of the snap force can reach almost the level of the normal wire tension during hanging in steady state. Following the consequent slack-wire cases, the snap load further increases until the tripod is successfully lifted off. The peaks of the fender force also increase following each slack-wire until the lift-off is completed.

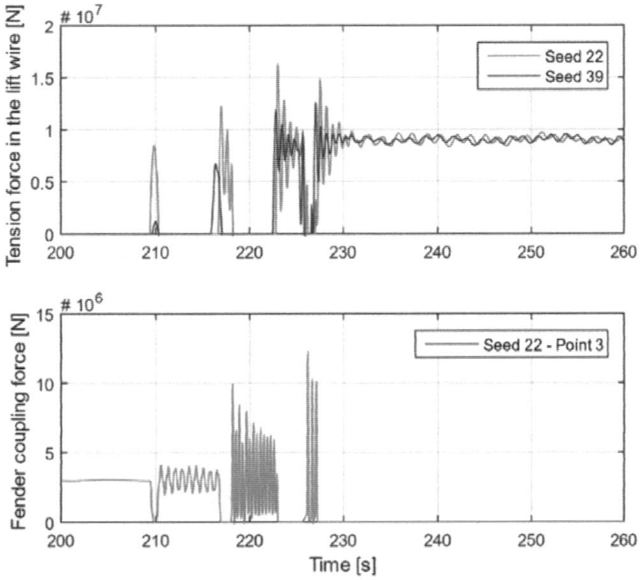

Figure 14. Time series of the coupling forces ($H_s = 2.5m, T_p = 7s$ and $Dir = 180°$).

Regarding the motions of the system during lift-off, the whole process can be divided into three phases. Phase one lasts until the lift wire gets taut (around 209s). The dynamic system can be treated as two separate systems, the barge with tripod seated on and the floating HLV. Phase two is the continuous lifting until the successful lift-off (until around 227s). The last phase refers to the consequent steady state hanging in air which also comprises of two separate systems, the free barge and the HLV with the lifted tripod. Slack and re-hit occurs during phase two.

Figure 15 shows the critical motions of the tripod during the lift-off from a barge. In the first phase, the motions of the tripod are similar as those of onboard lift-off. In this phase, little roll or sway can be observed due to the head sea condition, which contributes to the near-zero tip motion in Y direction. On the other hand, small pitch and surge motions occur which leads to few tip motions in X direction. During phase two, tripod motions become bigger but still moderate. Attention is brought to the motions immediately after the successful lift-off. Large tripod motions can be observed, followed by noticeable tip motions in both X and Y direction. Even with time dependent ballasting system, significant motions would be involved on the HLV due to the transfer of the heavy load (the tripod). In this case, more attention shall be paid to verify the safety clearance to avoid the collision.

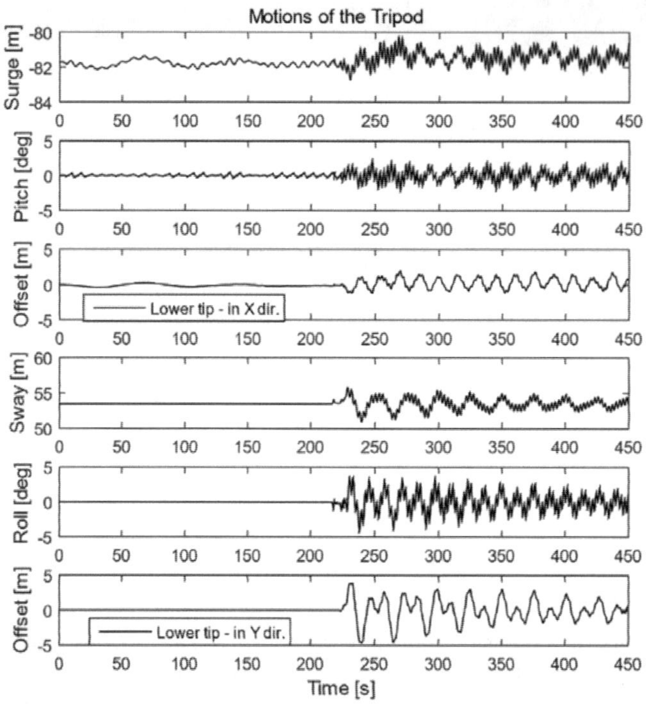

Figure 15. Critical motions of the tripod ($H_s = 2.5m, T_p = 7s$ and Dir=180°).

Compared to the onboard lift-off with minor motion responses, lift-off from a barge is much more demanding, accompanied by critical snap load and significant motion responses.

7. Conclusions

The installation of a tripod foundation using a heavy lift vessel (HLV) was investigated in this study. The limiting sea states were firstly calculated in the frequency domain and later verified in the time domain. In frequency domain analysis, case studies were included using different wave directions and crane tip positions. Then, numerical modelling and time-domain simulations of the coupled dynamic system during the lowering phase and the lift-off phase were performed. For the lowering phase, comparative studies using two types of installation vessel, the HLV and the Jack-up, were investigated. For the lift-off phase, two scenarios, lift-off from the own deck of the HLV and lift-off from a transport barge were studied. Besides, case study on various hoisting speeds during the onboard lift-off was also presented.

Based on the investigations, some essential findings can be summarized in the following:

1) The crane tip vertical motion can be greatly mitigated by locating the crane tip toward the mid-ship and consequently severer sea states can be tolerated. Rougher sea states can also be tolerated by aligning the lifting vessel in the head sea rather than in oblique waves.
2) The predicted limiting sea states using frequency domain method is found inappropriate due to the neglecting of the coupling effects of the lifting system. The calculated limiting H_s were overestimated in short waves but underestimated in long waves compared to the time-domain results with tripod submerged in 40m draft.
3) The vessel type has significant impact on the system responses. Both the wire tension and the tripod motions are smaller by using the jack-up. Especially during the lowering in air, barely any tripod motions occur when using the jack-up.

4) Compared with the onboard lift-off, there are much more challenges for the lift-off from a barge. Under the considered wave condition, the onboard lift-off operation can be smoothly performed while the lift-off from the barge experiences critical snap load and violent tripod motions.

Taking into consideration the simplifications in the present numerical models, recommendations for future work are discussed as follows:

1) Shielding effects from the HLV on the responses of the tripod should be taken into consideration. The simulations in this study were carried out in the undisturbed waves. Referring to the investigation [7], shielding effects may be significant especially in short waves.
2) The effects from wave spreading (short crested wave) should be considered. This study was conducted in long crested, head sea wave condition for the nonstationary time domain simulations. It may result in non-conservative results, especially the roll related motions.

References

[1] WindEurope 2017 *The European offshore wind industry – key trends and statistics 2016* (windeurope.org)

[2] IEA 2013 *Technology roadmap: Wind energy.* URL: http://www.iea.org/publications/freepublications/publication/Wind_2013_Roadmap.pdf. [Accessed 02 02 2017]

[3] Gao Z, Acero W G, Li L, Zhao Y, Li C and Moan T 2016 Numerical simulation of marine operations and prediction of operability using response-based criteria with an application to installation of offshore wind turbine support structures *Proc. Marine Operations Specialty Symp. (MOSS2016)* singapore

[4] Haver S 2007 *Prediction of characteristic response for design purposes report (Draft version)*

[5] Li L, Acero W G, Gao Z and Moan T 2016 Assessment of allowable sea states during installation of offshore wind turbine monopiles with shallow penetration in the seabed *Journal of Offshore Mechanics and Arctic Engineering* vol 138(4) 041902

[6] Xu D 2016 *Numerical modelling and simulations for lowering of an offshore wind turbine tripod (Master thesis)* (Trondheim: Norwegian University of Science and Technology).

[7] Li L, Gao Z, Moan T and Ormbrg H 2014 Analysis of lifting operation of a monopile considering vessel shielding effect *Marine Structures* vol 39 pp 287-314

[8] Li L, Gao Z and Moan T 2015 Joint distribution of environmental condition at five European offshore sites for design of combined wind and wave energy devices *Journal of Offshore Mechanics and Arctic Engineering* vol 137 no 3

[9] Faltinsen O M 1990 *Sea loads on ships and offshore structures* (Cambridge: Cambridge University Press)

[10] DNV 2011 *Recommended Practice DNV-RP-H103, Modelling and analysis of marine operations* (Det Norske Veritas)

[11] Haver S 2016 *Marine Operations (lecture notes)* (Stavanger: University of Stavanger)

[12] Li L, Gao Z and Moan T 2016 Operability analysis of monopole lowering operation using different numerical approaches *Int. Journal of Offshore and Polar Engineering (ISSN 1053-5381)* vol 26 no 2 pp 88-89

[13] MARINTEK 2016 *SIMO Theory Manual (Version 4.8.2)* (MARINTEK)

[14] MARINTEK 2016 *SIMO User Guide (Version 4.8.2)* (MARINTEK)

[15] Li L, Gao Z and Moan T 2013 Numerical simulations for installation of offshore wind turbine monopiles using floating vessels *ASME 2013 32nd Int. Conf. on Ocean, Offshore and Arctic Engineering* Nantes

[16] DNV 2011 *Recommended Practice DNV-RP-C205, Environmental conditions and environmental loads* (Det Norske Veritas)

[17] Birkeland F M 2016 *Numerical Simulation for Installation of XL Monopile for Offshore Wind Turbine (master thesis)* (Trondheim: Norwegian University of Science and Technology)

[18] Acero W G, Gao Z and Moan T 2015 Assessment of the dynamic responses and allowable sea states for a novel offshore wind turbine installation concept based on the inverted pendulum principle *Energy Procedia* vol 94 pp 61-71

[19] Sarkar A and Gudmestad O T 2010 Splash zone lifting analysis of subsea structures *ASME 2010 29th Int. Conf. on Ocean, Offshore and Arctic Engineering* pp 303-312

[20] Mavrakos S 1988 Hydrodynamic coefficients for a thick-walled bottomless cylindrical body floating in water of finite depth *Ocean Engineering* vol 15 no 3 pp 213-229

[21] Baar J J M, Pijfers J G L and Santen J A 1992 Hydromechanically coupled motions of a crane vessel and a transport barge *Proc.Offshore Technology Conf.* (Houston:Offshore Technology Conferene)

[22] Li L, Gao Z and Moan T 2015 Response analysis of a nonstationary lowering operation for an offshore wind turbine monopile substructure *Journal of Offshore Mechanics and Arctic Engineering* vol 137 no 051902

IOP Conf. Series: Materials Science and Engineering **276** (2017) 012013 doi:10.1088/1757-899X/276/1/012013

Vortex-induced vibration (VIV) effects of a drilling riser due to vessel motion

R S Joseph, J Wang[*], M C Ong and J B Jakobsen

Department of Mechanical and Structural Engineering and Materials Science, University of Stavanger, Norway

* Contact author: jungao.wang@uis.no

Abstract. A marine riser undergoes oscillatory motion in water due to the vessel motions, known as global dynamic response. This to-and-fro motion of the riser will generate an equivalent flow that can cause Vortex-Induced Vibrations (VIVs), even in the absence of the ocean current. In the present work, full-scale measurement data of a drilling riser operating in the Gulf of Mexico are analysed. The VIV occurrences for the riser are identified from the data and the possible excitation sources are discussed. The oscillatory flow due to vessel motion is compared with the ocean current and its possibility to excite VIV is analysed. The full-scale data analysis provides an insight into the vessel motion-induced VIV of marine risers in the actual field environment.

1. Introduction

When slender marine structures like risers, free spanning pipelines and mooring lines are exposed to a current flow, they may experience oscillations or vibrations caused by the shedding of vortices around the structure. These are called Vortex Induced Vibrations (VIVs). A cylinder in still water will have many natural frequencies (f_n), for the different modes of vibrations. When the cylinder is exposed to a flow, vortex shedding occurs with a shedding frequency f_s. When the shedding frequency approaches one of the natural frequencies of the cylinder, there is resonance and the cylinder vibrates with a larger amplitude. The frequency of the response is approximately equal to the shedding frequency and the natural frequency. This is called as "lock-in". Once we have a "lock-in", the cylinder is said to experience VIV.

When a marine riser attached to a vessel moves back and forth in water due to the motion of the vessel, an equivalent oscillating current is generated and is experienced by the riser. This equivalent current flow can cause VIV similar to that caused by the ocean current. The vessel motion-induced VIV is influenced by the Keulegan-Carpenter (KC) number values along the riser.

The vessel motion-induced VIV was first reported in STRIDE, a Joint Industry Project focused on compliant risers [1]. It was further studied by Gonzalez [2], Le Cunff et al. [3] and Rateiro et al. [4]. Many investigations based on model tests were later conducted to study the VIV of a free-hanging riser under oscillatory motion. Kwon et al. [5] conducted experiments on a riser model under oscillatory motion for KC numbers as low as 2.24. Wang et al. [6] did similar a study with an 8 m long free-hanging riser subjected to pure vessel motion. The equivalent current velocity, effect of KC number and the characteristics of the VIV responses were investigated. The Cross-Flow (CF) VIV was observed for a KC number as low as 12. Small KC number cases are important given the fact that they can cause considerable fatigue to the riser.

An empirical method to predict the vessel motion-induced VIV of a riser under small KC numbers (KC<40) is proposed by Wang et al. [7]. This model takes into account the KC number distribution along the riser, the vessel motion frequency and the Strouhal relationship to predict the VIV response frequency. Under small KC numbers, the VIV response is no longer governed by the Strouhal relationship using St=0.2, which is valid for the steady current, but by the ratio N between the response frequency, f_{resp} and the vessel motion frequency, f_{im}. The values of the integer N for various KC regimes are documented by Sumer [8] and are shown in equation (1). It was observed that the value of N increases step-wise with the KC number. N is found to increase by 1 with an increase of 8 in KC number.

$$N = \frac{f_{resp}}{f_{im}} = \begin{cases} 2; & 7 < KC < 15 \\ 3; & 15 < KC < 24 \\ 4; & 24 < KC < 32 \\ \cdots & \end{cases} \tag{1}$$

If the KC regime and the vessel motion frequency are known, the response frequency can be estimated using the value of N corresponding to that KC regime.

Wu et al. [9] proposed an empirical model for the prediction of heave-induced VIV. This model considers several time windows within one motion period. The flow velocities over the length of the riser are calculated and a number of velocity snapshots at equal intervals over a period of oscillation are obtained. Each velocity snapshot, taken at a particular time instant over the length of the riser, is the equivalent flow profile for the VIV analysis. Responses for each flow profile are obtained and the average of all these responses is taken to describe the time-varying response of the whole riser.

Many of the investigations focusing on the vessel motion-induced VIV were based on the scaled model tests, performed in controlled laboratory conditions. To the authors' knowledge, no attempts have been made so far to study this phenomenon in an actual field environment. In the present paper, the full-scale measurements from a drilling riser are analysed and instances of VIV are identified. The possible excitation sources including ocean current and vessel motion are investigated. Furthermore, the uncertainties associated with the data analysis are highlighted.

2. Full-scale measurements of drilling riser

Full-scale measurements of a drilling riser were taken by British Petroleum during a drilling campaign in the Gulf of Mexico between 13th April 2007 and 11th July 2007. The dataset is donated to the VIV Data Repository hosted by the Center for Ocean Engineering at MIT, for the purpose of calibration and benchmarking of a VIV software [10] and is publicly accessible. The main parameters in the data set are the accelerations measured by the accelerometers at various points along the span of the riser. The dataset contains the following:

- Riser configuration, dimensions and riser weight.
- Tension and mud weight data.
- Current data at different water depth.
- Acceleration data at various points along the riser.

2.1. Riser configuration and instrumentation

Two wells were drilled during the drilling period. The configuration of the riser during drilling of well-1 is illustrated in Figure 1. The instrumentation consists of 13 standalone loggers. One of the loggers is located on the drill floor and measures the vessel acceleration. The remaining loggers are placed at various locations along the riser. The measurement loggers are named as S01, S02, ..., S13 with S01 situated on the drill floor. Their locations are indicated by the squares in Figure 1. It can be seen that the majority of the loggers are concentrated towards the lower end of the riser. Such an arrangement is to capture all the expected modes with a minimum possible number of loggers. An optimum placement of loggers should capture at least the quarter wavelength of the lowest mode expected [11]. The logger contains the sensors, batteries, memory card and all the associated electronics encased within a cylindrical casing [10]. The loggers are strapped to the structure. Typically, motion sensors are made of tri-axial accelerometers and tri-planar angular rate sensors [12], but the angular rate sensors are not active in this experiment.

The top tension applied to the riser is 8408 kN and the average density of mud used during the drilling period is 1580 kg/m^3. The properties of the riser sections are provided in table 1. The modulus of elasticity of the material is 2.07 x 1011 N/m^2 and the density is 7850 kg/m^3. The eigenmodes of the riser are shown in Figure 2.

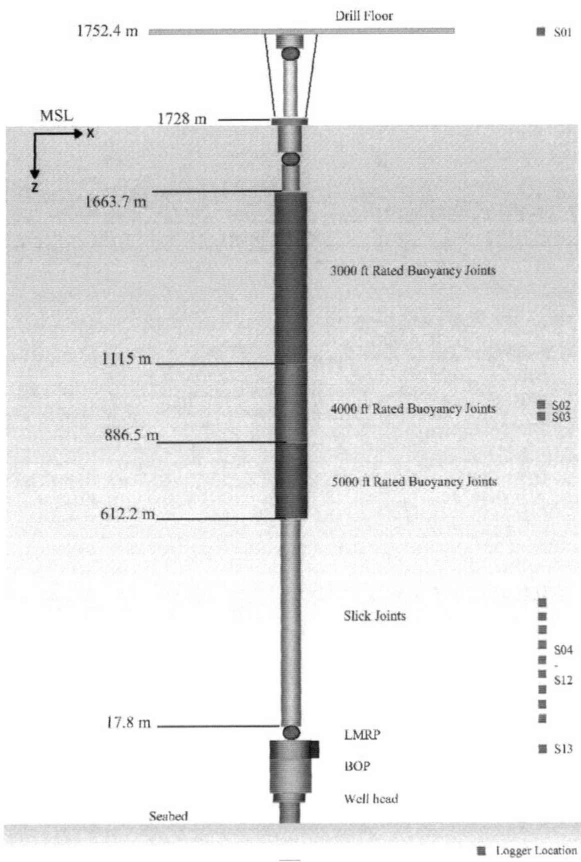

Figure 1. Configuration of the drilling riser for well-1 [10].

2.2. *Environment and current data*

The water depth at the location of well-1 is 1728 m and well-2 is 1729 m. The current at the location was measured using three ADCPs (Acoustic Doppler Current Profiler), two of which are mounted on the vessel. The current is sampled at every 10 minutes [10]. A constant density of 1025 kg/m^3 is assumed for the sea water.

2.3. *Accelerations and events*

The translational acceleration data are recorded continuously for a duration of 15 minutes at each 2 hour interval. This limited duration of measurements is adopted in order to distribute a limited battery life over the whole drilling period. The sampling frequency is 10 Hz. Each 15-minute-long recording period is termed as an event. There are a total of 1078 events captured during the drilling period, spread over 3 operational conditions – (1) drilling of well-1, (2) hang-off and transport and (3) drilling of well-2.

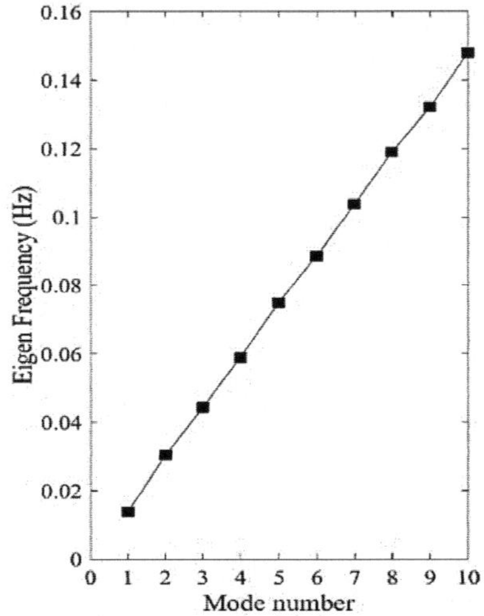

Figure 2. Eigenmodes and eigen frequencies for the riser.

Table 1. Riser sections and properties.

Description	L (m)	Stress diameter (m)	Thickness (m)	Hydrodynamic diameter (m)	Unit dry weight (kg/m)	Unit submerged weight (kg/m)	Bending stiffness EI (Nm^2)	Axial stiffness EA (N)
Diverter	4.1	0.5334	0.02	0.5334	3114.72	-	2.27×10^8	6.88×10^9
16.7 ft pup joints	5.1	0.5334	0.02	0.5334	706.76	614.88	2.27×10^8	6.88×10^9
Inner barrel	9.2	0.5334	0.02	0.5334	787.95	-	1.79×10^8	5.34×10^9
Outer barrel	30.4	0.6604	0.04	0.6604	863.10	750.89	7.49×10^8	1.54×10^{10}
Intermediate FJ	2.8	1.1938	0.35	1.1938	1904.01	1656.52	2×10^{10}	1.92×10^{11}
Termination joint	17.3	0.5334	0.02	0.5334	735.95	640.28	2.27×10^8	6.88×10^9
5 ft pup joints	1.5	0.5334	0.02	0.5334	1391.46	1210.49	2.27×10^8	6.88×10^9
20 ft pup joints	6.1	0.5334	0.02	0.5334	740.75	644.46	2.27×10^8	6.88×10^9
40 ft pup joints	12.2	0.5334	0.02	0.8636	614.44	534.56	2.27×10^8	6.88×10^9
3000 ft buoyancy joints	548.6	0.5334	0.02	1.2827	861.72	17.62	2.27×10^8	6.88×10^9
4000 ft buoyancy joints	228.6	0.5334	0.02	1.3081	904.98	27.54	2.27×10^8	6.88×10^9
5000 ft buoyancy joints	274.3	0.5334	0.02	1.3335	957.82	15.04	2.27×10^8	6.88×10^9
Slick joints	594.5	0.5334	0.02	0.8636	586.74	510.47	2.27×10^8	6.88×10^9
Lower FJ	2.7	1.4732	0.49	1.4732	2485.25	933.04	4.73×10^{10}	3.13×10^{11}
LMRP	3.4	5.6515	2.58	5.6515	29495.64	25661.20	1.04×10^{13}	5.15×10^{12}
BOP	7.2	5.6515	0.02	5.6515	26597.42	23139.76	1.04×10^{13}	5.15×10^{12}
Wellhead	4.4	0.9652	0.06	0.9652	1854.08	1613.09	3.49×10^9	3.37×10^{10}

3. Identification of VIV from the data

The VIV and the corresponding response frequency can be identified by performing a spectral analysis of the accelerations or displacements at each logger location. Figure 3 shows the acceleration spectrum at a particular logger. The responses seen on the spectrum can also be due to sources other than VIV such as drill string rotation or wave-induced motion [13]. VIV occurs in a narrow range of low frequencies. Higher frequency vibrations caused by drill string rotation can be identified based on the rotation speeds and can be discarded by passing the signal through a low-pass filter.

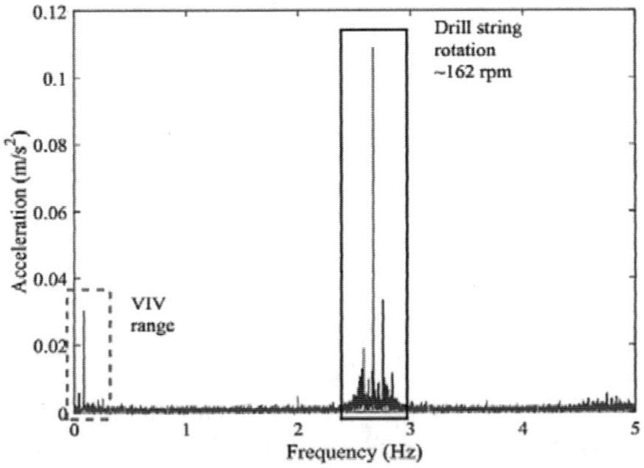

Figure 3. Acceleration spectrum in X direction at a logger showing VIV and other high frequency vibrations (Event dated 17/04/2007 1800 hrs).

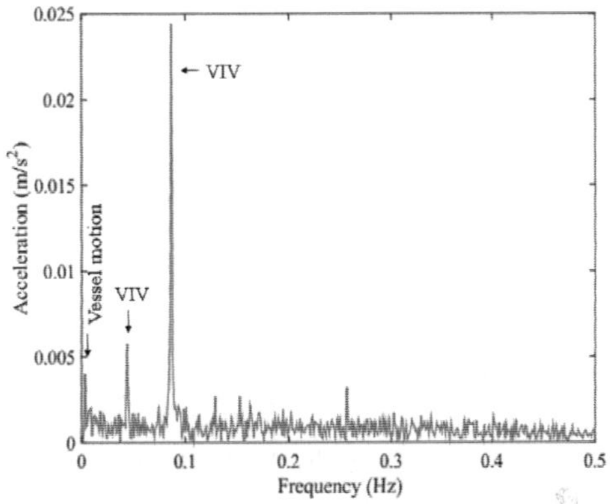

Figure 4. Acceleration spectrum in X direction at a logger showing peaks in the VIV range (Event dated 17/04/2007 1800 hrs).

Figure 4 shows the acceleration spectrum at low frequencies. The VIV response can be found by correlating the spectral peaks in the VIV range across all the loggers on the riser. The vessel motion occurs at very low frequencies and can be easily identified from a displacement spectrum. The displacement spectrum gives a very large peak for the vessel motion frequency at all the loggers. This can be correlated with logger S01 which measures the vessel acceleration and hence vessel motion frequency can be singled out.

In this section, a representative event from the full-scale data, measured on 17/04/2007 at 1800 hrs, is analysed. This corresponds to the riser configuration for the drilling of well-1. The acceleration signals are passed through a low-pass filter to filter-in signals in the VIV range and lower frequencies (< 0.5 Hz).

Figure 5 represents the spectra of accelerations in X direction across all the 13 loggers. Figure 6 represents those in the Y direction. In both directions, it can be seen that all the loggers except S01 (S01 measures vessel acceleration) have a peak response at a frequency of 0.0866 Hz. This is identified to be the response frequency of the VIV. S01 has a peak at a lower frequency which corresponds to the frequency of the vessel motion. Figure 7 shows the displacement amplitude spectrum of logger S01 which is attached to the vessel. Since it measures purely the motion of the vessel and not the VIV, the peak of the spectrum should provide us with the frequency of the vessel motion. In this way, the frequency of the vessel motion can be identified. From the figure, we can see that the vessel motion frequency is 0.0033 Hz, which is reasonable for a MODU's slow varying motion due to the second order differential frequency wave loads.

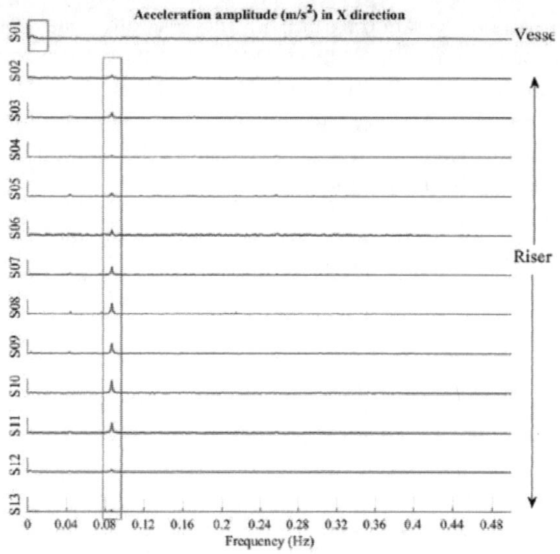

Figure 5. Spectra of acceleration in X direction across all loggers.

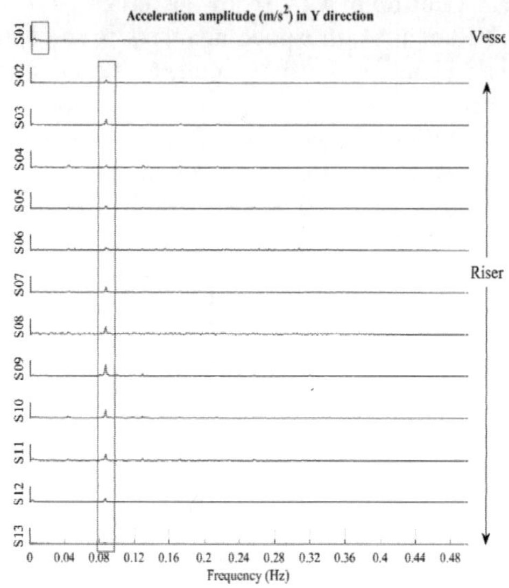

Figure 6. Spectra of acceleration in Y direction across all loggers.

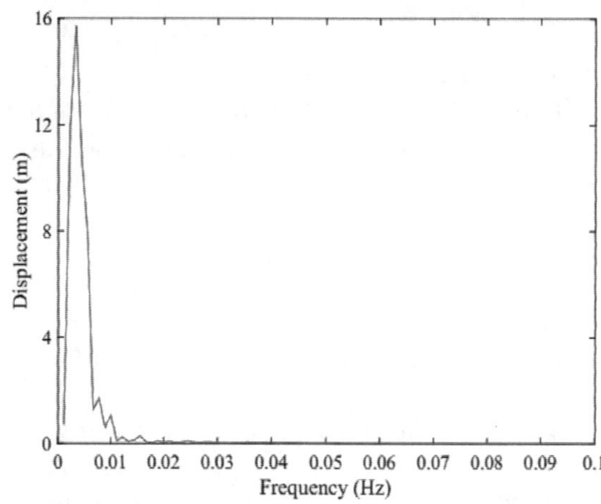

Figure 7. Displacement spectrum in X direction of logger S01.

The source of excitation of the VIV can be either the ocean current or the oscillatory flow due to the vessel motion. From the data, the ocean current details for this case are available up to a depth of 800 m. The direction of the current is at 136-196° w.r.t vessel bow.

3.1. Comparison of current profiles

3.1.1. Equivalent current profile: In order to describe the vessel motion-induced VIV, it is necessary to define the equivalent current that causes it. Equivalent current profile is the one that the riser "sees" due to its relative motion in the water. Since the vessel motions are oscillatory in nature, the movement of the riser results in time-varying velocity series at each point along its span. The representative maximum of the velocity at each point can be used to create a current profile.

In case of irregular vessel motions, Wang et al. [14] have proposed a method to generate the equivalent current profile in which the standard deviation of the velocity time series at each point along the length of the riser is multiplied by √2 to get the representative maximum. Hence the time-varying velocity can be simplified into an equivalent current profile according to equation (2).

$$V_e(z) = \sqrt{2} \times \sigma_v(z) \tag{2}$$

where $V_e(z)$ is the equivalent current profile and $\sigma_V(z)$ is the standard deviation of the time-varying velocity series at each point along the length.

Following steps are used to formulate the equivalent current profile in this case:

- The acceleration data is passed through a low-pass filter to filter-in the low frequencies (<0.04 Hz) which are associated with the vessel motion.
- The velocities in X and Y directions at each logger are derived by the integration of accelerations in X and Y respectively.
- The derived velocities are detrended after the integration to keep the dynamic part.
- The resultant of the velocities in X and Y is obtained. The standard deviation of the resultant velocity time series is found out at each logger and used in equation (2) to obtain the equivalent current profile.

It should be mentioned that the vessel and the riser are moving/vibrating in both X and Y directions. It is also hard to figure out the correlation among different sensors, partly because of the misalignment of the sensors. Therefore, the resultant of the velocities from both directions are taken when estimating the equivalent current profile, which should be on the conservative side which helps to negate any errors in the data due to the possible logger misalignments.

In Figure 8, the obtained equivalent current profile due to the motion of the vessel is compared with the ocean current. It is seen that the ocean current velocities are very low compared with the equivalent velocities from the vessel motion. The ocean current has a maximum value of 0.2 m/s, whereas the current speeds in the equivalent current profile are much higher owing to the higher velocities of the vessel's lateral motions. At the top of the riser, the velocity is as high as 0.95 m/s. The bottom part of the riser experiences a higher velocity due to the high accelerations.

Figure 9 shows the resultant KC number distribution for the drilling riser in this case. KC number is defined as follows:

$$KC(z) = \frac{2\pi A_n(z)}{D} \tag{3}$$

where $A_n(z)$ is the resultant displacement of the riser at a node and D is the diameter of the riser. The displacements are obtained by integrating the velocities. It is seen that the KC numbers are high for this case owing to the high amplitudes of vessel drifts.

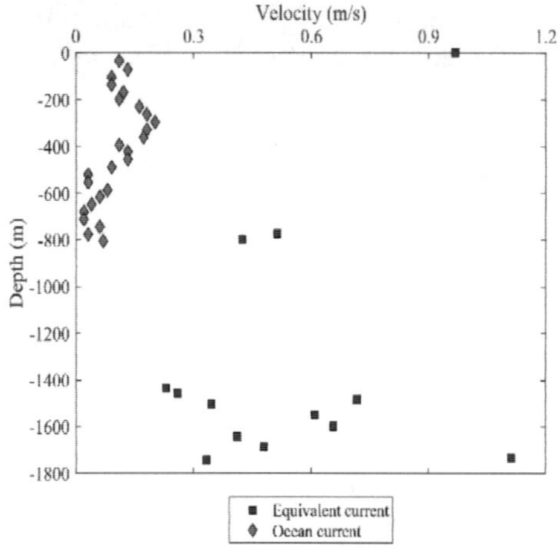

Figure 8. Generated equivalent current profile and the ocean current profile.

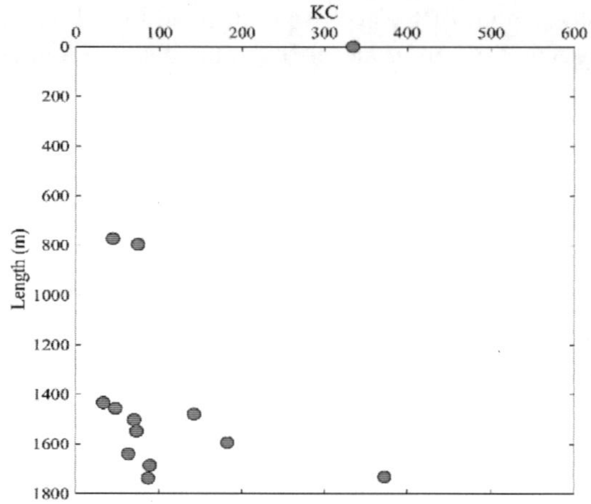

Figure 9. Resultant KC number distribution along the riser.

3.2. Numerical analysis

Both the current profiles are used as input in VIVANA to predict the VIV response frequency. Since the ocean current data is not available below a depth of 800 m, the profile is extrapolated to zero velocity at the bottom. A constant Strouhal number of 0.2 is used in the analysis. Figure 10 shows the comparison of the excitation frequencies along the riser due to the ocean current and the equivalent current due to vessel motion. This, in turn, is compared with the averaged spectrum of acceleration in X direction from all the loggers. It can be seen that the response frequency due to ocean current lies in a low range of 0-0.04 Hz, whereas that due to the equivalent current lie in a range of 0.06-0.11 Hz. The latter frequency range is closer to the peak of the observed acceleration spectrum. This is an indication that the equivalent current could excite the VIV. Although not dominant, a smaller peak is also visible at around 0.04 Hz in the spectrum.

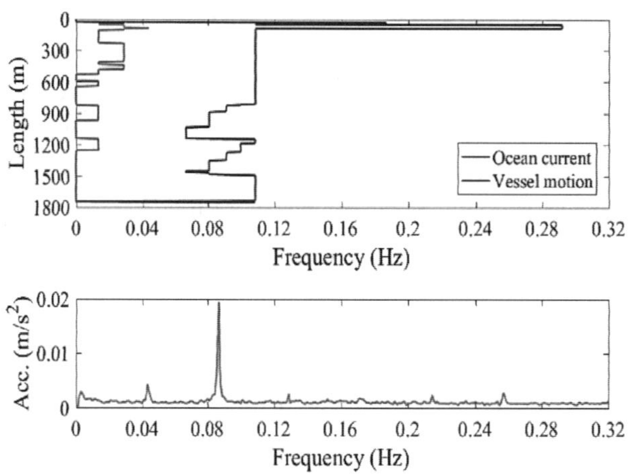

Figure 10. Comparison of excitation frequencies from VIVANA analysis with averaged acceleration spectrum in X direction.

3.3. Major uncertainties in the analysis

The complexity of the measurements and lack of information makes it difficult to interpret the data and ascertain whether the observed VIV is due to ocean current or vessel motion. Some of the uncertainties associated with the full-scale measurements and analysis are presented in this section.

A major uncertainty here is in the identification of the vessel motion frequency. The frequency resolution of the measurements is 0.001 Hz. Due to lack of vessel data or mooring details, it was difficult to ascertain the reasonableness of the obtained vessel motion frequencies. Also, the fact that the duration of measurements is only 15 mins adds to uncertainty in the obtained results. Prior works by Tognarelli et al. [13] and Thethi et al. [12] based on riser monitoring data were focused only on finding the response in the VIV frequency range. Typically, the accelerometers are not accurate at low frequencies, but the vessel displacements obtained by integration is within the range of typical vessel drifts.

The response obtained from the measurements is at a frequency of 0.0866 Hz. This falls in the typical wave frequency range. Information regarding the waves is not available and hence it was not possible to determine the effect of waves on the vessel and the riser.

The alignment of the axes of the loggers with the global axes is not guaranteed and hence there is uncertainty in the data. Also, it could be possible that the measurements are contaminated by gravity due to the bending motion of the riser.

The ocean current velocities are available only up to a depth of 800 m and had to be interpolated to zero at the bottom. The effect of ocean current on the vessel motion-induced VIV has not been considered here owing to the large difference in speeds between it and the equivalent current. A relative speed is only negligibly more/less than the speed of the equivalent current, especially at greater depths. For other cases when the equivalent current and the ocean current are closer in amplitude, it is hard to draw the same conclusion. More controlled model test should be designed and performed in order to find the effect of the ocean current on vessel motion-induced VIV, especially when the vessel motion is irregular and three-dimensional as have been observed from the full-scale measurement

4. Conclusion and scope of future work

Full-scale measurement data from a drilling riser was analysed in this study. VIV was identified from the data. The ocean current and the equivalent current caused by the vessel motion were used to predict the VIV. The equivalent current was found to excite VIV with a response frequency closer to the observations than the ocean current, which points out to the possibility of occurrence of vessel motion-induced VIV. The uncertainties associated with the field measurements and analysis makes it difficult to ascertain the actual source of excitation.

There remains scope for future work where model tests could be designed and performed for irregular vessel motion cases to obtain a thorough understanding of the phenomenon involved.

Acknowledgments

This work has been partly funded by the Program Area for Research "Computational Methods in Offshore Technology (COTech)" at Faculty of Science and Technology, University of Stavanger, Norway. The authors thank the Faculty of Science and Technology for supporting this work.

References

[1] Willis N R T and Thethi K S 1999 Stride JIP: Steel risers in deepwater environments – progress summary *Proc. of Offshore Technology Conf. (Houston).*
[2] Gonzalez E C 2011 High frequency dynamic response of marine risers with application to flow induced vibration *Massachusetts Institute of Technology.*
[3] Cunff C, Biolley F and Damy G 2005 Experimental and numerical study of heave- induced lateral motion (HILM) *Proc. of 24th Int. Conf. on Ocean, Offshore and Arctic Engineering (Halkidiki).*

[4] Rateiro F, Goncalves R T, Pesce C P, Fujarra A L C, Franzini G R and Mendes P 2013 A model scale experimental investigation on vortex-self induced vibrations (VSIV) of catenary risers *Proc. of 32nd Int. Conf. on Ocean, Offshore and Arctic Engineering (Nantes).*

[5] Kwon Y J, Kim H J and Jung D H 2015 A study for forced oscillation experiment for OTEC riser under current *Proc. of 25th Int. Ocean and Polar Engineering Conf. (Kona).*

[6] Wang J, Fu S, Ong M C and Li H 2016 Experimental investigation on vortex- induced vibration of a free-hanging riser under vessel motion *Proc. of 35th Int. Conf. on Ocean, Offshore and Arctic Engineering (Busan).*

[7] Wang J, Joseph R S, Ong M C and Jakobsen J B 2017 Numerical investigation on vessel motion-induced VIV for a free hanging riser under small Keulegan–Carpenter numbers *Proc. of 36th Int. Conf. on Ocean, Offshore and Arctic Engineering (Trondheim).*

[8] Sumer M B and Fredsøe J 2006 *Hydrodynamics Around Cylindrical Structures* (World Scientific Publishing Company).

[9] Wu J, Lie H, Larsen C M and Baarholm R J 2015 An Empirical Heave Induced VIV Prediction Model *Proc. of 34th Int. Conf. on Ocean, Offshore and Arctic Engineering (St. John's)* .

[10] BP and 2H Offshore 2008 VIV benchmarking repository–drilling riser without VIV suppression Data P*ackage Technical Note.*

[11] Natarajan S, Howells H, Deka D and Walters D 2006 Optimization of sensor placement to capture riser VIV response *Proc. of 25th Int. Conf. on Ocean, Offshore and Arctic Engineering (Hamburg).*

[12] Thethi R, Howells H, Natarajan S and Bridge C 2005 A fatigue monitoring strategy and implementation on a deepwater top tensioned riser *Proc. of 2005 Offshore Technology Conf. (Houston).*

[13] Tognarelli M A, Taggart S and Campbell M 2008 Actual VIV fatigue response of full scaledrilling risers: with and without suppression devices *Proc. of 27th Int. Conf. on Ocean, Offshore and Arctic Engineering (Estoril).*

[14] Wang J, Jaiman R K, Adaikalaraj P F, Shen L, Tan S B and Wang W 2016 Vortex-induced vibration of a free-hanging riser under irregular vessel motion *Proc. of 35th Int. Conf. on Ocean, Offshore and Arctic Engineering (Busan).*

Large eddy simulation of the tidal power plant deep green using the actuator line method

S T Fredriksson[1]*, G Broström[1], M Jansson[2], H Nilsson[3], and B Bergqvist[2]

[1]Department of Marine Sciences. University of Gothenburg, Gothenburg, Sweden
[2]Minesto AB, Gothenburg, Sweden
[3]Department of Mechanics and Maritime Sciences, Chalmers University of Technology, Gothenburg, Sweden

*Contact author: sam.fredriksson@gu.se

Abstract. Tidal energy has the potential to provide a substantial part of the sustainable electric power generation. The tidal power plant developed by Minesto, called Deep Green, is a novel technology using a 'flying' kite with an attached turbine, moving at a speed several times higher than the mean flow. Multiple Deep Green power plants will eventually form arrays, which require knowledge of both flow interactions between individual devices and how the array influences the surrounding environment. The present study uses large eddy simulations (LES) and an actuator line model (ALM) to analyze the oscillating turbulent boundary layer flow in tidal currents without and with a Deep Green power plant. We present the modeling technique and preliminary results so far.

1. Introduction

There are several new technologies emerging for extracting power from tidal currents. They span from turbines mounted on the bottom to devices that operates in mid-depth or at the surface [1]. The fact that the power plants will be mounted in regions with strong tidal currents implies that robust design of the equipment becomes an important issue. The tidal current in itself is quite well-known and easy to observe, but the small-scale turbulent fluctuations are less known and more demanding to observe. One way to describe the turbulence fields and their impact on the structure is to use computational fluid dynamic modelling.

There are some studies that have targeted the turbulence characteristics of tidal currents [2-4]. There are also some studies focusing on how bottom mounted turbines operate in a tidal flow [5-7]. The main part of the studies have focused on turbulence intensity and length scale of the turbulent eddies, but other quantities such as structure functions, probability density functions, intermittency, coherent turbulence kinetic energy, anisotropy invariants, and a scalar measure of anisotropy to characterize the turbulence have been suggested as well [8].

Power plants reduce the velocity behind the plants, creating a wake in the flow field. The wakes for atmospheric wind mills and wind farms have been extensively studied [9, 10]. It has been found that the return to normal flow conditions, e.g. turbulent fluctuations, depends on e.g. how rough the ground is and that it is faster over rough land than over smooth seas. For land-based wind farms it has become evident that they increase the mixing in the lower atmosphere [11, 12]. For offshore wind farms it has been found that the wind wake may force local up- and down-welling [13]. It can be argued that horizontal mixing is more important than vertical mixing in shallow water conditions as compared to the atmospheric counterpart. This fundamental difference between the (shallow and coastal) ocean and the atmosphere probably increases the importance of how tidal power plants should be placed in relation to each other and how they affect the surrounding area, as compared to wind farms. Today there are few studies, e.g. [14-17], of those aspects of the energy extraction from ocean currents.

1.1. Deep Green technology

The Deep Green power plant is a novel marine energy technology that produces electricity from tidal and/or ocean currents. The main components of the Deep Green, shown in Figure 1, are the wing, the axial turbine, the nacelle that comprises the generator and power electronics, and the rudders. The Deep Green is attached to a foundation on the seabed via three struts and a long tether. The control system steers the power plant in a predefined trajectory. In its current design, the length of the wing is 12 meters and the rated power of the generator is 500 kW. Multiple Deep Green devices will be placed together to form an array.

The Deep Green technology has the same working principle as a wind kite. The wing hydrodynamics enables the power plant to accelerate several times the speed of the flow, moving almost perpendicular to the flow. Accordingly, the Deep Green power plant can operate in low-flow stream and use less material in construction compared to other similar technologies, in relation to the installed capacity.

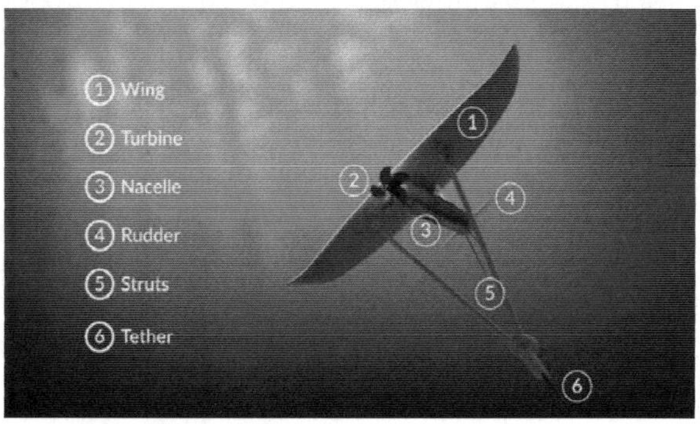

Figure 1. Illustration of the Deep Green device.

Figure 2. Sketch of the trajectory of the Deep Green.

The Deep Green is preprogrammed to fly in an 8-shaped trajectory, see Figure 2, while the water is pressed through the rotating turbine. A generator in the nacelle converts the turbine rotation to electricity which is transferred via the tether and the internal network to the grid.

Minesto is using a combination of in-house, open source and commercially available computational codes to perform CFD, rigid body and multi body simulations in the design work of the Deep Green technology. The present study gives an opportunity to better understand the underlying energy resource, i.e. the tidal flow of water. With a more realistic turbulent flow as input to the simulations, the dimensioning of components can be made with better accuracy. Studies of how the Deep Green device affects the turbulent flow downstream will be a valuable input to coming array designs.

1.2. The present study

The present paper consists of a methods section where the set-up of the simulations and the actuator line method is described. This is followed by a results section that focuses on the preliminary results, comparing the flow fields and turbulent structures with and without the Deep Green. In the last section we present our conclusions and future work.

2. Methods

This section presents the work process and the numerical set-up, followed by a description of the actuator line model (ALM).

2.1. Work process

The site in the presents study is approximately 80 m deep with a maximum temporal peak of the vertically averaged flow of about 1.6-2.4 ms^{-1} depending on the tidal acceleration variability. Initial studies show that the tidal flow is a strongly forced accelerating current, and that the turbulence fields are not close to be in a quasi-stationary state with the mean flow, but rather depends on the phase in the tidal cycles.

The present study is therefore divided in three steps:

1. Full tidal cycle large eddy simulations (LES) to both give information of the free turbulence characteristics and to produce initial conditions for precursor simulations. This is done without Deep Green, with horizontally cyclic boundary conditions and a time-varying force term that drives the flow.
2. Precursor LES to produce initial flow conditions and inlet boundary conditions for Deep Green simulations. This is done without Deep Green, with initial conditions from step 1 and a time-varying source term that drives the flow. The simulation is started at the instance in time, t_0, where the flow corresponds to a prior dynamics analysis in Dymola.
3. LES using precursor initial and inlet boundary conditions, including Deep Green where the wing and the turbine are modeled using an actuator line model technique. The precursor inlet boundary condition here determines the flow rate by setting the velocity, the sub-grid scale turbulent kinetic energy, and the turbulent viscosity. A Neumann boundary condition is used at the outlet.

Figure 3 and Table 1 show the computational domain and details about the three steps. The tidal cycle simulation in step 1 is performed using a pseudo-spectral method (pseudo-spectral in the horizontal direction and finite differences in the vertical direction). The method is available in a code herein referred to as NCAR-LES which targets geophysical turbulence [18-20]. The pseudo-spectral method allows accurate and fast execution but requires horizontally cyclic conditions. Initial studies in the present work show that the length scale of the turbulence is on the order of 100-150 m in the flow direction. In order to avoid locking of the large-scale turbulence structures due to the horizontally cyclic conditions we use a model domain spanning some 10 times this value in step 1 [21].

The turbulent structures are resolved down to a fraction of the length of the Deep Green structure, which has a wing span of 12 m. The aim of the precursor study (step 2) is to deliver inlet boundary conditions for the Deep Green simulation (step 3). It is performed using the OpenFOAM CFD solver and is an intermediate step that has the purpose of preparing step 3. OpenFOAM is a general-purpose finite volume solver to which an actuator line model can be attached [14, 22, 23]. It is not as efficient as NCAR-LES for the simulations in step 1, but it does not require cyclic conditions and it can be used for the arbitrary geometries that will be of interest in future work. In step 2 the results from step 1 at the time instance t_0 is used as the initial conditions, see Figure 6. t_0 is in the accelerating tidal phase, starting just before the peak flow rate. It is chosen at the time when the vertically averaged velocity over the depth of the trajectory of the Deep Green corresponds to the plug flow of 1.6 ms^{-1} that was used in a previous Dymola simulation that was performed to find e.g. the trajectory path. The step 2 simulation is driven by the time-varying tidal body force for 400 s while sampling the fluctuating velocity and sub-grid scale properties at the inlet boundary for every time step of the simulation. These temporally and spatially varying fields are then used as upstream boundary conditions for the step 3 simulation. The step 3 simulation starts at the same instance in time, t_0 , as that of step 2, using the same initial conditions as in step 2 but the sampled time-varying velocity, sub-grid scale turbulent kinetic energy, and the turbulent viscosity of step 2 are applied as the upstream boundary conditions.

The Deep Green plant is in step 3 modeled using the actuator line method, in which source terms are added to the momentum equations. Those source terms are based on separate studies of lift and drag coefficients, and the flight path is taken from a previous simulation using the Dymola software. At this stage it is thus only the flow that is affected by Deep Green, and not vice versa. The previous Dymola simulations are the opposite – the flight path is determined by the forces from the flow, but for a constant plug flow. It should be noted that the simulations in steps 2 and 3 are only performed for a short duration of real time, in the accelerating phase of the tidal flow, close to the maximum velocity. Further, the domain size is smaller than in step 1, however using the same mesh resolution. It is nevertheless assumed that the turbulent flow field of step 3 is representative enough for the purpose of that simulation at this stage. All such aspects will be addressed in future work. We have used the same time step in the step 2 and step 3 simulations.

2.2. Numerical set-up
The filtered Navier-Stokes equations used in the large eddy simulations are written as

$$\frac{\partial \tilde{u}_i}{\partial x_i} = 0, \tag{1}$$

$$\frac{\partial \tilde{u}_i}{\partial t} + \frac{\partial \tilde{u}_i \tilde{u}_j}{\partial x_j} = -\frac{1}{\rho}\frac{\partial \tilde{p}}{\partial x_i} + \nu \frac{\partial^2 \tilde{u}_i}{\partial x_j \partial x_j} - \frac{\partial \tau_{ij}}{\partial x_j} + \frac{F_{T,i}}{\rho} + \frac{F_{DG,i}}{\rho} \tag{2}$$

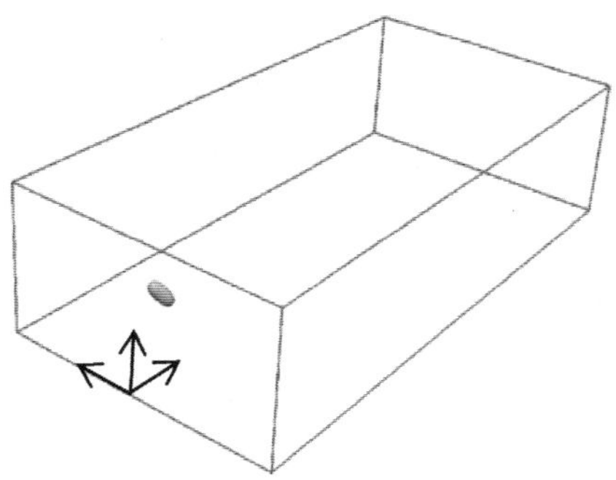

Figure 3. Sketch of computational domain where the Deep Green is indicated at the center of its trajectory, close to the upstream boundary.

Table 1. Code, computational domain, mesh resolution and forces

Step	Description	Code	Domain[1]	Resolution[2]	Force
1	Tidal cycle	NCAR-LES	16, 4, 1	2048, 512, 128	F_T
2	Precursor	OpenFOAM	4, 2, 1	512, 256, 128	F_T
3	Deep Green	OpenFOAM	4, 2, 1	512, 256, 128	F_{DG}

[1] Size relative to depth, $H = 80\ m$, in x, y, and z directions.

[2] In x, y, and z directions (equidistant).

where $\tilde{u}$ and $\tilde{p}$ stand for the filtered velocity and pressure respectively and ρ is the density assumed to be constant. t is time, x_i is the $x, y,$ and z directions. τ_{ij} is the sub-grid scale stress tensor modeled via the one-equation eddy-viscosity concept, using a transport equation for the sub-grid scale kinetic energy and a local length scale [24, 25]. $F_{T,i}$ and $F_{DG,i}$ are body forces to mimic the tidal forcing (T) and the actuator line model describing the Deep Green effect (DG), respectively. We have in equation (2) not included the effects of buoyancy, which can enhance or depress turbulence depending on its sign, as well as the effects of Coriolis forces caused by the rotation of the earth. The body force

$$F_{T,i} = f_T \cos(\omega t)\mathbf{e}_x \tag{3}$$

is activated during the tidal cycle and precursor simulations (steps 1 and 2) and deactivated during the Deep Green simulations (step 3). Here f_T is the amplitude, $\omega = 2\pi/t_0$ with the period t_0 set to 12 hours, and $\mathbf{e}_x$ denotes the unit vector in the x direction (general flow direction). The force from Deep Green, $F_{DG,i}$, (further discussed below) is only activated during the Deep Green simulations (step 3).

We apply cyclic boundary conditions in both horizontal directions for the tidal cycle and precursor simulation (steps 1 and 2), but only in the y-direction for the Deep Green simulation (step 3). In step 3, Neumann and Dirichlet boundary conditions are applied for the pressure at the upstream ($x = 0$) and downstream boundary ($x = 4H$), respectively. For the velocity and sub-grid scale properties we apply a temporal and spatial varying boundary condition, given by the precursor simulation, at the upstream boundary, and Neumann boundary condition at the downstream boundary. The bottom boundary condition is given as a rough wall condition with a bottom roughness parameter z_0=0.01 m in all simulations.

2.3. Actuator Line Model

The Deep Green plant is taken into account in step 3 using the actuator line model [26], where the body forces, see Figure 4, that arise due to the Deep Green are determined using a blade-element approach. An actuator line model is used since it is not possible to resolve the components of Deep Green and the resulting small-scale structures in a domain that is large enough to also cover the largest scales. Here the wing and the axial turbine is modeled. The wing is discretized in the spanwise direction using 10 elements, in combination with two-dimensional airfoil characteristics. The turbine is modeled with one element with a small size compared to the mesh size resulting in a source that can be considered to be a point source. In the following the word "foil" will be used as the general term for the modeling of both the wing and the turbine. Denoting the velocity (in vector notation) of a foil as $\mathbf{v}_f$ and the flow field velocity at the leading edge of the foil as $\mathbf{u}_{LE}$, the local flow velocity relative to the foil is given by

$$\mathbf{u}_{rel} = \mathbf{u}_{LE} - \mathbf{v}_f. \tag{4}$$

The angle of attack is found as

$$\alpha = sin^{-1}\left(\mathbf{e}_{wingPlaneNormal} \cdot \mathbf{e}_{rel}\right) \tag{5}$$

where $\mathbf{e}_{wingPlaneNormal}$ and $\mathbf{e}_{rel}$ denote the unit vector along the normal to the wing plane defined by the chord line and the span direction of the foil and the unit vector along the relative velocity, respectively, and · denotes the dot product. The three-dimensional actuator line force per spanwise unit length is calculated as

$$F_{foil,i} = \frac{1}{2}\rho U_{rel}^2 c(C_L\mathbf{e}_L + C_D\mathbf{e}_D) \tag{6}$$

where c is the chord length, $C_L = C_L(\alpha, Re)$ and $C_D = C_D(\alpha, Re)$ are the lift and drag coefficients respectively given in lookup tables. $\mathbf{e}_L$ and $\mathbf{e}_D$ denote the unit vectors in the lift and drag directions.

$\mathbf{v}_f$ and the position, and the orientation of the foil are given as an input from a prior simulation performed in the software Dymola that models the complete Deep Green including the control system, but for a plug flow only. This input is given and the forces are projected back to the flow at a position $1/4$ chord downstream the leading edge along the chord. We have chosen to have a distance between this point and the sampling point of the field velocity (leading edge) to decrease the influence of the source term on the sampled velocity. In order to avoid numerical oscillations due to steep gradients, the body force (source term) is projected by means of a spherical Gaussian function as [26]

$$F_{DG,i}(\mathbf{r}) = \frac{F_{foil,i}}{\varepsilon^3\pi^{3/2}} exp\left[-\left(\frac{|\mathbf{r}|}{\varepsilon}\right)^2\right] \tag{7}$$

where $\mathbf{r}$ is the vector between the cell where the source is to be applied and the actuator point, and ε controls the Gaussian width. Here the Gaussian widths are individually determined for each actuator element at each time step as the largest value out of three alternative formulations based on the 1) lift as a function of the chord length as $\varepsilon_{chord} = c_{chord}C$ where $c_{chord} = 1/4$, 2) mesh size as $\varepsilon_{mesh} = c_{mesh}2\Delta x$ where $c_{mesh} = 2$ and the cell length $\Delta x = \sqrt[3]{V_{cell}}$ is related to the cell volume, and 3) drag as $\varepsilon_{drag} = c_{drag}C_dC/2$ [23]. The thrust of the turbine is modeled similar to (6) using the thrust coefficient C_T instead of C_D, applying an element spanwidth smaller than the local mesh size.
The lift and drag coefficients have been determined by steady state analyses for a complete wing while varying the angle of attack. The coefficients are determined for ten sections as presented in Figure 5.

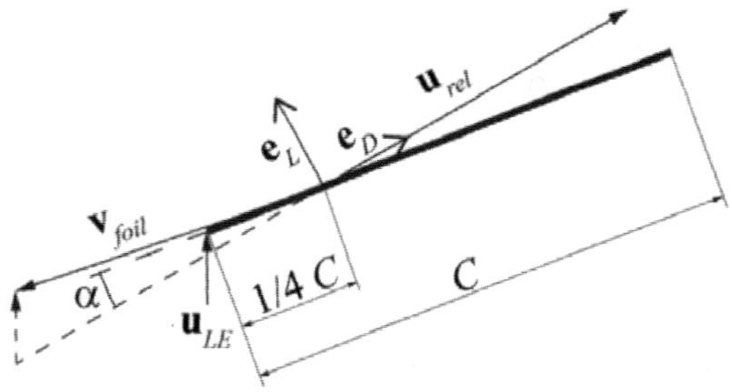

Figure 4. Sketch of a foil with chord length C.

Figure 5. Sections of the wing where the forces were determined in the steady state analysis.

Here the $k-\omega$ turbulence model was used and the flow velocity was $12\,ms^{-1}$. The Reynolds number, $Re = U_{rel}c/\nu$, is fairly stable during the trajectory of Deep Green, and it is known [27] that the lift and drag coefficients are fairly insensitive to Re for high enough Re. Hence, the possibility to express the lift and drag coefficients as a function of Re has not been used in this study. When two-dimensional airfoil data is used, it needs in general to be corrected for three-dimensional effects such as infinite span width and skew attack velocity in relation to the chord direction [26]. No correction for a finite-span is however needed in this study since the lift and drag coefficients have been determined using the actual three-dimensional finite wing. A correction due to skew inflow is a subject for future improvement once the data is available.

This work is based on the libraries of turbinesFoam [23, 28], which is distributed as a stand-alone user-contributed module for OpenFOAM. The libraries allow the ALM to be used through the fvOptions functionality in OpenFOAM. Our work has generalize the ALM to make it possible to prescribe an arbitrary path of the foils, not restricted to circular as typical for axial wind and tidal turbines [14]. In turbinesFoam there are a few features that we have chosen not to use or to investigate later on. These are

1) dynamic stall which isn't used since there is no risk for stall with the attack angles used in this study,
2) added mass which is used in conjunction to the dynamic stall correction,
3) flow curvature effect that arises due to a varying angle of attack along the chord direction which has not been taken into account in the present study since it is believed to only have a minor effect of the general behavior of the flow downstream of the foil following the prescribed path, and
4) end effects where we just started some sensibility studies that so far points in the direction of only minor influence of the downstream field.

IOP Conf. Series: Materials Science and Engineering **276** (2017) 012014 doi:10.1088/1757-899X/276/1/012014

3. Results

The main emphasis here is to present some typical results for the Deep Green simulation. Studies of the domain size, resolution sensitivity and code comparison will therefore be presented elsewhere. We start, however, by briefly presenting the results for the tidally driven simulation since this simulation gives the initial fields for the precursor study, which then in turn gives the initial conditions and inlet boundary conditions for the Deep Green simulation.

3.1. Tidal forcing

The volume averaged velocity during a tidal cycle simulation is shown as a function of time in Figure 6. The maximum volume averaged velocity at the second tidal peak is about 2.1 ms^{-1}. This specific case has not been run longer than shown in the figure, but simulations with a smaller domain with the same resolution show that the velocity at the third maximum is close to that at the second maximum, which shows that the second maximum is close to "fully developed". In a friction-free flow we would have a phase shift of 3 hours. Because of the bottom friction the phase shift becomes about 2 hours. Furthermore, we can see that the current response is skewed from the "non-frictional" sinus function.

The velocity profile, derived from the horizontally averaged velocity as a function of depth at time, t_0, is shown in Figure 7. We normalize the velocity with the volume averaged velocity $\langle \tilde{u}_1 \rangle_{z,trajectory} = 1.58\, ms^{-1}$, which is averaged over the depths that enclose the trajectory of the wing. This instance in time is chosen for the analysis since it is used for the initialization of the Deep Green simulation. The mean friction velocity at the bottom for this time is $0.081\, ms^{-1}$. There is a clear vertical gradient in the horizontally averaged velocity and it resembles to some extent the velocity profiles of a logarithmic layer. This shows that the bottom friction is an important process in the simulations. The velocity gradient over the depths that enclose the wing trajectory is about $0.1\, ms^{-1}$.

Snapshots of the flow field at a given depth are shown in Figure 8. This is a small part of the full domain that stretches $4H$ in the y direction and $16H$ in the x direction. We see the elongated scales in the x direction and that the turbulent fluctuations are stronger in $\tilde{u}_1$ than for $\tilde{u}_2$and $\tilde{u}_3$. In the w-velocity we clearly see that much of the vertical transport seems to be in bursts with high vertical velocities in a small part of the domain.

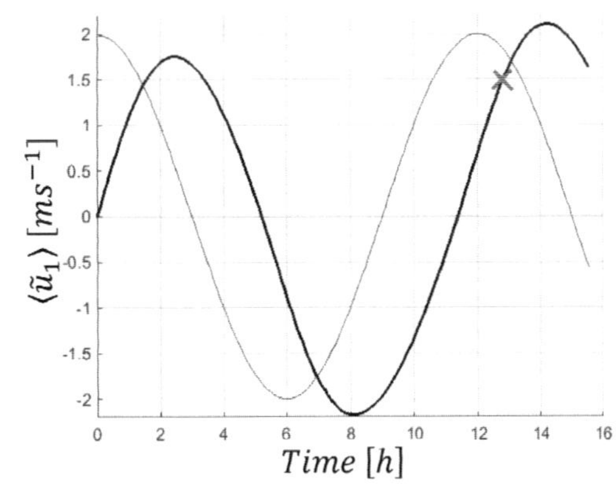

Figure 6. Black shows the volume averaged velocity from the cosines type of body forcing used to mimic the tidal forcing at a local position. The red "x" is the time, t_0, when the precursor and Deep Green simulation starts (here the volume avereged velocity is 1.47 ms^{-1}, while the volume averaged velocity for the vertical region the wing spans is 1.58 ms^{-1}). Blue shows the form of the tidal body forcing. Note that the body forcing has another dimension and magnitude, and that it is included here for comparison of the shapes and phase-shift.

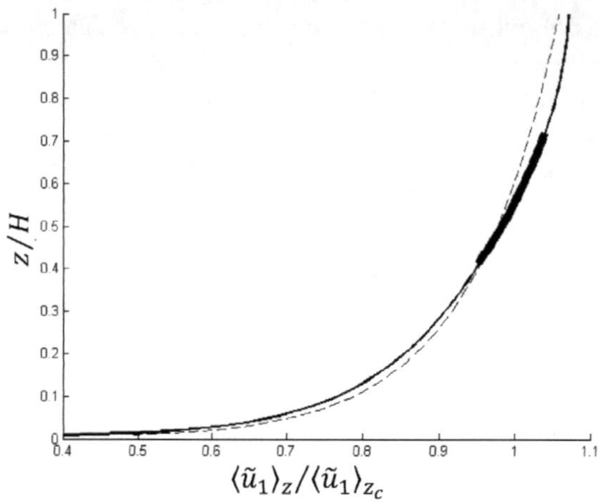

Figure 7. The horizontally averaged velocity as a function of the depth (thin line). The region of the wing trajectory is indicated by a thicker line. The dotted line shows the velocity profile of a logarithmic boundary layer flow.

3.2. Deep Green
A snap-shot of the flow field from a Deep Green simulation is shown in Figure 9. This is 300 s after the start of the Deep Green simulation, t_0, which means that approximately 15 full trajectories have passed since the start and that quasi-stationary conditions prevails. The vortices are visualized by isosurfaces of the second invariant of the velocity gradient tensor. This quantity is positive in regions of rotating flow, such as vortices, and is sometimes referred to as the "Q-criterion". The Q criterion is good for visualizing the vortices but the number given for the isosurfaces affects how long downstream the vortices seem to persist before breakdown and is therefore misleading. The strength of the vortices can, however, be compared to the vortices produced by the rough bottom and it is seen that the tip vortices are both stronger and longer than the bottom induced ones and persist fairly long. Another way of discussing the downstream effect is to study the wake in the velocity fields. It is customary to relate the downstream conditions to the distance normalized with a turbine diameter.

It is, however, for the Deep Green most likely the wing and its trajectory that dominate the downstream conditions rather than its turbine and therefore some other length scales to be considered are 1) span width S of the wing, 2) full width D_y of the trajectory, and 3) full height $D_z$of the trajectory. Here the coordinates of the position $1/4$ of the center line chord is used to determine D_y and D_z although the tips of the wing sometimes are further away during the trajectory. In Figure 9, we present slices of velocity at 1, 2, 3, and 4 times D_y downstream of the point (x_c, y_c, z_c) where the path trajectory crosses its way. It can be seen that the velocity field is affected at all these distances. Since $4D_y \approx 12D_z \approx 21S$ it is most likely the D_y rather than D_z or S that determines how long downstream the flow is affected. Future simulations with longer computational domain will show how long the actual flow field impact is.

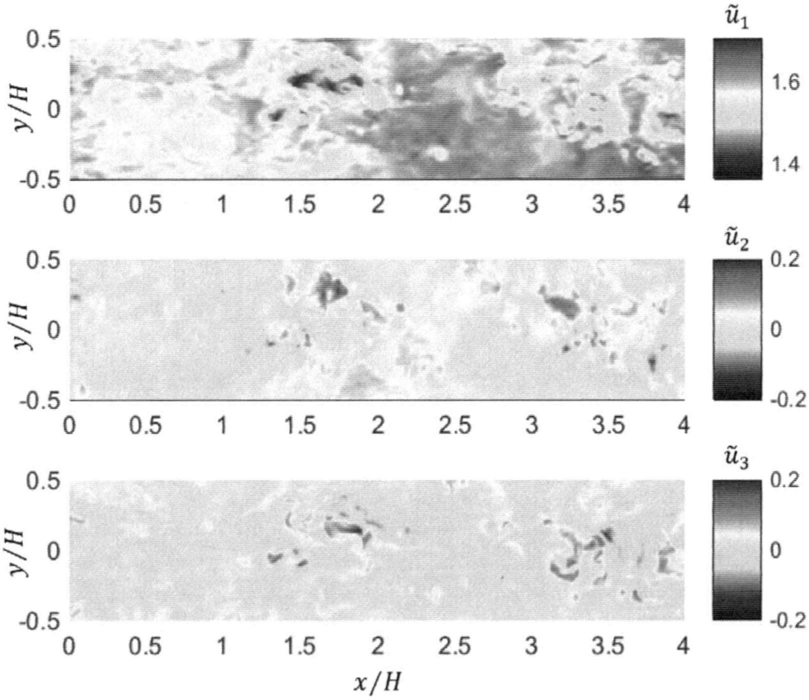

Figure 8. Instantaneous flow fields at the xy-plane at the center of the trajectory of Deep Green i.e., $z_c = 47.3\ m$ (from bottom), at the time t_0. Note the different color scales in the figures.

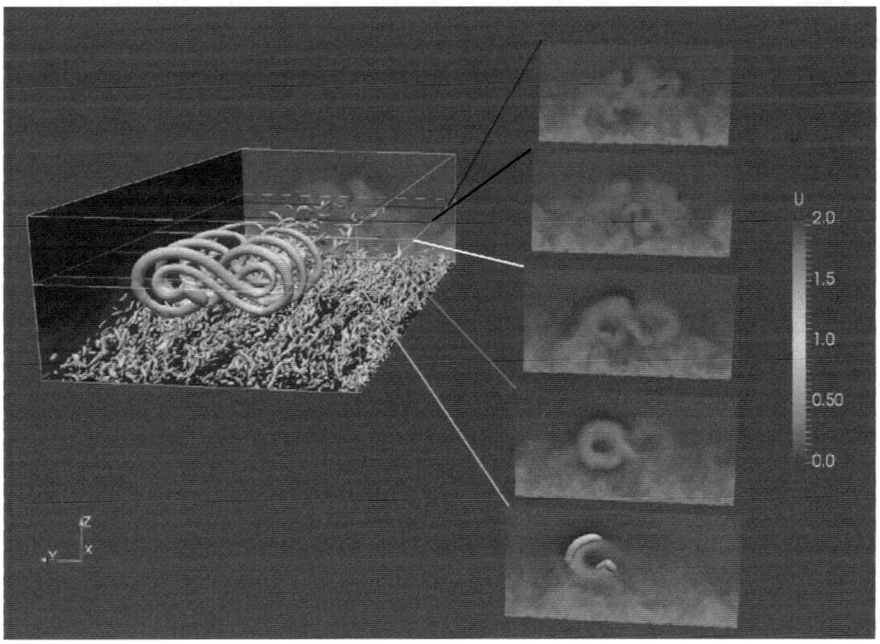

Figure 9. Instantaneous velocity fields 300 s (15 trajectories) after t_0. Velocities given at domain boundaries at $y = H, x = 4H$, and at yz-planes at $x = x_c, x_c + D_y, x_c + 2D_y, x_c + 3D$, and $x_c + 4D_y$. The grey isosurfaces mark a positive value of the second invariant of the velocity-gradient tensor which indicate vortices. The position of the Deep Green is visualized by the green isosurface of the force field.

Figure 10. Instantaneous normalized velocity fields $\underline{\tilde{u}_1}$, 300 s (15 trajectories) after t_0. The velocitiy field is given at the xy-plane at $z = z_c$ and the xz-plane at the y position where the Deep Green trajectory is closest to the surface.

The normalized velocity fields with Deep Green are shown in Figure 10 for the same instance as in 0 9. The horizontal plane is here chosen as $z = z_c$ (which is approximately 2 m above the mean of the max and min of the vertical position of the wing). Here the deviation between the instantaneous velocities and the spatially averaged velocity has been normalized with the spatially averaged velocity as $\underline{\tilde{u}_1} = \left(\tilde{u}_1 - \langle\tilde{u}_1\rangle_{z_c}\right)/\langle\tilde{u}_1\rangle_{z_c}$. It can be seen that the maximum normalized velocity deviations are approximately 40 %. It can also be seen that the deviations are fairly large even at $4D_y$ downstream of the trajectory crossing point.

The spatial velocity gradients are an important measure for the control system and the applied fluid forces on the Deep Green. One way of estimating these gradients is to use the magnitude of the vorticity. The normalized magnitude of vorticity $\underline{\tilde{\Omega}} = (\tilde{\Omega} - \langle\tilde{\Omega}\rangle_{z_c})/\langle\tilde{\Omega}\rangle_{z_c}$ is presented in Figure 11. It can be seen that $\tilde{\Omega}$ is affected all the domain (more than $4D_y$ downstream of x_c).

The time-averaged streamwise velocity profiles shown in Figure 12 are useful for the understanding of the wake evolution downstream of the Deep Green. These are taken in the centerline, and downstream of the Deep Green at various locations, similar to Figure 9. It is seen in Figure 12a) that the wake is asymmetric in the vertical direction, which is due to the vertical shear in the main flow. It can also be seen in the plot in the horizontal direction, see Figure 12b) that the wake persists further downstream (at least $4D_y$) for the off-center positions close to the xz-planes at the y position where the Deep Green trajectory is closest to the surface and bottom.

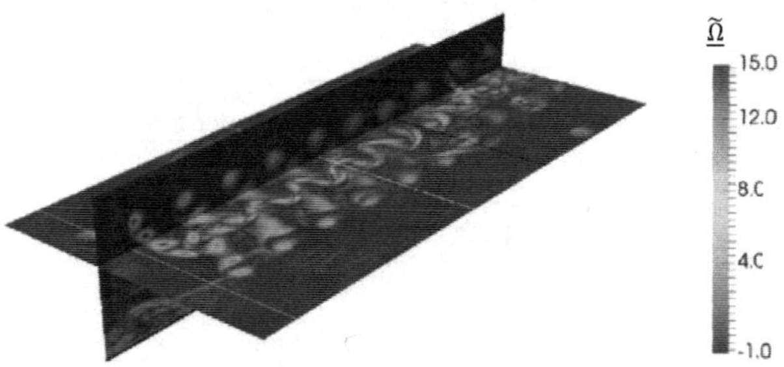

Figure 11. The magnitude of the vorticity for the instance 300 s after t_0. The vorticity field is given at the xy-plane at $z = z_c$ and the xz-plane at the y position where the Deep Green trajectory is closest to the surface.

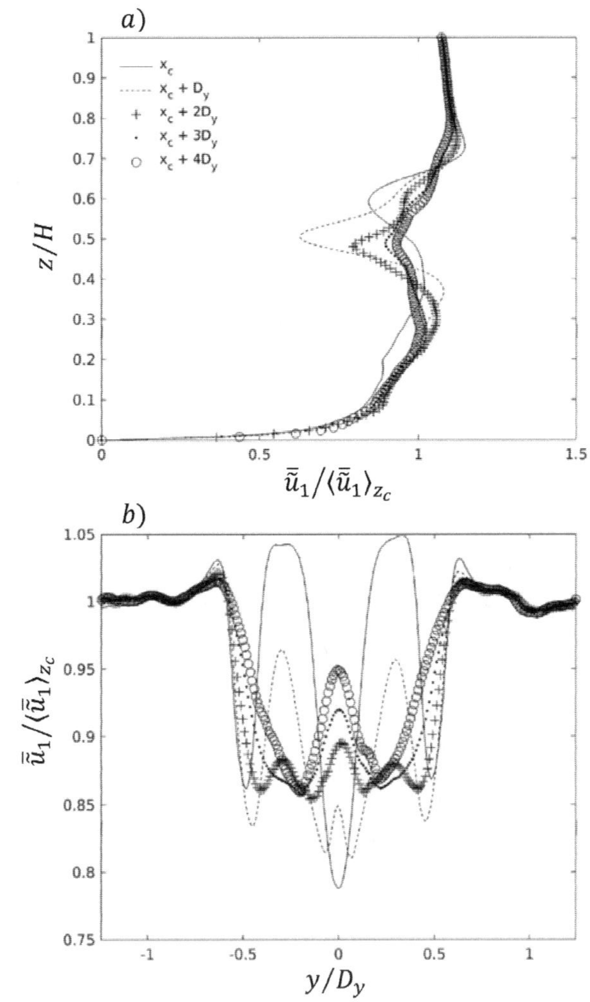

Figure 12. Velocity deficit comparison at locations downstream of the Deep Green trajectory center (x_c, y_c, z_c). Time averaged over 200-400 s after t_0. a) vertical profiles along the xz-plane at $y = y_c$. b) horizontal profiles along the yz-plane at $x = x_c$

4. Conclusions

The overall aim for this work is to find to what extent and how long downstream the Deep Green affect the flow field. This information is e.g., important to be able to optimize the packing of eventual arrays of power plants. The main focus of the present work is to design a numerical framework for studies of the interaction between the Deep Green tidal power plant and the tidal flow, under realistic conditions. A work process is designed, in which a the turbulence of the oscillating tidal flow is predicted using an efficient pseudo-spectral method, followed by finite volume simulations and the actuator line method to include the effects of the Deep Green power plant on the flow.

This ensures that the studies are performed in a realistic turbulent and oscillating tidal flow field. It is found that the numerical framework and work process are appropriate for the present studies. The numerically predicted undisturbed boundary layer (without the power plant) is comparable to the theory of the law of the wall for a rough bottom boundary condition. Initial studies in the present work show that the turbulence is dependent on the phase of the tidal cycle. The studies with the Deep Green power plant are performed during accelerating flow, close to the maximum flow. The results show that the tip vortices, their residues and the increased velocity fluctuations persist the complete computational domain in the streamwise direction. A preliminary analysis indicates that the trajectory width can be used as the characteristic length scale to estimate the affected distance, and the flow is significantly affected at least four trajectory widths downstream the power plant. This needs to be considered when optimizing and packing eventual arrays of power plants.

Some of the steps in the future are to 1) evaluate the need of a correction for lift and drag coefficients for skew inflow to the power plant wing, 2) evaluate the sensitivity to mesh density, 3) increase the computational domain in the streamwise direction in order to find how far downstream the flow is affected by the power plant.

Acknowledgement
This project is financed by the Swedish Energy Agency.

References

[1] Magagna D, Monfardini R, and Uihlein A 2017 JRC Ocean Energy Status Report 2016 Edition.
[2] Lewis M, Neill S, Robins P, Hashemi M, and Ward S *Renewable Energy*, **114** A, pp. 258-72.
[3] Mason-Jones A, O'doherty D, Morris C, O'doherty T, Byrne C, Prickett P, et al. 2012, *Energy*, **44**, pp. 820-29.
[4] Milne I A, Sharma R N, Flay R G, and Bickerton S 2013, *Phil. Trans. R. Soc.* A, **371**, pp. 20120196.
[5] Ahmed U, Apsley D, Afgan I, Stallard T and Stansby P 2017 *Renewable Energy*, **112**, pp. 235-46.
[6] Blackmore T, Myers L E and Bahaj A S 2016 *Int. J. Mar. Energy* **14**, pp. 1-26.
[7] Mason-Jones A, O'doherty D, Morris C and O'doherty T 2013 *Renewable Energy*, **52**, pp. 23-30.
[8] McCaffrey K, Fox-Kemper B, Hamlington P E and Thomson J 2015 *Renewable Energy*, **76**, pp. 441-53.
[9] Barthelmie R J and Jensen J 2010 *Wind Energy*, **13**, pp. 573-86.
[10] Barthelmie R J, Pryor S, Frandsen S T, Hansen K S, Schepers J, Rados K, et al., 2010 *J. Atmos. Oceanic Technol.*, **27**, pp. 1302-17.
[11] Roy S B 2011 *J. Wind Eng. Ind. Aerodyn.* **99**, pp. 491-8.
[12] Zhang W, Markfort C D and Porté-Agel F 2013 *Environ. Res. Lett.* **8**, p. 015002.
[13] Broström G 2008 *J. Mar. Syst.* **74**, pp. 585-91.
[14] Churchfield M J, Li Y and Moriarty P J 2013 *Philos. Trans. R. Soc.* A, **371**.
[15] Funke S W, Farrell P E, and Piggott M D 2014 *Renewable Energy*, **63**, pp. 658-73.
[16] Turnock S R, Phillips A B, Banks J and Nicholls-Lee R 2011 *Ocean Eng.*, **38**, pp. 1300-07.
[17] Malki R, Masters I, Williams A J and Nick Croft T 2014 *Renewable Energy*, **63**, pp. 46-54.
[18] McWilliams J C, Huckle E, Liang J-H and Sullivan P P 2012 *J. Phys. Oceanogr.* **42**, pp. 1793-816.
[19] Sullivan P P, McWilliams J C and Moeng C-H 1994 *Boundary Layer Meteorol.* **71**, pp. 247-76.
[20] Moeng C-H 1984 *J. Atmos. Sci.* **41**, pp. 2052-62.
[21] Fredriksson S T, Arneborg L, Nilsson H and Handler R A 2016 *J. Geophys. Res. C: Oceans*, **121**(10), pp. 7369–7389.
[22] Jha P K, Churchfield M J, Moriarty P J and Schmitz S 2014 *J. Sol. Energy Eng.* **136**, p. 031003.
[23] Bachant P, Goude A and Wosnik M 2016 *Actuator line modeling of vertical-axis turbines*, available at: https://arxiv.org/abs/1605.01449.
[24] Moeng C H and Sullivan P P2015 *Numerical models | Large-Eddy Simulation A2 - North, Gerald R, Encyclopedia of Atmospheric Sciences* (2nd Edition), J. Pyle and F. Zhang, Eds., Academic Press, Oxford, pp. 232-240.
[25] de Villiers E, 2007 *The potential of large eddy simulation for the modelling of wall bounded flows* Department of Mechanical Engineering, Imperial College London.
[26] Sorensen J N and Shen W Z 2002 *J. Fluids Eng.*, ASME Transactions, **124**, pp. 393-9.
[27] Abbott I H and von Doenhoff A E 1959, *Theory of wing sections, including a summary of airfoil data*, Courier Corporation.
[28] Bachant P, Goude A and Wosnik M 2016 *TurbinesFoam:* v0.0.7, available at https://zenodo.org/record/49422/export/dcite3#.WhNIFU2WzL8.

IOP Conf. Series: Materials Science and Engineering **276** (2017) 012015 doi:10.1088/1757-899X/276/1/012015

Fluid flow in steady and oscillatory lid-driven square cavities

[a]J Zhu, [a]L E Holmedal, [a]D Myrhaug, [a]H Wang

[a]*Dept. of Marine Technology, Norwegian University of Science and Technology*, Trondheim, Norway

E-mail: jianxun.zhu@ntnu.no

Abstract. This paper presents numerical simulations of steady and oscillatory lid-driven cavity flow at different Reynolds numbers with a fixed aspect ratio of 1:1. A projection method (P2 pressure correction method) is applied to solve the incompressible Navier-Stokes equations. The code is validated by comparison with published works of steady lid-driven flow at Re = 100, 400 and 1000. Oscillatory lid-driven cavity flow at different Reynolds numbers (100, 400 and 1000) at a fixed oscillation frequency has been investigated. It is observed that the oscillatory lid-driven cavity flow is substantially affected by the Reynolds number.

1. Introduction

Steady lid-driven cavity flow is one important benchmark for Navier-Stokes solvers due to its simple geometry and can also serve as a simplified model for industrial applications like e.g. wave-induced flow in sandpits (see e.g. [1]). Flow phenomena such as corner vortices and longitudinal vortices exist within the cavity. Flow structures in the cavity vary significantly at different Reynolds numbers $Re = Uh/\nu$ and aspect ratios $D = h/l$ (where ν, U, h and l are the kinematic viscosity of the fluid, the steady lid motion velocity, the depth and width of the cavity, respectively). Flow in a square cavity was investigated extensively by [2]. The predictions, limited up to $Re = 400$, show that there are three vortices existing in the cavity (one primary vortex occupying the central core and two other relatively small vortices located at the bottom corners of the cavity). Steady square lid-driven cavity flow at $Re \leq 10000$ was presented by [3] and [4]. The primary vortex and two corner vortices grow in strength as the Reynolds number increases while an upper left wall vortex is observed when the Reynolds number increases to 3200.

A few studies have been carried out for flow in a square cavity driven by an oscillating wall. The vortex dynamics have been investigated by [5] at different Reynolds numbers and Stokes numbers (here, the Stokes number is $St = \omega A^2/\nu$, where ω and A are the frequency and amplitude of the lid motion) for equal values ($Re = St$) ranging from 0 to 600. It is observed that one counter-rotating vortex pair and two corner vortices occur within the cavity during one cycle. Similar results have also been presented by [6], [7] and [8].

In the present paper, the steady lid-driven cavity flow at different Reynolds numbers has been investigated and compared with corresponding results obtained from [4]. Simulations of the oscillatory lid-driven cavity flow at a fixed oscillation frequency have been conducted at different Reynolds numbers. The results show good agreement with the previous data presented

IOP Conf. Series: Materials Science and Engineering **276** (2017) 012015 doi:10.1088/1757-899X/276/1/012015

by [8]. This approach facilitates a detailed study of the oscillatory lid-driven cavity flow covering a wide range of physical parameters in terms of Stokes numbers and Reynolds numbers.

2. Problem formation

A definition sketch of the flow in a square lid-driven cavity is shown in Figure. 1. The flow is considered to be incompressible and homogeneous. The continuity equation and Navier-Stokes equations are given as:

$$\nabla \cdot \vec{U} = 0 \tag{1}$$

$$\frac{\partial \vec{U}}{\partial t} + \vec{U} \cdot \nabla \vec{U} = -\frac{1}{\rho} \nabla p + \nu \Delta \vec{U} \tag{2}$$

where $\vec{U}$ is the velocity vector with the horizontal velocity component u and vertical component w; $Re = U_{lid}h/\nu$ for steady lid-driven cavity flow and $Re = U_{max}h/\nu$ for oscillatory lid-driven cavity flow (where U_{lid} is the steady velocity and U_{max} is the velocity amplitude, respectively). Moreover, the velocity presented in this paper is scaled by U_{lid} and U_{max} for the steady and the oscillatory lid-driven flow, respectively. No-slip velocity and Neumann pressure conditions are imposed on all the walls.

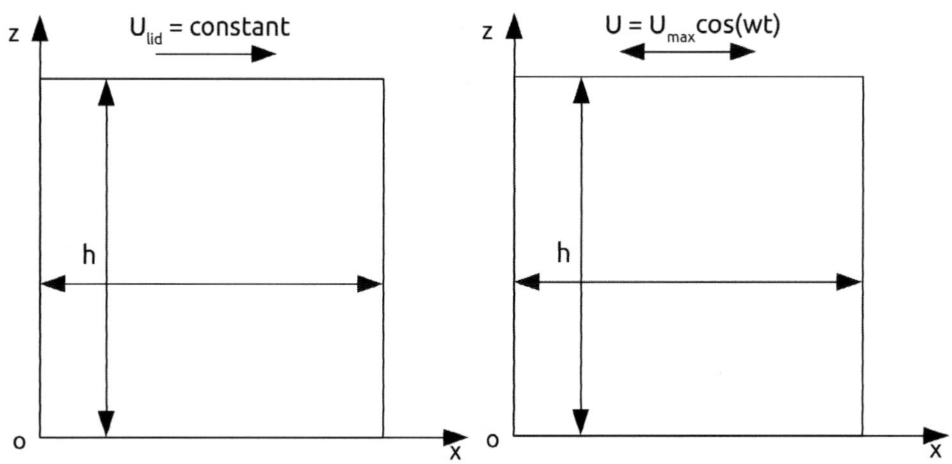

Figure 1. Schematic of the steady and oscillatory lid-driven cavity flow.

3. Numerical method

Equations (1) and (2) are discretized by second order central differences in space with a staggered grid arrangement. The impulse equations have been integrated in time using a semi-implicit second order scheme in conjunction with a projection method, where Adams-Bashforth and Crank-Nicolson are applied to the convective terms and diffusive terms, respectively.

A P2 projection method has been applied in a standard manner to solve the Navier-Stokes equations. Here an intermediate velocity field (which does not satisfy the continuity equation) is obtained by using the pressure gradient from the last time step (P2 method). A Poisson equation for the pressure is solved in such a way that the resulting pressure gradients are applied to correct the intermediate velocity field such that the resulting velocity satisfies the continuity equation.

4. Results of test cases and examples

4.1. Grid convergence test

The grid convergence test is carried out by using the predictions of the steady lid-driven flow in a square cavity at Re = 100, 400 and 1000. Five different uniform grid resolutions have been applied for each Reynolds number. The minimum values of the velocities $u(0.5, z/h)$ and $w(x/h, 0.5)$ and the maximum values of $w(x/h, 0.5)$ are shown in Table 1. At $Re = 100$, these values obtained from 64×64 grid cells deviate less than 0.1% from those obtained from 48×48 grid cells. As the Reynolds number increases to 400, the deviation between the values from 96×96 and 128×128 grid cells is less than 0.1% while for $Re = 1000$, this deviation is obtained between 128×128 and 160×160 grid cells.

Table 1. Results of cases with different grid resolutions

Test Cases	Velocity at the central line $min_z u(0, z/h)$	$min_x w(0, x/h)$	$max_x w(0, x/h)$
32	-0.2072	-0.2461	0.1734
48	-0.2099	-0.2488	0.1757
64	-0.2111	-0.2499	0.1765
96[a]	-0.2117	-0.2506	0.1771
128	-0.2120	-0.2508	0.1773
		Re = 100	
48	-0.3118	-0.4352	0.0.2871
64	-0.3179	-0.4417	0.2933
96	-0.3225	-0.4472	0.2976
128[a]	-0.3242	-0.4493	0.2991
160	-0.3250	-0.4498	0.2998
		Re = 400	
64	-0.3634	-0.4970	0.3524
96	-0.3753	-0.5117	0.3636
128	-0.3797	-0.5172	0.3679
160[a]	-0.3818	-0.5197	0.3699
192	-0.3825	-0.5207	0.3703
		Re = 1000	

[a]The case for investigation.

4.2. Two-dimensional steady lid-driven flow in a square cavity

The steady flow predictions with $Re = 100$, 400 and 1000 are presented in this section. Figure. 2 depicts the velocity profiles on the center-line of the cavity showing good agreement with the corresponding results obtained from [4]. The slope of the velocity profiles $u(0.5, z/h)$ and $w(x/h, 0.5)$ is nearly constant in the core region (i.e. away from the wall) for a Reynolds number of 1000. Moreover, the region with near uniform vorticity grows with increasing Reynolds number as shown in Figure. 3. It is demonstrated that the thickness of the boundary layers at the walls decreases as the Reynolds number increases.

Figure. 4 shows streamlines for the primary vortex and corner vortices at Re = 100, 400 and 1000. The primary vortex core moves towards the centre of the cavity and increases in strength while the corner vortices grow both in size and strength as the Reynolds number increases. The right corner vortex is smaller and weaker than the left vortex. Moreover, the stream-function and the vorticity contours are in good agreement with the results in [4].

4.3. Two-dimensional oscillatory lid-driven flow in a square cavity

Here results for the periodic solutions on the oscillatory lid-driven cavity flow at $Re = 100$, 400 and 1000 with a fixed oscillation frequency parameter $\omega' = \pi/3$ (here, $\omega' = h\omega/U_{max} = h/A$)

are given. The results are presented for 10 intervals of each oscillation cycle.

The oscillatory lid-driven cavity flow is expected to give a time-periodic flow. The criterion for the flow being fully developed is:

$$\frac{max|\vec{U}^n(x,z,t+T) - \vec{U}^{n-1}(x,z,t)|}{max|\vec{U}^n(x,z,t)|} \leq \epsilon \tag{3}$$

where n denotes the time step number; $\epsilon = 1 \times 10^{-6}$ and T is the period of the lid motion.

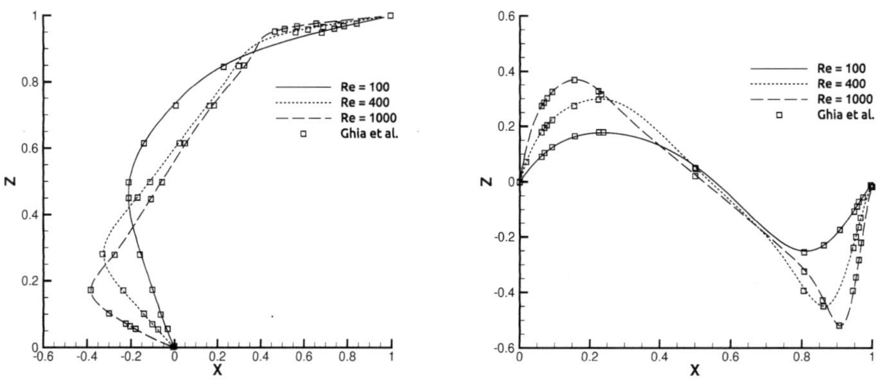

Figure 2. Comparisons of $u(0.5, z/h)$ and $w(x/h, 0.5)$ between predictions of the steady lid-driven cavity flow and reference data from Ghia et al. [4] at various Reynolds numbers

Figures 5 and 6 show the velocity profiles $u(0.5, z/h)$ and $w(x/h, 0.5)$, respectively. It is observed that the velocity profiles at t and $t+T/2$ are mirror images of each other. Comparison with the results in Fig. 2 reveals that the velocity profiles for the oscillatory lid-driven cavity flow are more complex than those for the steady lid-driven cavity flow. For instance, the velocity $u(0.5, z/h)$ at $t = 0.3T$ changes from a negative value to a positive maximum, and then decreases again to a negative value. Thereafter, it increases gradually to zero as it approaches the bottom wall. In other words, the direction of the velocity changes twice along the center-line indicating that at least one counter-rotating vortex pair emerge in the central region of the cavity. The boundary layer thickness beneath the moving lid decreases as the Reynolds number and consequently the Stokes number increase. This is consistent with laminar boundary layer theory (i.e. Stokes second problem described in [9]); the boundary layer thickness increases as the oscillation period (i.e. Stokes number) increases. Furthermore, the number of local extrema of $w(x/h, 0.5)$ increases as the Reynolds number increases. This indicates, as shown in Figure. 7 that the number of vortices near the central plane of the cavity increases.

Figure. 7 shows streamlines for different Reynolds numbers at $t = 0.4T$. The location and number of vortices within the cavity change for different Reynolds numbers. At $Re = 100$, two counter-rotating vortices occur in the cavity (one primary counterclockwise vortex and one small clockwise vortex are formed during the initial phase of the lid motion). As the Reynolds number increases from 100 to 400, an additional counterclockwise vortex appears beneath the clockwise vortex, and the size of the clockwise vortex increases. At $Re = 1000$, four vortices appear in the cavity, comprising two counter-rotating vortex pairs. Thus it appears that the flow structures in the cavity become more complex as the Reynolds number increases. The results for $Re =$ 100, 400 and 1000 with $\omega' = \pi/3$ are in good agreement with results obtained by [8].

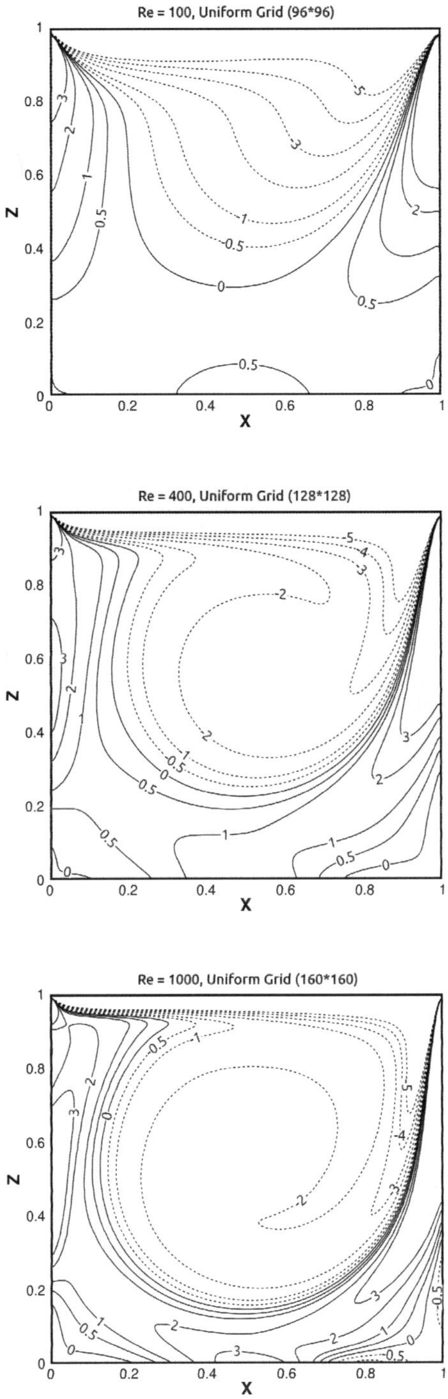

Figure 3. Contours of y-vorticity for the steady lid-driven cavity flow at different Reynolds numbers

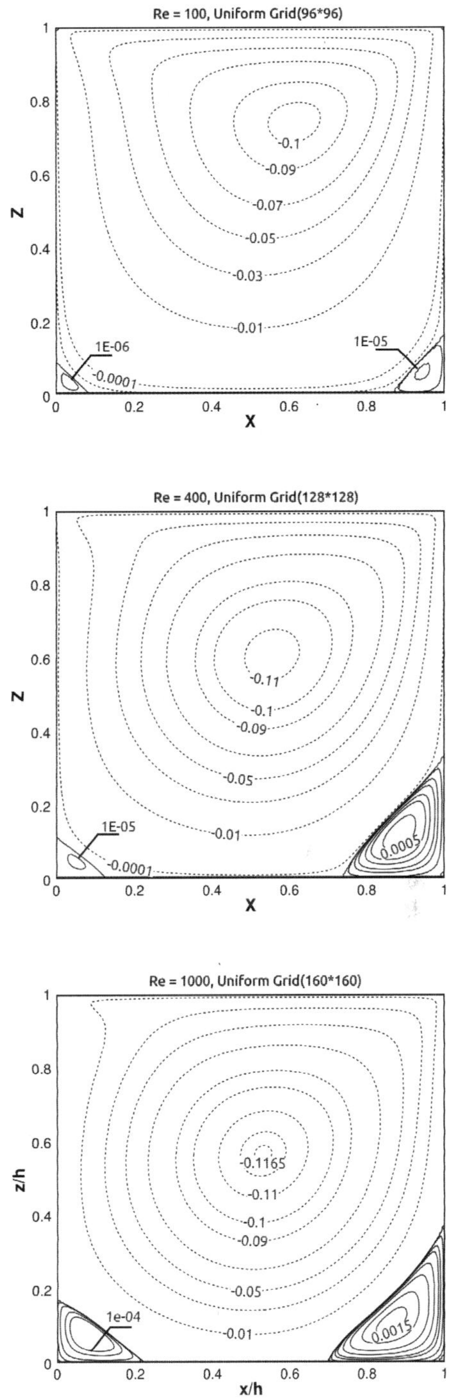

Figure 4. Contours of stream-function for the steady lid-driven cavity flow at different Reynolds numbers

IOP Conf. Series: Materials Science and Engineering **276** (2017) 012015 doi:10.1088/1757-899X/276/1/012015

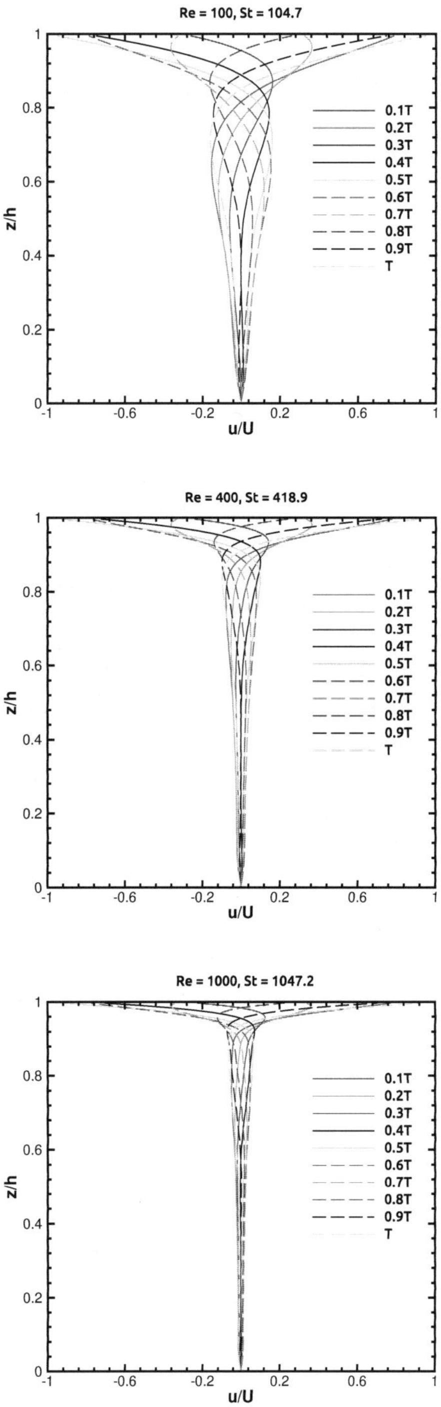

Figure 5. Profiles of $u(0.5, z/h)$ for the oscillatory lid-driven cavity flow at different Reynolds numbers with a fixed $\omega' = \pi/3$

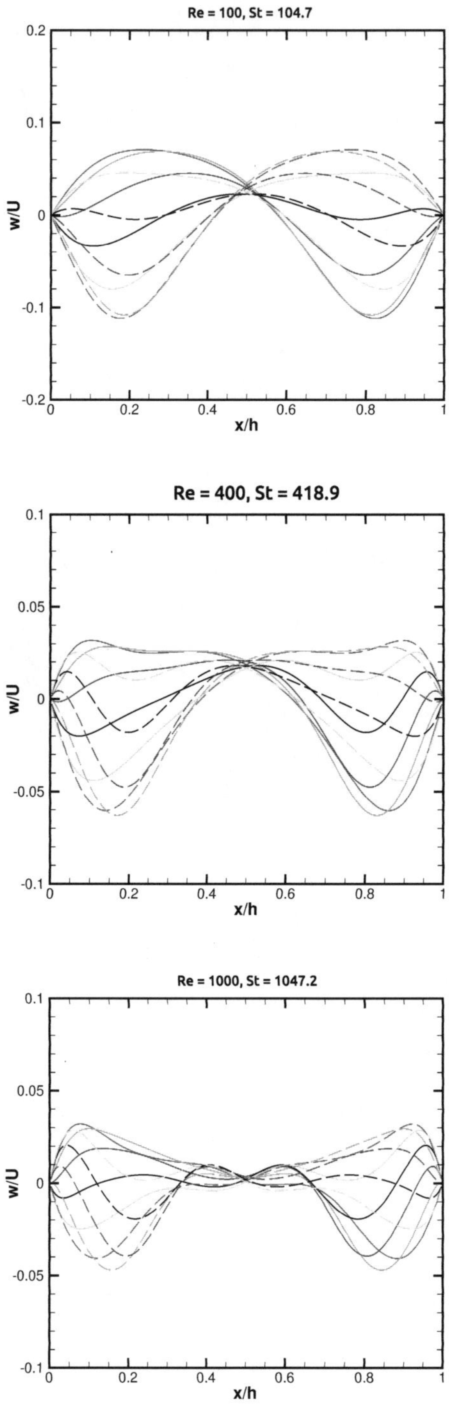

Figure 6. Profiles of $w(x/h, 0.5)$ for the oscillatory lid-driven cavity flow at different Reynolds numbers with a fixed $\omega' = \pi/3$. Velocities at different time instants are marked by the same color as in Figure. 5.

IOP Conf. Series: Materials Science and Engineering **276** (2017) 012015 doi:10.1088/1757-899X/276/1/012015

5. Summary and Conclusions

The flow features in the steady and oscillatory lid-driven cavity have been investigated. It appears that the oscillatory lid-driven flow exhibits more complex fluid dynamical behaviors than the steady lid-driven flow. For the chosen oscillation frequency, the boundary layer thickness at the top lid decreases as the Reynolds number increases. Finally, the number of vortices appearing in the central part of the cavity increases with increasing Reynolds number.

Acknowledgment

The first author would like to thank the support from the Department of Marine Technology, NTNU and the China Scholarship Council (Grant no. 201506680058).

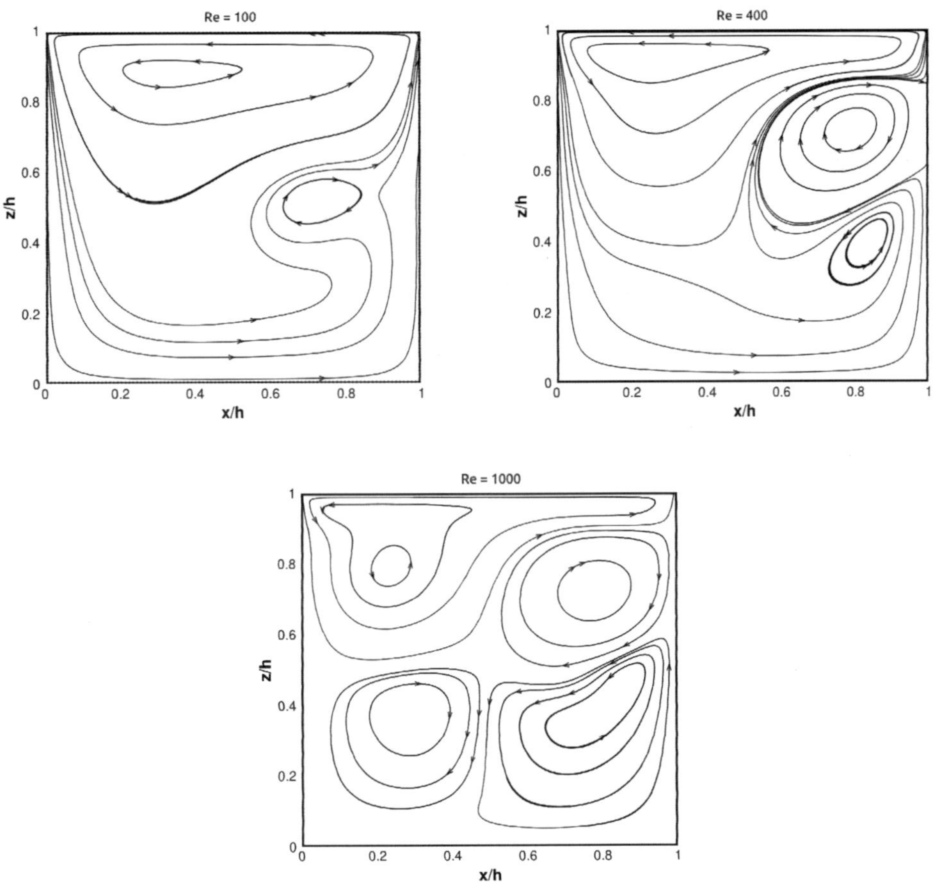

Figure 7. Streamline patterns of the oscillatory lid-driven flow with different Reynolds numbers at $t = 0.4T$.

References

[1] Van Rijn L, Soulsby R, Hoekstra P and Davies A 2005 *SANDPIT, Sand Transport and Morphology of Offshore Mining Pits.*
[2] Burggraf O R 1966 *J. Fluid Mech.* **24(1)** 113–151
[3] Benjamin A and Denny V 1979 *J. Comput. Phys.* **33(3)** 340–358
[4] Ghia U, Ghia K N and Shin C 1982 *J. Comput. Phys.* **48(3)** 387–411

[5] Duck P 1982 *J. Fluid Mech.* **122** 215–234
[6] Liu C H 2001 *Int. J. Numer. Methods Fluids* **35(5)** 533–557
[7] Iwatsu R, Hyun J M and Kuwahara K 1992 *J. Fluids Eng.* **114(2)**
[8] Mendu S S and Das P 2013 *Eur. J. Mech. B. Fluids* **39** 59–70
[9] Hermann S *et al.* 1979 *McGarw-Hill Book Company* 135–149

IOP Conf. Series: Materials Science and Engineering **276** (2017) 012016 doi:10.1088/1757-899X/276/1/012016

Limitations related to marine operations in the Barents Sea

O T Gudmestad
Department of Mechanical and Structural Engineering and Materials Science, University of Stavanger, Stavanger, Norway

E-mail: otgudmestad@gmail.com

Abstract. Some marine activities in the Barents Sea are normally ongoing year-round; others are dependent on limited weather windows. The limitations for the marine operations are the special weather conditions characterized by unpredictable Polar Low situations during the fall, winter and spring seasons, as well as cold temperatures that also are causing sea spray icing and the potential for drifting ice in certain parts of the Sea. It must also be realized that large distances combined with challenging meteorological and oceanographic criteria as well as darkness during the winter period represents a concern for evacuation and rescue, should it be necessary to abandon ships and platforms. The long distances to the locations farthest away from shore are, furthermore, out of reach of helicopter assistance. These aspects make it necessary to conduct hazard identification studies and to include all relevant historical knowledge in the hazard identification session, prior to the execution of marine operations in the Barents Sea.

1. Introduction

Marine operations in the Barents Sea are carried out by the oil and gas industry, by the fishing industry through activities for collecting cods and crabs and other kinds of fish, by the commercial transport industry as well as by personnel transport companies, including the cruise traffic.

Unpredictable Polar Low situations during the fall, winter and spring seasons characterize the Barents Sea as well as cold temperatures, causing sea spray icing and the potential for drifting ice in certain parts of the Sea. The Polar Lows limit the predictability of suitable weather windows, necessitating planning for emergency disconnection of marine operations. The cold temperatures give rise to needs for winterization of vessels and equipment. Sea spray icing limits the possibility for operability of equipment and raises concerns as to stability of certain kinds of smaller vessels. Furthermore, possible drifting ice represents a danger to any operations towards the north of the Barents Sea and needs to be monitored carefully. Particularly valuable areas in the Svalbard – Barents Sea – Lofoten area are shown in Figure 1 [1].

IOP Conf. Series: Materials Science and Engineering **276** (2017) 012016 doi:10.1088/1757-899X/276/1/012016

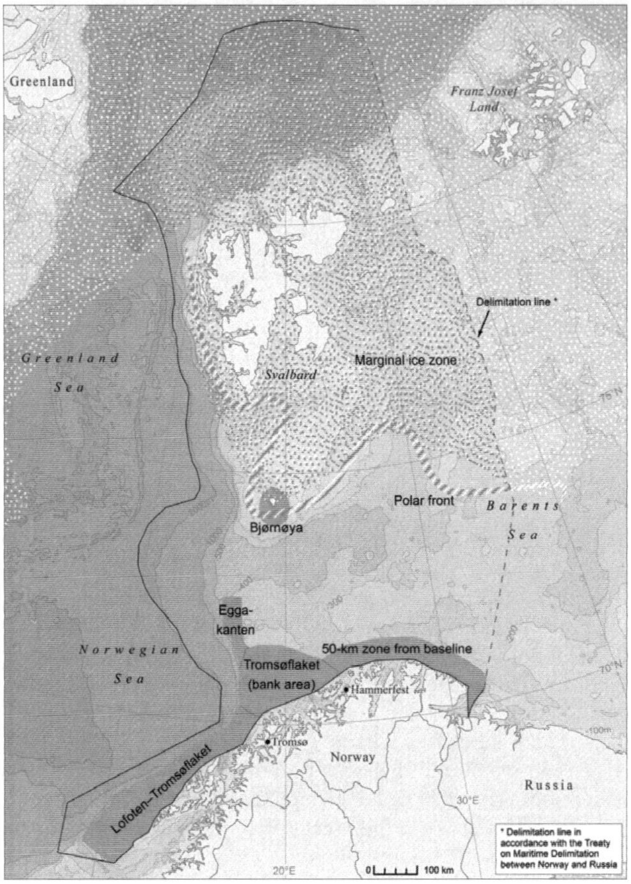

Figure 1. Particularly valuable and vulnerable areas in the Barents Sea–Lofoten area (shown in green), From [1]

2. Activities in the oil and gas industry

All activities in the oil and gas industry are vulnerable to the physical environmental conditions, in particular to the wave conditions. Both wave heights and wave periods are critical parameters [2] to ensure that resonant motions with rigs and offshore support vessels are avoided. Orimolade [3] and coauthors [4], [5] have discussed weather limitations for safe marine operations in the Barents Sea. An important finding is the potential for waves in Polar Low situations growing very quickly to large waves, confirming a model for waves in Polar Lows as proposed by Dysthe and Harbitz [6].

Without doubts the low temperatures in the Barents Sea (Figure 2, from reference [7]) will also influence the marine operations in the Barents Sea. Limitations due to combination of low temperatures and winds have to be considered due to the strong wind chill effects in low temperatures. In very low effective temperatures, personnel should not work outdoors.

Furthermore, winterization [8] of facilities and equipment is important where safe operations and safe escape be given highest priority, i.e. safety critical equipment must be prioritized. It will, however, be necessary to limit the amount of energy spent to keep the facilities free of ice, therefore, the consequences of heavy icing situations may be limitations in operational uptime for the facilities. Such situations may cause large pressure on the operating staff and the management of Barents Sea operations must be aware of the potential for lower production uptime in Barents Sea operations as compared to work in more temperate climate [9].

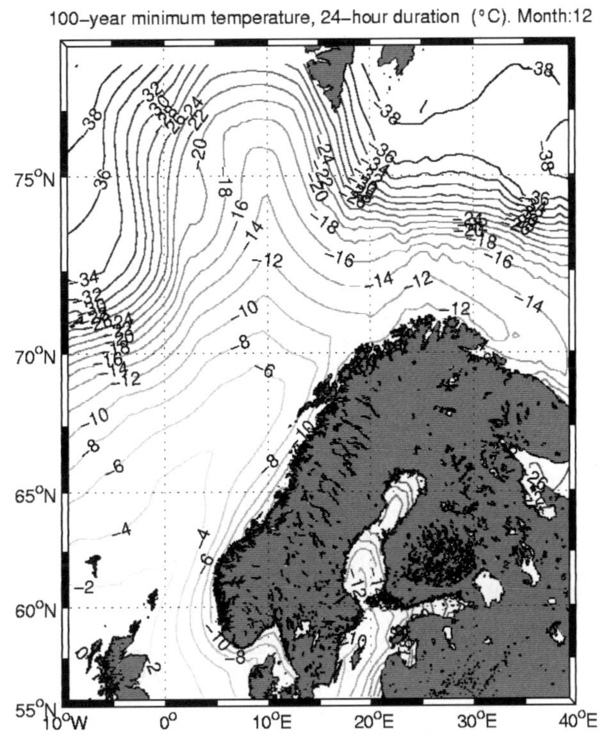

Figure 2. December: 24-hour duration minimum temperature at 2 meters above SWL on the Norwegian Continental Shelf (= minimum in data base -2°), data period 1958 – 2011. Figure 27 from [7]

There is an ongoing discussion in Norway regarding the limit for sea ice, "the ice edge", where year around operations could be halted by sea ice in certain years, see Figure 3, [10]. The discussion relates to whether only recent years or a longer period shall be taken for the estimate of the expected ice limit. Ice conditions in extreme years are shown in Figure 4, [11].

Limitation to year-round operations will be encountered in years with ice. Of particular concern is drifting bergy-bits or smaller icebergs, drifting near to facilities. Norsok N003 [7] and the relevant International Standard [12] call for ice management and evaluation of the possible need for disconnection should dangerous situations be threatening the facilities. The monthly maximum number of icebergs in May within a cell of 100 x 100 km is given in Figure 5 according to reference [13].

3. The fishing industry

The fishing industry has been operating in the Barents Sea for hundreds of years. In the past, the lack of models to forecast Polar Lows have caused considerable concern with the loss of a large number of persons.

As the fishing industry is one of Norway's most important industries, much attention has been paid to provide rescue possibilities for the fishermen. There are also situations when Norwegian authorities close fishing ports in the north to limit fishing vessels from leaving the ports for the fishing grounds.

The main concerns regarding technical issues are still the reliability of weather forecasts and the possibility that sea spray icing will lead to lack of stability, [14]. Also in recent years loss of fishermen and fishing vessels are reported. It should be noted that the fishing fleet follows the ice retraction and that the fleet now is active in the waters around Svalbard and that new seafood resources are explored when opportunities arise [15]. Figure 6 shows the area where snow-crabs can be found. The catching of snow-crabs is very profitable; this area represents challenging meteorological and oceanographical challenges [15].

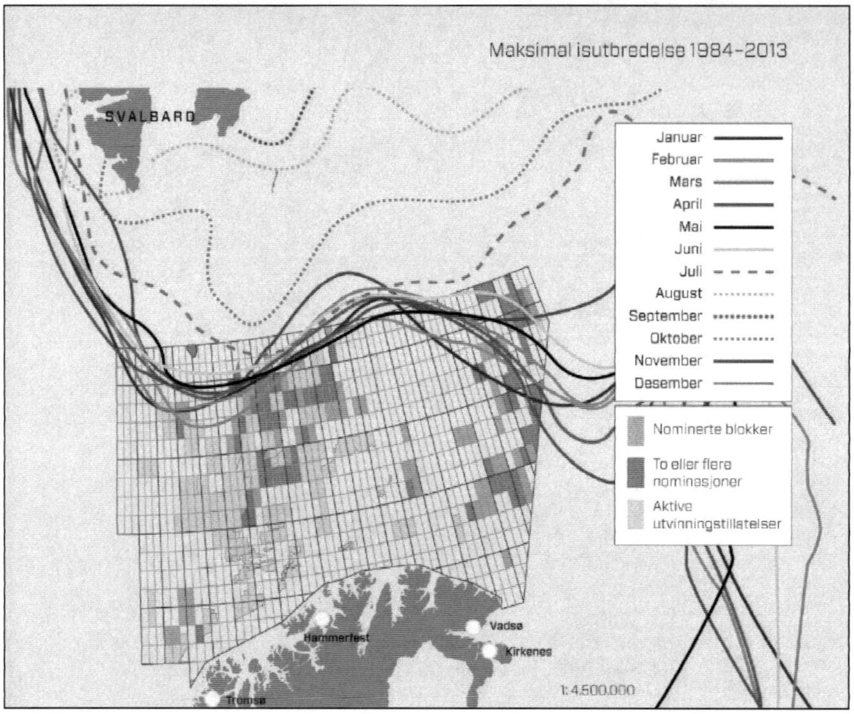

Figure 3. Ice in the Barents Sea during 1984 to 2013, [9].

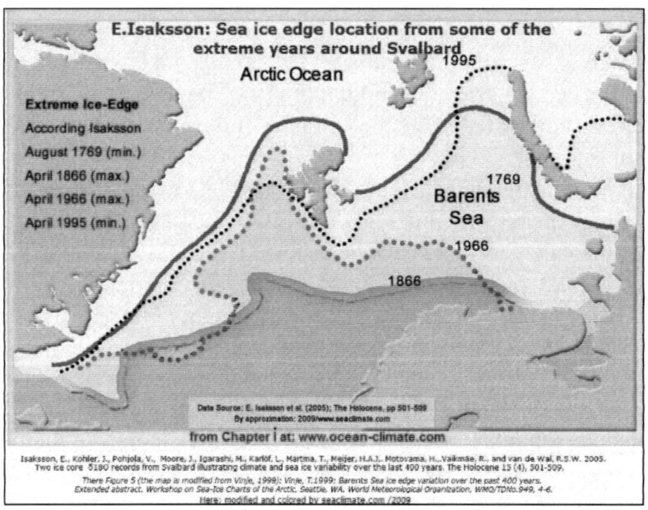

Figure 4. Sea ice edge location from some of the years with extreme ice [10], see http://www.arctic-warming.com/wp-content/uploads/2014/02/91.jpg.

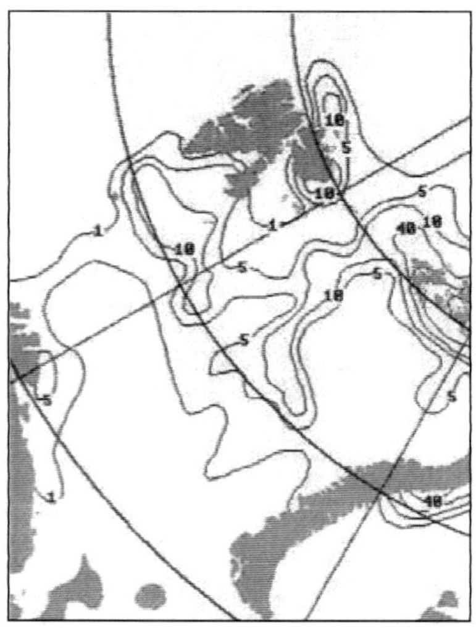

Figure 5. Monthly maximum number of icebergs in May within a cell of 100 x 100 km according to reference [13].

In order to ensure that agreed fishing quotas and agreed regulations are respected, the Norwegian Coast Guard is present in the Barents Sea and in the waters of the Svalbard archipelago. The Coast Guard vessels, furthermore, ensure a rescue backup to the rescue helicopters stationed at Banak airport.

4. Transport in the Barents Sea

Transport along the coast and across the Barents Sea is limited by weather conditions. Some days every year, ships from the coastal express fleet leaving from Bergen in the south for Kirkenes in the north have to seek shelter.

Furthermore, collisions between south-going and north-going vessels are of concern. Separate transport corridors have been established for southbound and for northbound vessels, respectively, Figure 7 [1]. More than 80 % of the total distance sailed in the Barents Sea–Lofoten area by vessels of gross tonnage over 10 000 within the areas covered by the traffic separation schemes between Vardø and Røst, and this includes almost 100 % of all tanker traffic. Tug boats are also stand by in harbours (including the LNG shipping harbour at Melkøya) to ensure that vessels losing propulsion can be towed so they will not be grounded.

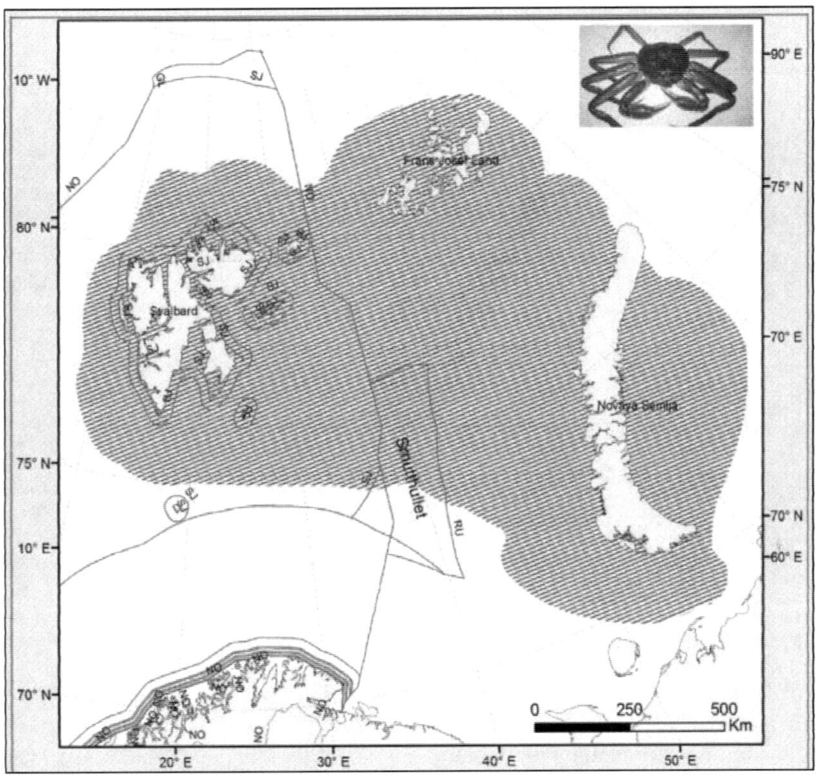

Figure 6. Area where snow-crabs can be, or is expected, to be found [15]

Measures implemented between 2005 and 2010 to improve maritime safety include [1]:

- An Automatic Identification System (AIS) for ships (information, tracking and collision prevention) has been introduced. It is estimated that the system has reduced the risk of collisions by 20 %.
- Satellite-based monitoring of sea areas has been further developed.
- The Vardø VTS Centre was established in 2007 and monitors high-risk traffic along the entire Norwegian coast, including Svalbard.
- The traffic separation schemes between Vardø and Røst were established in 2007.
- The emergency tugboat services have been further developed and three new tugboats have been hired on short-term contracts.
- The new Act relating to ports and navigable waters entered into force in 2010. It applies in Svalbard, and the earlier Harbour Act was made applicable to Svalbard in 2008 in regulations.

IOP Conf. Series: Materials Science and Engineering **276** (2017) 012016 doi:10.1088/1757-899X/276/1/012016

- The Norwegian Coastal Administration has developed a procedure for coordinating the actions of the authorities in situations where a vessel needs to be brought to a port of refuge. A prior assessment of suitable geographical areas has been made.
- To limit the damage in the event of a spill, a provision was introduced in 2007 forbidding ships calling at the nature reserves in eastern Svalbard from carrying or using any fuel other than light marine diesel. Since 1 January 2010 a corresponding provision has applied in the three large national parks in western Svalbard as well. A temporary exception until 2015 has been made for the approach to Ny-Ålesund and the Magdalenefjorden.

The Barents Sea is also the gateway to the North-Eastern Passage, The Northern Sea Route, Figure 8. The Norwegian city close to the Russian border, Kirkenes, is a good harbour for repair of vessels, for shelter in case of challenging weather conditions and for ensuring that Russian travel permits are obtained.

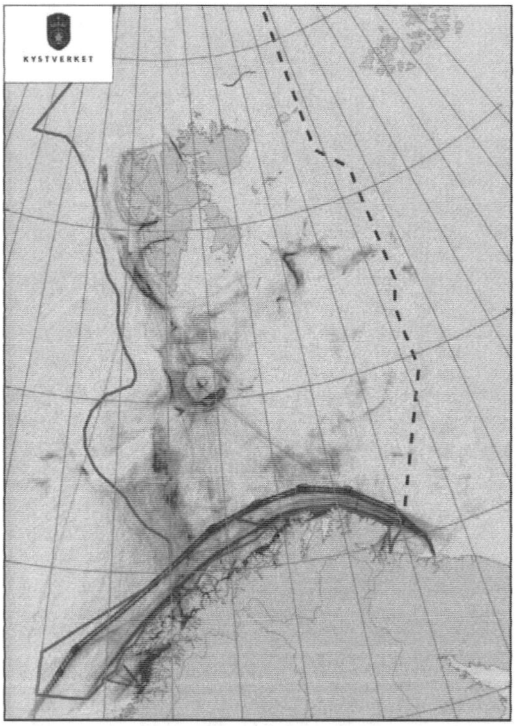

Figure 7. Traffic density in the management plan area, the traffic separation schemes between Vardø and Røst (thick red line) and near coast areas in the second half of 2010. The highest traffic density is indicated by the red shading [1].

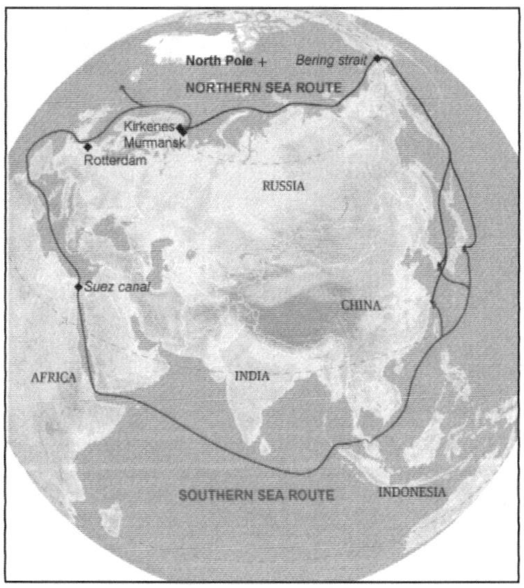

Figure 8. Kirkenes is the gateway to Northern Sea Route

5. Other commercial activities

Commercial activities have in the past included seal hunting and whaling. The risks involved to the participants were extreme, in some years a large percentage of the vessels and the hunters were lost, [16]. Figure 9 shows the situation with large ice floes riding on big waves.

Figure 9. Outside the ice edge, large ice floes surfing on waves [16]. Photo: W. Warholm

Recently, the cruise industry has shown considerable interests in the Barents Sea and in particular the coast of Svalbard. The cruise ships are normally travelling in the summer season; however, the Norwegian Coastal Express takes passengers all year along the Norwegian coast.

Of main concerns are the pollution from the vessels in the harbours and at sea and the search and rescue operations in case of accidents. The rescue capacity is limited in this region to a limited number of persons. For a discussion of evacuation, search and rescue see below.

6. Evacuation, search and Rescue

6.1. *Helicopter rescue*

When examining the feasibility of providing long-range search and rescue for personnel in the Barents Sea, we should consider a helicopter accident while en-route to or from an offshore petroleum installation in the Barents Sea or to a maritime accident. In [17] a combination of a SAR helicopter and multipurpose emergency response vessels is proposed, whereby improved search and rescue capacity is obtained both for personnel involved in the petroleum industry and others i.e., fisheries, maritime transport and tourism.

The basis for the paper [17], is the petroleum exploration activity in the far North-Eastern area of the Norwegian sector of the Barents Sea, Figure 10. The area is currently opened for oil and gas exploration. There is currently little or no infrastructure in the area beyond the coast. A method to provide SAR coverage over a distance of 260 nautical miles with a minimum rescue capacity of 21 persons within two hours is considered in [17]. Issues related to survival in cold water, immersion survival suits and performance requirements for search and rescue resources are challenging in order to provide an optimum combination and enhanced probability of survival if an incident should occur.

Operational considerations involving departure criteria for helicopter transport should be developed in order to ensure that persons travelling on a helicopter to remote locations in the Barents Sea have a reasonable prospect of surviving a helicopter ditching and subsequently being rescued. In [17] it is documented that Multipurpose Emergency Response Vessels, ERVs, equipped with dual Fast Recovery Daughter Craft, FRDC, capable of operating in an Arctic climate deployed at the remote location and en-route, together with an onshore based search and rescue, SAR, helicopter may provide a rescue capacity for 21 persons within 120 minutes.

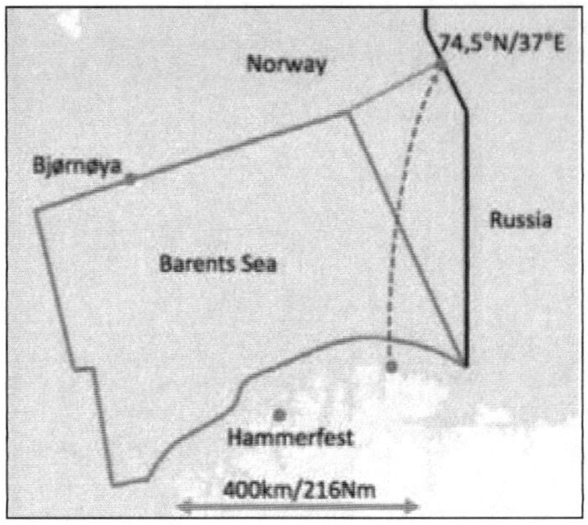

Figure 10. Map of the Barents Sea (Source: npd.no), from [17].

6.2. Rescue of persons from a cruise ship

The cruise industry's efforts to provide customers with visits to even more remote cruise locations may represent a large challenge in case there is a need to evacuate a cruise ship. Of particular concern is survivability in poorly insulated rescue means in Polar Regions. A search and rescue exercise (SARex 1) north of Spitzbergen during April 2016 [18] was, therefore, conducted to assess whether standard SOLAS rescue equipment would satisfy the five days survivability requirement of the IMO Polar Code [19], which was implemented in January 2017. Another exercise was conducted in May 2017 (SARex 2) with improved equipment, Figure 11.

The polar conditions experienced during the survival exercises arranged by University of Stavanger, GMC, Stavanger and the Norwegian Coast Guard at north Spitzbergen in April 2016 and May 2017 generated polar-specific challenges for the exercise participants and for the lifesaving equipment. It was planned to simulate relevant polar conditions, incorporating sea ice, sea swell, low air and water temperatures and remoteness.

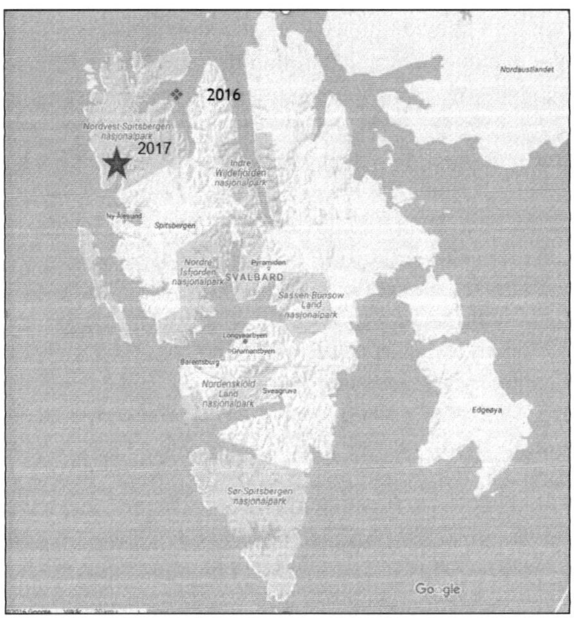

Figure 11. Locations for SARex 1 (2016) and 2 (2017)

The detailed objectives of the exercises and the associated research programs were to:

- Assess the adequacy of the lifesaving appliances (lifeboats and life rafts) as required by the IMO Polar Code.
- Identify the gaps between SOLAS [20] approved rescue craft (lifeboats and life rafts) and the requirements defined in the IMO Polar Code.
- Identify the gap between SOLAS approved personal protective equipment (PPE) and the requirements defined in the IMO Polar Code.
- Assess the personal/group survival kits as defined by the IMO Polar Code.
- Train Norwegian Coast Guard personnel in emergency procedures in ice-infested waters, with reference to evacuation and rescue from cruise ships.
- Identify the effects of improved equipment during the 2017 exercise as compared to the 2016 exercise.

The following topics were addressed in the exercises:

- Functionality of life raft/lifeboat under polar conditions.
- Functionality of personal protective equipment (PPE) (e.g. thermal protection/survival suits).
- Additional training requirements for crew and passengers.
- Evaluation of Coast Guard's search and rescue procedures, including handling of mass evacuations in Polar Regions.
- Evacuation to sea ice (2016).
- Evacuation by helicopter (2017).
- Identification of the functionality of search equipment in Polar Regions (2017).

We concluded after the 2016 exercise (SARex 1) that the IMO 5-day survival requirement poses strain on all rescue means and causes the needs for improved PPE (personal protection equipment, for example improved survival suites), insulated lifeboats and insulated life rafts. The follow-up search and rescue exercise (SARex 2) was thereafter conducted with improved rescue means. We saw relatively large effects of insulation of rescue means, heating of lifeboat, use of improved personal protection equipment and of limiting the number of personnel onboard the rescue means (because of the bulkier PPE needed for survival in Polar Regions).

7. Hazard identification and risk analysis

In order to ensure safe marine operations [21] in the Barents Sea, a thorough risk analysis is suggested. The risk analysis hazard identification phase will be most important where the following aspects are of main concern [22]:

- experience from operations in the area
- experience with the technology being applied
- weather forecast's reliability during the operation
- human factors relevant for the participants
- the company's experience with activities in the area
- the company management's ability to handle activities in the high north [9].

As relevant quantitative data will be sparse, the use of qualitative risk analysis schemes may be very useful to identify risks of concern and their possible mitigating measures. Operations which can be set on hold or be abandoned quickly have much lesser risks than operations which last for longer periods and which cannot be stopped. Regarding the complexity of risk analysis for Arctic operations, see [23].

The Arctic Operational Handbook [24] reviews different themes related to operations in the Arctic and concludes that:

- It is noted from the gap analysis that there is limited guidance in International Standardization Documents (ISO documents) for pipe lay, trenching and dredging operations, let alone for the arctic areas.
- Site specific operations should be considered when planning and carrying out operations.

- Considerable effort was performed to align the knowledge on weather conditions and in particular on the requirements for monitoring and forecasting as well as the requirements for decision based tools.
- For the transportation & logistic aspects, input to the report relied heavily on the existing guidance for arctic shipping which is further developed and was evaluated and transferred to recommendations for the specific services of this guide.
- The report provides guidance as required specifically for contractors expecting to work in arctic areas on the aspects of health, safety, training and also stakeholder mapping.
- A frame work is provided to perform environmental impact assessments both in early as well as detailed stages of design in order to ensure that impacts can be managed and mitigated. Specific attention was given to the evaluation of the loads on and the operation of disconnectable floating production units.

8. Conclusions

There exist considerable more limitations to marine operations in the Barents Sea compared to more temperate areas. The activities should be handled with extreme care to avoid damage to the environment and to assists and to avoid fatalities. Risk identification and risk analysis are important tools if properly conducted. All relevant historical information about the area must be included in the hazard identification [16]. It seems, however, that “waiting on weather” should be considered to ensure safe operations in the Barents Sea.

Acknowledgments

The author would like to acknowledge the cooperation with colleagues who have worked with him on issues related to marine operations in the Barents Sea, in particular students at the University of Stavanger, and scientists at NTNU, Trondheim and at University of Tromsø, the Arctic University of Norway. He will suggest that the theme is of much importance for Norway and he will encourage further work on the subject. It would be seen as a waste of research opportunities in case the theme is abandoned at the University of Stavanger.

The author will also challenge the oil and gas industry and the Petroleum Safety Authority of Norway to ensure that all aspects of marine activities are considered prior to giving consent to perform marine activities in the Barents Sea.

References

[1] Norwegian Gouvernment 2010 Integrated Management Plan for the Marine Environment of the Barents Sea–Lofoten Area, *Meld. St. 10 to the Parliament Oslo* https://www.regjeringen.no/en/dokumenter/meld.-st.-10-20102011/id635591/sec4 accessed 5th November 2017

[2] Natskår A, Moan T and Alvær P Ø 2015 Uncertainty in forecasted environmental consitions for reliability analyses of marine operations *Ocean Engineering* **108**, pp 636 -647

[3] Orimolade, A P 2017 Weather limitations for marine operations in the Barents Sea *PhD Thesis, University of Stavanger*

[4] Orimolade A P, Furevik B R, Noer G, Gudmestad O T, Samelson R M 2016 Waves in polar lows *J. Geophys. Res.: Oceans* **121**(8), pp 6470-6481

[5] Orimolade A P, Gudmestad O T 2017 On weather limitations for safe marine operations in the Barents Sea *COTech conference,* Stavanger, Norway

[6] Dysthe K B, Harbitz A 1987 Big waves from polar lows? *Tellus A* 39A:500-508

[7] Standards Norway 2016 Actions and action effects, Norsok N-003 *Standards Norway Oslo*

[8] DNV GL 2013 Winterization for Marine Operations Offshore Standard DNV-OS-A201 *DNV GL, Oslo* https://www.norskoljeoggass.no/Global/HMS-utfordringer%20i%20nordomr%C3%A5dene/Seminar%204%20-

%20Risikostyring%20og%20design/1530%20Fossheim%20DNV%20OS-A201.pdf, assessed 5th November 2017

[9] Gudmestad O T 2015 Ledelsesutfordringer ved leting og produksjon i Barentshavet *Arctic Safety Summit arranged by Petroleum Safety Authority, PSA* Tromsø

[10] Teknisk Ukeblad January 2015 Regjeringen med ny grense for iskanten *Teknisk Ukeblad, Oslo* https://www.tu.no/artikler/regjeringen-med-ny-grense-for-iskanten/223704 assessed 5th November 2017

[11] Isaksson E et al. 2005 Two ice core records from Svalbard illustrating climate and sea ice variability over the last 400 years *The Holocene* 15 (4), pp. 501 -509, 2005. http://www.arctic-warming.com/wp-content/uploads/2014/02/91.jpg, accessed 5th November 2017

[12] International Standards Organization ISO 2015 ISO19906, Arctic Offshore Structures *ISO Geneva Switzerland*

[13] Abramov V 1996 Atlas of Arctic Icebergs *Backbone Publishing Company*

[14] Orimolade A P, Gudmestad O T, Wold L 2017 Vessel stability in polar low situations *Ships and Offshore Structures*, Volume 12, Supplement 1, pages S82 – S87 DOI: 10.1080/17445302.2016.1259954, http://dx.doi.org/10.1080/17445302.2016.1259954 assessed 5th November 2017

[15] Jørgensen A 2016 Utfordringer ved snøkrabbefangst med hensyn på risiko ved operasjoner på sjøen Kartlegging av risiko og risikoreduserende tiltak *MSc thesis, University of Tromsø*, file:///C:/Users/otgud/AppData/Local/Packages/Microsoft.MicrosoftEdge_8wekyb3d8bbwe/TempState/Downloads/thesis.pdf,, accessed 5th November 2017

[16] Alme J B 2011 Ishavsfolk si erfaring *Tapir, Trondheim* (with an extensive summary in English)

[17] Jacobsen S J, Gudmestad O T 2013 Long-range Rescue Capability for Operations In the Barents Sea Paper OMAE2013-10616 Proceedings of *OMAE, Nantes* ISBN: 978-0-7918-5536-2

[18] Solberg K E, Gudmestad O T, Kvamme B O (Editors) 2016 SARex Spitzbergen, Exercise report, search and rescue exercise conducted off North Spitzbergen University of Stavanger http://hdl.handle.net/11250/2414815, accessed 5th November 2017

[19] IMO, The marine environment protection committee of the International Maritime Organization 2016 International code for ships operating in polar waters (polar code) *IMO Geneva Switzerland*

[20] IMO, International Convention for the Safety of Life at Sea (SOLAS) 1974, Internet page: http://www.imo.org/en/About/Conventions/ListOfConventions/Pages/International-Convention-for-the-Safety-of-Life-at-Sea-(SOLAS),-1974.aspx, assessed 3rd November 2016

[21] Gudmestad O T 2015 Marine Technology and Operations: Theory & Practice *WIT Press, Southhampton, UK.*

[22] Quale C, Gudmestad O T 2011 Technology and Operational challenges for the High North *IRIS and University of Stavanger*

[23] Kämpf M, Haley S 2011 Risk Management in the Arctic Offshore: Wicked Problems Require New Paradigms ISER Working Paper 2011.3 *University of Alaska* http://www.iser.uaa.alaska.edu/Publications/2011_10-riskmanagement.pdf assessed 10th November 2017

[24] Arctic Marine Operations Challenges & Recommendations, Joint International Project, Report no MAR11908-E/1133-RP01, 2013

IOP Conf. Series: Materials Science and Engineering **276** (2017) 012017 doi:10.1088/1757-899X/276/1/012017

Implications caused by SARex on the implementation of the IMO polar code on survival at sea

K E Solberg

GMC Maritime AS
University of Stavanger, Stavanger, Norway

E-mail: knut.espen.solberg@gmc.no

Abstract. The International Code for Ships Operating in Polar waters goes into effect on 01 January 2018 for all ships. This puts additional strain on vessel owners and operators as they will have to comply with an additional set of requirements. This includes the functional requirement of a minimum of 5 days survival time. The SARex exercise has elaborated on the issue of survival in close cooperation with the different stakeholders associated with the marine industry. Being an objective third party is important when organizing and executing these activities as all of the stakeholders has different agendas and priorities. Developing sustainable solutions is a balancing act, incorporating economic and political aspects as well as technology and requires a mutual common understanding of the mechanism involved.

1. Introduction

The extent of maritime activities taking place in the polar regions is increasing. Much of this activity is related to passenger transfer/cruise ship activity or offshore activity. Traditionally there were no extra requirements for vessels operating in cold climate environments, despite the additional challenges represented. As a result, many operators did not take into account the added risk associated with their cold climate operation.

2. The International Code for Ships Operating in Polar Waters - Regulatory rationale

The International Code for Ships Operating in Polar Waters is also known as the *International Maritime Organization (IMO) Polar Code* [1], and will go into effect on 01 January 2018 for all ships.

The International Code for Ships Operating in Polar Waters is a functional set of requirements, and utilizes a risk-based approach. It aims to mitigate the additional risks associated with marine activities in Polar waters. Having a risk-based approach induces additional strain on the chain involved in regulating, designing and conducing marine operations in a polar environment, as the risks are to be identified, assessed and mitigated through a holistic approach incorporating all aspects of the system.

Functional requirements are utilized in the offshore oil and gas industry on the Norwegian shelf. The success experienced with in this field can come as a result of a stable and close relationship between major oil operators, employees and authorities. Very few of the current marine regulations are based on a risk-based approach, with the exception of the International Safety Management (ISM) code. The process of assessing and mitigating the risk the identified in the IMO Polar Code requires in depth knowledge in relevant fields, and the outcome of the assessment is never any better than the knowledge available during the process.

The IMO Polar Code requires additional equipment to be carried, e.g. equipment that enables survival on ice/land and equipment that enables a minimum of 5 days survival time. 5 days have been defined by IMO due to the lack of infrastructure and remoteness present in the polar region. This represents a challenge with regards to the capacity in the rescue craft, and downsizing, operating with a reduced number of passengers, can be a possible solution.

Currently there is no common understanding of the interpretation of the code. As a result, there are variations between flag states and classification societies on how to achieve compliance. For vessel operators/transportation providers the lack of consistency, predictability and transparency represent not only a practical challenge. It also induces an economic risk as downsizing, can be a result of the implementation of the IMO Polar Code. Downsizing or operating with a reduced number of passengers will greatly affect the profit of a marine operation.

3. SARex

An increased activity within the field of expedition cruises is experienced in the Arctic. This increase is expected to persist in the coming years. The tourist industry in areas like Svalbard is already preparing to meet the expected increase in visitors to the area [2].

Much of the focus involves increasing the tourist activities and raising the revenue generated by the visitors. Few questions the safety and the risks involved in these types of activities. Limited understanding of the complex multi-discipline challenge of surviving in a rescue craft in a cold climate environment has troubled the marine industry since the first draft of the IMO Polar Code was released. Our goal has been to investigate and quantify survivability related to a major marine accident in the Arctic/Antarctic, and assess if our results are in line with the global societal expectations and the 5 day requirement as stated in the IMO Polar Code.

To increase the understanding related to the complex multi-discipline challenge of surviving in a rescue craft in an Arctic environment full-scale experiments have been conducted. The project has been called SARex [3] and the objective has been to increase the level of maritime safety through quantification of survivability in relation to Safety of Lives At Sea (SOLAS) approved equipment. The full-scale experiments have been conducted utilizing the Coast Guard vessel KV Svalbard in close cooperation with regulators (Norwegian Maritime Authority, Petroleum Safety Authority, ABS and DNV GL), equipment manufacturers (Norsafe and Viking-Life), in addition to experts within their respective fields (e.g. medical personnel and risk expert).

To fully grasp the concept of survival at sea, following an abandoning ship incident, there is a need for improved understanding of the mechanisms involved. An increased understanding of the mechanism involved, in combination with a functional based rule set, enables sub-suppliers to design their safety systems in a sustainable manner, incorporating not only the functionality required to supply adequate survival, but also at the same time take into account the economic demands enforced by the industry.

The results from our experiments have been communicated to the relevant major stakeholders in the global marine industry, including the Norwegian Government, maritime administrations, IMO, vessel operators/transportation providers and equipment suppliers.

4. The path of regulatory development

Much of the current regulatory regime enforced by IMO [4] has been developed as a result of a retroactive processes preceding a major accident, e.g. the Titanic accident triggered the development of the SOLAS convention.

It is however important to note that the implementation processes taking place in IMO requires extensive time and consensus among the member states. As almost every country in the world is a member state of IMO. Considerable divergence among cultures, financial situation and involvement is experienced among the different members. As a result, common consensus among the member states can at times be hard to obtain. It is at times experienced that political agendas overrule scientific facts in the voting processes.

As there are many stakeholders and agendas, regulatory development can be regarded as a process of causation, where the process focus on predictable aspects of an uncertain future. We humans have limited control of the future. We retrieve knowledge to handle uncertainties. Both the regulatory goal and the regulatory development path can be unclear, just like economic decision-making. The regulatory development process has strong analogies with the theories described in Saras D. Sarasvathy "*Causation and Effectuation: Toward a theoretical shift from economic inevitability to entrepreneurial contingency"* [5].

Traditionally full scale experiments or "reduced full scale" experiments, e.g. towing tank experiments, have been utilized in design processes. In cases where the problem is complex and cross-disciplined, where all the individual mechanisms and interactions between the mechanisms have not been properly understood a holistic or "black box" approach can be utilized.

5. Stakeholders

Working with regulatory development within the marine industry on an international level require an in-depth understanding of both the evident and at times hidden agendas of the different stakeholders. The main stakeholders affecting work related to marine safety can be summarized as follows:

- *IMO* – The International Maritime Organization is a special agency under the UN. It has currently 172 member states, usually represented by their maritime administration. IMO is organized through 5 committees, each with several sub-committees. The work associated with life saving appliances is anchored in the legal instrument, the SOLAS Convention, which is administered by the Maritime Safety Committee (MSC).

  Many of the decisions made in IMO are based on finding common denominators and reaching a consensus among the member states. This process is time consuming and often involves taking into account political and national interests.
- *National interests* – In Norway the marine industry is governed by the "Norwegian Ministry of Trade, Industry and Fisheries" and the national interests are administered by Norwegian Maritime Authority [6]. The Norwegian Maritime Authority (NMA) is not only administering and enforcing our national requirements, but is also administering our maritime registers (NIS/NOR registers). The vessels registered in our national registers are to comply with our maritime regulations. In most cases, the vessel owners are companies registered in Norway. Due to the income generated by the taxes imposed on the vessel owners, the individual nations strive to have commercially competitive regulations, both within the maritime regime and the taxation scheme.
- *Petroleum Safety Authority (PSA)* – The responsibility of the Petroleum Safety Authority [7] is to ensure an adequate safety level on offshore installations on the Norwegian Shelf. It is administered by the "Ministry of Labor and Social Affairs". PSA is only concerned with our national interest and no international consensus is required with regards to regulatory development/ implementation. PSA has no formal legal connection to the maritime industry (ships/vessels registered under NIS/NOR or any other maritime administrations), unless they are drilling on the Norwegian Shelf. They will however enforce requirements on offshore drilling operators, including their sub-suppliers like offshore supply vessels. Traditionally the requirements enforced by the PSA have been more conservative than the ones enforced by the NMA, addressing weaknesses in maritime regulatory regime.
- *Classification societies* – Classification societies are interpreting the regulations defined by the costal administrations. In some cases, they act on behalf of the costal administrations. Other times they act as objective third parties. It is however important to note that having vessels registered in a classification society generates income for the society. This mechanism forces the societies to compete against each other in an aggressive marked. As a result, the societies

have to balance the need for conservative interpretation of the regulations with the cost implied on vessel owner/operator to keep a fleet registered under their rules.

- *Vessel owners/operators* - The vessel owners/ operators have to cover the cost associated with the regulatory requirements. The owners/operators also have to pay insurance, which again is only valid if the vessel complies with the flag state requirements, typically enforced by class. In general, you are regarded as a responsible owner/operator if you operate in compliance with the flag/port state requirements.
- *Equipment producers* – the equipment producers provide the vessel owners/operators with equipment that enables them regulatory compliance. The safety equipment is usually evaluated on regulatory compliance, price, capacity, weight and volume, where regulatory compliance has to be in place, and where price is the key most important parameter determining the sales volume.
- *Ship officers/crew* – The training of the vessel crew is defined in the IMO STCW convention and their interest are safeguarded through unions, e.g.
- Norwegian Seafarers' Union [8]. The unions enforce strong interest in the safety of the officers/crew, and have representatives present in IMO.
- *Passengers* – The safety of passengers is safeguarded by no individual organization. Usually their safety is the responsibility of the cruise operator/transportation provider. Their motivation of safeguarding their passengers is the risk of economic implications caused by an incident/accident. It is however important to note that the cruise operator/transportation provider main motivation is to generate a profit, which involves keeping the cost at a minimum level. To stay commercially competitive, they are often forced to keep the cost related to safety equipment at a minimum, but still within the levels defined by the regulatory regime.

6. Societal perspectives

In the marine industry the decision processes is seldom “black and white”, and it involves many considerations that interact on the different stakeholders in different ways. This is influenced by culture, economic robustness, global politics and facts. As described in “Technologies of Humility” [9] science only offers part of the picture. It is important to understand that reaching consensus with regards to regulatory development and regulatory interpretation is as much a balancing act, incorporating political, economical and cultural aspects as well as objective scientific data.

There are 3 main stakeholders affecting the design of safety equipment for the maritime industry, the IMO requirements (enforced through the flag state and class rules), the equipment suppliers and the vessel operators. Each party has their own agenda and much of the challenges are associated with finding the right compromise between cost and functionality. The different agendas can be summarized as follows:

- Regulators – ensuring a safety-level that is globally considered acceptable, within the frames defined in IMO and giving their flag state register no commercial dis-advantage.
- Equipment suppliers – supplying equipment that is fulfilling regulatory requirements and at the same time is commercially attractive.
- Vessel operators – providing the safety as required by the regulators at the lowest possible cost.

As defined in Social Construction of Technology (SCOT) [10] there is not just one way or one best way to design an artifact. Development of new technology or utilization of new combinations of existing technology could increase the safety levels considerably. This would reveal many unique opportunities with regards to development of new commercial products. However, the marked will not purchase these products unless they or their functionality is defined as a compulsory requirement in the governing regulations, or the products can be regarded as cost efficient solutions.

A result of the above mechanisms is that increased maritime safety is mainly accomplished through regulatory development and regulatory interpretation. This will later will be followed by development of new products. “Proof” that the current situation is not in line with the global societal expectations typically initiates regulatory development. This "proof" can be obtained through a major accident, and traditionally the marine regulations are retroactive and major regulator changes has emerged in the wake of major accidents.

"Proof" can also emerge because of scientific documentation. Through the SARex exercises, we have investigated the survivability to be expected if a real accident occurred in an Arctic/Antarctic region. The survivability figures obtained does not meet the global societal expectations or the 5 day requirement as defined by the IMO Polar Code.

As the maritime administrations are responsible for defining the acceptable risk levels associated with maritime activity. It is therefore important that the results from SARex have to be communicated to the maritime administrations, which further will communicate these finding to the international community through the IMO regime.

For the stakeholders represented as regulators, our results have to be incorporated in their organizations. The vessel operators/transportation providers are in general skeptical to findings that will induce additional costs that is to be carried by them. On the contrary to what many expect, the equipment manufacturers have very few opinions on the issue, as they only provide equipment according to regulatory requirements and have no responsibility beyond that with regards to the functionality or survivability provided by their equipment.

The international maritime industry is a complex structure with many stakeholders. Michael Gibson addresses this type issue in "Science new social contract with society" [11] where he states that the price for increased complexity in society is a pervasive uncertainty. The same can be the case in the marine industry where ownership and responsibility can be hard to identify [12].

7. Responsible Research and Innovation Challenges

A responsible research and innovation approach [13] continuously seeks to:

- Anticipate – evaluate the impacts induced by the research activity
- Reflect – reflect on the implications of the results from the research activity
- Engage – opening up for relevant discussions in a broader audience
- Act – utilizing the above processes to influence the direction of the research process.

One way of fulfilling the above principles is that all societal actors are to cooperate and work together to align both expectations and results to societal needs. The SARex project has incorporated representatives from all major stakeholders within the maritime industry. This induces continuously discussions and dialog on the purpose, direction and implications of our findings.

The SARex project and its findings have also been present at several academic conferences, in addition to industry conferences. As the project has broad public interest, the results have also been communicated through media, in addition to several closed industry seminars.

Through our work on communication we have obtained dialog with multiple stakeholders that otherwise would have been difficult.

Among the maritime industry, in addition to all project participants, there is no disagreement on the importance of the issue of survivability on rescue crafts, and further knowledge is required. There is however conflict of interests among the different stakeholders. This is mainly due to the potential cost induced by our findings.

One of the main principles of the regulatory regime imposed by IMO is that there is to be no discrimination among the member states. From a vessel operator/transportation providers point of view this means that all competitors are to compete on "equal grounds". To our knowledge, there is currently no common consensus with regards to interpretation of the IMO Polar Code. If SARex can contribute to help the global marine industry reach a consensus that would be beneficial for all parties involved, despite the fact that there will be a higher cost associated with the solution. IMO processes are essential to reach this consensus. This takes time and a closure cannot be expected for several years.

There are currently no indications that disruptive effects will reduce the need for cold climate marine activities in the future. However, there might be some unforeseen and unpredicted societal effects of our work. There are many examples where societal impacts and effects have not been adequately considered in the early phases of the project [14]. Unforeseen impacts of our work remains speculations. If it turns out that survival along the lines defined in the IMO Polar Code (a minimum of 5 days) is not achievable within the limits of the industry, a combination the following effects are to be expected:

- IMO will have to reverse the implementation of the Polar Code. Reversing the implementation of the regulations is extremely difficult. It is a process that will take years and involve high-level political discussions in the IMO.
- Investment in the expedition cruise industry is dependent on predictability and transparency. As the Polar Code represents a possible high cost, and potentially a loss of income due to downsizing measures, investments can be regarded as a high risk venture.
- The nearest vessel in case of a cruise ship incident in the high north is most likely another cruise ship. With lower economic margins there is to be expected a reduced activity. Fewer vessels will again result in a longer response time. As a result, the passengers on expedition cruises to the high Arctic will expose themselves to an increased risk.
- The polar states will have to substantially increase their budgets set aside for Search & Rescue capacities to ensure a sustainable Arctic development.

The societal impacts and effects will most likely not reveal it selves until a major accident occurs, typically involving loss of life. Not only will the accident have to occur, but the accident will also have to be communicated to the public through the public media channels. Based on historical events it is a paradox that lives have to be lost before a major effort is put into regulatory development trying to safeguard lives.

8. Concluding remarks

Closure of the challenge related to survival in rescue crafts is achieved when the global marine industry reaches a consensus and perceives the problem as solved. This process will be influenced by not only the objective facts generated by academia, but also political and economic agendas. Science can only indicate that a potential disaster can occur and the effect of the different mechanisms at play.

It is evident that an increasing part of the marine industry is moving its operation into polar regions. Limited activity has been observed with regards to development of infrastructure in the relevant areas. This includes communication, SAR, and oil spill preparedness. This lack of infrastructure put a larger strain and responsibility on the industry which has to mitigate this lack of infrastructure. The costs associated with these mitigation-measures have to be covered by the industry.

As the global maritime industry is a complex and competitive industry where almost every party has a separate agenda. It is highly important that we, as a scientific institution, stay clear of commercial economic opportunities related to this work. Our credibility within the maritime sector is dependent maintaining the status as an objective third-party. The moment we, as a scientific institution, contribute in driving a political process in a direction where our motivation can be linked to our own economic gain; industry, regulators, governments and the IMO will question our credibility. This will terminate our role as a leading knowledge provider for the maritime sector.

References

[1] International Maritime Organization 2016 *POLAR Code-International Code for Ships Operating in Polar Waters* (London-IMO Publishing, ISBN 978-92-801-1628-1)

[2] Brunvoll R 2015 *MASTERPLAN Svalbard mot 2025-Sysselmannen Svalbard.* Retrieved from http://www.sysselmannen.no/Documents/Svalbard_Miljovernfond_dok/Prosjekter/Rapporter/2015/14-%2020%20Masterplan%20Svalbard%20mot%202025.pdf

[3] Solberg K E, Gudmestad O T and Kvamme B O 2016 *Exercise Report-SARex Spitzbergen, Search and rescue exercise conducted north of Spitzbergen* (University of Stavanger, ISSN 0806-7031/ISBN 978-82-7644-677-7) Retrieved from https://brage.bibsys.no/xmlui/handle/11250/2414815

[4] International Maritime Organization 2017 Retrieved from http://www.imo.org/en/Pages/Default.aspx

[5] Sarasvathy S D 2001 *Academy of management Review* **26(2)** pp. 243-63.
[6] Sjøfartsdirektoratet 2017 Retrieved from Sjøfartsdirektoratet: https://www.sjofartsdir.no/
[7] Petroleum Safety Authority 2017 Retrieved from http://www.ptil.no/?lang=en_US
[8] Norsk Sjømannsforbund 2017 Retrieved from http://www.sjomannsforbundet.no/?gclid =CL2h8Yr989MCFZbJsgodigQKJA
[9] Jasanoff S 2007 *Nature* **450(7166)** pp. 33-33.
[10] Bijker, W 1984 *Social Construction of Technology. In Social Construction of Facts and Artifacts* (Sage Publications).
[11] Gibson M 1999 *Nature* **402** pp. C81-4.
[12] NTB 2017 Februar 23 Dagbladet. *Aner ikke hvem som eier Jæren-havarert skip. Skipet har heller ikke godkjenning til å seile:* Retrieved from http://www.dagbladet.no/nyheter/aner-ikke-hvem-som-eier-jaeren-havarert-skip-skipet-har-heller-ikke-godkjenning-til-a-seile/67327657
[13] EPSRC's AREA framework for Responsible Research and Innovation 2017 Retrieved from https://www.epsrc.ac.uk/research/framework/area/
[14] van den Hoven J ed 2013 *Options for strengthening Responsible Research and Innovation* (Brussels-Publications Office of the European Union).

On weather limitations for safe marine operations in the Barents Sea

A P Orimolade* and **O T Gudmestad**
Department of Mechanical and Structural Engineering and Materials Science, University of Stavanger, Stavanger, Norway

*Contact author: aporimolade@hotmail.com

Abstract. The Barents Sea wave conditions and the weather conditions that often accompany fully developed polar lows are presented in this paper. The studies on the extreme waves showed that the extreme waves in the Barents Sea decreases as we move further north. The Barents Sea wave conditions are found to be more accessible for marine operations in the summer months compared to the North Sea and the Norwegian Sea wave conditions. However, the quality of actual weather forecasts and the contribution of long periodic swells to the total sea in the Barents Sea could negate the longer weather windows observed in the area. Further, the weather conditions that often accompany fully developed polar lows represent limitations to marine operations in the Barents Sea. It was found that the significant wave height in polar lows could be up to 9 m. In addition, snow and ice accretion in polar lows are deemed operation and safety hazards. Overall, marine operation may not be carried out within the period when a polar low is likely to occur.

1. Introduction

With increasing activities in the Barents Sea, the improved knowledge of the various metocean conditions (meteorology and oceanography) that are peculiar to this area of the Arctic sea, will aid in planning and execution of safe marine operations, and in designing reliable offshore structures. The main objective of this paper is to contribute to the knowledge of the wave conditions, and the occurrences of polar lows in the Barents Sea. The findings presented in the paper are an abridge version of the summary of the PhD thesis titled "*Weather limitations for safe marine operations in the Barents Sea*" [1].

We have studied the Barents Sea wave conditions from two perspectives: the extreme wave conditions for design purposes, and the relatively calm wave conditions for marine operations. The studies on the extreme wave conditions are driven mainly by the availability of wave hindcast data. Here, we have considered 57 years wave hindcast data for four locations, the locations are shown in Figure 1 as BS1 – BS4. The hindcast data are from the NORA10 (Norwegian Reanalysis 10 km) database, and are made available courtesy of Birgitte Furevik of MET Norway (Norwegian Meteorological Institute).

The studies on the wave conditions for marine operations are sub-divided into three categories. These are; study of the probable weather windows in the Barents Sea using wave hindcast data, study of the quality of wave forecasts in the Barents Sea comparing wave forecasts and wave observations, and study of the contribution of swell waves to the total sea in the Barents Sea using wave hindcast data. For the probable weather windows, we have considered the 57 years wave hindcast data for BS1 – BS5, Heidrun in the Norwegian Sea, and Ekofisk in the North Sea, see Figure 1. Consideration is given here to Heidrun and Ekofisk for comparison. For the quality of wave forecasts, we have considered wave forecasts and wave observations at BS6 (Figure 2), the results are compared with the quality of wave forecasts at Heidrun. The contribution of swell waves to the total sea is studied using the 57 years wave hindcast for BS1 – BS4 (Figure 1). The wave observations, wave forecasts, and wave hindcast are also made available courtesy of Birgitte Furevik of MET Norway.

With regards to the occurrences of polar lows, our focus was on the weather conditions that often accompany the fully developed polar lows, and how such weather conditions may limit marine operations in the Barents Sea. Here, we studied the probable significant wave heights that may accompany the fully developed polar lows, the probable additional snow and ice (from sea spray icing) loads that a vessel caught up in polar lows may experience, and the effects of the additional loads on the stability of different vessels.

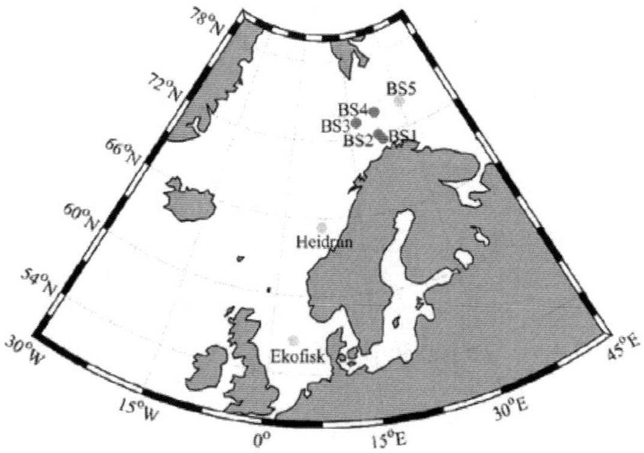

Figure 1. Locations of wave hindcast data for extreme wave studies (red dots), and locations of wave hidncast data for weather windows studies (green dots) [1].

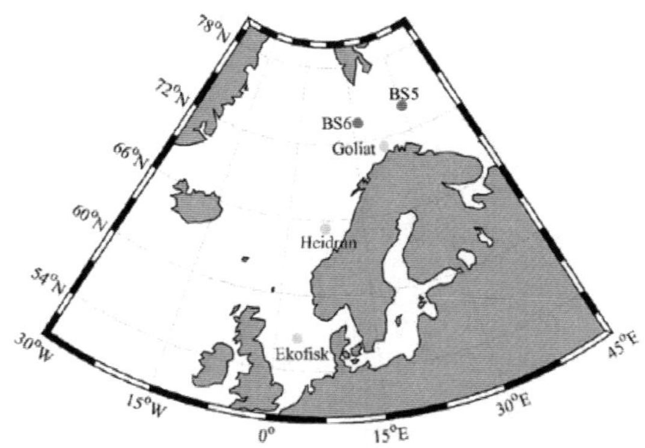

Figure 2. Locations of wave observations from buoys (red dots), and locations of wave observations from wave radar (green dots) [1].

The study of the probably significant wave heights is driven by the records of wind speeds in polar lows from 1999 – 2015, and a wave prediction model proposed by [2]. Also studied is the accuracy of wave forecasts during the passage of polar lows, using wave forecasts and wave observations during the passage of two polar low events near the location Goliat (Figure 2). The studies on the probable additional snow and ice loads is driven by few metocean observations and previously published papers.

Overall, the studies summarized in this paper showed that the extreme waves in the Barents Sea decreases as we move further north, this is in agreement with [3]. The 57 years wave hindcast suggested good possibility of carrying out marine operations in the summer months, here defined as the periods from May – September. The quality of wave forecasts for the Barents Sea is also found to be good at lead times less than 30 hours. However, the contribution of swell waves to the total sea may further limit the observed weather windows. In addition, it was found that the significant wave heights in polar lows would limit most marine operations, particularly in the winter months (October – April) when polar lows usually occur. However, when compared with the 100-year wave conditions in the Barents Sea, the significant wave height in polar lows is not as large. The additional snow and ice loads that a vessel at sea may experience during polar low occurrences are deemed both operation and safety hazards.

2. The extreme wave conditions

The sea state may be characterized by a set of integrated parameters, including the significant wave height (H_s), the spectral peak period (T_p), and the direction of propagation (θ). Among these parameters, the H_s is commonly used to describe the severity of the sea state. Often, the sea state is a combination of wind sea and swell waves. In the study of the extreme wave conditions for the Barents Sea, we have considered the omni-directional total sea. The extreme waves are analyzed by considering the marginal distribution of H_s, and the joint distribution of H_s and T_p.

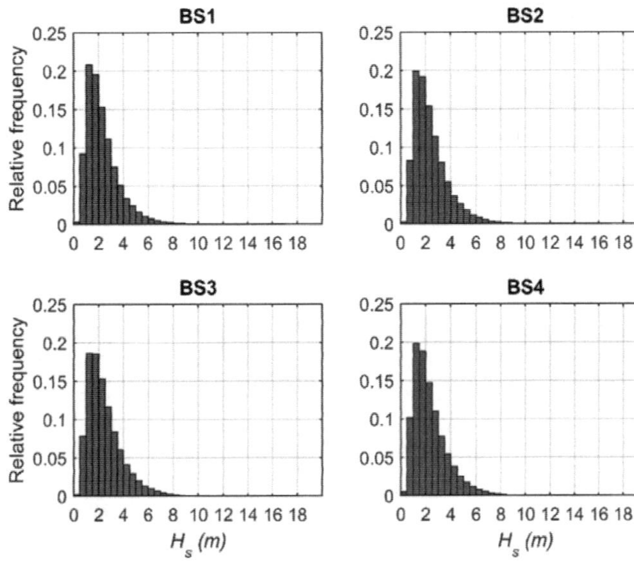

Figure 3. Relative frequencies of the significant wave heights (H_s) for locations BS1 – BS4 (1957 – 2014): the data is bin using the class interval of 0.5 m [1].

2.1 Marginal distribution of H_s

The relative frequencies of the H_s for the locations BS1 – BS4 is shown in Figure 3 based on a class interval of 0.5 m. The values of the H_s are from 0.1 – 17.0 m for BS1 and BS4, 0.2 – 17.0 m for BS2, and 0.2 – 17.9 m for BS3.

To predict the 100-year H_s at the locations, three commonly used extreme wave prediction methods are adopted. These are, the initial distribution (ID) method, the peak over threshold (POT) method, and the annual maximum (AM) method. For the ID method, the 3-parameter Weibull distribution is selected as the candidate probability distribution [4] that is fitted to all the 3-hourly wave hindcast data at each location. The parameters of the 3-parameter Weibull distribution are estimated using the method of moments (MOM).

For the POT method, peak excesses above a threshold of 7 m, and above a threshold of 9 m are considered. The threshold selection is based on two of the threshold selection techniques described by [5]. The generalized Pareto distribution (GPD) [5], the exponential distribution, and the 2-parameter Weibull distribution [4] are considered as probable candidate probability distributions for the peak excesses. The parameters of the GPD are estimated using the maximum likelihood estimation (MLE) method [5], while the parameters of the exponential distribution and the parameters of the 2-parameter Weibull distribution are estimated using the MOM. A comparison of the estimated parameters and the visual goodness of fit of the probability distributions to the peak excesses suggested that the 2-parameter Weibull is sufficient.

For the AM method, annual is defined as the period from September to August. The generalized extreme value (GEV) distribution [5], and the Gumbel distribution [4], are considered as probable candidate probability distributions for the annual maxima wave hindcast data. The parameters of the GEV distribution are estimated using the MLE method, while the parameters of the Gumbel distribution are estimated using the MOM. The Gumbel distribution is adopted in this study, as the shape parameter of the GEV distribution does not deviate significantly from zero.

The 100-year H_s at the locations BS1 – BS4 are estimated from the parameters of the adopted probability distributions. A summary of the estimated 100-year H_s at the four locations is shown in Table 1. Generally, the estimated values suggested a decrease in the extreme wave further north. The differences in the estimated 100-year H_s when using the different methods give an indication of the probable uncertainty that may be associated with the estimates.

It is difficult to select the best estimate from the three extreme wave prediction methods used in this study. This is because, each of the methods has its limitations. We find the estimates from the POT method to be dependent on the selected threshold, and the selection of the suitable threshold to be rather challenging. Further, we find both the POT method and the AM method to be sensitive to the few extreme significant wave heights in the wave hindcast data. We therefore adopted the estimates from the ID method.

2.2 Joint distribution of H_s and T_p

The joint frequencies of H_s and T_p based on the 57 years wave hindcast data for the four locations is shown in Figure 4. The figure shows that the T_p at the locations can be up to 24 s.

From the joint frequencies presented in Figure 4, the long-term joint distribution of H_s and T_p are established using the Conditional Modelling Approach (CMA) [4, 7 – 9]. In this approach, the marginal distribution of H_s is modelled by a 3-parameter Weibull probability density function, and the distribution of T_p conditional on H_s is modelled by a lognormal probability density function.

Extrapolating the parameters of the lognormal distribution (mean and variance) to extreme sea states require fitting of continuous functions to the point estimates. Over the years, a 3-parameter continuous function or a 4-parameter continuous function has been found to be a good fit to the variance point estimates [4, 8, 10, 11]. In this study, both the 3-parameter and the 4-parameter continuous functions are compared, and the 3-parameter continuous function is found to be sufficient.

For the four locations, the estimated parameters of the joint distribution of H_s and T_p are summarized in Table 2. These parameters can be used in estimating the joint H_s-T_p for any return period, in generating contour plots of long-term joint distribution, or in generating environmental contour lines.

Using the inverse first-order reliability method (IFORM) described by [11], environmental contour lines are generated for the four locations based on the parameters in Table 2, the contour lines are shown in Figure 5. These can be used in selecting short-term sea states for an approximate long-term response analysis for the locations.

3. Wave conditions for marine operations

Marine operations can be classified into *weather restricted* and *unrestricted* marine operations [12] described weather restricted marine operations as operations with planned operation period of less than 72 hours (3 days), while unrestricted marine operations are of longer duration. Safe planning and execution of weather restricted marine operations are dependent on reliable weather forecasts, while unrestricted marine operations could be planned using statistical data.

In this study, we have examined the wave conditions for marine operations in the Barents Sea from three perspectives; the probable weather windows for an operation that required that the H_s is less than 3 m, the quality of the forecasted H_s, and the contribution of swell waves to the total sea in the area.

Table 1. Estimated 100-year H_s for locations BS1 – BS4 using the three extreme wave prediction methods, the maximum value at each location is shown in italics; Hs (m). (Table from [6])

Method	**BS1**	**BS2**	**BS3**	**BS4**
ID	*16.09*	15.83	15.11	14.46
POT (7 m)	15.98	15.89	15.25	14.24
POT (9 m)	15.96	*16.26*	*16.47*	*15.95*
AM	16.03	16.09	15.92	14.95

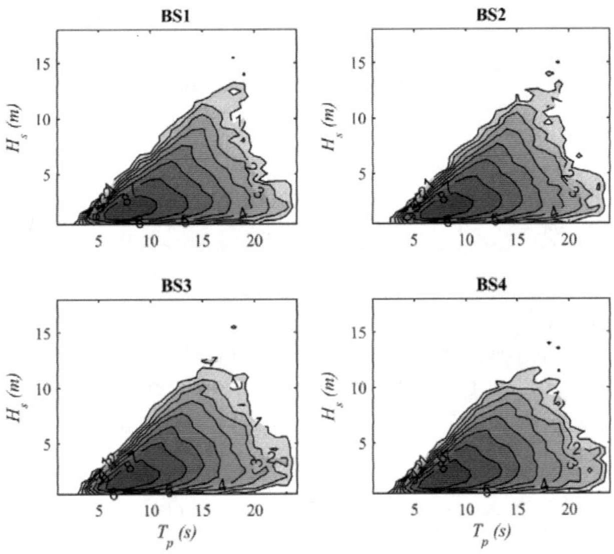

Figure 4. Logarithmic contour plots of H_s and T_p over the 57 years period (BS1 – BS4) [1].

Table 2. Estimated parameters of the joint probability distribution, the variance coefficient b_1 is fixed at 0.001 [1])

	Marginal H_s 3-parameter Weibull			**Conditional T_p** Mean coefficients			**Conditional T_p** Variance coefficients	
	α	β	γ	a_1	a_2	a_3	b_2	b_3
BS1	1.12	1.58	0.77	1.46	0.58	0.34	0.134	0.318
BS2	1.16	1.69	0.76	1.38	0.64	0.31	0.113	0.275
BS3	1.24	1.85	0.70	1.42	0.60	0.32	0.092	0.239
BS4	1.24	1.79	0.73	1.42	0.58	0.34	0.107	0.252

α - shape parameter, β - scale parameter, γ - location parameter
a_1, a_2, a_3 – coefficients of the fitted mean continuous function
b_1, b_2, b_3 – coefficients of the fitted variance continuous function

3.1 The weather windows

The probable weather windows for marine operations is investigated using the 57 years wave hindcast data. It should be noted that the presented weather windows are only useful for planning purposes, giving the planner an idea of what is likely from historical data. Actual weather forecasts at the locations are required for safe execution of operations.

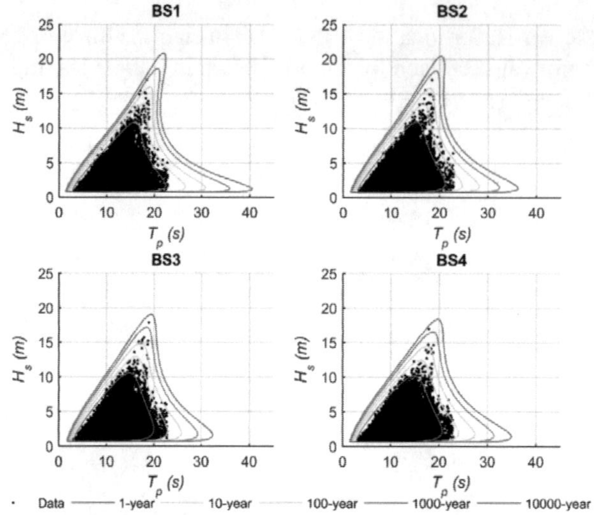

Figure 5. Environmental contour lines of H_s and T_p based on the 57 years wave hindcast data (black dots) [1].

Persistence analysis is used to estimate the probable weather window in each month, here, we defined persistence as a continuous duration in which the significant wave height (H_s) remains below (calm duration) or above (storm duration) the defined limiting significant wave height (e.g. $H_{s(\lim)} = 3$ m) in each month. The monthly average calm duration (τ_c) is defined as:

$$\tau_c = T_c / N_c \quad (1)$$

where T_c is the total calm hours in a month < $H_{s(\lim)}$, N_c is the number of calm windows in the month < $H_{s(\lim)}$.

For the limiting H_s of 3 m, the monthly average calm duration estimated from (1) is presented in Figure 6 for BS1 – BS4. The figure shows that July is the month with the probable longest weather window for a limiting H_s of 3 m. As indicated by the blue line in the figure, it may be difficult to find a 3-day weather window for a weather restricted marine operation in these areas in the winter months (October – April).

A comparison of the weather windows at BS5, Heidrun, and Ekofisk suggested that there is a higher probability of finding a 3-day weather window in the Barents Sea in the summer months, compared to the North Sea and the Norwegian Sea. Details of this comparison is presented in [13].

3.2 Wave forecast accuracy

As noted in Section IIIA, reliable weather forecasts are very important for safe marine operations. Considering the limited infrastructures in the Barents Sea, and its Arctic conditions, it may be more difficult to find reliable weather forecasts for the Barents Sea compared to the Norwegian Sea and the North Sea. In this section, we present the accuracy of wave forecasts for a location in the Barents Sea (BS6), compared to Heidrun in the Norwegian Sea.

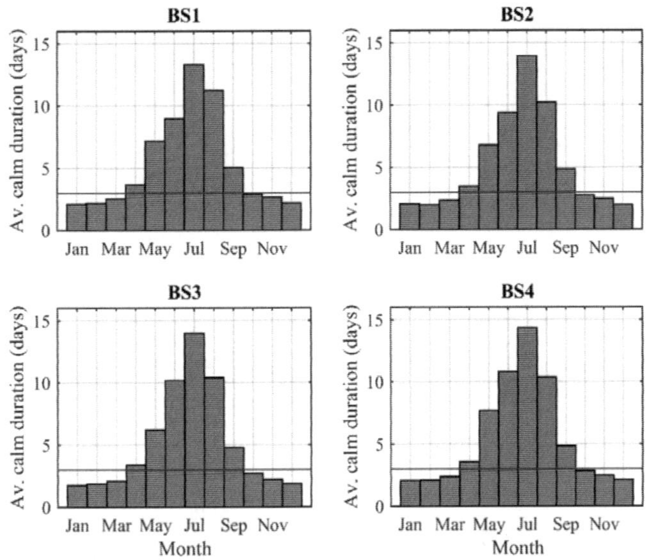

Figure 6. Monthly average calm duration for a limiting H_s of 3 m, using the 57-year wave hindcast for locations BS1 – BS4; a 3-day weather window is difficult to find in the winter months indicated by the blue line [1].

We have considered the available buoy wave observations from February – September 2010, for the Barents Sea location. Wave observations from a radar aboard the Heidrun platform have been obtained for the same period. Wave forecasts from the WAM10 model at 10 km resolution are obtained for the two locations.

The wave forecasts are then compared with the wave observations for each location, to estimate the accuracy of the forecasted waves. The measures of accuracy adopted are the mean absolute error (MAE) and the root mean square error (RMSE). Generally, the lower the estimated values of these measures of accuracy, the higher the accuracy of the forecast.

The estimated values of MAE and RMSE for the two locations are presented in Figure 7 when considering all the available wave observations. In Figure 8, we presented these values when considering only significant wave heights less than or equal to 3 m. A comparison of Figure 7 and Figure 8 shows that the accuracy of the forecasts do not change significantly when using only wave observations less than or equal to 3 m.

From the figures, it is observed that the accuracy of the wave forecasts at the Barents Sea location is higher in the summer months (May – September), compared to the winter months (February – April). For the Norwegian Sea location, no particular trend is observed in the forecast accuracy, and the accuracy appears even lower in July compared to some winter months. It is not straightforward to explain the observed trend for the Norwegian Sea location; however, it is noted that the wave observations are from different sources.

The inferences that can be drawn from the wave forecast accuracy are: the wave forecasts for the Barents Sea are of good quality at lead times less than 30 hours, the accuracy then decreases rapidly beyond this lead time. However, these findings may not be generalized, considering that the analysis is based on only one year data, which does not reflect inter-annual variability in the wave observations.

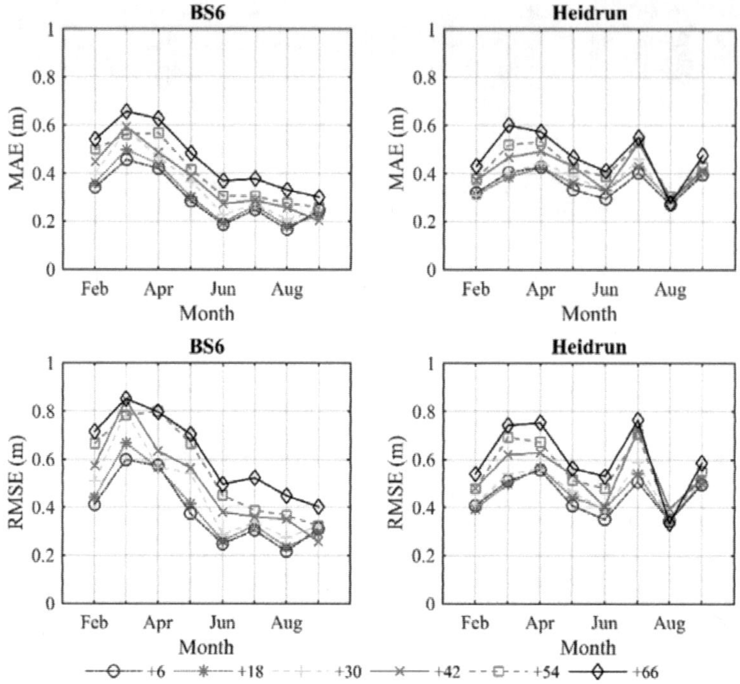

Figure 7. Comparison of wave forecasts accuracy at the locations BS6 and Heidrun using all the available wave observations in 2010 [1].

MAE – mean absolute error, RMSE – root mean square error. The legend represents the different forecast lead time.

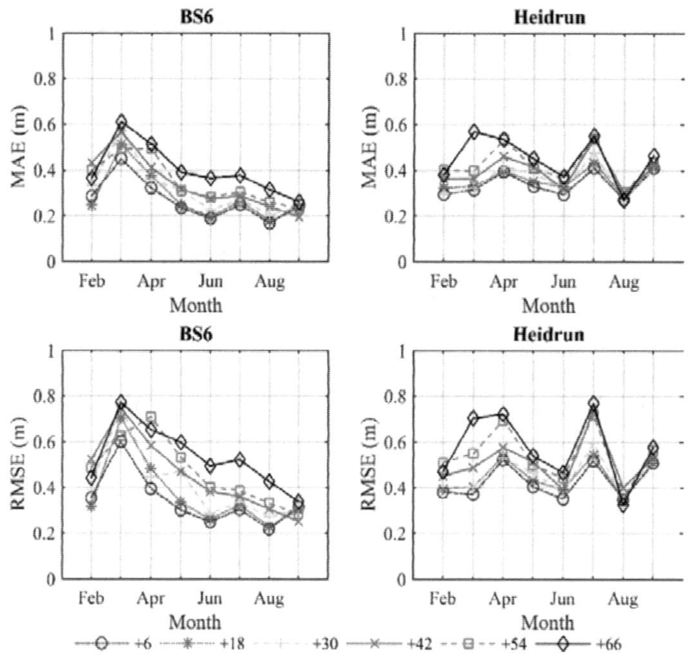

Figure 8. Comparison of wave forecasts accuracy at the locations BS6 and Heidrun using only wave observations less than or equal to 3 m in 2010 [1].

3.3 Contribution of swell waves to the total sea

Swell waves are long periodic waves generated by distant storms. Compared to a young wind sea generated by local storm, a "swell is rather regular and long-crested", and therefore has a narrow banded spectrum [14 - 16]. On the other hand, a "young wind sea is irregular and short-crested", having a much broader spectrum [15, 16].

The knowledge of the swell waves at a location for marine operations is of importance. This is because, even for a very calm sea state (H_s < 1 m), the presence of swell waves may limit the operation, and lead to long waiting on weather (WOW). A practical example is given by [16]. In addition, the knowledge of the swell waves at a location will aid in selecting a suitable wave spectrum for the planning of marine operations.

For locations in the Nordic Seas (North Sea, Norwegian Sea, and Barents Sea), [17] showed that swells are more prevalent in the Norwegian Sea, followed by the Barents Sea, with the least swell occurring in the North Sea. In this study, we investigated the contribution of swells to the total sea in the Barents Sea, using the 57 years wave hindcast data for BS1 – BS4.

For the four locations, the contribution of wind waves and swell waves to the total sea, based on their direction of propagation are shown in Figure 9 and Figure 10. It can be inferred that the contribution of southwesterly/westerly swells to the total sea in both winter and summer is significant. Therefore, swell waves should be well accounted for when planning for marine operations in the Barents Sea.

On the selection of wave spectra for marine operations: When planning for marine operations, the foregoing suggests that a double peaked wave spectrum is better suitable to describe the wave environment in the Barents Sea. The Torsethaugen wave spectrum [18] is therefore recommended.

However, it should be noted that the choice of the Torsethaugen spectrum is not a rigid one, the JONSWAP wave spectrum [19] may also be applicable for certain classes of H_s and T_p combination as shown in [20]. Further, the Torsethaugen wave spectrum does not account for wave directionality, in detailed analysis therefore, one must model the wind sea and the swell components separately. In such analysis, a combination of the Torsethaugen and JONSWAP wave spectra could be considered.

4. Polar lows

The polar low is one of the meteorological phenomena that are considered as additional challenges to marine operations in the northern Norwegian Sea and in the Barents Sea. The definition of a polar low that is commonly adopted is "*a small, but fairly intense maritime cyclone that forms poleward of the main baroclininc zone*" [21]. They have horizontal extents approximately between 150 and 600 km, and are referred to as mesoscale. They have been likened to hurricanes due to their appearances on satellite images [22]. However, polar lows are of smaller scales and are less intense compared to hurricanes. A satellite image of a fully developed polar low over the Barents Sea is shown in Figure 11.

An average of 12 fully developed polar lows have been observed in the Nordic Seas yearly, with a majority of these occurring in the northern Norwegian Sea and in the Barents Sea. Figure 12 shows the locations where fully developed polar lows are first observed in the Nordic Seas from December 1999 – October 2015. It is seen from the figure that a majority of the polar lows are observed in northern Norwegian Sea and Barents Sea.

Due to their mesoscale nature, and the limited infrastructures, polar lows are difficult to forecast. In many instances, the meteorologists do not identify a polar low until fully developed. See [23] for an example of a polar low event where the meteorologist did not identify the polar low until it hits a semi-submersible rig in the Barents Sea.

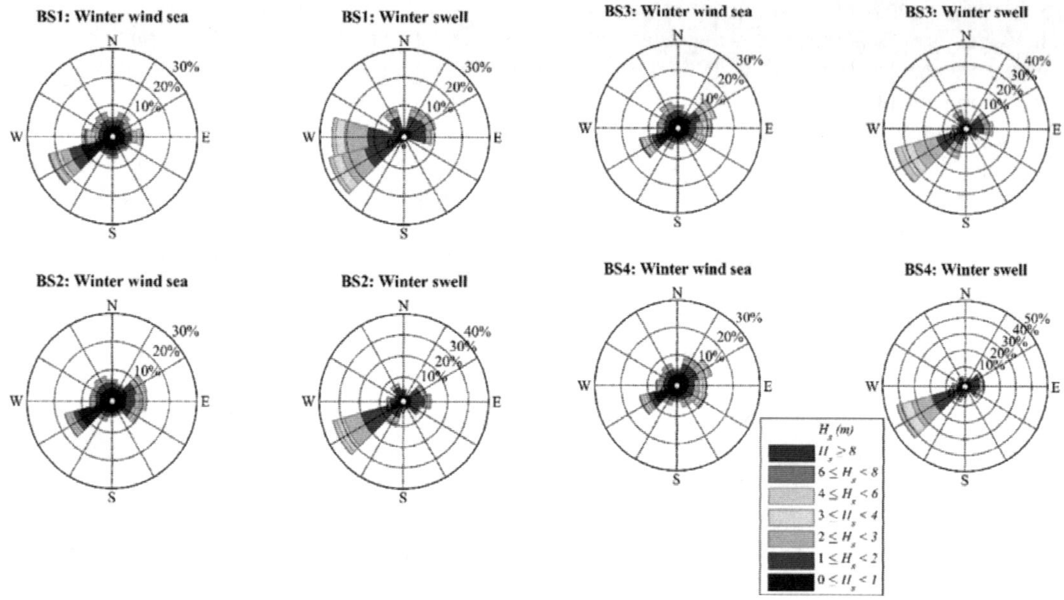

Figure 9. Winter wave roses for the four locations, based on the 57-year wave hindcast data: the length of each rose petal shows the frequency of occurrence of waves (wind sea/swell) from different directions [1].

The polar lows are considered a challenge to marine operations because of the weather conditions that often accompany them, including high wind speeds, heavy snowfall, poor visibility, and possible large waves. In this section, we focused on the wave growth in polar lows, the probable snow and ice accretion that a vessel caught up in polar lows may experience, and their effects on the stability of vessels of different sizes.

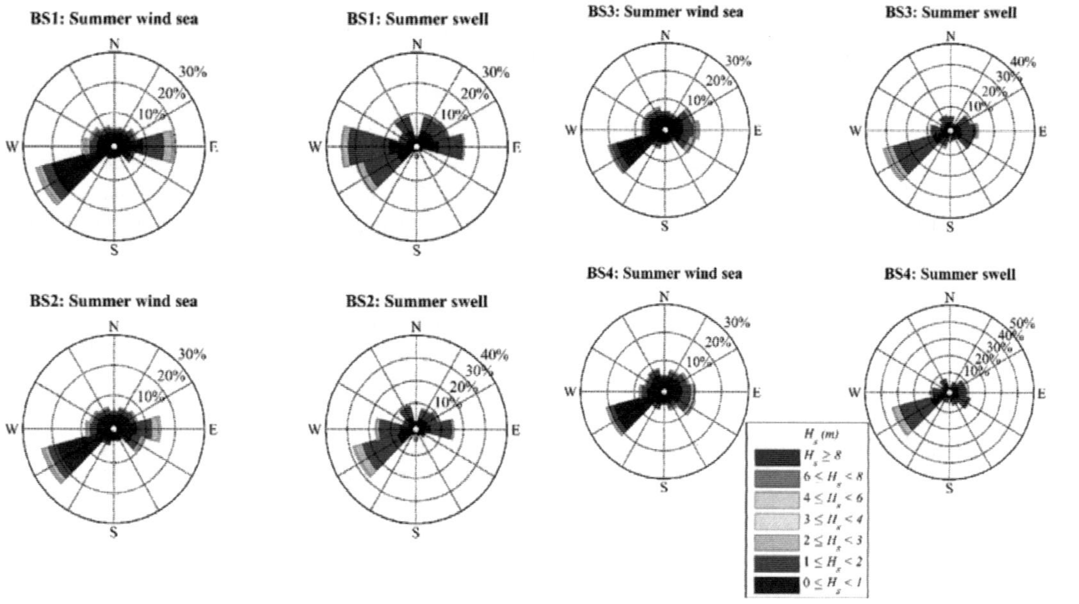

Figure 10. Summer wave roses for the four locations, based on the 57-year wave hindcast data: the length of each rose petal shows the frequency of occurrence of waves (wind sea/swell) from different directions [1].

4.1 Waves in polar lows

Generally, wave growth is associated with wind speed and fetch. Considering the limited horizontal extent of polar lows, one may not expect large waves in polar lows. However, a wave moving along the path of a translating polar low can experience the fetch winds for a longer duration, *if the wave group velocity matches the polar low velocity*. Such a wave may therefore grow larger than a wave in a rather stationary fetch. The described phenomenon is termed group velocity quasi-resonance, effective fetch, dynamic fetch, trapped fetch, or extended fetch [2, 24 - 27].

Figure 11. A satellite image of a fully developed polar low over the Barents Sea from 0418 UTC 27 February 1987 (Courtesy Neil Lonie: NERC Satellite Receiving Station, Dundee University, Scotland, http://www.sat.dundee.ac.uk/).

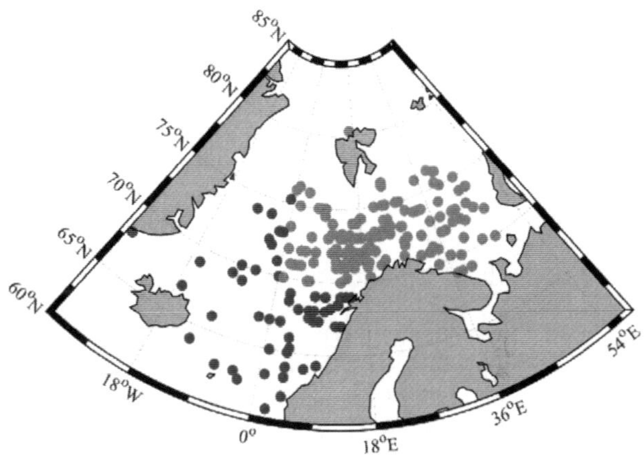

Figure 12. Locations where fully developed polar lows are first observed in the Nordic Seas, from December 1999 – October 2015; the red dots show the concentration of the polar lows in northern Norwegian Sea and the Barents Sea [1]. (The data is obtained courtesy of G. Noer and T. Lien, MET Norway).

Using the effective fetch concept and the solution to the fetch limited empirical relations by [19, 28], [2] proposed a one-dimensional wave model for the prediction of wave growth in polar lows. According to [2], a wave in polar low could experience an extended spectral evolution if the following conditions are satisfied:

$$l_{\max} >> l \tag{2}$$

$$0.25 < V / U < 0.50 \tag{3}$$

$$l > 8.13 \cdot 10^4 \cdot (V^4 / gU^2) \tag{4}$$

where l is the fetch associated with the polar low, l_{max} is the maximum fetch along a relatively linear path, V is the polar low velocity, and U is the wind speed associated with the polar low.

By adopting the wave prediction model by [2], we find that the waves in a majority of the fully developed polar lows in the Nordic Seas could have experienced an extended spectral evolution, if the polar lows velocities, V, have been between 8 – 10 m/s [29]. We also find that the significant wave heights at the end of the spectral evolution could be greater than 9 m [29]. As earlier noted, such large waves could not have developed if the polar lows are rather stationary.

4.2 Wave forecast accuracy in polar lows

In Section IIIB, we presented the accuracy of wave forecasts in the Barents Sea using the data for one location. No polar lows were observed at the location in the periods covered by the data. In this section, we examined the accuracy of wave forecasts in polar lows, using two specific polar low events from December 2015. The polar lows under consideration have passed near the Goliat platform in the Barents Sea.

The wave forecasts used in the study are from two sources, the WAM10 model earlier mentioned, and a recent model, AROME-Arctic at 4 km resolution. We compared the wave forecasts to the buoy wave observations at Goliat, using data from one day before the arrival of the polar lows, up to one day after the polar lows have passed the location of the observations (Goliat), see [29].

The comparison showed that both wave forecast models under predicted the observed waves in the two polar low events, especially at the time when then waves are observed to reach their maximum values. This observation is not surprising, as wave forecasts in polar lows is considered a complex problem, considering the nonlinear relationship between the stationary storm factors of wind speed, duration, and fetch, and the translating speed of a wave generating polar low.

4.3 Snow and ice accretion in polar lows

The amount of snowfall in polar lows are usually referred to qualitatively, mainly as *heavy, moderate, or light snowfall.* Heavy snowfall often accompany fully developed polar lows; see for example [21]. According to [30], a warning of heavy snow signifies the possibility of snowfall of 15 cm or more in 12 hours, or 20 cm or more in 24 hours.

In the study by [31], we concluded that up to 30 cm snow could accrete on the horizontal surfaces of vessels at sea during polar low occurrences. This is based mainly on reviewed papers, as there are very sparse information about snowfall at sea. The 30 cm snowfall is equivalent to the recommended 10000-year characteristic snow load by [3] if the snow density is 400 kg/m^3.

The accretion of sea spray icing on vessels at sea during polar low occurrences is attributed to the high wind speeds that often accompany polar lows in combination with low air temperatures. It is noted that severe ice accretion is often associated with sea spray icing resulting from wave-structure interaction, compared to sea spray icing resulting from wind. Therefore, a vessel in transit during polar lows may experience more ice accretion than a rather stationary vessel.

Using a few metocean observations, and a review of past icing experiences, [31] concluded that up to 25 cm ice could accrete on a vessel at sea during polar low occurrences. For a ship-shaped vessel, the 25 ice is considered in the vessel's maximum ice accretion zone, with a reduction factor considered at the other zones.

We have examined the effect of the additional 30 cm snow and 25 cm ice on the stability of three ship-shaped vessels of similar geometries but different sizes. We find that the stability of the smallest vessel is the most affected by the additional snow and ice loads. In our study, we have considered only symmetrical snow and ice distribution on the horizontal surfaces of the vessel. However, as shown by [32], asymmetrical snow and ice load distributions could be of higher concern.

In the study by "in press" [33], we also examined the effect of the additional snow and ice loads on the stability of semi-submersibles, using a sixth generation DP3 and moored harsh environment semi-submersible drilling rig as a case study.

By considering the snow load on all the semi-submersible horizontal surfaces, we find that the snow load could be very significant. For snow thicknesses of 15 and 30 cm, the resulting snow loads are 369.56 and 739.18 tons, respectively. However, we find that such symmetrically distributed snow loads will have little effect on the intact stability of a semi-submersible.

Ice loads resulting from sea spray icing are considered to accrete mainly on the windward side of the semi-submersible. We find that the ice load may not exceed 300 tons in any of the semi-submersible loading conditions (transit, operation, survival), even for ice thickness up to 25 cm. However, the ice load may exceed 600 tons if the sea spray icing event extends to the main deck of the semi-submersible, this is possible when the semi-submersible is on deep draft.

The case with the sea spray icing extending to the main deck is found to be the most dangerous, with the semi-submersible heeling about 12 degrees. It should be noted that heeling above 5 degrees will be of significant concern to personnel aboard a semi-submersible, and such heeling will generally hinder drilling operations.

5. Summary of findings

The estimated 100-year H_s from the four Barents Sea locations indicate a decrease in the extreme H_s further north, this is in agreement with the expected spatial trend in the wave climate in the Barents Sea.

Longer weather windows are observed for marine operations in the Barents Sea in the summer months, compared to the North Sea and the Norwegian Sea. Further, the wave forecasts for the Barents Sea are found to be of good accuracy, especially at forecast lead times less than 30 hours. However, in polar low situations, the wave forecast accuracy is deemed poor.

Although, waves in polar lows are not as large as the 100-year design wave for the Barents Sea, however, waves in polar lows will limit most marine operations. The probable snowfall and vessel icing in polar lows are additional limitations to marine operations in the area, generally snow and ice accretion are deemed both operation and safety hazards.

Overall, the weather conditions that often accompany fully developed polar lows represent limitations to marine operations in the Barents Sea. Marine operations should not be carried out within the periods when polar lows are likely to occur, even if the probability of occurrence is small.

Acknowledgment

The author would like to acknowledge the Research Council of Norway and the University of Stavanger, Norway (IN-10604), for providing the resources to prepare this paper.

The author would also like to acknowledge the contributions from all the co-authors in the papers used in writing this summary of the PhD findings.

References

[1] Orimolade A P 2017 *Weather limitations for marine operations in the Barents Sea* (Stavanger: University of Stavanger)

[2] Dysthe K B and Harbitz A 1987 Big waves from polar lows? *Tellus A*: **39** 500-08.

[3] Norsk Sokkels Konkuranseposisjon 2016 *Actions and action effects* (Lysaker: Standards Norway) NORSOK Standard N-003

[4] Det Norske Veritas 2014 *Environmental conditions and environmental loads* (Oslo: Det Norske Veritas) DNV-RP-C205

[5] Coles S 2001 *An introduction to statistical modeling of extreme values* (London: Springer)

[6] Orimolade A P, Haver S and Gudmestad O T 2016 Estimation of extreme significant wave heights and the associated uncertainties: A case study using NORA10 hindcast data for the Barents Sea *Mar. Struct.* **49** 1-17

[7] Bitner-Gregersen E M and Haver S 1991 *Joint environmental model for reliability calculations Proc. 1st Int. Offshore and Polar Engineerin Conference* (Edinburg: ISOPE)

[8] Bitner-Gregersen E M 2005 *Joint probabilistic description for combined seas Proc. ASME 2005 24th Int. Conference on Offshore Mechanics and Arctic Engineering* (Halkidiki: ASME)

[9] Bitner-Gregersen E M 2015 Joint met-ocean description for design and operations of marine structures *Appl. Ocean Res.* **51** 279-292

[10] Haver S and Nyhus K A 1986 *A wave climate description for long term response calculations Proc. 5th Int. Offshore Mechanics and Arctic Engineering Symposium* (Tokyo: ASME)

[11] Winterstein S R, Ude T C, Cornell C A, Bjerager P and Haver S 1993 *Proc. 6th Int. Conf. on Structural Safety and Reliability* (Innsbruck: ICOSSAR)

[12] Det Norske Veritas 2011 *Marine operations General* (Høvik: Det Norske Veritas) DNV-OS-H101

[13] Kvamme B O, Orimolade A P, Haver S K and Gudmestad O T 2016 *Marine operation windows offshore Norway Proc. ASME 2016 35th Int. Conference on Ocean Offshore and Arctic Engineering* (Busan: ASME)

[14] McCreight K K 1998 *A note on the selection of wave spectra for design evaluation* (Bethesda Maryland: Naval Surface Warfare Center, Carderock Division)

[15] Holthuijsen L H 2007 *Waves in oceanic and coastal waters* (Cambridge: Cambridge University Press)

[16] Gudmestad O T 2015 *Marine technology and operations: Theory & practice* (Southhampton: WIT Press)

[17] Semedo A, Vettor R, Breivik Ø, Sterl A, Reistad M, Soares C G and Lima D 2014 The wind sea and swell waves climate in the Nordic Seas *Ocean Dynam.* **65** 223-240

[18] Torsethaugen K 1996 *Model for double peaked wave spectrum* (Trondheim: SINTEF Civil and Environmental Engineering)

[19] Hasselmann K *et al.* 1973 Measurements of wind-wave growth and swell decay during the Joint North Sea Wave Project (JONSWAP) *Dtsch. Hydrogr. Z. Suppl. A 8*

[20] Orimolade A P and Gudmestad O T 2016 *Wave spectra for weather restricted marine operations with emphasis on operations in the Barents Sea Proc. 26th Int. Ocean and Polar Engineering Conference* (Rhodes: ISOPE)

[21] Rasmussen E A and Turner J 2003 *Polar lows: Mesoscale weather systems in the polar regions* (Cambridge: Cambridge University Press)

[22] Emanuel K A and Rotunno R Polar lows as arctic hurricanes 1989 *Tellus A* **41** 1-17

[23] Olsen A-M, Noer G, Øiestad M, and Magnusson A K 2012 *Validation of weather reports sent to Transocean Barents operated by NORECO at the EIK location in the Barents Sea* (Bergen/Tromsø: Metocean forecasting, Norwegian Meteorologial Institute)

[24] Swamp Group 1985 *Ocean wave modeling* (New York: Springer Science & Business Media)

[25] Young I R *A parametric hurricane wave prediction model* [Online]," Available: https://icce-ojs-tamu.tdl.org/icce/index.php/icce/article/view/ 4289 (Last accessed 26-05- 2017).

[26] MaCafee A W and Bowyer P J 2005 The modeling of trapped-fetch waves with tropical cyclones - A desktop operational model *Weather Forecast.* **20** 245-263

[27] Bowyer P J and MaCafee A W 2005 The theory of trapped-fetch waves with tropical cyclones – An operational perspective *Weather Forecast.* **20** 229–244

[28] Phillips O M 1977 *The dynamics of the upper ocean* (Cambridge: Cambridge University Press)

[29] Orimolade A P, Furevik B R, Noer G, Gudmestad O T and Samelson R M 2016 Waves in polar lows *J. Geophys. Res.: Oceans* **121** 6470-6481

[30] National Oceanic and Atmoshperic Administration website:
https://forecast.weather.gov/glossary.php?letter=h (Last accessed 20-06-2017)

[31] Orimolade A P, Gudmestad O T and Wold L E 2017 Vessel stability in polar low situations *Ships Offshore Struc.* **12** S82-S87

[32] Chung K K 1995 *Ship icing and stability* (Alberta: University of Alberta)

[33] Orimolade A P, Larsen S, and Gudmestad O T 2017 Vessel stability in polar low situations: Case study for semisubmersible drilling rigs *Ships Offshore Struc.*

A practical guideline for human error assessment: A causal model

Y Z Ayele[1, 2*], **A Barabadi**[2].

[1] Faculty of Engineering, Østfold University College, Fredrikstad, Norway
[2] Department of Engineering and Safety, UiT The Arctic University of Norway, Tromsø, Norway

*Contact author: yonas.z.ayele@uit.no

Abstract. To meet the availability target and reduce system downtime, effective maintenance have a great importance. However, maintenance performance is greatly affected in complex ways by human factors. Hence, to have an effective maintenance operation, these factors needs to be assessed and quantified. To avoid the inadequacies of traditional human error assessment (HEA) approaches, the application of Bayesian Networks (BN) is gaining popularity. The main purpose of this paper is to propose a HEA framework based on the BN for maintenance operation. The proposed framework aids for assessing the effects of human performance influencing factors on the likelihood of human error during maintenance activities. Further, the paper investigates how operational issues must be considered in system failure-rate analysis, maintenance planning, and prediction of human error in pre- and post-maintenance operations. The goal is to assess how performance monitoring and evaluation of human factors can effect better operation and maintenance.

1. Introduction

To meet the availability target and reduce system downtime, effective maintenance operation have a great importance. Maintenance performance is greatly affected in complex ways by human factors, since these factors play an important role in the pre- and post-maintenance operations [1, 2]. Further, there are several factors, which influences human error, such as internal (psychological and physiological) or external (technological and social) human performance shaping factors [2-4]. Therefore, in order to have an effective maintenance operation, these factors needs to be assessed and quantified. Further, all factors affecting human performance must be recognized and quantified throughout the various phases of the maintenance operation. The evaluations of these factors can help to have an effective maintenance operation with high level of safety.

Human factors analysis is rooted in the concept that the frequency and consequences of human errors are related to the maintenance work environment, culture, and maintenance procedures. This can be accounted for in the design of equipment, structures, maintenance processes and procedures. In general, the human error analysis concentrates on the effects of human performance on the maintenance operation, recognising that pre- and post-maintenance operations are influenced by human actions. To evaluate the risk of human error during maintenance activities, several studies have been carried out; see e.g. Apostolakis, et al. [5], Noroozi, et al. [1], Deacon, et al. [6], Noroozi, et al. [7], and Khan, et al. [8]. For instance, Noroozi, et al. [1] proposed a risk-based methodology for pre- and post-maintenance and, illustrated how to calculate the human error probabilities (HEP) for both maintenance activity. Apostolakis, et al. [5] critically reviewed the human cognitive reliability (HCR) and the success likelihood index methodology–multiattribute utility decomposition (SLIM-MAUD) model.

For a long time, traditional HEA strategies have been preoccupied with estimating the frequency of an undesirable event and its magnitude to give an overall measure of human error or risk. This type of human error or risk measure is quite useful for prioritising risks (the larger the number, the greater the risk); however, it is normally impractical and can be irrational when applied blindly [9]. One immediate problem with expressing the risk as a product of frequency and consequence is that, usually, one cannot directly obtain the numbers needed to calculate the risk without complete knowledge of the relation between the causes and effects of risk of events [10]. Hence, to avoid the inadequacies of traditional human error assessment approaches and to provide solutions to the problems mentioned above, the application of BN is gaining popularity and has been discussed in several pieces of the literature; see e.g. Fenton and Neil [9], Røed, et al. [10], Ayele, et al. [11] and Lee and Lee [12]. BNs are particularly useful in the HEA of maintenance activities, as they allow us to understand the causal relationship as well as combine any historical data that is available with qualitative data and subjective judgements about the risk of events.

Most of the conventional BN based risk assessment approaches are, however, broad, holistic, practical guides or roadmaps, developed for off-the-shelf nuclear systems, for non-maintenance operations. Moreover, current issues regarding the traditional HEA methodologies are lack of an explicit causal model linking observed behaviour to personal and situational factors–theoretical &empirical basis; lack of consistency or reproducibility; lack of traceability/ transparency; and variability in results [2]. Furthermore, there is a lack of causal methodology to quantify the HEPs of maintenance activities. A BN model offers a way to use both limited data and expert judgment to estimate HEPs. Data can be used to inform both the BN structure and the BN quantification.

The main purpose of this paper is thus to propose a BN based HEA (BNB-HEA) framework for maintenance operation. The proposed framework can help to assess the effects of human performance influencing factors (PIFs) on the likelihood of human error during maintenance activities. Further, this paper investigate how operational issues must be considered in system failure-rate analysis, maintenance planning, as well as prediction of human error in pre- and post-maintenance operations. The goal is to assess how performance monitoring and evaluation of human factors can effect better operation and maintenance.

The rest of the paper is organised as follows. The basic concepts of static BN are described in Section 2. Thereafter, the proposed BNB-HEA framework for maintenance activities is presented in Section 3. Finally, some concluding remarks are presented in Section 4.

In this paper, human error (HE) is taken to mean the failure to implement a definite task (or performance of a not allowed action) that could result in disruption of planned tasks or damage to equipment and property. The human error with which this paper is concerned are all in some way related to 'maintenance activity'. They are the actual or potential threat of adverse effect of inadequate training and skill, poor maintenance instructions and operating procedures, poor work layout, poor equipment design and improper work tools [13]. Further, risk is roughly considered as a function of the probability of particular occurrences and, the expected losses (lives lost, persons injured, damage to property, and disruption of economic activity) caused by a particular phenomenon [14, 15].

2. Bayesian network – a bird's eye view

Static BN are a probabilistic graphical model consists of a qualitative part, an acyclic directed graph (DAG), where the nodes represent random variables and a quantitative part, a set of conditional probability functions [16]. The nodes can be discrete or continuous, and may or may not be observable and the arcs (from parent to child) represent the conditional dependencies or the cause-effect relationships among the variables [16]. Parent nodes are nodes with links pointing towards the child nodes. Nodes that are not connected represent variables, which are conditionally independent of each other. Further, when BN contain discrete and continuous variables (nodes) generally it is called a Hybrid BN.

The quantitative part of a BN structure can be represented as a product of conditional distribution of each node N_i given its parents nodes $parents(N_i)$. Each node is described by the conditional probability function of that variable. Then, the joint probability distributions, considering discrete variable, can be expressed as:

$$\Pr(N_1, N_2, \dots, N_M) = \prod_i^M \Pr(N_i|parents(N_i)) \quad (1)$$

where: $\Pr(N_i|parents(N_i)$ is the conditional distribution mass function of node, N_i.

Moreover, to understand the complex interaction between PIFs and estimate the consequence of human error, a causal approach can be used. Figure 1 illustrates the causal relationships between a human error and other main events.

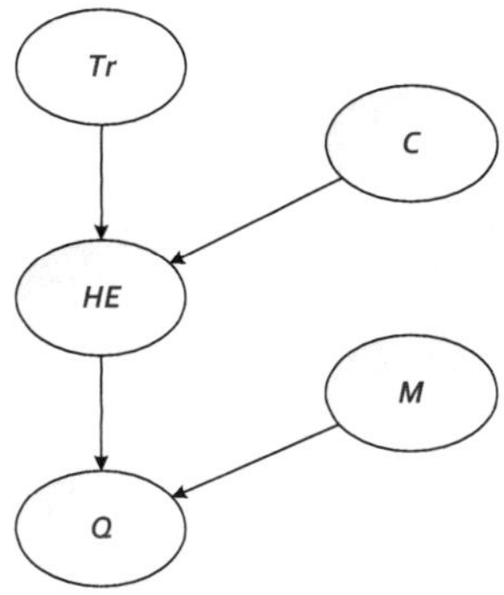

Figure 1. Causal relationships between a human error and other main events.

$\boldsymbol{Tr}$ denotes a set of any initiating (trigger) events, which are the performance influencing factors and $\boldsymbol{C}$ denotes the control measure, such as barriers. $\boldsymbol{HE}$ denotes the main risk event – human error, which shows that the cause of the trigger $\boldsymbol{Tr}$ produces the effect $\boldsymbol{HE}$. $\boldsymbol{Q}$ denotes the consequence of the human error. $\boldsymbol{M}$ denotes the mitigating event that prevents any cause of $\boldsymbol{Q}$, such as any response that avoids or reduces the consequence event.

In the causal relationships, a human error, which is an event can be characterised by causal chain involving [9]:

- The event itself (i.e. the error),
- At least one consequence event that characterizes the impact,
- One or more initiating (trigger) events,
- One or more control events, which can stop the trigger event, and
- One or more mitigating events, which avoid or reduce the consequence event.

3. Proposed BNB-HEA framework

The proposed BNB-HEA framework consists of two parts: qualitative and quantitative. The main aim of qualitative part is to investigate the interaction of the predominant PIFs and their negative synergy effect on the maintenance activities. On the other hand, the focus of quantitative part is to estimate the posterior probabilities of the human error and, quantify the risk by employing causal analysis, i.e. BN.

3.1. Qualitative assessment

Figure 2 illustrates specific steps that should be followed to construct the BN for maintenance operations and, to estimate the main query – the posterior probability of the human error.

Step 1.1 – Define the problem and perform evaluation of human PIFs: The purpose the initial stage of the proposed framework is to study and investigate the effect of the predominant PIFs on the pro- and post– maintenance activities. Further, the interaction of the PIFs, the dependability of these factors on various variables, their negative synergy effect on the human perfomance needs to be assessed and specified. During this process, the consultation of all relevant stakeholders should be corroborated.

Step 1.2 – Perform classification of the maintenance task: In the next step, the assessment of the main repair or maintenance tasks during system failure should be carried out. For instance, the repair or maintenance tasks during gearbox failure are as follows:

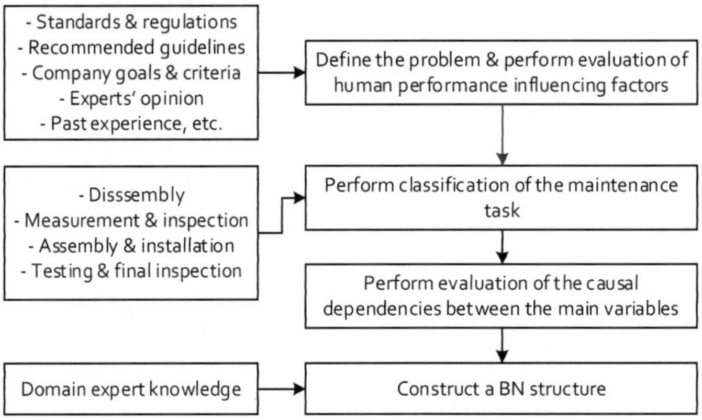

Figure 2. Qualitative part of the proposed BNB-HEA

- Disassembly of gear box;
- Measurement and inspection;
- Corrective maintenance;
- Assembly and installation; and
- Testing and final inspection.

Step 1.3 – Perform evaluation of the causal dependencies between the main variables: In the next stage, the interactions or causal dependencies between the main variables, i.e. the PIFs, maintenace operation, and the human error, needs to be understood and then the structure of the BN has to be decided. In general, BNs can be used for three kinds of reasoning [17]: *i*) causal reasoning – from known causes to unknown effects, *ii*) diagnostic reasoning – from known effects to unknown causes, and *iii*) a combination of casual and diagnostic reasoning.

Step 1.4 – Construct a BN structure: The final stage in the qualitative evaluation is to construct the BN. Intrinsically, the static BN is a solution for estimating the potential human error probabilities with maintenance operations. The main aim of this step is thus to build the Bayesian structure that captures the main variables, which comprises both discrete and continuous variables. During this stage, the key is to focus on the causal relationships between the main variables.

3.2. Quantitative assessment

Figure 3 describes the quantitative part of the proposed BNB-HEA framework and, illustrates the specific steps that should be followed to determine the posterior probability of the human error.

Step 2.1 – Define the state of the discrete nodes: The first step in the quantitative assessment is to define the state of each discrete node. A discrete node (variable) is one with a well-defined finite set of possible values, called states [18]. The state can take binary values (such as true or false) or ordered values (such as low, medium, or high) [19].

Step 2.2 – Assign MPT for each root discrete node and CPT for other discrete nodes: After specifying the states of discrete nodes, then the next step is to quantify the relationships between the connected nodes (variables). In this step, marginal probabilities of each root nodes should be assigned. For other discrete node (other than the root nodes), conditional probability tables (CPT's) needs to be defined. For each particular discrete node, all possible combinations of values of those parent nodes needs to be observed; and such combination is called instantiation of the parent [19]. For a Boolean network, for instance, a variable with n parents requires a CPT with 2^{n+1} probabilities [19]. These probabilities can be estimated or assigned using direct elicitation and/ or machine-learning techniques [19].

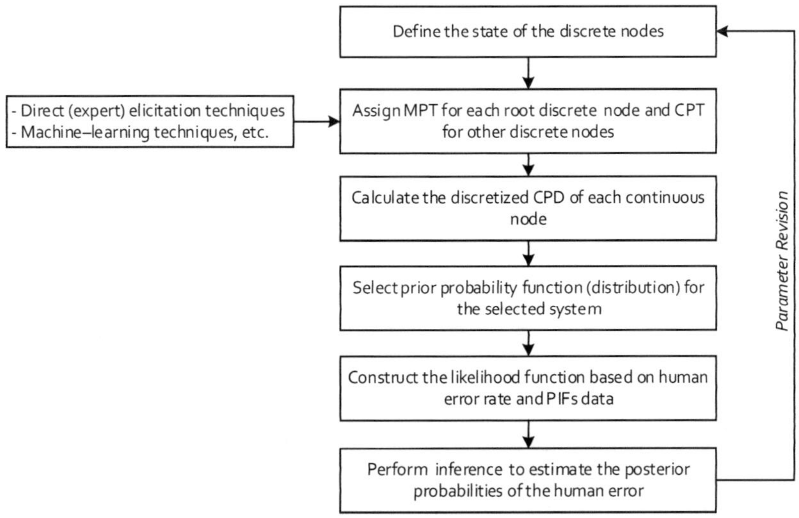

Figure 3. Quantitative part of the proposed BNB-HEA

Step 2.3 – Calculate the discretized CPD of each continuous node: The next stage is defining the conditional probability distributions (CPD's) for each continuous variable. A continuous variable (node) is one which can take on a value between any other two values [18]. Typically, there are two approaches to handle the continuous variable: static and dynamic discretization. Both approaches try to specify the states of the continuous nodes. Basically, a static discretization requires the breakup of the total range of the continuous variable into a number of intervals [18]. However, this process is cumbersome and error prone, and where a model contains numerical nodes having a potentially large range, results are necessarily only crude approximations [16]. To overcome this problem, the dynamic discretization has been developed, by Neil, et al. [20].

Step 2.4 – Select prior probability function (distribution) for the selected system: In this stage, a prior probability function or distribution needs to be asserted, for the defined system or component. This function is the representation of the failure rate of the system or component; and failure rate is the measure of frequency of system or component failure. The prior function describes the probability of *n* or fewer failures during a time interval of $(0, t)$, when all PIFs are equal to zero or absent (i.e. ''ideal'' operating environment), during maintenance operations. For instance, by assuming that the components fail according to a Poisson process, the probability of *n* or fewer failures, can be estimated by the following equation [20]:

$$P(S_f) = \sum_{i=0}^{n} \frac{(\lambda t)^n}{n!} e^{(-\lambda t)} \tag{2}$$

where $P(S_f)$ is the probability of n or fewer failures and λ is a failure rate of the system/ component.

Step 2.5 – Construct the likelihood function based on human error rate and PIFs data: After defining the prior probability function and observing the PIFs data, then the likelihood function has to be constructed. Likelihood function generally is the joint probability function and, it can be expressed as a product of conditional probabilities [21]. Hence, by considering discrete PIFs, the likelihood function of the system failure, based on Glickman and van Dyk [21] approach, can be expressed as follows:

$$L(S_f|Tr) = p(Tr_1, Tr_2, \ldots, Tr_{r-1}, Tr_r|S_f) \tag{3}$$

where $Tr_1, Tr_2, \ldots, Tr_{r-1}, Tr_r$ is a set of PIFs or triggers.

Then, by grouping the PIFs into vectors of size R, (3), can be re-written as follows:

$$L(S_f|Tr) = \prod_{i=1}^{M} p(Tr_i|S_f) \tag{4}$$

where M is a vector of size.

By following the same approach, the likelihood function of the human error can be expressed as, by considering the discrete PIFs variables:

$$L(HE|Tr) = \prod_{i=1}^{M} p(Tr_i|HE) \tag{5}$$

where HE is representing the human error.

Step 2.6 – Perform inference to estimate the posterior probabilities of the human error: The final stage is to perform inference for estimating the posterior probabilities of the human error. Probabilistic inference is the task of computing the probability of each node in BN, according to the most recent PIFs to provide posterior probabilities. The posterior distribution combines prior PIFs information with actual observed data to predict the future potential human error probabilities. That means the current information about the PIFs will be used to continuously update the potential human error relating to maintenance activity. Simply, the distribution describes the probability that the error will occur, given the predominant PIFs has observed. The posterior distribution of the human error, considering discrete PIFs (variables), based on Glickman and van Dyk [21] approach, can be expressed as:

$$P(HE|Tr) = \frac{P(HE)P(HE|Tr)}{\int p(HE)p(Tr|HE)d_{HE}} \tag{6}$$

By substituting the likelihood function and applying Bayes' theorem, Equation (6) can be re-written as:

$$P(HE|Tr) = \frac{P(HE)L(HE|Tr)}{P(Tr)} \propto P(HE)L(HE|Tr) \tag{7}$$

To solve, Equation (6) and Equation (7), we can first multiply the prior distribution by the likelihood, and then determine the marginal constant that forces the expression to integrate to 1.

4. Concluding remarks

This work introduced a framework for a human error assessment of maintenance activities based on BN. The methodology consists of two parts: qualitative and quantitative. The qualitative analysis involves the following steps:

(i) evaluation of the PIFs (to investigate the influence of the PIFs on the maintenance),
(ii) performing maintenance task identification (to investigate the main repair or maintenance tasks),
(iii) evaluating the causal dependencies between the main variables (to understand the interactions between the main variables), and
(iv) constructing a BN structure.

The quantitative part illustrates the specific steps that should be followed to estimate the posterior probability of the human error and involves the following steps: (i) defining the state of each discrete node, (ii) assigning a MPT for root discrete nodes and a CPT for other discrete nodes, (iii) calculating the discretized CPD of each continuous node, (iv) selecting the prior probability distribution for the selected system, (v) constructing the likelihood function, based on the system failure rate data, (vi) computing the posterior distribution or probabilistic inference.

The findings are as follows:

- The proposed BNB-HEA framework is beneficial as it outlines a set of steps that assist the maintenance manager to estimate the probabilities of the human error due to the negative impact of performance influencing factors.
- By employing the proposed BNB-HEA framework, decision maker can analyse different maintenance strategies for operational costs, risk, flexibility, and resource constraints.

Nomenclature

BN	Bayesian networks
HE	Human error
HEA	Human error assessment
HEPs	Human error probabilities
HCR	Human cognitive reliability
SLIM-MAUD	Success likelihood index methodology–multiattribute utility decomposition
BNB-HEA	Bayesian networks based human error assessment
PIFs	Performance influencing factors
$\Pr(N_i \| parents(N_i)$	The conditional distribution mass function of node, N_i.
N_i	Conditional distribution of a node i
$P(S_f)$	The probability of n or fewer failures
λ	Failure rate of the system/ component.
$L(S_f \| Tr)$	Likelihood function of the system failure,
$Tr_1, Tr_2, \ldots, Tr_{r-1}, Tr_r$	A set of PIFs.
$L(HE \| Tr)$	Likelihood function of the human error
$P(HE \| Tr)$	Posterior distribution of the human error

References

[1] Noroozi A, Khan F, MacKinnon S, Amyotte P and Deacon T 2014 Determination of human error probabilities in maintenance procedures of a pump *Process Saf. Environ. Prot.,* **92(2)** pp 131-141.

[2] Aalipour M, Ayele Y Z and Barabadi A 2016 Human reliability assessment (HRA) in maintenance of production process: a case study *Int. J. Sys. Assurance Eng. Manage.* **7(2)** pp. 229-238.

[3] McLeod R W 2015 Designing for Human Reliability: Human Factors Engineering in the Oil, Gas, and Process Industries. Gulf Professional Publishing.

[4] De Felice F, Petrillo A, Carlomusto A and Ramondo A 2012 Human Reliability Analysis: A review of the state of the art *IRACST–Int. J. Res. Manage. Technol. (IJRMT)* **2(1)**.

[5] Apostolakis G, Bier V and Mosleh A 1988 A critique of recent models for human error rate assessment *Reliab. Eng. Syst. Safe.* **22(1-4)** pp 201-217.

[6] Deacon T, Amyotte P, Khan F and MacKinnon S 2013 A framework for human error analysis of offshore evacuations *Safety Science,* **51(1)** pp 319-327.

[7] Noroozi A, Khakzad N, Khan F, MacKinnon S and Abbassi R 2013 The role of human error in risk analysis: Application to pre- and post-maintenance procedures of process facilities *Reliab. Eng. Syst. Saf.* **119** pp 251-258.

[8] Khan F I, Amyotte P R and DiMattia D G 2006 HEPI: A new tool for human error probability calculation for offshore operation *Saf. Sci.* **44(4)** pp 313-334.

[9] Fenton N and Neil M 2012 Risk assessment and decision analysis with Bayesian networks CRC Press.

[10] Røed W, Mosleh A, Vinnem J E and Aven T 2009 On the use of the hybrid causal logic method in offshore risk analysis *Reliab. Eng. Syst. Saf.* **94(2)** pp 445-455.

[11] Ayele Y Z, Barabady J andDroguett E L 2015 Risk assessment of Arctic drilling waste management operations based on Bayesian Networks *Proc. Int. Conf. in Safety and Reliability of Complex Engineered Systems: ESREL 2015* (Zurich, Switzerland) pp 1907-1915.

[12] Lee C-J and Lee K J 2006 Application of Bayesian network to the probabilistic risk assessment of nuclear waste disposal *Reliab. Eng. Syst. Saf.* **91(5)** pp 515-532.

[13] Ayele Y Z and Barabadi A 2016 Risk based inspection of offshore topsides static mechanical equipment in Arctic conditions *Proc. Int. Conf. in Industrial Engineering and Engineering Management (IEEM)* pp 501-506.

[14] Blanchard W 2006 Select Emergency Management-Related Terms and Definitions. Vulnerability Assessment Techniques and Applications (VATA).

[15] Ayele Y Z, Barabady J and Droguett E L 2016 Dynamic Bayesian Network-Based Risk Assessment for Arctic Offshore Drilling Waste Handling Practices *J. Offshore Mech. Arct. Eng.* **138(5)** p 051302.

[16] Marquez D, Neil M and Fenton N 2010 Improved reliability modeling using Bayesian networks and dynamic discretization *Reliab. Eng. Syst. Saf.* **95(4)** pp 412-425.

[17] Mihajlovic V and Petkovic M 2001 Dynamic bayesian networks: A state of the art. ***TR-CTIT-34*****(TR-CTI).**

[18] Norsys 2015 Discrete vs. Continuous. Available: http://www.norsys.com/WebHelp/NETICA/X_CY_discrete_vs_continuous.htm

[19] Korb K B and Nicholson A E 2010 Bayesian artificial intelligence CRC press.

[20] Hassan J, Khan F and Hasan M 2012 A risk-based approach to manage non-repairable spare parts inventory *J. Qual. Maint. Eng.***18(3)** pp 344-362.

[21] Glickman M E and van Dyk D A 2007 Basic Bayesian Methods in Topics in Biostatistics Springer pp 319-338.

IOP Conf. Series: Materials Science and Engineering **276** (2017) 012020 doi:10.1088/1757-899X/276/1/012020

Limit cycle oscillations at resonances

For systems subjected to nonlinear damping or external forces

K Hellevik* **and O T Gudmestad**

Department of Mechanical and Structural Engineering and Material Science, University of Stavanger, 4036 Stavanger, Norway

*Contact author: karina.hellevik@gmail.com

Abstract. This paper deals with limit cycles in one degree of freedom systems. The van der Pol equation is an example of an equation describing systems with clear limit cycles in the phase space (displacement-velocity 2 dimensional plane). In this paper, it is shown that a system with nonlinear loading, representing the drag load acting on structures in an oscillatory flow (the drag term of the Morison equation), will in fact exhibit limit cycles at resonance and at higher order resonances. These limit cycles are stable, and model self-excited oscillations. As the damping in the systems is linear and constant, the drag loading will to some degree work as negative damping. The consequences of the existence of these limit cycles are that systems starting at lesser amplitudes in the phase plane will exhibit increased amplitudes until the limit cycle is obtained.

1. Introduction

The phase plane is a plane where the system's position and velocity are plotted for an increasing time, t. The phase plane method, which evolves around finding limit cycles, is adapted from Struble and Martin [1]. We will investigate the phase plane motions of solutions of nonlinear equations considered to model physical phenomena of importance in engineering. We are in particular interested in resonance phenomena between the natural frequency of the system and the imposed load on the system.

2. Limit cycles

A closed trajectory in the phase plane, where closed trajectories spiral either towards or away, is termed a *limit cycle*. The trajectories around the limit cycle are not closed. Limit cycles are either stable, half-stable or unstable. When the surrounding trajectories all spiral towards the limit cycle, it is said to be stable. If all the near trajectories spiral away, it is unstable, and if some spiral towards while other spiral away, the limit cycle is half-stable. Half-stable limit cycles are uncommon. The stable limit cycles are of importance as they model systems with self-excited vibrations or oscillations. Systems with a stable limit cycle will, independent of the starting conditions, settle into a steady trajectory in the phase plane, where there is a balance between generation and dissipation of energy.

Limit cycles only occur in systems with nonlinear terms. A linear system may have a closed trajectory, but it is then surrounded by other closed trajectories when we select other initial conditions. Nonlinear systems may be described by ordinary differential equations, containing nonlinearity in the stiffness, damping or loading terms.

For more information about limit cycles and self-excited oscillations, see Hagedorn [2] from page 117.

$$\frac{d^2x}{dt^2} - \mu(1-x^2)\frac{dx}{dt} + x(t) = 0 \tag{1}$$

The van der Pol equation (1) is an example of a system with nonlinear damping, and is well known for exhibiting limit cycles. In the van der Pol Equation, the damping term will, for some time, work as negative damping, and pumping energy into the system. The phase plane diagram for the van der Pol equation with μ=1 is shown in Figure 1. The starting point, decided by the initial conditions (ICs) of each trajectory is marked by a square on the figure, with an arrow indicating the direction of the trajectories. All trajectories spiral towards the limit cycle. The van der Pol equation has been used in a wide variety of sciences, ranging from electricity, to engineering to medicine [3]. The burst like oscillations emerging towards the limit cycles (when starting from low initial values) have been used as a model to explain tremors in Parkinson disease. The van der Pol equation is further discussed in Bogoliubov and Mitropolsky [4] from page 186 and in Guckerheimer and Holmes [8] from page 67.

3. Drag loading

The Morison Equation is used to describe a system with a nonlinear loading term. It is used to describe the force on a cylinder in oscillatory flow. The Morison loading consists of a drag force and a mass force.

In this paper, the following equation is considered:

$$m\frac{d^2x}{dt^2} + c\frac{dx}{dt} + kx(t) = F_0 \sin(\omega t)\,|\sin(\omega t)| \tag{2}$$

Where the loading term represents the drag force from the Morison loading. F_0 is a constant, ω is the loading frequency, t is the time, and m, c and k represents the systems mass, damping and stiffness, respectively. Notice that the forcing term is representing a term proportional to the velocity squared, representing turbulent flow past a cylinder (an obstacle in the flow path). In this paper, only one degree of freedom systems (as given by (2)) will be investigated.

It should be noted that F_0 is proportional to the absolute value of the water particle velocities in case of forces caused by fluid flow. For more discussion about the Morison equation, see Gudmestad [5].

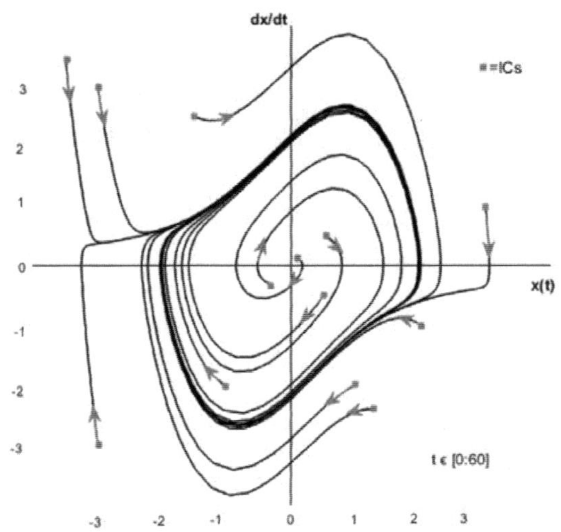

Figure 1: Phase plane diagram for the unforced van der Pol equation exhibiting a clear limit cycle.

IOP Conf. Series: Materials Science and Engineering **276** (2017) 012020 doi:10.1088/1757-899X/276/1/012020

3.1. Resonance

Resonance occurs when the frequency of the loading is equal to the frequency of the system, i.e. when $\omega = \omega_0$. The systems natural frequency ω_0 is given as:

$$\omega_0 = \sqrt{\frac{k}{m}} \tag{3}$$

During resonance, the energy from the external source is directly fed into the system, which will have amplitude growth only limited by the damping of the system.

3.2. Limit cycles for system with nonlinear loading

This paper considers equation (2) where examples of solutions are given for linear, constant damping, c = 0,5.

The Morison equation is not typically associated with limit cycles, and the solutions are rarely represented in the phase plane. As the drag force consists of a term proportional to the water velocity squared, and the damping term, $c\frac{dx}{dt}$ is connected to the velocity of the structural motion, it is natural to investigate if the drag force can in fact work as damping in the system, and more importantly, if it could exhibit apparent negative damping. Note that the term $\sin(\omega t)\,|\sin(\omega t)|$ can be linearized using Fourier series transformation:

$$\sin(\omega t)\,|\sin(\omega t)| = c_0 + c_1 \sin(\omega t) + c_2 \cos(2\omega t) + c_3 \sin(3\omega t) + c_4 \cos(4\omega t) + \cdots \tag{4}$$

In this paper, the following values are chosen (notice that other values could be chosen as well) to illustrate the behaviour at resonance):

- $F_0 = 50$
- k = 2
- m = 2 (giving $T_0 = 2\pi$ s)
- c = 0,5
- ω = 1,0 (at resonance)
- ω = 0,5 (at second order resonance)
- ω = 0,333 (at third order resonance)
- ω = 0,75 (out of resonance).

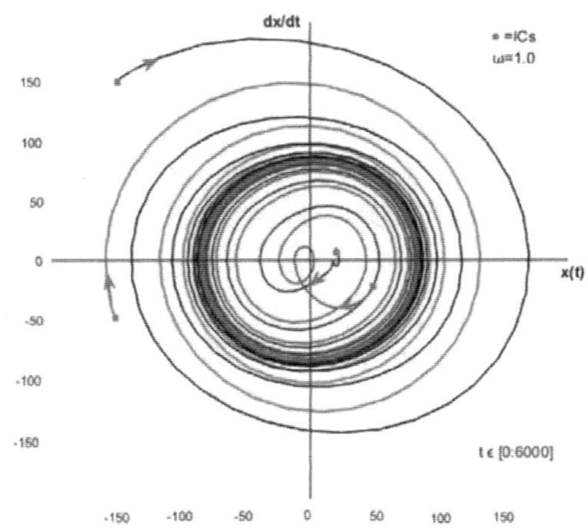

Figure 2: Phase plane diagram for a system with nonlinear drag force at resonance, ω=1, Four different trajectories (with different initial conditions, IC) are shown, which all spiral towards the limit cycle marked in red.

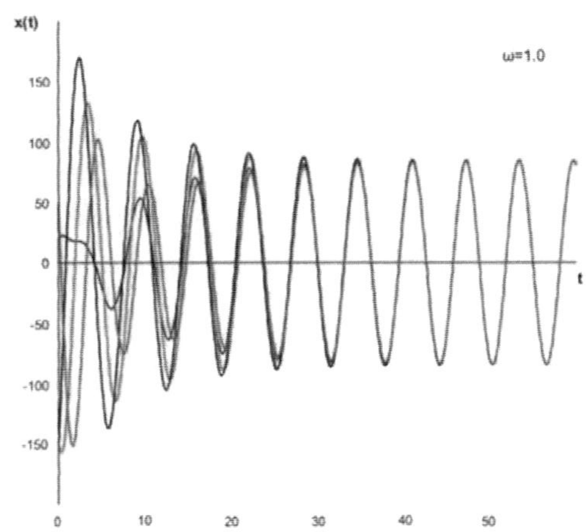

Figure 3: Position vs time curve for a system subjected to nonlinear drag force at resonance, ω = 1,0. Four different trajectories are shown, overlapping after approximately 35 s.

Figure 2 shows the phase plane diagram for a system at resonance, i.e. when ω = 1. It exhibits a clear limit cycle, marked red in the figure, and all near trajectories spiral towards it. The figure shows four different trajectories, marked by starting points and arrows indicating their directions.

Figure 3 shows the same trajectories in the traditional displacement vs time plot. The trajectories are shown for the first 50s. Around t = 30 s, the trajectories are all overlapping, continuing along the same path. Notice the growth of the displacement for the system starting at low initial values.

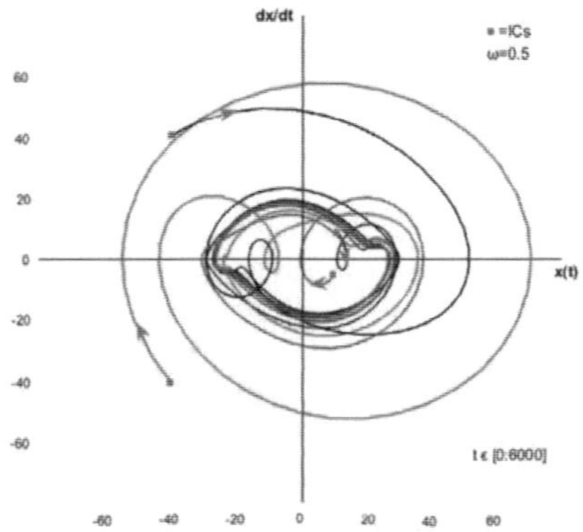

Figure 4: Phase plane diagram for a system with nonlinear drag force at higher order resonance, ω = 0,5. Four different trajectories are shown, which all spiral towards the limit cycle marked in red.

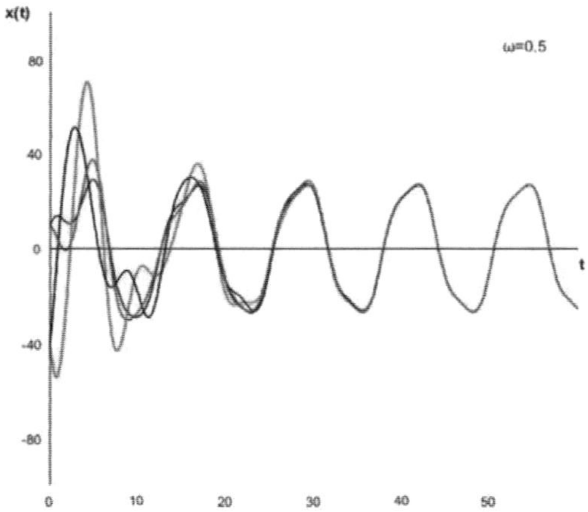

Figure 5: Position vs time curve for a system subjected to nonlinear drag force at higher order resonance, ω = 0,5. Note that for this system, the trajectories overlap after 25s approximately.

Figure 4 shows the phase plane diagram for a system with ω = 0,5, i.e. second order resonance. Four different trajectories are shown, and for this system, as well, a clear limit cycle emerges. Figure 5 shows the position vs time plot for the first 35 s for the system at higher order resonance. This system's trajectories use less time to reach the limit cycle, and are overlapping from approximately t = 25 s.

3.3. Trajectories experiencing apparent negative damping

Both Figure 2 and Figure 4 show limit cycles in the phase plane. The figures show four trajectories starting at different ICs, all are spiralling towards the limit cycle. Some of the trajectories have ICs making them start outside the limit cycle, spiralling inwards towards the limit cycle. However, some trajectories start on the inside of the limit cycle, experiencing a growth in amplitudes for both position and velocity until they reach the limit cycle. These trajectories are of interest as they experience apparent negative damping, "speeding the system up".

For clarity of the negative damping effect, Figure 6 shows the phase plane diagram for the system at resonance, with trajectories starting at ICs inside the limit cycle. All the trajectories experience a growth in amplitudes until reaching the limit cycle. Figure 7 shows the same trajectories in the position vs time plane.

IOP Conf. Series: Materials Science and Engineering **276** (2017) 012020 doi:10.1088/1757-899X/276/1/012020

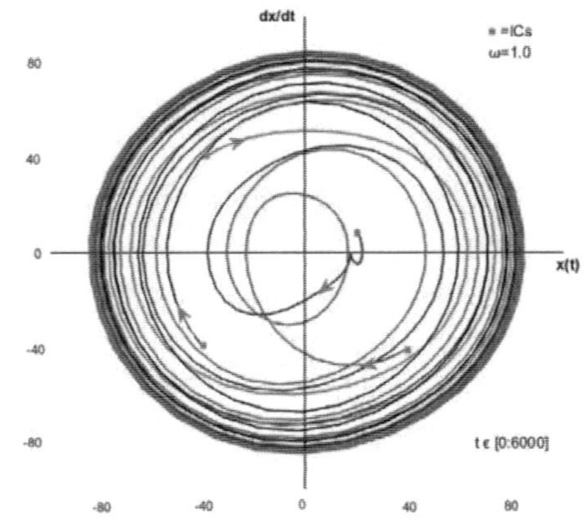

Figure 6: Phase plane diagram for the system at resonance, ω=1,0 with four different trajectories, all starting from ICs inside the limit cycle.

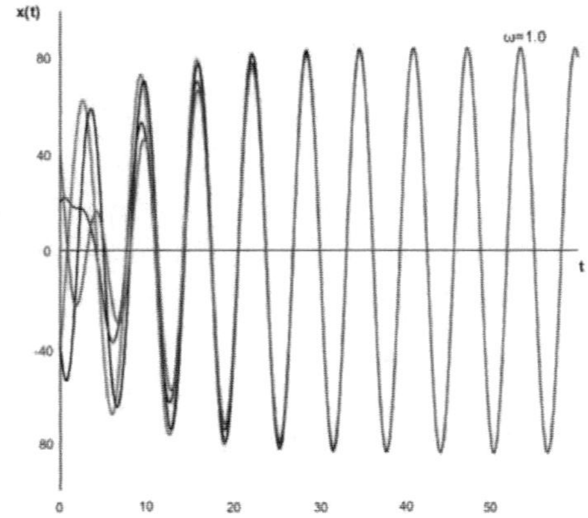

Figure 7: Position curve for the system at resonance, ω=1,0 with trajectories starting inside the limit cycle

3.4. Drag loading with current

The loading term in Equation (2) is representing drag loading for water waves without current. Adding a current u_0 to the system results in the following equation:

$$m\frac{d^2x}{dt^2} + c\frac{dx}{dt} + kx(t) =$$
$$F_0(u_0 + a\sin(\omega t))\,|(u_0 + a\sin(\omega t))| \tag{5}$$

Where u_0 and a are constants. In Figure 8 and Figure 9 the values are set to $u_0 = 0,1$ and $a = 1,0$, respectively, i.e. a small value of the current is added. We obtain the same form of the solutions as discussed before for the case of no current.

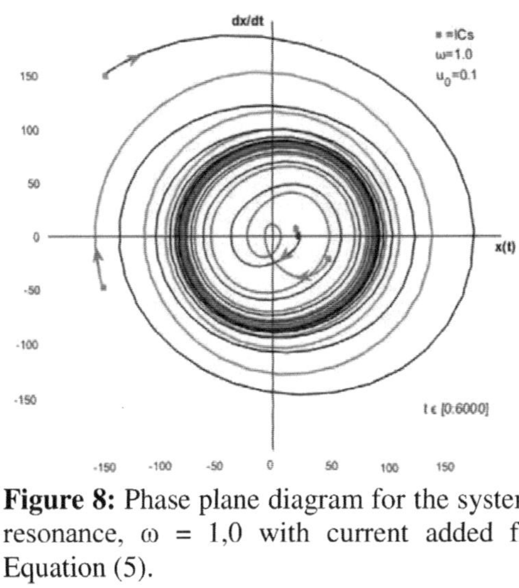

Figure 8: Phase plane diagram for the system at resonance, ω = 1,0 with current added from Equation (5).

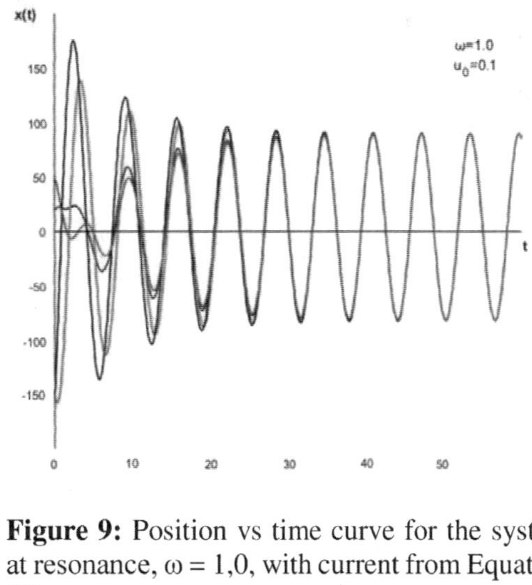

Figure 9: Position vs time curve for the system at resonance, ω = 1,0, with current from Equation (5).

3.5. Drag loading on a flexible structure

For a flexible structure, the drag loading term in sinusoidal waves will, according to [6], take the form:

$$F_0(b\frac{dx}{dt} + a\sin(\omega t))\left|b\frac{dx}{dt} + a\sin(\omega t)\right| \quad (6)$$

i.e. as the structure moves, the force will be adjusted by the velocity, dx/dt, of the structural motion.

The velocity of a wave particle is here given as $F_0 a\sin(\omega t)$where the wave velocity amplitude is $F_0 a$. The velocity of the structure is given by $F_0 b\frac{dx}{dt}$. When the structure is fixed to the bottom, the velocity of the structure is a lot less than the velocity of the wave particles. In order to achieve this, the values of a and b in Equation (6) are set to 1,0 and 0,002 respectively. Figure 10 shows the phase plane diagram for four different trajectories, which all spiral towards a limit cycle.

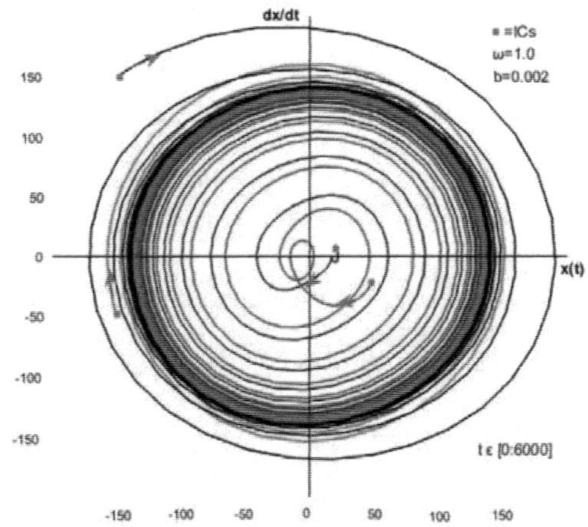

Figure 10: Phase plane diagram for a system at resonance subjected to loading from Equation (6). Four different trajectories are shown, all spiralling towards the limit cycle.

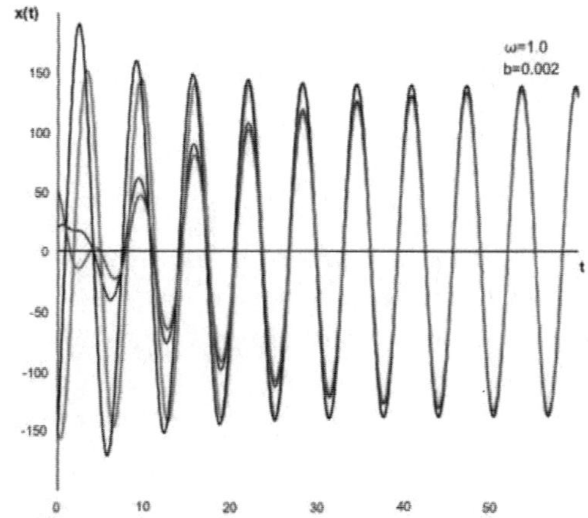

Figure 11: Position vs time plot for the system at resonance subjected to loading from Equation (6). Note that all the trajectories overlap at approximately t = 50 s.

Figure 11 shows the position vs time plot for the system at resonance with forcing term from Equation (6). The trajectories overlap after approximately 50s, reaching the limit cycle.

Figure 12 shows the limit cycles for different values of b. As b increases, the velocity and position amplitudes of the limit cycle increases, as the structure becomes less stiff, resulting in more structural movements. When *b* exceeds approximately 0,029, the phase plane trajectories spiral out from the limit cycle.

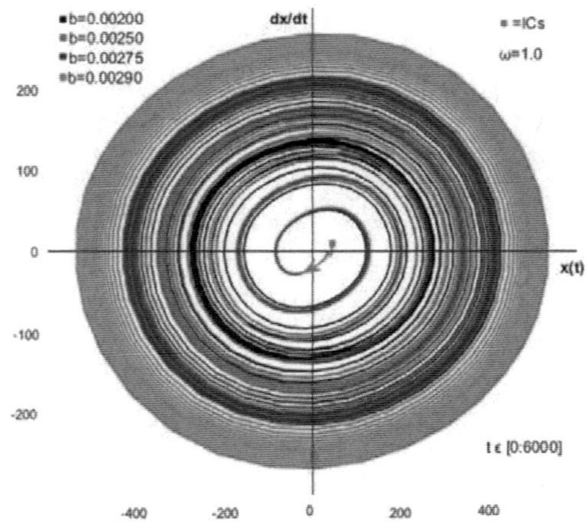

Figure 12: As b increases, the stiffness of the structure decreases, allowing more movement of the structure. This results in larger amplitude limit cycles.

3.6. Comparison of limit cycles for different degrees of resonance

Figure 13 shows the limit cycles for the system subjected to Equation (6) at resonance (red), at second order resonance (blue), at third order resonance (magenta) and out of resonance (black). Figure 14 shows the position vs time plot for the same trajectories.

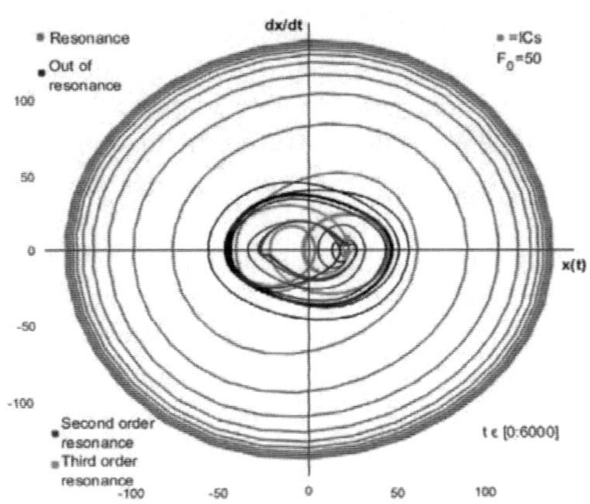

Figure 13: Systems with different degree of resonance subjected to loading from Equation (6) in the phase plane.

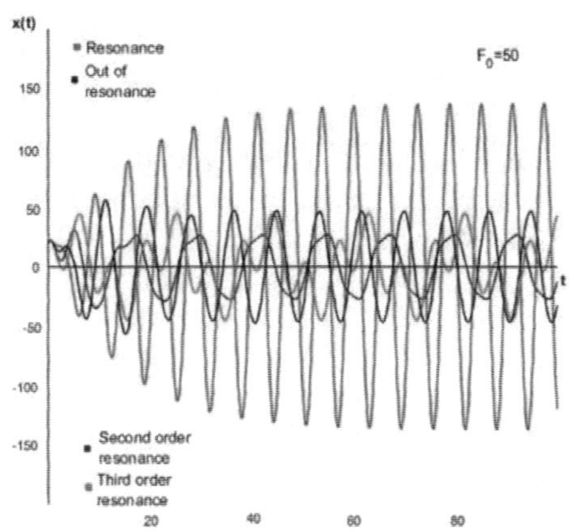

Figure 14: Position vs time plot for systems with different degrees of resonance, subjected to loading from Equation (6).

Figure 15 shows the phase plane diagram for systems at varying degrees of resonance, with F_0 adjusted to a realistic value of the nonlinear forcing. The position vs time plot for the same systems are shown in Figure 16. It should be noted that the adjustment of F_0 reflects the estimation of the relative loading from waves with the associated periods (at resonance the wave period is 2π and at second order resonance the wave period is 6π).

In this case, we see that the contributions to the displacement from the higher order resonances are more pronounced than from the main resonance due to the larger absolute value of the forcing term in larger amplitude waves (normally associated with larger wave periods).

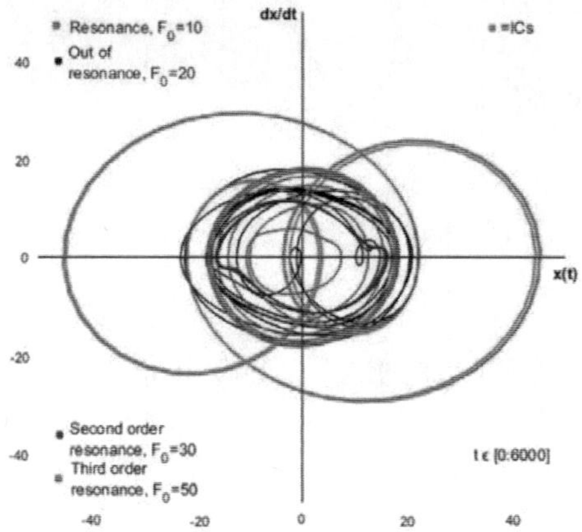

Figure 15: Phase plane diagram for systems at various degrees of resonance and various value of F0, subjected to loading from Equation (6).

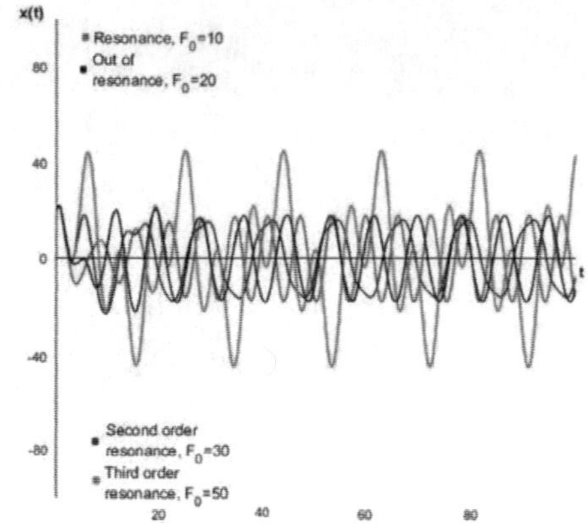

Figure 16: Position vs time plot for systems of varying degrees of resonance and various values of F0, subjected to loading from Equation (6).

4. Conclusions

A one degree of freedom system subjected to drag loading and linear, constant damping will have limit cycles near and at resonance. As the damping is constant and linear, the drag loading will to some degree work as negative damping, making the generation and dissipation of energy in the system balanced.

Limit cycles are seen in cases both with and without current, as well as for more flexible systems subjected to drag load on a flexible structure. For higher order resonances, $\beta = 1/2$ and $\beta = 1/3$, similar behaviour is observed, thereby energy from waves with higher periods, for example 15 seconds and 10 seconds, respectively, are fed into a system with natural period of 5 seconds. In storm situations, the energy spectrum has high values for higher periodic waves and considerable displacements and velocities can be expected. The drag load from the Morison equation is used to describe oscillatory flow past an obstacle. Typically, the oscillatory flow is representing sea waves, and the obstacles fixed offshore structures in the sea.

It is suggested that this application may not be the only use for this type of loading model. The heart pumps blood at a certain frequency, so the oscillatory flow could be the blood flowing and the obstacle could be a blood clot. Using the drag load from the Morison equation in this regard would describe the flow in blood veins past obstacles as for example blood clots. The research question suggested is therefore if certain frequencies of the heartbeat would cause resonance in the blood system and the possible movements of the clots in case of changing heart rhythm and more forced blood circulation during physical efforts, a situation most undesirable for the patient.

References

[1] Struble R A and Martin W T 1962 Nonlinear differential equations *McGraw-Hill Book Company*
[2] Hagedorn P 1981 Non-linear oscillations *Clarendon Press.*
[3] Beuter A, Glass L, Mackey M C, Titcombe M S 2003 Nonlinear dynamics in physiology and medicine *Springer-Verlag.*
[4] Bogoliubov N N and Mitropolsky Y A 1961 Asymptotic methods in the theory of non-linear oscillations *Hindustan Publishing Corp.*
[5] Gudmestad O T 2015 Marine technology and operations. Theory and practice *WIT Press.*
[6] Gudmestad O T and Connor J J 1983 Linearization methods and the influence of current on the nonlinear hydrodynamic drag force *Applied Ocean Research* **5** 4.
[7] Gudmestad O T 2011 Transient motions of an oscillating system caused by forcing terms proportional to the velocity of the structural motion *Philosophical Transactions of The Royal Society* **369.**

[8] Guckerheimer J.and Holmes P 1983 Nonlinear oscillations, Dynamical Systems, and Bifurcations of Vector Fields *Applied Mathematical Sciences 42, Springer-Verlag.*

Boundary element method solution for large scale cathodic protection problems

D C Rodopoulos[1], T V Gortsas[1], S V Tsinopoulos[2] and D Polyzos[1]*

[1]Department of Mechanical Engineering & Aeronautics, University of Patras, Greece
[2]Department of Mechanical Engineering, Technical Research Institute, Greece

Corresponding author: polyzos@upatras.gr

Abstract. Cathodic protection techniques are widely used for avoiding corrosion sequences in offshore structures. The Boundary Element Method (BEM) is an ideal method for solving such problems because requires only the meshing of the boundary and not the whole domain of the electrolyte as the Finite Element Method does. This advantage becomes more pronounced in cathodic protection systems since electrochemical reactions occur mainly on the surface of the metallic structure. The present work aims to solve numerically a sacrificial cathodic protection problem for a large offshore platform. The solution of that large-scale problem is accomplished by means of "PITHIA Software" a BEM package enhanced by Hierarchical Matrices (HM) and Adaptive Cross Approximation (ACA) techniques that accelerate drastically the computations and reduce memory requirements. The nonlinear polarization curves for steel and aluminium in seawater are employed as boundary condition for the under protection metallic surfaces and aluminium anodes, respectively. The potential as well as the current density at all the surface of the platform are effectively evaluated and presented.

1. Introduction

Damage due to corrosion is the major material degradation factor for underwater and underground metallic structures. The most effective solution to that problem is the combination of painting and the installation of a sacrificial or impressed current cathodic protection (ICCP) system. Painting provides a protective coating that isolates the metal from the surrounding medium, while cathodic protection (CP) system facilitates the metallic structure with an external source of current that protects the areas where the paint is damaged and/or its adhesion with those metallic surfaces has been lost. The sacrificial CP system is a passive protection method where metals with a more negative electrochemical potential than the structure we have to protect are placed on the metallic surface playing the role of anode. On the other hand ICCP system can be considered as an active CP system since an external power supply is used as a current source so that to increase the potential level of the protective surface and thus minimizing corrosion.

In both cases of the aforementioned CP systems, their efficiency is based on the accurate evaluation of the electric potential and protection current density distribution on the surface of the protected metallic structure. This is a potential problem defined in an infinite space surrounding the metallic place where the electrochemical process of corrosion takes place. A robust numerical tool for solving such problems is the Boundary Element Method (BEM) [1-4]. Two remarkable advantages it offers as compared to the Finite Element Method are the reduction of the dimensionality of the problem by one and its high solution accuracy. Thus, although the electrochemical process occurs at the metallic surface and the surrounding electrolytic medium, the just described CP problems can be solved by discretizing only the under protection metallic surface. However, despite its advantages, the brutal application of BEM to large-scale problems suffers from very time-consuming computations and high demands for computer memory capacity. On the other hand, the design of an efficient CP system demands to model the exact geometry of very large metallic structures, optimize the anodes and impressed current source location, examine the interference with nearby CP systems, predict the life span of the installed CP system, evaluate the efficiency of the CP system in different environments, simulate the transient dynamic response of the CP system under working condition and check the performance of CP system under different damage scenarios [5]. Consequently, the need for an advanced BEM that is able to solve quickly and economically large-scale cathodic protection engineering problems is apparent.

In the present work, a large-scale problem dealing with the protection of an offshore oil steel platform via a sacrificed CP system is treated with the aid of PITHIA [6], an advanced BEM package able to solve large scale potential problems by exploiting hierarchical matrices and adaptive cross approximation technics in conjunction with GMRES iterative solver that accelerates the solution process. The methodology is explained in brief in the section after next and demonstrated in the last section where the abovementioned CP problem is solved and presented.

2. Description of the CP problem

Consider a closed metallic surface S surrounded by an infinitely extended, homogeneous electrolyte and a large sphere with surface S_∞ where the system structure-electrolyte is embedded (Figure 1). The surface S consists of the area of anode S_c, the area of cathode S_a and the painted surface S_p, while the three dimensional (3D) domain of the electrolyte is denoted by Ω.

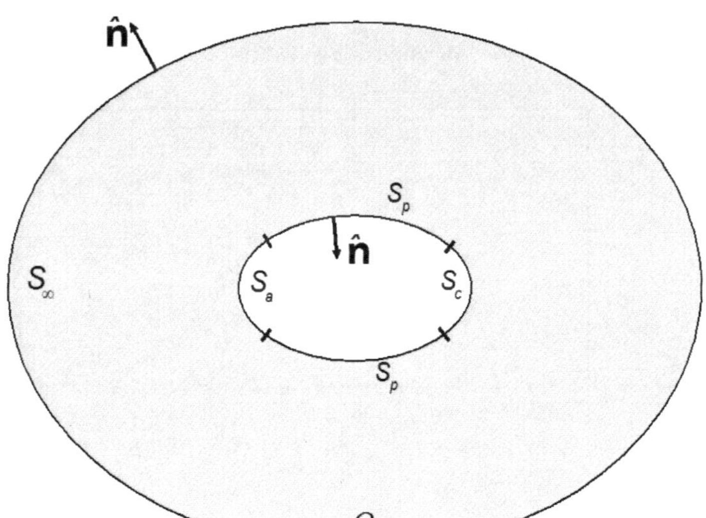

Figure 1. Illustration of a sacrificed CP system.

The electric potential field φ developed in the electrolyte satisfies the Laplace equation [7]:

$$\nabla^2 \varphi = 0 \tag{1}$$

with ∇ denoting the gradient operator.

The solution of Equation (1) at a field point **x** admits an integral representation of the form [7]:

$$c(\mathbf{x})\varphi(\mathbf{x})+\int_S \partial_n G(\mathbf{x},\mathbf{y})\varphi(\mathbf{y})dS_y=\int_S G(\mathbf{x},\mathbf{y})\partial_n\varphi(\mathbf{y})dS_y \tag{2}$$

where $S \equiv S_\infty \cup S_p \cup S_c \cup S_a$, $\partial_n \equiv \hat{\mathbf{n}}\cdot\nabla$ is the directional derivative with respect to the normal unit vector $\hat{\mathbf{n}}$ defined at all surfaces and externally to domain Ω, **y** represents source points lying across the boundaries of the problem, $c(\mathbf{x})$ is a coefficient obtaining the values 1 and 0.5 when point **x** lies in space Ω and its boundaries, respectively and $G(\mathbf{x},\mathbf{y})$ stands for the fundamental solution of 3D Laplace Equation (1) written as

$$\begin{aligned} G(\mathbf{x},\mathbf{y}) &= \frac{1}{4\pi r} \\ \partial_n G(\mathbf{x},\mathbf{y}) &= -\frac{1}{4\pi r^2}\hat{\mathbf{n}}\cdot\mathbf{r} \\ r &= |\mathbf{y}-\mathbf{x}| \end{aligned} \tag{3}$$

At the boundaries of domain Ω, three types of boundary conditions can be assigned. At the perfectly painted surfaces the current density is vanished, while for the exposed metal surfaces, which play the role of a cathode in order to be protected, a nonlinear Robin boundary condition is imposed, usually derived experimentally. At the anodic surface S_α the potential can be considered either constant and known for standard material properties and geometry or can be treated as an anodic material obeying to a Robin nonlinear relation between potential and current density, as in the case of cathodic surfaces. For the case of sacrificial CP system that utilises aluminium anodes to protect steel in seawater, the nonlinear polarization curves are depicted in Figure 2 [8].

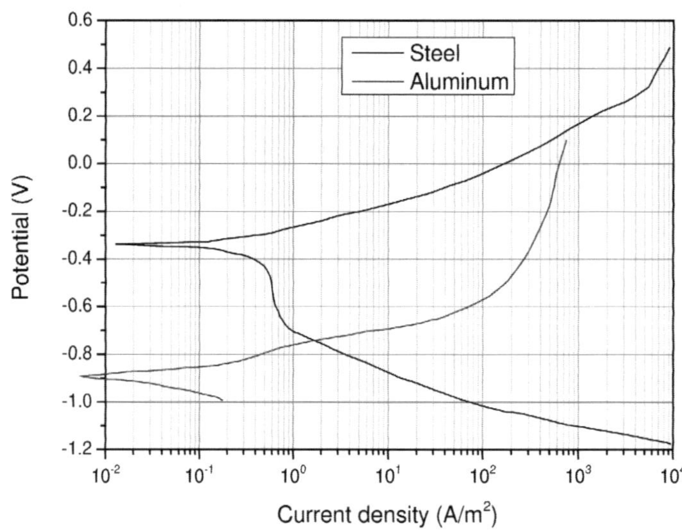

Figure 2. Polarization curves for steel and aluminum anodes in sea-water [8].

For the problem of Figure 1, the just described boundary conditions read:

$$i(\mathbf{x}) \equiv -\sigma\partial_n\varphi(\mathbf{x}) = 0, \quad \mathbf{x}\in S_p \tag{4}$$

$$\left.\begin{array}{c}\varphi_a(\mathbf{x})=\varphi_0\\ \text{or}\\ i_a(\mathbf{x})=g(\varphi_a(\mathbf{x}))\end{array}\right\}\mathbf{x}\in S_a \tag{5}$$

$$i_c(\mathbf{x})=f(\varphi_c(\mathbf{x})),\quad \mathbf{x}\in S_c \tag{6}$$

$$\partial_n\varphi(\mathbf{x})=0,\quad \mathbf{x}\in S_\infty \tag{7}$$

where σ is the conductivity of seawater, i is the current density and g, f represent the upper branch of the blue polarization curve and the lower branch of steel polarization curve shown in Figure 2, respectively.

The boundary condition (7) implies that the constraint equation

$$\int_{S_p\cup S_c\cup S_a}\partial_n\varphi(\mathbf{y})dS_{\mathbf{y}}=0 \tag{8}$$

which express the conservation of charge on the considered structure, is automatically fulfilled.

3. BEM solution

In the present section, an advanced BEM used for the numerical solution of the above-described cathodic protection problem is described in brief. To this end the boundaries of the problem of Figure 1 are discretized into 3-noded triangular or 4-noded quadrilateral linear elements. Then the integral Equation (2) written for the surface node i, obtains the discrete form:

$$\frac{1}{2}\varphi^i+\sum_e\sum_a H_{iea}\varphi^a=\sum_e\sum_a G_{iea}q^a \tag{9}$$

where e,α denote the number of elements and nodes per element, respectively, $q\equiv\partial_n\varphi$ is the potential flux, while the coefficients H_{iea}, G_{iea} have the form:

$$H_{iea}=\int_{-1}^{1}\int_{-1}^{1}\partial_n G\left(\mathbf{x}^i,\mathbf{y}^e(\xi_1,\xi_2)\right)N^a(\xi_1,\xi_2)\left|\mathbf{J}^e\right|d\xi_1 d\xi_2 \tag{10}$$

$$G_{iea}=\int_{-1}^{1}\int_{-1}^{1}G\left(\mathbf{x}^i,\mathbf{y}^e(\xi_1,\xi_2)\right)N^a(\xi_1,\xi_2)\left|\mathbf{J}^e\right|d\xi_1 d\xi_2 \tag{11}$$

where $N^a(\xi_1,\xi_2)$ represents the functions used for the interpolation of potentials and potential fluxes in a local coordinate system ξ_1,ξ_2, while **J** stands for the Jacobian of the transformation from the global Cartesian coordinate system to the local one.

Collocating Equation (9) at all nodes one obtains the following system of algebraic equations:

$$[\mathbf{H}]\cdot\{\boldsymbol{\varphi}\}=[\mathbf{G}]\cdot\{\mathbf{q}\} \tag{12}$$

where the elements of matrices [**H**] and [**G**] are integrals of the form of Equations (10) and (11), respectively, evaluated numerically via highly accurate, direct integration techniques illustrated in [9], while the vectors {**φ**} and {**q**} contain all the known and unknown nodal values of potential and potential fluxes, respectively.

Finally, applying a Newton-Raphson iterative procedure for the nonlinear boundary condition (6) and satisfying all the boundary conditions of the problem, Equation (11) can be written as

$$[\mathbf{A}]\cdot\{\mathbf{x}\}=\{\mathbf{b}\} \tag{13}$$

with the vectors $\{\mathbf{x}\}$ and $\{\mathbf{b}\}$ containing all the unknown and known nodal values of the CP problem, respectively.

The just described conventional BEM formulation produces full populated and non-symmetric collocation matrices [A] increasing thus the computational cost and confining the method to the solution of relatively small problems with no more than 100.000 degrees of freedom (DOFs). A very efficient method that circumvents that problem and accelerates drastically the solution process of a BEM code is that of the Fast Multipole BEM (FM/BEM). The FM/BEM accelerates the computation of matrix [A] of the final system of algebraic equations [A]{x}={b} to O(N) operations and reduces the memory requirements to O(N) as well. This is accomplished, first, by analytically expanding the fundamental solutions of the differential operator of the problem around the centres of a hierarchical cell structure through fast multipole expansions, second, by performing analytical integrations utilizing constant elements and third, by using an iterative solver (e.g. GMRES) instead of a direct one. The use of FM/BEM for the solution of elastic and acoustic problems is explained in the book of Liu [7].

An alternative to the FM/BEM is the use of a BEM enhanced by Hierarchical Matrices (HM) and Adaptive Cross Approximation (ACA) techniques that accelerate drastically the computation of matrix [A] and also reduce the memory requirements. That acceleration is possible due to the nature of the fundamental solutions, which are functions of the distance between the source and field points and thus only a small number of elements of the collocation matrix [A] are calculated, while the rest of them are approximated via the already evaluated elements. According to ACA/BEM, the matrix [A] is organized into a hierarchical structure of blocks depended on the geometry of the problem. Applying a geometrical criterion the blocks are characterized either as non-admissible, where the ACA algorithm is inefficient and thus the conventional BEM is employed or admissible where ACA is effective and is used to calculate only a small number of their rows and columns. Each admissible block is represented in a low rank matrix format via the product of two matrices formed by the previously calculated rows and columns, respectively. This low rank format in conjunction with an iterative solver, like GMRES, leads to significant reductions in memory requirements and CPU time due to the acceleration of the matrix vector multiplication.

Table 1 presents the gain coming from the application of ACA/BEM to large-scale elastic problems solved by Gortsas *et al.* [10]. Although the FMM/BEM seems to be faster than ACA/BEM, it appears some disadvantages as it is compared to the ACA/BEM, namely, it requires the knowledge of the multipole expansions of the fundamental solution of the problem and significant and complex modifications in a conventional BEM code in order to be implemented. On the other hand, ACA/BEM is a black box algorithm applied upon the collocation matrix [A] and thus its implementation is the same regardless of the differential operator of the problem. More details one can find in the work of Gortsas *et al.* [10].

Table 1. CPU time and memory requirements for conventional BEM and ACA/BEM appearing in [9]

Degrees of Freedom x 10^6	Memory requirements (GB)		CPU time (sec)	
	Conventional BEM	ACA/BEM	Conventional BEM	ACA/BEM
0.1	78	2	30000	2000
0.5	-	23	-	52000
1.0	-	49	-	11700
1.2	-	68	-	14500

Depending on the problem, ACA/BEM can be applied either to the final matrix [A] or to the matrices [H] and [G] of Equation (12) and before the satisfaction of the boundary conditions of the problem. The later solution is adopted in the present work because of the nonlinear nature of the boundary conditions (5) and (6), where a Newton-Raphson iterative procedure is applied according to the scheme:

$$\left.\begin{array}{c} \varphi_a(\mathbf{x}) = \varphi_0 \\ or \\ \partial_n \varphi_a^k(\mathbf{x}) = \frac{1}{\sigma}\tilde{g}(\varphi_a^{k-1}(\mathbf{x})) + \frac{1}{\sigma}\frac{\partial \tilde{g}(\varphi_a^{k-1}(\mathbf{x}))}{\partial \varphi}\Delta\varphi_a^k(\mathbf{x}) \end{array}\right\} \mathbf{x} \in S_a \tag{14}$$

$$\partial_n \varphi_c^k(\mathbf{x}) = \frac{1}{\sigma}\tilde{f}(\varphi_c^{k-1}(\mathbf{x})) + \frac{1}{\sigma}\frac{\partial \tilde{f}(\varphi_c^{k-1}(\mathbf{x}))}{\partial \varphi}\Delta\varphi_c^k(\mathbf{x}), \quad \mathbf{x} \in S_c \tag{15}$$

where k is the step where the boundary element method machinery is applied, while $\tilde{g}$, $\tilde{f}$ represent the best fitting functions of the polarization curves g and f, respectively.

4. Numerical solution

In order to demonstrate the applicability of the above described ACA/BEM to CP engineering problems, a typical offshore oil steel platform, galvanically protected by a number of sacrificial anodes, is analysed. The structure of the platform is a frame consisted by hollow circular columns and beams, shown in Figure 3. The dimensions of the platform are (55m) x (36m) x (41m), with 47m of its height being immersed in the seawater. Thus, the total surface of the platform lying below the waterline is 3694 m^2 and 75 identical sacrificial anodes are utilized for its corrosion protection. Sacrificial anodes are rectangular parallelepiped in shape with dimensions (4m)x(0.3m)x(0.3m) and total surface area of 4.98m^2, mounded at columns and beams as it is shown in Figure 4. The conductivity of the seawater was taken equal to 4 $\Omega^{-1}m^{-1}$, while on the surface of the anodes first the constant potential of -1.1V has been assigned and next the polarization curve of aluminium anodes has been considered. The immersed metallic surface of the platform is treated as a cathode obeying to the polarization curve of bare steel in sea-water depicted in Figure 2.

According to the BEM described in the previous section, only the surface of the electrolyte touching the outer surface of the frame, is needed to be discretized. Additionally, in order to ensure the conservation of charge on the structure, the semi-infinite medium of the problem is bounded by a large semi-sphere, with a radius of 100 m. The surface of the semi-sphere is also discretized and a zero current density boundary condition is imposed. The surface of the electrolyte being in contact with the platform is discretized by 20594 eight-noded quadratic quadrilateral and six-noded quadratic triangular elements, shown in Figure 4, while for the interpolation of both potential and current density fields, linear four-noded quadrilateral and three-noded triangular elements has been employed, thus resulting to a problem with 24716 DOFs.

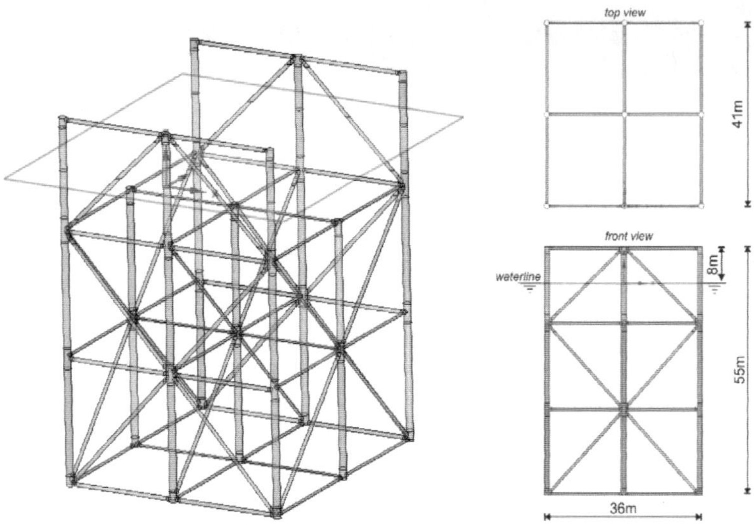

Figure 3. Geometry and dimensions of the offshore oil steel platform analysed in the present work.

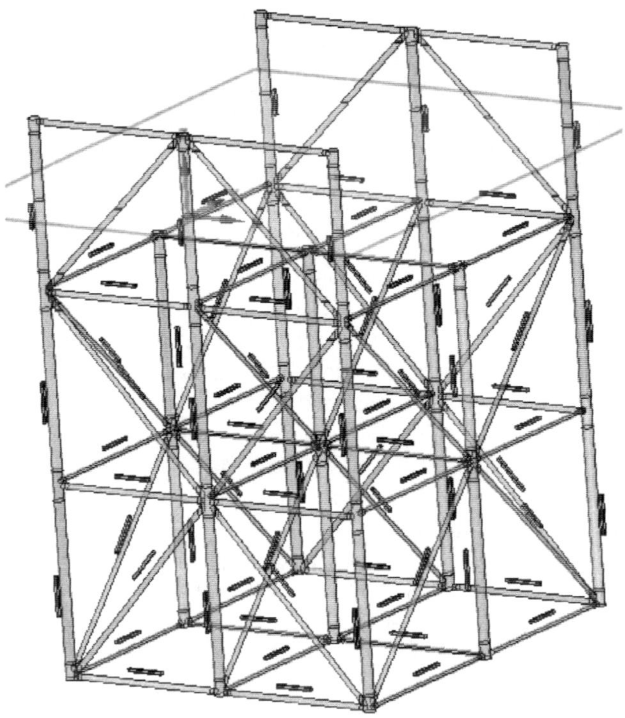

Figure 4. Placement of rectangular parallelepiped sacrificial anodes for the protection of the platform of Figure 3.

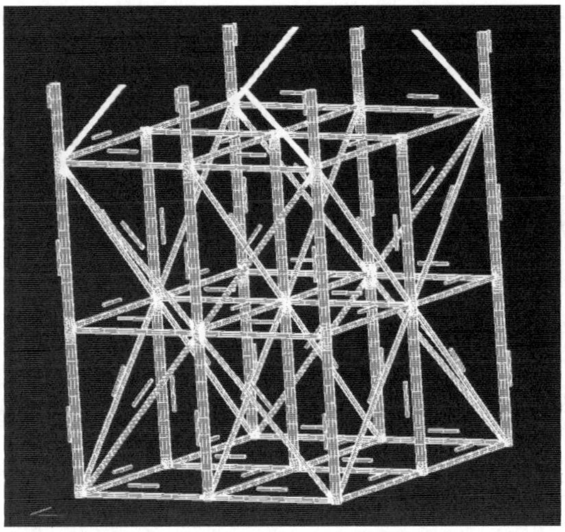

Figure 5. Boundary element discretization of the electrolyte surface being in contact with the outer surfaces of the platform, used for the ACA/BEM solution of the exterior potential CP problem.

For comparison reasons, the problem has been solved via both the conventional BEM and the ACA/BEM version of PITHIA package. The solution of the problem via the ACA/BEM code was accomplished in 35 minutes, almost 50 times faster than in the case of conventional BEM code. The difference here between conventional BEM and ACA/BEM is more pronounced than that of Table I due to the iterations required for the treatment of the nonlinear boundary condition valid at the cathodic surface of the platform.

The calculated potential and current density fields on the sunken platform surface are depicted in Figures 6 and 7, respectively. The minimum potential value of -0.55V and the maximum value of -0.2V are observed at platform points being very close and away from the considered anodes, respectively. On the other hand, the current density varies in the range of 2 mA/m^2 to 50 mA/m^2. Although the above CP problem has been solved for demonstration purposes, from engineering point of view the examined platform is not protected adequately since the steel in seawater is protected when it is polarized below to -0.58V. Thus, the jacket of seventy five sacrificial anodes is not enough and a larger number of distributed anodes, perhaps two anodes per beam/column, should be tested.

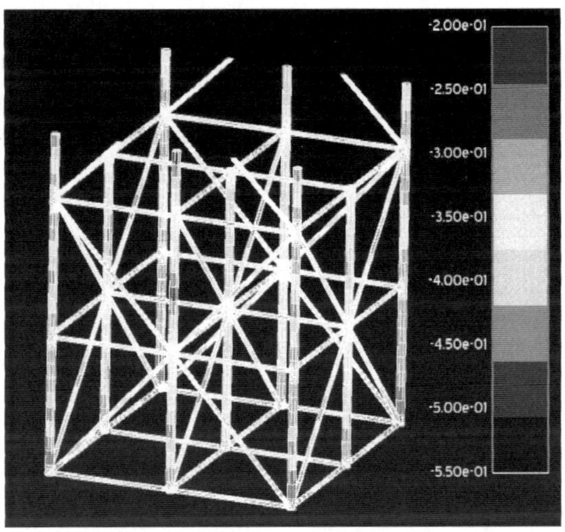

Figure 6. Potential distribution (V) at the submerged surface of the metallic platform

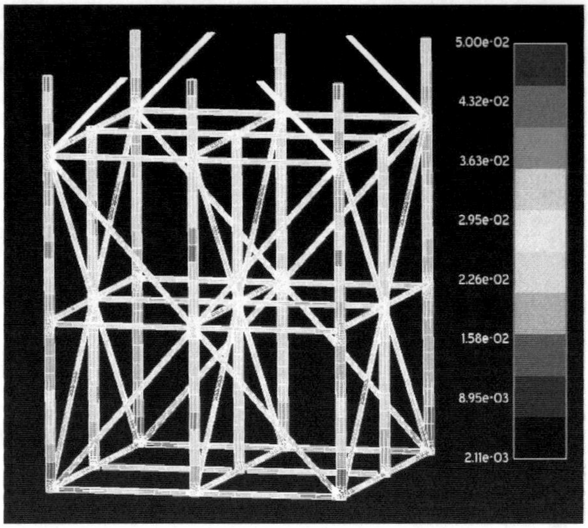

Figure 7. Current density distribution (A/m^2) at the submerged surface of the metallic platform

5. Conclusions

PITHIA, an advanced Boundary Element Method (BEM) enhanced by Adaptive Cross Approximation (ACA) and Hierarchical Matrices (HM) techniques, has been employed for the solution of a potential problem dealing with the sacrificial Cathodic Protection (CP) of a large metallic oil-platform. The nonlinear polarization curve for steel in sea-water was employed as boundary condition for the metallic surfaces we need to protect. Due to the nonlinear form of polarization curve the problem has been solved iteratively via a Newton-Raphson scheme. The solution process of the final BEM system of linear algebraic equations was accelerated via an iterative GMRES solver and the potential as well as the current density at all the surface of the platform were effectively evaluated and presented. Due to the efficiency of the utilized ACA/BEM, the problem was solved 50 times faster than in the case of using conventional BEM, while the gain in storage memory was significant so that the problem to be solved in a desktop PC with 64 GB RAM. The convergence, the efficiency and the accuracy of the obtained results indicate that PITHIA is a robust numerical tool for simulating large-scale CP problems.

References

[1] Zamani N G, Chuang J M and Porter J F 1987 *Int. J. Numer. Method Eng.* **24** pp. 605–620.

[2] Santiago J A F and Telles J C F 1997 *Int. J. Numer. Method Eng.* **8** pp. 76–78.

[3] DeGiorgi V G and Wimmer S A 2005 *Eng. Anal. Boundary Elem.* **29** pp. 15–28.

[4] Wang X, Hou B, Wang Z, Song J and Chen S 2012 *Eng. Anal. Boundary Elem.* **36** pp. 903-906

[5] Mathiazhagan A 2010 *Int. J. Chem. Eng. Appl.* **1(3)** pp. 217-221.

[6] FEAC website: http://www.feacomp.com/pithia, visited: 25.10.2017.

[7] Liu Y 2009 *Fast Multipole Boundary Element Method: Theory and Applications in Engineering* (Cambridge University Press).

[8] Tsai T-M 1996 *Journal of the Marine Science and Technology* **4(1)** pp. 17-21.

[9] Polyzos D, Tsinopoulos S V and Beskos D E 1998 *European Journal of Mechanics A/Solids*, **17(3)** pp. 515-536.

[10] Gortsas T, Tsinopoulos SV and Polyzos D 2015 *CMES: Computer Modeling in Engineering & Sciences*, **107(4)** pp. 321-343.

Finite element simulation of long wave impact on floating breakwaters with variable stiffness

T K Papathanasiou[1]*, A Karperaki[2] and K A Belibassakis[3]

[1]Department of Mechanical, Aerospace and Civil Engineering, Brunel University London, Uxbridge UB8 3PH, UK
[2]School of Naval Architecture and Marine Engineering, National Technical University of Athens, Zografos, 15773, Greece
[3]School of Naval Architecture and Marine Engineering, National Technical University of Athens, Zografos, 15773, Greece

*Contact author: theodosios.papathanasiou@brunel.ac.uk

Abstract. The hydroelastic response of flexible, floating breakwaters is a subject of interest for coastal engineering applications. In this study, a higher order hydroelastic finite element is applied to the simulation of floating breakwaters of variable stiffness undergoing long wave impact. The main aim is the evaluation of breakwater efficiency in terms of transmitted and reflected wave characteristics. It is established that, for the wave-lengths examined, the maximum amplitude and wave-length of the transmitted pulse are strongly dependent on the breakwater stiffness. Finally it is shown that for case of a periodic stiffness profile the transmitted energy is minimised when the modulation wavelength is comparable to the wavelength of the incoming excitation.

1. Introduction

Slender flexible floating bodies have been studied by several authors using both analytical [1] and numerical techniques [2]. In the literature, the response of geophysical formations such as ice floes [2-3] and even specific types of floating breakwaters are modelled under thin plate theory assumptions [4-5]. Kirchhoff's classical plate theory becomes relevant as the horizontal dimensions of the structure in question are significantly larger than its thickness, rendering the hydroelastic responses of the floating structure dominant over rigid body motions. A structure complying with the aforementioned assumption of negligible thickness compared to horizontal dimensions is expected to flex under wave excitation, leaving only fractions of the incident wave energy to be transmitted. In turn part of the remaining energy will either reflect or dissipate due to friction and damping [6]. A floating breakwater structure is expected to shelter marinas and harbours in an environment of limited fetch [4-5]. The principal aim of floating breakwater configurations is to mitigate the transmitted wave energy and the transmitted wave amplitude. The investigation of reflection and transmission characteristics of a coupled hydroelastic system yields valuable information for the optimal design of a floating breakwater structure, aiming in wave action attenuation.

Commonly, the linear water wave theory is employed for the study of regular wave reflection and transmission characteristics of a structure resting on a layer of inviscid and irrotational fluid. In this line of thought, notable are the works of Meylan and Squire [7] and Montiel [8] who study reflection-transmission characteristics for the case of a thin ice floe, with the latter work accounting for the Archimedean draft. In the given investigations the transmission coefficient of the energy conserving system was correlated with the incident wavelength. Bennets et al. [9] and Smith and Meylan [10] explored the effects of plate thickness in the computed transmitted energy.

As already mentioned the Kirchoff thin plate assumptions are employed for structures with small thickness-to-length ratio. Moreover, as floating breakwaters are positioned near-shore the study in a shallow depth environment is relevant. In that note, Sturova [11] studied the transient response of a thin, heterogeneous plate in shallow water conditions by means of an eigenfunction expansion method. Additionally, Praveen et al [12] considered in their work, the hydroelastic response of a thick plate under the long wave assumptions. However, in nearshore environments the abrupt variability of the bathymetry must also be taken into account. Papathanasiou et al. [13] presented a higher order finite element scheme for the solution of the transient hydroelastic problem featuring a thin, heterogeneous plate, floating over variable, shallow bathymetry.

In the present contribution, we focus on the determination of reflection-transmission coefficients for the coupled hydroelastic system involving a thin, elastic strip featuring inhomogeneities, which interacts with incoming long waves. The motivation of the present study lays in the analysis and design of novel breakwater configurations. In addition, the present investigation is able to provide a simple tool for determining regions of excessive wave energy transmission for harvesting purposes. A similar study was conducted by the authors in [14] focusing on the effects of bathymetry on the reflected and transmitted wave characteristics. In this contribution we focus on the effect of the elastic modulus of the floating structure. A parametric study, involving several modulation types of the stiffness value along the breakwater is performed and optimal stiffness variations are determined. The optimization criteria involve the minimization of energy and maximum wave amplitude in the transmission region. Locally increased breakwater stiffness properties can also be achieved through the use of elastic connections to the seabed [15]. The response of flexible breakwaters, elastically connected to the seabed is studied in the same framework too. These, very local in nature, stiffness enhancements are found to induce significant alterations in the wave reflection properties of the breakwaters.

2. Organization of paper

The present paper is structured as follows: Section III briefly presents the governing mathematical equations for the problem of calculating the flexural, hydroelastic response of a floating, thin plate over shallow water bathymetry. Section IV explores the variational form of the presented initial boundary value problem and presents the implementation of the proposed finite element scheme. Section V presents a series of numerical results exploring the reflected and transmitted wave characteristics, firstly for the case of a breakwater with constant stiffness and then for the case of a breakwater featuring variable Young's modulus. Finally, in Section VI key findings are summarized and further research is proposed.

3. Governing equations

In the present section, the governing equations for a coupled hydroelastic problem featuring a thin, floating plate over shallow waters will be briefly presented. The reader is directed to relevant works in the literature for a more in depth discussion [2,11,13]. A two dimensional Cartesian coordinate system is employed, with the horizontal axis x coordinate system coinciding with the mean water level, and the vertical axis z pointed upwards. The plate is assumed to extend indefinitely in the direction vertical to the xz plane. The case of a flexible elastic strip is therefore considered. The respective 1D domain $\Omega:(-\infty < x < \infty)$ is occupied by a layer of inviscid and irrotational fluid. The floating elastic strip is located at the 1D domain extending from $x=0$ to $x=L$. The thickness distribution of the flexible strip is $\tau(x)$ and its density ρ_p. The density of the fluid (water) is denote by ρ_w.

The symbols S_0, S_1, S_2 will be used to denote the three subregions $0 < x < L$, $-\infty < x < 0$ and $L < 0 < +\infty$, respectively (see also Figure 1). The three regions will be termed 'region of hydroelastic interaction', 'region of transmission' and 'region of reflection' respectively. In S_0 the free water surface elevation $\eta(x,t)$ is assumed to coincide with the plate deflection. The velocity potential functions $\varphi_i, i = 0,1,2$ will be used for the three abovementioned domains. The bathymetry in the domain of 'hydroelastic interaction' is given by $s(x) - d(x)$, where $s(x)$ is the variable depth of the seabed with respect to the undisturbed free surface and $d(x) = \rho_p \rho_w^{-1} \tau(x)$ is the plate draft (according to Archimedes principle). The semi-infinite, thin strip assumption allows for the modelling of the plate by means of the Euler-Bernoulli beam theory. For the hydrodynamic modeling, the linearized Shallow Water Equations are employed. The following non-dimensional variables are employed for the derivation of the resulting 1-D hydroelastic system,
$\tilde{x} = x / L, \tilde{\eta} = \eta / L, \tilde{t} = g^{1/2} L^{-1/2} t$ and $\tilde{\varphi}_i = g^{-1/2} L^{-3/2} \varphi_i$, for $i = 0,1,2$

After dropping tildes for simplicity the non-dimensional system is written as,

$$M\ddot{\eta} + \left(K\eta_{xx}\right)_{xx} + \eta + \dot{\varphi}_0 = 0, \ x \in S_0, \tag{1}$$

$$\dot{\eta} + (H\varphi_{0x})_x = 0, \ x \in S_0, \tag{2}$$

$$\ddot{\varphi}_1 - (H\varphi_{1x})_x = 0, \ x \in S_1, \tag{3}$$

$$\ddot{\varphi}_2 - (H\varphi_{2x})_x = 0, \ x \in S_2, \tag{4}$$

with the coefficients $M(x) = m(x)\rho_w^{-1}L^{-1}$, $K = D\rho_w^{-1}g^{-1}L^{-4}$, $H(x) = \left[s(x) - d(x)\right]/L$ in the 'hydroelastic interaction' region, $H(x) = s(x)/L$ outside the 'hydroelastic interaction' region and g the acceleration of gravity. The flexural rigidity of the plate is $D = E\tau^3(12(1-v^2))^{-1}$, where E is the Young's modulus and v the Poisson's ratio of the plate material. The non-dimensionalised quiescence conditions at infinity are given as

$$\varphi_{1x} = 0(x \to -\infty) \text{ and } \varphi_{2x} = 0(x \to \infty). \tag{5}$$

At the interfaces between subregions, mass and energy conservation dictated the following matching conditions

$$H(0^-)\varphi_{1x}(0^-,t) = H(0^+)\varphi_{0x}(0^+,t), \tag{6a}$$

$$H(1^-)\varphi_{0x}(1^-,t) = H(1^+)\varphi_{2x}(1^+,t), \tag{6b}$$

$$\dot{\varphi}_1(1^-,t) = \dot{\varphi}_0(1^+,t) \text{ and } \dot{\varphi}_0(1^-,t) = \dot{\varphi}_2(1^+,t). \tag{7}$$

For a freely floating plate, the non-dimensional boundary conditions at the edges, corresponding to zero shear force and bending moment, are

$$K\eta_{xx}\big|_{x=0} = 0, \ K\eta_{xx}\big|_{x=1} = 0 \text{ and} \tag{8}$$

$$K\eta_{xxx}\big|_{x=0} = 0, \quad K\eta_{xxx}\big|_{x=1} = 0. \tag{9}$$

The initial- boundary value problem is completed with appropriate initial conditions of the form

$$\eta_0(x,0) = \phi_0(x,0) = 0, \ x \in S_0, \tag{10}$$

$$\phi_1(x,0) = \partial_t\phi_1(x,0) = 0, \tag{11}$$

$$\phi_2(x,0) = 0, \ \partial_t\phi_2(x,0) = -G(x), \tag{12}$$

The above initial conditions imply that an initial upper surface disturbance, denoted by $G(x)$, is allowed to propagate at the offshore region S_2 at the beginning of time, while the free water surface in region S_1 and the upper surface in the hydroelastic coupling region S_0 are at rest.

4. Finite element implementation

The weak form of the above initial-boundary value problem is subsequently derived by the employment of appropriate weight functions. The procedure is described in detail in [13] and briefly outlined below for completeness in presentation. The presented Equations (1)-(4), valid in given subregions of $\Omega:(-\infty<x<\infty)$, are multiplied with the corresponding weight functions: $v\in H^2(S_0)$, $-w_0\in H^1(S_0)$, $w_1\in H^1(S_1)$, $w_2\in H^1(S_2)$, where H^k is the space of functions with square integrable k order derivative, and finally integrated over the respective subdomains . After performing integration by parts,

$$\int_0^1 Mv\ddot{\eta}dx+\int_0^1 Kv_{xx}\eta_{xx}dx+\int_0^1 v\eta dx+\int_0^1 v\dot{\varphi}_0 dx=0 \tag{13}$$

$$-\int_0^1 w_0\dot{\eta}dx+\int_0^1 w_{0x}H\varphi_{0x}dx-\left[w_0H\varphi_{0x}\right]_0^1=0 \tag{14}$$

$$\int_{-\infty}^0 w_1\ddot{\varphi}_1 dx+\int_{-\infty}^0 w_{1x}H\varphi_{1x}dx-\left[w_1H\varphi_{1x}\right]_{-\infty}^0=0 \tag{15}$$

$$\int_1^{\infty} w_2\ddot{\varphi}_2 dx+\int_1^{\infty} w_{2x}H\varphi_{2x}dx-\left[w_2H\varphi_{2x}\right]_1^{\infty}=0 \tag{16}$$

Adding Equations (13)-(16) and the employment of the quiescent conditions at infinity and the matching conditions imposed by material continuity at the interfaces results in the formulation of the following variational problem,

Find $\eta(x,t)$, $\varphi_0(x,t)$ $\varphi_1(x,t)$ and $\varphi_2(x,t)$ such that for all $v\in H^2(S_0)$, $-w_0\in H^1(S_0)$, $w_1\in H^1(S_1)$ and $w_2\in H^1(S_2)$ it is

$$\begin{aligned}\int_0^1 Mv\ddot{\eta}dx-\int_0^1 w_0\dot{\eta}dx+\int_{-\infty}^0 w_1\ddot{\varphi}_1 dx+\int_1^{\infty} w_2\ddot{\varphi}_2 dx\\ +a(\eta,v)+b_0(\varphi_0,w_0)+b_1(\varphi_1,w_1)+b_2(\varphi_2,w_2)=0\end{aligned} \tag{17}$$

where the bilinear functionals appearing in (17), are defined as: $a(\eta,v)=\int_0^1\left(Kv_{xx}\eta_{xx}+v\eta\right)dx$, $b_0(\varphi_0,w_0)=\int_0^1 w_{0x}H\varphi_{0x}dx$, $b_1(\varphi_1,w_1)=\int_{-\infty}^0 w_{1x}H\varphi_{1x}dx$ and $b_2(\varphi_2,w_2)=\int_1^{\infty} w_{2x}H\varphi_{2x}dx$.

In Papathanasiou et al. [13], it is shown that for the examined hydroelastic system the following quantity, which is actually the energy of the system, is conserved,

$$\begin{aligned}\mathcal{E}(t)=\left(\int_0^1 M\dot{\eta}^2dx\right)+\left(\int_{-\infty}^0 \dot{\varphi}_1^{\,2}dx\right)+\left(\int_1^{\infty}\dot{\varphi}_2^{\,2}dx\right)+a(\eta,\eta)\\ +b_0(\varphi_0,\varphi_0)+b_1(\varphi_1,\varphi_1)+b_2(\varphi_2,\varphi_2)\end{aligned}. \tag{18}$$

The total energy of the system is proven to be conserved in time and equal to the energy provided form the initial free water surface disturbance, hence it holds that $\mathcal{E}(t)=\mathcal{E}(0)$ for every $0\le t\le T$, where T is the examined time interval.

The above variational problem will be solved by means of the higher order finite element scheme developed in [13]. In particular, we are interested for the solution of the initial boundary value problem corresponding to a long wave pulse generated in the 'transmission region', propagating and interacting with the floating plate. The computational spatial domain that will be used is set such that x extends from $-x_A$ to x_B, where these values define a spatial domain large enough so that the pulses do not reach the computational domain boundaries in the examined time interval. Thus the conditions for the velocity potential spatial derivative (velocity) at infinity are also valid at the edges of the computational domain.

In the free surface regions, i.e. the 'transmission' and the 'reflection' regions, the velocity potential function is approximated by fourth order Lagrange polynomial shape functions. In the region of hydroelastic coupling a special hydroelastic element introduced in [13] is used. The special element features quantic Herminte polynomials for the approximation of the plate deflection and fourth order Lagrange polynomials for the velocity potential function in the region. Hence, the approximate solution in each element is given as,

$$\eta^h = \sum_{i=1}^{6} H_i(x)\eta_i^h(t) \text{ and } \varphi_j^h = \sum_{i=1}^{5} L_i(x)\varphi_{ij}^h(t), j = 1,2 \tag{19}$$

Substituting the above expressions in the variational problem defined by Eq. (17) results in a second order system of the form $\mathbf{M}\partial_{tt}u + \mathbf{C}\partial_t u + \mathbf{K}u = 0$, where the vector u contains the nodal unknowns. The implication in this otherwise standard dynamical system is that each matrix is singular. The stability results in [13] however guarantee a unique solution of this discrete system. That is, an appropriate linear combination of the matrices appearing in this dynamic system must be invertible. This fact enforces the application of an implicit time integration procedure. Subsequently, a Newmark time integration scheme (see [13]) is employed in order to calculate the solution of the ordinary differential equation system. The particular values for the Newmark method $\gamma = 1/2$ and $\beta = 1/4$ have been used in all the numerical results that follow. The number of time steps employed is in all cases sufficient to ensure high accuracy.

5. Numerical results

In this section, the efficiency of flexible breakwaters, in terms of reflection and transmission characteristics, will be analysed with respect to the stiffness of the flexible elastic component. More specifically, several types of stiffness distribution along the flexible elastic strip will be considered and the portion of energy transmitted to the near-shore region will be derived. In addition, the distorted form of the main pulse, after interaction with the breakwater, will be calculated and the reduction in amplitude of the transmitted pulse will be studied. Throughout the following analysis the length of breakwater is taken as $L = 250$ m and its thickness is set to $\tau(x) = 0.5$ m. The density of the elastic material of the breakwater is $\rho_p = 900$ kg/m^3 while the water density is $\rho_w = 1025$ kg/m^3. All the numerical results involve an incoming wave created by an initial water surface elevation at the offshore region. This initial upper surface elevation has the general form

$$G(x) = 2A_0 \exp\left(-\mu\left(x - x_0\right)^2\right) \tag{20}$$

where the non-dimensional pulse amplitude is $A_0 = 3.3 \cdot 10^{-4}$, while the non-dimensional, positive parameter controlling the smoothness of the pulse is set to $\mu = 250$. Finally, x_0 denoted the point of origin for the initial disturbance located in region S_2. The initial profile splits into a left and a right propagating wave. The right propagating wave travels to the offshore region and is just plotted as a reference wave profile. It is the left propagating wave that interacts with the floating breakwater. Due to the large wavelength of the pulses, compared with the depth of the basin considered, and the related assumption of shallow water hydrodynamics, the pulses remain unaltered when propagating in constant bathymetry regions.

At the hydroelastic region, the bathymetry varies creating a shoal approximated by the function,

$$H(x) = 4 + 2\tanh(10x). \tag{21}$$

At this point it is noted that for presentation purposes, and with no loss in the generality of the presented theoretical analysis, the origin of Cartesian axis is horizontally translated so as to include the entire computational domain in the right half-plane. In that line of thought, the employed bathymetry

(21) corresponds to a smooth shoaling profile given by a hyperbolic tangent with the point of inflection coinciding with the middle of the floating breakwater, as seen in Figure 1. In the same figure the initial disturbance is seen to split into two identical pulses with half the initial amplitude travelling is opposite directions and featuring no signs of dispersion as previously mentioned.

5.1. Constant stiffness

At a first step, the case of constant stiffness will be examined. For that, three different examples with breakwaters of constant stiffness will be presented. The first examined case concerns a very flexible floating strip. In particular, the Young's modulus value $E = 5 \cdot 10^7 \ N/m^2$ is considered. Note that the Young's modulus values selected do not represent any specific material but rather the macroscopic elastic behaviour of the floating component. The second constant stiffness case involves a flexible breakwater of intermediate stiffness. The value of Young's modulus selected for this second case is $E = 5 \cdot 10^9 \ N/m^2$. Finally, a very stiff breakwater is introduced and assigned the value $E = 5 \cdot 10^{11} \ N/m^2$.

In Figure 2 snapshots of the propagating pulse impacting the floating breakwater and subsequently transmitting in region S_1 are shown for the case featuring a 'soft' breakwater. The upper surface elevation at the same time instants are depicted in Figure 3 and 4 for the intermediate stiffness and the 'stiff' breakwater cases respectively.

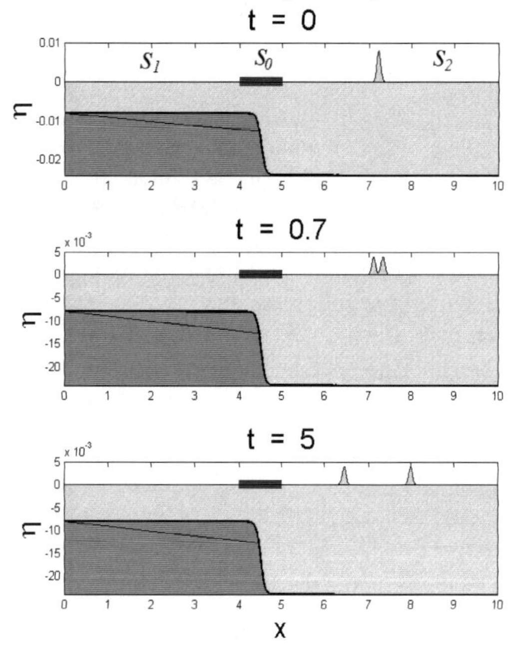

Figure 1. Initial Surface elevation and long wave propagation in the offshore region

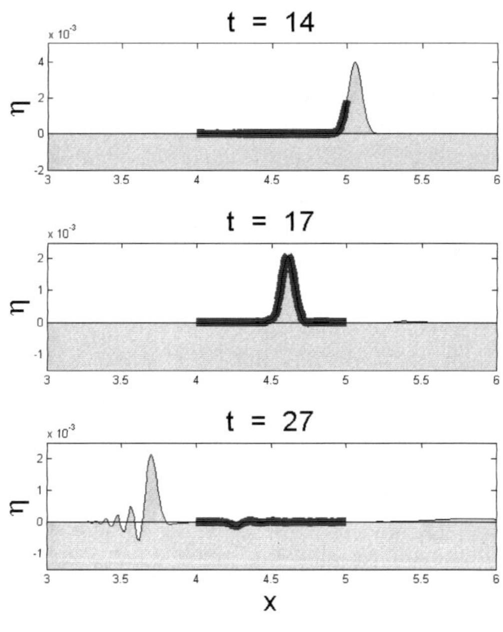

Figure 2. Hydroelastic interaction and transmitted wave profile in the case of a 'soft' flexible breakwater with $E = 5 \cdot 10^7 \ N/m^2$.

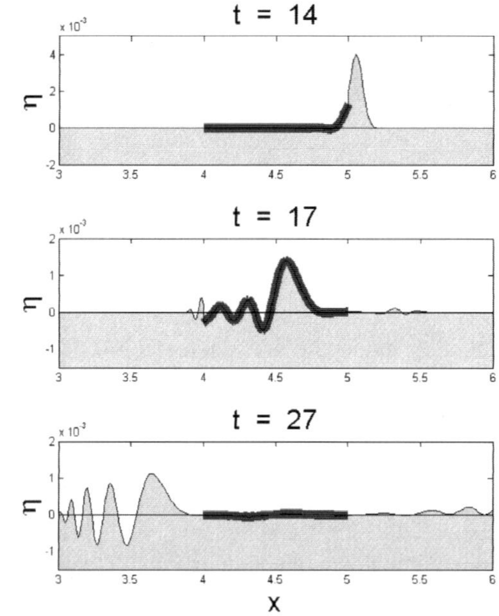

Figure 3. Hydroelastic interaction and transmitted wave profile in the case of a relatively 'stiff' flexible breakwater with $E = 5 \cdot 10^9\ N/m^2$.

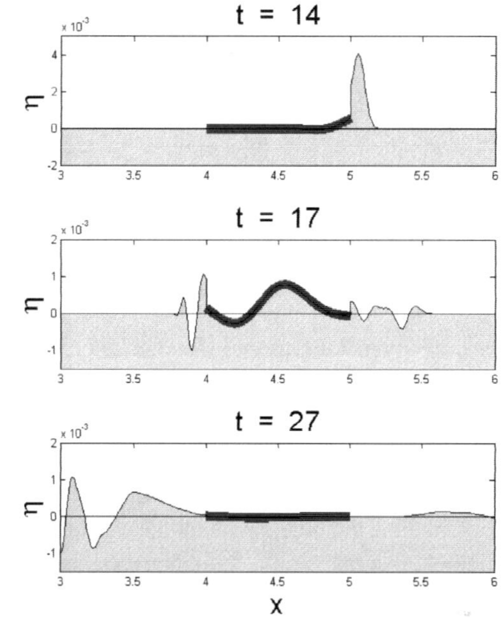

Figure 4. Hydroelastic interaction and transmitted wave profile in the case of a very 'stiff' breakwater with $E = 5 \cdot 10^{11}\ N/m^2$.

Comparing Figures 2, 3, 4 it is readily deduced that as stiffness increases, the amplitude of the reflected pulse is seen to increase while the transmitted wave amplitude is successfully attenuated. Additionally, the 'softer' case in Figure 2 leads to a transmitted pulse with wavelength similar to the initial disturbance that impacts the structure, along with a small dispersive trail. In Figures 3 and 4 however the transmitted pulse appears significantly altered. In Figures 5, 6 and 7 the energy of the system, defined in (18) is illustrated for all the considered constant stiffness cases. The nondimensional total energy of the system $\mathcal{E}$, denoted by the thick black line, is found to be constant in time.

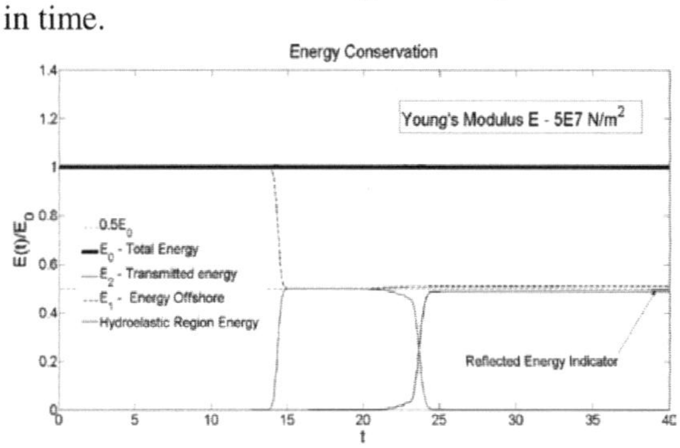

Figure 5. Energy of the system before, during and after the Hydroelastic interaction and portion of the reflected pulse energy in the case of a 'soft' flexible breakwater with $E = 5 \cdot 10^7\ N/m^2$.

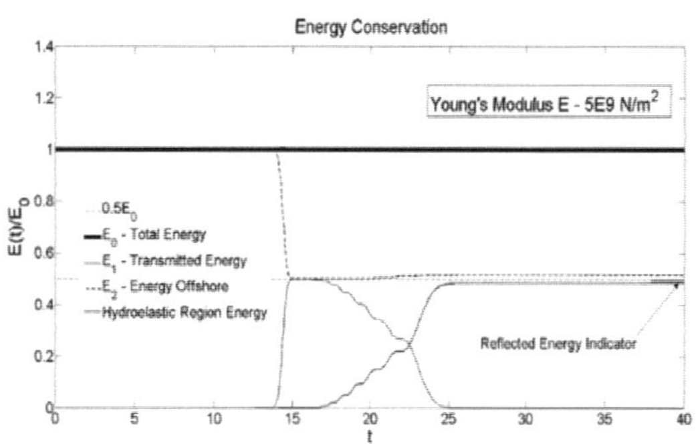

Figure 6. Energy of the system before, during and after the Hydroelastic interaction and portion of the reflected pulse energy in the case of a relatively 'stiff' flexible breakwater with $E = 5 \cdot 10^9\ N/m^2$.

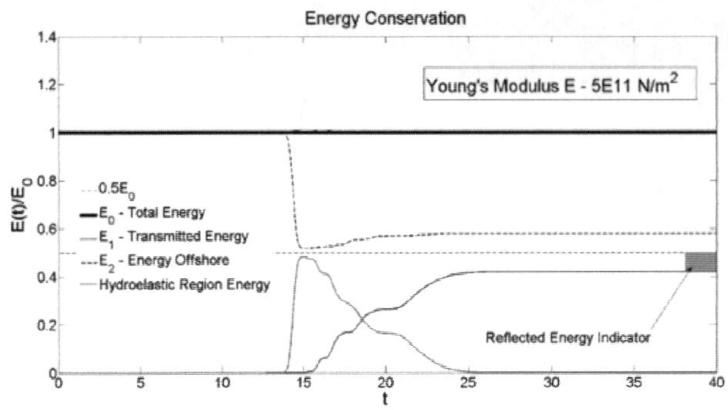

Figure 7. Energy of the system before, during and after the Hydroelastic interaction and portion of the reflected pulse energy in the case of a very 'stiff' breakwater with $E = 5 \cdot 10^{11}\ N/m^2$.

The total energy of the system includes the energy of the pulse that propagates in the offshore region, away from the breakwater. Hence the energy that reaches the structure is half the initial energy of the system. The pulse propagating towards the nearshore region is partly reflected back in S_2 due to the presence of the elastic body and partly transformed into hydroelastic energy before finally being transmitted into S_1. This implies that the energy of the pulse interacting with the breakwater is $0.5\mathcal{E}$. This energy level is depicted in the figures by a thin, dashed line. The energy against time in every subregion is also depicted in the figures. The difference between the transmitted energy in region S_2, given by the continuous blue line and half of the energy that impacts the breakwater is an indicator of the pulse energy that is reflected back in the offshore region. It is clearly seen that the reflected energy increases with increasing elastic stiffness. The given outcome is of interest since maximization of reflection corresponds to the optimal attenuation of transmitted energy.

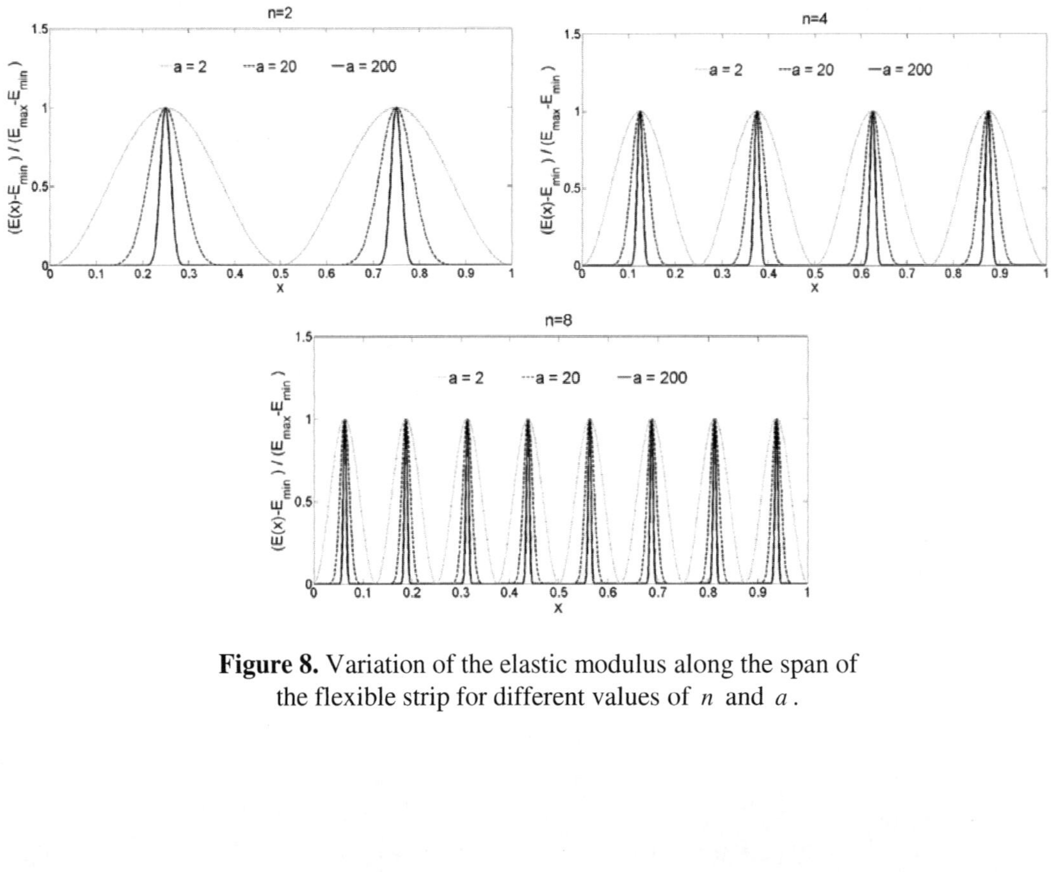

Figure 8. Variation of the elastic modulus along the span of the flexible strip for different values of n and a.

5.2. *Variable stiffness*

For the analysis of flexible breakwaters with variable stiffness, the following profile for Young's modulus will be adopted

$$E(x) = E_{\min} + \left(E_{\max} - E_{\min}\right)\sin^a\left(n\pi(x - x_0)\right). \tag{22}$$

The above distribution suggests that Young's modulus is allowed to vary between a minimum and a maximum value. Parameters a and n that appear in expression (22) affect the shape of the stiffness profile in terms of periodic variation frequency and stiffness of the peaks, as shown in Figure 8. The minimum value of the Young's modulus selected is $E = 5\cdot 10^9\ N/m^2$ while the maximum value selected is $E = 5\cdot 10^{11}\ N/m^2$. It is evident that significantly increasing the exponent a results in a localized elastic stiffening which is equivalent to the presence of an elastic connection of the breakwater with the seabed at n locations.

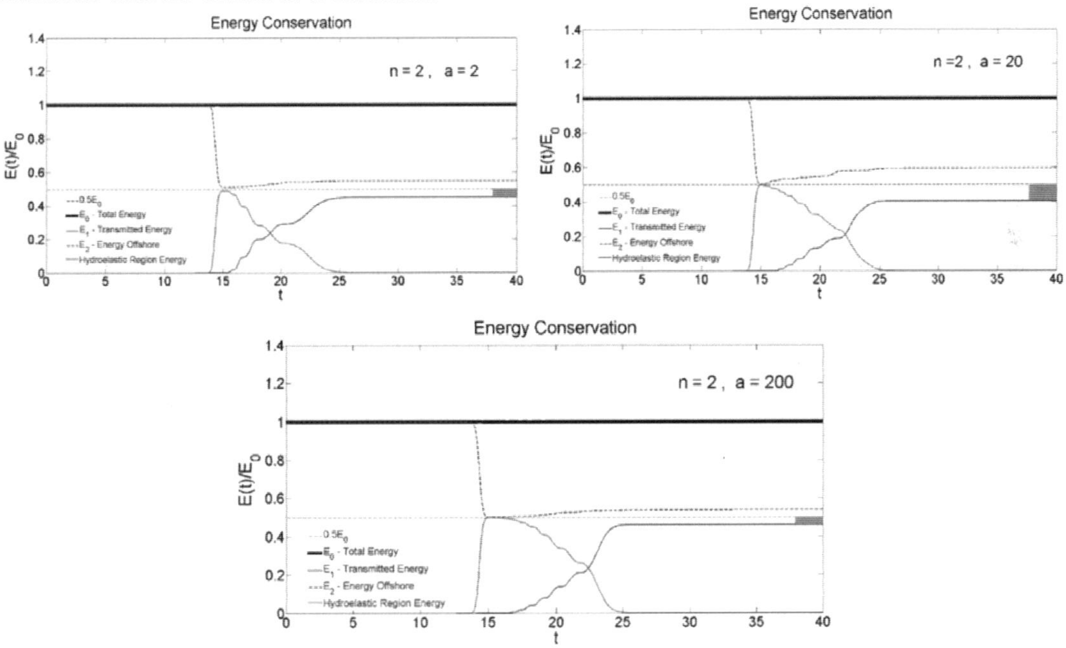

Figure 9. Energy balance and transmitted energy indicator for $n = 2$ and different values of the exponent a.

The reader is referred to the work of Karperaki et al. [15] for an in depth analysis of the hydroelastic response of a thin floating structure, elastically connected to the seabed. The energy balance principle is again explored for several variable stiffness profiles featuring different values of parameters a and n. In Figure 9 the case of a constant number of peaks $n = 2$ in the periodic fluctuation of the Young's modulus is employed. An increasing steepness parameter a is considered. It is seen that the reflected pulse is maximized for the intermediate case of $a = 20$. It is clearly seen however in Figures 9, 10, 11 that the optimum transmitted energy attenuation is achieved for a steepness parameter that depends on the number of peaks in the periodic variation for profile (22). This suggests an optimization problem. More specifically, in Figure 10, where $n = 4$, the maximization of reflected energy is achieved for $a = 2$. For an increased number of periodic fluctuations, as seen in Figure 11, the optimal case is achieved for $a = 20$.

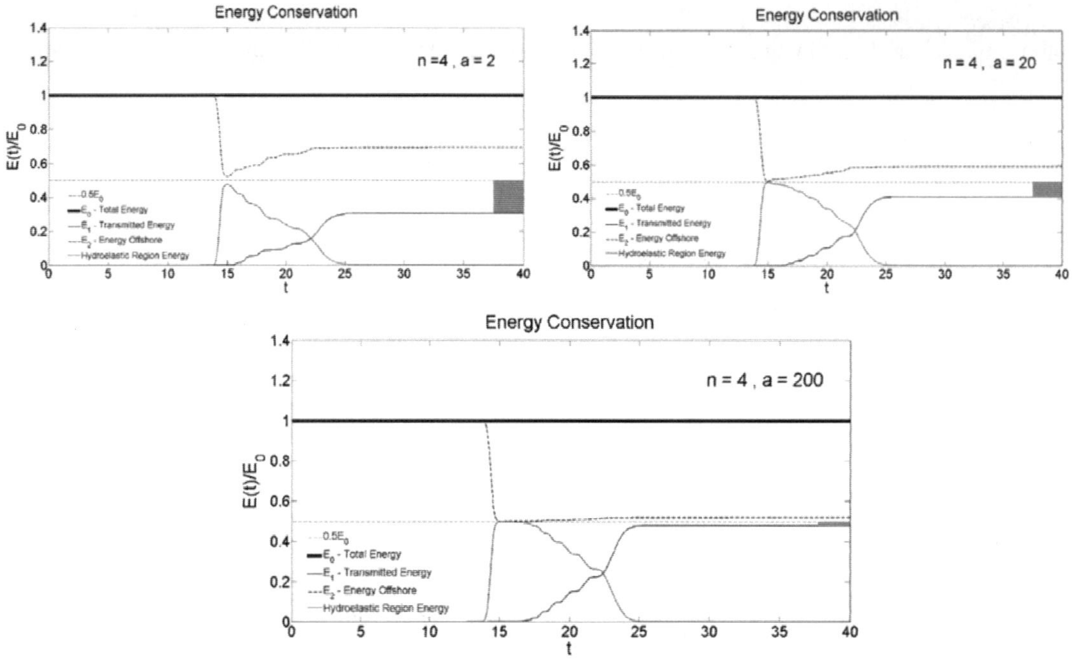

Figure 10. Energy balance and transmitted energy indicator for $n = 4$ and different values of the exponent a.

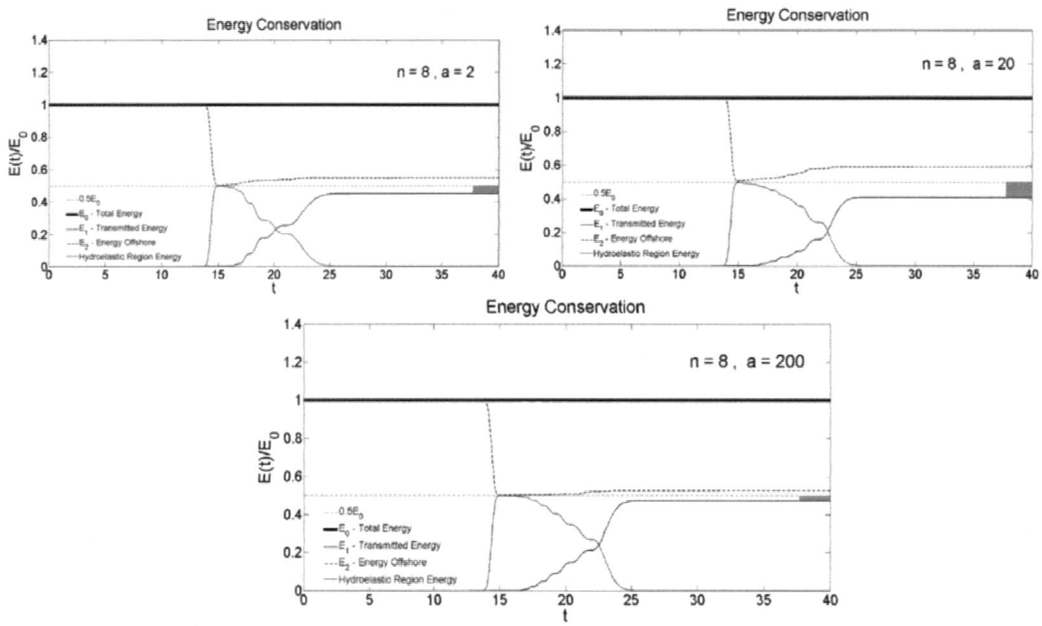

Figure 11. Energy balance and transmitted energy indicator for $n = 8$ and different values of the exponent a.

The given result is fundamentally different from the previous observations for a constant stiffness profile, where increasing the Young's modulus results in decreasing transmitted energy. For the examined stiffness profile, maximization of the reflected energy is achieved when the wavelength of the periodic Young's modulus distribution is comparable with the incoming pulse wavelength. In such cases the hydroelastic pulse suffers multiple reflections that increase the total reflected energy. When the periodicity of the Young's modulus assumes a wavelength that is much larger than the one of the excitation pulse, the reflected wave energy reduces again.

In order to further investigate the correlation between the parameters a and n a parametric study was conducted for a constant steepness parameter $a = 2$ and a varying number of fluctuations. The results of the study shown in Figure 12 suggest that the minimization of the transmitted energy is deduced for $n = 5$. In Figure 13, snapshots of the upper surface deflection for the above optimal scenario are shown. The amplitude of the transmitted wave appears significantly reduced when compared to the amplitude of the incoming pulse suggested by the amplitude of the hydroelastic response in $t = 15$, just moments after impact.

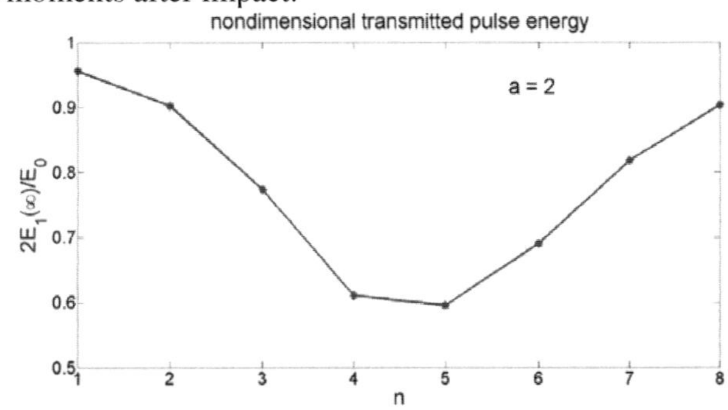

Figure 12. Energy transmitted in the nearshore region for $a = 2$ and different values n.

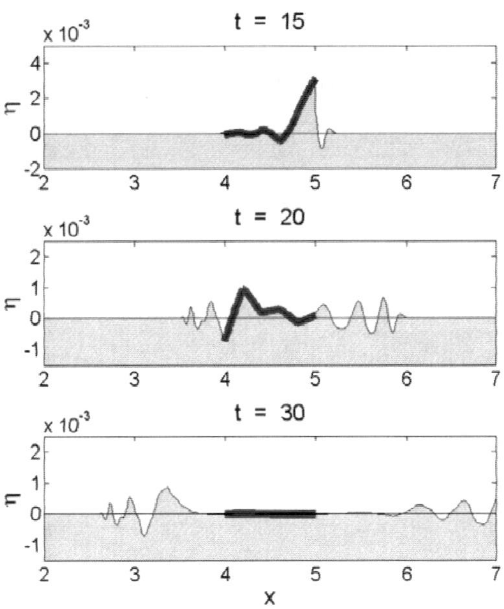

Figure 13. Snapshots of the upper surface elevation for the case of minimum transmitted energy $n = 5$, $a = 2$.

6. Conclusions

In the present contribution the hydroeastic response of a floating elastic breakwater in shallow water conditions is studied in terms of the characteristics of the reflected and transmitted waves. After a brief presentation of the governing equations, the equivalent variational problem is presented and a higher order finite element scheme is proposed for the calculation of the breakwater response. Next, the case of a breakwater of constant stiffness has been considered, where a clear correlation between the employed Young's modulus and the amplitude and form of the transmitted pulse is shown. The transmitted wave energy was shown to be attenuated for increasing stiffness. In order to examine a case of variable elastic modulus a profile featuring periodic fluctuations was employed. In the employed profile, the Young's modulus is allowed to vary between a chosen set of minimum and maximum values while the number of periodic fluctuations and the steepness of the modulation peaks are controlled by parameters. Minimisation of the reflected energy was achieved when the wavelength of the periodic structure employed for the Young's modulus profile was comparable with the incoming wave excitation. A parametric study to further investigate the tuning of the employed stiffness profile parameters was carried out. Future developments involve incorporation of nonlinearities in the hydroelastic breakwater model.

References

[1] Stoker J J 1957 *Water Waves: the mathematical theory with applications* (New York: Interscience Publishers, Inc).

[2] Papathanasiou T K, Karperaki A E, Theotokoglou E E and Belibassakis K A 2015 Hydroelastic analysis of ice shelves under long wave excitation *Nat. Hazards Earth Syst. Sci.* **15** pp 1821-57.

[3] Squire V A 2008 Synergies between VLFS Hydroelasticity and Sea Ice Reasearch *Int. J. offshore polar* **18** pp 1-13.

[4] Wang C M and Tay Z Y 2015 Very Large Floating Structures: Applications, Research and Development *Proc Engineering* **14** pp 62-72.

[5] Wang C M, Watanabe E and Utsunomiya T 2008 *Very large floating structures*, (London: Taylor and Francis).

[6] Koutandos E, Prinos P and Gironella X 2005 Floating breakwaters under regular and irregular wave forcing: reflection and transmission characteristics *J. Hydraul. Res.* **43** pp 174-188.

[7] Meylan M H and Squire V A 1994 The response of ice floes to ocean waves *J. Geophys Res* **99** pp 891-900.

[8] Montiel F, Bennetts L G and Squire V A 2012 The transient response of floating elastic plates to wavemaker forcing in two dimensions *J. Fluids Struct.* **28** pp 416-433.

[9] Bennetts L G, Biggs N R T and Porter D A 2007 A multi-mode approximation to wave scattering by ice sheets of varying thickness *J. Fluid Mech.* **579** pp 413-43.

[10] Smith M J A and Meylan M H 2011 Wave scattering by an ice floe of variable thickness, *Cold Reg. Sci. Technol.* **67** pp 24-30.

[11] Sturova I V 2009 Time-dependent response of a heterogeneous elastic plate floating on shallow water of variable depth *J. Fluid Mech.* **637** pp 305-25.

[12] Praveen K M, Karmakar D, and Nasar T 2016 Hydroelastic analysis of floating elastic thick plate in shallow water depth *Perspect. in Sci.* **8** pp 770-72.

[13] Papathanasiou T K, Karperaki A, Theotokoglou E E and Belibassakis K A 2015 A higher order FEM for time-domain hydroelastic analysis of large floating bodies in an inhomogeneous shallow water environment *Proc. R. Soc. A* **471** 20140643.

[14] Papathanasiou T K and Karperaki A E 2017 Transmission and reflection of long waves over steep bathymetry variations using large floating strips of shallow draft *Proc of 7th Int. Conf.* on Computational Methods for Coupled Problems in Science and Engineering, COUPLED 2017 (12-14 June, Rhodes).

[15] Karperaki A E, Belibassakis K A and Papathanasiou T K 2016 Time-domain, shallow-water hydroelastic analysis of VLFS elastically connected to the seabed *Mar. Struct.* **48** pp 33-51.

IOP Conf. Series: Materials Science and Engineering **276** (2017) 012023 doi:10.1088/1757-899X/276/1/012023

An efficient structural finite element for inextensible flexible risers

T K Papathanasiou[1]*, S Markolefas[2], P Khazaeinejad[1] and H Bahai[1]

[1]Department of Mechanical, Aerospace and Civil Engineering, Brunel University London, UK
[2]Department of Mechanical Engineering, Technological Educational Institute of Sterea Ellada, Chalkida, Greece

*Contact author: theodosios.papathanasiou@brunel.ac.uk

Abstract. A core part of all numerical models used for flexible riser analysis is the structural component representing the main body of the riser as a slender beam. Loads acting on this structural element are self-weight, buoyant and hydrodynamic forces, internal pressure and others. A structural finite element for an inextensible riser with a point-wise enforcement of the inextensibility constrain is presented. In particular, the inextensibility constraint is applied only at the nodes of the meshed arc length parameter. Among the virtues of the proposed approach is the flexibility in the application of boundary conditions and the easy incorporation of dissipative forces. Several attributes of the proposed finite element scheme are analysed and computation times for the solution of some simplified examples are discussed. Future developments aim at the appropriate implementation of material and geometric parameters for the beam model, i.e. flexural and torsional rigidity.

1. Introduction and literature review

As oil exploration and production progresses into deeper waters the demand for efficient simulation and design of flexible riser systems, functioning in harsh subsea environments and operating at high internal and external pressures, increases. Numerical models used for analysis of flexible riser heavily rely on the structural component representing the main body of the riser as a slender beam [1]. This structural component is subjected to several characteristic loading conditions including self-weight, buoyant and hydrodynamic forces, and internal pressure [2]. The complexity of the geometry, deformations, material behaviour and loads acting on the riser leads to computationally intensive modelling problems. An appropriate selection of the riser structural model can significantly facilitate the analysis.

Several researchers have proposed models for flexible riser simulation and analysis. The basic component of the proposed models is a slender beam-type structural element representing the riser body. In the late 70s, Kirk et al. [3] presented a frequency domain normal mode solution for an unbuoyed riser under periodic excitation. The excitation originated formed a surface vessel motion in the direction of wave propagation. Their structural analysis was based on a normal mode solution (free vibration modes and rigid body motions) of a variable tension, beam-column equation. The model utilised Morison's formula for drag forces and included the effects of a linearly varying current as well. Subsequently, Kirk [4] employed a modal/Galerkin solution for the linearised single wave and linearised spectral frequency analysis of tension-leg-platform risers.

The importance of nonlinearities in the riser response was subsequently pointed out by several authors. Bernitsas et al. [5] presented a static, 3D, small strain-large deformation analysis of deep water risers, based on an incremental finite element predictor-corrector procedure. These authors considered both extensible and inextensible risers. Coupling of torsion and bending, as well as deformation dependent boundary conditions and external loads were included in the model. In a following study, Bernitsas and Kokarakis [6] demonstrated that differences as high as 28% in the riser maximum displacement may appear in the presence of nonlinearities. At the same time, differences of 31% in the lower ball joint angle and 40% in the maximum equivalent stress were reported by the same authors. Three dimensional nonlinear dynamic analysis has been extended to more involved configurations and applied to riser bundles by Vlahopoulos and Bernitsas [7]

A different modelling approach, based on the lumped mass discretisation method was presented by Ghadimi [8]. A time domain solution using tangent stiffness increments and Wilson-theta numerical integration was utilised by this author. The proposed analysis featured also a simple seabed contact model for catenary risers. The simulation strategy was reported to produce results in good agreement with experiments. Trim [9] retained the small displacement structural model using finite elements and the modal form but supplemented it with nonlinearities originating from the fluid forcing and stochastic attributes of some parameters.

Regarding analytical or semi-analytical techniques, Cheng et al. [10] analysed linear vibrations of marine risers using perturbation techniques to derive asymptotic expansions based on the dynamic stiffness methods. In their analysis they employed and extended previous results reported by Kim and Triantafyllou [11]. A very simple model, admitting analytic solution, for the axial dynamic response of marine risers in installation was proposed by Wang et al. [12]. The same authors used also the finite difference method to analyse statically the combined effect of axial and lateral forces of marine risers [13].

More recent advances in the analysis and simulation of flexible risers include several studies focusing on complex dynamic analysis or the analysis of structural behaviour. Rahmati, Norouzi et al. [14] experimentally and numerically analysed the behaviour of flexible risers under bending, taking into account the internal structure of the component. Klaycham et al. [15] employed the finite element method to analyse the nonlinear vibrations of marine risers accounting for large displacements and taking into account the strain energy associated with the rotational restraint at the bottom end of the riser. Zhu et al. [16] studied that possibility of energy harvesting using a flexible riser and attached free-to rotate impellers. Finally, Cuamatzi-Melendez et al. [17] discussed the bisymmetric collapses in flexible risers in high external pressure environments.

The above literature review is not exhaustive, but rather indicative of the developments in the analysis and simulation of flexible risers within the last decades. Of major importance is the observation that the structural component representing the flexible riser is typically considered to have nonlinear response and very complex dynamics, while at the same time simplicity in the formulation of the governing equations and efficiency in the solution are desired.

In this study, a structural finite element for an inextensible riser, based on the approach developed in [1] regarding the application of the inextensibility constraint, is presented. In particular, the inextensibility constraint is applied only at the nodes of the meshed arc length parameter. Among the virtues of the proposed approach is the flexibility in the application of boundary conditions and the easy incorporation of dissipative forces. Several attributes of the proposed finite element scheme are analysed and CPU times for the solution of many simplified examples are computed. Future developments aim at the appropriate implementation of material and geometric parameters for the inextensible beam model, i.e. flexural and torsional rigidity. These data will be generated by the multi-scale model for flexible riser systems developed in [2].

2. Organisation of paper

The paper is organised as follows: in Section 3 the governing equations for an inextensible beam featuring a nonlinear bending moment-curvature relation are derived from the variation of the energy functional. Based on the variational form of the problem a finite element procedure, similarly to the one proposed in [1] is presented in Section 4. The main result, i.e. the formulation of a numerical scheme that features a linearization of the nonlinear flexural modulus and preserves the convergence rate characterising the method of Bartels [1] is contained in this section. Finally, Section 5 is devoted to the presentation of relevant numerical results. The study ends with a concluding paragraph summarising the current findings and mentioning future developments. In particular, the proposed scheme refers to the flexural response of the beam and does not include the effects of torsion. This addition, is very important and is targeted as a near future step towards the development of an integrated methodology for the efficient and robust simulation of flexible risers.

3. Governing equations

In this section, the equation governing the large deflection of inextensible beams will be introduced and briefly analysed. Subsequently, the finite element procedure introduced by Bartels [1] is adopted for the simulation of elements representing flexible riser components. The analysis presented here is focused on the riser body itself and does not incorporate fluid-structure interaction phenomena. The aim is to first obtain an efficient model, and an associated robust solution methodology for the inextensible structural member undergoing large deformations and featuring some degree of nonlinearity in the moment-curvature constitutive relation.

Let L and T be positive constants representing the length of inextensible beam and the total time that the phenomenon under consideration lasts, respectively. The beam domain is $I=[0,L]$. For any $s \in I$, the vector valued function

$$u(s)=\begin{bmatrix}x(s) & y(s) & z(s)\end{bmatrix}^T, \tag{1}$$

defines the location in the Euclidian space of each point along the length of the inextensible beam. The curvature at each point s representing the arc-length parameter is $\kappa(s)=u_{ss}$, where a subscript denotes differentiation with respect to the respective variable.

The kinetic energy of a flexible riser with variable properties is

$$K(\partial_t u)=\frac{1}{2}\int_I m\left|\partial_t u\right|^2 ds, \tag{2}$$

where m is the mass distribution along the riser body and $\left|\partial_t u\right|$ denotes the measure of the velocity vector. In the following analysis, the strain energy functional is assumed to have the form

$$U(u)=\frac{1}{2}\int_I EI\left|u_{ss}\right|^2 ds+\frac{1}{4}\int_I b\left|u_{ss}\right|^4 ds, \tag{3}$$

where $EI=EI(s)$ represents the flexural rigidity of the riser and $b=b(s)$ is a parameter introducing nonlinear elastic behaviour in the response. This expression of the strain energy is aimed to simulate phenomena related to relatively small values of the curvature, by introducing a small degree of nonlinearity to the moment-curvature relation.

Assuming now an energy functional of the form

$$J(\partial_t u,u,\lambda)=K(\partial_t u)+U(u)+\frac{\lambda}{2}\int_I\left(\left|u_s\right|^2-1\right)ds, \tag{4}$$

in order to incorporate the inextensibility constraint through a Lagrange multiplier λ representing the action that enforces the inextensibility and setting the first variation, using the Gâteaux derivative, to be zero, it is

$$\lim_{\varepsilon\to 0}=\frac{J(\partial_t u+\varepsilon\partial_t\chi,u+\varepsilon\chi,\lambda+\varepsilon\mu)-J(\partial_t u,u,\lambda)}{\varepsilon}=0. \tag{5}$$

Equation (5) implies the variational form:

$$\int_I m\partial_{tt}u\cdot\chi ds+\int_I\left(EI+b\left|u_{ss}\right|^2\right)u_{ss}\cdot\chi_{ss}ds+\int_I\lambda u_s\cdot\chi_s ds=0, \tag{6}$$

and

$$\left|u_s\right|^2-1=0 \text{ in } I, \tag{7}$$

almost everywhere in $(0,T)$ for all admissible functions χ.

Introducing now the nondimensional quantities $\tilde{s}=s/L$, $\tilde{u}=u/L$ and $\tilde{t}=tL^{-2}\sqrt{EI/m}$, the variational form becomes (after dropping tildes)

$$\int_I\partial_{tt}u\cdot\chi ds+\int_I\left(1+B\left|u_{ss}\right|^2\right)u_{ss}\cdot\chi_{ss}ds+\int_I\Lambda u_s\cdot\chi_s ds=\int_I Q\cdot\chi ds, \tag{8}$$

where $B=bL^{-2}/EI$, $\Lambda=\lambda L^3/EI$ and $Q=qL^3/EI$ and $I=[0,1]$.

In the nondimensional form, the strain energy density as a function of the curvature $\kappa=u_{ss}$ reads

$$V(\kappa)=\frac{1}{2}\left|\kappa\right|^2+\frac{1}{4}b\left|\kappa\right|^4. \tag{9}$$

A plot of the V as a function of κ, for different values of the nonlinearity parameter b is depicted in Figure 1. The bending moment for the above model, defined as the energy conjugate quantity of the curvature is

$$\mathrm{M}(\kappa)=\left(1+B\left|\kappa\right|^2\right)\kappa. \tag{10}$$

A plot of the bending moment as a function of the curvature is depicted in Figure 2 for different values of the nonlinearity parameter B. Finally, a plot of the bending modulus, denoted here as $\mathrm{dM/d}\kappa$ is shown in Figure 3. The feasible domains for the above mentioned quantities are highlighted in green color. For $B=0$ the linear bending moment-curvature model is retrieved. Positive values of B correspond to a type of elastic hardening in the behaviour of the beam, while negative values produce a reduction in the flexural rigidity, with increasing curvature.

This particular model can be employed up to values of curvature that retain the positive definiteness of the strain energy in order to simulate the small curvature region of the more involved behaviour that has been proposed in [2].

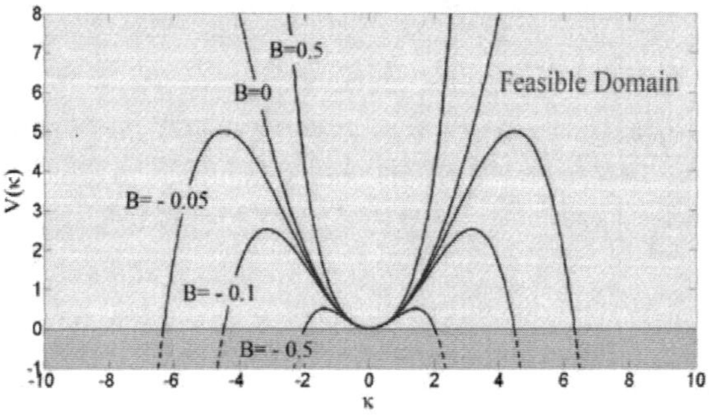

Figure 1. Strain energy density for different values of the nonlinearity parameter *B*.

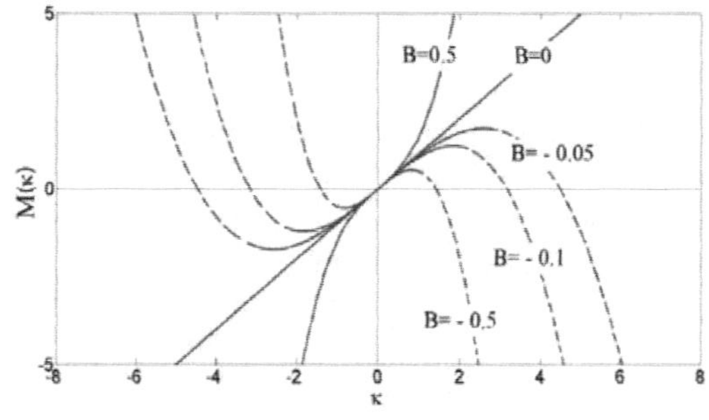

Figure 2. Bending moment for different values of the nonlinearity parameter *B*.

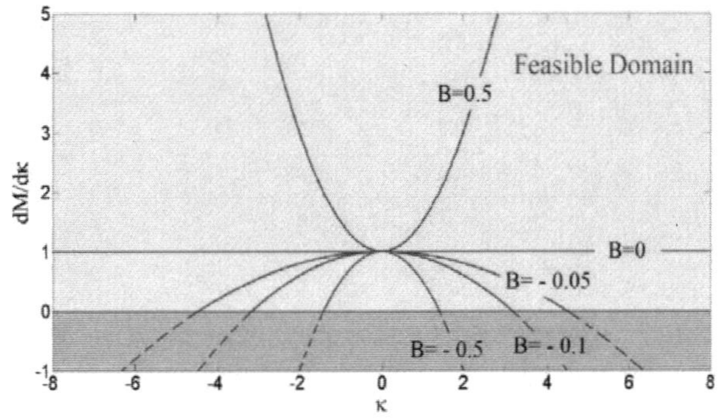

Figure 3. Flexural rigidity for different values of the nonlinearity parameter *B*.

4. The proposed finite element scheme

For the efficient solution of the inextensible beam model equations presented in the previous section, the numerical procedure proposed by Bartels [1] for the vibration of inextensible curves of constant flexural rigidity is utilised. The efficient method proposed by Bartels, based on the satisfaction of the inextensibility constraint only on the nodes of the finite element mesh, is extended here to the case of an inextensible beam that has a curvature dependent flexural modulus and in particular a bending moment-curvature relation as defined in Equation (10). It should be noted that the scheme proposed here includes only the flexural response of the riser and for realistic applications it should be coupled with axial and torsional effects as well. In the following, basic steps for the derivation of an integrated methodology for the efficient and robust simulation of flexible risers are described. Initially, following Bartels [1], for the time discretisation of the system the backward difference quotient D^{-} is introduced in Equation (11)

$$D^{-}v^{n+1} = \frac{v^{n+1} - v^{n}}{\tau}, \tag{11}$$

In the above expression v^{n} denotes the velocity of the system $\partial_t u$ at time instant t_n and $\tau = t_{n+1} - t_n$ is the time step (assumed constant in this study but in general variable). Furthermore, the update (12) of the deflection is introduced:

$$u^{n+1} = u^{n} + \tau v^{n+1}, \tag{12}$$

The scheme involving spatial discretisation using standard Hermite interpolation for each finite element and expressions Equation (11) and Equation (12) for discretisation with respect to time is:

Given $u^{j}, v^{j} \in H^{2}(I;\mathrm{R}^{3})$, $j = 0,1,2,...,n$ *and* $\tau > 0$, *find* $v^{k+1} \in H^{2}(I;\mathrm{R}^{3})$ *such that*

$$\int_I D^{-}v^{n+1} \cdot \chi ds + \int_I \left(1 + B\left|u_{ss}^{n} + \tau v_{ss}^{n+1}\right|^{2}\right)\left[u_{ss}^{n} + \tau v_{ss}^{n+1}\right] \cdot \chi_{ss} ds = 0, \tag{13}$$

and $v_s^{n+1} \cdot u_s^{n} = 0$ in I, for all $\chi \in H^{2}(I;\mathrm{R}^{3})$, with $\chi_s \cdot u_s^{n} = 0$ in I. Update the deflection through $u^{n+1} = u^{n} + \tau v^{n+1}$, where $H^{2}(I;\mathrm{R}^{3})$ denotes the space of functions with square-integrable second derivatives defined in I, with values in the Euclidian space R^{3}. The value $j = 0$ corresponds to the definition of appropriate initial conditions for the deflection and velocity of the inextensible beam.

For the approximate solution of the above problem, using the definition of $D^{-}v^{n+1}$, multiplying the discrete variational form by τ and expanding the terms in the second integral we obtain

$$\begin{aligned}&\left(v^{n+1}-v^{n},\chi\right)+\tau\left(\left[1+B\left|u_{ss}^{n}\right|^{2}\right]u_{ss}^{n},\chi_{ss}\right)\\&+\tau^{2}\left(\left[1+B\left|u_{ss}^{n}\right|^{2}1\right]v_{ss}^{n+1},\chi_{ss}\right)+2\tau^{2}\left(B\left[u_{ss}^{n}\cdot v_{ss}^{n+1}\right]u_{ss}^{n},\chi_{ss}\right)\\&+\tau^{3}\left(B\left|v_{ss}^{n+1}\right|^{2}u_{ss}^{n},\chi_{ss}\right)+2\tau^{3}\left(B\left[u_{ss}^{n}\cdot v_{ss}^{n+1}\right]v_{ss}^{n+1},\chi_{ss}\right)\\&+\tau^{4}\left(B\left|v_{ss}^{n+1}\right|^{2}v_{ss}^{n+1},\chi_{ss}\right)=0\end{aligned}, \tag{14}$$

where $(\ ,\ )$ denotes the $L^2(I)$ inner product defined by the integrals in the definition of the discrete problem. Using now the identity

$$\left(u_{ss}^{n}\cdot v_{ss}^{n+1}\right)u_{ss}^{n}=\left(u_{ss}^{n}\otimes u_{ss}^{n}\right)v_{ss}^{n+1}, \tag{15}$$

the following result is easily derived from Equation (14).

PROPOSITION Let $\mathbf{I}_3$ denote the 3×3 identity matrix. Then it is

$$\begin{aligned}&\left(v^{n+1},\chi\right)+\tau^{2}\left(\left[\mathbf{I}_{3}+B\left|u_{ss}^{n}\right|^{2}\mathbf{I}_{3}+2B\left(u_{ss}^{n}\otimes u_{ss}^{n}\right)\right]v_{ss}^{n+1},\chi_{ss}\right)=\\&\qquad\left(v^{n},\chi\right)-\tau\left(\left[1+B\left|u_{ss}^{n}\right|^{2}\right]u_{ss}^{n},\chi_{ss}\right)+O(\tau^{3})\end{aligned}, \tag{16}$$

The above proposition suggests that the simper form Equation (16) can be used instead of Equation (14), while the remainder is of the order $O(\tau^3)$. This is a higher order term compared to the order of approximation in the finite element procedure employed by Bartels [1] and thus the use of Equation (16) instead of Equation (14) does not affect the rate of convergence.

The linearized updated procedure defined by Equation (16) has certain similarities with the *secant modulus* method (method of Kačanov), however it is not identical with the secant modulus method obtained by linearizing the problem and then introducing the backward difference quotient for the time derivative approximation. For the approximation of the displacement field either 2- node Hermite elements or 3-node Hermite elements [18] can be used.

5. Numerical results and discussion

In this section, the efficiency of the proposed model and the numerical solution strategy is investigated. The possible use of the proposed finite element for applications involving the large deflection of flexible marine risers (Figure 4), featuring large displacements in the 3D space, various types of boundary conditions and possible large masses with degrees of freedom attached at one end (representing possibly a vessel motion due to surface gravity wave fields) is discussed.

Regarding the effectiveness of the proposed methodology and the effect of the nonlinearity in the moment-curvature relation, one of the examples studied in [1] will be analysed for different values of the nonlinearity parameter B. In particular, the case of an unwinding helix is selected. Following Bartels [1], we select $I=[0,2\pi]$, $T=30$ and

$$u(s,0)=\left[\sin(\sqrt{0.99}s)\quad\cos(\sqrt{0.99}s)\quad\sqrt{0.01}s\right],\ v(s,0)=\left[0\quad0\quad0\right],$$

while at $s=0$ it is

$$u(0,t)=\left[0\quad1\quad0\right],\ u_s(0,t)=\left[\sqrt{0.99}\quad0\quad\sqrt{0.01}\right],$$

representing clamped boundary conditions.

Figures 5 and 6 present snapshots of the unwinding helix evolution for the linear bending moment-curvature model (blue line) and the nonlinear case (red line).

In Figure 5, the case B = -0.025 is examined, corresponding to a 'softening' flexural modulus with increasing curvature, while in Figure 6, it is $B = 0.1$ (red line), corresponding to a flexural modulus that increases with increasing values of the curvature measure. The results for $B = 0$ are also included. This case is presented for two reasons. First, it provides a means of validation since it has already been analysed in [1]. The results obtained in this study for $B = 0$ are identical to those obtained by Bartels [1]. The second reason is to demonstrate the effect that the nonlinearity in the bending behaviour of the beam has on the deflections. To that end, the linear case is retained as a reference.

The violation of the inextensibility constraint (that is the total increase in the length of the beam) in the examples presented in Figures 5 and 6 is at the most 0.05 in a total (nondimensional) length of 6.28. As reported also in [1], if the time increment is appropriately selected for a given spatial mesh size, the present numerical scheme should approximate the inextensibility feature very effectively. The time needed for the solution of the above presented examples is only a few minutes. The number of 2-node elements used is 30 and the time steps less than 2000.

As a second example, a helix with two concentrated masses attached at its ends is studied. The mass at the lower end is assigned a very large value, leading to virtually zero motion of the lower end. This case would approximately correspond to a fixed end. Such an edge condition would be appropriate for the simulation of the riser connection to the seabed. The mass attached at the upper end is significantly lower and thus the upper end undergoes significant displacements. This type of edge condition might be relevant to the attachment of a flexible riser onto a vessel or a floating platform. In the nondimensional setting, the mass attached to the lower end is 10^6, while the one at the upper end is 100. The remaining parameters are $I = [0, 4\pi]$, $T = 30$ and

$$u(s,0) = \left[\sin(\sqrt{0.25}s) \quad \cos(\sqrt{0.25}s) \quad \sqrt{0.75}s\right], \; v(s,0) = \left[0 \quad 0 \quad 0\right],$$

while at $s = 0$ it is

$$u(0,t) = \left[0 \quad 1 \quad 0\right], \; u_s(0,t) = \left[\sqrt{0.27} \quad 0 \quad \sqrt{0.75}\right].$$

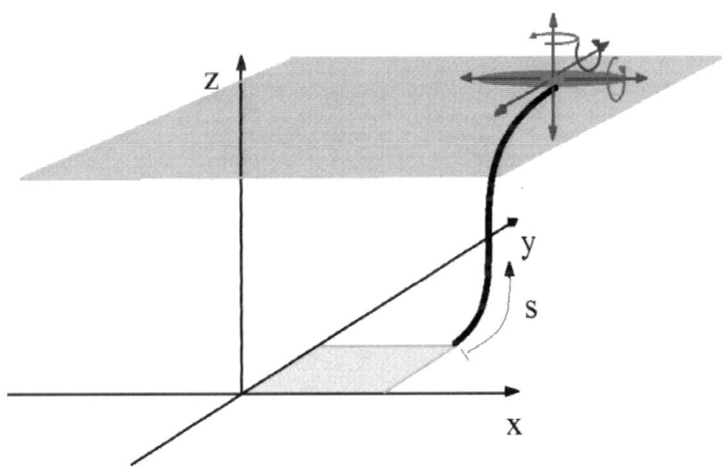

Figure 4. Schematic representation of a flexible marine riser.

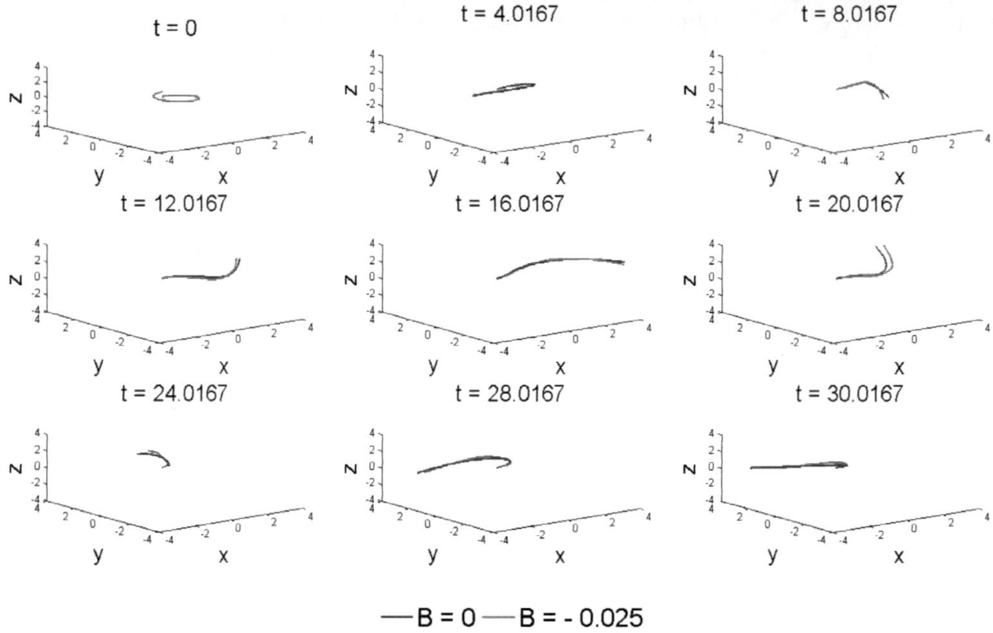

Figure 5. Evolution of the unwinding helix for $B = 0$ and B = -0.025.

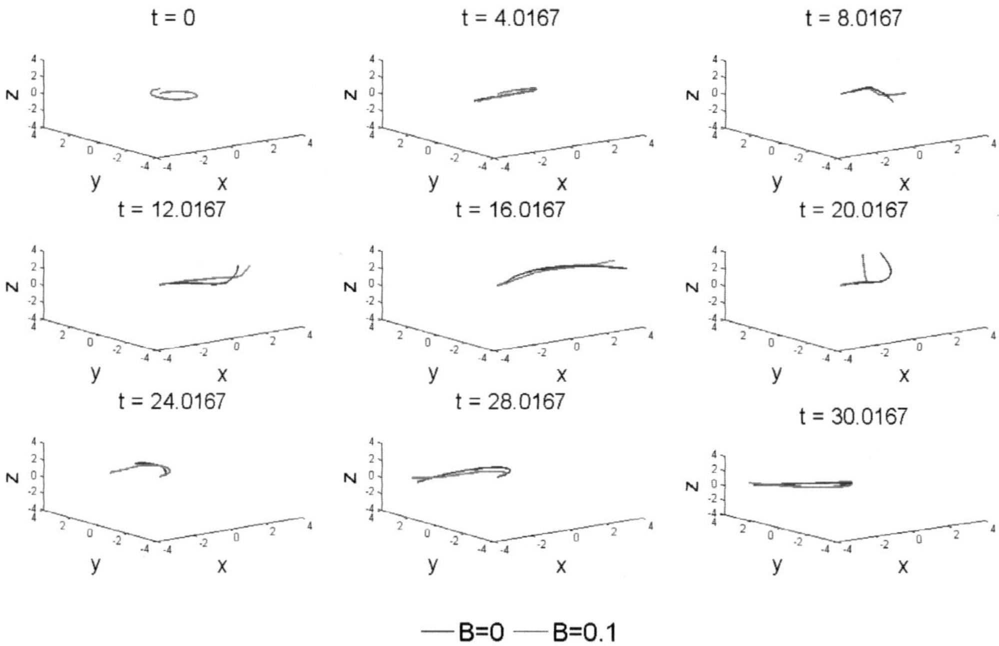

Figure 6. Evolution of the unwinding helix for $B = 0$ and $B = 0.1$.

Again the violation in the inextensibility constraint leads to an insignificant variation in the total length of the beam, even in cases where very few elements and time steps are used. In all the examined cases, the total time needed for the evaluation of the evolution was only a few minutes. This suggests that the efficient solution methodology proposed in [1], and applied to the case of riser like structures in the present work, might be of significant interest when multiple simulations for the purpose of parametric studies are needed. Snapshots of the evolution in the beam deflection are depicted in Figure 7. The velocity along the length of the beam is also depicted in the form of arrows. It should be noted that the selected example corresponds to the case $B = 0$.

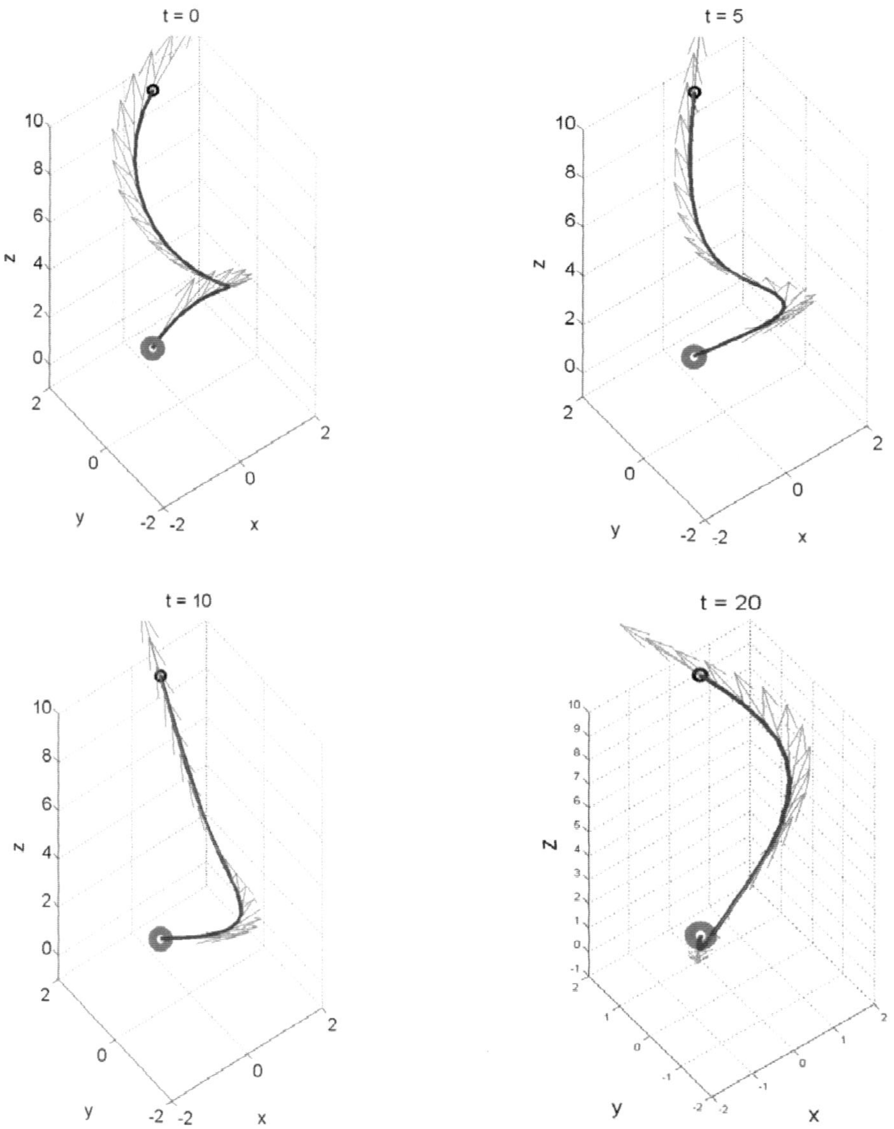

Figure 7. Snapshots of the evolution (example 2) for $B = 0$.

6. Conclusions

In this study, a structural finite element for an inextensible riser, based on a point-wise application of the inextensibility constraint is presented. In particular, the inextensibility constraint is applied only at the nodes of the meshed arc length parameter. The method is extended to include some degree of nonlinearity in the curvature-bending moment relation. The specific form adopted corresponds to the small curvature regime of more sophisticated riser mechanical behaviour models proposed in the literature. Future research focuses on the incorporation of the torsion effects while retaining the robustness of the numerical solution technique developed. Among the virtues of the proposed approach is the flexibility in the application of boundary conditions and the easy incorporation of dissipative forces.

References

[1] Bartels S 2016 A simple scheme for the approximation of elastic vibtations of inextensible curves *IMA J. Numer. Anal.* **36** pp 1051-71.

[2] Rahmati M T, Bahai H and Alfano G 2016 An accurate and comptutationally efficient small-scale nonlinear FEA of flexible risers *Ocean Eng.* **121** pp. 382-91.

[3] Kirk C L, Etok E U and Cooper M T 1979 Dynamic and static analysis of a marine riser *Appl. Ocean Res.* **1** pp 125-35.

[4] Kirk C L 1985 Dynamic response of marine risers by single wave and spectral analysis methods *Appl. Ocean Res.* **7** pp 2-13.

[5] Bernitsas M M, Kokarakis J E and Imron A. 1985 Large deformation three-dimensional static analysis of deep water marine risers *Appl. Ocean Res.* **7** pp. 178-87.

[6] Bernitsas M M and Kokarakis J E 1988 Importance of nonlinearities in static riser analysis *Appl. Ocean Res.* **10** pp 2-9.

[7] Vlahopoulos N and Bernitsas M.M. 1991 Three-dimensional nonlinear dynamics of nonintegral riser bundle *J. Ship Res.* **35** pp. 40-57.

[8] Ghadimi R. 1988 A simple and efficient algorithm for the static and dynamic analysis of flexible marine risers *Comput. Struct.* **29** pp. 541-55.

[9] Trim A D 1990 Time-domain random dynamic analysis of marine risers and estimation of non-Gaussion bending stress statistics *Appl. Ocean Res.* **12** pp. 162-74.

[10] Cheng Y, Kim Vandiver J. and Moe G. 2002 The linear vibration analysis of marine risers using the WKB-based dynamic stiffness method *J. Sound Vib.* **251** pp. 750-60.

[11] Kim Y C and Triantafyllou M S 1984 The nonlinear dynamics of long slender cylinders *J. Energy Resour. Technol.* **106** pp 250-56.

[12] Wang Y, Gao D and Fang J 2014 Axial dynamic analysis of marine riser installation *J. Nat. Gas Sci. Eng.* **21** pp. 112-17.

[13] Wang Y, Gao D and Fang J 2014 Static analyis of deep-water marine riser subjected to both axial and lateral forces in its installation *J. Nat. Gas Sci. Eng.* **19** pp 84-90.

[14] Rahmati M T, Norouzi S, Bahai H and Alfano G 2017 Experimental and numerical study of strucutral behaviour of a flexible riser model *Appl. Ocean Res.* **67** pp 162-168.

[15] Klaycham K, Athisakul C and Chucheepsakul S 2017 Nonlinear vibration of marine riser with large displacement *J. Marine Sci. Technol.* **22** pp. 361-75.

[16] Zhu H and Gao Y 2017 Vortex induced vibration response and energy harvesting of a marine riser attached by a free to rotate impeller *Energy* **134** pp 532-44.

[17] Cuamatzip-Melendez R, Castillo-Hernandez O, Vazquez-Hernandez A O and Vaz M A 2017 Finite element and theoretical analyses of bisymmetric collapses in flexible risers for deepwaters developments *Ocean Eng.* **140** pp 195-208.

[18] Papathanasiou T K, Karperaki A, Theotokoglou E E and Belibassakis K A 2014 A higher order FEM for time-domain hydroelastic analysis of large floating bodies in an inhomogeneous shallow water environment Proc. Royal Soc. A 471 20140643.

[19] Ong M C, Bachynski E E and Økland O D 2017 Dynamic Responses of Jacket-Type Offshore Wind Turbine Using Decoupled and Coupled Models *J. Offshore Mech. Arct. Eng.* **139** 041901.

IOP Conf. Series: Materials Science and Engineering **276** (2017) 012024 doi:10.1088/1757-899X/276/1/012024

Influence of extreme low temperature conditions on the dynamic mechanical properties of carbon fiber reinforced polymers

S P Zaoutsos* and M C Zilidou

Laboratory of Advanced Materials and Constructions, Department of Mechanical Engineering, Technological Educational Institute of Thessaly, Ring Road Larissas-Trikalon, Larissa, Greece.

* Contact Author: szaoutsos@teilar.gr

Abstract. In the current study dynamic mechanical analysis (DMA) is performed in CFRPs that have been exposed for certain periods of time to extreme low temperatures. Through experimental data arising from respective DMA tests the influence of low temperature exposure (-40 °C) on the dynamic mechanical properties is studied. DMA tests were conducted in CFRP specimens in three point bending mode at both frequency and thermal scans in order to determine the viscoelastic response of the material in low temperatures. All experimental tests were run both for aged and pristine materials for comparison purposes. The results occurred reveal that there is deterioration both on transition temperature (Tg) and storage modulus values while there is also a moderate increase in the damping ability of the tested material as expressed by the factor tanδ as the period of exposure to low temperature increases.

1. Introduction

Carbon fiber reinforced polymer composites (CFRPs) mainly based on thermoset polymers are being increasingly used in a variety of modern engineering applications such as aviation and space industry or in offshore infrastructures because of their advantage to high stiffness and strength values over weight ratio, corrosion resistance and ease of fabrication.

However, as the use of composite materials expands into the field of oil pipeline networks and aircraft transportation, their components are subjected to more hostile environments, including exposure to extreme low or cryogenic temperature conditions.

In these conditions, such as very low temperatures and moisture environment, the performance of the CFRPs still remains under consideration and extensive study [1 - 4]. Thus, it is crucial to understand the response of advanced composite materials to such conditions in order to produce a well-designed, structurally efficient component using the minimum amount of material and understanding of how such materials behave at cryogenic temperatures.

Under service conditions, CFRP bars are subjected to a wide range of very low temperatures. The exposure to these temperatures may have radical effects on fiber-matrix bond deterioration and matrix cracking compared to any exposure. Sub-zero temperatures can cause changes in mechanical properties and create additional microcracks in CFRP materials.

Additionally at lower temperatures microcracking can further result in the increase of water absorption at higher temperatures, leading to an increased matrix plasticization or delaminations. Frozen water present in cracks and voids may also lead in debonding and transverse microcrack growth resulting in material degradation, increased brittleness and changes in damage mechanisms. The effect of low-temperature thermal cycling on mechanical properties of carbon-vinylester composites and E-glass/vinyl ester composites has been investigated by a number of researchers [5-10]. In their studies they pointed out that freeze-thaw events could cause significant degradation of the mechanical properties and glass-transition temperatures of the FRP specimens immersed in an aqueous environment.

In order to predict the mechanical response of the material in these environments it is very important to know their performance for longer periods of time, thus the viscoelastic properties must be determined. To the author's knowledge limited or no data exist for the dynamic viscoelastic response of CFRPs under severe low temperature conditions.

The study aims to determine the effect of extreme low temperature in the dynamic viscoelastic properties of CFRPs and the change in their rheological behavior through the technique of dynamic mechanical analysis.

2. Theoretical background

Dynamic mechanical analysis has been used extensively in literature to study and characterize materials, mainly polymers and composites, that exhibit time dependent behaviour. Through a sinusoidal stress applied to the specimen, the strain of the material is measured, allowing thus to determine the storage modulus E', the loss modulus E'', tanδ and the glass transition temperature Tg of the material as well as to identify transitions corresponding to other molecular motions by varying the temperature or the frequency of the applied stress.

In a more extensive manner, by observing that the creep compliance vs. log (time) or log (frequency) curves for different temperatures retain the same shape, a technique of shifting these curves and superimposing them to a unique reference master curve can be applied [11-13]. This is the Time Temperature Superposition Principle (TTSP) that has been extensively used in literature [14-20] for polymers and composite materials.

In general, the steps followed when TTSP is applied so as to generate a creep master curve can be described by the methodology below.

- A material specimen is subjected to a constant load at a pre-set temperature level as in conventional creep testing, and the evolution of the creep strain of the specimen is recorded vs the log (time).
- Parallel experiments are performed for different specimens or even the same specimen, in modern DMA equipment, at different temperature levels and the relevant creep curves are obtained.
- A desired reference temperature is selected (TR).
- All the individual creep curves corresponding to different temperature levels are shifted along the log (time) scale to be superimposed to a master or mother curve.

If it is possible to generate a smooth master curve by applying a horizontal shift along the log (time or temperature) axis, the material can be then characterized as a thermorheologically simple material (TSM). However, for some materials a vertical shift factor may be needed to obtain a smooth master curve; they are then classified as thermorheologically complex materials (TCM).

TTSP has been used for polymeric materials, as it is governed by the solid rules of linearity and superposition that are applied in order to obtain the master curves. Usually, TTSP results exhibit the success of the method. Apart from polymer composites of various types [21], it has been successfully applied by researchers to investigate a variety of materials when time-related or temperature-activated procedures are considered. For example, it has been successfully employed for the characterization of asphalt concrete [22], viscous metallic liquids [23], solid rocket propellants [24] and wood [25].

3. Experimental procedure

3.1. Materials and manufacturing

Manufacturing of composite laminates was performed by Vacuum Assisted Resin Infusion Molding (VARIM) (Figure 1). Stacked woven carbon fiber layers, were cured for 12h at room temperature. The carbon fabric used was a C400P (TR50S) Plain Weave, (E= 240 GPa, σ_{ult} =4.9 GPa). Resin type was R2940/Infusion RTM (E=3.6 GPa, $\sigma_{ult} \approx$ 70-85 MPa, Tg $\approx$ 67-80 °C). Both room temperature curing resin and carbon fabric were supplied by FIBERMAX Composites Ltd. The fabric was hand laid in 6 layers of a stacking sequence $[(0/90)_2/\pm45]_s$ resulting in quasi isotropic Plates of 400x400x2.50 mm^3. The layers were vacuum-bagged and readily mixed resin/accelerator mixture infiltrated them by a flexible tube placed in one side of the vacuum bag. Specimens prior and after low temperature exposure were cut from these plaques by a diamond saw blade and polished with polishing paper in order to eliminate surface microcracks and local heterogeneities (Figure 2).

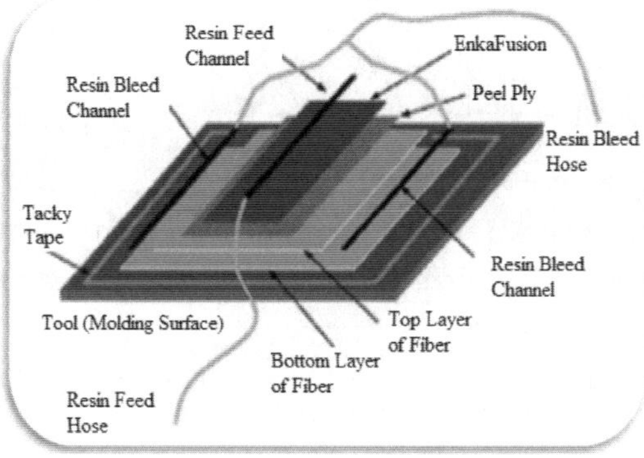

Figure 1. Schematic representation of the Vacuum Assisted Resin Infusion Molding technique used for the fabrication of the composite plates.

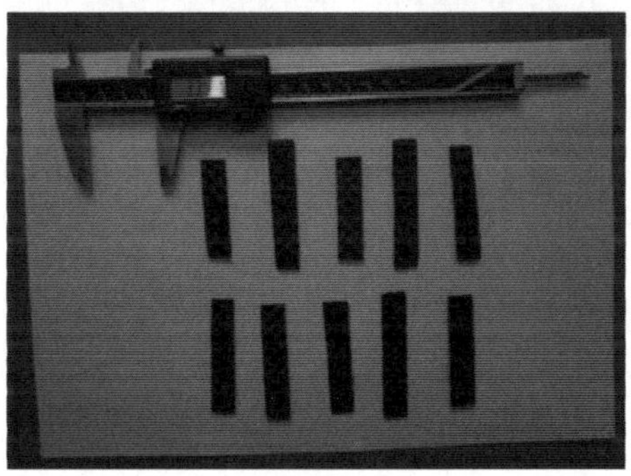

Figure 2. Specimens of CFRPs used for mechanical testing.

3.2. Composite plaques freezing procedure

In order to investigate the effect of low temperature on composites an experimental program applied included exposure of the composite plaques in a climatic chamber Instron mod.3119-409 (High Wycombe, UK) at -40 °C for three different periods of time, namely 30, 45 and 60 days. Composite plaques were then immediately afterwards removed and placed in a dry air freezer for 8 hours.

3.3. Testing

Three point bending testing (3PB) was performed on an Instron 3382 Universal Testing Machine as per EN ISO 178:1996, equipped with a specially manufactured high-stiffness bending rig (Figure 3). Specimens' nominal dimensions were Length = 50 mm, Width = 12 mm, Thickness = 2.50 mm and crosshead speed of the three point bending test was set at 2 mm/min. Support span was 40mm. All specimens were set in isothermal conditions prior to testing in order to avoid discrepancies occurring from the variation of temperature. Five specimens per sample were used for property averaging.

A similar specimen (50 mmx12 mmx2.5 mm) was used for the dynamic mechanical analysis (DMA) in three point bending mode on a DMA Q800 system of TA Instruments (Figure 4). Temperature scans were performed at a frequency of 1 Hz while the temperature was varied from T = 25-150 °C and heat rate was set at dT/dt =3 °C/min. Multi-frequency tests were performed at a range of 1 to 200 Hz and a temperature that varied from T = 25-150 °C.

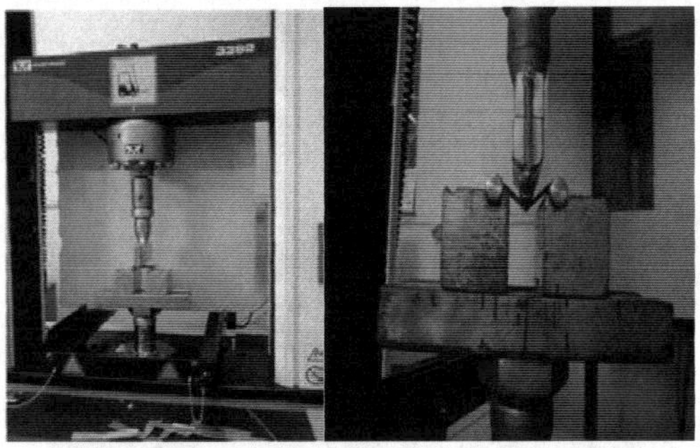

Figure 3. Three point bending tests procedure in Instron 3382 UTM.

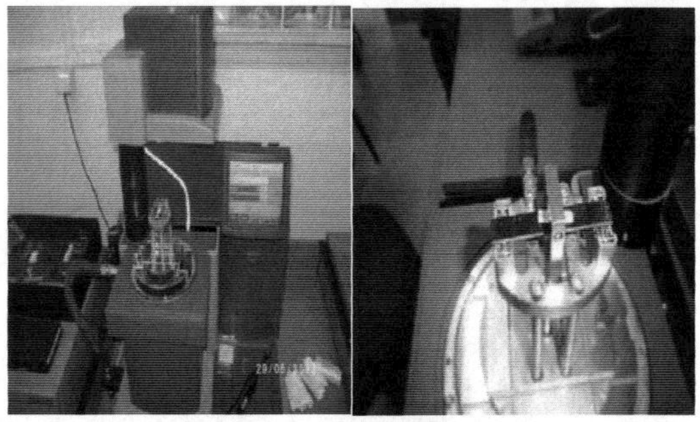

Figure 4. The type of the three point bending clamp used in mechanical testing in TA Q800 DMA.

4. Results and discussion

4.1. Three point bending static tests

Three point bending strength and modulus values determined from respective three point bending tests deteriorate as freezing time increases as seen in Table 1. This is a result confirming the general trend that environmental factors such as alternating temperature and humidity, accounts for mechanical property loss.

Table 1. Flexural properties for all composite specimens

Specimen Type/ Aging Procedure	Bending Modulus [GPa]	Bending Strength [MPa]
Pristine Specimens	$21.32^{\pm4.3}$	$693.15^{\pm78}$
30 days at -40°C	$18.16^{\pm3.5}$	$657.31^{\pm59}$
45 days at -40°C	$14.20^{\pm2.7.}$	$614.46^{\pm61}$
60 days at -40°C	$12.87^{\pm2.1}$	$579.17^{\pm69}$

4.2. Dynamic mechanical analysis tests-Temperature dependence

DMA spectra of dynamic mechanical properties i.e. storage modulus G', loss modulus G'' and tanδ parameter as a function of temperature are shown in Figures5-7 for the carbon fiber epoxy specimens. As it can be observed, the effect of the period of freezing temperature of -40°C on the values of the dynamic properties is obvious. Dynamic storage modulus (Figure 5) tends to decrease with rising testing temperature while its value drops by almost 5000 MPa confirming the results from static three point bending testing. A typical sigmoidal curve is observed in all cases. So, the tested CFRP exhibits a common "viscoelastic" time-temperature dependent response. The glass transition temperature (T_g), as it may be determined from the onset of the slope of the Dynamic Storage Modulus E'-Temperature T curve, also deteriorates.

Damping coefficient as expressed by the tanδ parameter appears to be increased with the variation of temperature (Figure 7). At the same time, an increase of the period of freezing temperature shifts the values of Tg at a lower level verifying the trend in the plots of Dynamic Storage modulus E'.

This is an important finding implying that the material loses part of its ability for energy distortion however gains in capability in vibrational dumping, which is a positive effect in the design of composites structures.

The mean shift of Tg from 0 days to 60 days exposure at -40 °C, is about -10 °C (from ~84 to ~74 °C) which is also an important part of restraint of the service temperature of a composite thus reducing his use capability.

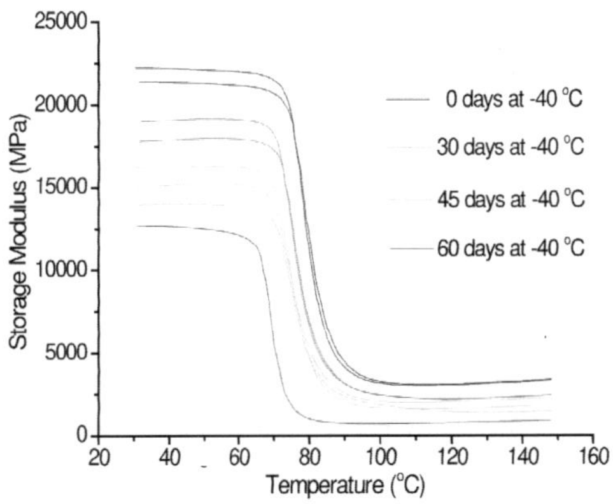

Figure 5. Storage modulus G' as a function of temperature and frequency.

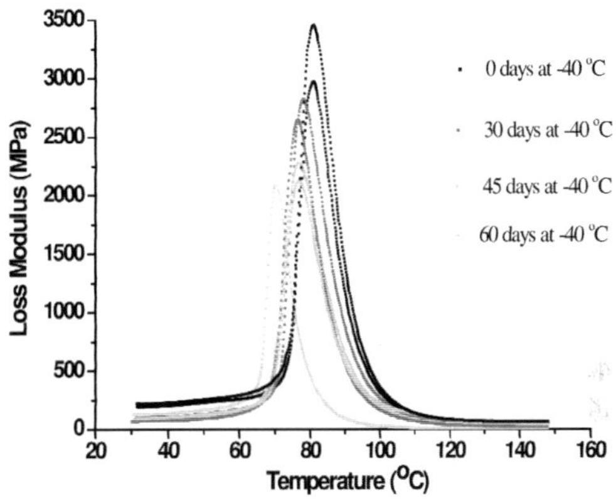

Figure 6. Loss modulus G'' as a function of temperature and frequency.

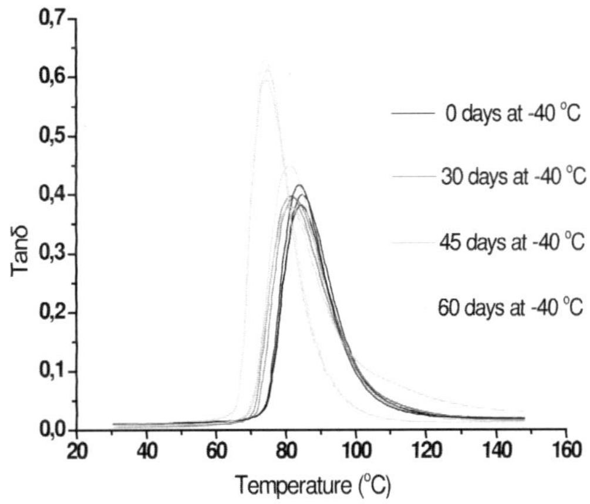

Figure 7. Tanδ as a function of temperature and frequency.

4.3. Dynamic mechanical analysis tests-Frequency dependence

Plots of storage modulus, loss modulus and tanδ for the three tested loading scenarios versus frequency occurring from scans of 35 °C up to 150 °C temperature range are appended in Figures 8-10. The curves that are to be superimposed are depicted in Figures 8a, 9a, 10a and 11a (in Appendix A), while the respective master curves are shown in Figures 8b, 9b, 10b, and 11b. It should be noted that all shifts in the curves that were to be superimposed in Figures 9a-11a were performed with respect to a reference curve that was experimentally determined at a temperature of 150 °C.

A comparison of the evolution of storage modulus versus frequency for specimens that have been exposed at -40 °C for three different periods of time is shown in Figure 12. As expected, an increase of the storage modulus as frequency per decade increases is observed, however, a tendency for decrease in the values of Storage Modulus as the exposure time to freeze increases is also obvious.

Similar behavior for loss modulus versus frequency is also depicted in Figure 13. The values of Loss Modulus tend to lower values as the composite remains for longer times in the reference freezing temperature. Finally, a view of the tanδ values is shown in Figure 14 confirming the previous trend of decrease versus freezing temperature. An important point should be also noted concerning the transition range of the stiffening of the material at around 2×10^5 rad/sec.

5. Conclusions

The study of the static and dynamic mechanical behavior of a fiber reinforced polymer composite exposed at a temperature of -40 °C proved the effect of freezing period on the mechanical properties of the material.

More precisely, deterioration of the values of static bending modulus and bending strength was observed while dynamic storage modulus was also reduced as the time of exposure at the tested low temperature was increased. On the contrary, an increase in the loss modulus versus time of exposure was noted. The above experimental observations were followed as expected by an increase of the dumping factor tanδ.

Glass transition temperature (Tg) was also affected by freezing time, moving to lower values thus limiting the service temperature of the material. Time temperature Superposition principle applied in CFRPs specimens denoted a moderate decrease in storage modulus, loss modulus and tanδ values versus freezing time at longer times, however, glass transition temperature seems not to be affected by this variation.

Appendix A

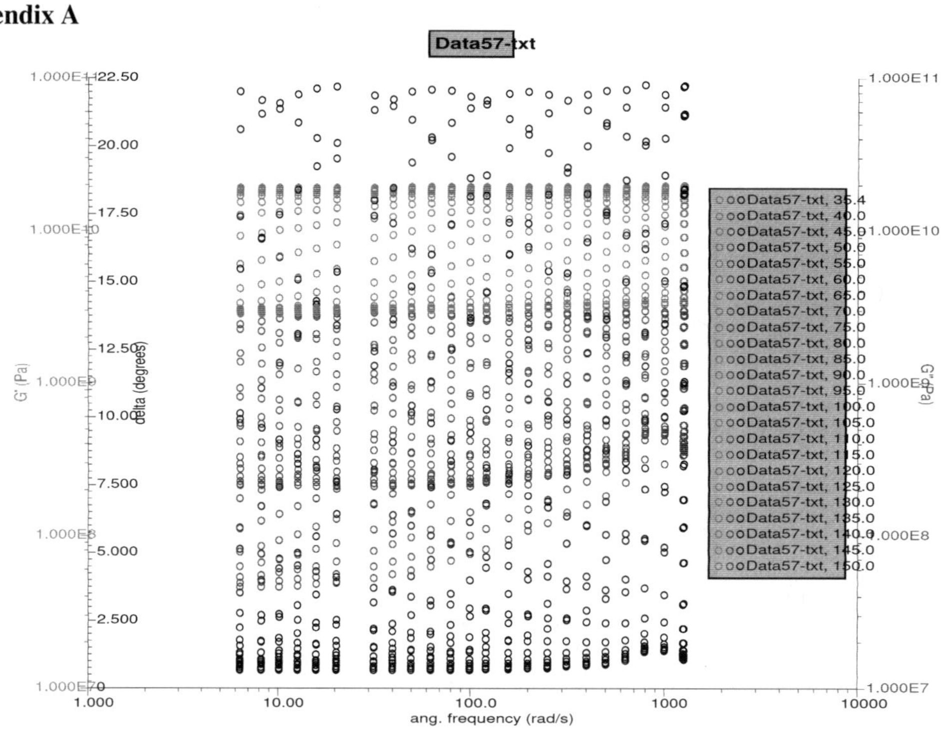

a)

IOP Conf. Series: Materials Science and Engineering **276** (2017) 012024 doi:10.1088/1757-899X/276/1/012024

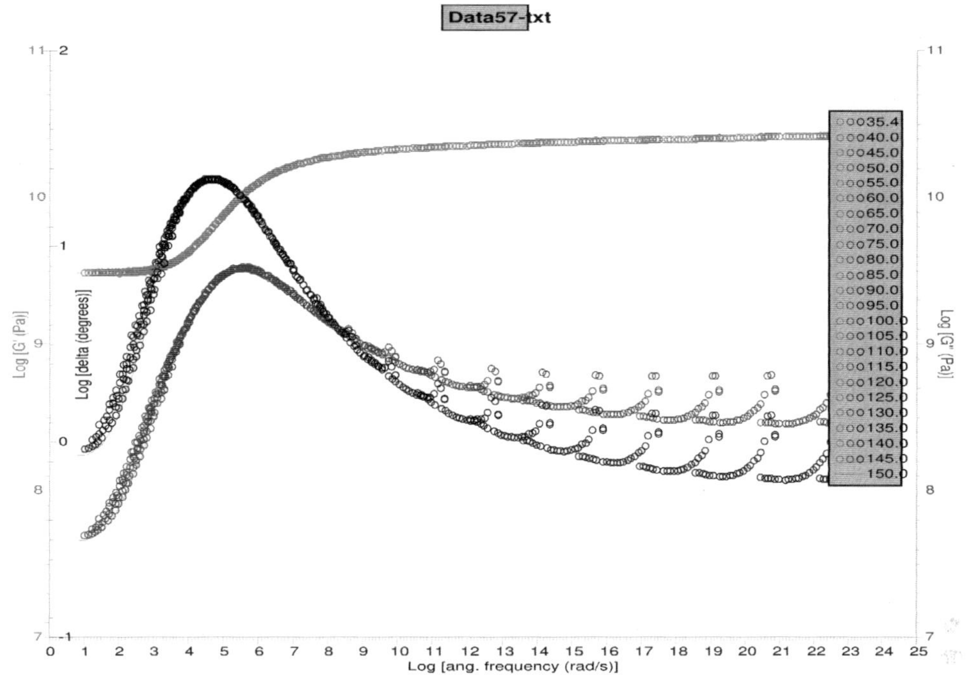

b)

Figure 8. a) G', G'' and tanδ curves versus fequency at different temperatures ranging from 35 to 150 °C that are to be superimposed and b) Master curves for G', G'' and tanδ as a function of frequency, for neat specimens.

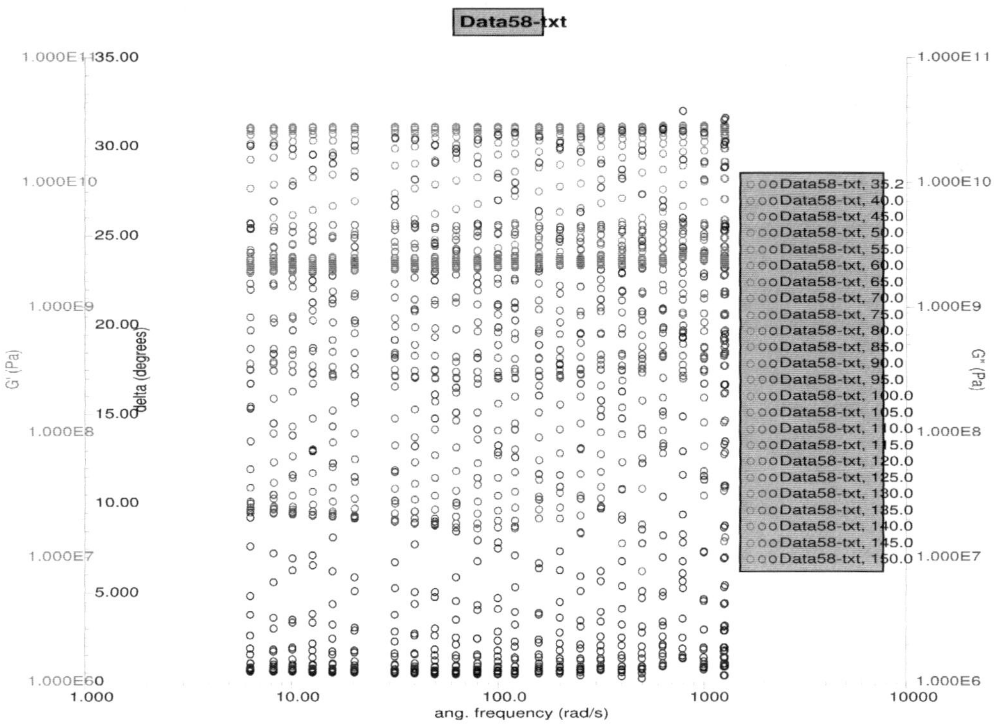

a)

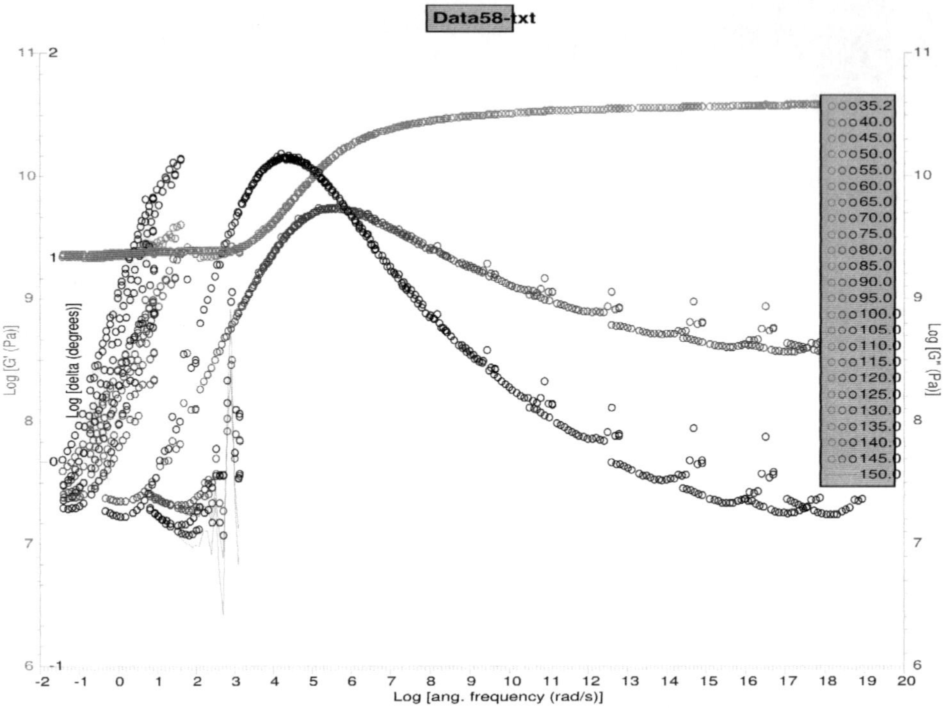

b)

Figure 9. a) G', G'' and tanδ curves versus fequency at different temperatures ranging from 35 to 150 °C that are to be superimposed and b) Master curves for G', G'' and tanδ as a function of frequency, for specimens for 30 days at -40 °C.

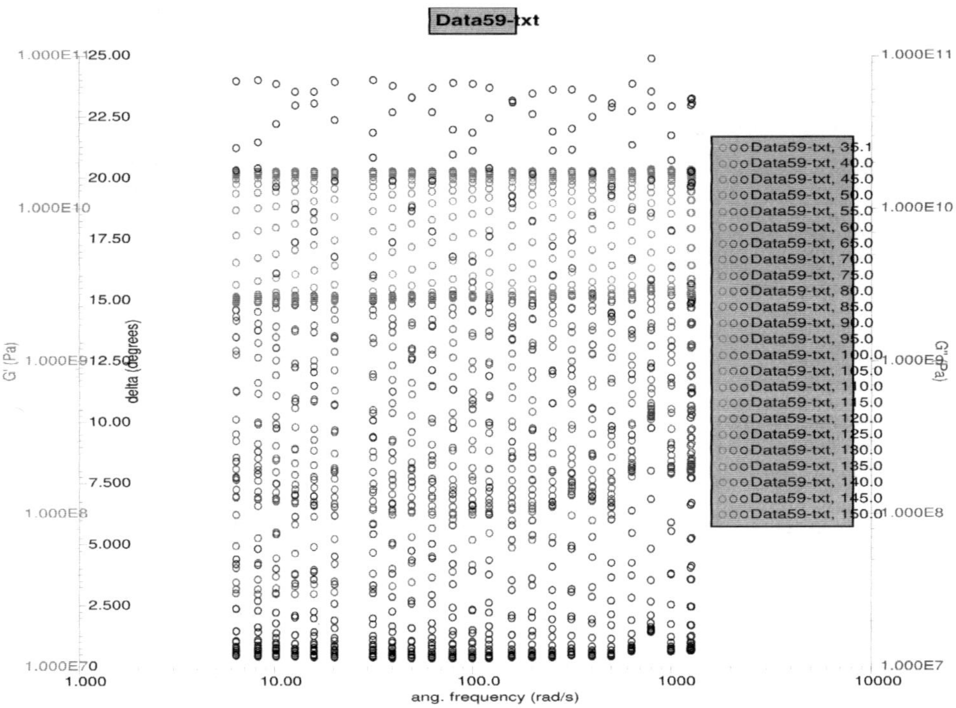

a)

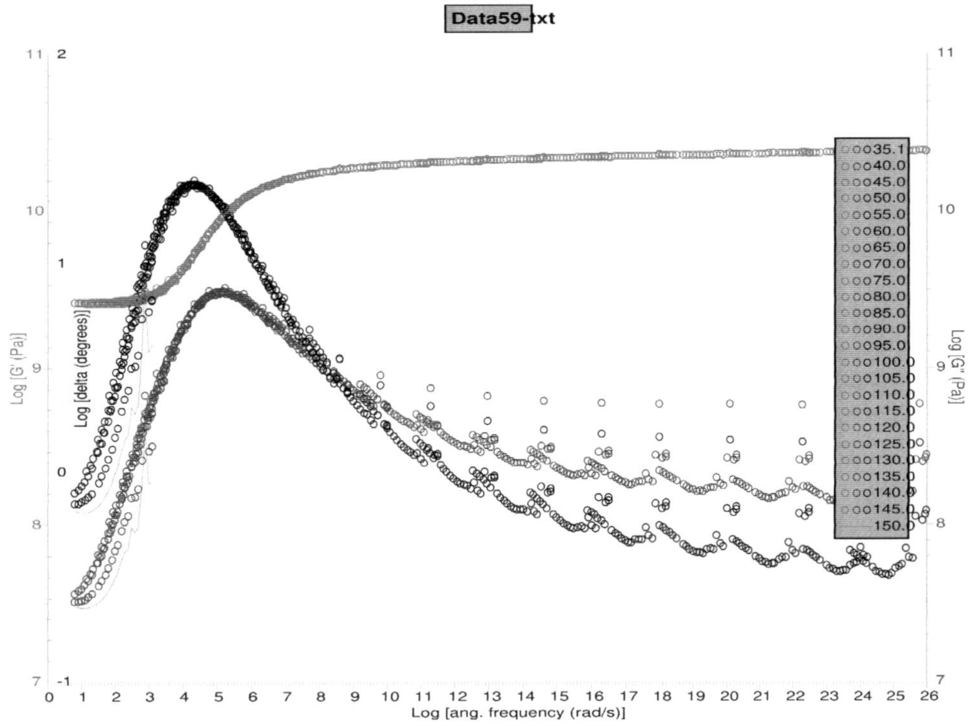

b)

Figure 10. a) G', G'' and tanδ curves versus fequency at different temperatures ranging from 35 to 150 °C that are to be superimposed and b) Master curves for G', G'' and tanδ as a function of frequency, for specimens for 45 days at -40 °C.

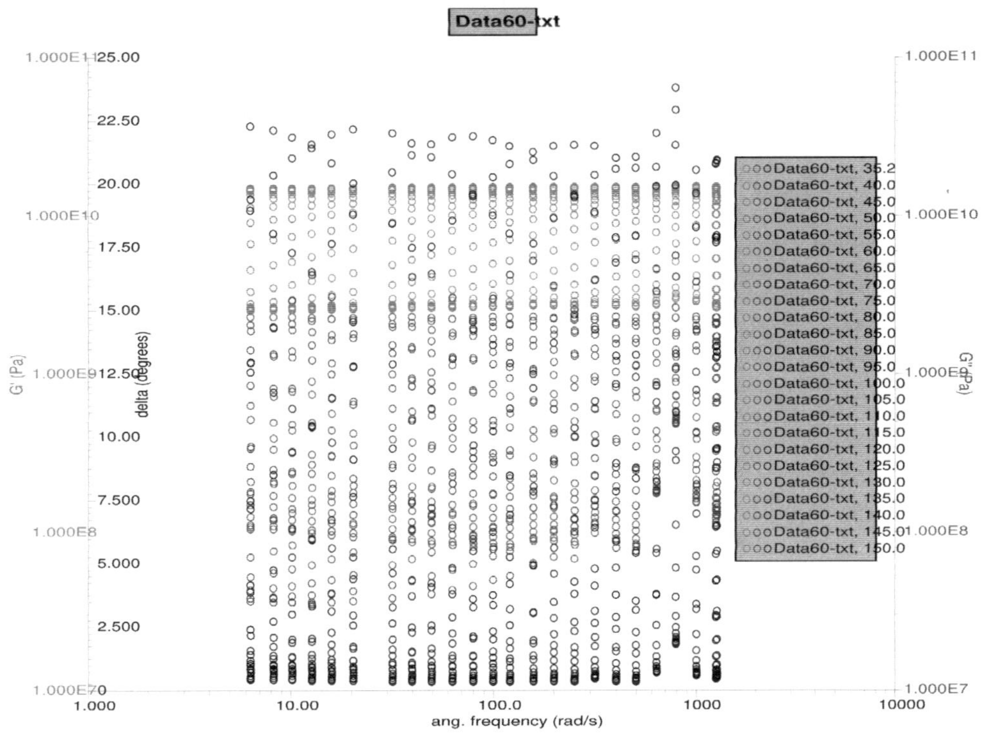

a)

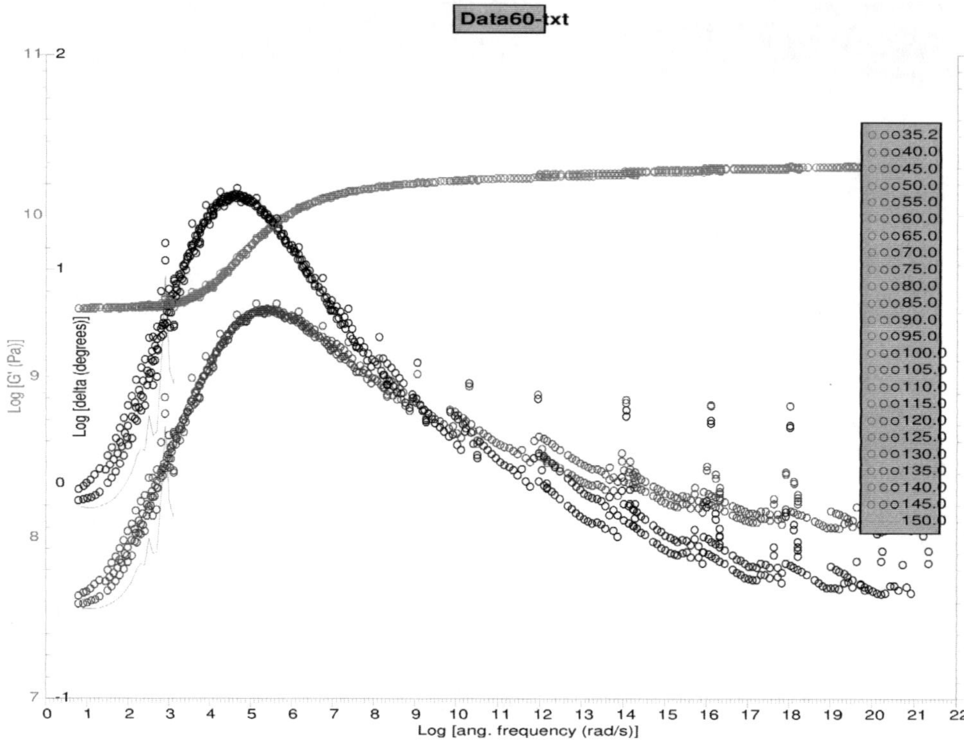

b)

Figure 11. a) G', G'' and tanδ curves versus fequency at different temperatures ranging from 35 to 150 °C that are to be superimposed and b) Master curves for G', G'' and tanδ as a function of frequency, for specimens for 60 days at -40 °C.

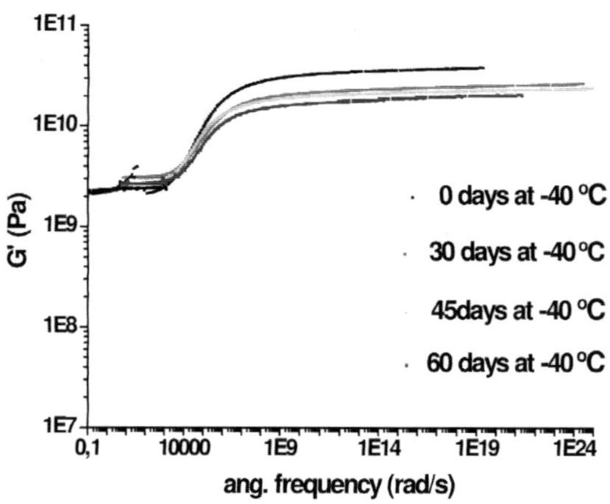

Figure 12. Comparison of superimposed storage modulus as a function of frequency for different exposure periods at -40 °C.

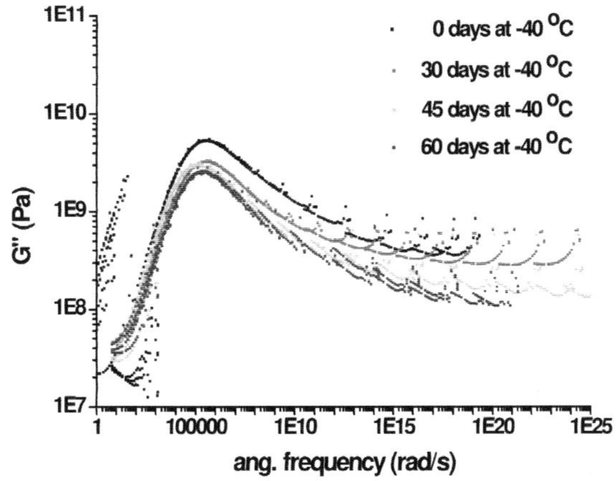

Figure 13. Comparison of superimposed loss modulus as a function of frequency for different exposure periods at -40 °C.

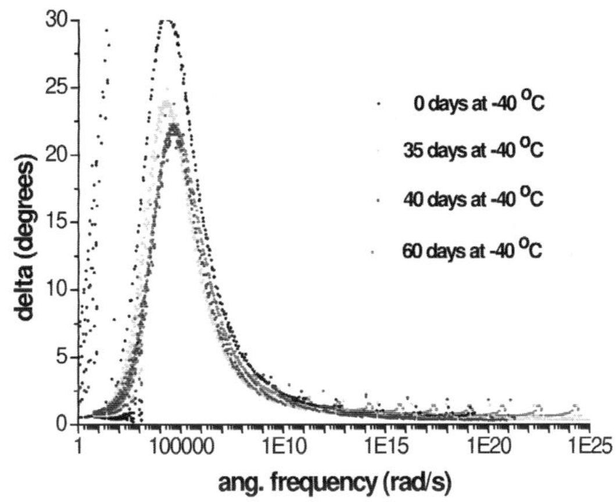

Figure 14. Comparison of superimposed tanδ as a function of frequency for different exposure periods at -40 °C.

References

[1]. Reed R P and Golda M 1998 Cryogenic Properties of Unidirectional Composites *Cryogenics* **34** pp. 909-28

[2]. Chu W and Karbhari V M 2005 Effect of water sorption on performance of pultruded e-glass/vinylester composites *J. Mater. Civ. Eng.* **17**(1) pp. 63–71

[3]. Shrivastava A K and Hussain M N 2008 Effect of Low Temperature on Mechanical Properties of Bidirectional Glass Fiber Composites J. Compos. Mater. **42**(22) 2407

[4]. Karbhari V M, Chin J W, Hunston D, Benmokrane B, Juska T, Morgan R, Lesko J J, Sorathia U, and Reynaud D 2003 Durability Gap analysis for fiber reinforced polymer composites in civil infrastructure *J. Compos. Constr.* **7**(3) pp. 238–241

[5]. Dutta P K 1992 Low temperature compressive strength of glass-fiber reinforced polymer composites *Proc. 11th Int. Conf. on Offshore Mechanics and Arctic Engineering, ASME* (New York)

[6]. Lord H W and Dutta P K 1988 On the design of polymeric composite structures for cold regions applications *J. Reinf. Plast. Compos.* 7(5) pp. 435–58

[7]. Haramis J 2003 *Freeze-thaw durability of polymeric composite materials for use in civil infrastructure* (Blacksburg, Va: Ph.D. dissertation, Dept. Of Civil Engineering, Virginia Polytechnic Institute and State Univ.)

[8]. Karbhari V M, Rivera J and Dutta P K 2000 Effect of short-term freeze-thaw cycling on composite confined concrete *J. Compos. Constr.* **4**(4) pp. 191–97

[9]. Karbhari V M, Rivera J and Zhang J 2002 Low-temperature hygrothermal degradation of ambient cured E-glass/vinyl ester composites *J. Appl. Polym. Sci.* **86** pp. 2255–60

[10]. Barton M and Greenfield D C L 1986 The use of Dynamic Mechanical Methods to Study the Effect of Absorbed Water on Temperature-dependent Properties of an Epoxy Resin-Carbon Fibre Composite *J. British Polymer* **18**(1)

[11]. Ferry J D 1980 *Viscoelastic properties of polymers. 3rd edition* (John Willey and Sons, Inc.) ISBN: 978-0-471-04894-7

[12]. Leaderman H 1943 *Elastic and Creep Properties of Filamentous Materials* (Washington: Textile Foundation)

[13]. Markovitz H 1975 Superposition in rheology *J Polym.Sci.: Polymer Symposia.* **50** pp. 431-56

[14]. Tobolsky A V and Andrews R D 1945 Systems manifesting superposed elastic and viscous behaviour *Chem. Phys.* **13** pp.3-27

[15]. Brinson H F, Morris D H and Yeow Y T 1978 A new experimental method for the accelerated characterisation and prediction of the failure of polymer-based composites laminates *6th Int. Conf. for Experimental Stress Analysis* (Munich, West Germany)

[16]. Griffith WI 1980 *The accelerated characterisation of viscoelastic composite materials* (USA Virginia Polytechnic Institute and State University: PhD Thesis)

[17]. Marsimov R D and Plume E 1982 Predicting the creep of unidirectional reinforced plastics with thermorheologically simple structural components *Mech. Comp. Mater.* **18**(6) pp. 737-44

[18]. Jeon H Y, Kim S H and Yoo H K 2002 Assessment of long-term performances of polyester geogrids by accelerated creep test *Polym. Test.* **21** pp. 489-95

[19]. Povolo F, and Fontelos M 1987 Time-temperature superposition principle and scaling behavior *J. Mater. Sci.* **22** pp. 1530-34

[20]. Brinson L C and Gates T S 1995 Effects of physical aging on long term creep of polymers and polymer matrix composites *Int J Solids and Struct* **32**(6/7) pp. 827-46

[21]. Georgiopoulos P P and Kontou E 2015 The effect of wood-fiber type on the thermomechanical performance of a biodegradable polymer matrix *J. Appl. Polym. Sci.* **132** pp. 42185-94

[22]. Schwartz C, Gibson N and Schapery R A 2002 Time-Temperature Superposition for Asphalt Concrete at Large Compressive Strains: Transportation Research Record *J. the Transportation Research Board* **1789** pp. 101-12

[23]. Meyer A, Busch R and Schober H 1999 Time-Temperature Superposition of Structural Relaxation in a Viscous Metallic Liquid *Phys. Rev. Lett.* **83** pp. 5027-29

[24]. Jeremic R 1999 Some Aspects of Time Temperature Superposition Principle Applied for Predicting Mechanical Properties of Solid Rocket Propellants *Propellants Explos. Pyrotech.* **24** pp. 221–3

[25]. Samarasinghe S, Loferski J R and Holzer S M 1994 Creep modeling of wood using Time-temperature superposition *Wood and Fiber Science.* **26**(1) pp. 122-30.

Numerical assessment of wind turbine blade damage due to contact/impact with tower during installation

[a,b]**Amrit Shankar Verma***, [d]**Nils Petter Vedvik**, [a,b,c]**Zhen Gao**

[a]Department of Marine Technology, Norwegian University of Science and Technology (NTNU), Trondheim, NO-7491, Norway
[b]Centre for Marine Operations in Virtual Environments (SFI MOVE), NTNU, Larsgardsvegen 2, 6009, Ålesund, Norway
[c]Centre for Autonomous Marine Operations and Systems (AMOS), NTNU, Trondheim, NO-7491, Norway
[d]Department of Mechanical and Industrial Engineering, NTNU, Richard Birkelandsvei 2B, NO-7491, Trondheim, Norway
amrit.s.verma@ntnu.no; nils.p.vedvik@ntnu.no; zhen.gao@ntnu.no

*Contact author: amrit.s.verma@ntnu.no

Abstract. The use of floating crane vessel for installation of offshore wind turbine blades presents a great challenge in terms of its random motions and is likely to increase the probability of the blade hitting the preassembled tower during lifting operation. To evaluate the consequences of such scenarios and to determine the allowable motions or sea states for such operations, it is very important to understand the damage development in the blade due to impact. The present paper employs the application of high fidelity finite element method to investigate the damage behavior in the blade when the leading edge of the blade hits the tower. A nonlinear time domain structural analysis using ABAQUS was conducted on the DTU 10 MW reference blade model which is based on shell elements. Damage assessment along with the nature of evolution of various energies is examined and presented for two different impact velocities with modified layup stacking sequence at the contact region.

1. Introduction

The current trend in offshore wind industry is bigger turbines and larger offshore wind farms which demands bigger and heavier components. Moreover, offshore wind industry is developing very fast, moving from shallow water to intermediate water [1]. However, one of the most challenging, risky and cost demanding phases for any offshore wind project is its installation phase [2]. The assembly and installation phase for any offshore wind project is a highly dynamic phase and contributes about 20% of the total CAPEX cost [3, 4]. The contribution of the installation phase to the CAPEX cost is expected to increase in the near future if the current state of the art methods and procedures are used for the planning and execution of marine operations related to offshore wind turbines, most of which are in principle adopted and based on the experiences from traditional oil and gas industry [2]. Thus, the planning and execution of marine operations related to offshore wind farms requires the study of risks and challenges explicitly dedicated to offshore wind turbines, which can further be employed to establish improved and optimized methods and procedures specifically for the installation phases related to the

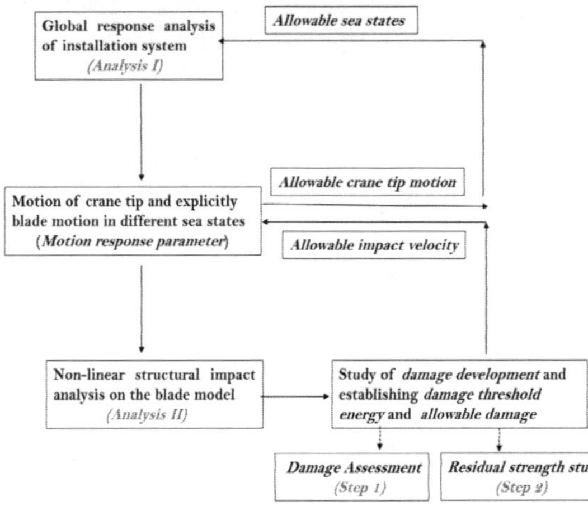

Figure 1. Explicit structural response based method

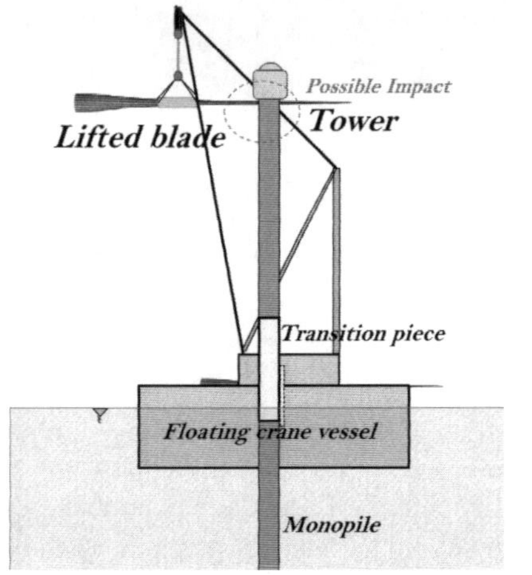

Figure 2. Illustration showing impact during lifting of the blade using floating crane vessel

offshore wind farms. Such methods should in principle reduce the CAPEX cost by increasing the weather window for installation as well as guarantee the safety of the components and stability of the installation systems by reducing the risk involved in such operations.

One such novel method called explicit response method was proposed by Guachamin Acero, et al. *2016* [3, 5]. This method is based on a very systematic approach that helps in assessment of operational limits for any offshore wind turbine installation operation in general by considering various risks, critical events, limiting response parameters and estimate operational limits [3, 5, 6]. The methodology is based on the numerical simulation of the actual marine operation and operational limits using response based criteria can be derived for a specific marine operation related to offshore wind turbine installation. The details of this generic approach can be found in [3, 5, 6]. In this way, a very practical operational limit can be derived which can be used for planning and execution of marine operation. Guachamin Acero, et al. 2016 [3, 5, 6] has successfully utilized this methodology to estimate the operational limit for mating operation of transition piece with monopile. Li, et al. 2016 [2, 7] has also used this approach to estimate operational limits for initial hammering process of the monopile using the heavy lift floating vessel (HLV). The present paper is motivated towards utilizing and extending this methodology to establish operational limits for the installation phase of a single blade being lifted using a jack up or a floating vessel while considering the risk of impact on the blade and finding damage threshold energy and allowable damage in the blade which will corresponds to allowable impact velocity (motion response parameter) obtained from *Analysis I* to derive back the allowable sea states (Figure 1). This allowable sea state will represent the safe response level for the blade and can be utilized for decision making during real time execution. However, derivation of these limits is out of the scope of this paper and the paper is focused explicitly on the study of damage resistance (*Step 1*) which is the study of damage assessment in the composite structure due to impact [8, 9].

At the moment, mainly jack-up vessels are used for the installations of offshore wind turbines and they provide a very stable platform to carry on operations in the sea but are mostly suitable for shallow water installation works up to the water depth of 30-60 m [5, 10]. However, in the near future, more and more floating installation vessels will be developed and used for installation of offshore wind turbines in intermediate or deep water [1]. The use of floating crane vessel for installation of offshore wind turbines presents great challenges in terms of its random motion mainly induced by wave loads [2, 3, 5, 7] and can lead to impact or contact events during installation owing to their excessive wave induced roll motions and wind and vessel induced crane tip motion (Figure 2). Such an undesirable event can be quite significant especially for the blades which are composite structures and are very much susceptible to impact induced damages during installation [11].

The sensitivity and the concern for the composite structures to impact induce damage is not new and has been a major concern in the military and aerospace sector. Offshore wind turbine blades are made of composite material as combinations of single skin laminates and sandwich configurations [12] and can also be in principle quite vulnerable to such undesirable impact/contact events when compared with other components made of steel structures getting installed [13, 14]. The property of the anisotropic composite materials is quite complex and has been found very sensitive to both very low and high velocity impact damages as compared to the isotropic steel structures. Steel structures in principle exhibit ductile behavior and the material continue to flow even after yielding in the plastic phase [15] and can dissipate a lot of absorbed energy in the plastic phase before fracture [15]. Composite material on the other hand are brittle in nature and can only dissipate absorbed energy either in the elastic deformation phase or through damage mechanism. These damage mechanisms are susceptible to grow and can cause significant strength reduction due to such impact events [12, 16-21] and thus such damage analysis on the blade is very important if the overall safety level of the installation phase is to be guaranteed. Moreover, the use of sandwich structures further complicates the understanding of blade under impact. There can be other reasons and challenges to do such an analysis on the composite blade. The loads arising due to such impact scenarios on the blade can be very large and questions the integrity of the entire wind turbine blade which are not designed to carry such large impact loads [13, 14]. Again, an offshore wind blade can exhibit several coupled and complex damage and failure modes which can affect the strength of the blade in different ways [22, 23]. The damage on the blade can be in principle localized or distributed [24]. The damage can be failure of fiber in tension (distributed damage), failure of fiber in compression (localized damage), matrix cracking, delamination between the layer of plies in the laminate (interlaminar failure mode), crack through the adhesive joints of trailing edge, leading edge or at the interior bond lines at laminate and beam assembly leading to adhesive failure [24, 25]. It has been shown that delamination of laminates and adhesive bond failure are the most critical damage modes for wind turbine blades [11, 22, 23, 26, 27]. Moreover, the size and location of the delamination also influences the residual compressive strength and reliability in blade [11, 22, 23, 26, 27]. Another challenge for the blade in such impact events is that it is not always feasible to visually inspect such damages, especially damages from small or low velocity impacts and are called barely visible impact damages (BVID) [18, 19, 28]. However, such damages on the blade could still lead to considerable reduction in the strength of the blade and thus it requires a careful study to predict damage mode in the blade under impact to understand the effect of such impact induced damage on the integrity of the blade [18, 19]. Moreover, such damaged blades could demand quite large repair and maintenance cost and could suffer premature failure during its design life if the damage levels are not analyzed and the blades are installed onto the turbine's hub and thus can cause major down time during the productivity hours offshore.

The present paper employs the application of high fidelity finite element method to investigate the damage behavior in the blade when the leading edge of the blade hits the tower. The DTU 10 MW model which is based on shell elements is utilized as the base case for studying such a scenario. The particulars of the blade, together with the parent and the modified composite layup at the impact region and the modeling details are explained in Section 2 followed by a detailed discussion on the damage models in Section 3. Finally, results and discussions followed by conclusions are presented in Section 4 and 5 of this paper respectively.

2. Impact scenario and numerical modeling strategy

The present work considers the contact scenario when the leading edge of the lifted blade (near the yoke region) impacts the tower during lifting which can occur owing to wave induced motion especially highly sensitive roll motion. The numerical model for this work is based on the DTU 10 MW reference blade model and will be referred as 'base model' or 'parent model' in the remaining paper.

2.1. Contact scenario and contact region

The contact scenario along with the contact region of the blade are very important parameters to be considered in a contact/impact analysis of the blade during installation. In principle, the contact scenario depends upon the type of lifting method considered (Figure 3).

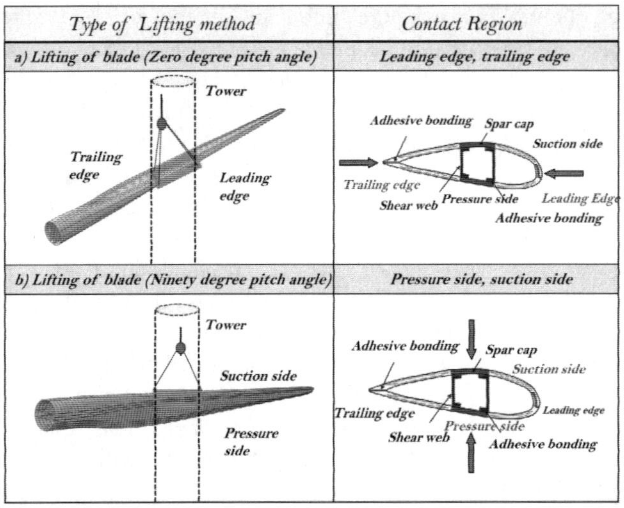

Figure 3. Illustration of type of lifting method and possible contact regions [42]

The often-used lifting method considered in the practice is the lifting of the blade either with a zero degree-pitch angle (horizontal lifting) or with a ninety degree-pitch angle (vertical lifting) (Figure 3) [29]. Horizontal lifting can cause the leading edge or the trailing edge side of the blade to impact the tower. On the other hand, the vertical lifting of the blade can cause the pressure or suction side of the blade vulnerable (Figure 3). Also, each scenario can have different contact regions and the criticality of the damage will vary as different sections of the blade are optimized with different composite layup. This paper investigates the scenario when the leading-edge side of the blade hits the tower during lifting operation under zero pitch angle (horizontal lifting) at a region (*'A'*) quite close to the yolk region of the lift ($0.44<r/R<0.48$) with two different impact velocity considered: 0.1 m/s and 0.5 m/s (Figure 6-7).

2.2. Composite layup at the impact region

The base model utilized in this work is a DTU 10 MW blade model made of GFRP plies with balsa as core material and is 86.4 m (R) long with 5.4 m root diameter [25, 31]. The outer surface of the blade model except for the shear web was used as the shell reference surface to take into advantage the smoothness property [30] as well as to aid in the contact formulation as the impacts occur between the exterior surfaces of blade and the tower. The structural design perspective of the blade is based on a conventional approach using a load carrying box girder with two shear webs along with a third shear web closer to the trailing edge of the blade [31].

As per DTU parent model, the whole blade was partitioned into 11 regions circumferentially and 100 regions radially for defining composite layup. However, the main motive of the DTU 10 MW blade model was to check the global strength of the blade under global aerodynamic loads which is not sensitive to details of the layup definition and thus the stacking sequence based on multidirectional ply formulation was considered along with the properties derived based on Classical Laminate Theory [31] [32]. The layup considered was based on triaxial, biaxial and uniaxial nomenclature. The layup for the contact region of the blade (*'A'*) considered in this paper based on this multidirectional nomenclature is presented in Figure 4. A triaxial ply in the layup is considered as a ply with [+45/-45/0] degree properties with 35%, 35% and 30 % contribution respectively. Similarly, a biaxial ply was considered as ply with [+45/-45] degree properties with 50-50 % contribution and a uniaxial ply was considered as the ply with [0/90] degree ply properties with 95% and 5% contribution respectively.

The same composite layup based on the above nomenclature representing ply, with equivalent properties of individual plies, cannot be used for the study of impact analysis on the blade at the impact region. This is due to the fact that the nature of impact is very much sensitive to the layup and it is not possible then to explicitly say which layer is most critical to damage. A typical offshore wind blade utilizes individual ply stacked with +45, -45 and 0-degree angle with each orientation contributing to cater specific loading condition. 0° ply is for the bending stiffness whereas +45 and -45 ply are used to cater for torsional and buckling resistance [9, 16, 33]. Moreover, it has been found that the delamination arising at the interface of the ply cause sub-laminate buckling and can be very critical [11, 22, 23, 26, 27] which in principle cannot be captured by multidirectional stacking sequence. Thus, to explicitly say which ply through the thickness will be more critical would require a layup in a unidirectional nomenclature representing individual ply with their specific orientation.

Thus, the composite layup for the impact region of the blade was modified and defined (Figure 5) in the terms of individual plies layup orientation keeping the basic parameters like thickness of the region (34.6 mm), mass (42.37 tons) and COG of the blade (26.2 m from the blade root) constant. Further, eigenvalue analysis for the whole blade model along with the impact analysis with the tower gave consistent results for both parent and modified layup. The material properties were derived from the DTU 10 MW database as well as from [11, 22, 23, 26] and are listed in Table I of section III.

2.3. Numerical modelling and assumptions

The numerical modelling of the laminates and composites for the finite element analysis very much depends upon the choice of element which is determined by the kinematic assumption enforced [34] for a particular type of element based on the composite shell theory. The choice of element for modelling the composite blade shell for the impact analysis in this study was considered as general purpose thick shell S4R element (Figure 7). S4R thick shell element is a 4-node element with finite strain and is suitable for large strain analysis. Further, the S4R element is enforced by first order shear deformation theory (FSDT) also called as Mindlin shear flexible theory and estimates transverse shear stresses using a shear correction factor and is well suited to model thick laminates or materials with low shear modulus as is the case with the composite materials. For Mindlin shear flexible theory, straight lines normal to the mid plane remain straight but not necessarily normal after deformation and thus considers interlaminar shear strains and was found suitable to model the blade. However, the use of shell elements for the blade could only capture intralaminar failure mode in the blade and could not analyze delamination between the interfaces of the ply which could be critical for the blade under impact.

However, the delamination can be qualitatively predicted as the matrix cracking in the ply could further lead to delamination. Moreover, to explicitly understand the progressive damage of the delamination, use of cohesive elements or fracture mechanics approach is required and then the blade must be modelled with solid continuum elements which makes the analysis very computationally expensive. However, the scope of this paper is to investigate the failure modes using the DTU 10 MW blade model which is based on shell element formulation. For simplicity, the core material in the form of balsa was considered to be part of the layup stacking sequence itself (Ply 5, Figure 8). The outer layer of the blade shell is taken as the reference shell surface. This allows contact of the blade outer shell with the tower which is physically true. Moreover, adhesive material along the trailing edge was modelled with standard hexahedral continuum solid element with eight nodes and reduced integrations scheme (C3D8R) which were surface to surface tie constrained (Figure 7) with the upper and lower aerodynamic shell. Moreover, adhesive connections along the spar-web were modelled with a simple edge-to-edge connection. The tower considered in this numerical study has a diameter 5 m with height 110m and was modelled as a discrete rigid body with (R3D4) element (Figure 7) and with fixed encastre boundary condition. The blade was given initial velocity in the x-direction (V_x) with no velocity in other two direction (V_y=V_z=0). The study does not include the numerical modelling of wire or yolk of the crane

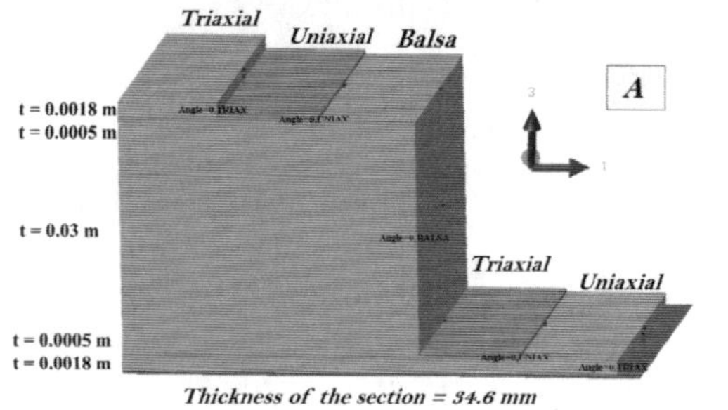

Figure 4. Composite Layup at the contact region (Parent DTU definition)

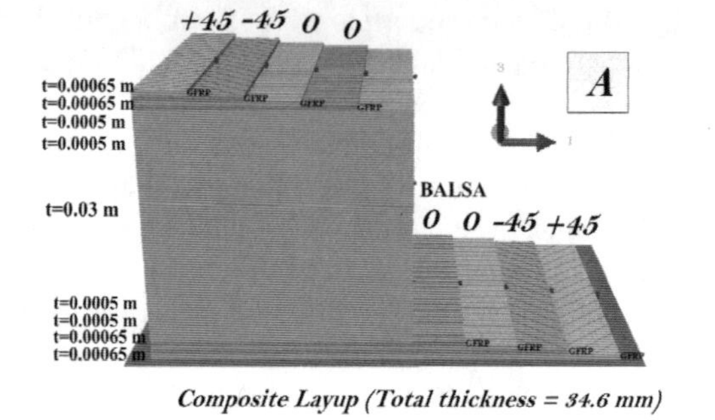

Figure 5. Composite Layup at the contact region (Modified layup at contact region 'A')

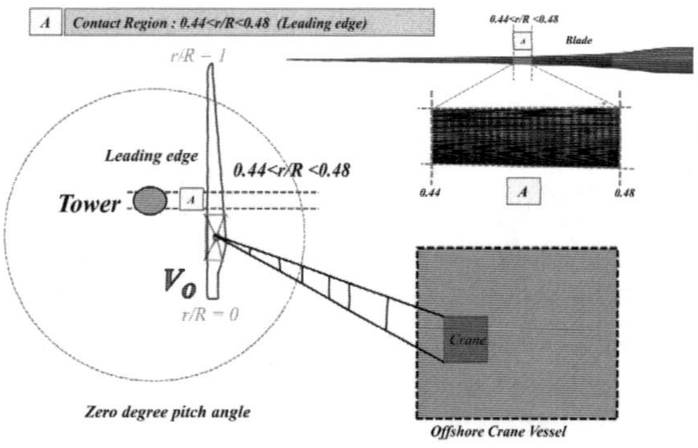

Figure 6. Bird view schematic representation of contact scenario and impact region of the blade marked with '*A*' (0.44 < r/R < 0.48)

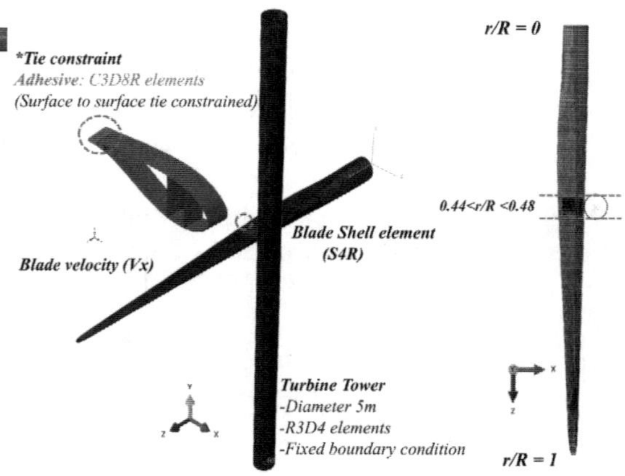

Figure 7. Numerical modelling detail for the blade and turbine tower

used for the lift and neglects the effect of gravity in the finite element study.

The contact between the blade reference surface and the tower outer surface was modelled using the general contact algorithm available in ABAQUS Explicit. The Penalty enforcement algorithm with a friction coefficient of 0.3 along with hard contact pressure over closure interaction behavior was defined. There was no mass scaling considered in any part of the model in the explicit analysis. Each ply was defined with 3 section points. Figure 8 shows section point's nomenclature considered while defining composite layup and it can be seen that there were 27 section points in total considered throughout the thickness out of which all section points other than 13, 14 and 15 were reserved for the composite material. The section points 13, 14 and 15 (Ply-5) were saved for outputting results for the balsa material. The post processing of results at section points makes it possible to get information for the individual ply in the shell based design of laminate composite structure. Figure 9 shows the mesh convergence study performed on the blade to calibrate result which will be independent of the mesh density. For this purpose, the maximum stress at the impact region was plotted with varying element sizes at the contact region for a case having velocity of impact as 0.1 m/s. The contact region finally had a very refined area with element dimension of 48 x 48 mm^2 (S4R) with other region having a coarse mesh size with elements varying from 50 mm to 0.2 m with total 87,456 elements chosen and the analysis takes almost about 26 hours to complete. This is owing to small stable time increment required by ABAQUS kernel to handle very small element size in contact.

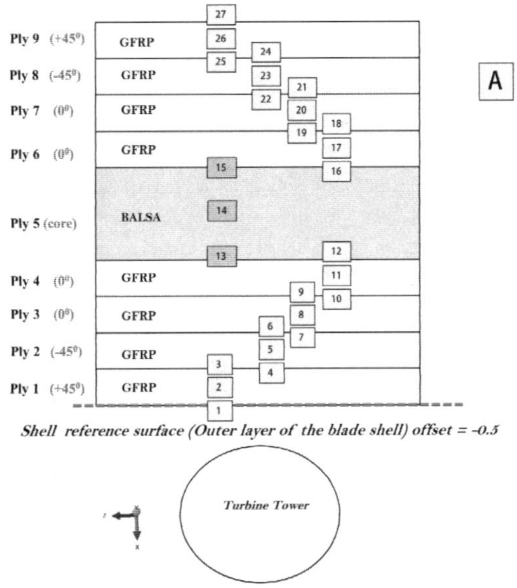

Figure 8. Composite layup with section points considered for each ply

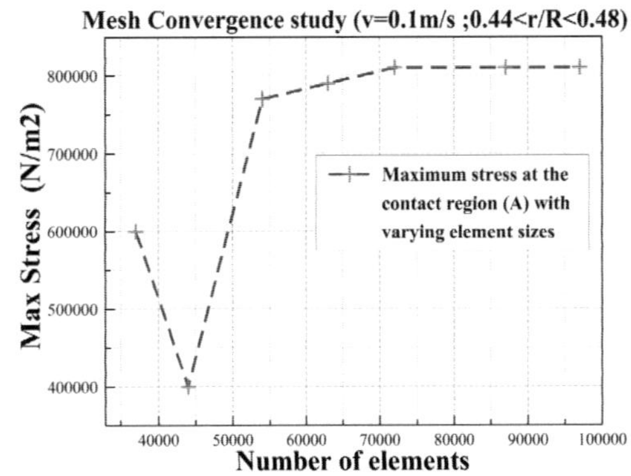

Figure 9. Mesh Convergence study for the global shell model

3. Constitutive damage models for impact analysis on blade

The present paper utilizes damage models explicitly for intralaminar damage modes in the composite ply and damage in the balsa which is utilized as the core material in the blade. They are explained below briefly:

3.1. Intralaminar damage model for composite ply

The damage quantification and assessment of composites at the ply level can be evaluated based on stress or strain based failure criteria or by an appropriate polynomial based failure criteria like Tsai-Wu or Tsai-Hill or by continuum damage mechanics approach [9]. The analysis in this paper utilizes a very well known strength based failure criteria to evaluate the onset and propagation of fiber failure and matrix cracks also referred to as progressive damage model in ABAQUS and other finite element solvers [18, 19, 30, 35]. The intra-ply damage model in this paper is based on Hashin failure criteria along with the energy-based damage evolution law which is based on the work of [36, 37] and can effectively describe the post damage initiation behaviour. This progressive damage model can distinguish different stages of the composite under impact loading history.

The first phase is based on undamaged constitutive behaviour where the material is assumed to behave linearly elastic followed by a damage initiation point described by Hashin's failure criterion and post damage Initiation behaviour given by a damage evolution law [36, 37]. *Hashin and Rotem, et al. 1973* [38] proposed a failure criteria which was further modified by *Hashin 1980* [35] and it utilizes the effective stress space (σ') as the failure surface and can be used to predict four major damage mode in the composite laminates- matrix tension, matrix compression, fiber tension and fiber compression. The initiation criteria have the following general form [30, 35]:

1. *Tensile fiber failure for* $\sigma'_{11} \geq 0$

$$F_f^t = \left(\frac{\sigma'_{11}}{X^T}\right)^2 + \alpha\left(\frac{\tau'_{12}}{S^L}\right)^2 \tag{1}$$

2. Compressive fiber failure for $\sigma'_{11} < 0$

$$F_f^C = \left(\frac{\sigma'_{11}}{X^C}\right)^2 \tag{2}$$

3. Tensile matrix failure for $\sigma'_{22} > 0$

$$F_m^t = \left(\frac{\sigma'_{22}}{Y^T}\right)^2 + \left(\frac{\tau'_{12}}{S^L}\right)^2 \tag{3}$$

4. Compressive matrix failure for $\sigma'_{22} < 0$

$$F_m^C = \left(\frac{\sigma'_{22}}{2S^T}\right)^2 + \left[\left(\frac{Y^C}{S^L}\right)^2 - 1\right]\left(\frac{\sigma'_{22}}{Y^C}\right) + \left(\frac{\tau'_{12}}{S^L}\right)^2 \tag{4}$$

Where σ'_{ij} denote the effective stress tensor; X^T, X^C, Y^T *and* Y^C denotes the allowable tensile and compressive strength of unidirectional laminate in longitudinal and transverse direction respectively. Further, S_L, S_T denote the allowable in-plane and out of plane shear strength of the composites in their respective principal material directions [30]. Moreover, here α denotes the contribution of shear stress in the tensile fiber failure mode (α =0, in this study). Once the F_f^t, F_f^c, F_m^t or F_f^t reaches 1 for an integration point, it implies that the damage initiation criteria have been met for any of the four-possible mode of failure explained above. The results can be processed at all the section points through the thickness of the laminate defined for this study using output request HSNMTCRT, HSNFCCRT, HSNMCCRT, and HSNFTCRT for tensile matrix failure, compressive fiber failure, compressive matrix failure and tensile fiber failure respectively except the section points 13, 14, 15 which are section points reserved for balsa material. Again, the energy based damage evolution is utilized, which is used to describe post damage initiation material behavior and calculate the stiffness of the damaged elements post damage initiation point by the relation given below:

$$\sigma = C_d \varepsilon \tag{5}$$

$$C_d = \frac{1}{D}\begin{bmatrix} (1-d_f)E_1 & (1-d_f)(1-d_m)\nu_{21}E_1 & 0 \\ (1-d_f)(1-d_m)\nu_{12}E_2 & (1-d_m)E_2 & 0 \\ 0 & 0 & (1-d_s)GD \end{bmatrix} \tag{6}$$

$$D = (1-d_f)(1-d_m)\nu_{12}\nu_{21} \tag{7}$$

$$d_f = \begin{cases} d_f^t \text{ if } \sigma'_{11} \geq 0 \\ d_f^c \text{ if } \sigma'_{11} < 0 \end{cases} \tag{8}$$

$$d_m = \begin{cases} d_m^t \text{ if } \sigma'_{22} \geq 0 \\ d_m^c \text{ if } \sigma'_{22} < 0 \end{cases} \tag{9}$$

$$d_s = (1 - d_f^t)(1 - d_f^c)(1 - d_m^t)(1 - d_m^c) \tag{10}$$

where C_d is the damaged material stiffness, d_f, d_m and d_s are damage variables estimated based on the evolution law given by stress-displacement relation with linear material softening [30] and represent the current damaged state for the fiber and matrix in different modes. This can be obtained by calling ABAQUS output card - DAMAGEMC (matrix compression), DAMAGEMT (matrix tension), DAMAGEFC (fiber compression), DAMAGESHR (shear) and DAMAGEFT (fiber tension) at the required section points of the ply.

3.2. Failure criteria considered for the core material :

The shear failure criteria for balsa material modelled as a part of shell composite layup in the blade section (Figure 9-13) is described with a very simple failure criterion and is based on von Mises stress criterion with equivalent plastic strain and considers the balsa as an elastic perfectly plastic material. The isotropic hardening is considered where the yield stress evolves as the plastic strain accumulates. Moreover, the damage model has been adopted and verified in the work by Nanami, et al 2012 [13, 14, 30]. Under this damage model, material yielding starts when von Mises stress reaches the allowable strength of balsa taken as 5.4MPa. This value is derived from [39]. Then, it is further assumed that failure occurs when the equivalent plastic strain corresponds to failure strain of balsa material which is taken as 0.8. The plasticity model is given by the following equation [30]:

$$\sigma_Y = \sqrt{\sigma_{11}^2 + \sigma_{22}^2 - \sigma_{11}\sigma_{22} + 3\tau_{12}^2} \tag{11}$$

where σ_y is the von Mises stress and σ_{11}, σ_{22}, τ_{12} are components of in-plane effective stress vector, and accumulated plastic strain is given by

$$\varepsilon^{pl} = \varepsilon_0^{pl} + \int_0^t \sqrt{(2/3)\dot{\varepsilon}^{pl} : \dot{\varepsilon}^{pl}}, \tag{12}$$

where ε^{pl}_0 is the initial equivalent plastic strain. The material properties for balsa as well as GFRP plies were obtained from [18, 19, 22, 23, 31] [31, 39-41] and are listed in Table 1.

Table 1. Material Properties implemented

S. No	Properties	Uniaxial	Biaxial	Triaxial	Balsa
1	ρ (Kg/m^3)	1915.5	1845.0	1683.0	110
2	E1 (GPa)	41.63	13.92	21.79	0.050
3	E2 (GPa)	14.93	13.92	14.67	0.050
4	E3 (GPa)	13.42	12.09	12.09	2.730
5	G12 (GPa)	5.047	11.50	9.413	0.0167
6	G23 = G13 (GPa)	5.0469	4.53	4.53	0.150
7	ν_{12}	0.241	0.533	0.478	0.5
8	ν_{23}	0.33	0.3329	0.3329	0.013
9	ν_{13}	0.2675	0.275	0.275	0.013
10	X_T (MPa)	903.6	214.2	472.06	-
11	X_C (MPa)	660.1	184.8	324.16	5.4
12	$Y_T = Y_C$ (MPa)	42.1	184.8	127.12	-
13	$S_L = S_T$ (MPa)	58.65	143.9	99.25	-
14	G_{FT} (J/m^2)	1200	1200	1200	-
15	$G_{FC} = G_{MT} = G_{MC}$ (J/m^2)	4000	4000	4000	-

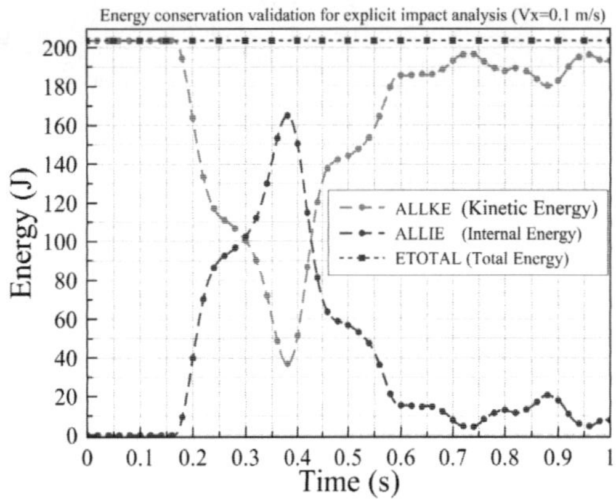

Figure 10. Variation of ALLKE, ALLIE and ETOTAL(Vx = 0.1 m/s and 0.44 <r/R < 0.48)

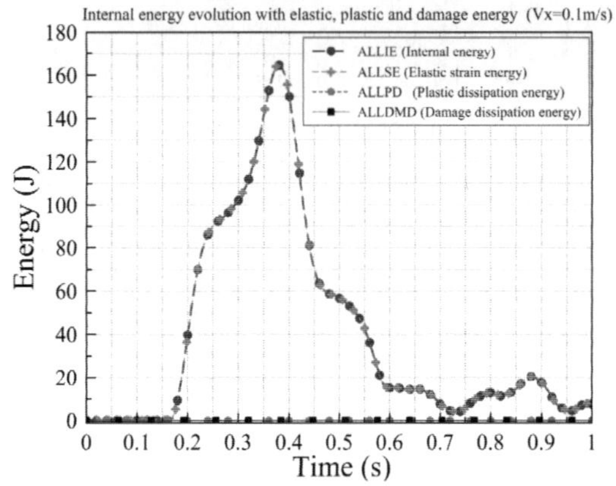

Figure 11. Evolution of ALLIE, ALLSE, ALLDMD and ALLPD (Vx = 0.1 m/s and 0.44 <r/R < 0.48)

4. Results and discussion

This section deals with the results and discussion for damage analysis on the blade including assessment of damage initiation and damage evolution for the impact with the tower for two different impact velocity (a) 0.1 m/s (b) 0.5 m/s at the contact region (*0.44 < r/R < 0.48*) of the blade based on Hashin's intralaminar failure model for fiber and matrix damage for GFRP plies and von-Mises yield stress criteria with equivalent plastic strain indicator for balsa material in the shell layup. Also, the contact-force time history for the cases along with damage criterion for the most critical ply is reported and discussed in this section. Moreover, the discussion on the nature of evolution of different types of energy relevant for the impact scenario is made which in principle can also be qualitatively used to indicate the development of any damage.

Initial impact velocity 0.1 m/s:

The most important result for an impact analysis based on ABAQUS explicit algorithm is the check for the validity of energy conservation [30] and the nature of evolution of impact energy with time. This is required to confirm that the considered numerical model is fit to estimate damage assessment in the blade under impact scenario. Figure 10 shows the variation of kinetic energy (ALLKE), internal energy (ALLIE) and total energy (ETOTAL) involved in the simulation for the entire model and it can be seen that the ETOTAL is almost constant throughout the simulation and the sum of ALLIE and ALLKE corresponds to ETOTAL. Thus, the conservation of energy during the impact analysis on the blade is obtained. The simulation time to investigate this impact study was taken as 1 second with the blade coming in contact with the tower at 0.18s of the simulation time and at this time kinetic energy of the blade starts reducing with the development of internal energy in the blade. Now, this internal energy (ALLIE) in principle is utilized either as the elastic strain energy (ALLSE), damage dissipation energy (ALLDMD) or plastic deformation energy (ALLPD) based on the mechanics of the material under impact. The contact with the tower lasts till 0.58s of the simulation and the blade rebounds.

Figure 11 shows the variation of internal energy (ALLIE) with strain energy (ALLSE), plastic dissipation energy (ALLPD) and damage dissipation energy (ALLDMD) and it further confirms the validation of energy conservation during the impact analysis as the sum of ALLSE, ALLDMD, and ALLPD corresponds to ALLIE of the structure. Moreover, it was observed that ALLIE is almost similar to ALLSE throughout the simulation which implies that there is no noteworthy damage in the impact area. This is a very qualitative indicator of deciding whether there is damage in the material or not based on energy results. The quantification of damage can further be done based on the damage model and failure criteria applied for specific materials which will be discussed later. Moreover, by analyzing the evolution of hourglass energy or the artificial energy (ALLAE), it was found that this energy was negligible. Thus, from all the above energy evolution results, it was confirmed that the considered numerical model can estimate evolution of damage in the blade during impact scenario.

Moreover, from the contact force time history (Figure 12), it can be seen that the curve is very smooth before and during the impact from 0.18 s to 0.21 s with maximum contact force reaching as around 33

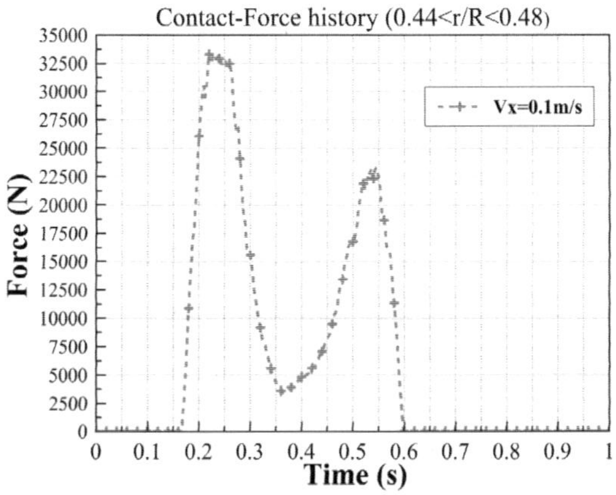

Figure 12. Contact force time history (Vx = 0.1 m/s and 0.44 <r/R < 0.48)

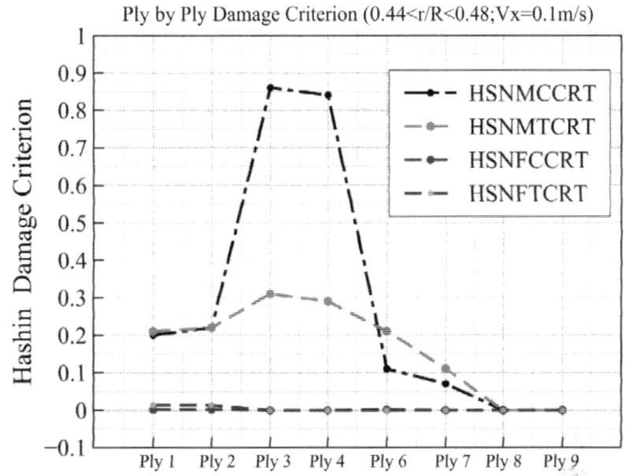

Figure 13. Ply by ply maximum Hashin damage criterion per element

kN at 0.21 s. After that, the contact-force curve displays a small vibration followed by decline and incline of contact area of the blade from 0.28 s to 0.34 s respectively (decline and incline of contact force). This is in--line with the observed edge wise vibration of the blade, together with a fluttering motion in the blade tip relative to the other region of the blade. This is due to the eccentricity of the contact region and a relatively lighter and tapered section of the blade towards the tip of the blade. The contact with the blade further lasts till 0.58 s and finally the blade rebounds. Further as per the implemented damage model, it was seen that none of the ply in the layup has damage criterion equal to one (Figures 13, 14), meaning that the damage initiation criteria was not met and thus there was no intralaminar damage initiation.

The maximum Hashin criterion was 0.83 in matrix compression damage mode (*HSNMCCRT)*. Moreover, since there were three section points reserved for each ply during the preprocessing set up (except section points 13,14,15), it was possible to output the Hashin damage criterion at specific section points for each ply. This can help in identifying which ply through the thickness of the laminate is critical and prone to failure initiation. Moreover, for the cases, where there is actual damage achieved (Hashin failure criterion>1), it is also possible to quantify which ply is damaged (Figure 18(a)). However, for this case since the Hashin criterion was not met, no damage is initiated in any mode of failure. However, it is still possible to compare which ply has the highest damage criterion. Figure 13 presents the Hashin criterion developed at each ply level and it can be seen that Ply 3 and Ply 4 has the highest Hashin damage criterion more than 0.8 for the damage in matrix compression. None of the ply had Hashin damage criterion more than 0.1 for damage mode in fiber tension (HSNFTCRT) or compression (HSNFCCRT). Again, for the shell section points (13, 14, and 15) reserved for Ply no 5 for the balsa material, it was possible to check the highest von Mises stress or any plastic strain developed. However, it was found that the maximum stress levels (0.6-1.8 MPa) observed were quite lower than the allowable stress level of balsa which is taken as 5.4 MPa and it was further seen that there was no plastic strain developed in the balsa as it has not reached the yield strength (the output card PEEQ was unchecked). Moreover, adhesive elements around the trailing edge and the composite elements along the spar flange

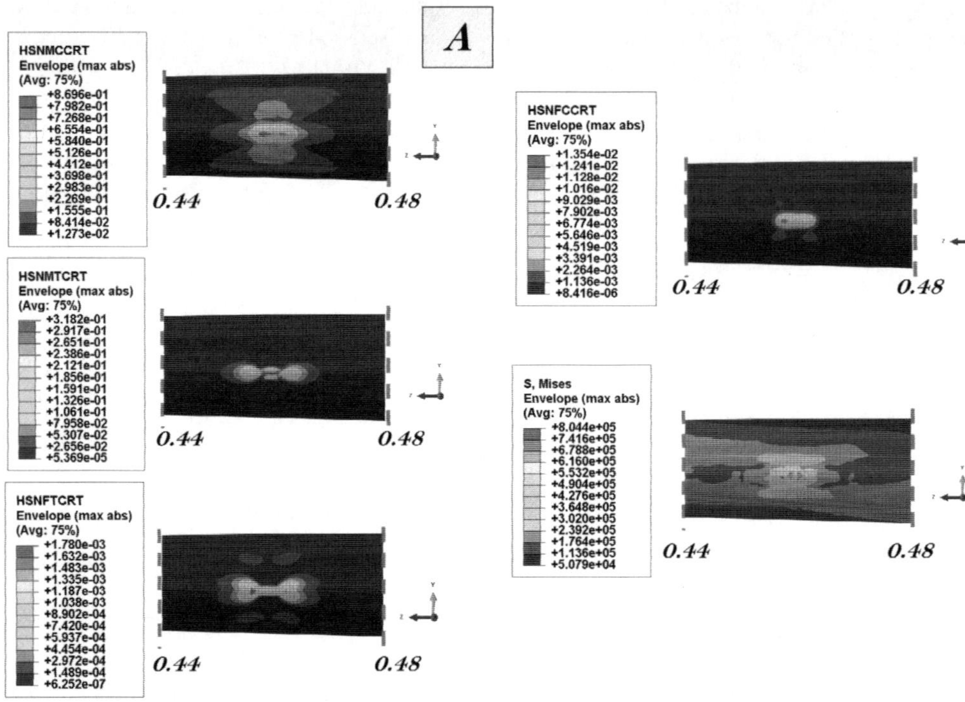

Figure 14. Hashin damage criterion for composite ply and von Mises stresses in Balsa (Vx = 0.1 m/s)

and shear web connection had very low stress levels which further can imply qualitatively that these regions were safe during the impact.

Initial impact velocity 0.5 m/s:

The validation of energy conservation was very well observed for ETOTAL as well as ALLIE for this case when the blade impacts the tower at the initial velocity of 0.5 m/s, and hence the numerical model considered can be used to estimate damage evolution in the blade (Figure 15). However, in this case ALLSE is similar to the ALLIE only until 0.061 s of the total simulation time and can be seen that at this time ALLDMD appears and the damage initiation in the blade has started. Moreover, there was a very small development of ALLPD (Figure 15) in the numerical model which implies qualitatively that the balsa material would have developed a very small plastic strain and the stress levels at the section points (13, 14, 15) would have exceeded allowable stress value. The clear picture of this will be explained later when the damage is explained with regard to the damage models implemented.
Also, the blade comes in contact with the tower at 0.05 s. Figure 16 shows the contact force time history along with the evolution of ALLDMD where it can be seen that the nature of contact force curve with the first contact force peak follows the development of damage energy (only the first impact at the first contact force peak induces the damage and not the second peak, Figure 16). Once the maximum contact force of 163 kN along with maximum damage energy is reached, there is a rapid decrease in the force peak curve which implies damage state representing failure mode in the impact region. Again, this is just a qualitative indication of impact induce damage.
From the damage model implemented for the composite material, it was seen that the blade at the impact region has developed Hashin damage criterion equal to one in matrix compression and matrix tension failure mode (Figure. 18(e)) which implies the initiation of damage. Once, the criterion was met, the damage evolution law implemented would determine the final damaged state of the composite ply at the

impact region. It was found that post damage initiation, the final damaged state for the composite ply was checked for shear damage variable (DAMAGESHR), matrix tensile damage (DAMAGEMT) variable and matrix compression damage variable (DAMAGEMC) which implies that the composite plies had failed in shear, matrix compression and matrix tension failure modes.

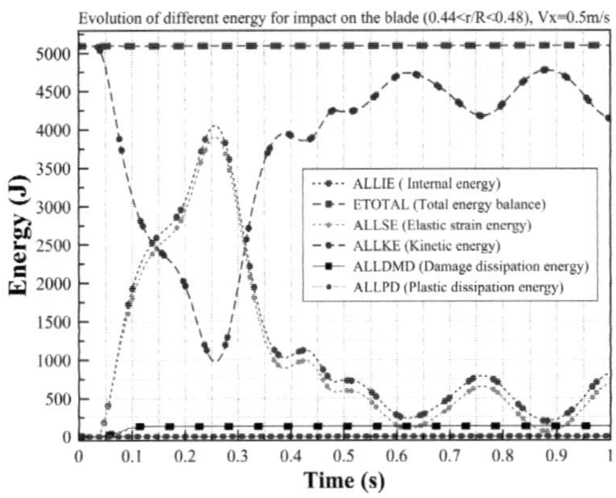

Figure 15. Evolution of various energy for the impact analysis (Vx = 0.5 m/s)

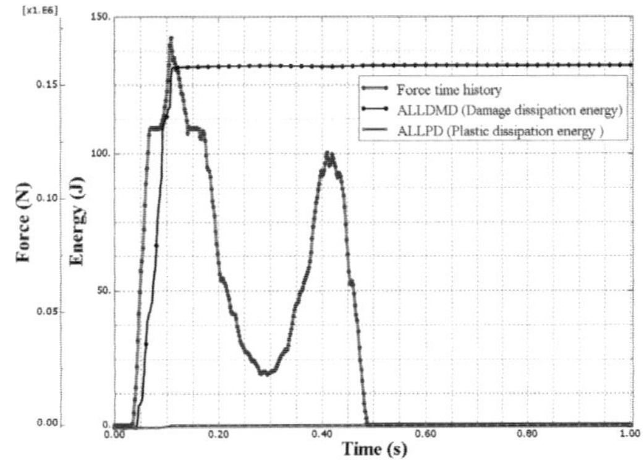

Figure 16. Contact force time history v/s ALLDMD (Vx = 0.5 m/s)

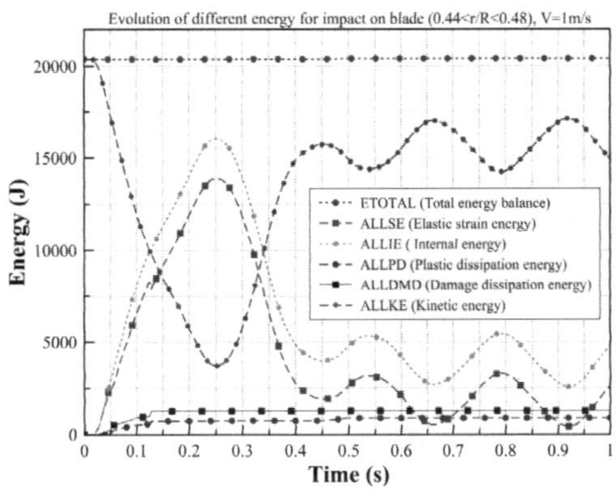

Figure 17. Contact force time history v/s ALLDMD (Vx = 0.5 m/s)

Figure. 18 (a) shows the envelope (max) for these variables as well as for the respective plies at the middle most section point which shows the maximum damages occurring in the impact area (ply wise) in different damage modes. It can be seen that the maximum damage occurs at the top most plies quite close to the impact region and they lie above the sandwich core balsa. However, Ply 6 which lies below the core, had some damage in matrix compression and shear. Moreover, different plies had different damage area given the fact the impact region had a curvature. The stress level and plastic strain output at the section points reserved for balsa material (13, 14 and 15) showed that there was development of plastic strain (Figure.18 (d)). However, it was well below the allowable plastic strain. Moreover, it can also be qualitatively argued that since Ply 1, Ply 2 and Ply 3 had developed matrix cracks, they have a high chance of developing delamination between them as it has been reported in the literature that matrix crack could lead to delamination in the ply with fibers of different orientations. The solid adhesive elements near the trailing edge had very low and acceptable stress level developed. However, some region far from the impact and along the spar flange and shear web connection developed considerable high stresses as seen from Figure. 18(c) and could failure of adhesive connections along the interface. This observation also develops the understanding that it is quite possible to see the effect of the impact or damage area, quite far from the contact region and thus for higher velocity impacts, in order to see all possible damages and failure modes, the whole blade model must be studied in the analysis and suitable failure criteria must be implemented. Also, since the spar cap region did not have balsa material in the layup definition, it could cause the failure of adhesive connection by causing the failure of adherend itself. However, this is just a qualitative inference. In order to quantitatively analyze the failure of adhesive material, it would require a proper numerical modelling technique using fracture mechanics approach such as VCCT (Virtual Crack Closure technique) or by CZM (Cohesive Zone Modelling) approach. Such an analysis to utilize the cohesive elements at the interface of the spar flange and shear web region would come at a high computation cost and would require an efficient numerical modelling strategy and modification/adjustment on the available DTU 10 MW blade which is primarily based on shell element formulation.

From the energy evolution history for a case with impact velocity 1 m/s (Figure 17), it was observed that the damage energy and plastic deformation energy increases further rapidly. This significant increase in ALLPD also implies development of further damage to the core in the sandwich structure and it would require modifications/adjustments to the existing DTU 10 MW numerical blade model to capture other detailed failure mode under impact study. These modifications would also demand dedicated experiments/ numerical study to an extent to validate blade behavior under impact evaluated with a numerical solver. This is planned to be addressed in the future research work.

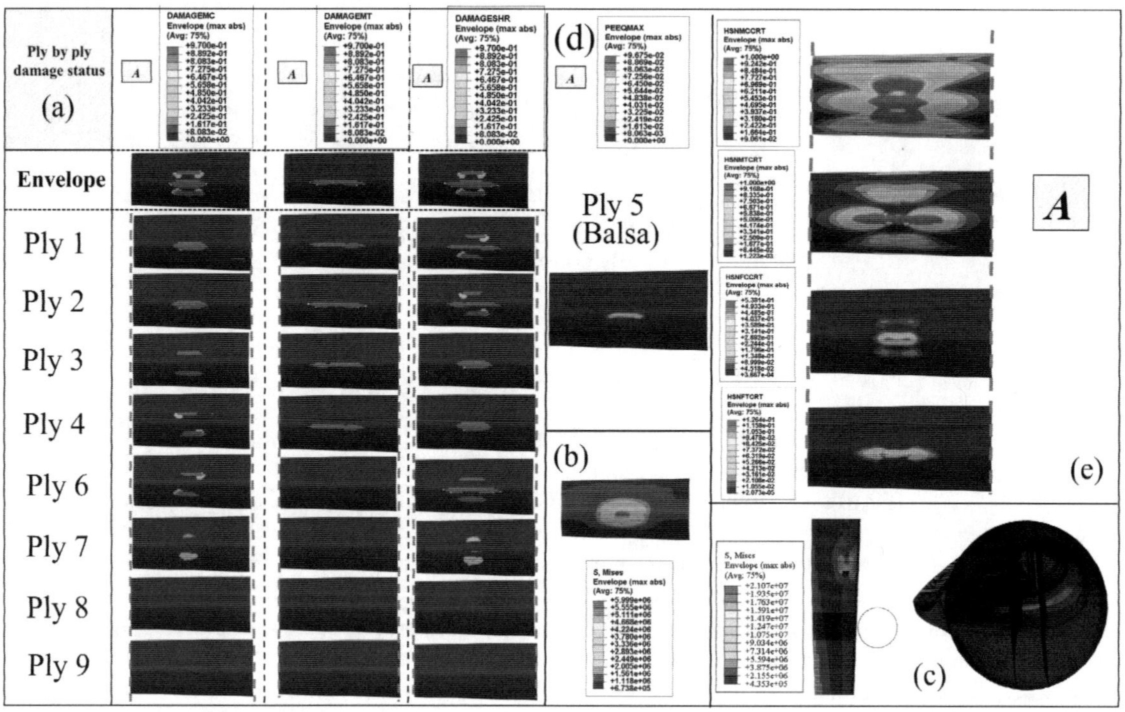

Figure 18. (a) Damage state of the composite ply (t = 1s) (b) Stress level developed in balsa (c) Stress level at spar flange and shear web region of the blade (d) Plastic strain developed (PEEQ) in balsa, Ply-5 (e) Hashin damage criterion developed at the impact region (Vx = 0.5 m/s; 0.44 < r/R < 0.48)

5. Conclusions

In this investigation, the numerical simulation of impact analysis of the wind turbine blade with the tower during installation was performed using advanced capability of finite element study in ABAQUS explicit environment. The emphasis was on the progressive damage modelling on the available DTU 10

MW wind turbine composite blade model based on shell element formulation under a scenario when the leading edge of the blade hits the tower at the contact region (0.44 < r/R < 0.48). The whole purpose of such an analysis to estimate allowable sea state based on explicit structural response method was briefly mentioned. The details for the numerical modelling were described. Finally, the impact analysis was performed for two different velocity of impact- 0.1 m/s, 0.5 m/s. The impact energy varies with the square of velocity and thus it was found that for the case of 0.1 m/s, there was no damage in the blade and all the kinetic energy in the blade transferred to the internal energy was lost by the motion of the blade tip and excessive vibration at the contact region. However, for the case of 0.5 m/s, significant damage was observed in the impact region (A) of the blade and damaged state was quantified at the ply level. It was also found that it is even possible to get damage in the composite plies on the opposite side of the impacted surface. It can also be concluded that the damage threshold energy (DTE) would lie somewhere in between the impact energy for 0.1 m/s and 0.5 m/s. The stress levels along the spar flange and shear web connection far from the impact region experiences high stresses which implies the requirement of the whole blade for the numerical analysis with appropriate modelling technique required and cannot just be studied on the local area of the blade. The extent of damage is considerable under such scenario and thus in order to evaluate allowable damage, which allows installing the blade with some level of damage would require strength analysis on the damaged blade. Overall, computationally to further study and extract other possible failure modes like delamination in the ply and the failure of adhesive connection, some modifications are required on the available DTU 10 MW blade numerical model especially for relatively high velocity of impact. This will be addressed in future research. However, still for low velocity impact, the current model gives good notion of energy evolution and damage modes relevant for blade under impact especially intra-ply failure modes.

Acknowledgement
This work was made possible through the financial support from the Norwegian Research Council under Centre for Marine Operations in Virtual Environments (SFI MOVE), NFR project no. 237929. The authors would also like to thank DTU for making the blade numerical model available in public domain.

References

[1] Rodrigues S, Restrepo C, Kontos E, Pinto RT, Bauer P 2015 Trends of offshore wind projects. *Renewable and Sustainable Energy Reviews*. 2015; **49** :pp 1114-35.
[2] Li L, Gao Z, Moan T 2015. Comparative Study of Lifting Operations of Offshore Wind Turbine Monopile and Jacket Substructures Considering Vessel Shielding Effects. *The Twenty-fifth International Ocean and Polar Engineering Conference*; 2015: *International Society of Offshore and Polar Engineers.*
[3] Gao Z, Wilson Guachamin Acero LL, Zhao Y, Li C, Moan T. 2016 Numerical simulation of marine operations and prediction of operability using response-based criteria with an application to installation of offshore wind turbine support structures. *Marine Operations Specialty Symposium (MOSS 2016).*
[4] Tegen S, Hand M, Maples B, Lantz E, Schwabe P, Smith A. 2012 *Cost of Wind Energy Review.* National Renewable Energy Laboratory (NREL), Golden, CO.; 2012.
[5] Acero WG, Li L, Gao Z, Moan T. 2016 Methodology for assessment of the operational limits and operability of marine operations. *Ocean Engineering*. 2016; **125** :pp 308-27.
[6] Acero WG, Gao Z, Moan T. 2017 Methodology for assessment of the allowable sea states during installation of an offshore wind turbine transition piece structure onto a monopile foundation. *Journal of Offshore Mechanics and Arctic Engineering* 2017.
[7] Li L, Gao Z, Moan T. Operability analysis of monopile lowering operation using different numerical approaches. *International Journal of Offshore and Polar Engineering*. 2016; **26(02)** :pp 88-99.
[8] Kim H, DeFrancisci G, Halpin J. *Impact Damage of Composite Structures*. Long-Term Durability of Polymeric Matrix Composites. 2012; **3**: pp (143-180)

[9] Abrate S. *Impact on composite structures*: Cambridge university press; 2005.
[10] El-Reedy MA. *Offshore structures: design, construction and maintenance:* Gulf Professional Publishing; 2012.
[11] Haselbach PU, Bitsche R, Branner K. The effect of delaminations on local buckling in wind turbine blades. *Renewable Energy*. 2016; **85** :pp 295-305.
[12] Feng D. *Simulation of low-velocity impact damage in sandwich composites*: Universita'degli Studi di Cagliari; (PhD Thesis), 2014.
[13] Nanami N, Ochoa O. Bird Impact Study of a Preloaded Composite Wind Turbine Blade. Paper Presented at the *The 19th International Conference on Composite Materials, Montreal, Canada.*
[14] Nanami N, Ochoa O Damage Assessment of a Large-Scale Hybrid Composite Wind Turbine Blade. *Journal of Mechanical Engineering and Automation*. 2016; **6(5)** :pp 117-27.
[15] Agrawal S, Singh KK, Sarkar P. Impact damage on fibre-reinforced polymer matrix composite– a review. *Journal of Composite Materials*. 2014; **48(3)** :pp 317-32
[16] Abrate S, Ferrero J, Navarro P. Cohesive zone models and impact damage predictions for composite structures. *Meccanica*. 2015; **50(10) :** pp 2587-620.
[17] Bernal AN. *Simulation of the low-velocity impact behavior of glass reinforced composite pipes*: The Petroleum Institute (United Arab Emirates); 2015.
[18] Perillo G, Jørgensen J. Numerical/Experimental Study of the Impact and Compression after Impact on GFRP Composite for Wind/Marine Applications. *Procedia Engineering*. 2016; **167** :pp 129-37.
[19] Perillo G, Vedivik N, Echtermeyer A. Numerical and experimental investigation of impact on filament wound glass reinforced epoxy pipe. *Journal of Composite Materials*. 2015; **49(6)** :pp 723-38.
[20] Reiner J, Torres JP, Veidt M, Heitzmann M. Experimental and numerical analysis of drop-weight low-velocity impact tests on hybrid titanium composite laminates. J*ournal of Composite Materials*. 2016; **50(26)** :pp 3605-17.
[21] Skaar MW. *Modeling and Testing of Impact Damage in Composite PressureVessels*: NTNU; (Master thesis), 2015.
[22] Haselbach PU, Branner K. Initiation of trailing edge failure in full-scale wind turbine blade test. *Engineering Fracture Mechanics*. 2016; **162** :pp 136-54.
[23] Haselbach PU, Eder MA, Belloni F. A comprehensive investigation of trailing edge damage in a wind turbine rotor blade. *Wind Energy*. 2016; **19(10) :** pp 1871-88.
[24] McGugan M, Pereira G, Sørensen BF, Toftegaard H, Branner K. Damage tolerance and structural monitoring for wind turbine blades. Phil Trans R Soc A. 2015; **373(2035)** :2014.
[25] Jensen FM. *Ultimate Strength of a Large Wind turbine Blade (PhD Thesis)*: Risø National Laboratory, Technical University of Denmark; 2008.
[26] Haselbach PU. *Ultimate Strength of Wind Turbine Blades under Multiaxial Loading:* DTU Wind Energy; (PhD Thesis), 2015
[27] Toft HS, Branner K, Berring P, Sørensen JD. Defect distribution and reliability assessment of wind turbine blades. *Engineering Structures*. 2011; **33(1)** :pp 171-80.
[28] Minakuchi S, Okabe Y, Mizutani T, Takeda N. Barely visible impact damage detection for composite sandwich structures by optical-fiber-based distributed strain measurement. *Smart Materials and Structures*. 2009; **18(8)** :085018.
[29] Zhao Y Cheng Z, Sandvikd PC, Gao Z, Moan T. An integrated dynamic analysis method for simulating installation of a single blade for offshore wind turbines (In press) *Ocean Engieering 2017.*
[30] *ABAQUS V. 6.14 Documentation*. Dassault Systemes Simulia Corporation. 2014.
[31] Bak C, Zahle F, Bitsche R, Kim T, Yde A, Henriksen LC, et al., editors. *The DTU 10-MW reference wind turbine. Danish Wind Power Research* 2013.

[32] Reddy JN. *Mechanics of laminated composite plates and shells: theory and analysis* : CRC press; 2004.

[33] Cox K, Echtermeyer A. Structural design and analysis of a 10MW wind turbine blade. *Energy Procedia*. 2012; **24** :pp 194-201.

[34] Barbero EJ. Finite element analysis of composite materials using AbaqusTM: *CRC press;* 2013.

[35] Hashin Z. Fatigue failure criteria for unidirectional fiber composites. ASME, Transactions, *Journal of Applied Mechanics*. 1981; **48** :pp 846-52.

[36] Camanho PP, Davila C, De Moura M. Numerical simulation of mixed-mode progressive delamination in composite materials. *Journal of composite materials*. 2003; **37(16)** :1415-38.

[37] Camanho PP, Dávila CG. *Mixed-mode decohesion finite elements for the simulation of delamination in composite materials*. (NASA) 2002.

[38] Hashin Z, Rotem A. A fatigue failure criterion for fiber reinforced materials. *Journal of composite materials*. 1973; **7(4)** :pp 448-64.

[39] Vural M, Ravichandran G. Microstructural aspects and modeling of failure in naturally occurring porous composites. *Mechanics of Materials*. 2003; **35(3)** :pp 523-36.

[40] Nanami N. *Structural and Damage Assessment of Multi-Section Modular Hybrid Composite Wind Turbine Blade* 2014 (PhD Thesis), Texas A&M University

[41] Weijermars W. *Mechanical behaviour of composite sandwich panels in bending after impact.* 2016 *(Master thesis)*, University of Twente.,Enschede, Netherlands

[42] Aymerich F. Composite materials for wind turbine blades: issues and challenges, SYSWIND Summer school-July 2012-University of Patras, Italy (Schematic of the cross-section of the blade with different parts).

IOP Conf. Series: Materials Science and Engineering **276** (2017) 012026 doi:10.1088/1757-899X/276/1/012026

A case study on topology optimized design for additive manufacturing

A W Gebisa* and H G Lemu
Department of Mechanical and Structural Engineering and Materials Science, University of Stavanger, Stavanger, Norway

*Corresponding author: aboma.w.gebisa@uis.no

Abstract. Topology optimization is an optimization method that employs mathematical tools to optimize material distribution in a part to be designed. Earlier developments of topology optimization considered conventional manufacturing techniques that have limitations in producing complex geometries. This has hindered the topology optimization efforts not to fully be realized. With the emergence of additive manufacturing (AM) technologies, the technology that builds a part layer upon a layer directly from three dimensional (3D) model data of the part, however, producing complex shape geometry is no longer an issue. Realization of topology optimization through AM provides full design freedom for the design engineers. The article focuses on topologically optimized design approach for additive manufacturing with a case study on lightweight design of jet engine bracket. The study result shows that topology optimization is a powerful design technique to reduce the weight of a product while maintaining the design requirements if additive manufacturing is considered.

1. Introduction

Topology optimization is a mathematical tool to make optimized material distribution in a part to be designed. By doing so, it gives us a part topology that is more natural and more complex. There are different topology optimization approaches developed during the last three decades. For instance, set level approach, homogenization method, SIMP method and density approach are among the main topology optimization techniques [1, 2]. The initial topology optimization developments considered the conventional manufacturing techniques that are either subtractive or formative. These conventional manufacturing systems have limitations in producing complex shape geometries as they have different manufacturing constraints. The birth of additive manufacturing gave another bright future opportunity for topology optimization as with additive manufacturing (AM) technologies; producing complex shape geometry is achievable. This is because, in AM, the systems do not require any tooling for producing a part. For this and other similar capabilities of the technologies, topology optimization and additive manufacturing are considered ideal couples.

To implement AM technology in production of functional parts, advances in materials technology and design optimization are considered as the key areas of current research. Regarding the last-mentioned research challenge, a design approach that can directly transfer the design concept of the engineer to a produced part without any due consideration for manufacturing constraint and enable optimum utilization of the part under loading is sought. If proper and efficient algorithms are developed, topology optimization techniques can play a key role in the future development of AM technology. This case study article is intended to explore the role of topology optimization in AM to design a lightweight product.

The article is structured as follows. Section 2 discusses the basics of topology optimization with its benefits and the formulation of topology optimization problems. In Section 3, topology-optimized design for additive manufacturing is discussed considering lightweight product design. Section 4, elaborates the case study considered in the project followed by Section 5 that ends the report by drawing some conclusions.

2. Topology optimization

In mechanical design problem, there came to exist a structural optimization concept in which the size, shape and topology of a product are optimized separately or simultaneously. These concepts are very important to find an optimal size, shape and material distribution during the product development process. As shown in Figure 1 [3], size optimization deals with the optimization of size of an object (length, width or depth); whereas shape optimization is to find optimal shape of a part or openings in a part; while topology optimization deals with optimizing the whole geometry of the part including both size and shape. In this section, the concept of topology optimization with its benefits and formulation of topology optimization problems are discussed.

2.1. What is topology optimization?

Topology optimization is one of the structural optimization techniques that optimizes the distribution of material within a specified design space for a given loading and boundary conditions while fulfilling the performance requirements of the product. Most of the topology optimization techniques are carried out by collective use of Computer Aided Design (CAD) concept, Finite Element Analysis (FEA) concept and different optimization algorithms in consideration of different manufacturing techniques as shown on the topology optimization process on Figure 2 [4]. The use of CAD in topology optimization is to make a rough/initial model of the product to be optimized, whereas FEA is used to see the distribution of stresses and displacements throughout the product. The topology optimization is performed to remove the areas of the part that are not sufficiently supporting the applied loads and not undergoing significant deformation and thus not contributing to the overall performance of the part.

Based on the design problem requirement, different optimization algorithms are used to remove the portion of the material in the product that are not supporting the applied load. Moreover, the topology optimization is done to satisfy certain design objectives and maintain the design constraints. Based on the problem at hand, the objectives might be to minimize the compliance of the part, i.e. to maximize the stiffness of the part, as compliance is the inverse of the stiffness; the constraints could be the maximum allowable deformation, the maximum mass fraction and so on. The topology optimization tools generate a complex natural shape that shows the removal of materials based on the objectives and constraints set in the design problem. The design is then finalized on CAD software to get smooth and manufacturable part following the shape generated from the topology optimization process. Ultimately, the final optimized design is validated using FEA tools to satisfy the design requirements so that the product meets its overall performance.

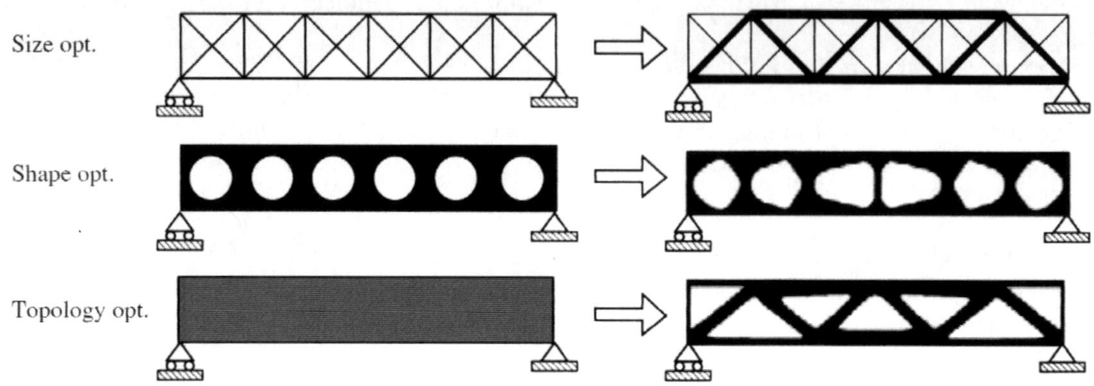

Figure 1. Comparative illustration of size, shape and topology optimization

Structural optimization, in general, has huge potential benefits in product development process. Topology optimization in particular has the following benefits in the design process

- Creating lightweight structures
- Creating ready to manufacture design
- Reducing time-to-market
- Saving huge amount of material
- Reducing physical test
- Saving large amount of processing energy
- Reducing physical prototype build

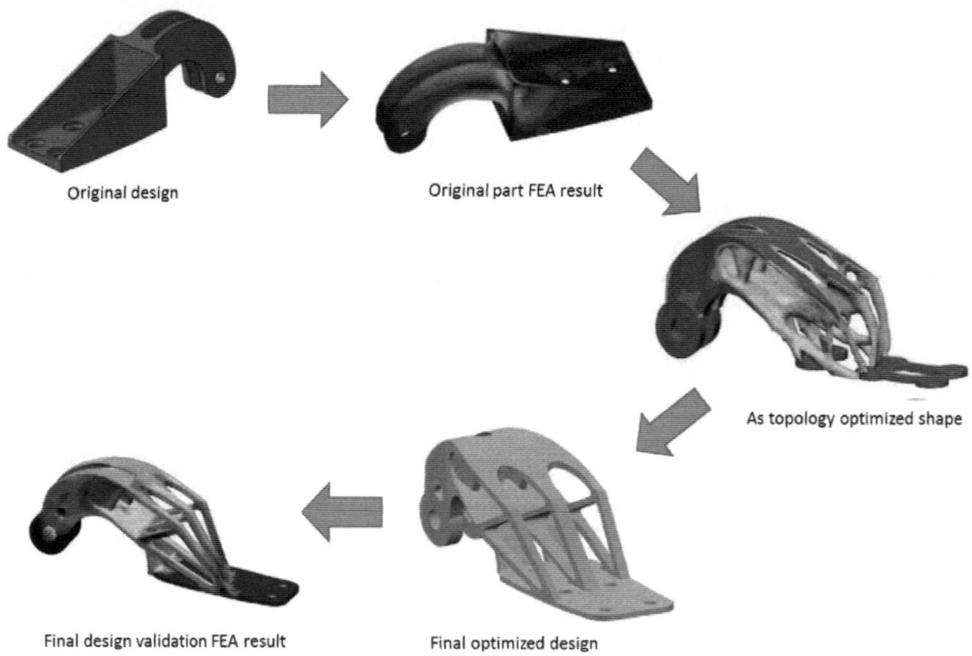

Figure 2. Topology optimization process

2.2. *Formulation of topology optimization problem*

The formulation of topology optimization problem has to be made very carefully as the best solution is obtained only if the problem is formulated in a proper way. The following five steps are used in the formulation of design optimization problem [5].

1. *Develop topology optimization problem statement:* In this step, the design goal, what to accomplish by the optimization and set of criteria will be clarified.
2. *Data and information collection:* In this step, all the necessary information are collected.
3. *Identification/definition of design variables*: In this step, the design variables that describe the system are identified and defined.
4. *Identification of a criterion to be optimized:* The criteria to evaluate and stop the optimization process has to be clarified. These criteria are called objective functions, which have to be maximized or minimized depending on problem requirements.
5. *Identification of constraints:* The restrictions on the problem are identified at this stage. They are extracted from the resources and performance requirements.

Mathematically the optimization problem can be formulated as in equation (1) [6]:

$$\begin{aligned} \text{Minimize :} \quad & f(\mathbf{x}) \\ \text{Subject to :} \quad & g_i(\mathbf{x}) \leq 0, \quad i = 1,\ m \\ & h_j(\mathbf{x}) = 0, \quad j = 1,\ p \\ & \mathrm{x}_k^{\ell} \leq \mathrm{x}_k \leq \mathrm{x}_k^{\mathrm{u}}, \quad k = 1,\ n \end{aligned} \tag{1}$$

where f(x) is the objective function to be minimized whereas g(x) and h(x) are inequality and equality constraints respectively. The m and p are the number of inequality and equality constraints respectively; x is the vector of design variables; n is the total number of design variables; and x^l_k and x^u_k are the lower and upper bounds of the kth design variable x_k, respectively.

3. Topology optimized design for additive manufacturing

Topology optimization is a powerful design approach to save time, material and energy that are not economically achievable with any other manufacturing process if explored to its maximum by considering additive manufacturing. Additive manufacturing (also referred to as 3D printing) is an emerging manufacturing technology that has the potential to outpace or replace the conventional manufacturing approaches. AM is defined as "process of joining materials to make objects from 3D model data, usually layer upon layer, as opposed to subtractive manufacturing methodologies " [7]. Additive manufacturing technology has unique capabilities; shape complexity (produces any shape that can be designed), materials complexity (processes multi-material products), hierarchical complexity (products internal structure ranging from mesoscale to macroscale) and functional complexity (produces multiple parts as a single functional product) [8]. Topology optimization has started long time ago with consideration of the existing manufacturing techniques. However, the existing techniques did not benefit sufficiently from this design approach because there are manufacturing constraints for the optimized designs. The optimized topology could not be produced, as there are limitations on the geometric complexity to produce with the existing manufacturing techniques.

In product development process, lightweight design has huge benefits including reduction in material consumption, reduction in processing and energy consumption. The potential approach to reduce the weight of a product is to use the topology optimization with additive manufacturing techniques as reported elsewhere [9-19]. Topology optimization can be employed for new product design or redesigning of an already existing product so that lightweight product can be obtained while maintaining its functional requirements. Topology optimized lightweight design can be realized by using different materials internal structures: solid structures or cellular structures. Topology optimization with additive manufacturing is considered to redesign a bracket to reduce its weight from 70 gm to around 42 gm, which is about 40% reduction [20]. A similar opportunity in re-designing a bracket with more lightweight that was manufactured using laser additive manufacturing is reported in [21]. These kinds of geometries are very difficult to manufacture using the traditional techniques but they can be realized with additive manufacturing.

Designers at ARUP in Netherlands have also explored the opportunity of topology optimization for additive manufacturing by redesigning a steel node for a tensegrity structure [22]. In their research, they found out that the topology-optimized design for additive manufacturing could enable the weight reduction of the node from 20 kg to 5 kg (75% reduction) without compromising the functional and structural performance of the product. Lightweight topology optimized automotive control arm design with lattice structure is investigated as another opportunity to be produced using additive manufacturing [23].

It is also reported [24] that, a hip joint implant interior part is topologically optimized and fabricated using additive manufacturing. The implant part has interior pore that makes it much complex and unachievable with conventional manufacturing techniques. However, the part is produced with additive manufacturing since the technology made the geometry complexity free. Moreover, it was not possible to find porous implants that allow the ingrowth of bone and cells if design optimization and additive manufacturing were not implemented together in the product development process. OptiStruct and Altair product design team helped RUAG Space Switzerland Zurich to topology optimize an existing antenna support arm to be produced using additive manufacturing that resulted in half weight reduction from the old version [25]. Topology optimized cargo sling with 15% volume reduction thus weight reduction is also another research that demonstrated the potential of topology optimization to make lightweight design [26].

Figure 3 shows topology optimized design process. First, the original CAD design is drawn in a 3D-CAD modeling software. The original design is then structurally analyzed with the given loading conditions to see the stress and displacement distribution. Based on the stress and displacement distribution, the topology optimization removes material from areas that does not significantly contribute to carry the applied loads. Based on the topology optimization result the part is remodeled in a CAD software. The new CAD model is then verified with FEA to carry the loads and to satisfy the design requirements. If the model satisfies the verification physical model verification is done using any of physical prototyping methods. If not the remodeling is done again until verification is done. The final design is then prepared for the final additive manufacturing. The process is employed in the next section, case study, to redesign a jet engine bracket to show the potential of topology optimized design for additive manufacturing.

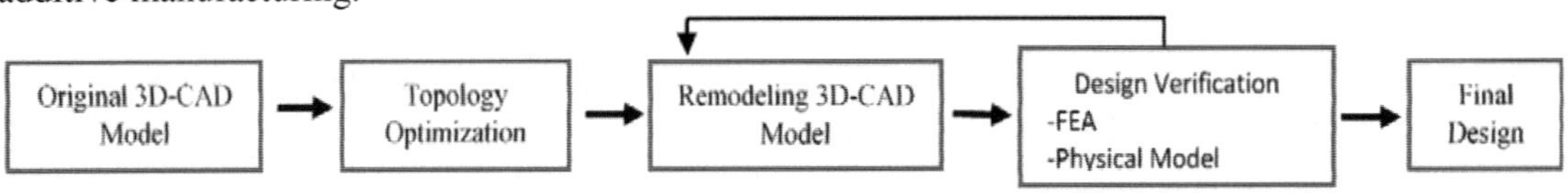

Figure 3. Topology optimized design process

4. Case study

In this study, a jet engine bracket shown in Figure 4 [27] is considered as a case study to show the potential of topology optimized design approach in reducing the weight of a product if coupled with additive manufacturing. The original bracket is based on General Electric (GE) design. The bracket is designed to sustain different loading conditions shown in Figure 5.

4.1. Analysis of the original bracket

Structural finite element analysis is carried out on the original bracket to see the stress distribution in the part. The contour plots shown in Figure 6 show the von Misses stress distribution for all the load cases in the part. The large areas of the part that are shown in blue color in the contour plot indicate inefficient use of material. It is very likely that these areas of the part need material removal as they have negligible effect on the performance of the part.

4.2. Design optimization

In this study, the bracket is redesigned using topology optimization design approach considering four loading conditions. The objective of the study is to reduce the weight of the bracket while satisfying all the design requirements. Altair Hypemesh 14 Optistruct commercial software is employed for the topology optimization. The design requirements for the engine bracket are listed below [27].

- The optimized geometry must fit within the original part envelope.
- The material was Ti-6Al-4V with an assumed yield strength of 904 MPa at the service temperature (23°C).
- Minimum material feature size (wall thickness): 1.13 mm
- Interface 1: 19 mm diameter pin. The pin was to be considered infinitely stiff for analysis purposes.
- Interfaces 2–5: 9.5-24 AS3239-26 machine bolt. Nut face 10.287 mm max ID and 14.173 mm min OD. The bolts were to be considered infinitely stiff.

The topology optimization is formulated as in equation (2):

Minimize the weighted compliance i.e. to maximize the stiffness

Subject to

Mass fraction < 30 % (2)

Minimum feature size >1.13 mm

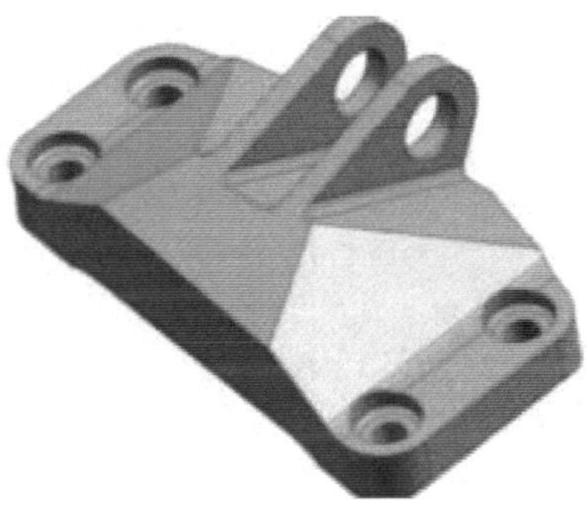

Figure 4. Original jet engine bracket.

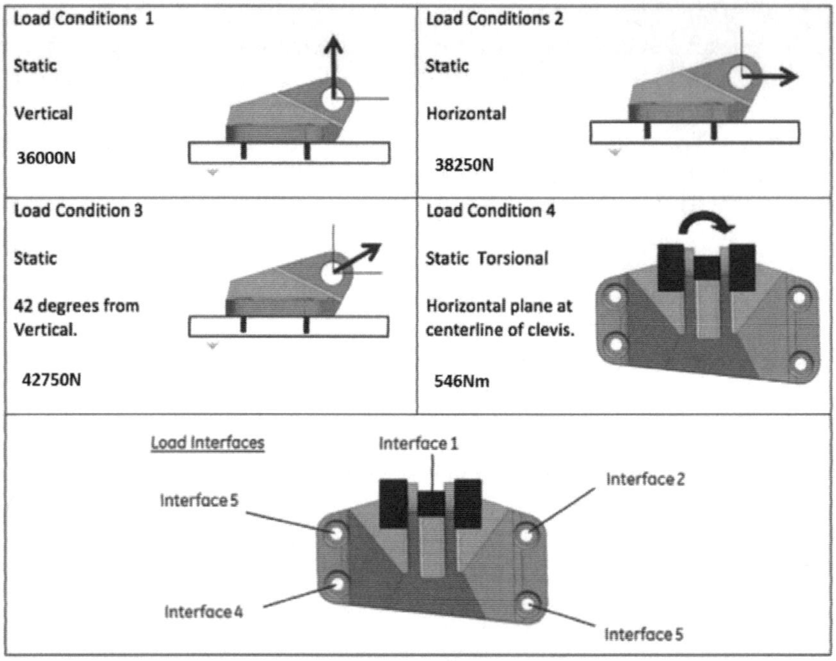

Figure 5. Loading conditions and function description of the engine bracket [27].

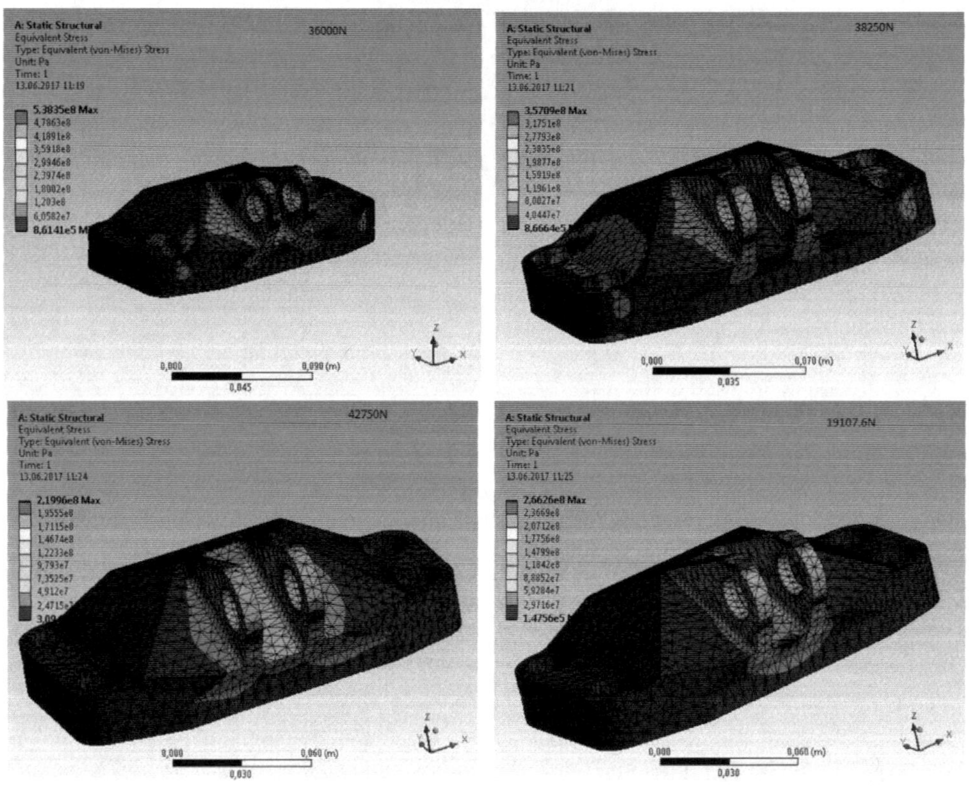

Figure 6. FEA of original engine bracket.

IOP Conf. Series: Materials Science and Engineering **276** (2017) 012026 doi:10.1088/1757-899X/276/1/012026

The initial stage of the topology optimization is to divide the part into design (green) and non-design (yellow) spaces as shown in Figure 7. This is to separate the part into two parts; design space, a space on which the topology optimization is carried out and non-design space, a space that has nothing to do with the topology optimization and is unchanged. Furthermore, the non-design space is the part of the object through which, it is connected to others parts. Both spaces are then meshed using the same tool, Altair Hypemesh 14 Optistruct commercial software as shown in Figure 8. The mesh is done very fine with element size of 0.25 mm to obtain an even and acceptable materials distribution throughout the part.

Figure 7. Design (green) and non-design (yellow) spaces

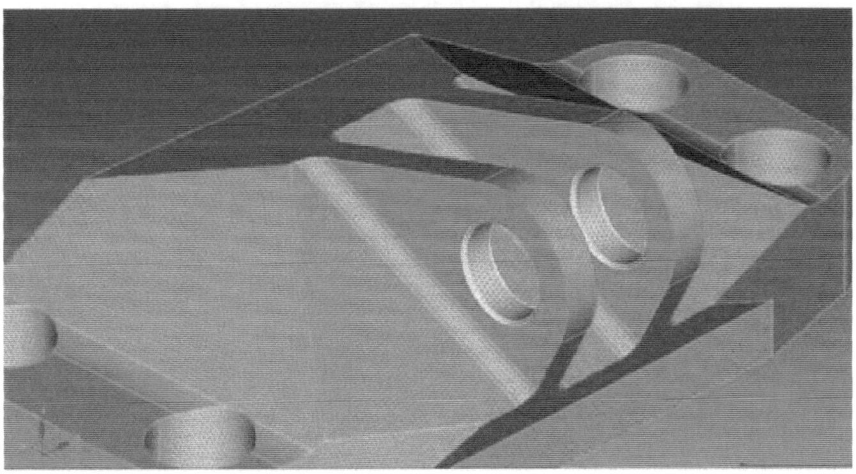

Figure 8. Meshing of the engine bracket.

All the loading conditions and the constraints are then applied on the part through the non-design spaces. The loads are applied separately on the part through the two horizontal hole surfaces defined as non-design spaces as shown on Figure 9 to make the part ready for optimization. The torque is applied as two equivalent force couples in different directions as there was a problem in the software to apply a torque.

The topology optimization tool computes the stress and displacement distribution on the part for all the loads separately. Based on the stress and displacement distribution, parts that are supporting less load or not significantly contributing to support the load are removed resulting in a very natural geometry shown in Figure 10 (left). As optimized the geometry is a very natural rough geometry showing the optimized material layout throughout the part. Since the initial optimized geometry from the tool is rough and does not look good aesthetically, remodelling of the part is important. The final design is then prepared on Autodesk Inventor Professional following the topology optimization as closely as possible as shown in Figure 10 (right).

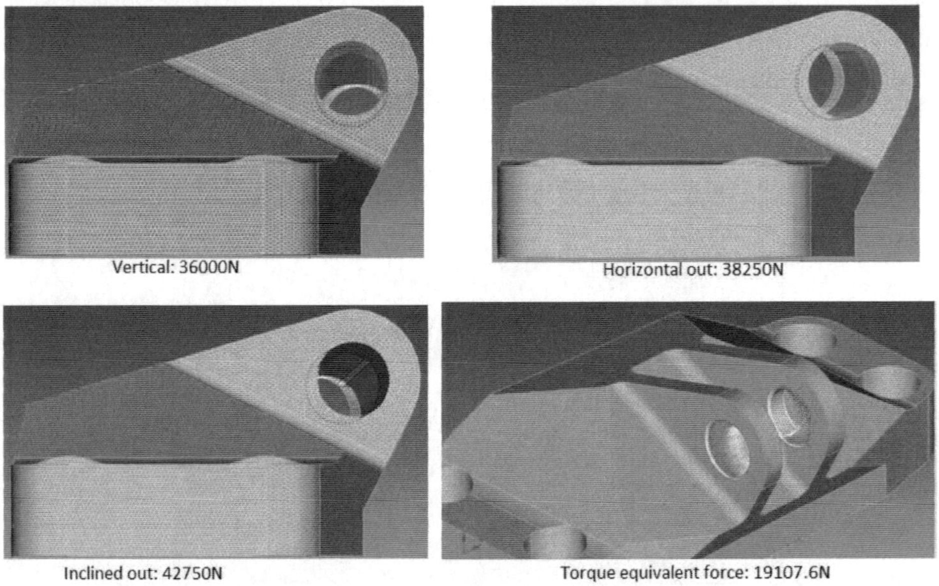

Figure 9. Loads applied on the engine bracket.

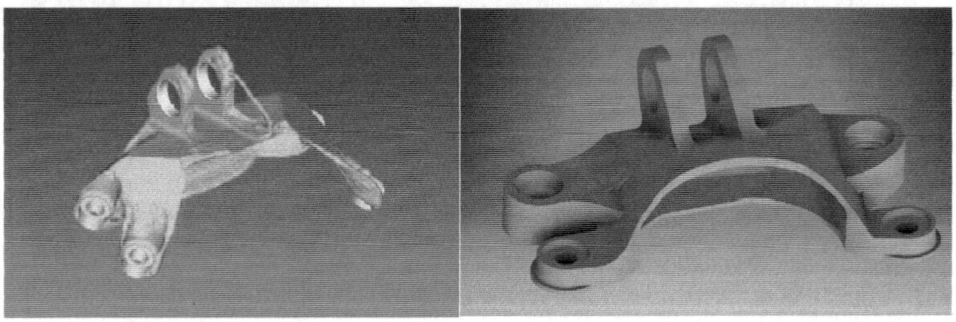

Figure 10. Topology optimized geometry: as optimized (left) and remodelled final part (right).

The newly designed part will have to sustain the same mechanical load while fulfilling the same design requirements. The final design has to be verified with the given design criteria that is the yield strength. The von Misses stress values for all the load cases should not exceed the yield strength. The structural verification analysis is done on ANSYS R17 Academic Educational tool. As shown in Figure 11 the von Misses stresses for all the cases are below the yield strength of the material. The final design satisfies the yielding condition with safety factor from 1.11 to 2.17. The final design is additive manufactured using fused deposition modelling (FDM) Fortus 450 from Stratasys for the physical design verification as shown in Figure 12. The final design resulted in 65% weight reduction, which is from 2.067 kg original part to 0.72 kg final one.

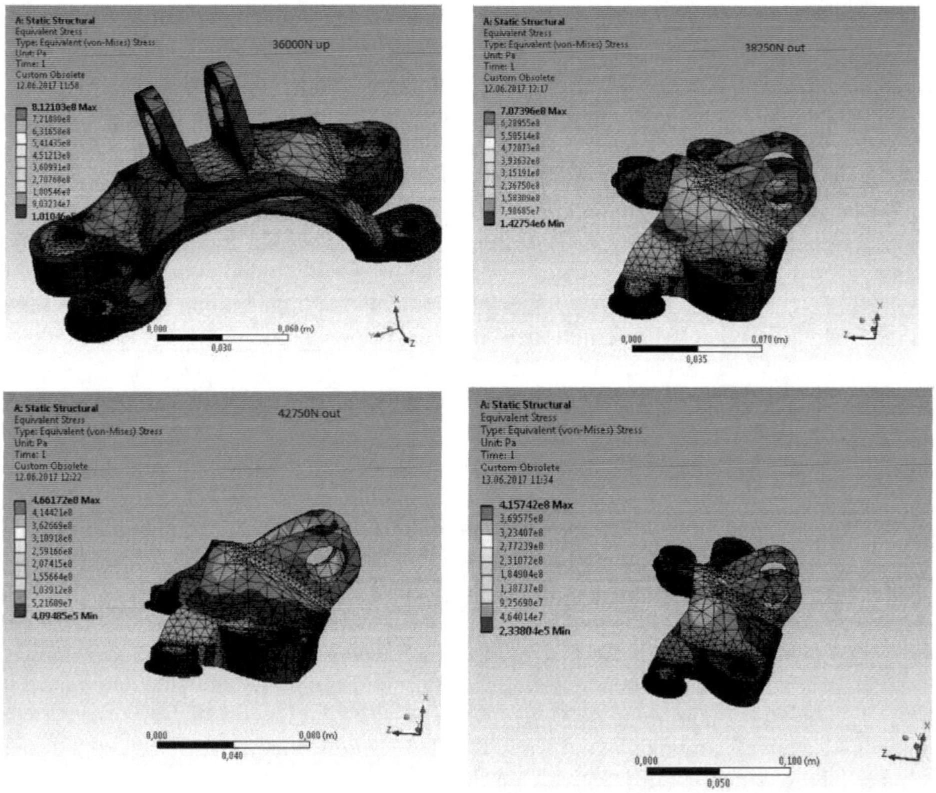

Figure 11. Final engine bracket design stress verification contour plots.

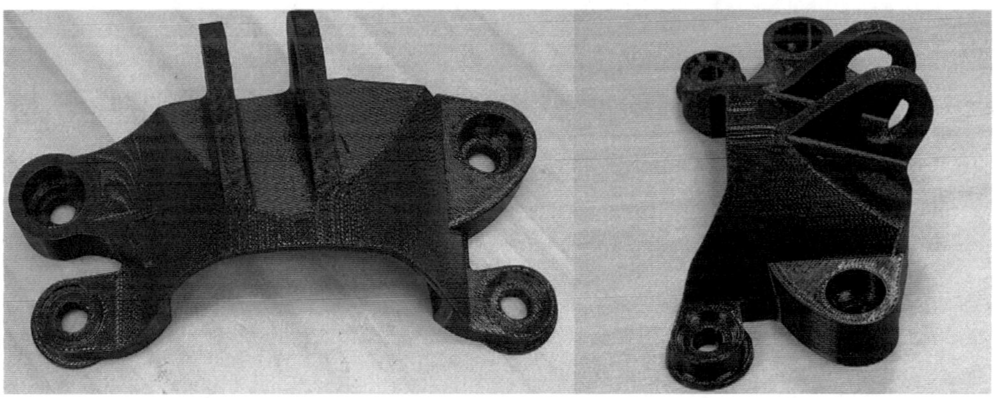

Figure 12. Additive manufactured final engine bracket design.

5. Conclusion

From the case study, it can be concluded that, topology optimization is a powerful design concept to reduce the weight of structural products. The reduction of weight saves huge amount of material and processing energy thus huge amount of money. It also shows that the capability of topology optimization can be fully utilized with additive manufacturing techniques, as the manufacturing constraints in the conventional methods are no longer available. From the case study result, which is 65% weight reduction, it can also be concluded that topology optimized design for additive manufacturing can reduce huge portion of the mass thus result in lightweight design.

IOP Conf. Series: Materials Science and Engineering **276** (2017) 012026 doi:10.1088/1757-899X/276/1/012026

References

[1] Rozvany G IN 2009 *Struct. Multidiscipl. Optim.* **37(3)** pp 217-237.

[2] Sigmund O and Maute K 2013 *Struct. Multidiscipl. Optim.* **48(6)** pp.1031-1055.

[3] Chau Le *Topshape* http://topshape.bm-rd.com/topology-optimization-overview/ Last accessed 13.06.17.

[4] Tomlin M and Meyer J 2011 *Proc. of the 7th Altair CAE technology conf.*

[5] Arora J 2004 *Introduction to optimum design* (San Diago- Elsevier Academic Press)

[6] Chang K H 2014 Design Theory and Methods Using CAD/CAE (London- Elsevier Academic Press).

[7] ASTM F2792-12a 2012 *Standard terminology for additive manufacturing technologies* (Withdrawn 2015) (ASTM International- West Conshohocken).

[8] Gibson I, Rosen D W and Stucker B 2014 *Additive manufacturing technologies: 3D printing, rapid prototyping, and direct digital manufacturing* (Springer-New York).

[9] Emmelmann C, Petersen M, Kranz J and Wycisk E 2011 *SPIE Eco-Photonics 2011, Int. Society for Optics and Photonics* **8065** pp. 80650L1-80650L12.

[10] Doubrovski Z, Verlinden J C and Geraedts J M 2011 *ASME 2011 Int. Design Engineering Technical Conf. and Computers and Information in Engineering Conf.* pp. 635-646.

[11] Nguyen J, Park S and Rosen D W 2013 *Int. J. Precis. Eng. Manuf.* **14.6** pp. 1071-1078.

[12] Strano G, Hao L, Everson R M and Evans K E 2013 The *Int. J. of Advanced Manufacturing Technology* **66** pp. 1247–1254.

[13] Zhu J-H, Zhang W-H and Xia L 2016 *Arch. Comput. Methods Eng.* **23 (4)** pp. 595-622.

[14] Brackett D, Ashcroft I and Hague R, 2011 *Proc of the solid freeform fabrication symp* (Austin).

[15] Klahn C, Leutenecke B and Meboldt M 2014 *Procedia CIRP* **21** pp. 138-143.

[16] Lindemann C, Reiher T, Jahnke U and Koch R 2015 *Rapid Prototyping J* **21(2)** pp. 216-227.

[17] Buchbinder D, Schleifenbaum H, Heidrich S, Meiners W and Bültmann J 2011 *Physics Procedia* **12** pp.271-278.

[18] Meisel N A, Williams C B and Druschitz A 2012 *Proc. of the Int. Solid Freeform Fabrication Symp* pp. 162-176.

[19] Moon S K, Tan Y E, Hwang J and Yoon Y J 2014 *Int. J. of Precision Eng. and Manufacturing-Green Technology* **1(3)** pp. 223-228.

[20] Salonitis K Al and Zarban S 2015 *Procedia CIRP* **36** pp. 193-198.

[21] Emmelmann C, Sander P, Kranz J and Wycisk E *Physics Procedia* **12** pp. 364-368, 2011.

[22] Ren S and Galjaard S 2015 *Springer Int. Publishing, Design Modelling Symp 2015* pp. 35-44.

[23] Dias W *Altair* http://insider.altairhyperworks.com/design-and-optimization-of-lattice-structures-for-3d-printing-using-altair-optistruct/ Last accessed 02.08.2017.

[24] Gardan N and Schneider A 2015 *J. Manuf. Syst.* **37** pp. 417-425.

[25] *Topologically Optimized and 3D Printed Component on its Way to Space.* http://www.altairproductdesign.com/CaseStudyDetail.aspx?id=52 Last accessed 02.08.2017

[26] Rosen D W 2014 *Virtual and Physical Prototyping*, **9(4)** pp. 225-232.

[27] Grabcad Community, https://grabcad.com/challenges/ge-jet-engine-bracket-challenge Last accessed 02.08.2017.

IOP Conf. Series: Materials Science and Engineering **276** (2017) 012027 doi:10.1088/1757-899X/276/1/012027

A review of fatigue crack propagation modelling techniques using FEM and XFEM

K Rege and H G Lemu

Department of Mechanical and Structural Engineering and Materials Science, University of Stavanger, Norway

kristen.rege@uis.no

Abstract. Fatigue is one of the main causes of failures in mechanical and structural systems. Offshore installations, in particular, are susceptible to fatigue failure due to their exposure to the combination of wind loads, wave loads and currents. In order to assess the safety of the components of these installations, the expected lifetime of the component needs to be estimated. The fatigue life is the sum of the number of loading cycles required for a fatigue crack to initiate, and the number of cycles required for the crack to propagate before sudden fracture occurs. Since analytical determination of the fatigue crack propagation life in real geometries is rarely viable, crack propagation problems are normally solved using some computational method. In this review the use of the finite element method (FEM) and the extended finite element method (XFEM) to model fatigue crack propagation is discussed. The basic techniques are presented, together with some of the recent developments.

1. Introduction

Components which are subjected to fluctuating loads are found virtually everywhere: Vehicles and other machinery contain rotating axles and gears, pressure vessels and piping may be subjected to pressure fluctuations (e.g. water hammer) or repeated temperature changes, structural members in bridges are subjected to traffic loads and wind loads, while those in ships and offshore structures are subjected to the combination of wind loads, wave loads and currents. If the components are subjected to a fluctuating load of a certain magnitude for a sufficient amount of time, small cracks will nucleate in the material. Over time, the cracks will propagate, up to the point where the remaining cross-section of the component is not able to carry the load, at which the component will be subjected to sudden fracture [1]. This process is called fatigue, and is one of the main causes of failures in structural and mechanical components [2]. In order to assess the safety of the component, engineers need to estimate its expected lifetime. The fatigue life is the sum of the number of loading cycles required for a fatigue crack to nucleate/initiate, and the number of cycles required for the crack to propagate until its critical size has been reached [2, 3]. In this paper, computational methods to estimate the lifespan of a propagating crack whose initial geometry is known will be considered.

Estimations of the fatigue crack propagation rate, da/dN, are normally based on a relation with the range of the stress intensity factor, ΔK, which is a linear elastic fracture mechanics (LEFM) parameter for quantifying the load and geometry of the crack. Paris, Gomez and Anderson [4] first proposed the existence of such a relation in 1961, and its simplest form is the Paris law [5]:

$$\frac{\mathrm{d}a}{\mathrm{d}N} = C\Delta K^m \tag{1}$$

where a is the length of an edge crack (or half the length of an internal crack), N is the number of loading cycles, and C and m are scaling constants. Due to its simplicity, the Paris law is only applicable for intermediate values of ΔK (region II in Figure 1), under constant amplitude cyclic loading [6]. Furthermore, the constants C and m are influenced by the applied stress ratio $R = K_{min}/K_{max}$, and crack closure effects are not taken into account. Therefore, a number of different crack propagation laws have been proposed, each one taking different factors into account [7]. One of the most detailed and commonly used is the so-called NASGRO equation [8]:

$$\frac{\mathrm{d}a}{\mathrm{d}N} = C\left[\frac{(1-f)}{(1-R)}\cdot\Delta K\right]^m \cdot \frac{\left(1-\frac{\Delta K_{\mathrm{th}}}{\Delta K}\right)^p}{\left(1-\frac{K_{\mathrm{max}}}{K_{\mathrm{c}}}\right)^q} \tag{2}$$

where ΔK_{th} is the threshold value for fatigue crack propagation, K_c is the critical stress intensity factor at which fracture occurs, $\Delta K = K_{max} - K_{min}$, $R = K_{min}/K_{max}$, f is a crack opening function and C, m, p and q are empirical constants. Note that for a given material, C and m do not have the same numerical values in different crack propagation laws.

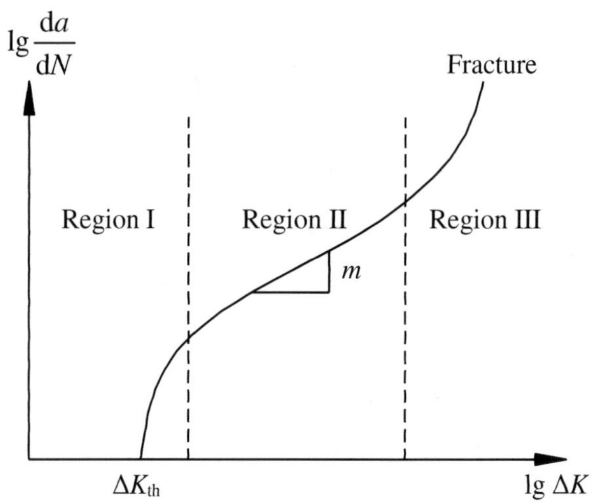

Figure 1. Typical fatigue crack growth curve.

The loading and displacement of a crack can be described by the three modes of fracture, each with its own stress intensity factor; mode I (tensile opening, K_I), mode II (in-plane sliding, K_{II}) and mode III (tearing/out-of-plane shear, K_{III}) [6]. The different modes require different values for the constants in the crack propagation law. In the case of mixed-mode fatigue, it may be necessary to use an effective mixed-mode stress intensity factor [9], for instance as given in [10], or a modified crack propagation law [11].

The crack propagation life can be estimated by integrating equation (1) or (2). However, the stress intensity factors K_{max} and K_{min} are normally functions of the crack length a, and depend on the geometry of the structure. Analytical integration of equations (1) and (2) is therefore rarely viable for complicated geometries. Instead, crack propagation problems are normally solved using some computational method, e.g. the Finite Element Method (FEM) [12]. The crack propagation process is then solved in a step-wise manner. For each step, the crack is advanced a small length, and the number of cycles required for the next crack increment is estimated using one of the crack propagation laws. In order to accomplish this, the computational method needs to perform the following tasks within each step [12, 13]:

1. Computation of the minimum and maximum stress and displacement fields within the cracked component.
2. Evaluation of the minimum and maximum stress intensity factors for the crack.
3. Evaluation of the direction for further crack propagation.
4. Generation of a representation of the crack advancement.

This process is repeated until the critical stress intensity factor is reached; $K_{max} = K_c$, and the number of experienced cycles are summed to obtain the crack propagation life.

The main challenge when using the finite element method to estimate the fatigue crack propagation, lies in the fourth task in the list above [12]. In order to evaluate stress intensity factors for the advanced crack, the finite element mesh needs to be updated, a process which has been shown to be challenging [6]. This has led to the development of alternative computational methods to handle propagating cracks, among them the eXtended Finite Element Method (XFEM) [14, 15], the Boundary Element Method (BEM) [16], hybrids between finite and boundary element methods, e.g. the Symmetric Galerkin Boundary Element Method – Finite Element Method (SGBEM-FEM) [17] and the Scaled Boundary Finite Element Method (SBFEM) [18], and meshless methods [19].

In this review, the use of the finite element method (FEM) and the extended finite element method (XFEM) to model fatigue crack propagation will be discussed. The basic techniques will be presented, together with some of the recent developments. The review will focus on the modelling techniques, and only to a less extent on their applications.

2. Crack propagation be the finite element method

2.1. *Computation of stress and displacement fields, and the stress intensity factor*

The first issue when estimating fatigue crack propagation rates using the finite element method is the computation of sufficiently accurate values for the stress intensity factor for the crack at the maximum and minimum applied loads during each cycle. In order to compute the stress intensity factor, the stress and displacement fields of the whole component are also needed. Three simple methods for computing the stress intensity factors of mode I cracks from a finite element stress field were presented by Chan, et al. [20] already in 1970. These methods were called the stress method, the displacement method and the line integral method, and were all computed using ordinary linear (constant-strain) triangle elements, with a high degree of refinement at the crack tip.

2.1.1. The stress method. In the stress method, nodal stress values are extrapolated to the crack tip. It is often easiest to do this extrapolation along the crack plane, in which case the stress intensity factor is related to the stress normal to the crack plane, σ_{yy}, by [6]:

$$K_{\mathrm{I}} = \lim_{r \to 0} \left[\sigma_{yy} \sqrt{2\pi r} \right] \tag{3}$$

where r is the distance from the crack tip. The nodal values of K_{I} are plotted as a function of r, and extrapolated to $r = 0$. As ordinary finite elements are not able to represent the stress singularity at the crack tip, the nodal values of K_{I} closest to the tip should be omitted when performing the extrapolation [20]. The singularity of the stress field also causes this method to be one of the least accurate methods available [6]. A similar approach is also possible for mode II and mode III cracks.

2.1.2. The displacement method. The second method is called the displacement method, and involves a relation between the stress intensity factor and the crack-opening displacement u_y [20]. For mode I loading, this relation is given as [6, 20]:

$$K_{\mathrm{I}} = \lim_{r \to 0} \left[\frac{E^* u_y}{4} \sqrt{\frac{2\pi}{r}} \right] \tag{4}$$

where:

$$E^* = \begin{cases} E, & \text{plane stress} \\ \dfrac{E}{1-\nu^2}, & \text{plane strain} \end{cases} \tag{5}$$

E is the elastic modulus, while ν is Poisson's ratio. As the opening displacement u_y is equal to zero at the crack tip, the nodal values of u_y near the crack tip must be disregarded when performing the extrapolation of K_{I} to $r = 0$. The displacement method does generally give more accurate estimations than the stress method [6].

2.1.3. Finite elements at the crack tip. In order to reduce the mesh quality required for the displacement method, special crack tip elements were developed in the 1970's, which were able to describe the singularity which exists in the near-crack stress field. One of them was the isoparametric bilinear rectangle (quadrilateral) 4-node element with special shape functions, proposed by Tracey [21] in 1971, which should be used at the crack tip. Four years later, Henshell and Shaw [22] declared in the title of their paper that "Crack tip finite elements are unnecessary". They could show that the crack tip singularity could be represented by 8-node isoparametric quadratic rectangle elements, if the mid-side nodes closest to the crack tip were moved to the "quarter points," i.e. ¼ of the element length away from the crack tip. Similar 2D elements, as well as three-dimensional elements, were independently developed by Barsoum, and presented in the same journal, two issues later [23].

When used at the crack tip, these so-called "quarter-point elements" [24] have a reasonably accurate stiffness, but the local values of stress and displacement within the quarter-point elements are poor [22]. The nodal displacements for these elements should therefore be omitted when calculating K. Recommendations for generating a suitable mesh with quarter-point elements for evaluating K_{I} by the displacement method have been given by Menandro, et al. [25] and Guinea, et al. [26], and a typical example of the crack tip mesh is shown in Figure 2. Note that the quarter-point (8-node quadratic rectangle) elements are collapsed down to triangles, where each element has three nodes located at the crack tip.

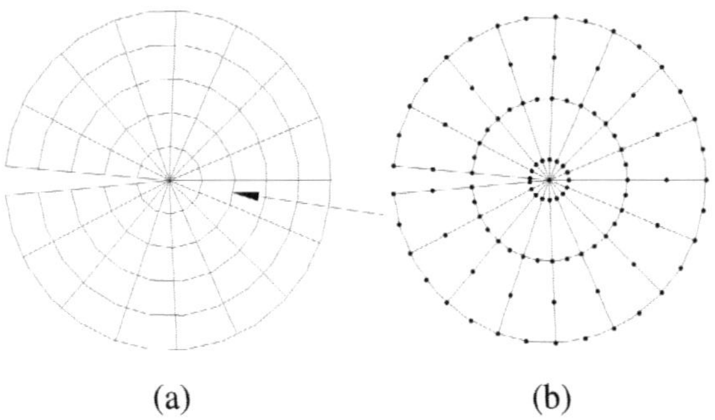

Figure 2. (a) Typical rosette pattern for the FEM mesh at a LEFM crack tip, and (b) detail of the two inner rings, with nodes shown. Note the use of quarter-tip elements only in the innermost ring.

2.1.4. Energy release rate methods. The potential energy decrease per unit crack advance is called the energy release rate, G, and may be used to characterise crack growth in a linear or nonlinear elastic body [27-29]. Rice [27] showed that the energy release rate could be computed by a path independent line integral, which for a two-dimensional problem is defined by:

$$J = \int_{\Gamma} \left(W\,\mathrm{d}y - \mathbf{T} \cdot \frac{\partial \mathbf{u}}{\partial x} \mathrm{d}s \right) \tag{6}$$

where Γ is a curve surrounding the crack tip, $W(x, y)$ is the strain-energy density field, and x, y are the Cartesian coordinates being parallel and normal to the crack tip, respectively. $\mathbf{T}$ is the traction vector associated to the outward normal to Γ, $\mathbf{u}$ is the displacement vector, and ds is an incremental arc length along Γ.

The third method to compute the stress intensity factors mentioned by Chan, et al. [20] is the line integral method. This method consists of numerically evaluating the J integral, equation (6), for the finite element solution over an arbitrary path surrounding the crack tip. In the case of small-scale yielding, the energy release rate may be related to the mode I stress intensity factor by the following equation [20, 27]:

$$K_{\mathrm{I}} = \sqrt{GE^{*}} = \sqrt{JE^{*}} \tag{7}$$

The J integral may be evaluated at a remote contour, like the outer boundary of the geometry [20], which improves its numerical accuracy, compared with the stress method and the displacement method. The application of this method to three-dimensional problems is more difficult, however, because the integral becomes a surface integral. It is difficult both to define the surface and to perform the numerical integration [29]. This led to the development of alternative methods to evaluate the energy release rate, e.g. the virtual crack extension methods developed by Parks [28], Hellen [30] and deLorenzi [29]. The most accurate and efficient [6] method for the numerical evaluation of the energy release rate seems to be the domain integral method, formulated by Shih, et al. in 1986 [31]. They compute the J integral by using the divergence theorem, in which the three-dimensional surface integral is transformed into a volume domain integral, which is evaluated using Gaussian quadrature.

For mixed mode I+II loading with small-scale yielding, the following relationship exists between the energy release rate and the stress intensity factors [14, 27]:

$$J = G = \frac{1}{E^{*}}\left(K_{\mathrm{I}}^{2} + K_{\mathrm{II}}^{2}\right) \tag{8}$$

In order to extract the stress intensity factors from the J integral, Yau, et al. [32] developed a technique using an interaction integral. The combination of the interaction integral with the domain integral method is shown in [14, 33].

The main advantage of the energy release rate methods is that accurate estimations for the stress intensity factor may be obtained even with a relatively coarse mesh [26]. A finer mesh, with quarter-point elements at the crack tip, is required if determination of the stress field is part of the objective. The domain integral is often recommended for practical use. Some find it easier to implement in certain FE codes (because quarter-point elements are not needed) [34], and others have found it to be more stable than the displacement method [24]. On the other hand, if quarter-point elements may easily be used in a FE code, the displacement method does not need any specialised post-processing routine to obtain the stress intensity factor. Both the displacement method and the domain integral method are able to accurately predict the stress intensity factor [24]. The displacement method is still widely used [9, 26, 35], and Guinea, et al. [26] question whether the domain integral actually is the most efficient.

2.2. Evaluation of the direction for further crack growth

If a crack is subjected to a mixed-mode loading, the propagating crack seeks the path of least resistance. Several theories have been proposed to choose this path. The three most used [35] are the criteria of maximum tangential (circumferential) stress [36], maximum energy release rate [37] and minimum strain energy density [38]. Other criteria include the criterion of maximum dilatational strain energy density [39] and the criterion of minimum accumulated strain energy [11]. There is no general agreement about which criterion should be used for a given material, but Bittencourt, et al. [24] have shown that the three former criteria predict basically the same crack growth trajectory for poly(methyl methacrylate) (PMMA), which is a brittle material. The criterion of maximum tangential stress is often applied in FEM simulations of fatigue crack growth, because it is simple to implement, as it has an approximate explicit solution for the crack growth direction θ as a function of K_{I} and K_{II} [9, 14, 35].

To implement the evaluation of the direction for further crack propagation in FE codes is generally not a problem [12].

2.3. Representation of the crack advancement

When the stress intensity range and the crack growth direction have been found, the number of cycles required for the crack to propagate a distance Δa may be estimated by the crack propagation law. As the crack propagation law is not linear with respect to Δa, the stress intensity factor range ΔK needs to be re-evaluated after the increment Δa. Reducing this increment will increase the accuracy of the solution, at the cost of increased computational effort.

In order to re-evaluate ΔK, the crack increment needs to be represented by the finite element mesh. As noted in section 2.1, some mesh refinement at the crack tip is normally needed. When the crack tip moves due to the crack increment, the focused region of the mesh should follow. The most common technique when modelling propagating fatigue cracks under LEFM conditions is to perform global or local re-meshing [9, 24, 35, 40]. Local re-meshing is generally preferred, due to the lower computational effort, compared to global re-meshing.

Local re-meshing was employed already in 1986 by Højfeldt and Østervig [40] to predict the shape and crack propagation life of fatigue cracks in shafts with a diameter transition. They employed the displacement method and quarter-point elements to estimate the stress intensity factors of the three-dimensional crack, and the criterion of minimum strain energy density to estimate the crack growth direction. Bittencourt, et al. [24] and Miranda, et al. [35] used the local re-meshing technique, together with quarter-point elements and the displacement technique, in their methodology for assessment of fatigue crack propagation in two-dimensional components. The re-meshing technique was also used by Alegre and Cuesta [9] to model mode I+II crack propagation in a valve.

The local re-meshing technique generally consists of four steps [24], as illustrated in Figure 3; (a) removing the existing mesh around the crack, (b) advancing the length of the crack, (c) applying quarter-point elements in a uniform rosette pattern at the crack tip, and (d) generating a new mesh in the open area.

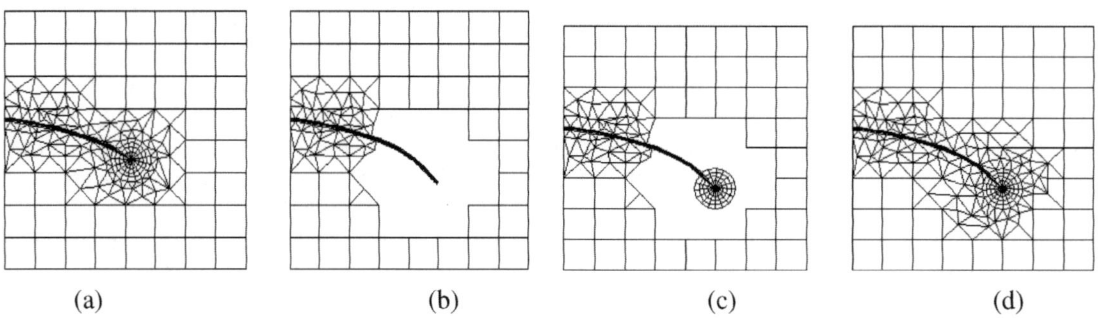

Figure 3. Local re-meshing technique. (a) Existing mesh. (b) The mesh around the crack tip is removed, and the crack is advanced. (c) Rosette of elements are applied at the crack tip. (d) The remaining area is meshed.

Recent research in crack propagation modelling by FEM often deals with plasticity induced crack closure, e.g. [41-43], in which the material plasticity is explicitly taken into account in the finite element analysis. Re-meshing is then highly cumbersome, with respect to both computational and programming effort, because the plastic strain history needs to be mapped from the old mesh to the new one [6]. Furthermore, accuracy is lost during this mapping process. As an alternative, it is possible to create a single mesh which accommodates crack growth, by being refined along the assumed crack propagation path. The disadvantage of this approach is that the crack path and shape are predetermined by the mesh [6, 12]. In addition, the crack increment Δa has to correspond to the element size, making the crack growth response mesh dependent. By employing micrometre-size elements, this approach is nevertheless widely used [42-43].

In order to advance the crack through the refined mesh, the following techniques may be used [6, 44]: (a) Removing elements along the crack front once a failure criterion is reached, (b) releasing the nodes at the crack tip at specific load steps or according to a failure criterion, so that these elements are no longer connected, or (c) using cohesive elements. The cohesive elements are zero-thickness elements which are placed in between the ordinary elements, and for which a certain force-displacement law is specified. When computing fatigue crack propagation with plasticity induced crack closure, the node release technique is the most common to use [45], but cohesive elements are also used by some researchers [42, 43, 46]. It should be noted that the plastic crack tip field does not contain the $1/\sqrt{r}$ singularity [6], making quarter point elements unnecessary when studying plasticity induced crack closure.

Colombo and Giglio [12] recently developed a crack advancement technique for LEFM conditions which is a combination of the local re-meshing and node release techniques. Their technique may be described by four steps: (a) the crack increment is projected on the FE mesh, (b) the elements touched by the crack increment are deleted, (c) the open area is re-meshed by debondable triangular elements which conform to the crack increment, and (d) the nodes along the crack increment are released. This technique limits the re-meshing to the absolute minimum, while still keeping the crack growth mesh-independent. The resulting mesh is inadequate for evaluating the stress intensity factors using the displacement technique, however. Therefore, the submodelling technique [47, 48] is used, in which the displacements from this mesh are imposed on an independent focused mesh which represents the crack tip using quarter-point elements (as shown in Figure 2(a)). The displacement field of this focused mesh is then used to estimate the stress intensity factor. The submodelling technique has also been used by Schöllmann, et al. [49] in their fatigue crack growth simulation software.

2.4. Current trends

In recent years the research focus seems to have shifted from simple geometries modelled under LEFM conditions towards three-dimensional cracks in complex geometries or with plasticity induced crack closure. A review of the literature on finite element analysis of three-dimensional fatigue cracks is found in [50], while an overview of the finite element analysis of plasticity induced crack closure is found in [45]. We will here limit our discussion to three of the latest works.

Aguilar Espinosa, et al. [41] have used four-node quadrilateral elements with the node release technique in Abaqus, in order to estimate the crack opening and closing stress intensity factors for a crack in a four-point bending specimen subjected to fatigue loading. The contact of the crack flanks was simulated by using Abaqus' surface contact boundary condition. However, even with a mesh refined to 4 μm elements at the crack tip, the estimated crack opening and closing loads are significantly lower than the experimentally determined values.

Solanki, et al. [45] stated that the physics of fatigue crack growth is not taken into account when the node release technique is used. García-Collado, et al. [42] illustrate some of the differences between the results obtained by using a physics-based cohesive element technique, and those obtained by using the node release technique, like differences in the plastic strain field around the crack tip. They used an element size of 15 μm at the crack tip, with 13 213 nodes in total to model a compact-tension specimen. In order to reduce the mesh refinement required at the crack tip, Hu, et al. [43] propose a special singular element to be used at the crack tip when considering plasticity induced crack closure in two-dimensional geometries. This single element covers the whole crack tip, and its radius should be equal to the plastic zone length. The element is based on a cohesive zone model, and is used to simulate variable amplitude propagation of a mode I fatigue crack.

3. Crack propagation by the extended finite element method

As indicated in section 2.3, the methods for evaluating fatigue crack growth by the ordinary finite element method do normally contain one of two shortcomings: The methods are either mesh-dependent, or requires re-meshing. In order to avoid both these shortcomings, Belytschko and Black published a mesh-independent method with minimal re-meshing in 1999 [14]. This method was further developed by Moës, Dolbow and Belytschko [51] into a mesh-independent method without any re-meshing. The method has later become known as the eXtended Finite Element Method (XFEM) [15, 44], and has become widely popular [52] for solving continuum mechanics problems containing discontinuities like cracks and material interfaces. In addition to crack propagation problems, XFEM has been employed to solve two-phase flows and fluid-structure interaction problems [15], and has even been proposed as an applicable tool to predict the deformation of a brain subjected to a surgeon's cut [53].

As illustrated in Figure 4, XFEM uses a non-conforming mesh to model the crack (or other discontinuities), i.e. the cracks are modelled independently of the mesh [44]. This is made possible by "enriching" the elements cut by the crack, by adding special shape functions to take care of the local discontinuities and singularities around the crack [15]. The mesh-independency makes it possible to use the same mesh for all stages of a growing fatigue crack. Even though the modelling of the crack is said to be independent of the mesh, Ren and Guan [54] clearly illustrate that the level of mesh refinement at the crack influences the accuracy of the representation of a three-dimensional crack. The mesh-independency does rather imply that the crack growth increment and orientation may be chosen independently of the mesh.

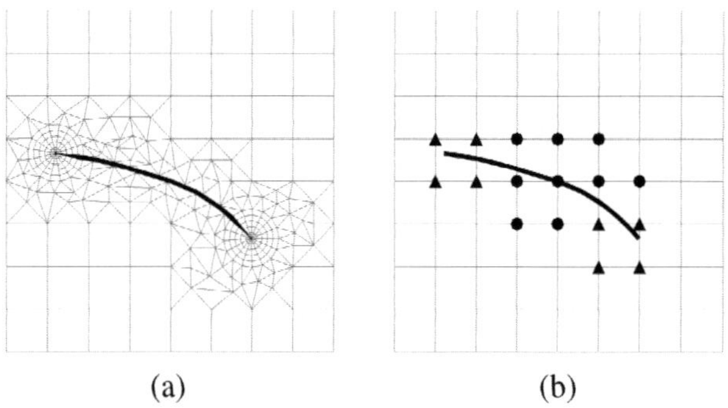

Figure 4. Illustrative sketches of (a) a conforming FEM mesh and (b) a non-conforming XFEM mesh. ▲ indicates tip enriched nodes, while ● indicates step enriched nodes. Element sizes not to scale.

We will here briefly review XFEM for the purpose of modelling fatigue cracks. An extensive review of XFEM in general is found in [15], and some of its numerous applications are reviewed in [55].

3.1. Representation of the crack

As the extended finite element method mesh does not normally conform to the crack, it is necessary to formulate the position and shape of the crack face and the position of the crack tips mathematically. This is normally (but not necessarily) done using the level set method [15, 44, 56, 57]. An arbitrary crack may be described by two level set curves. The first level set curve, $\phi(\mathbf{x},t)$, is the signed-distance function, which is equal to zero along the whole crack surface, as well as at its tangential extensions at its ends, as shown in Figure 5(a) [15, 56]:

$$\phi(\mathbf{x},t) = \pm \min_{\mathbf{x}_\Gamma \in \Gamma(t)} \left\| \mathbf{x} - \mathbf{x}_\Gamma \right\|, \ \forall \mathbf{x} \in \Omega \tag{9}$$

where the sign is positive and negative on opposite sides of the crack. $\mathbf{x}_\Gamma$ is the closest point to $\mathbf{x}$ on the crack face $\Gamma(t)$. Ω represents the domain of the solid.

While $\phi(\mathbf{x},t)$ represents the position and geometry of the crack face, it does not specify where the crack ends. This is accomplished by a second level set curve, $\gamma(\mathbf{x},t)$, which is orthogonal to $\Gamma(t)$ and equal to zero at the crack front [15], as shown in Figure 5(b). The crack front corresponds to the crack tip for a two-dimensional crack, or to the perimeter of a three-dimensional crack [57].

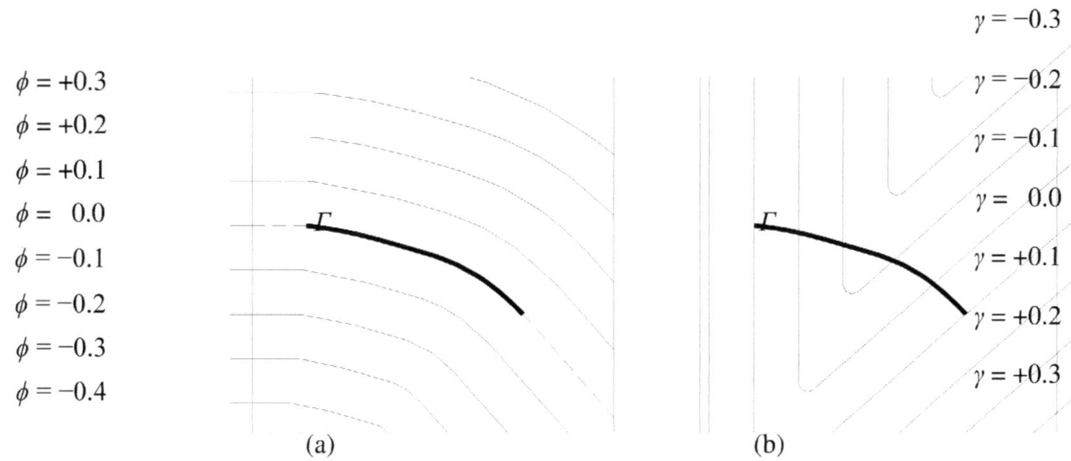

Figure 5. Level set curves (a) $\phi(\mathbf{x},t)$ and (b) $\gamma(\mathbf{x},t)$ for a 2D crack.

Given these two level sets, the geometry and position of the crack is given by [15, 56]:

$$\Gamma(t)=\{\mathbf{x}:\ \phi(\mathbf{x},t)=0\wedge\gamma(\mathbf{x},t)\leq 0\} \tag{10}$$

As the crack grows, $\Gamma(t)$ is updated, based on the crack increment length and direction. One of the possible ways to update the crack representation is the fast marching method used by Sukumar, et al. [57]. Alternative methods are mentioned in [15].

The values of the level sets are normally saved at the element nodes, and interpolated within the finite elements using standard shape functions [15]. It should be noted that this interpolation makes the accuracy of the crack representation dependable on the element size and the number of nodes per element, as illustrated by Ren and Guan [54].

3.2. *The stress and displacement fields*

The fundamental requirement of any finite element method is its ability to represent the stress and displacement fields of the loaded solid. With a crack being defined by the level set method, the extended finite element method needs to be able to compute these fields with sufficient accuracy. In order to take the discontinuity and the singularity around the crack into account, the XFEM introduces an enrichment to the finite elements which are cut by the crack [15, 44]. More precisely, "a node is enriched if its support is cut by the crack" [51]. The nodes are enriched by the introduction of an additional set of degrees of freedom, **q**. The nodes do still contain their traditional degrees of freedom, **u**, meaning that the enriched nodes have a higher number of degrees of freedom than the ordinary nodes. The displacement field within the elements is then approximated by the following expression:

$$\mathbf{u}^h(\mathbf{x})=\underbrace{\sum_{i\in I}N_i(\mathbf{x})\mathbf{u}_i}_{\text{standard FE approx.}}+\underbrace{\sum_{i\in I^*}N_i^*(\mathbf{x})\cdot[\psi(\mathbf{x})-\psi(\mathbf{x}_i)]\mathbf{q}_i}_{\text{enrichment}} \tag{11}$$

which is the standard XFEM approximation [15]. The enriched nodes I^* are a subset of all the nodes I, $N_i(\mathbf{x})$ are the standard FEM shape functions and $\psi(\mathbf{x})$ is called the enrichment function. $N_i^*(\mathbf{x})$ are some functions which have the partition of unity property [15]:

$$\sum_{i\in I^*}N_i^*(\mathbf{x})=1 \tag{12}$$

As the standard FEM shape functions $N_i(\mathbf{x})$ do have the partition of unity property [44], $N_i^*(\mathbf{x})$ are usually chosen to be equal to the $N_i(\mathbf{x})$. The term $-\psi(\mathbf{x}_i)$ is called shifting, and is used to remove the effects of $\mathbf{q}_i$ on the nodes and ensure compatibility across elements, i.e. to ensure that $\mathbf{u}^h(\mathbf{x}_i) = \mathbf{u}_i$.

The enrichment function is responsible for introducing discontinuities and singularities to the displacement field. Belytschko and Black [14] proposed the enrichment functions that describe the singular displacement field around the crack tip under LEFM conditions in 1999:

$$\psi_{\text{tip}}(\mathbf{x}) = \sqrt{r}\left\{\cos\frac{\theta}{2},\ \sin\frac{\theta}{2},\ \sin\frac{\theta}{2}\sin\theta,\ \cos\frac{\theta}{2}\sin\theta\right\} \tag{13}$$

Here, r and θ are the polar coordinates with the origin at the crack tip, and $\theta = 0$ along the parallel extension of the crack into the material.

The above enrichment is based on linear elastic fracture mechanics (LEFM), and is therefore only representative for cases where the crack tip plastic zone is considered to be small [44]. For the case of significant plasticity, an alternative set of crack tip enrichment functions was developed by Elguedj, et al. [58] in 2006. They still assume that the plasticity is confined to a region near the crack tip, and apply the Hutchinson-Rice-Rosengren (HRR) solution for a power-law hardening material, proposing the following crack tip enrichment functions:

$$\psi_{\text{tip}}(\mathbf{x}) = r^{\frac{1}{n+1}}\left\{\sin\frac{\theta}{2},\ \cos\frac{\theta}{2},\ \sin\frac{\theta}{2}\sin\theta,\ \cos\frac{\theta}{2}\sin\theta,\ \sin\frac{\theta}{2}\sin 3\theta,\ \cos\frac{\theta}{2}\sin 3\theta\right\} \tag{14}$$

n is here the hardening exponent of the material.

Even though the enrichment functions, equations (13) & (14), are discontinuous along $\theta = \pm\pi$ [14], i.e. throughout the crack, they are not readily applicable to describe long, severely curved or three-dimensional cracks [14, 51]. Hence, Belytschko and Black stated their method to require "minimal re-meshing." Moës, et al. [51] removed the need of re-meshing altogether, by introducing an additional enrichment, to be used for the enriched nodes along the crack, away from the crack tip. This enrichment models the discontinuous displacement field over the crack by using the sign function [15, 51]:

$$\psi_{\text{step}} = \operatorname{sign}\left(\phi(\mathbf{x},t)\right) = \begin{cases} -1 \text{ if } \phi(\mathbf{x},t) < 0, \\ 0 \ \text{ if } \phi(\mathbf{x},t) = 0, \\ 1 \ \text{ if } \phi(\mathbf{x},t) > 0. \end{cases} \tag{15}$$

This enrichment function is equal to −1 and +1 on opposite sides of the crack, which is normally represented by the level set $\phi(\mathbf{x},t)$. Moës, et al. [51] used the symbol $H(\mathbf{x})$ for this enrichment function, and subsequent authors have often referred to it as the Heaviside function, e.g. [53, 55, 57, 59], even though the Heaviside step function formally has the property $\text{H}(\phi(\mathbf{x})) = 0$ if $\phi(\mathbf{x}) \leq 0$ [15]. However, the sign and the Heaviside step functions do actually lead to identical results because they span the same approximation space [15].

By using the crack tip and step enrichments, an appropriate approximation for the stress and displacement fields may be obtained, both for two-dimensional [51, 60] and three-dimensional [57, 59] cracks. If we divide the additional degrees of freedom, $\mathbf{q}$, into those around the crack tip, $\mathbf{b}$ (indicated by ▲ in Figure 4(b)), and those remaining, $\mathbf{a}$ (indicated by ● in Figure 4(b)), we can rewrite equation (11) [15]:

$$\mathbf{u}^h(\mathbf{x}) = \sum_{i\in I} N_i(\mathbf{x})\mathbf{u}_i + \sum_{i\in I^*_{\phi,\gamma}} N_i^*(\mathbf{x})\cdot\left[\psi_{\text{step}}(\mathbf{x}) - \psi_{\text{step}}(\mathbf{x}_i)\right]\mathbf{a}_i + \sum_j \sum_{i\in I^*_{\text{tip}}} N_i^*(\mathbf{x})\cdot\left[\psi^j_{\text{tip}}(\mathbf{x}) - \psi^j_{\text{tip}}(\mathbf{x}_i)\right]\mathbf{b}_i^j \tag{16}$$

It should be noted that integration of the weak form of the finite element formulation is not straight forward when the elements contain singularities and discontinuities, because the standard Gauss quadrature requires a smooth integrand and a finite order polynomial [44]. Special integration techniques have therefore been developed for this purpose [15, 44, 51, 58].

3.3. Evaluation of the stress intensity factor and the direction for further crack growth
The stress intensity factor is commonly extracted from the extended finite element solution by employing the domain form of the interaction integral [14, 51, 54, 57, 58], as explained in section 2.1.4. It is likely that this method is preferred over the displacement method, because the crack tip displacements are less readily available in the XFEM results than in corresponding FEM results. Moës, et al. [51] used a proper mesh and demonstrated computation of the stress intensity factors for two-dimensional mode I+II cracks within 1–2 % deviation from the analytical values. The XFEM enrichment takes care of the near-crack singularity, and linear elements can therefore be used even for the LEFM analyses.

The direction for further crack growth is often evaluated using the criterion of maximum tangential stress [36], e.g. in [14, 51, 54, 60], just like in fatigue analyses using the ordinary FEM. The angle of propagation is estimated using a function of K_I and K_{II}, and the level sets describing the crack are updated accordingly. The relation between the angle of propagation and the level set updates are illustrated in [44].

3.4. Fatigue crack propagation and current trends
The first application of the extended finite element method to fatigue crack growth propagation was illustrated already by Moës, et al. [51] in 1999. They modelled the propagation of mode I+II fatigue cracks propagating from two holes in a two-dimensional plate with 2650 nodes. Two different values for the crack increment size are used, but the numerical results are not compared to any experimental ones.

Sukumar, et al. [57] presented the first XFEM simulation of planar fatigue cracks propagating through a three-dimensional solid (i.e. pure mode I cracks) in 2003. They use the level set method to track the position of the crack front, and the fast marching method to advance the crack front. Using three-dimensional meshes of 24x24x24 eight-node (linear) hexahedral elements, they compute the stress intensity factors for a planar penny-shaped crack with errors between 0.4 % and 2.9 %, while for a planar elliptical crack the errors are between 0.6 % and 3.7 %. The fatigue simulations correctly predicted an initially elliptical crack to develop into a penny-shaped crack, but the crack propagation life is not compared to any experimental results.

Comparisons of XFEM fatigue simulations to experimental results have recently been published by Bergara, et al. [59]. In this work, they have used the XFEM-based LEFM approach which is implemented in Abaqus to simulate the growth of a semi-elliptical crack located at the side of a beam specimen subjected to four-point bending. The specimen is completely modelled in three dimensions by 8-node hexahedral elements, with refinements around the crack and the boundary condition sites. With 173 892 nodes in total, approximately one week was required to run one fatigue crack propagation simulation using three 3.40 GHz processors. Bergara, et al. report excellent correspondence between the numerical and experimental crack propagation histories, as well as good agreement for the evolution of the crack geometry. However, there are large differences between the experimental and numerical stress intensity factor ranges; up to approximately 40 %. This seems remarkable given the fine mesh and good correspondence in the crack propagation histories.

Zhan, et al. [3] have proposed an entire framework for the fatigue life prediction of metallic components, where the crack initiation life is evaluated using FEM with continuum damage mechanics, while the crack propagation life is evaluated using XFEM. The initiated crack is set to be 0.1 mm long. Computational and experimental results have been compared for a mode I crack in a fuselage structure, modelled by 8-node hexahedral elements, using 20 106 nodes for half the model. The predicted crack propagation life was 28 % shorter than the experimental one, indicating a potential for improving the technique.

A new method for reducing the required structural mesh density near the crack has recently been proposed by Ren and Guan [54]. Their method is intended for the crack growth analysis of three-dimensional and arbitrarily shaped cracks subjected to mixed mode loading. They note that the level set method is not able to fully represent the crack front surface if the crack has a complicated geometry or if the mesh is relatively coarse. It is therefore proposed to use an individual mesh for the crack, and let the crack geometry in the XFEM model of the structure be described by this mesh, instead of the level set method. It is shown that this method makes the values for the stress intensity factor range more uniform along the crack front, and they are also less dependent on the structural mesh. However, the crack mesh is shown to be very fine in the illustrations, and necessarily needs to be updated (re-meshed) as the crack propagates. It has not been shown whether the computation time saved by using a coarser overall mesh makes up for the time required to re-mesh the crack for each propagation increment. Still, it is possible that this technique is more efficient for analysing fatigue crack propagation in large and complex components.

4. Discussion

The studies reviewed in section 3.4 indicate that even though XFEM is capable of accurately predicting stress intensity factors for simple "handbook" geometries, further development is required in order to make accurate predictions of stress intensity factors and fatigue crack propagation lives for more complex geometries. It should be noted that the accuracy of the predicted crack propagation life will always be limited by the fracture mechanics assumptions and the crack propagation law which is used, however. On the other hand, even though the XFEM was developed to reduce the computational effort required to simulate fatigue crack propagation, none of the studies reviewed do actually compare computation times between FEM and XFEM. This is remarkable.

In 1986 Højfeldt and Østervig [40] were able to predict the fatigue crack propagation life of shafts with shoulder fillets, with errors below 20 % in just 20 400 s (CPU-time). These results were obtained using the ordinary FEM with a three-dimensional mesh and the local re-meshing technique. By exploiting symmetry, only half the shaft was modelled, using only 1160 nodes. One would expect that after 30 years of research, either the accuracy of the computational results would increase, or the computation time would decrease, especially with the introduction of XFEM. The works considered in this review are inconclusive in this regard, mainly because computation times (with processor frequencies) are rarely stated. In order to choose between the various computational techniques for modelling fatigue crack propagation, and their variants, such information is essential.

Yazid, et al. state that "about the only drawback" of the present XFEM "is the need for a variable number of degrees of freedom per node" [55]. This is mainly a challenge when XFEM is incorporated into existing FEM codes [15]. A more important drawback, as indicated above, for both FEM and XFEM, seems to be the large number of nodes and long computation time required. FEM and XFEM are challenged by methods requiring significantly fewer nodes, like the boundary element method (BEM) [16] and the SGBEM-FEM by Nikishkov, Park and Atluri [17]. Recent comparisons by Dong and Atluri [52, 61] indicate that the SGBEM-FEM is both more accurate and more efficient than XFEM to analyse propagating fatigue cracks under LEFM conditions.

5. Conclusions

Both the ordinary Finite Element Method (FEM) and the eXtended Finite Element Method (XFEM) are able to provide good predictions for the fatigue crack growth propagation in simple geometries. While the FEM mesh conforms to the fatigue crack, the crack is only implicitly modelled in XFEM, independently of the mesh.

In FEM, the domain integral method is often recommended for evaluating the stress intensity factors, while the displacement method is still often used in practice. The domain integral method is commonly used also in XFEM. The criterion of maximum tangential stress is chosen for determining the crack propagation path in the majority of the analyses. Under linear elastic fracture mechanics (LEFM) conditions, local re-meshing is used to advance the crack in FEM, whereas either the node release technique or cohesive elements are used under elastic-plastic fracture mechanics conditions. In XFEM, it is common to use the level set method to describe the crack geometry.

As the complexity of the problem increases, the accuracy of both methods decreases. The latest research therefore focuses on applying and improving these techniques to complex problems, like three-dimensional crack propagation under mixed-mode loading. In the current literature, there is also an increasing focus on modelling plasticity induced closure and fatigue propagation under variable amplitude loading using FEM.

For both methods, there is also a focus on improving their efficiency, by introducing new techniques. Even though the XFEM was developed to reduce the computational effort required for crack propagation problems, none of the studies reviewed do actually compare computation times between FEM and XFEM. This is remarkable, and should be considered for later studies.

Acknowledgements

This work is part of a PhD study supported by the Norwegian Ministry of Education and Research. The financial support is acknowledged.

References

[1] Boyer H E ed 1986 *Atlas of Fatigue Curves* (Ohio: ASM International)
[2] Suresh S 1998 *Fatigue of Materials 2nd ed* (Cambridge: Cambridge University Press)
[3] Zhan Z, Hu W, Li B, Zhang Y, Meng Q and Quan Z 2017 Continuum damage mechanics combined with the extended finite element method for the total life prediction of a metallic component *Int. J. Mech. Sci.* **124-125** 48-58
[4] Paris P C, Gomez M P and Anderson W E 1961 A rational analytic theory of fatigue *The Trend in Engineering* **13** 9-14
[5] Paris P and Erdogan F 1963 A critical analysis of crack propagation laws *J. Basic Eng.* **85** 528-33
[6] Anderson T L 2005 *Fracture Mechanics: Fundamentals and Applications 3rd ed* (Boca Raton: CRC Press)
[7] Kłysz S and Leski A 2012 Good practice for fatigue crack growth curves description *Applied Fracture Mechanics* ed A Belov (InTech) chapter 7 pp 197-228
[8] Forman R G and Mettu S R 1992 Behavior of surface and corner cracks subjected to tensile and bending loads in Ti-6Al-4V alloy *Fracture Mechanics: 22nd Symp. (Vol. I) ASTM STP 1131* ed H A Ernst, et al. (Philadelphia: American Society for Testing and Materials) pp 519-46
[9] Alegre J M and Cuesta I I 2010 Some aspects about the crack growth FEM simulations under mixed-mode loading *Int. J. Fatigue* **32** 1090-5
[10] Tanaka K 1974 Fatigue crack propagation from a crack inclined to the cyclic tensile axis *Eng. Fract. Mech.* **6** 493-507
[11] Pavlou D G, Labeas G N, Vlachakis N V and Pavlou F G 2003 Fatigue crack propagation trajectories under mixed-mode cyclic loading *Eng. Struct.* **25** 869-75
[12] Colombo D and Giglio M 2006 A methodology for automatic crack propagation modelling in planar and shell FE models *Eng. Fract. Mech.* **73** 490-504
[13] Peng X, Atroshchenko E, Kerfriden P and Bordas S P A 2017 Isogeometric boundary element methods for three dimensional static fracture and fatigue crack growth *Comput. Methods Appl. Mech. Eng.* **316** 151-85
[14] Belytschko T and Black T 1999 Elastic crack growth in finite elements with minimal remeshing *Int. J. Numer. Methods Eng.* **45** 601-20
[15] Fries T-P and Belytschko T 2010 The extended/generalized finite element method: An overview of the method and its applications *Int. J. Numer. Methods Eng.* **84** 253-304
[16] Mi Y and Aliabadi M H 1994 Three-dimensional crack growth simulation using BEM *Comput. Struct.* **52** 871-8
[17] Nikishkov G P, Park J H and Atluri S N 2001 SGBEM-FEM alternating method for analyzing

3D non-planar cracks and their growth in structural components *Comput. Model. Eng. Sci.* **2** 401-22

[18] Yang Z 2006 Fully automatic modelling of mixed-mode crack propagation using scaled boundary finite element method *Eng. Fract. Mech.* **73** 1711-31

[19] Bordas S, Rabczuk T and Zi G 2008 Three-dimensional crack initiation, propagation, branching and junction in non-linear materials by an extended meshfree method without asymptotic enrichment *Eng. Fract. Mech.* **75** 943-60

[20] Chan S K, Tuba I S and Wilson W K 1970 On the finite element method in linear fracture mechanics *Eng. Fract. Mech.* **2** 1-17

[21] Tracey D M 1971 Finite elements for determination of crack tip elastic stress intensity factors *Eng. Fract. Mech.* **3** 255-65

[22] Henshell R D and Shaw K G 1975 Crack tip finite elements are unnecessary *Int. J. Numer. Methods Eng.* **9** 495-507

[23] Barsoum R S 1976 On the use of isoparametric finite elements in linear fracture mechanics *Int. J. Numer. Methods Eng.* **10** 25-37

[24] Bittencourt T N, Wawrzynek P A, Ingraffea A R and Sousa J L 1996 Quasi-automatic simulation of crack propagation for 2D LEFM problems *Eng. Fract. Mech.* **55** 321-34

[25] Menandro F C M, Moyer jr E T and Liebowitz H 1995 A methodology for crack tip mesh design *Eng. Fract. Mech.* **50** 713-26

[26] Guinea G V, Planas J and Elices M 2000 K_I evaluation by the displacement extrapolation technique *Eng. Fract. Mech.* **66** 243-55

[27] Rice J R 1968 A path independent integral and the approximate analysis of strain concentration by notches and cracks *J. Appl. Mech.* **35** 379-86

[28] Parks D M 1974 A stiffness derivative finite element technique for determination of crack tip stress intensity factors *Int. J. Fract.* **10** 487-502

[29] deLorenzi H G 1985 Energy release rate calculations by the finite element method *Eng. Fract. Mech.* **21** 129-43

[30] Hellen T K 1975 On the method of virtual crack extensions *Int. J. Numer. Methods Eng.* **9** 187-207

[31] Shih C F, Moran B and Nakamura T 1986 Energy release rate along a three-dimensional crack front in a thermally stressed body *Int. J. Fract.* **30** 79-102

[32] Yau J F, Wang S S and Corten H T 1980 A mixed-mode crack analysis of isotropic solids using conservation laws of elasticity *J. Appl. Mech.* **47** 335-41

[33] Shih C F and Asaro R J 1988 Elastic-plastic analysis of cracks on bimaterial interfaces: Part I – small scale yielding *J. Appl. Mech.* **55** 299-316

[34] Courtin S, Gardin C, Bézine G and Hamouda H B H 2005 Advantages of the *J*-integral approach for calculating stress intensity factors when using the commercial finite element software ABAQUS *Eng. Fract. Mech.* **72** 2174-85

[35] Miranda A C O, Meggiolaro M A, Castro J T P, Martha L F and Bittencourt T N 2003 Fatigue life and crack path predictions in generic 2D structural components *Eng. Fract. Mech.* **70** 1259-79

[36] Erdogan F and Sih G C 1963 On the crack extension in plates under plane loading and transverse shear *J. Basic Eng.* **85** 519-25

[37] Hussain M A, Pu S L and Underwood J 1974 Strain energy release rate for a crack under combined mode I and mode II *Fracture Analysis: Proc. of the 1973 National Symp. on Fracture Mechanics, Part II ASTM STP 560* ed G R Irwin (Philadelphia: American Society for Testing and Materials) pp 2-28

[38] Sih G C 1974 Strain-energy-density factor applied to mixed mode crack problems *Int. J. Fract.* **10** 305-21

[39] Theocaris P S and Andrianopoulos N P 1982 The T-criterion applied to ductile fracture *Int. J. Fract.* **20** R125-30

[40] Højfeldt E and Østervig C B 1986 Fatigue crack propagation in shafts with shoulder fillets *Eng. Fract. Mech.* **25** 421-7

[41] Aguilar Espinosa A A, Fellows N A, Durodola J F and Fellows L J 2017 Development of numerical model for the determination of crack opening and closure loads, for long cracks *Fatigue Fract. Eng. Mater. Struct.* **40** 571-85

[42] García-Collado A, Vasco-Olmo J M and Díaz F A 2017 Numerical analysis of plasticity induced crack closure based on an irreversible cohesive zone model *Theor. Appl. Fract. Mech.* **89** 52-62

[43] Hu X, Bui T Q, Wang J, Yao W, Ton L H T, Singh I V and Tanaka S 2017 A new cohesive crack tip symplectic analytical singular element involving plastic zone length for fatigue crack growth prediction under variable amplitude cyclic loading *Eur. J. Mech. A/Solids* **65** 79-90

[44] Belytschko T, Liu W K, Moran B and Elkhodary K I 2014 *Nonlinear Finite Elements for Continua and Structures 2nd ed* (Chichester: Wiley)

[45] Solanki K, Daniewicz S R and Newman jr J C 2004 Finite element analysis of plasticity-induced fatigue crack closure: an overview *Eng. Fract. Mech.* **71** 149-71

[46] Nguyen O, Repetto E A, Ortiz M and Radovitzky R A 2001 A cohesive model of fatigue crack growth *Int. J. Fract.* **110** 351-69

[47] Cook R D, Malkus D S, Plesha M E and Witt R J 2002 *Concepts and Applications of Finite Element Analysis 4th ed* (Hoboken: Wiley)

[48] Knight jr N F, Ransom J B, Griffin jr O H and Thompson D M 1991 Global/local methods research using a common structural analysis framework *Finite Elements Anal. Des.* **9** 91-112

[49] Schöllmann M, Fulland M and Richard H A 2003 Development of a new software for adaptive crack growth simulations in 3D structures *Eng. Fract. Mech.* **70** 249-68

[50] Branco R, Antunes F V and Costa J D 2015 A review on 3D-FE adaptive remeshing techniques for crack growth modelling *Eng. Fract. Mech.* **141** 170-95

[51] Moës N, Dolbow J and Belytschko T 1999 A finite element method for crack growth without remeshing *Int. J. Numer. Methods Eng.* **46** 131-50

[52] Dong L and Atluri S N 2013 Fracture & fatigue analyses: SGBEM-FEM or XFEM? Part 1: 2D structures *Comput. Model. Eng. Sci.* **90** 91-146

[53] Vigneron L M, Duflot M P, Robe P A, Warfield S K and Verly J G 2009 2D XFEM-based modeling of retraction and successive resections for preoperative image update *Comput. Aided Surg.* **14** 1-20

[54] Ren X and Guan X 2017 Three dimensional crack propagation through mesh-based explicit representation for arbitrarily shaped cracks using the extended finite element method *Eng. Fract. Mech.* **177** 218-38

[55] Yazid A, Abdelkader N and Abdelmadjid H 2009 A state-of-the-art review of the X-FEM for computational fracture mechanics *Appl. Math. Model.* **33** 4269-82

[56] Stolarska M, Chopp D L, Moës N and Belytschko T 2001 Modelling crack growth by level sets in the extended finite element method *Int. J. Numer. Methods Eng.* **51** 943-60

[57] Sukumar N, Chopp D L and Moran B 2003 Extended finite element method and fast marching method for three-dimensional fatigue crack propagation *Eng. Fract. Mech.* **70** 29-48

[58] Elguedj T, Gravouil A and Combescure A 2006 Appropriate extended functions for X-FEM simulation of plastic fracture mechanics *Comput. Methods Appl. Mech. Eng.* **195** 501-15

[59] Bergara A, Dorado J I, Martin-Meizoso A and Martínez-Esnaola J M 2017 Fatigue crack propagation in complex stress fields: Experiments and numerical simulations using the extended finite element method (XFEM) *Int. J. Fatigue* **103** 112-21

[60] Nasri K and Zenasni M 2017 Fatigue crack growth simulation in coated materials using X-FEM *Comptes Rendus Mecanique* **345** 271-80

[61] Dong L and Atluri S N 2013 Fracture & fatigue analyses: SGBEM-FEM or XFEM? Part 2: 3D solids *Comput. Model. Eng. Sci.* **90** 379-413

Experimental and numerical study of Bondura® 6.6 PIN joints

I Berkani[1], Ø Karlsen[2]*, H G Lemu[1]

[1]Faculty of Science and Technology, University of Stavanger, Stavanger, Norway
[2]Bondura® Technology, Bryne, Norway

*Contact Author: oyvind@bondura.no

Abstract. Pin joints are widely used in heavy-duty machinery such as aircrafts, cranes and offshore drilling equipment to transfer multi-dimensional shear forces. Their strength and service life depend on the clamping force in the contact region that is provided by interference fits. Though the interference fits provide full contact at the pin-hole interface under pretension loads, the contact interface reduces when the pin is subjected to an external load and hence a smaller contact surface leads to dramatic increase of the contact stress. The PIN joint of Bondura® Technology, investigated in this study, is an innovative solution intended to reduce the slack at the contact surface of the pin joint of heavy-duty machinery by using tapered sleeves on each end of the PIN. The study is aimed to better understand the contact pressure build-up and stress distribution in the supporting contact surface under pre-loading of the joint and the influence of temperature difference between part assembly and operation conditions. Numerical simulation using finite element method and diverse experimental tests were conducted. The numerical simulation and the test results, particularly the tests conducted with lubricated joints, show good conformance.

1. Introduction

Several components of heavy-duty machinery are usually assembled using interference fit that is one of the low cost joining methods and widely used in the industry [1]. Interference fit can be seen as a semi-permanent joint and, if properly designed, the assembled system can resist the relative movement of the parts. Thus, the high radial pressure created at the interface can transmit forces or torques without a relative motion at the contact interface. If improperly designed, i.e. too much slack is created, for instance, between bolts or pins and the mating surface, operation of heavy-duty machinery with slack can be a key issue with serious consequences. Wear and tear in the joints can increase maintenance costs and can reduce the service life of the whole system, but the most serious consequence is loss of human life. Slack occurs because of the bolt being smaller than the mating surface due to either contact surface wear during operation or due to too small interference fit between the parts. In a conventional pin-support connection the force is applied to one single point, this will result in the deformation of the support (host surface) and continues until the support is worn out and thus has to be replaced.

The risk of having slack in a pin-joint depends on the level of the clamping force at assembly and the behaviour and size of the external load. The clamping force is mostly introduced by press or shrink fit where a pin with a larger diameter than the corresponding hole is pressed into the hole or the assembly is done by freezing the support or freezing the bolt using liquid nitrogen respectively. When the whole assembly returns to operating temperature, the pressure created at the interface allows transmission of forces or torques [2]. This might appear as an acceptable solution, but disassembly of the parts can be demanding in certain cases. For instance, it may demand that the pin can only be detached from the support by torch cutting, which has a high associated cost and does not guarantee the safety of equipment. This is where the Bondura® Technology, which introduces rather simple but impressive concept, i.e. the Bondura® PIN joint comes in. It is an innovative solution intended to reduce the slack at the contact surface of the pin joint of heavy-duty machinery such as in offshore installations, among others, drilling and hoisting/lifting machinery by using tapered sleeves on each end of the PIN. Figure 1 illustrates the components involved in one of the Bondura® PIN joints i.e. Bondura® 6.6. In this assembly, clamping force is induced by tightening bolts that push a conical sleeve along the pivot pin and creating a press load on the internal surface of the assembly.

Compared with the conventional press or shrink fit where the load is focused on a line along the mating surface, the Bondura® PIN joint offers 360° distribution of the load in the joint while preloading due to the fact that the tapered sleeves on each end of the PIN are wedged into place and eliminate any slack. When external force is acting, this load is distributed over 180° of the contact surface. The tapered sleeves of the PIN assembly can also eliminate the wear problem due to ovality. The Bondura® bolt is found in usage in a variety of areas on heavy machinery and they are also certified for usage offshore by DNV GL (Det norske veritas) and ABS PDA (American Bureau of Shipping Product Design Assessment).

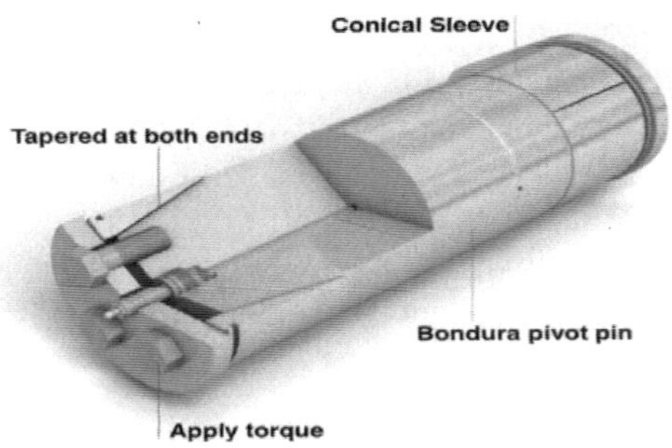

Figure 1. Illustratve picture of Bondura® 6.6 product series

Calculation of interference fit in the design process is normally based on the classical theory of Timoshenko [3] that assumes the interface as the thick-walled tube theory with internal pressure. This assumption may be valid for simple cylindrical parts and it does not allow for the calculation and simulation of the behaviour in most industrial cases [4]. In addition, the application of the traditional methods to design of interference fit is limited to 2D stress analysis under loading within the elastic region. On the other hand, recent studies are mostly based on simulations using finite element method (FEM) that facilitates study of the stress and deformation distribution at the interface [5, 6] using a virtual model. This study asserts that 3D analysis of interference fitted connections are best obtained by using FEM approach when compared with the traditional design based on the thick-wall cylinder theory. One clear advantage of the approach is that the actual geometry and working conditions can be accounted for in the model. FEM based approaches to design of interference fits are also reported using finite load incremental theory [7] for frictional contact analysis and stress pattern analysis of the shrink fit of liner and sleeve in presence of high temperature and pressure [8]. Predicting the contact pressure, which highly depends on the contact geometry, is one of the challenges in the design of interference fits. A study of the mechanics of interference fits in dental implants [9] shows that tapered fits provide better reliable connection.

Following this introduction section, the article is divided into six main sections. Section 2 presents the motivations for conducting the study. The main research parts of the work, i.e. experimental and numerical study are presented in Section 3 and 4 respectively. Then the results obtained from the two aspects of the research are compared and discussed in Section 5. Finally, the concluding remarks are given in Section 6.

IOP Conf. Series: Materials Science and Engineering **276** (2017) 012028 doi:10.1088/1757-899X/276/1/012028

2. Motivation and Problem Description

This article reports a study conducted as part of a master thesis by the first author [10]. The study was initiated by Bondura® Technology Company to better understand performance of its Bondura® 6.6 product series. Among others, the company wants to investigate the performance of the Bondura® PIN joints in terms of

1. the relation between the tightening torque and the radial stress/strain generated at the contact surface and
2. the influence of temperature variation between location of assembly and operation of the system, for instance in cases where the product is assembled in a low temperature region such as in Norway and operated in high temperature region such as the Middle East, and vise-versa.

Thus, the study focuses on developing numerical simulation models for two products of the Bondura® 6.6 PIN series, namely Bondura® 6.6 Ø88,9 and Bondura® 6.6 Ø120, conducting experimental tests on both products and compare the analysis results with experimental investigation. The study was conducted by varying tightening torque of bolts, lubrication conditions and temperature variations.

3. Experimental Study

In this study, a total of two Bondura® 6.6 PINs were tested. The test samples have diameters of 88,9 mm and 120 mm, and both samples are 201 mm long. The mechanical properties of the components of the two product series, as obtained from the material certificates and standards (EN 10083-3 & EN 10269) are given in Appendix A (Table 3) together with illustration of the components in the assembly. As stated earlier, the objective of the tests was to find the relationship between the torque generated upon tightening the screws, and the radial stress/pressure generated due to reactions in the supports. To begin with, the tests were carried out under room temperature, and later on, the ambient temperature was varied while documenting the stress effect observed at the supports.

3.1. The Experimental Methods and Materials

The two Bondura® 6.6 PINs are assembled and connected to the Test Jig on one end, as recommended in the Installation manual, i.e. the pivot pin was inserted so that the end surface is flush with outer surface of the support of the jig. Then, the components (conical sleeves, plate and screws) were assembled and the screws were tightened alternately to the specified torque on this end.

A test boss with cylindrical form, which acts as a second support, was inserted at the other end of the test PINs. The test boss has an insert with a Distance Indicator-X, as shown in Figure 2, so that Bondura® Multi Tool can be used for gentle and easy removal of conical sleeve. Figure 3 illustrates a fully assembled Bondura® 6.6 PINs that are ready for testing.

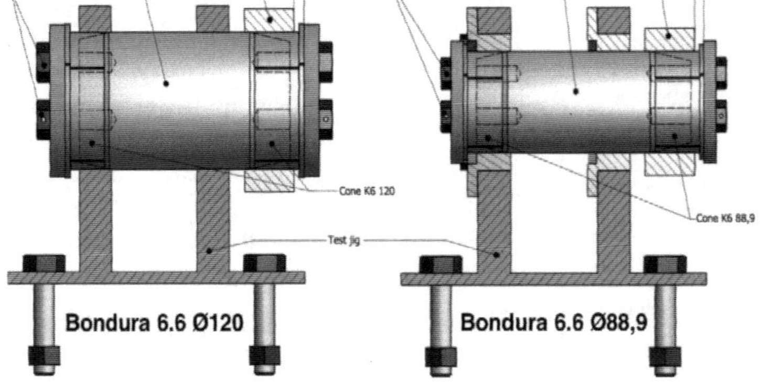

Figure 2. Test boss with a distance indicator-x

Figure 3. Illustration of the test assembly for the two test setups

To examine the loading and make mechanical stress analysis, measurements were performed with strain gauges. Four strain gauges were bonded to the test bosses with a cylindrical form, which will act as a support. Several tests at different angles were executed at room temperature. The first test boss position was fixed in such a way that the measuring points are parallel with the center of the surface of the conical sleeve. The test boss was then tilted 45° from the first position so that the measuring points become parallel to the cutting opening of the conical sleeve. The location of strain gauges around the test bosses is illustrated in Figure 4.

Before starting the test, the friction surfaces of the PIN joints and conical sleeves of both test cases were cleaned, dried and made free of any oil and grease substances. This was to make the samples thoroughly cleaned and free of any unwanted particles that might interfere in the measurement process.

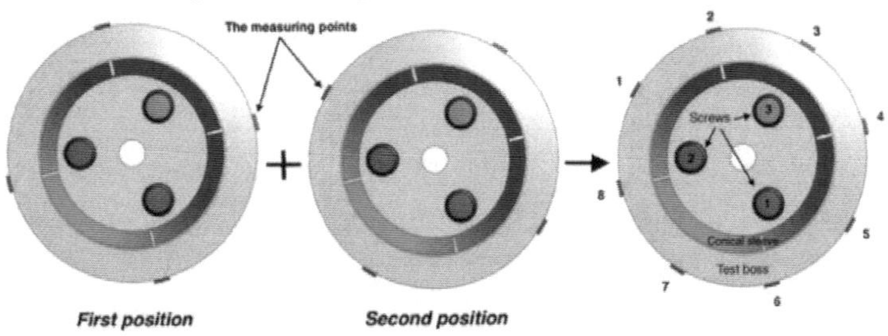

Figure 4. Illustration of the location of strain gauges

To ensure that the conical sleeves do expand correctly according to the installation manual, it was found important to tap with a hammer on the plates. This step had to be skipped during testing because the test boss is not fixed, and because of the strain gages' sensitivity, which can result in incorrect measurements.

The screws were tightened using a torque wrench. Table 1 shows the applied torque values under measurement. The measurement process was repeated alternatively until all indicated torque has been applied to the screws, and all corresponding reaction forces have been registered.

Table 1. List of measured torques

Bondura® 6.6	**Applied torque [Nm]**						
Ø88,9	120	140	160	180	200	--	--
Ø120	290	310	330	350	360	370	380

To measure at higher temperatures, the screws were first tightened to 180 Nm and 350 Nm torque for Bondura® 6.6 Ø88,9 and Bondura® 6.6 Ø120 respectively. The whole structure is then moved into the oven, and the temperature of the oven was increased stepwise to 40 °C and then to 60 °C. The reaction forces were registered at room temperature, and the change in the reaction forces due to temperature change to 40 °C and 60 °C were registered.

With the application of the transient heat conduction theory and air velocity in the oven, the time that heat uses to get inside the core of pivot pins was calculated. To avoid any measurement errors and make sure that correct measurements were taken, the structure was left one hour more in the oven.

IOP Conf. Series: Materials Science and Engineering **276** (2017) 012028 doi:10.1088/1757-899X/276/1/012028

3.2. The Experimental Results

Tangential and axial strain on the external surface of the test bosses was measured with strain gauges, while radial stress/pressure on internal surface of test bosses was calculated using Catman software based on the thick-walled cylinder theory for all positions. This was done by increasing the temperature starting from room temperature and the screw threads were lubricated.

The test was repeated at least three times in each position at room temperature and twice at increased temperature. The experimentally obtained results of radial stresses on internal surface (contact surface between test boss and conical sleeve) are illustrated graphically (Figure 5). Each of the differently coloured graphs in this figure represents one of the eight measuring points.

3.2.1 Test results of Bondura® 6.6 PINs at room temperature: For test conducted on Bondura® 6.6 Ø88.9, as the torque increases (i.e., when the screws are tightened), the compressive radial stress continues to increase. The plots in Figure 5 show the development of the radial stress at the different eight measuring points. It is worthy to point out that the pressure varies in the different measuring points, as the screws continue to be tightened alternately. There appears to be intervals where the radial stress remains stable, this is when the tightening is paused so that the value of the radial stress generated will be more apparent for each value of the applied torque. The pills along the applied torque direction indicate the transition from a torque value to another, i.e. while tightening.

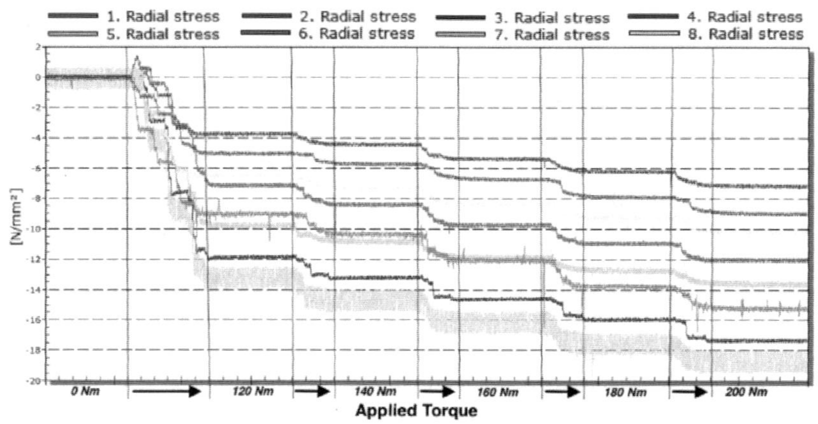

Figure 5. Variation of radial stresses with torque for Bondura® 6.6 Ø88,9

Comparably, for the Bondura® 6.6 Ø120, the compressive radial stress increases as the screws are tightened (the torque increases) as shown in Figure 6. This trend does not apply to all the points of measurement as it is the case for the Bondura® 6.6 Ø88.9. At measuring point 7, which is located between the first and the second screw, the radial stress remains positive throughout the whole tightening procedure. The positive radial stress value indicates the absence of any pressure existing in this point. Another particular observation is the case of point 3 (the blue graph) that is characterized by frequent noise throughout the tightening process. The noise is reduced as the tightening is stopped and a value becomes readable in the graph. This noise might be resulting due to the sensitivity of the strain gauges near the measurement point.

While tightening bolted connections, the tightening torque goes partly to preload the bolt and partly consumed in overcoming frictional resistance in the threaded contact between the male and female parts and the bearing/collar friction at the flat surface of the fastener or nut head.

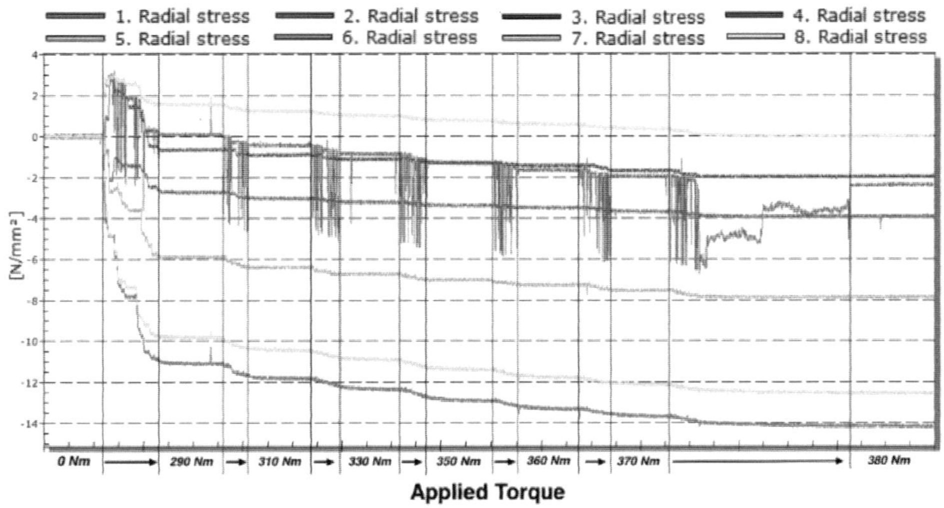

Figure 6. Variation of radial stresses with torque for Bondura® 6.6 Ø120

Many studies indicate that about 90% of the tightening torque is consumed by friction, while only 10% is used as a useful torque to preload the bolt [11]. In other words, the torque–preload relationship is highly sensitive to friction that is influenced by factors such as joint geometry [12], physical property of bolt material [13], lubrication [14], surface finish [15], etc.

Though the accuracy of the method in terms of controlling the amount of the preload of the bolt is seriously limited, the common method of tightening bolted connections is using torque wrench. A simplified torque-preload relationship for this method is given by [14, 16];

$$T = k \cdot F_0 \cdot d \tag{1}$$

where T = tightening torque by torque wrench, k = torque coefficient (where k = 0.2 is recommended for standard screw threads), F_0 = bolt preload and d = the nominal diameter of the bolt. The bolt sizes used for the two Bondura® PIN sizes, namely Bondura® 6.6 Ø88,9 and Bondura® 6.6 Ø120, are M16 and M20 respectively.

To study the influence of friction while applying tightening torque and record the changes in stresses/pressure applied to the test boss, the Bondura® 6.6 PINs were again tested after lubricating the screw threads and data on the contact pressure (Table 2) and radial stress (Figures 7 and 8) were registered.

It is observable from Table 2 that the average pressure in the contact surface, for both dry friction and lubricated threads, increases with increasing torque. Comparing the percent increase for each applied torque, higher pressures are generated when the threads are lubricated implying that friction loss is reduced and more torque is used to preload the bolt. For Bondura® 6.6 Ø88,9, the percent increase of the average pressure, in general increases with increasing torque, while the reverse is observed for Bondura® 6.6 Ø120. In this case, though the average pressure in dry friction noticeably increases with the torque, the average pressure under lubricated thread condition show insignificant change.

From the plots of the radial stresses in Bondura® 6.6 Ø88,9 and Bondura® 6.6 Ø120 given in Figures 7 and 8 respectively, the most significant observed changes are described as follows:

- Radial stress/pressure curves approach each other in the four measuring points, giving better pressure distribution.
- The order of radial stresses (from minimum to maximum values) in those four measuring points changed.
- Bondura® 6.6 Ø120 gets higher average pressure increase compared with the dry friction case.
- The radial stresses are higher for lubricated screw threads of Bondura® 6.6 Ø88.9, while they are almost stable for Bondura® 6.6 Ø120.

Table 2. Average pressure while tightening at dry friction and lubricated threads

Bolt	Applied torque [Nm]	Average pressure [MPa]		Pressure increase (%)
		p_{av_dry}	$p_{av\_lubricated}$	
Bondura® 6.6 Ø88,9	120	6.06	6.93	14.3
	140	6.96	8.19	17.7
	160	8.14	9.82	20.6
	180	9.28	10.20	9.9
	200	10.45	12.72	21.7
Bondura® 6.6 Ø120	290	4.79	9.47	97.7
	310	5.30	10.67	101.3
	330	5.70	10.93	91.8
	350	6.20	11.01	77.6
	360	6.53	11.07	69.5
	370	6.82	11.11	62.9
	380	7.27	11.15	53.4

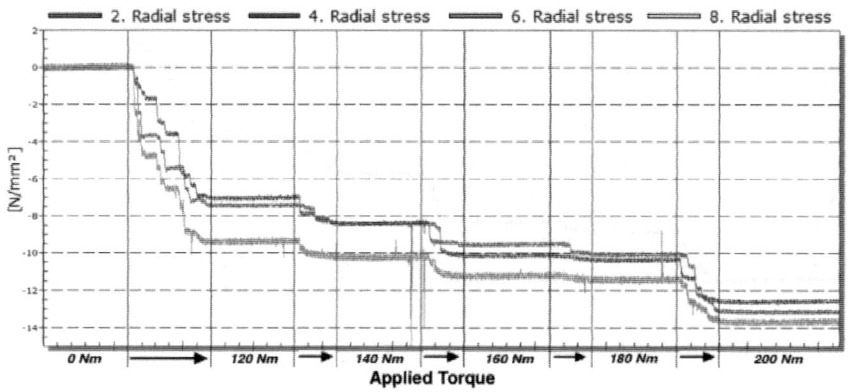

Figure 7. Plot of radial stresses for Bondura® 6.6 Ø88,9

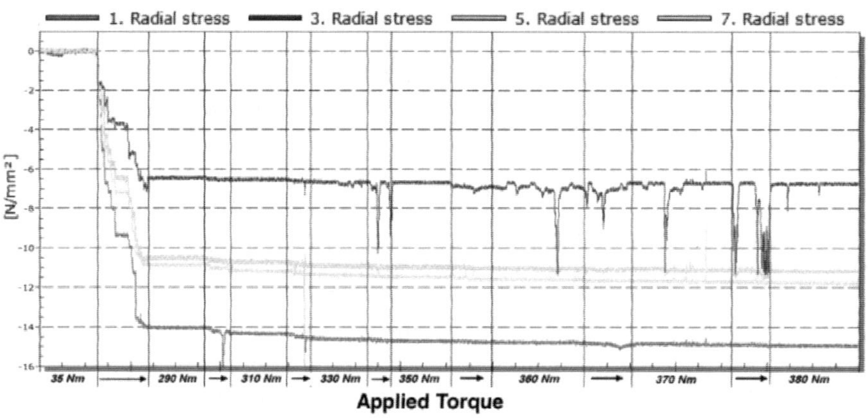

Figure 8. Plot of radial stresses for Bondura® 6.6 Ø120

3.2.2 Test results of Bondura® 6.6 PINs at higher temperature: The temperature was varied throughout the system after assembly at room temperature. With increasing test temperature, higher compressive radial stresses are observed for both pin joints, as depicted in Figure 9 for Bondura® 6.6 Ø88,9 and Figure 10 for Bondura® 6.6 Ø120. Compared with room temperature tests, the results of higher temperature tests show more stable trend with increased temperature.

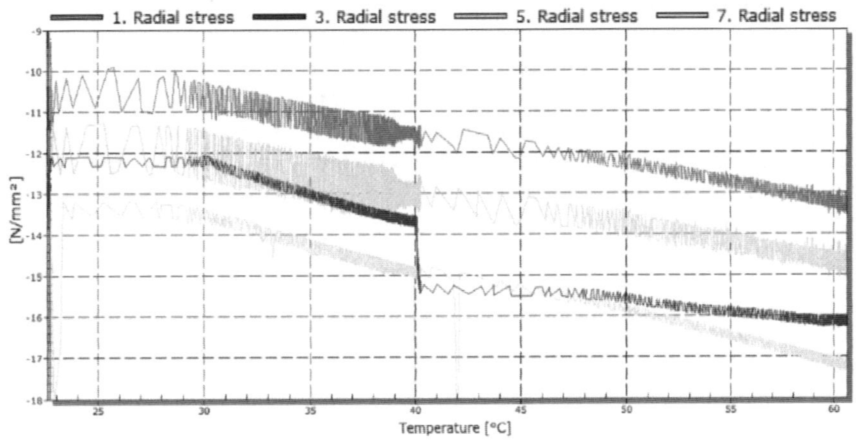

Figure 9. Plot of radial stresses for Bondura® 6.6 Ø88,9 at higher temperature

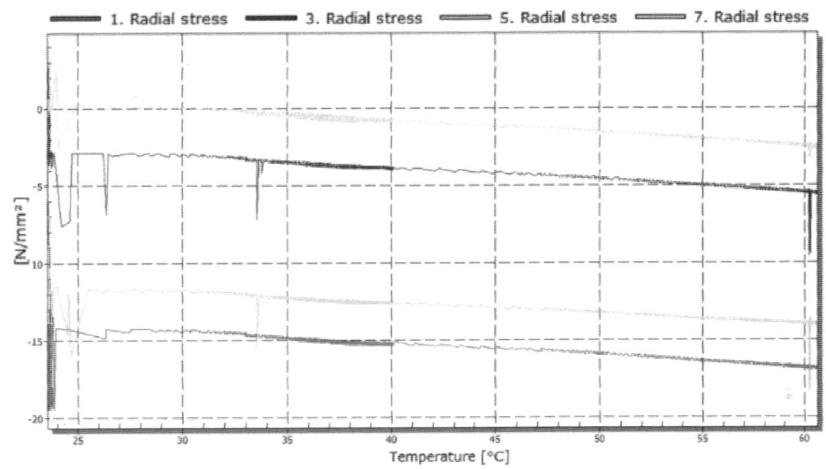

Figure 10. Plot of radial stresses for Bondura® 6.6 Ø120 at higher temperature

4. Numerical Study

Finite Element Method is, nowadays, an approach widely used for stress analysis of mechanical systems. Though the method is used to get approximate solutions, the approach leads to better understanding of the stresses and strains in the mechanical system at lower cost, and near accurate results can be obtained provided that the system is properly modelled with relevant boundary conditions. In this study of Bondura® 6.6 PIN joints, the radial stresses/pressure distribution on internal surface of test bosses was investigated using finite element analysis in ANSYS Workbench. In addition, the stress distribution, axial deformation and pressure change with respect to temperature variation are calculated for both test bosses. The goal is to obtain approximate solutions of Bondura® 6.6 test PINs and compare the results with data from the experimental, which can used as a validation of the simulation model.

4.1. Simulation methodology

The Computer-aided Design (CAD) models of Bondura® 6.6 PINs have been developed using Inventor CAD application and imported the solid geometry in to the FEM analysis tool in ANSYS Workbench. To perform the simulation, the necessary Pre-processing steps including adding material properties to Engineering Data, creating a Coordinate System, defining contact surfaces and setting up loads and supports have been done.

The Bolt Pretension load option available in ANSYS Workbench was employed to allow pretension loads produced by the installation torque of tightening the bolts (from Table 1) to be easily analyzed. Preload was applied gradually on the screws.

For the temperature fluctuation analysis, Thermal Condition was added to the simulation, where the temperature is defined at 22 °C under applied preload, and the thermal condition is ramped up step-by-step to 60 °C for high temperature analysis, then ramped down through the same steps to -40 °C for low temperature analysis.

4.2. Simulation results

To explore the effect of Bondura® 6.6 PINs pressure on the support, the radial stress in the test boss's contact surface against the conical sleeve was computed. In the following section, radial stress affecting contact surface of test bosses from conical sleeve of pins are presented graphically.

4.2.1 Simulation of Bondura® 6.6 PINs at room temperature: The simulation results of the computed radial stresses for the two PINs, Bondura® 6.6 Ø88,9 and Bondura® 6.6 Ø120 are plotted in Figures 11 and 12 respectively. In those figures, the minimum and maximum radial stresses that are calculated for both test bosses with every tightening step are shown. The results show that radial stress distribution is different almost around 360° contact surface of the test boss.

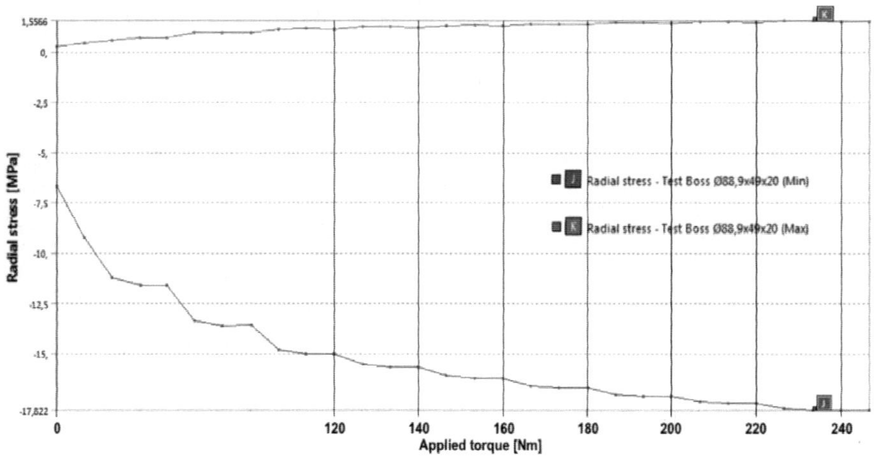

Figure 11. Computed radial stresses (max. & min) for Bondura® 6.6 Ø88,9

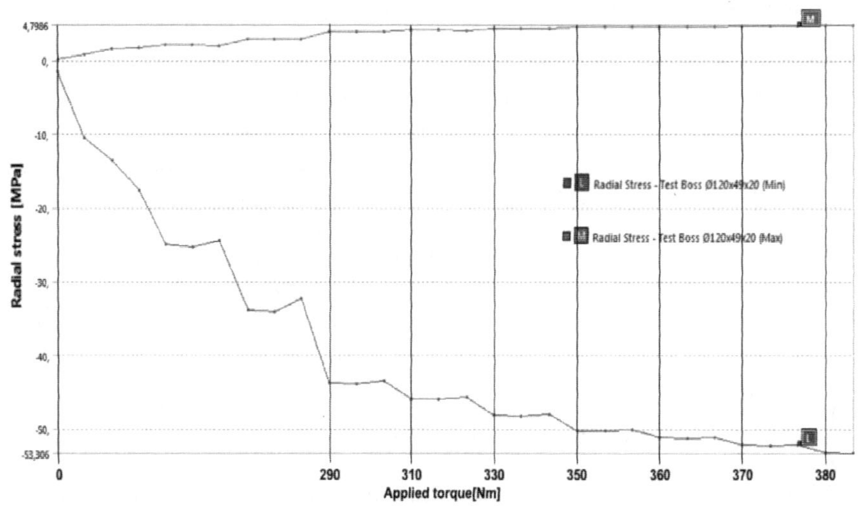

Figure 12. Computed radial stresses (max. & min) for Bondura® 6.6 Ø120

In other words, the stress distribution on the contact surface of the test bosses is not uniform as can be observed from the stress patterns given in Figure 13. The maximum radial stress experienced in the area close to the screw are all within the frustum area for Bondura® 6.6 Ø88,9, and it also centers the test boss's contact surface to conical sleeve as seen Figure 13(a), (b) and (c). This distribution of radial stress/pressure is more optimal, as it is more circularly symmetrical. The highest stress in the area close to the first screw for test boss of Bondura® 6.6 Ø120 is within frustum area, as seen in Figure 13(a'), while the second tight screw gets less stress as shown in Figure 13(b'). In the cross section of the third tight screw, a minimum stress is in the frustum area. Furthermore, the numerical analysis shows that the highest stress is experienced near the edge of the conical sleeve for the third screw, Figure 13(c´). This is due to the test boss deformation resulting from alternate screws tightening.

Deformation of test bosses in axial direction (z-direction) is calculated during simulation and the results are shown in Figures 14 & 15 for Bondura® 6.6 Ø88,9 and Bondura® 6.6 Ø120 respectively. The test boss of Bondura® 6.6 Ø88,9 (Figure 14) experiences little differences between maximum and minimum deformation, throughout the whole preload procedure. On the contrary, the test boss for Bondura® 6.6 Ø120 (Figure 15) experiences major difference between maximum and minimum deformation under preload at the same step. The most significant deformation happens in boss area that is close to the third screw.

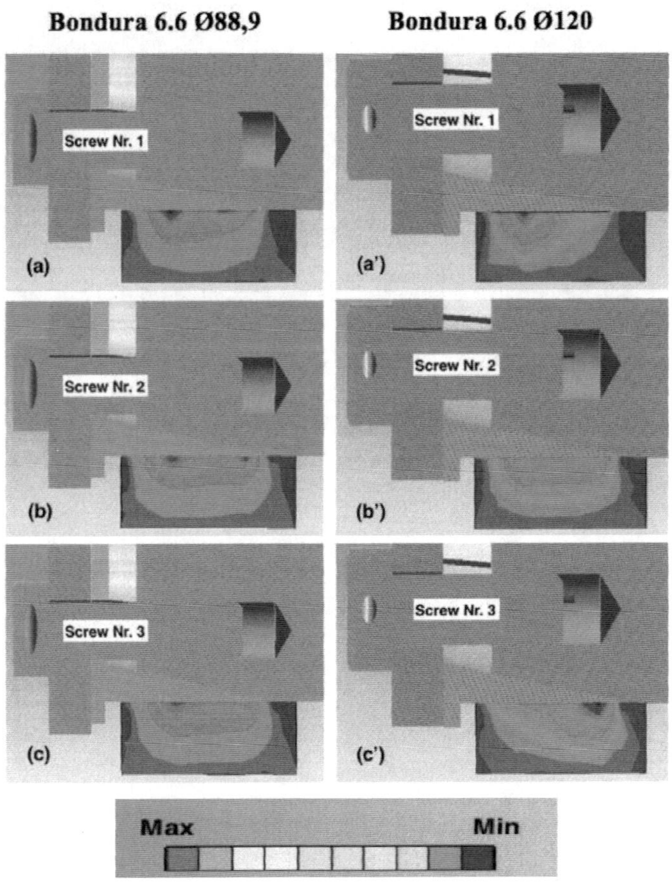

Figure 13. Simulated stress patterns of Bondura 6.6.Ø88,9 (Left) and Bondura 6.6 Ø120 (Right)

IOP Conf. Series: Materials Science and Engineering **276** (2017) 012028 doi:10.1088/1757-899X/276/1/012028

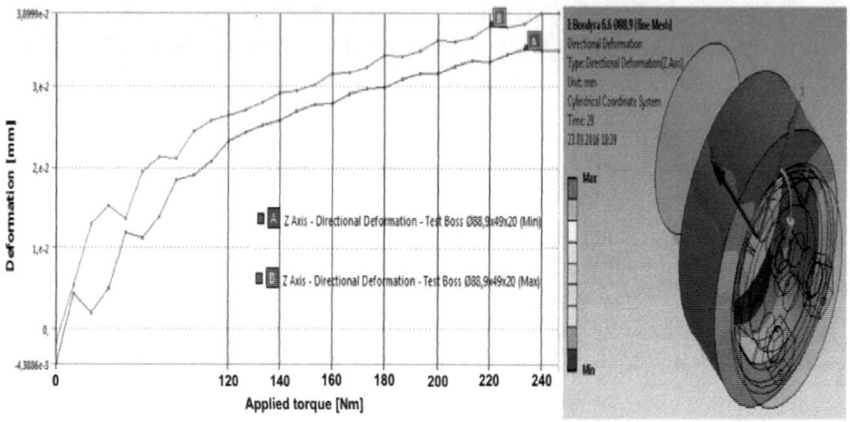

Figure 14. Axial deformation of Bondura® 6.6 Ø88,9 as afunction of applied torque

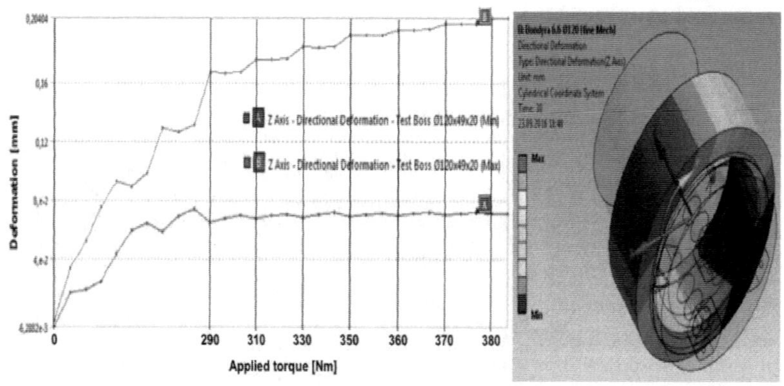

Figure 15. Axial deformation of Bondura® 6.6 Ø120 as afunction of applied torque

Simulation of Bondura® 6.6 PINs at temperature fluctuation: Upon variation of the temperature, the minimum radial stress distribution was studied using ANSYS and the result for both test bosses is shown in Figure 16. In this case, there appears to be very little change in radial stress for both Bondura® 6.6 PINs.

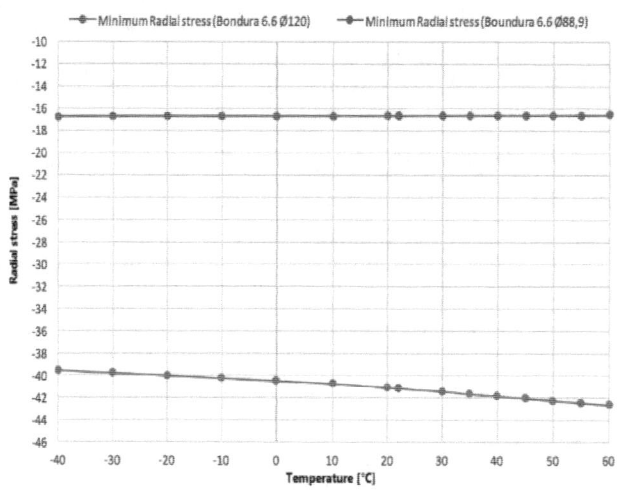

Figure 16. Minimum radial stress of Bondura® 6.6 PINs as a function of temperature

IOP Conf. Series: Materials Science and Engineering **276** (2017) 012028 doi:10.1088/1757-899X/276/1/012028

5. Discussion and Comparison of Results

The distinct feature of the Bondura® PINs is its ability to distribute the pressure 360° in its envelope. The experimental work has shown that this can be achieved in the case of Bondura® 6.6 Ø88,9 at room temperature. On the other hand, the experiment was not able to confirm the same results for Bondura® 6.6 Ø120. There appears to be an absence of pressure in a certain measurement point inside the test boss, which might discredit the initial claim. These results, however, are not definitive and might have occurred due to several factors, including the following:

- *Not hammering of the assembly*: It is instructed that the plate structure connected to the conical sleeve is to be hammered during mounting. This is to ensure that the latter expands correctly. A decision to neglect this step was made since the hammering would have resulted in considerable noise readings from the strain gauges.

Unfixed test boss: This means the test boss moves with conical sleeve when torque is applied. This is not the case under industry applications where the surrounding structure is fixed.

The analysis performed at room temperature mainly indicates an uneven distribution of pressure over the 360° contact surface between the conical sleeve and the test boss for both Bondura® 6.6 PINs. When the simulation plots were further examined, potential factors causing this were identified. these are size of the screw used and its resulting frustum area, the positioning of these screws relative to the test boss and the deformation of the test boss occurring when tightening.

Theoretically, the clamping force increased to a value surpassing 10% when the screws are lubricated and, in turn, the average pressure measures increased. Figure 17 shows the average pressure of Bondura® 6.6 PIN joints with and without-lubricated screw threads. The experimental test results show that the average pressure increases by approx. 17% for Bondura® 6.6 Ø88,9 while the average pressure of Bondura® 6.6 Ø120 gets an increase about 80%. Another noticeable thing is that the pressure distribution is more even when mounting the structure after lubrication.

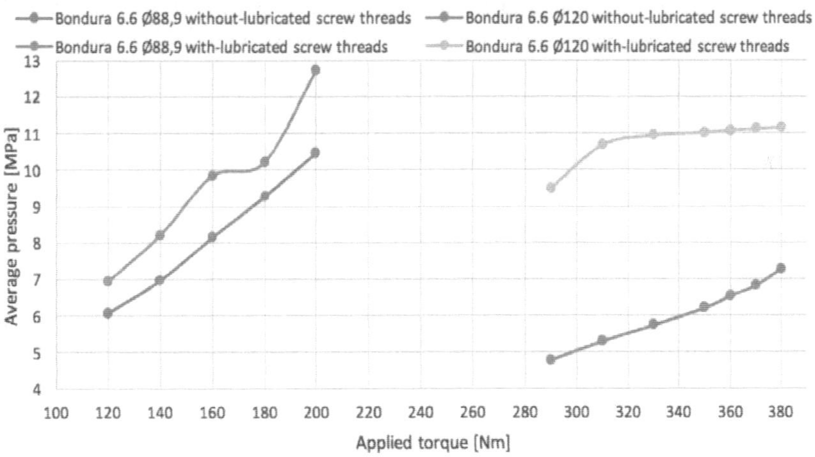

Figure 17. The average pressure of Bondura® 6.6 PINs with & without-lubricated screw threads

Under experimental testing, measurements are taken at only four points positioned on the external surface of the test boss. Using theoretical calculations, pressure on the inside surface was then calculated. Four points measurement may not be sufficient to get a clear idea of the behavior of the pins. This is because a very limited area of the surface of the test boss is covered by these points. The numerical analysis on the other hand, provides a lot more data. The entire contact surface to the test boss towards Bondura® 6.6 PINs can be checked, and even stress distribution into the material, deformation and status of the contact surfaces could be animated and plotted.

When the results from the experimental testing and numerical analysis are compared, it is observed that the pressure of the experimental testing is within the maximum and minimum pressure interval produced by the numerical analysis. The modelled pressure values in the mid contact surface of test boss to Bondura® 6.6 Ø88,9 are within close range of the values obtained through the experimental work, while the test boss of Bondura® 6.6 Ø120 had an average pressure that was higher in the numerical analysis than the ones measured under the experimental testing. It is possible to state that stress distribution is similar for both the experimental testing and the numerical analysis. There appears to be more pressure close to the first tight screw and less close to the last, this applies mainly to the area where the measurement points were set under the experiment.

Since the test bosses get different pressure in each of the measuring points, shear stress is generated. Moreover, the numerical analysis shows that the test bosses get some axial stress in the contact surfaces due to friction of deformation of the test bosses. The existence of shear stress and axial stress that was disregarded in the equation used for the calculation of the experimental results might be the error source in this case, causing a deviation in the results between the experimental testing and numerical analysis.

Another issue that should be pointed out regarding the numerical analysis is that the design requirements for FEM model with the tightening option "bolt pretension" are often obtained by a simple clamping force (axial load). During experimental work, the applied torque generates not only the axial loading but also the collar friction and thread friction. With lubricated thread, axial load is increased and hence a more "pure" force transition is achieved. This is why the pressure with lubricated threads during experimental testing is very similar to pressures in simulation.

In the case of temperature fluctuation, it is expected that the pressure is going to be constant for Bondura® 6.6 Ø88,9, given the fact that the thermal expansion coefficient is equal in all its components. However, the experimental results depict a drop in radial stress. This contrast might be due to the measurement being taken on the outside, and the software calculating the desired pressure inside. These assumptions might not be very accurate in determining the changes occurring along the thickness separating the placement of the strain gauges, and the desired measurement point inside when temperature is involved.

For Bondura® 6.6 Ø120, there appears to be a pressure variation in the form of a rising trend as it is depicted. This is in accordance with the initial expectation given that the thermal expansion coefficient varies from the pivot pin to the rest of the components. Since the coefficient for the pivot pin is larger than the surrounding components, it is deemed to expand more drastically compared to the components enveloping it.

Pressure is held constant with temperature variation in the numerical analysis of Bondura® 6.6 Ø88,9, this is expected due to the thermal expansion coefficient in all parts of the PIN all being too similar. In the experimental testing, pressure has increased about 9% with temperature increase. Regarding Bondura® 6.6 Ø120, pressure in both the numerical and experimental analysis is observed to be increasing with temperature increase. The average increase in pressure is about 13% during the experimental study and approximately 1.8% in the numerical analysis. Differences in the results for Bondura® 6.6 Ø88,9 may be due to the thermal expansion coefficient being the same in all parts. Since the strain gauges are on the external surface of test bosses, the material expansion might also be due to temperature increase, and not pressure increase on the internal surface. In this situation, calculations of pressure using the measured values during testing can possibly be a source of error.

6. Conclusion

The study conducted on two product variants of Bondura® technology (as a master thesis project) at University of Stavanger is partly presented in this article. Both experimental investigations and numerical simulation & analysis approaches were implemented. Both studies concluded that a 360° pressure distribution occurs throughout the cylindrical support. The pressure distribution in both pivot joints is documented to be circular asymmetrical, the highest pressure being registered in the area closer to the first tightened screw.

From the two product variants, extremely uneven pressure distribution is registered for Bondura® 6.6 Ø120 under the experiment. The numerical analysis showed similar results for the same area, but it has also disclosed that although the pressure was minor in the measured point, most of the pressure occurs at the edge of the conical sleeve.

Lubricating the screw threads has resulted in an improved pressure distribution for both Bondura® 6.6 assemblies. Consequently, the average pressure increases relative to pressure registered for non-lubricated screw threads indicating the fact that the frictional loss is reduced and part of the torque that could have lost is converted into preloading of the bolts. In general, the results from the experimental investigation and numerical analysis show similar trends, particularly for Bondura® 6.6 Ø88,9, though non-conclusive results are observed for the experimental work with temperature variation.

This study has revealed a number of potential future research areas on the product series. The test boss, in this study, was not fixed against axial movement, and hence constraining its motion while preloading the screws and using several tangential strain gauges can provide better control of the stress/pressure distribution under the experiment. Other possible further works include; low temperature test with proper instrumentation, post installation investigation of the pivot joints and simulation of the numerical model with high performance FEA tools.

Acknowledgement
The authors would like to acknowledge the support of Yaaseen A. Amith in setting up the laboratory tests in the Metrology and Solid Mechanics Lab of the Department of Mechanical & Structural Engineering and Materials Science, University of Stavanger.

References

[1] Truman C E and Booker J D 2007 Analysis of a shrink-fit failure on a gear hub/shaft assembly *Eng. Fail. Anal.* **14** pp. 557–72

[2] Wick C and Veilleux R F 1987 *Quality control and assembly – tool and manufacturing engineers handbook vol. 4.* (MI, Dearborn: Society of Manufacturing Engineers)

[3] Timoshenko S P 1956 *Strength of materials Part II: advanced theory and problems. 3rd ed.* (Krieger Pub. Co) pp. 205-13

[4] Boutoutaou H, Bouaziz M and Fontaine J F 2013 Modelling of interference fits with takingi nto account surfaces roughness with homogenization technique *Int. J. Mech. Sci.* **69** pp. 21–31

[5] Zhang Y, McClain B and Fang X D 2000 Design of interference fits via finite element method *Int. J. Mech. Sci.* **42** pp.1835–50

[6] Sen S and Aksakal B 2004 Stress analysis of interference fitted shaft–hub system under transient heat transfer conditions *Mater. Des.* **25**(5) pp. 407-17

[7] Okamoto N and Nakazawa N 1979 Finite element incremental contact analysis *Int. J. Numer. Methods Engi.* **14** pp. 337 - 57

[8] Oh J H, Lee S J and Lee D G 1997 Manufacturing of an alumina linear-steel sleeve under high pressure and temperature *Int. J. Mach. Tools Manuf.* **37** pp. 1511 - 23

[9] Bozkaya D and Müftü S 2003 Mechanics of the tapered interference fit in dental implants *J. Biomech.* **36** pp. 1649–58

[10] Berkani I 2016 *Contact stress and temperature variation analysis of Bondura bolts: Numerical and experimental study* (Master Thesis: University of Stavanger)

[11] Zhu L, Hong J and Jianga X 2016 On controlling preload and estimating anti-loosening performance in threaded fasteners based on accurate contact modeling *Tribol. Int.* **95** pp. 181–91

[12] Bickford J H 2007 *Introduction to the design and behavior of bolted joints: non- gasketed joints* (NewYork: Marcel Dekker)

[13] Nassar S A, El-Khiamy H, Barber G C, Zou Q and Sun T S 2005 An experimental study of bearing and thread friction in fasteners *Trans ASME J. Tribol.* **127** pp. 263–72

[14] Croccolo D, De Agostinis M and Vincenzi N 2011 Failure analysis of bolted joints effect of friction coefficients in torque-preloading relationship *Eng. Fail. Anal.* **18** pp. 364–73

[15] Nassar S A and Sun T S 2007 Surface roughness effect on the torque–tension relationship in threaded fasteners *Proc. IMechE Part J: Eng. Tribol.* **221** pp. 95–103

[16] Juvinall R C and Marshek K M 2006 *Fundamental of machine component design. 4th ed.* (Asia: Wiley & Sons Pt. Ltd.)

APPENDIX A

Table 3 contains mechanical properties of materials of parts the Bondura® 6.6 PIN joint assembly reported in this article. These material data are obtained from material certificates and Standards EN 10083-3 & EN 10269. The components of the assembly are also illustrated in Figure 18.

Table 3. Mechanical properties of components of Bondura® 6.6 product series

	PIN Ø88.9	PIN Ø120	Accessories *	Screws
Material type	34CrNiMo6 (SS2541)	42CrMo4V (SIS2244)	S355J2+N	8.8
Young's modul. [GPa]	190-210	190-210	200	210
Poisson's ratio	0.27-0.30	0.27-0.30	0.303	0.3
Yield strength [MPa]	962	784	403	640
Tensile strength [MPa]	1074	914	547	800
Thermal expan. [$10^{-6}.K^{-1}$]	11.1	12.1	11.1	10.8

*ACCESSORIES: BOSS, CONE SLEEVE AND PLATE

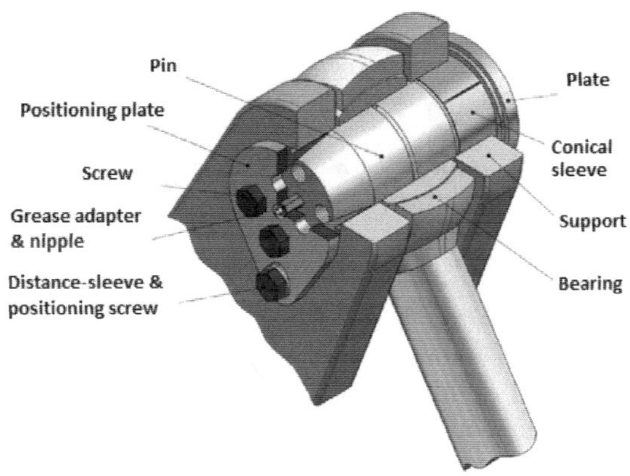

Figure 18. Illustration of components of Bondura® 6.6 assembly

IOP Conf. Series: Materials Science and Engineering **276** (2017) 012029 doi:10.1088/1757-899X/276/1/012029

Recursive thoughts on the simulation of the flexible multibody dynamics of slender offshore structures

J Schilder*, M Ellenbroek, and A de Boer
Faculty of Engineering Technology, University of Twente
P.O. Box 217, 7500AE Enschede, The Netherlands

*Corresponding author: j.p.schilder@utwente.nl

Abstract. In this work, the floating frame of reference formulation is used to create a flexible multibody model of slender offshore structures such as pipelines and risers. It is shown that due to the chain-like topology of the considered structures, the equation of motion can be expressed in terms of absolute interface coordinates. In the presented form, kinematic constraint equations are satisfied explicitly and the Lagrange multipliers are eliminated from the equations. Hence, the structures can be conveniently coupled to finite element or multibody models of for example seabed and vessel. The chain-like topology enables the efficient use of recursive solution procedures for both transient dynamic analysis and equilibrium analysis. For this, the transfer matrix method is used. In order to improve the convergence of the equilibrium analysis, the analytical solution of an ideal catenary is used as an initial configuration, reducing the number of required iterations.

1. Introduction

In offshore engineering, many long and slender structures are encountered, such as pipelines and risers. These structures are subject of study in a wide variety of applications such as design optimization, operability studies, real time control and so on. The fact that such studies require many design evaluations, long transient analyses or iteration times that are faster than real time, emphasizes the importance of efficient and accurate structural models.

The most important contribution of this paper is in providing an overview of how various methods, that individually are all well-developed and well-documented, can be combined to form a powerful solution strategy for the dynamic simulation of slender offshore structures specifically. These various methods will be briefly explained here, together with their suitability for offshore applications.

For the modelling of slender offshore structures, different approaches can be pursued. In particular for the purpose of pipe-laying, different methods are explained in textbooks such as [1]. For the equilibrium analysis of pipelines, analytical methods based on catenary theory are well-documented. The ideal catenary solution can be found in standard textbooks on statics such as [2] and can be expanded to an elastic catenary in which the effects of bending stiffness are taken into account [3]. It is still common to compare more elaborate models that do not possess an analytical solution, with catenary-like solutions, as for example shown in [4].

In order to study the dynamic behavior of more complicated structural geometries and topologies, and the effects of additional phenomena such as contact mechanics, plasticity and hydro-elastic coupling, a broad range of numerical models is developed. In rigid multibody dynamics (sometimes referred to as the rigid finite element method) the structure is modeled as a series of rigid bodies connected by ideal elastic elements in the form of (torsion) springs. The details of this formulation are explained excellently in the standard work by Haug [5] and an application of this method in offshore engineering is for instance found in [6].

Departing from linear finite element formulations, co-rotational finite element formulations have been developed to cope with the geometric nonlinearities that result from the large displacements [7]. A benefit of this method is the strong relation with linear finite element formulation. However, in particular for three-dimensional problems and systems that move at high velocities, many elements may be required in order to produce accurate results. This is due to the fact that the effects of quadratic velocity terms and elastic deformation are not taken into account in the inertia forces.

Flexible multibody dynamics is a natural extension of rigid multibody dynamics. Within the framework of flexible multibody dynamics, the floating frame of reference formulation is used in this work, which is described clearly in the standard work by Shabana [8]. The key concept is that for each body the rigid and flexible motion are separated and described by distinct generalized coordinates: a floating frame attached to each flexible body describes the body's rigid motion with respect to an inertial coordinate frame, whereas a linear combination of elastic modes describes the local deformations with respect to the floating frame. These elastic modes can be obtained directly from a linear finite element model of the body, which allows the possibility of well-developed model order reduction techniques.

The decision to use the floating frame of reference formulation in this work is a fundamental one, as it emphasizes the natural possibility of including a pipeline or riser model in a multibody dynamics model of the vessel and its on-board equipment and a possible seabed model. How to efficiently include hydrodynamic forces into this flexible multibody model using potential theory is described in [9].

The efficiency of recursive solution procedures in systems that have a chain-like or tree-like topology lies in the fact that the number of operations to solve the equations of motion at a certain iteration is only of $O(n)$, with n the size of the system matrix. The majority of the recursive methods is based on the successive elimination of bodies from the chain by condensation. An example of this method for the purpose of dynamic simulations of space structures is found in [10]. Alternatively, the transfer matrix method can be used as a recursive method. The fundamentals of this method for linear elastic problems is well explained in [11]. How to extent this method for geometrically nonlinear problems is described in [12], in which also a comparison has been made with the condensation method. It is found that the transfer matrix method seems to be advantageous for closed-loop systems, as it does not require the use of Lagrange multipliers for loop-closure constrains.

The outline of this paper is as follows: In Sections 2 and 3, the details of the floating frame of reference formulations that are necessary for our present purposes are given. This concerns the kinematics of an elastic body and the development of the constrained equations of motion of a flexible multibody system. In Section 4, it is described how for the particular case of a chain-like structure that contains many bodies with two interface points only, the equations of motion can be rewritten in terms of the absolute interface coordinates. In this way, the Lagrange multipliers can be eliminated from the equations of motion and a formulation in terms of minimal coordinates is obtained. In Sections 5 and 6, it is shown how these equations of motion can be solved recursively using the transfer matrix method for transient dynamic analysis and equilibrium analysis, respectively. To this end, the equation of motion of a single flexible body is rewritten to transfer matrix form after which it is explained how to couple it with neighboring bodies. Section 7, summarizes the basics of catenary theory and explains how the analytical solution of the ideal catenary can be used in the initialization of equilibrium analysis. A numerical example is developed for the demonstration of the equilibrium analysis. The paper is closed with the conclusions in section 8.

2. Kinematics of an elastic body

Consider a material point A located on a flexible body. A coordinate frame is rigidly attached to this material point. The combination of the location of A and the orientation of the coordinate frame located in A defines a coordinate system. In this work, this will be referred to as the position of A. The position of A can be expressed relative to a coordinate frame in another point B by the (3×1) position vector $\mathbf{r}_A^{B,B}$ and the (3×3) rotation matrix $\mathbf{R}_A^B$. The position vector $\mathbf{r}_A^{B,B}$ defines the location of A (lower index) with respect to the location of B (second upper index) and its components are expressed in the coordinate system of B (first upper index). The rotation matrix $\mathbf{R}_A^B$ defines the orientation of A (lower index) with respect to the orientation of B. The rotation matrix is an orthogonal matrix of the proper type, i.e. its determinant is equal to 1 and its transpose equals its inverse:

$$[\mathbf{R}_A^B]^T = [\mathbf{R}_A^B]^{-1} = \mathbf{R}_B^A. \tag{1}$$

From the orthogonality property, it follows that the time derivative of the rotation matrix can be expressed as:

$$\dot{\mathbf{R}}_A^B = \tilde{\boldsymbol{\omega}}_A^{B,B} \mathbf{R}_A^B, \tag{2}$$

where the overdot operator $(\dot{\ })$ denotes partial differentiation with respect to time. $\tilde{\boldsymbol{\omega}}_A^{B,B}$ is the skew symmetric matrix constructed from the angular velocity vector $\boldsymbol{\omega}_A^{B,B}$. Here, the tilde operator $(\tilde{\ })$ is introduced such that when applied on a (3×1) vector $\mathbf{a}$, it yields a skew symmetric matrix $\tilde{\mathbf{a}}$:

$$\tilde{\mathbf{a}} = \begin{bmatrix} 0 & -a_3 & a_2 \\ a_3 & 0 & -a_1 \\ -a_2 & a_1 & 0 \end{bmatrix}. \tag{3}$$

Similarly as above, the virtual change or variation in the rotation matrix can be expressed as:

$$\delta \mathbf{R}_A^B = \delta \tilde{\boldsymbol{\pi}}_A^{B,B} \mathbf{R}_A^B, \tag{4}$$

where delta operator $\delta(\ \ )$ denotes the variation of its argument. $\delta \boldsymbol{\pi}_A^{B,B}$ is referred to as the virtual rotation of A with respect to B with its components expressed in frame B.

In the floating frame of reference formulation, the absolute position of any material point A on a flexible body with respect to a fixed inertial reference frame O is decomposed in the absolute position of the body's floating frame, denoted here by P, with respect to the inertial frame O and the relative position of A with respect to the floating frame P. This decomposition is shown graphically in Figure 1.

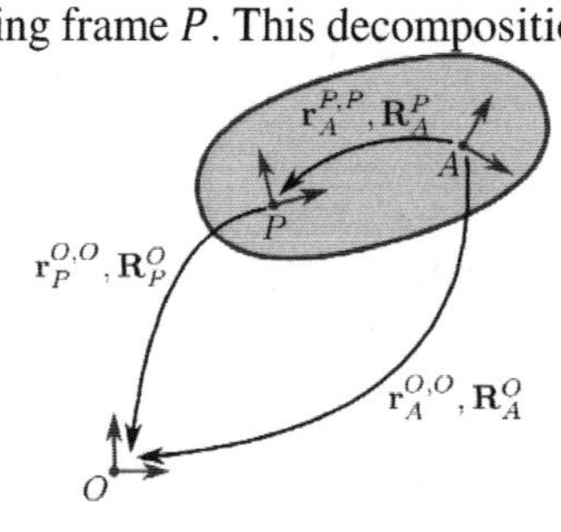

Figure 1. Decomposition of the absolute position of a material point on a flexible body using the floating frame of reference formulation.

With the use of the floating frame, the location of A with respect to O is decomposed as:

$$\mathbf{r}_A^{O,O} = \mathbf{r}_P^{O,O} + \mathbf{R}_P^O \mathbf{r}_A^{P,P}. \tag{5}$$

The orientation of A with respect to O can be decomposed as:

$$\mathbf{R}_A^O = \mathbf{R}_P^O \mathbf{R}_A^P. \tag{6}$$

The body's flexible behavior is described locally, with respect to the floating frame. Hence, the position vector $\mathbf{r}_A^{P,P}$ can be expressed as the sum of the position vector of A relative to P on the undeformed body $\mathbf{x}_A^{P,P}$ and an elastic displacement due to flexibility $\mathbf{u}_A^{P,P}$:

$$\mathbf{r}_A^{P,P} = \mathbf{x}_A^{P,P} + \mathbf{u}_A^{P,P}. \tag{7}$$

In case of small elastic deformations, the rotation matrix $\mathbf{R}_A^P$ can be linearized:

$$\mathbf{R}_A^P = \mathbf{I} + \tilde{\boldsymbol{\theta}}_A^{P,P}, \tag{8}$$

in which only the terms up to the first order in the Taylor series of the rotation matrix are retained. For this, the elastic rotation $\boldsymbol{\theta}_A^{P,P}$ that depends on the gradient of the displacement field must be small.

Following the theory of linear elasticity, the displacement field $\mathbf{u}_A^{P,P}$ can be expressed as a linear combination of a finite number of modes:

$$\mathbf{u}_A^{P,P} = \boldsymbol{\Psi}(\mathbf{x}_A^{P,P})\mathbf{q}_e, \tag{9}$$

where $\boldsymbol{\Psi}$ is the matrix of mode shapes and $\mathbf{q}_e$ the vector of generalized elastic coordinates.

The floating frame of reference formulation allows in principle for any combination of modes. For bodies with a complex geometry, these modes could be for instance the body's natural modes as obtained using finite element software. Modal order reduction techniques, such as the Guyan-Iron method, Craig-Bampton method, etc. can be applied to reduce the number of modes. For bodies with a simple beam-like geometry, standard cubic Hermitian interpolation functions can be used, similarly to the finite element method. In fact, it can be shown that in this case the floating frame of reference formulation for a body is equivalent to the co-rotational formulation for a finite element [13].

In order to obtain an expression for the linear velocity of A with respect to O, Equation (5) is differentiated with respect to time. Using the appropriate mathematical properties of the rotation matrix and skew symmetric matrix, this can be rewritten to:

$$\dot{\mathbf{r}}_A^{O,O} = \dot{\mathbf{r}}_P^{O,O} - \mathbf{R}_P^O\tilde{\mathbf{r}}_A^{P,P}\mathbf{R}_O^P\boldsymbol{\omega}_P^{O,O} + \mathbf{R}_P^O\dot{\mathbf{r}}_A^{P,P}. \tag{10}$$

For the angular velocity of A with respect to O holds:

$$\boldsymbol{\omega}_A^{O,O} = \boldsymbol{\omega}_P^{O,O} + \mathbf{R}_P^O\boldsymbol{\omega}_A^{P,P}. \tag{11}$$

Equation (10) and (11) are combined to a compact notation:

$$\mathbf{v}_A^{O,O} = [\mathbf{R}_P^O][\tilde{\mathbf{r}}_A^{P,P}]^T[\mathbf{R}_O^P]\mathbf{v}_P^{O,O} + [\mathbf{R}_P^O]\mathbf{v}_A^{P,P}, \tag{12}$$

in which the (6×1) velocity vectors $\mathbf{v}$ contain both the linear and angular velocity vectors, $[\mathbf{R}]$ denotes a (6×6) compound rotation matrix and $[\tilde{\mathbf{r}}]^T$ is defined as:

$$[\tilde{\mathbf{r}}]^T \equiv \begin{bmatrix} \mathbf{1} & \tilde{\mathbf{r}}^T \\ \mathbf{0} & \mathbf{1} \end{bmatrix}, \tag{13}$$

where sub- and superscripts are omitted above for brevity. The expression for the virtual displacement of A relative to O can be written in a similar form as Equation (12).

3. Equations of motion in the floating frame of reference formulation

The equation of motion of a flexible body is derived using Hamilton's principle:

$$\int_{t_1}^{t_2}(\delta T - \delta V + \delta W)\, dt = 0, \tag{14}$$

where δT is the virtual kinetic energy, δV is the virtual internal elastic energy and δW is the virtual work by externally applied body forces and surface tractions. Substitution of the appropriate kinematic relations in the expressions for the virtual energies in Equation (14) and using the argument that Hamilton's principle must hold for arbitrary intervals $[t_1, t_2]$ and for all virtual displacements, the floating frame of reference equation of motion in standard form is obtained:

$$\begin{bmatrix} \mathbf{M}_r & \boldsymbol{\Gamma}^T \\ \boldsymbol{\Gamma} & \mathbf{M}_e \end{bmatrix}\begin{bmatrix} \mathbf{a}_P^{O,O} \\ \ddot{\mathbf{q}}_e \end{bmatrix} + \mathbf{Q}_v + \begin{bmatrix} \mathbf{0} & \mathbf{0} \\ \mathbf{0} & \mathbf{K}_e \end{bmatrix}\begin{bmatrix} \mathbf{0} \\ \mathbf{q}_e \end{bmatrix} = \mathbf{Q}_a, \tag{15}$$

where $\mathbf{a}_P^{O,O}$ is the absolute acceleration of the floating frame. $\mathbf{M}_r$ is the rigid body mass matrix, $\mathbf{M}_e$ the mass matrix due to elastic modes and $\boldsymbol{\Gamma}$ are the modal participation factors that couple rigid and elastic motion. $\mathbf{K}_e$ is the generalized stiffness matrix due to elastic modes. The vector $\mathbf{Q}_v$ contains the inertia forces that are quadratic in the velocity and $\mathbf{Q}_a$ contains the externally applied generalized forces and modal forces.

In a multibody system, the equations of motion of all individual bodies can be combined and written in the following compact form:

$$\mathbf{M}\ddot{\mathbf{q}} = \mathbf{Q}, \tag{16}$$

in which the quadratic velocity forces and elastic forces are contained in $\mathbf{Q}$, as well as the unknown constraint forces that result from kinematic constraints between distinct bodies or between a body and the fixed world. In order to obtain a solvable set of equations, the corresponding kinematic constraint equations need to be taken into account. Holonomic constraint equations are considered in the following form:

$$\boldsymbol{\Phi} = \mathbf{0}. \tag{17}$$

Differentiation with respect to time twice yield the acceleration equation of the constraints:

$$\boldsymbol{\Phi}_q\ddot{\mathbf{q}} = \boldsymbol{\gamma}, \tag{18}$$

where $\boldsymbol{\Phi}_q$ is the Jacobian matrix of the constraint equations. The combination of Equation (16) and (18) forms the constraint equations of motion in Lagrange multiplier form, which is of the differential-algebraic type:

$$\begin{bmatrix} \mathbf{M} & \boldsymbol{\Phi}_q^T \\ \boldsymbol{\Phi}_q & \mathbf{0} \end{bmatrix} \begin{bmatrix} \ddot{\mathbf{q}} \\ \boldsymbol{\lambda} \end{bmatrix} = \begin{bmatrix} \mathbf{Q} \\ \boldsymbol{\gamma} \end{bmatrix}, \tag{19}$$

in which the constraint forces are removed from $\mathbf{Q}$ and $\boldsymbol{\lambda}$ are the Lagrange multipliers.

4. Transformation to interface coordinates

The constrained equations of motion in Lagrange multiplier form Equation (19) are valid for any flexible multibody system subjected to holonomic constraints. In co-rotational finite element formulations, no Lagrange multipliers are present. This is due to the fact that the equations of motion are written in terms of the absolute nodal coordinates. Since constrains are applied at the nodes, they result in a direct relation between nodal coordinates or they prescribe nodal coordinates to have a certain value.

Based on co-rotational finite element formulations, it is observed that it is possible to eliminate the Lagrange multipliers from the floating frame of reference equation of motion, if it is possible to establish a coordinate transformation from the floating frame of reference coordinates $\mathbf{q}$, to the interface coordinates $\mathbf{q}_I$ corresponding to the interface points at which the kinematic constraint equations are applied.

In general, such a coordinate transformation is not straightforward. Previously, it was established by locating the floating frame in an interface point [14] or by assuming the position of floating frame to be the weighted average of the interface coordinates [15]. More generally, this coordinate transformation can be established by demanding zero elastic deformation at the location of the floating frame [13].

However, the bodies that form slender offshore structures in a typical chain-like topology, often have two interface points only: connecting the body to its two immediate neighbors. In this specific case, the desired coordinate transformation can be established easily for the case that the floating frame coincides with one of the interface points, provided that the elastic behavior is described by an appropriate set of modes. To this end, consider a flexible body with interface points i and j. Let the body's floating frame coincide with interface point i. Using Equation (12), the absolute velocity of interface point j can be expressed as:

$$\dot{\mathbf{q}}_j^{O,O} = [\mathbf{R}_i^O][\tilde{\mathbf{r}}_j^{i,i}]^T[\mathbf{R}_O^i]\dot{\mathbf{q}}_i^{O,O} + [\mathbf{R}_i^O]\mathbf{v}_j^{i,i}. \tag{20}$$

In the specific case that the elastic displacement field is described using the 6 static interface modes of a Craig-Bampton reduction, the local velocities $\mathbf{v}_j^{i,i}$ equal the time derivative of the generalized elastic coordinates $\dot{\mathbf{q}}_e$. This is an immediate consequence of the fact that a static interface mode evaluated at the corresponding interface point equals 1. For this case, it is possible to use Equation (20) to express the time derivative of the local in terms of the absolute interface velocities:

$$\dot{\mathbf{q}}_e = -[\tilde{\mathbf{r}}_j^{i,i}]^T[\mathbf{R}_O^i]\mathbf{v}_i^{O,O} + [\mathbf{R}_O^i]\mathbf{v}_j^{O,O}. \tag{21}$$

Since the absolute coordinates of the floating frame of reference coincide with the absolute nodal coordinates of node i, the following coordinate transformation is obtained on the velocity level:

$$\dot{\mathbf{q}} = \mathbf{A}\dot{\mathbf{q}}_I, \quad \mathbf{A} \equiv \begin{bmatrix} \mathbf{1} & \mathbf{0} \\ -[\tilde{\mathbf{r}}_j^{i,i}]^T[\mathbf{R}_O^i] & [\mathbf{R}_O^i] \end{bmatrix}. \tag{22}$$

Differentiating Equation (22) with respect to time yields the coordinate transformation on the acceleration level:

$$\ddot{\mathbf{q}} = \mathbf{A}\ddot{\mathbf{q}}_I + \dot{\mathbf{A}}\dot{\mathbf{q}}_I. \tag{23}$$

Applying this coordinate transformation on the equations of motion Equation (16) results in:

$$\widehat{\mathbf{M}}\ddot{\mathbf{q}}_I = \widehat{\mathbf{Q}}, \tag{24}$$

in which $\widehat{\mathbf{M}}$ is the transformed mass matrix. $\widehat{\mathbf{Q}}$ is the corresponding generalized force vector in which also the squared velocity term due to the transformation Equation (23) is included.

With the equations of motion in the form of Equation (24), constrains between bodies can be applied direcly, eliminating dependent generalized coordinates from the equations. The resulting equations of motion are expressed in the minimal coordinates, i.e. the number of equations of motion equals the number of independent generalized coordinates.

For the sake of completeness, it is mentioned that $\widehat{\mathbf{M}}$ is not constant, but depends on both the rotation matrix as well as the body's elastic deformation. For small deformations, it is often reasonable to neglect the dependency on the elastic deformation. In [12] it is shown that under this assumption $\widehat{\mathbf{M}}$ can be related to the body's finite element mass matrix $\mathbf{M}_{FEM}$ that is condensed to the interface points:

$$\widehat{\mathbf{M}}\big|_{\mathbf{u}=\mathbf{0}} = [\mathbf{R}_i^O]\mathbf{M}_{FEM}[\mathbf{R}_O^i]. \tag{25}$$

Note that in co-rotational finite element formulations, the global mass matrix is also assembled by rotating the local finite element mass matrices, which are constant, to the global frame. In the floating frame of reference formulation, as presented in this work, the dependency of the mass matrix on the elastic deformation can, but does not need to be neglected: the terms follow naturally from the formulation. Also the quadratic velocity terms are still included in $\widehat{\mathbf{Q}}$, these terms are typically not taken into account in a co-rotational finite element formulation.

5. Transfer matrix method for dynamic analysis

As a consequence, of the chain-like topology of many offshore structures, recursive solution procedures are well-suited for dynamic simulations. In this work, the transfer matrix method is used for solving the geometrically nonlinear equations of motion. To this end, consider the equation of motion Equation (16) for a single body, expressed in its interface coordinates, in the following partitioned form:

$$\begin{bmatrix} \mathbf{M}_{ii} & \mathbf{M}_{ij} \\ \mathbf{M}_{ji} & \mathbf{M}_{jj} \end{bmatrix}\begin{bmatrix} \ddot{\mathbf{q}}_i \\ \ddot{\mathbf{q}}_j \end{bmatrix} = \begin{bmatrix} \mathbf{F}_i \\ \mathbf{F}_j \end{bmatrix} + \begin{bmatrix} \mathbf{Q}_i \\ \mathbf{Q}_j \end{bmatrix}, \tag{26}$$

where $\mathbf{F}_i$ and $\mathbf{F}_j$ denote the constraint forces at interface points i and j respectively and are separated from the remaining right hand side forces. In the transfer matrix method, Equation (26) is rewritten such that the acceleration and interface force at j are expressed in terms of the acceleration and interface force at i:

$$\begin{bmatrix} \ddot{\mathbf{q}}_j \\ \mathbf{F}_j \end{bmatrix} = \begin{bmatrix} \mathbf{M}_{ij} & \mathbf{0} \\ \mathbf{M}_{jj} & -\mathbf{1} \end{bmatrix}^{-1}\begin{bmatrix} -\mathbf{M}_{ii} & \mathbf{1} \\ -\mathbf{M}_{ji} & \mathbf{0} \end{bmatrix}\begin{bmatrix} \ddot{\mathbf{q}}_i \\ \mathbf{F}_i \end{bmatrix} + \begin{bmatrix} \mathbf{M}_{ij} & \mathbf{0} \\ \mathbf{M}_{jj} & -\mathbf{1} \end{bmatrix}^{-1}\begin{bmatrix} \mathbf{Q}_i \\ \mathbf{Q}_j \end{bmatrix}. \tag{27}$$

This is of the form:

$$\mathbf{z}_j = \mathbf{T}_i^j \mathbf{z}_i + \hat{\mathbf{z}}_i^j, \tag{28}$$

where $\mathbf{z}_i$ and $\mathbf{z}_j$ are the so-called state vectors of interface points i and j respectively, $\mathbf{T}$ is the transfer matrix from j with respect to i and $\hat{\mathbf{z}}$ is an internal state vector due to elastic and quadratic velocity forces.

When the equation of motion of each flexible body in the chain is written in the form Equation (28), the bodies can be coupled by considering the appropriate coupling conditions. For example, in case of a rigid coupling, the accelerations of the connected interface points are equal and the interface forces are equal but of opposite sign, following Newton's third law. Mathematically, such a coupling can be realized by a permutation matrix $\mathbf{P}$:

$$\mathbf{z}^{(k+1)} = \mathbf{P}\mathbf{z}^{(k)}, \quad \mathbf{P} \equiv \begin{bmatrix} \mathbf{1} & \mathbf{0} \\ \mathbf{0} & -\mathbf{1} \end{bmatrix}, \tag{29}$$

where $\mathbf{z}^{(k)}$ and $\mathbf{z}^{(k+1)}$ are the state vectors of the interface point that bodies k and k+1 have in common. In this way, the state vector of last interface point of the chain, say $\mathbf{z}_N$ can ultimately be related to the state vector of the first interface point $\mathbf{z}_1$:

$$\mathbf{z}_N = \mathbf{T}_1^N \mathbf{z}_1 + \hat{\mathbf{z}}_1^N. \tag{30}$$

At this point, the boundary conditions at both ends of the chain can be applied. In the case of pipe-laying one could for instance prescribe zero deformation at the location of the sea bed, prescribe the water depth and the force applied by the tensioner system. Then, from Equation (30) the unknown displacements and constraint forces at both ends of the chain are determined. All intermediate states can be solved recursively by back substitution. The solving of all state vectors is then followed by numerical time integration to the next time step.

Because the transfer matrix method is a recursive solution procedure that requires $O(n)$ computations, it is potentially a very efficient algorithm. A benefit of the transfer matrix method as presented in this work over many other recursive solution procedures, is that no knowledge of boundary conditions is required to set up the recursive equations. In particular, no loop-closure constraints are required in order to solve structures with a closed-loop topology. Such a loop-closure constraint would be required in a condensation-based recursive method if one prescribes the water depth.

6. Transfer matrix method for equilibrium analysis

Analyses that are performed with the purpose of e.g. operability studies, require transient dynamic simulations. In order to prevent unwanted start-up behavior, the transient simulation is started from a static equilibrium configuration. The equilibrium equations of a single body are obtained directly by removing all inertia related terms from the left hand side of Equation (15). Also for this static case, the assembly of the equilibrium equations of the entire system can be done using Lagrange multipliers. Due to the geometric nonlinearities on the position level, it is not possible to establish an explicit coordinate transformation to interface coordinates. As a consequence, it is not possible develop a state transfer matrix on the position level.

It is common however to solve nonlinear equilibrium equations using a total Lagrangian or updated Lagrangian formulation, in which the equations are solved iteratively using a Newton-Raphson scheme. Within each iteration, the incremental change in the degrees of freedom $\Delta\mathbf{q}$ is solved from a set of linearized equations in the tangent space of the equilibrium equations. For increment $t + \Delta t$ these equations can be written as [16]:

$$\mathbf{K}_t \, \Delta\mathbf{q} = \mathbf{Q}^{(t+\Delta t)} - \mathbf{Q}_e^{(t)}, \tag{31}$$

where $\mathbf{K}_t$ is the tangential stiffness matrix, which consists of the material stiffness matrix, as determined in Equation (15), as well as the geometric stiffness matrix that follows from taking the variation of the nonlinear equilibrium equations. The right hand side is the difference between the externally applied forces of the next iteration $\mathbf{Q}^{(t+\Delta t)}$ and the internal elastic forces of the current iteration $\mathbf{Q}_e^{(t)}$. If at the current iteration, the nonlinear equations are solved exactly, the right hand side equals the incremental change in the externally applied forces $\Delta\mathbf{Q}$. However, in the notation used in Equation (31), a possible residual at the current iteration is taken into account during the next iteration.

The linearized equilibrium equation of the entire structure can be assembled using Equation (31) for each body in the chain. At this point it is possible to set up a transfer matrix method for solving the equilibrium equations. In this procedure, a state vector would consists of the position of an interface point and the interface forces. The remainder of the solution procedure is identical as the procedure for a transient dynamic simulation.

In order to determine the equilibrium configuration of an offshore pipeline that is suspended from a vessel and supported by the sea bed, an initially straight pipe can be considered. This pipe can be incrementally loaded, solving Equation (31) at each load increment, until equilibrium is reached.

To demonstrate this procedure, a numerical example is performed on pipe that is subjected to its own weight in downward direction and a vertical tip force equal to its own weight in upward direction. The pipe is divided into 10 bodies. The total length of the pipe is 1 (m). The cross section has radius 0.01 (m) and the wall thickness is 0.001 (m). The Young's modulus is 70 (GPa). The applied weight per unit length is 100 (kN/m). This load is such that the deformation of the pipe will be large, such that geometrical nonlinear effects need to be included. The pipe is initially straight and the equilibrium configuration is obtained using 5 equal load increments. Within each load increment, iterations are performed until convergence is reached. Figure 2 shows the converged equilibrium configurations of each load increment.

7. Equilibrium analysis using catenary theory

To solve the equilibrium configuration of a suspended pipeline from a pipe that is initially straight, large displacements are required. As a consequence, the use of load increments and the Newton-Raphson procedure are required, which may need quite some iterations before convergence is reached. It could be said that an initially straight pipe is a very poor initial estimation of the equilibrium configuration.

It is found that the equilibrium configuration of a pipe line is reasonable well approximated by the analytical solution from catenary theory. For instance, in [4] results are presented that in case for a J-lay process show the striking similarity between the ideal catenary solution and the pipeline. In order to benefit from the fact that the catenary theory yields an analytical solution, many references can be found in which this theory is expanded to include effects such as longitudinal straining, finite bending stiffness and inertia effects. In this work, this semi-analytical approach is not followed on purpose, in order to emphasize the possibility to include the pipeline into a multibody simulation.

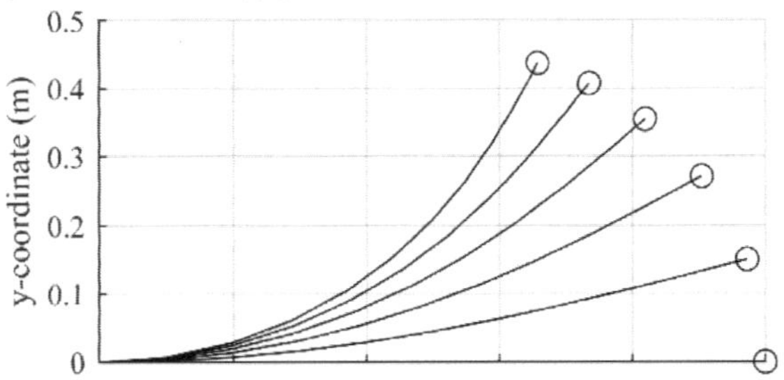

Figure 2. Equilibrium configuration of the pipe using 5 load increments.

However, using the ideal catenary solution for the initialization of the equilibrium analysis could greatly reduce the effort required in this stage. The nonlinear differential equation of a standard catenary is [2]:

$$\frac{d^2y}{dx^2} = \frac{\rho Ag}{T_0}\sqrt{1 + \left(\frac{dy}{dx}\right)^2}, \tag{32}$$

in which $y(x)$ is the shape of the catenary, ρAg the weight of the cable per unit length and T_0 the constant horizontal component of the pretension in the cable. The solution of Equation (32) is:

$$y(x) = \frac{T_0}{\rho Ag}\left[\cosh\left(\frac{\rho Ag}{T_0}x\right) - 1\right]. \tag{33}$$

At every position, the slope of the catenary can be determined from:

$$\tan(\theta) = \frac{dy}{dx}. \tag{34}$$

The horizontal component of the tension force equals T_0 at every position. The vertical component of the tension force equals the weight of the cable up to that point, which can be expressed as:

$$T_y(x) = T_0 \sinh\left(\frac{\rho Ag}{T_0}x\right). \tag{35}$$

The internal bending moment in the catenary is zero by definition. For a given water depth and pretension T_0, the solution of the catenary is determined. Then, Equation (32) to (34) are evaluated at the interface points to establish the state vectors at these interface points. These state vectors are used as an initial guess in the iterative Newton-Raphson procedure that solves the equilibrium configuration of the actual multibody system.

Also in the case of the numerical example used in the previous section, it is found that the catenary solution is close to the equilibrium configuration of the pipe. To demonstrate this, Figure 3 shows the final equilibrium configuration of the pipe again (solid line), together with the analytical catenary solution that passes through the end point (dashed line). The close resemblance of the catenary to the pipe line suggests that it would indeed be beneficial to use the catenary solution as an initial guess in the equilibrium analysis of the pipeline. In this way, the use of load increments could be prevented and only few Newton-Raphson iterations are required.

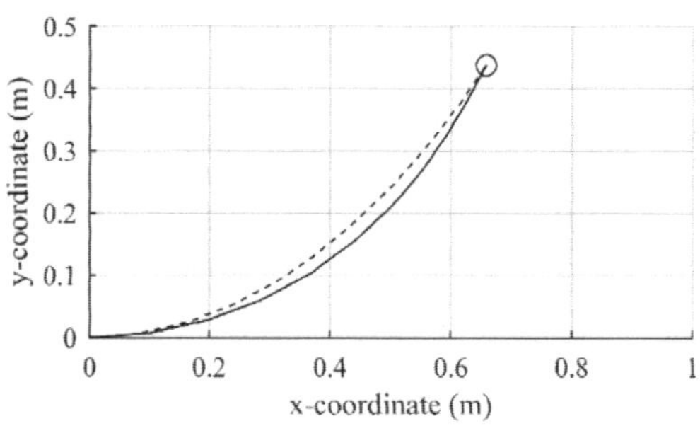

Figure 3. Resemblance of the ideal catenary (dashed line) with the pipe's equilibrium configuration (solid line).

8. Conclusion

In this work, thoughts have been presented on how to model slender offshore structures using the floating frame of reference formulation, suitable for the fast simulation of the structure's flexible multibody dynamics. Starting with the standard floating frame of reference formulation, a coordinate transformation to absolute interface coordinates is applied in order to remove the Lagrange multipliers from the equations of motion. This allows for the use of potentially efficient recursive solution procedures, such as the transfer matrix method. For the equilibrium analysis, it is explained how the use of the ideal catenary solution could be used to prevent the use of load increments. In this way, the number of iterations can be reduced and convergence improved.

Future work is aimed on the validation of the proposed approach with simulations published in literature. In this, the effect of using a recursive scheme in combination with the use of the ideal catenary for the equilibrium analysis on computation times will be investigated. Once reliable results have been obtained, a pipeline model can be coupled to a multibody dynamics model of the pipe-laying vessel. At this point, full hydrodynamic coupling can be included. In this way the transient dynamic behavior of the combined system subjected to various sea-states can be studied with great numerical efficiency.

References

[1] Palmer A C and Kind R A, 2008, *Subsea Pipeline Engineering: 2nd Ed.* (Penn Well Books)
[2] Meriam J L and Kraige L G, 2008 *Engineering Mechanics Statics, 6th Ed.* (John Wiley & Sons)
[3] Wang C Y and Watson L T 1982 The elastic catenary *Int. J. Mech. Sci.* **24(6)** pp. 349-57
[4] Jensen G A, Säfström N, Nguyen T D and Fossen T I 2010 A nonlinear PDE formulation for offshore vessel pipeline installation *Ocean Eng.* **37** pp. 365-77
[5] Haug E J 1989 *Computer Aided Kinematics and Dynamics of Mechanical Systems: Basic Methods* (Prentice Hall)
[6] Szczotka M 2010 Pipe laying simulation with an active reel drive *Ocean Eng.* **37** pp. 539-48
[7] de Vries F H, Geijselaers H J M and van den Boogaard A H 2017 A nonlinear dynamic corotational finite element model for submerged pipes *1st Conf. of Computational Methods in Offshore Technology, COTech (Stavanger)*
[8] Shabana A A 1998 *Dynamics of Multibody Systems* (Cambridge University Press)
[9] Schilder J P, Ellenbroek M H M, Hagmeijer R and de Boer 2016 A Hydro-elasticity in flexible multibody dynamics *Int. Conf. on Noise and Vibration Engineering (Leuven)*, ISMA
[10] Ellenbroek M H M 1994 *On the fast simulation of multibody dynamics of flexible space structures* (PhD thesis: University of Twente)
[11] Pestel E C and Leckie F A 1963 *Matrix Methods in Elasto Mechanics: 1st Ed.*(McGraw-Hill)
[12] Schilder J P, Ellenbroek M H M and de Boer A 2017 Recursive solution procedures for flexible multibody systems: comparing condensation and transfer matrix methods *ECCOMAS Thematic Conf. on Multibody Dynamics (Prague)*
[13] Ellenbroek M H M and Schilder J P 03-05-2017 On the use of absolute interface coordinates in the floating frame of reference formulation for flexible multibody dynamics *Multibody Sys. Dyn.*
[14] Cardona A and Géradin M 1991 Modelling of super elements in mechanism analysis *Int. J. Numer. Methods Eng.* **32** pp. 1565-96
[15] Cardona A 2000 Super elements modelling in flexible multibody dynamics *Multibody Sys. Dyn.* **4** pp. 245-66
[16] de Borst R, Crisfield M A, Remmers J J C and Verhoosel C V 2012 *Non-linear Finite Element Analysis of Solids and Structures: 2nd Ed.* (John Wiley & Sons.

A nonlinear dynamic corotational finite element model for submerged pipes

F H de Vries[1]*, H J M Geijselaers[1], A H van den Boogaard[1] and A Huisman[2]
[1]University of Twente, Faculty of Engineering Technology, P.O. Box 217, 7500 AE Enschede, The Netherlands
[2]Allseas Engineering B.V., Delft, The Netherlands

*Corresponding author: f.h.devries@utwente.nl

Abstract. A three dimensional finite element model is built to compute the motions of a pipe that is being laid on the seabed. This process is geometrically nonlinear, therefore co-rotational beam elements are used. The pipe is subject to static and dynamic forces. Static forces are due to gravity, current and buoyancy. The dynamic forces exerted by the water are incorporated using Morison's equation. The dynamic motions are computed using implicit time integration. For this the Hilber-Hughes-Taylor method is selected. The Newton-Raphson iteration scheme is used to solve the equations in every time step. During laying, the pipe is connected to the pipe laying vessel, which is subject to wave motion. Response amplitude operators are used to determine the motions of the ship and thus the motions of the top end of the pipe.

1. Introduction

Typical numerical simulations of offshore pipelaying tend to have high computational costs. This is due to the long time period of a simulation in combination with several variables, such as wave direction and water depth. Commercial software is available for these type of simulations, such as Offpipe [1], PipeLay [2] and Orcaflex [3]. Each of these software packages has its possibilities as well as limitations. Due to increased analysis demands and market trends there is a need for quicker and more efficient software. The current work brings existing methods together in order to build an efficient dynamic model for pipe laying simulations.

Offshore pipelay analyses are typically performed using a Finite Element approach. In these analyses large rotations of the elements occur. In literature several methods for large rotations can be found. Reissner [4] and Simo and Vu Quoc [5], [6] developed the geometrically exact beam method. This method is named exact because no assumptions have been made on the size of the deformations. However, since discretization will be used, this method is still an approximation. This method converges in a small number of iterations for small beam lengths. However, to analyze offshore pipe laying efficiently, long elements are required. The geometrically exact method requires a large number of iterations to converge when using long elements.

An alternative method for large rotations is the absolute nodal coordinate formulation, presented by Shabana [7]. An advantage of this method is that it results in a constant mass matrix. However when using a time integration method from the Newmark-family [8], the mass matrix is not inverted in every iteration. Thus the only advantage of a constant mass matrix is that is does not need to be rotated. Disadvantages of this method are the more complex stiffness matrix and the larger number of degrees of freedom compared to the co-rotational method. Furthermore Romero [9] and Bauchau and Han [10] compared the absolute nodal coordinate and geometrically exact formulations. In both papers, no improvements of accuracy or computational costs are found for the absolute nodal coordinate formulation.

Shabana [11] suggested a substructuring technique with a floating frame of reference. The deformations of the substructures are described by elastic vibration modes. This method is often used in multibody dynamics. Disveld [12] investigated in his MSc thesis if this method would be more efficient for offshore pipelaying application. Additionally, he investigated a recursive method with two fixed ends. These methods were shown to be less efficient compared to the co-rotational finite element method.

The chosen method for large rotations in the current numerical model is therefore the co-rotational method, presented in Section 2. The co-rotational method has been well established in literature [13], [14] and more recently [15], [16]. The formulation used in the current model is the method of Crisfield [17][18].

General distributed loads are added in the global coordinate system in Section 3. Thereafter two special kinds of distributed loads are considered. In Section 4 buoyancy is added using the method of Yazdchi [19]. New in this paper is the addition of equivalent moments for the buoyancy due to a distributed pressure. Furthermore, when the pipe moves through the water, forces are exerted by the water on the pipe. These hydrodynamic forces are included in the model using Morison's equation [20]. This also accounts for the forces due to the velocity and acceleration of the water, as can be seen in Section 5.

To be able to use large time steps in the numerical time integration, implicit time integration is chosen. A well-known method is the Newmark-β method [8], which requires parameters β and γ. If these parameters are chosen at $\beta = 1/4$ and $\gamma = 1/2$, the method is second order accurate and unconditionally stable for linear systems. However the current model is nonlinear. To maintain stability numerical damping can be added by increasing parameter γ, but then the second order accuracy is lost. For this reason, for the Hilber-Hughes-Taylor method, or HHT-α method [21], is selected, which is an expansion of the Newmark-β method. The HHT-α method uses a parameter α to include damping and is second order accurate independent of its parameter. Further details of this method can be found in [21]. In Section 6 both static and dynamic examples using the presented method are presented.

2. Three dimensional co-rotational method

This section is an introduction to the co-rotational beam element in a three dimensional space. This element has two nodes. Each node has six degrees of freedom, three translations and three rotations. The foundation of the method is that it uses a local coordinate system to calculate small strains. This is presented in Section 2.1. This local coordinate system is rotated to the global coordinate system. The corresponding rotation matrix is shown in Section 2.2.

Because the co-rotational method is geometrically nonlinear, an iterative solution procedure is needed. The Newton-Raphson method is selected.

$$\boldsymbol{u}_{sys}^{n+1} = \boldsymbol{u}_{sys}^{\mathrm{n}} - \mathrm{K}_{sys}^{-1}\boldsymbol{F}_{sys} \tag{1}$$

Here $\boldsymbol{F}_{sys}$ is the system force vector. This vector is built using the global force vector and the external force vector: $\boldsymbol{F} = \boldsymbol{F}_{int} - \boldsymbol{F}_{ext}$. The internal force vector is derived in section 2.3. K_{sys} is the system stiffness matrix. This matrix is assembled from the tangent stiffness matrices of all elements. The tangent stiffness matrix is presented in section 2.4.

2.1. Local coordinate system

In a co-rotational formulation a local 'co-rotational' coordinate frame is introduced. The origin of this frame is placed on the first node of the element. The local coordinate frame is placed such that the x-axis runs through the first and the second node. In other words the displacements in local y- and z-direction are always zero at the nodes. This placement of the coordinate system results in only seven local degrees of freedom.

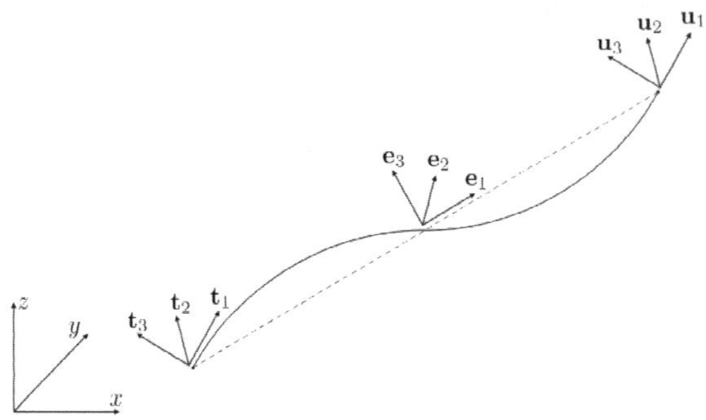

Figure 1. Co-rotational beam element with orthogonal unit vectors to define rotations

$$\boldsymbol{p}_l = [\psi_1, \theta_1, \phi_1, u, \psi_2, \theta_2, \phi_2]^T \tag{2}$$

An important factor when rotating in three dimensions is that the rotations are in general not commutative. This means that a rotation cannot be added to another rotation. The three dimensional rotations are implemented using nodal triads. Detailed derivations can be found in [18]. For each element, two nodal triads, T and U, and one element triad, E, are defined. This can be seen in Figure 1. Each triad contains three orthogonal unit vectors. The unit vectors of the three triads are named $\boldsymbol{t}_i$, $\boldsymbol{e}_i$ and $\boldsymbol{u}_i$.

$$\mathrm{T} = [\boldsymbol{t}_1, \boldsymbol{t}_2, \boldsymbol{t}_3] \quad \mathrm{E} = [\boldsymbol{e}_1, \boldsymbol{e}_2, \boldsymbol{e}_3] \quad \mathrm{U} = [\boldsymbol{u}_1, \boldsymbol{u}_2, \boldsymbol{u}_3] \tag{3}$$

These unit vectors define the local rotations of the element nodes. The middle triad defines the element coordinate system. As shown by Crisfield [18], these unit vectors can be used to find the local rotations.

$$\begin{aligned} 2\sin\psi_1 &= -\boldsymbol{t}_3^T\boldsymbol{e}_2 + \boldsymbol{e}_3^T\boldsymbol{t}_2 \\ 2\sin\theta_1 &= \boldsymbol{t}_3^T\boldsymbol{e}_1 - \boldsymbol{e}_3^T\boldsymbol{t}_1 \\ 2\sin\phi_1 &= -\boldsymbol{t}_2^T\boldsymbol{e}_1 + \boldsymbol{e}_2^T\boldsymbol{t}_1 \\ 2\sin\psi_2 &= -\boldsymbol{u}_3^T\boldsymbol{e}_2 + \boldsymbol{e}_3^T\boldsymbol{u}_2 \\ 2\sin\theta_2 &= \boldsymbol{u}_3^T\boldsymbol{e}_1 - \boldsymbol{e}_3^T\boldsymbol{u}_1 \\ 2\sin\phi_2 &= -\boldsymbol{u}_2^T\boldsymbol{e}_1 + \boldsymbol{e}_2^T\boldsymbol{u}_1 \end{aligned} \tag{4}$$

After each iteration, nodal triads T and U are updated as follows.

$$\begin{aligned} \mathrm{T}_{n+1} &= \Delta\mathrm{T}(\Delta\boldsymbol{\alpha})\mathrm{T}_n \\ \mathrm{U}_{n+1} &= \Delta\mathrm{U}(\Delta\boldsymbol{\beta})\mathrm{U}_n \end{aligned} \tag{5}$$

Here $\Delta\boldsymbol{\alpha}$ and $\Delta\boldsymbol{\beta}$ are the iterative spin variables. These variables are the update increments of the angles at the end of an iteration. They are non-additive and represent the vector about which the nodal system is rotated. The length of the vector is the angle. ΔT and ΔU are rotation matrices. Here it is chosen to use Rodrigues rotation matrix [22].

The element triad E is updated differently. For this matrix first a rotation matrix from the first node to the second node is determined.

$$\Delta\mathrm{R}(\boldsymbol{\gamma}) = \mathrm{UT}^T \tag{6}$$

Spurrier's algorithm [23] is applied to ΔR to find spin variable $\boldsymbol{\gamma}$. This is used to find the mean rotation $\bar{\mathrm{R}}$.

$$\bar{\mathrm{R}} = \Delta \mathrm{R}_m \left(\frac{\boldsymbol{\gamma}}{2}\right) \mathrm{T} \tag{7}$$

Here $\Delta \mathrm{R}_m(\boldsymbol{\gamma}/2)$ is obtained using the Rodrigues rotation matrix. The mean rotation $\bar{\mathrm{R}}$ is used to find the matrix E. First the vector $\boldsymbol{e}_1$ is obtained by

$$\boldsymbol{e}_1 = \frac{1}{L}(\boldsymbol{x}_{21} + \boldsymbol{d}_{21}) = \frac{1}{\mathrm{L}} \begin{bmatrix} x_2 - x_1 + u_2 - u_1 \\ y_2 - y_1 + v_2 - v_1 \\ z_2 - z_1 + w_2 - w_1 \end{bmatrix} \tag{8}$$

Here x_i, y_i and z_i are the coordinates of node i and u_i, v_i and w_i are the displacements of node i. These are all in the global coordinate system. Using Crisfield's mid-point rule approximation and $\bar{R} = [\boldsymbol{r}_1, \boldsymbol{r}_2, \boldsymbol{r}_3]$, $\boldsymbol{e}_2$ and $\boldsymbol{e}_3$ are found.

2.2. Rotation matrix

A rotation matrix is derived to rotate from the local coordinate system to the global coordinate system. This is done by taking the variation of the local variables.

$$\delta \boldsymbol{p}_l = [\delta\psi_1, \delta\theta_1, \delta\phi_1, \delta u, \delta\psi_2, \delta\theta_2, \delta\phi_2]^T = \mathrm{B}\delta\boldsymbol{p} \tag{9}$$

This derivation can be found in [18]. Note that a different coordinate system is used here. The resulting rotation matrix is

$$\mathrm{B} = \begin{bmatrix} \frac{1}{2\cos\psi_1}(\boldsymbol{t}_2^T \mathrm{Q}(\boldsymbol{r}_3) - \boldsymbol{t}_3^T \mathrm{Q}(\boldsymbol{r}_2) + [\boldsymbol{0}, \boldsymbol{e}_2^T \tilde{\boldsymbol{t}_3} - \boldsymbol{e}_3^T S \tilde{\boldsymbol{t}_2}, \boldsymbol{0}, \boldsymbol{0}]) \\ \frac{-1}{2\cos\theta_1}(\boldsymbol{t}_1^T \mathrm{Q}(\boldsymbol{r}_3) + [\boldsymbol{t}_3^T \mathrm{A}, \boldsymbol{e}_1^T \tilde{\boldsymbol{t}_3} - \boldsymbol{e}_3^T \tilde{\boldsymbol{t}_1}, -\boldsymbol{t}_3^T A, \boldsymbol{0}]) \\ \frac{1}{2\cos\phi_1}(\boldsymbol{t}_1^T \mathrm{Q}(\boldsymbol{r}_2) + [\boldsymbol{t}_2^T \mathrm{A}, \boldsymbol{e}_1^T \tilde{\boldsymbol{t}_2} - \boldsymbol{e}_2^T \tilde{\boldsymbol{t}_1}, -\boldsymbol{t}_2^T \mathrm{A}, \boldsymbol{0}]) \\ [-\boldsymbol{e}_1^T, \boldsymbol{0}, \boldsymbol{e}_1^T, \boldsymbol{0}] \\ \frac{1}{2\cos\psi_2}(\boldsymbol{u}_2^T \mathrm{Q}(\boldsymbol{r}_3) - \boldsymbol{u}_3^T \mathrm{Q}(\boldsymbol{r}_2) + [\boldsymbol{0}, \boldsymbol{0}, \boldsymbol{0}, \boldsymbol{e}_2^T \widetilde{\boldsymbol{u}_3} - \boldsymbol{e}_3^T \widetilde{\boldsymbol{u}_2}]) \\ \frac{-1}{2\cos\theta_2}(\boldsymbol{u}_1^T \mathrm{Q}(\boldsymbol{r}_3) + [\boldsymbol{u}_3^T \mathrm{A}, \boldsymbol{0}, -\boldsymbol{u}_3^T \mathrm{A}, \boldsymbol{e}_1^T \widetilde{\boldsymbol{u}_3} - \boldsymbol{e}_3^T \widetilde{\boldsymbol{u}_1}]) \\ \frac{1}{2\cos\phi_2}(\boldsymbol{u}_1^T \mathrm{Q}(\boldsymbol{r}_2) + [\boldsymbol{u}_2^T \mathrm{A}, \boldsymbol{0}, -\boldsymbol{u}_2^T \mathrm{A}, \boldsymbol{e}_1^T \widetilde{\boldsymbol{u}_2} - \boldsymbol{e}_2^T \widetilde{\boldsymbol{u}_1}]) \end{bmatrix} \tag{10}$$

This is a 7x12 matrix. Here $\tilde{\boldsymbol{t}}_i$ is the skew symmetric cross product matrix of vector $\boldsymbol{t}_i$. Furthermore Q is the following 3x12 matrix.

$$\mathrm{Q}(\boldsymbol{r}_i) = [-\mathrm{Q}_1(\boldsymbol{r}_i), \mathrm{Q}_2(\boldsymbol{r}_i), \mathrm{Q}_1(\boldsymbol{r}_i), \mathrm{Q}_2(\boldsymbol{r}_i)] \tag{11}$$

with

$$\begin{aligned} \mathrm{Q}_1(\boldsymbol{r}_i) &= -\frac{1}{2}\boldsymbol{r}_i^T \boldsymbol{e}_1 \mathrm{A} - \frac{1}{2}(\boldsymbol{e}_1 + \boldsymbol{r}_1)\boldsymbol{r}_i^T \mathrm{A} \\ \mathrm{Q}_2(\boldsymbol{r}_i) &= -\frac{1}{2}\tilde{\boldsymbol{r}_\iota} + \frac{1}{4}\boldsymbol{r}_i^T \boldsymbol{e}_1 \tilde{\boldsymbol{r}_\iota} + \frac{1}{4}(\boldsymbol{e}_1 + \boldsymbol{r}_1)\boldsymbol{e}_1^T \tilde{\boldsymbol{r}_\iota} \end{aligned} \tag{12}$$

and

$$\mathrm{A} = \frac{1}{L}(\mathrm{I} - \boldsymbol{e}_1^T \boldsymbol{e}_1). \tag{13}$$

Here I is a 3x3 identity matrix.

2.3. Internal forces

The three dimensional stiffness matrix can be found in many textbooks like [24]. It is presented as a 12x12 matrix. However the size of this matrix can be reduced when using the co-rotational method. Since the number of local degrees of freedom is reduced to seven, the local stiffness matrix is reduced to 7x7.

$$\mathrm{K}_l = \frac{1}{L_0}\begin{bmatrix} GJ & 0 & 0 & 0 & -GJ & 0 & 0 \\ 0 & 4EI & 0 & 0 & 0 & 2EI & 0 \\ 0 & 0 & 4EI & 0 & 0 & 0 & 2EI \\ 0 & 0 & 0 & EA & 0 & 0 & 0 \\ -GJ & 0 & 0 & 0 & GJ & 0 & 0 \\ 0 & 2EI & 0 & 0 & 0 & 4EI & 0 \\ 0 & 0 & 2EI & 0 & 0 & 0 & 4EI \end{bmatrix} \tag{14}$$

The stiffness matrix multiplied by the local displacements vector yields the local element forces.

$$\boldsymbol{f}_l = \mathrm{K}_l \boldsymbol{p}_l \tag{15}$$

The forces in the global coordinate system are found by rotating the local forces.

$$\boldsymbol{F}_{int} = \mathrm{B}^T \boldsymbol{f}_l \tag{16}$$

2.4. Consistent tangent stiffness

The tangential stiffness matrix is found by taking the variation of the forces in the global coordinate system.

$$\begin{aligned} \delta\boldsymbol{F}_{int} = \mathrm{K}\delta\boldsymbol{p} &= \mathrm{B}^T\delta\boldsymbol{f}_l + \delta\mathrm{B}^T\boldsymbol{f}_l \\ &= \mathrm{B}^T\mathrm{K}_l\mathrm{B}\delta\boldsymbol{p} + \mathrm{K}_{geo}\delta\boldsymbol{p} \end{aligned} \tag{17}$$

The result is the rotated local stiffness matrix plus the geometric stiffness matrix.

$$\mathrm{K} = \mathrm{B}^T\mathrm{K}_l\mathrm{B} + \mathrm{K}_{geo} \tag{18}$$

The derivation and details of the geometric stiffness matrix can be found in Crisfield [18]. Note that he uses a different coordinate system.

3. Gravity

This section describes how distributed loads such as gravity can be added to the co-rotational model. The method used assumes the load is distributed uniformly over the element. This can also be used for other distributed loads such as water current.

3.1. Equivalent forces and moments

The global forces at the first node are a function of the distributed force vector q, which contains the distributed forces in the global coordinate system.

$$\overline{\boldsymbol{F}}^q = \frac{1}{2}L_0\boldsymbol{q} \tag{19}$$

The global forces at the second node are equal to the global forces at the first node. The moments at the first node are first calculated in the local coordinate system. The moment around the local x-axis is zero since the external forces have no influence on torsion. The moment around the local y-axis is due to all forces in $\boldsymbol{e}_3$-direction. The minus sign is due to the right-hand-rule. The moment around the local z-axis is due to all forces in $\boldsymbol{e}_2$-direction. The moments at the first node are represented by:

$$\boldsymbol{M}_l^q = \frac{1}{12} L_0^2 \begin{bmatrix} 0 \\ -\boldsymbol{q}^T \boldsymbol{e}_3 \\ \boldsymbol{q}^T \boldsymbol{e}_2 \end{bmatrix}. \tag{20}$$

These moments can be rotated to the global coordinate system using element triad E from (3).

$$\boldsymbol{M}^q = \mathrm{E}\boldsymbol{M}_l^q \tag{21}$$

The moments at the second node are minus the moments at the first node. Now the element force vector due to distributed forces becomes as follows:

$$\boldsymbol{F}^q = \begin{bmatrix} \overline{\boldsymbol{F}}^q \\ \boldsymbol{M}^q \\ \overline{\boldsymbol{F}}^q \\ -\boldsymbol{M}^q \end{bmatrix} \tag{22}$$

3.2. Consistent tangent stiffness

Since these forces are a function of the displacements, a tangent stiffness matrix needs to be derived in order to maintain quadratic convergence using the Newton-Raphson method. This matrix is derived by taking the variation of $\boldsymbol{F}^q$. The variation of the forces at each node $\overline{\boldsymbol{F}}^q$ is zero since $\delta \boldsymbol{q} = 0$. Therefore the first three rows and rows 7-9 of the stiffness matrix are zero. The variation of $\boldsymbol{M}^q$ is calculated as follows.

$$\delta \boldsymbol{M}^q = \delta \mathrm{E} \boldsymbol{M}_l^q + \mathrm{E} \delta \boldsymbol{M}_l^q \tag{23}$$

This yields the following contribution to the stiffness matrix:

$$\mathrm{K}_q = \begin{bmatrix} 0 \\ \mathrm{K}_{sub} \\ 0 \\ -\mathrm{K}_{sub} \end{bmatrix}. \tag{24}$$

Note that 0 on the first and third row represents a 3x12 zero matrix. The matrix K_q has size 12x12. K_{sub} is given by the following 3x12 matrix.

$$\mathrm{K}_{sub} = \frac{1}{12} L_0^2 \left(-\boldsymbol{q}^T \boldsymbol{e}_3 \mathrm{Q}(\boldsymbol{r}_2) + \boldsymbol{q}^T \boldsymbol{e}_2 \mathrm{Q}(\boldsymbol{r}_3) - \boldsymbol{e}_2 \boldsymbol{q}^T \mathrm{Q}(\boldsymbol{r}_3) + \boldsymbol{e}_3 \boldsymbol{q}^T \mathrm{Q}(\boldsymbol{r}_2) \right) \tag{25}$$

Here $\mathrm{Q}(\boldsymbol{r}_i)$ is from (11).

4. Buoyancy

The treatment of buoyancy forces is based on Yazdchi [19]. In his work the buoyancy forces are directly related to the finite element method. Yazdchi divides the distributed loads in three sections. First the forces due to distributed pressure, second the forces due to curvature of the pipe and lastly the forces due to capped ends.

4.1. Distributed pressure

The forces due to distributed pressure can be calculated as follows. On a horizontal pipe section the distributed force in N/m is

$$q_b = (A_0\rho_w - A_i\rho_i)g. \tag{26}$$

Here A_0 is the outer section area of the pipe, A_i is the inner section area of the pipe, ρ_w is the density of the sea water, ρ_i is the density of the fluid in the pipe and g is the standard acceleration due to gravity.

This equation can be found by integrating the pressure over the surface area of the pipe. The result is determined by the pressure difference in the water. However if the pipe is not horizontal, the pressure difference between top and bottom of the pipe is smaller. Therefore the factor $\boldsymbol{e}_n^T\boldsymbol{k}$ is added.

$$\bar{q}_b = q_b\boldsymbol{e}_n^T\boldsymbol{k} \tag{27}$$

Here $\boldsymbol{k}$ is a unit normal in the global z-direction. $\boldsymbol{e}_n$ is a unit normal defined by $\boldsymbol{k}$ and $\boldsymbol{e}_1$, which is the unit normal in the direction of the axis of the pipe. The derivation of $\boldsymbol{e}_n$ can be found in [19].

If an element is only partially submerged, the forces are distributed equivalently over both nodes. This is done by introducing factors k_A and k_B. If the pipe is completely submerged, k_A and k_B are $1/2$. If only the first node (node A) is submerged, the parameters k are determined from the position of the first and second node with respect to the water line. These distances h_A and h_B are positive when the node is submerged and negative when the node is above water.

$$\begin{aligned} k_A &= \bar{h} - k_B \\ k_B &= 1/2\bar{h}^2 \end{aligned} \tag{28}$$

Here $\bar{h}$ is calculated by

$$\bar{h} = \frac{h_A}{h_A - h_B} \tag{29}$$

Up to here the buoyancy is identical to [19]. However Yazdchi neglects all equivalent moments. These are presented here.

The moments are also distributed equivalently using factors h_{AM} and h_{BM}. If the element is completely submerged, these terms are equal to $1/12$ and $-1/12$ respectively. In order to find the size of the factors when the element is partly submerged, the following steps are taken.

If only the first node is submerged, the moment at the first node can be calculated using

$$M_A = \int_0^a q\frac{\partial w}{\partial \theta_1}dx. \tag{30}$$

Here a is the distance between the first node and the water surface. The local displacement w within the element is determined by a qubic Hermite interpolation function.

By solving this integral factor k_{AM} is found. The same can be done for the moment at the second node to find k_{BM}.

$$\begin{aligned} k_{AM} &= \frac{1}{4}\bar{h}^4 - \frac{2}{3}\bar{h}^3 + \frac{1}{2}\bar{h}^2 \\ k_{BM} &= \frac{1}{4}\bar{h}^4 - \frac{1}{3}\bar{h}^3 \end{aligned} \tag{31}$$

The equivalent nodal forces can now be found.

$$\boldsymbol{F}_1^b = \bar{q}_b \begin{bmatrix} k_A L \boldsymbol{e}_n \\ k_{AM} L^2 (\boldsymbol{e}_1 \times \boldsymbol{e}_n) \\ k_B L \boldsymbol{e}_n \\ k_{BM} L^2 (\boldsymbol{e}_1 \times \boldsymbol{e}_n) \end{bmatrix} \tag{32}$$

Here $\boldsymbol{e}_n$ is the direction of the equivalent forces and $(\boldsymbol{e}_1 \times \boldsymbol{e}_n)$ is the direction of the equivalent moments.

4.2. Curvature effects

The forces due to curvature effects are also implemented as given by Yazdchi. Note that he uses a different coordinate system.

$$\boldsymbol{F}_2^b = q_b \begin{bmatrix} h_A \tan\theta_1 \, \boldsymbol{e}_3 - h_A \tan\phi_1 \, \boldsymbol{e}_2 \\ \boldsymbol{0} \\ -h_B \tan\theta_2 \, \boldsymbol{e}_3 + h_B \tan\phi_2 \, \boldsymbol{e}_2 \\ \boldsymbol{0} \end{bmatrix} \tag{33}$$

The third row is the negative of the first row because of the direction of rotation at the second node. Note that the force $\boldsymbol{F}_2^b$ is dependent on the water depth. The force is only applied on submerged nodes.

4.3. Capped ends

Thirdly Yazdchi also gives a buoyancy term due to capped ends. This term is applied to an end of the pipe.

$$\boldsymbol{F}_2^b = q_b \begin{bmatrix} h_A \boldsymbol{e}_1 \\ \boldsymbol{0} \\ -h_B \boldsymbol{e}_1 \\ \boldsymbol{0} \end{bmatrix} \tag{34}$$

5. Hydrodynamic forces

Hydrodynamic forces are calculated using Morison's equation [20]. The following equation is Morison's equation for a moving pipe in moving water.

$$\boldsymbol{q}_{Morison} = \frac{1}{2} \rho C_d D \dot{\boldsymbol{d}}_{rn} |\dot{\boldsymbol{d}}_{rn}| + \rho C_a \frac{\pi}{4} D^2 \ddot{\boldsymbol{d}}_{rn} + \rho \frac{\pi}{4} D^2 \ddot{\boldsymbol{d}}_{wn} \tag{35}$$

Here ρ is the density of the water, C_d is the drag coeficient, D is the pipe diameter, $\dot{\boldsymbol{d}}_{rn}$ and $\ddot{\boldsymbol{d}}_{rn}$ are the relative velocity and relative acceleration in normal direction of the pipe. C_a is the added mass coefficient.

The first term of (35) is the drag force, the second term is the added mass force. The last term is due to the pressure gradient, called the Froude-Krylov force. This force is a function of the acceleration of the water in normal direction $\ddot{\boldsymbol{d}}_{wn}$.

The relative velocity is calculated from the difference between the water velocity and the pipe velocity. The same can be done for the relative acceleration.

$$\begin{aligned} \dot{\boldsymbol{d}}_r &= \dot{\boldsymbol{d}}_w - \dot{\boldsymbol{d}} \\ \ddot{\boldsymbol{d}}_r &= \ddot{\boldsymbol{d}}_w - \ddot{\boldsymbol{d}} \end{aligned} \tag{36}$$

The relative velocity in the normal direction of the pipe is calculated by subtracting the tangential relative velocity from the relative velocity. Again the same can be done for the relative normal acceleration.

$$\begin{aligned} \dot{\boldsymbol{d}}_{rn} &= \dot{\boldsymbol{d}}_r - \dot{\boldsymbol{d}}_r^T \boldsymbol{e}_1 \boldsymbol{e}_1 \\ \ddot{\boldsymbol{d}}_{rn} &= \ddot{\boldsymbol{d}}_{rn} - \ddot{\boldsymbol{d}}_r^T \boldsymbol{e}_1 \boldsymbol{e}_1 \end{aligned} \tag{37}$$

The velocities and acceleration of the pipe are determined by the HHT-α time integration method. The velocities and acceleration of the water are determined by the Airy wave theory.

6. Results

The numerical model formed from the equations described in this paper is implemented in C++ using the FeaTure framework for finite element codes [25]. This section shows static and dynamic results from several examples.

6.1. Static: Buoyancy

To validate the buoyancy forces, the reaction forces of three examples are compared to analytical results. The analytic results are obtained using Archimedes Law.

$$F = A_o \rho_w g L \tag{38}$$

Here $A_o L$ is the volume of the pipe and ρ_w is the density of the water.

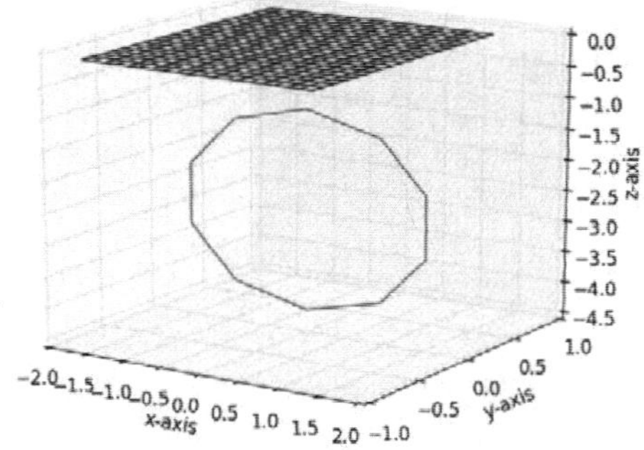

Figure 2. Pipe inclined at $x = 0$ and deformed to a circle using prescribed deformations

Table 1. Buoyancy forces [N]

Example	Analytical	Numerical
Vertical	315.89	315.89
Horizontal	157.95	157.95
Circle	3158.9	3158.9

In the three buoyancy examples the outer diameter is set to $d = 0.2$ m. The density of the water is set to $\rho_w = 1025\ \mathrm{kg/m^3}$. The first example is a vertical pipe of length $L = 1$ m inclined at the water surface. Numerically the buoyancy force is determined by the capped end term. The second example is a horizontal pipe of length $L = 1$ m supported at both ends. The third example is a pipe inclined at $x = 0$ and deformed to a circle using prescribed deformations. This example is depicted in Figure 2. The blue area is the water surface area.

In Table 1 the analytical and numerical solutions of the three examples are given. It can be seen that the numerical solutions are equal to the analytical solutions.

IOP Conf. Series: Materials Science and Engineering **276** (2017) 012030 doi:10.1088/1757-899X/276/1/012030

In the example with the horizontal pipe, the pipe is slightly curved due to the buoyancy forces. Thus curvature forces are acting on the pipe. The curvature forces are a function of the water depth. Thus if the pipe is moved downward, the curvature forces become larger and the total buoyancy force becomes smaller.

6.2. Static: Convergence

In the next example a pipe with a diameter of 457 mm and wall thickness 31 mm is considered. Other properties can be found in Table 2. This pipe placed 10 m above water. A distributed load of -3000 N/m is applied to the beam in the first two steps. In the third step the buoyancy forces are added. The result is shown in Figure 3. The blue area in this figure is the water surface.

Table 2. Beam properties

Parameter	Size
Length	100 m
E-modulus	207 GPA
Poisson ration	0.3
Density pipe	7700 kg/m^3
Density water	1025 kg/m^3

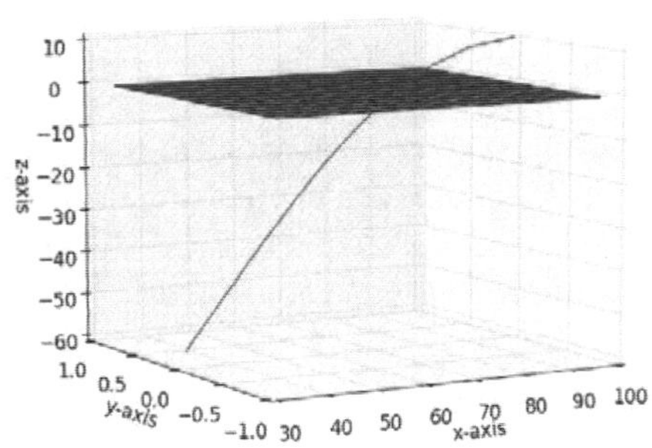

Figure 3. Deformed pipe resulting from three load steps; two steps of distributed load and one step of buoyancy

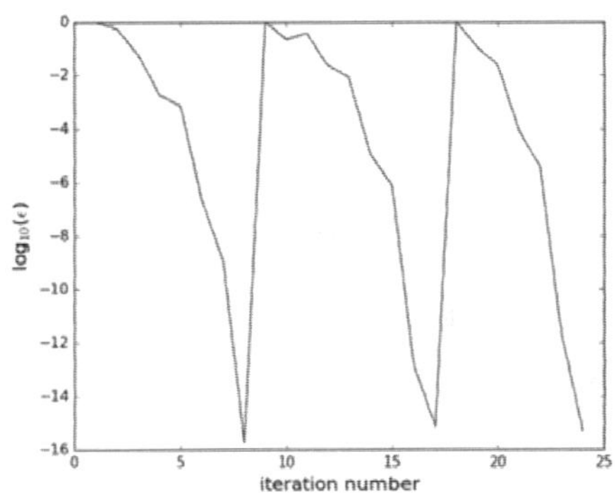

Figure 4. Convergence using displacement based error

The convergence using the tangential stiffness matrix is depicted in Figure 4. As expected for the Newton-Raphson iteration method, this convergence is quadratic. Furthermore the convergence is checked using a numerical stiffness matrix instead of the tangential stiffness matrix. No difference in convergence was observed.

6.3. Dynamic: Small deformations by force

In this example a cantilever beam subject to small deformations is considered. The length of this beam is $L = 100$ m and it is divided in 10 elements. Width and height of the beam are 0.1 m. Only in the first time step a load of $F = 1000$ N is applied at the free end of the beam. This example is solved using the HHT-α method with $\alpha = 0$. This is equal to the Newmark-β method with parameters $\alpha = 1/2$ and $\beta = 1/4$. The results of the y-displacement of the end of the beam is shown in Figure 5.

Table 3. Eigenfreciencies [Hz] of the cantilever beam

n	Analytical	Numerical
1	0.00838	0.00834
2	0.0525	0.0525
3	0.147	0.147
4	0.288	0.288
5	0.476	0.473
6	0.711	0.703

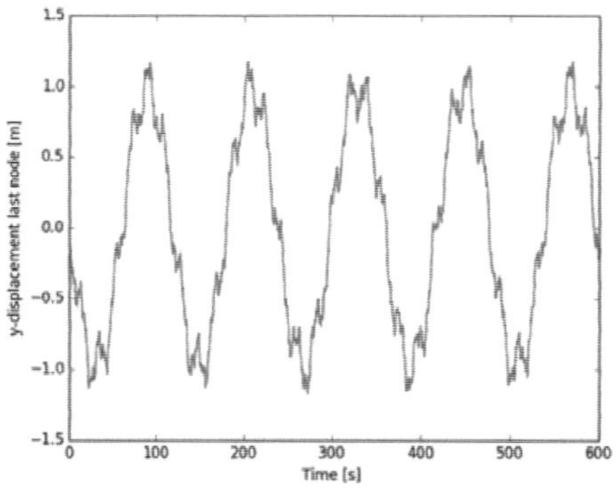

Figure 5. Response of beam tip to impulse load

In Table 3 the eigenfrequencies of the result shown in Figure 5 are compared to the analytically calculated eigenfrequencies of a cantilever beam. It is observed that these eigenfrequencies correspond very well to the numerical results.

6.4. Dynamic: Numerical damping

In this section the effect of the numerical damping in the HHT-α method is checked using an example problem. In this problem a cantilever beam is loaded with two equally large moments at the tip. The first moment is around the y-axis, the second moment is around the z-axis. This is depicted in Figure 6

The diameter of the pipe is 559 mm (22 inch) and the wall thickness is 21 mm. The beam is divided into 10 elements. Other pipe properties are given in Table 2. Using a time step of $\Delta t = 0.1$ s the displacements are calculated until total time of 60 s is reached.

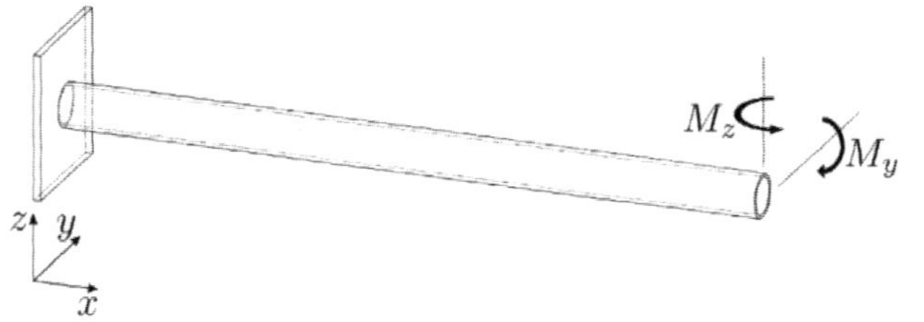

Figure 6. Example problem numerical damping

Two cases are considered. The first case is without numerical damping, thus $\alpha = 0$. The second case is with $\alpha = 0.03$. The total displacement of the tip of the pipe is calculated. The Fourier transform of the total displacement is depicted in Figure 7 and Figure 8. In these figures it can be observed that the HHT-α method shows numerical dissipation for the 3rd eigenfrequency, around 1 Hz, and higher. The calculation time and required number of iterations to converge is depicted in Table 4.

IOP Conf. Series: Materials Science and Engineering **276** (2017) 012030 doi:10.1088/1757-899X/276/1/012030

The HHT-α method is not unconditionally stable for nonlinear systems. This is observed by increasing the time period calculated. When using no numerical damping the system becomes unstable after approximately 300s. This problem is solved by including the numerical damping by setting $\alpha = 0.03$.

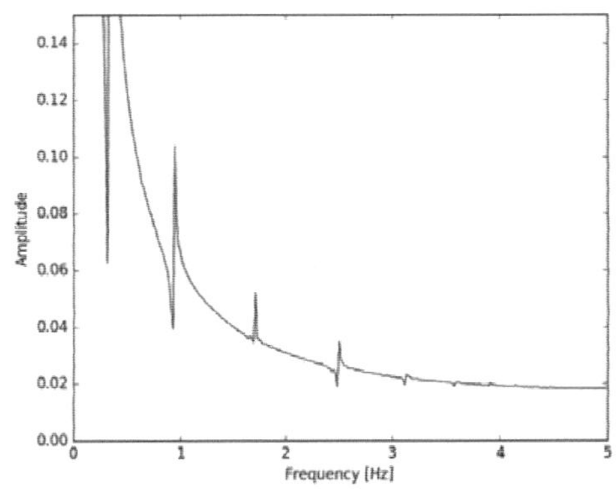

Figure 7. Fourier transform of example without numerical damping

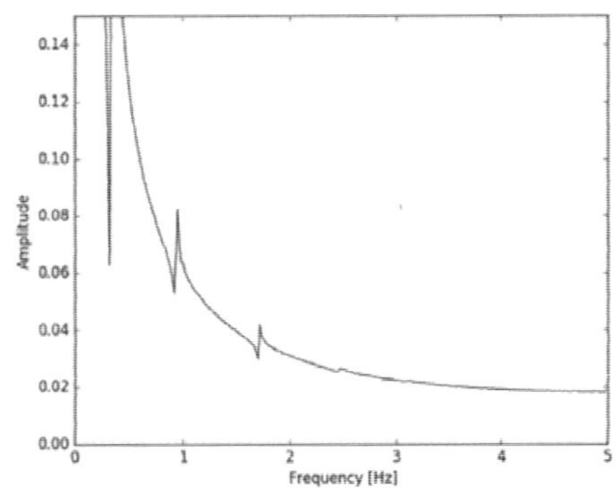

Figure 8. Fourier transform of example including numerical damping

Table 4. Number of iterations and calculation time of HHT method

α	**Iterations per step**	**Calculation time**
0	3-4	2.3 s
0.03	2-3	1.7 s

6.5. Dynamic: Hydrodynamic forces

A massless pipe with a length of 100 meters is placed horizontally 100 meter below the water surface. The pipe has a diameter of 0.2 m and a wall thickness of 50 mm. The mesh consists of 10 elements and the water density is 1025 kg/m^3. At time $t = 0$ the pipe is horizontal and the displacements are zero. Figure 9 shows the result of the dynamic calculation without Morison forces. Figure 10 shows the results including Morison forces. It can be seen that the displacements u and w in x- and z-direction damp out quickly.

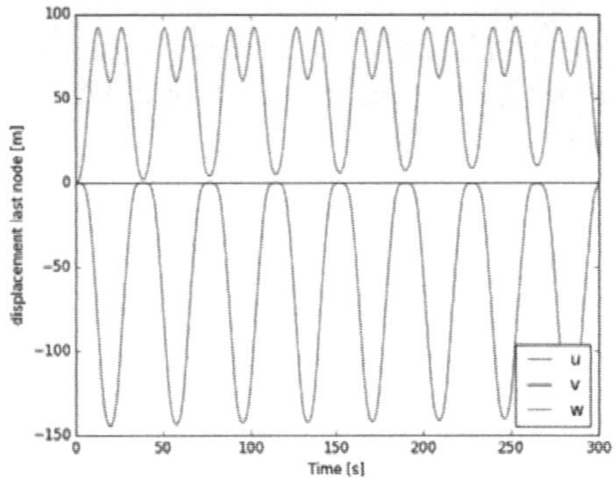

Figure 9. Displacement of the tip of a massless pipe under buoyancy load without Morison forces

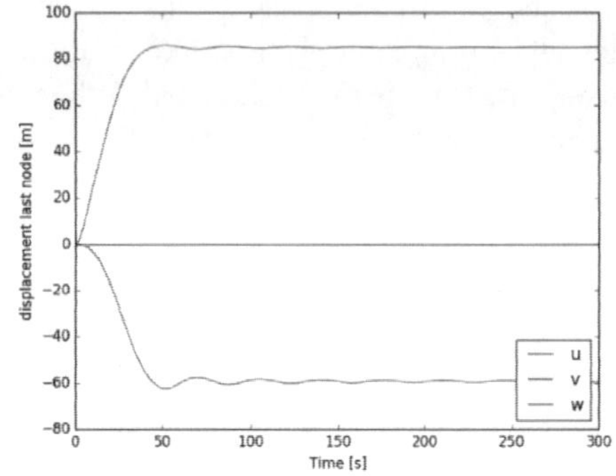

Figure 10. Displacement of the tip of a massless pipe under buoyancy load including Morison forces

6.6. Dynamic: Pipe response to vessel movement

A pipe with diameter of 457 mm (18 inch) with a wall thickness of 31 mm is considered. The pipe has a length of 200m. A section of length 100 m is supported by the pipe laying vessel. The centre of motion of the vessel is placed at the middle of the supported section. The unsupported part of the pipe is divided in 10 elements. Other pipe properties can be found in Table 2.

A distributed load of -3000 N/m is exerted on the pipe in vertical direction. When the pipe is below the water surface at $z = 0$ buoyancy forces act on the pipe. The static solution of this example can be found in Figure 11.

The movement of the centre of motion of the vessel is a response to the wave by Response Amplitude Operators (RAO) [26][27]. The wave motion is determined by a Pierson-Moskowitz spectrum [28] based on significant wave height H_s and mean zero crossing up period T_z [29]. This spectrum represents a fully developed sea. Figure 12 shows the response of the vessel's centre of motion to an irregular wave with significant wave height $H_s = 3$ m and mean zero crossing up period $T_z = 7$ s. Figure 13 shows the displacement of the free end of the pipe with respect to the static solution.

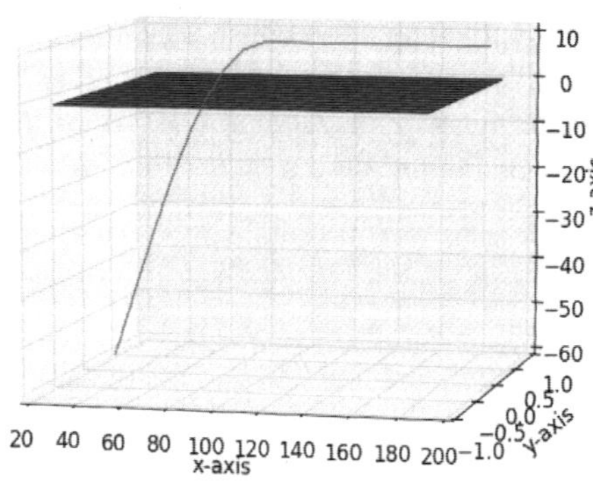

Figure 11. Static solution of pipe partly supported by vessel

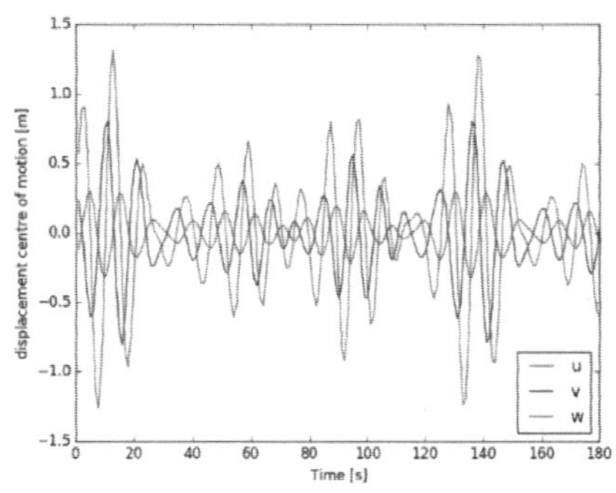

Figure 12. Displacement of vessel's centre of motion

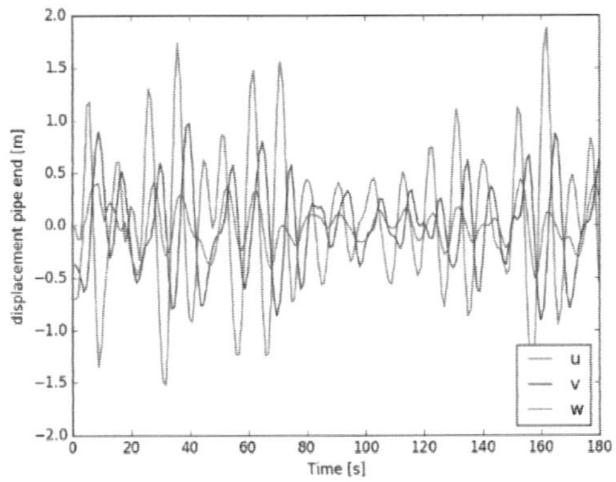

Figure 13. Displacement with respect to static solution of free pipe end

7. Conclusion

The nonlinear co-rotational finite element model presented in this paper shows good results for buoyancy forces and gravity forces. Furthermore, the static convergence is quadratic due to the Newton-Raphson method in combination with a consistent tangent stiffness matrix, resulting in an efficient calculation.

The implicit HHT-α method used for numerical time integration works well with buoyancy load and gravity. Numerical damping prevents instability problems. Furthermore, the Morison forces result in damping in the system.

References

[1] Malahy R C 1985 A nonlinear finite element method for the analysis of the offshore pipe laying problem (Rice University)
[2] Wood Group *PipeLay* https://www.woodgroup.com/ [Accessed: 16-Aug-2017].
[3] Orcina Ltd *OrcaFlex* https://www.orcina.com/SoftwareProducts/ OrcaFlex/ [Accessed: 16-Aug-2017]
[4] Reissner E 1972 On one-dimensional finite-strain beam theory: the plane problem *J. Appl. Math. Physic* **23(5)**
[5] Simo J C 1985 A finite strain beam formulation. The three-dimensional dynamic problem. Part I *Comput. Methods Appl. Mech. Eng.*, **49(1)**, pp. 55–70
[6] Simo J C and Quoc L V 1986 A three-dimensional finite-strain rod model. Part II: computational aspects *Comput. Methods Appl. Mech. Eng.* **58(1)** pp. 79–115
[7] Shabana A A, Hussien H A and Escalona J L 1998 Application of the Absolute Nodal Coordinate Formulation to Large Rotation and Large Deformation Problems *J. Mech. Des.* **120(2)**
[8] Newmark N M 1959 A method of computation for structural dynamics *J. Eng. Mech. Div.* **85(3)** pp. 67–94
[9] Romero I 2008 A comparison of finite elements for nonlinear beams: the absolute nodal coordinate and geometrically exact formulations *Multibody Syst. Dyn.* **20** pp. 51–68
[10] Bauchau O A, Han S, Mikkola A and Matikainen M K 2014 Comparison of the absolute nodal coordinate and geometrically exact formulations for beams *Multibody Syst. Dyn.* **32(1)** pp. 67–85
[11] Shabana A A 1983 A coordinate reduction technique for dynamic analysis of spatial substructures with large angular rotations *J. Struct. Mech.* **11(3)** pp. 401–31
[12] Disveld H J 2017 *Simulation of nonlinear pipeline dynamics development of a recursive method* MSc thesis: University of Twente CTW.16/TM-5788

[13] Crisfield M A 1990 A consistent co-rotational formulation for non-linear, three-dimensional, beam-elements *Comput. Methods Appl. Mech. Eng.*, **81(2)** pp. 131–50
[14] Nour-Omid B and Rankin C C 1991 Finite rotation analysis and consistent linearization using projectors *Comput. Methods Appl. Mech. Eng.* **93(3)** pp. 353–84
[15] Le T 2013 *Nonlinear Dynamics of Flexible Structures Using Corotational Beam Elements* (KTH Royal Institute of Technology)
[16] Li Z X 2007 A co-rotational formulation for 3D beam element using vectorial rotational variables *Comput. Mech.* **39(3)** pp. 309–22
[17] Crisfield M A 1991 *Non-linear Finite Element Analysis of Solids and Structures: Volume 1* (John Wiley & Sons)
[18] Crisfield M A 1997 *Nonlinear Finite Element Analysis of Solids and Structures, Volume 2* (John Wiley & Sons)
[19] Yazdchi M and Crisfield M A 2002 Non-linear dynamic behaviour of flexible marine pipes and risers *Int. J. Numer. Methods Eng.* **54(9)** pp. 1265–308
[20] Morison J, Johnson J and Schaaf S 1950 The force exerted by surface waves on piles *J. Pet. Technol.* **2(5)**
[21] Hilber H M, Hughes T J R and Taylor R L 1977 Improved numerical dissipation for time integration algorithms in structural dynamics *Earthq. Eng. Struct. Dyn.* **5(3)** pp. 283–92
[22] Rodrigues O 1840 Des lois géométriques qui régissent les déplacements d'un système solide dans l'espace: et de la variation des cordonnées provenant de ces déplacements considérés indépendamment des causes qui peuvent les produire *J. Math. Pures Appl.* **5** pp. 380–440,
[23] Spurrier R A 1978 Comment on 'Singularity-free extraction of a quaternion from a direction-cosine matrix *J. Spacecr. Rockets* **15(4)** p. **255**
[24] Cook R D, Malkus D S, Plesha M E and Witt R J 2002 *Concepts and applications of finite element analysis, 4th Ed.* (John Wiley & Sons)
[25] van den Boogaard A H 2006 *FeaTure: A finite element analysis toolkit for use in a research environment* [Online]. Available: www.utwente.nl/en/et/ms3/research-chairs/nsm/-facilities/software/ feature_/. [Accessed: 16-Aug-2017].
[26] Dean R G and Dalrymple R A 1991 *Water wave mechanics for engineers and scientists (Vol. 2).* (World Scientific Publishing Co Inc.)
[27] Faltinsen O M 1993 *Sea loads on ships and offshore structures* (Cambridge University Press.)
[28] Pierson W J and Moskowitz L 1964 A proposed spectral form for fully developed wind seas based on the similarity theory of SA Kitaigorodskii J. *Geophys. Res.* **69(24)**
[29] Barltrop N D P and Adams A J 1991 *Dynamics of fixed marine structures, 3th ed.* (Butterworth-Heinemann)

IOP Conf. Series: Materials Science and Engineering **276** (2017) 012031 doi:10.1088/1757-899X/276/1/012031

Application of foam-extend on turbulent fluid-structure interaction

K Rege and B H Hjertager

Department of Mechanical and Structural Engineering and Materials Science, University of Stavanger, Norway

kristen.rege@uis.no

Abstract. Turbulent flow around flexible structures is likely to induce structural vibrations which may eventually lead to fatigue failure. In order to assess the fatigue life of these structures, it is necessary to take the action of the flow on the structure into account, but also the influence of the vibrating structure on the fluid flow. This is achieved by performing fluid-structure interaction (FSI) simulations. In this work, we have investigated the capability of a FSI toolkit for the finite volume computational fluid dynamics software foam-extend to simulate turbulence-induced vibrations of a flexible structure. A large-eddy simulation (LES) turbulence model has been implemented to a basic FSI problem of a flexible wall which is placed in a confined, turbulent flow. This problem was simulated for 2.32 seconds. This short simulation required over 200 computation hours, using 20 processor cores. Thereby, it has been shown that the simulation of FSI with LES is possible, but also computationally demanding. In order to make turbulent FSI simulations with foam-extend more applicable, more sophisticated turbulence models and/or faster FSI iteration schemes should be applied.

1. Introduction

Fatigue is responsible for the majority of failures in structural and mechanical components [1]. One of the many causes of fatigue loading is the action of a turbulent flow field on a structure, e.g. turbulent flow through water turbines [2], wind turbines [3], or around flexible structures like marine risers [4-7]. In order to assess the safety and reliability of these structures, it is necessary to estimate their fatigue life, which depends on the actual stress history experienced at their critical spots.

While being subjected to the turbulent flow, the structures will vibrate. If the structure is very stiff compared to the fluid, the influence of its vibration on the fluid flow will be negligible [8]. Then, the stress history may be found by first evaluating the flow field around or through a rigid model of the structure, and then imposing the resulting fluid load history to a structural dynamics analysis of the structure [9]. This is called an uncoupled simulation. However, slender structures like marine risers are flexible. Therefore, they will be subjected to significant deformations, thereby influencing the fluid flow. In these cases, it will be necessary to perform coupled fluid-structure interaction (FSI) simulations, in which the deformed shape of the structure is explicitly taken into account when calculating the fluid flow, in order to obtain realistic estimations for the stress history [10, 11].

Because of their complexity, FSI problems are normally solved using computational techniques. These techniques may be divided into two groups: Monolithic approaches and partitioned approaches. In monolithic approaches [12, 13], a unified mathematical model is formulated for the entire regime (fluid and structure), and solved as a single large system of equations [14] for each time step. Stability and convergence rate may be better in a monolithic approach, but the system matrices may become ill-conditioned, requiring appropriate preconditioners [12].

In partitioned approaches [14-17] on the other hand, the structure and fluid regimes are each solved independently by separate mathematical models, i.e. computational solvers. At the interface between the fluid and the structure, information is interchanged between the two solvers, thereby enabling the fluid-structure interaction to be taken into account. The FSI problem is solved sequentially, as shown in Figure 1. The partitioned approach makes it possible to use existing, well-developed solvers to compute each of the two regimes, e.g. to use a finite element solver to compute the structural vibrations and a finite volume solver to compute the fluid flow. Because of this flexibility, the partitioned approach is very popular, and is employed in the majority of the FSI research reported [15].

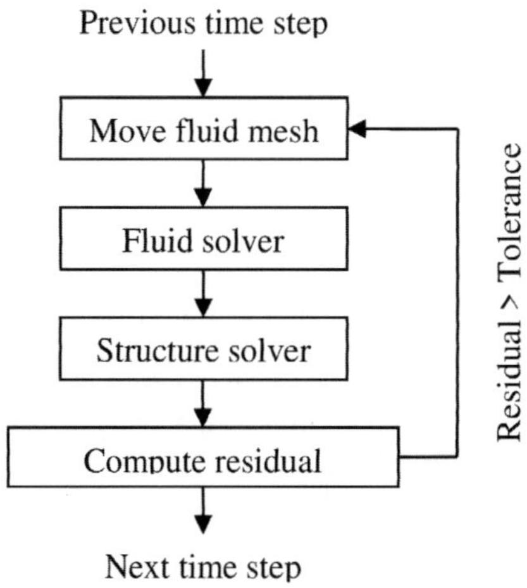

Figure 1. Partitioned FSI approach with strong coupling.

For the partitioned approaches, it is possible to employ either weak or strong coupling [12, 15]. When weak coupling is employed, the fluid and structure regimes are each solved once at each time step. However, this scheme does not guarantee that the instantaneous equilibrium between the fluid and structure is fulfilled. In strong coupling, this is ensured by iteratively computing the fluid and structure fields until instantaneous equilibrium is satisfied, as indicated in Figure 1. According to Münsch and Breuer [18], weak coupling leads to severe stability problems in most applications, and strong coupling should therefore be employed, especially for flexible structures.

Fluid-structure interaction has been studied for a number of flexible structures. Holmes, et al. [5] have simulated the vortex-induced vibration of marine riser models using the AcuSolve finite element (FE) computational fluid dynamics (CFD) solver. In these simulations, the motions of the riser are computed by combining the CFD results with the eigenvalues and eigenvectors of the riser. Holmes, et al. propose to use a detached eddy simulation (DES) or an unsteady Reynolds averaged Navier-Stokes (URANS) turbulence model, but do not specify which turbulence model they have used. The same riser model has also been simulated by Kamble and Chen [6], using the finite volume (FV) CFD solver FANS3D, together with the chimera (overset) grid technique, in which a very fine mesh around the riser overlaps the mesh for the rest of the domain. Using a large eddy simulation (LES) model and about two millions cells, together with a finite difference scheme to simulate the curvature of the riser, good correlation with experimental results has been obtained.

Vortex-induced vibration of marine risers has also been simulated by Menter, et al. [19], by using the element-based finite volume software ANSYS-CFX to compute a laminar simulation of the flow, and the ANSYS structural FE software to compute the structural vibrations. Nguyen and Nguyen [7] used the FV CFD solver OpenFOAM to simulate vortex-induced vibration of rigid, elastically mounted circular cylinders, using a DES turbulence model.

Other simulations of offshore structures include the uncoupled simulation of wave actions on a simplified platform deck structure by Marzban, et al. [20], and the strongly coupled simulation of the movement of a closed riser filled with mud by Paczkowski, et al. [21].

In the case of turbulent FSI, several test cases have been simulated by Münsch and Breuer [18], using strong coupling between a FV CFD solver and a FE structural solver, employing LES. The simulation results have only been compared to other computational results or laminar experimental data, however. Nilsson, et al. [22] have employed weak coupling and LES to simulate structural excitation of thin plate structures subjected to acoustics and gas flow in OpenFOAM, commenting on some short-comings. Recently, Cesur, et al. [16] studied the flow past a flexible cantilever beam, employing implicit large eddy simulation and strong FSI coupling, providing a thorough discussion of the results.

As indicated by the above literature review, the majority of FSI studies employ a finite element solver for the structural dynamics. However, recently a FSI toolkit was developed for foam-extend, a branch of the open source finite volume CFD code OpenFOAM [23], by Tuković, Cardiff, Karač, Jasak and Ivanković [24]. The toolkit employs the finite volume method (FVM) to solve both the fluid and structure regimes, in a partitioned approach, and is based on their previous works [25-30]. The toolkit supports parallel computation of FSI problems with weak or strong coupling between a laminar or turbulent incompressible flow and an elastic structure subjected to large deformations. Ponweiser, et al. [31] have investigated the scalability of this toolkit for multiple processors, and Šekutkovski, et al. [32] have used the toolkit together with a hybrid RANS-LES turbulence model, in order to simulate flow around aircraft wings.

The objective of this paper is to investigate whether the FSI toolkit for foam-extend 4.0 is readily applicable to predict the fatigue life of flexible structures subjected to a turbulent flow. Available test cases for FSI are limited. Turek and Hron's test case [33] is for laminar flow, while Pereira Gomes and Lienhart's turbulent test case [34] employs two structural materials of different densities, which is currently not supported in the FSI toolkit. Therefore, it is chosen to implement a LES turbulence model to Richter's laminar test case of a flexible wall in an enclosed channel flow [13], which also is included as a tutorial in the toolkit. This test case has also been simulated by Mehl, et al. [14] and Gillenbaart, et al. [17].

LES is chosen as the turbulence model, because Reynolds-averaged Navier-Stokes (RANS) models do not resolve the turbulent fluctuations which may be important for fatigue evaluation, while direct numerical simulation (DNS) has a high computational cost [16]. While previous FSI simulations with LES have often employed the Smagorinsky model [6, 18, 22], the one-equation k_{SGS} eddy viscosity model is used in the current work. The goal is to obtain the stress history for a critical spot on the flexible structure. The stress history should be recorded for a sufficient amount of time, so that it is representative for the full fatigue life of the flexible structure, and be usable for fatigue life evaluation.

2. Governing equations

The governing equation for a solid structure is normally stated in the Lagrangian formulation, in which the reference frame moves and deforms with the structural deformations. On the other hand, the governing equations for fluid flow are normally stated in the Eulerian formulation, in which the reference frame remains stationary, independently of the fluid flow. Like the majority of partitioned FSI solvers [15], the FSI toolkit for foam-extend employs the Arbitrary Lagrangian-Eulerian (ALE) formulation for the fluid flow. This means that the reference frame for the fluid domain is independent of the fluid flow (Eulerian), but deforms according to the deformations of the structure (Lagrangian). In practice, this also means that the fluid mesh deforms according to, and conforms to, the deformed shape of the structure.

2.1. Governing equations for the fluid flow

The isothermal flow of an incompressible and Newtonian fluid may be fully described by the continuity equation and the Navier-Stokes equations. However, in order to solve these equations accurately with even the smallest turbulent eddies and fastest fluctuations resolved, an extremely fine spatial mesh and very small time steps would be required. Instead, we use large eddy simulation (LES), where only the large-scale turbulent eddies are resolved. This makes it possible to use a coarser mesh and larger time steps, decreasing the computational effort.

In LES, the small-scale turbulent eddies, which are smaller than the cutoff width Δ (normally equal to the cell size), are not resolved [35]. Instead, they are modelled by sub-grid-scale (SGS) stresses, τ_{ij}, which represent the effects of the unresolved motions. This spatial filtering decomposes the variables ϕ (pressure and velocity components) into a sum of a filtered component, $\overline{\phi}$, and a sub-grid-scale component, ϕ', i.e. $\phi = \overline{\phi} + \phi'$ [36]. The governing equations for the fluid flow are solved for the filtered components, in which case the incompressible LES continuity equation in the ALE formulation becomes [15]:

$$\operatorname{div}\left(\overline{\mathbf{u}} - \mathbf{u}_{\mathrm{m}}\right) = \frac{\partial\left(\overline{u}_i - u_{\mathrm{m}i}\right)}{\partial x_i} = 0 \tag{1}$$

where $\overline{\mathbf{u}} = \left[\overline{u}_1, \overline{u}_2, \overline{u}_3\right]$ is the filtered fluid velocity field, $\mathbf{u}_{\mathrm{m}}$ is the velocity of the reference frame (i.e. the mesh), and $\mathbf{x} = [x_1, x_2, x_3] = [x, y, z]$ are the Cartesian coordinates. The filtered and incompressible Navier-Stokes equations in the ALE formulation are given as [16, 25, 28, 35, 36]:

$$\rho_{\mathrm{f}} \frac{\partial \overline{u}_i}{\partial t} + \rho_{\mathrm{f}} \frac{\partial}{\partial x_j}\left(\overline{u}_i\left(\overline{u}_j - u_{\mathrm{m}j}\right)\right) = -\frac{\partial \overline{p}}{\partial x_i} + \mu_{\mathrm{f}} \frac{\partial^2 \overline{u}_i}{\partial x_j^2} - \frac{\partial \tau_{ij}}{\partial x_j} \tag{2}$$

where ρ_{f} is the density of the fluid, t is time, $\overline{p}$ is the filtered fluid pressure field, μ_{f} is the dynamic viscosity of the fluid and τ_{ij} are the SGS stresses. For simplicity, the effects of gravitation are neglected in the simulation.

The SGS stresses are given as [35]:

$$\tau_{ij} = \rho_{\mathrm{f}} \overline{u_i u_j} - \rho_{\mathrm{f}} \overline{u}_i \overline{u}_j \tag{3}$$

The first term in this equation cannot be calculated from the solution of the filtered governing equations. Instead, it has to be modelled. The simplest SGS model is the Smagorinsky-Lilly model, which is often used, e.g. [6, 18, 22, 36]. However, this model contains a constant which needs to be adjusted on a case-to-case basis [35]. Therefore, the one-equation k_{SGS} eddy viscosity model will be used in the current work. This model includes one additional transport equation, which is solved for the SGS turbulent kinetic energy, k_{SGS} [35]:

$$\frac{\partial\left(\rho_{\mathrm{f}} k_{\mathrm{SGS}}\right)}{\partial t} + \operatorname{div}\left(\rho_{\mathrm{f}} k_{\mathrm{SGS}}\left(\overline{\mathbf{u}} - \mathbf{u}_{\mathrm{m}}\right)\right) = \operatorname{div}\left(\frac{\mu_{\mathrm{SGS}}}{\sigma_k} \operatorname{grad}\left(k_{\mathrm{SGS}}\right)\right) + 2\mu_{\mathrm{SGS}} \overline{S}_{ij} \cdot \overline{S}_{ij} - \rho_{\mathrm{f}} \varepsilon_{\mathrm{SGS}} \tag{4}$$

where the SGS viscosity is given as $\mu_{\mathrm{SGS}} = \rho_{\mathrm{f}} C'_{\mathrm{SGS}} \Delta \sqrt{k_{\mathrm{SGS}}}$, the local strain rate of the resolved flow is given as $\overline{S}_{ij} = \frac{1}{2}\left(\partial \overline{u}_i / \partial x_j + \partial \overline{u}_j / \partial x_i\right)$, and the rate of dissipation is given as $\varepsilon_{\mathrm{SGS}} = C_\varepsilon k_{\mathrm{SGS}}^{3/2} / \Delta$. C'_{SGS} and C_ε are constants, while Δ is the filter cutoff width. With the SGS viscosity known, the SGS stresses may be calculated from:

$$\tau_{ij} = -2\mu_{\mathrm{SGS}} \overline{S}_{ij} + \frac{1}{3} \tau_{ii} \delta_{ij} \tag{5}$$

where δ_{ij} is the Kronecker delta. The van Driest damping function [37] has been used in the simulations to reduce the near-wall eddy viscosity.

2.2. Governing equation for the structure

The governing equation for the structure is the conservation of linear momentum. For an isothermal structure in the total Lagrangian formulation, taking large deformations into account, this equation may be written as [26, 38]:

$$\int_{\Omega_0} \rho_{s0} \frac{\partial \mathbf{u}_s}{\partial t} \mathrm{d}\Omega_0 = \int_{\Gamma_0} \mathbf{n}_0 \cdot \left(\mathbf{S} \cdot \mathbf{F}^{\mathrm{T}} \right) \mathrm{d}\Gamma_0 + \int_{\Omega_0} \rho_{s0} \mathbf{f}_b \, \mathrm{d}\Omega_0 \tag{6}$$

where ρ_{s0} is the initial density of the structure, $\mathbf{u}_s$ is its velocity vector, Ω_0 is its initial volumetric domain, Γ_0 is its initial surface, and $\mathbf{n}_0$ is its outward pointing unit normal vector. $\mathbf{S}$ is the applied second Piola-Kirchhoff stress, $\mathbf{f}_b$ is any body force acting, and $\mathbf{F}$ is the deformation gradient, given as:

$$\mathbf{F} = \mathbf{I} + \left(\nabla_0 \mathbf{d}_s \right)^{\mathrm{T}} \tag{7}$$

where $\mathbf{d}_s$ is its displacement vector and ∇_0 represents the gradient operator at the undeformed geometry.

The second Piola-Kirchhoff stress $\mathbf{S}$ may be related to the more commonly used Cauchy stress $\boldsymbol{\sigma}$ (current force acting on current geometry) by the following equation:

$$\boldsymbol{\sigma} = \frac{1}{\det \mathbf{F}} \mathbf{F} \mathbf{S} \mathbf{F}^{\mathrm{T}} \tag{8}$$

For an elastic St. Venant-Kirchhoff material, the constitutive equation is:

$$\mathbf{S} = 2\mu_s \mathbf{E} + \lambda \operatorname{tr}(\mathbf{E}) \mathbf{I} \tag{9}$$

where $\mathbf{E}$ is the Green-Lagrangian strain and μ_s and λ are the Lamé parameters [38]:

$$\mathbf{E} = \frac{1}{2} \left(\mathbf{F}^{\mathrm{T}} \mathbf{F} - \mathbf{I} \right) \tag{10}$$

$$\mu_s = \frac{E}{2(1+\nu_s)}, \quad \lambda = \frac{\nu_s E}{(1+\nu_s)(1-2\nu_s)} \tag{11}$$

E is the elastic modulus and ν_s is Poisson's ratio for the material.

2.3. Fluid-structure interface

At the fluid-structure interface, the following conditions must be met [15]:

$$\begin{gathered} \mathbf{u}_f = \mathbf{u}_s \\ \boldsymbol{\sigma}_s \cdot \mathbf{n} = \boldsymbol{\sigma}_f \cdot \mathbf{n} \end{gathered} \tag{12}$$

Here, $\mathbf{u}_f$ is the fluid velocity field, $\mathbf{u}_s$ is the structural velocity, $\mathbf{n}$ is the unit normal vector of the interface, and $\boldsymbol{\sigma}_s$ and $\boldsymbol{\sigma}_f$ are the Cauchy stress matrices for the solid and fluid, respectively.

3. Discretisation of the equations

3.1. Discretisation of the fluid equations

The FSI toolkit for foam-extend 4.0 is used to solve the governing equations numerically, using the finite volume method. For the fluid flow, the PISO (Pressure Implicit with Splitting of Operators) scheme is used in the present study.

The SGS stresses calculated as part of the LES model are similar in magnitude to the numerical truncation errors [35]. It is therefore important to reduce these errors, so that they do not swamp the SGS stresses. This is done by using second-order or higher-order discretisation schemes in time and space.

The recommended temporal discretisation scheme for LES is the Crank-Nicolson scheme [39], while the backward scheme may give unacceptable dispersion [22]. The Crank-Nicolson scheme is an implicit scheme, which is second-order accurate in time, by weighting field variables at two succeeding time steps equally [35]. OpenFOAM (and foam-extend) allows the Crank-Nicolson scheme to be mixed with the first-order implicit Euler scheme. It is recommended to use a mix of 90 % Crank-Nicolson and 10 % Euler scheme in OpenFOAM, in order to bound/stabilise the scheme [40]. However, investigations by the authors have shown that using 80 % Crank-Nicolson and 20 % Euler works better in foam-extend, as illustrated by pure CFD simulations in Figure 2.

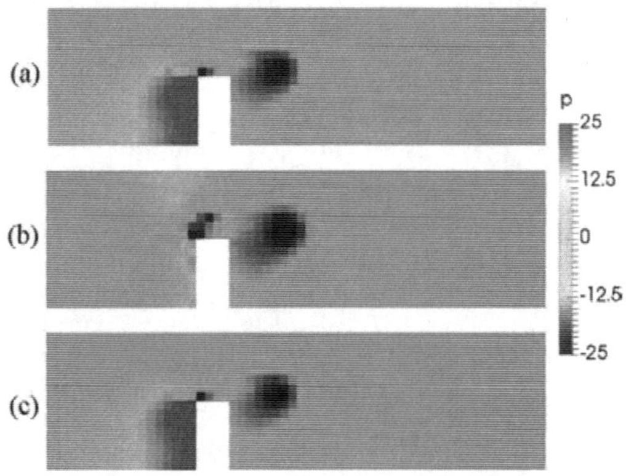

Figure 2. Pressure fields [Pa] for identical cases at t = 0.09 s, computed by (a) OpenFOAM 4.0, Crank-Nicolson 0.9, (b) foam-extend 4.0, Crank-Nicolson 0.9 and (c) foam-extend 4.0, Crank-Nicolson 0.8. For (b), the pressure field in front of the obstruction suddenly disappears between t = 0.08 s and t = 0.09 s.

The Courant number is given as $Co = |\mathbf{u}_f| \cdot \Delta t / \Delta x$, where $|\mathbf{u}_f|$ is the magnitude of the local velocity vector, Δt is the length of the time step and Δx is the length of the finite volume cell. The maximum Courant number should be kept below 1.0 when using the transient PISO scheme [41]. Here, a constant time step length has been used to keep the maximum Courant number around 0.53.

For the spatial discretisation of the convection and diffusion terms, and the gradient, central differencing with Gaussian integration is used. More specifically, OpenFOAM's scheme "Gauss filteredLinear2 1 0" was used for the div(phi,U) term, in order to remove staggering caused by pressure-velocity decoupling. Further details of the discretisation schemes (in the OpenFOAM branch) may be found in [40].

3.2. Discretisation of the structure equations

The finite volume discretisation of equation (6) contains one temporal term, one diffusive term and two source terms [30].

The first-order fully implicit Euler temporal discretisation scheme was first added to the FSI toolkit, and has been used in the current work. The second-order backward scheme has later been added to the FSI toolkit [24], and should be used for increased accuracy [17].

The diffusive term is discretised using central differencing with Gaussian integration, and gradients are computed using the least-squares approach. Further details of the discretisation of the governing equation for the structure may be found in [30].

3.3. Discretisation of the fluid-structure interaction

The fluid-structure interaction is solved in a partitioned approach, with strong coupling, as illustrated in Figure 1. The displacement increment and velocity of the structure are transferred to the fluid, by deforming the fluid mesh, while the pressure and viscous forces are transferred from the fluid to the structure. This is the most common procedure, as the reversed exchange of information might lead to instability [15]. Hence, the fluid forces are applied on the structure as natural boundary conditions.

3.3.1. Under-relaxation. As part of the iterative strong coupling, the fluid mesh is deformed at each iteration step. However, in order to stabilise the iterations and enforce convergence [14], the fluid mesh is not moved the same distance as the predicted structural displacement at the last iteration step. Instead, the iteration is under-relaxed. This means that the displacement of the fluid mesh at iteration step i+1 is calculated as:

$$\mathbf{d}_{i+1} = \omega_i \tilde{\mathbf{d}}_{i+1} + (1-\omega_i)\mathbf{d}_i \tag{13}$$

where $\mathbf{d}_i$ is the displacement of the fluid mesh at iteration step i, $\tilde{\mathbf{d}}_{i+1}$ is the structural displacement predicted in iteration step i, and ω_i is called the relaxation factor.

Three relaxation schemes are available in the FSI toolkit for foam-extend [24]: Fixed relaxation (i.e. ω_i = constant), Aitken dynamic relaxation and IQN-ILS. The fixed relaxation often leads to slow convergence [14], and dynamic relaxation is therefore employed. The convergence rate of both the dynamic relaxation techniques has briefly been tested, but no significant difference was observed. The Aitken dynamic relaxation method is one of the most popular methods [42], and has been employed in the current work. Details of the method may be found in [43]. In the current work, an initial relaxation factor of 0.4 is used at each time step, and iterations are made until a relative tolerance of 10^{-5} is reached. If this tolerance is not reached within 200 iterations, the solver simply moves on to the next time step.

3.3.2. Mesh movement. The motion of the fluid mesh is computed by a Laplacian approach, in which the points are moved according to the Laplace equation [15, 25]:

$$\nabla \cdot (\gamma \nabla \mathbf{u}_{\mathrm{m}}) = 0 \tag{14}$$

where γ is the diffusion coefficient, which is chosen to be proportional to the squared inverse distance from the moving boundary. The solver "refVelocityLaplacian" is chosen, with 2 non-orthogonal correctors.

4. Modelling approach

4.1. Geometry, properties and boundary conditions

The geometry of the problem chosen for this study is shown in Figure 3. It is based on Richter's laminar test case [13], but has been extended to full three-dimensionality, in order to allow turbulence modelling. The geometry of the problem consists of a flexible wall (structure) inside a rectangular duct. The duct has one inlet and one outlet. Its width is 0.8 m and its height is 0.4 m. The flexible wall is 0.4 m wide, 0.1 m thick and 0.2 m high, and is fixed to the floor of the duct, as shown in Figure 3. While Richter's case had an inlet length, IL, of 0.45 m, preliminary studies have shown that this is too short for the turbulent case, and therefore it has been increased to IL = 0.95 m (units are here added to Richter's non-dimensional measures).

As in Richter's case, the duct conveys an artificial fluid of density ρ_{f} = 1000 kg/m^3 and kinematic viscosity $\nu_{\mathrm{f}} = 10^{-3}$ m^2/s. In Richter's case, the fluid enters the duct with mean velocity U = 0.15 m/s. This velocity is here increased to U = 5 m/s, in order to get turbulence in the wake of the flexible wall. In this case, the Reynolds number, Re, for the flow entering the duct is 2667, which is within the critical zone between fully laminar (Re < 2000) and fully turbulent (Re > 4000) flow [44]. This value should be sufficient to give turbulence in the wake of the flexible wall, however.

For simplicity, a laminar inlet velocity profile, i.e. a parabolic velocity profile, has been used as an approximation. The peak velocity in the middle of the cross-section is $u_{\mathrm{max}} = 2U$ = 10 m/s. The no-slip condition is applied for the bottom, top and sides of the fluid domain, as well as for the interface to the flexible wall. At the outlet, the zero gradient condition is imposed on the velocity, and the pressure has a Dirichlet boundary condition with zero pressure.

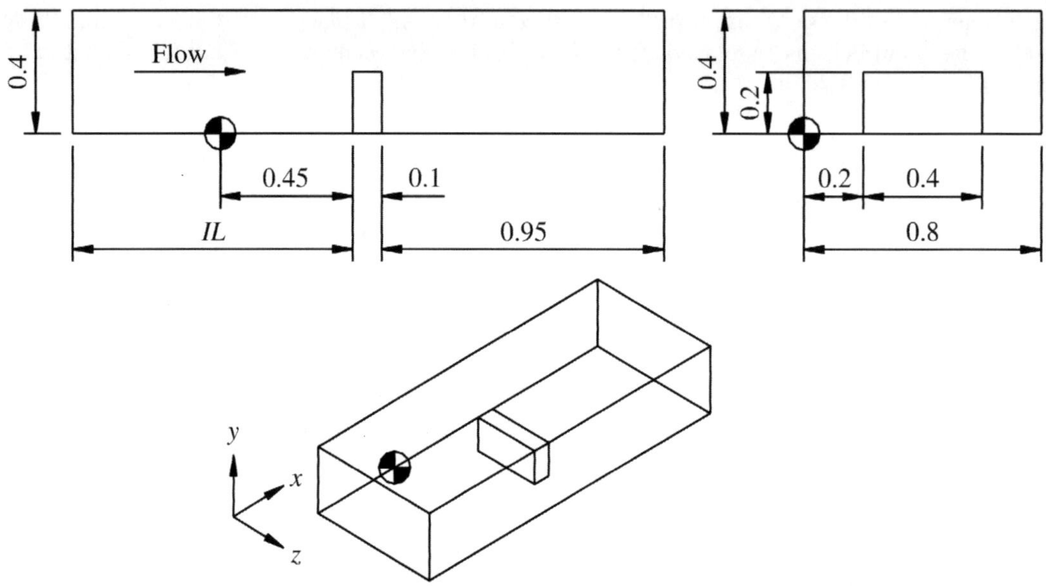

Figure 3. Geometry of the problem. All dimensions are given in meters.

The properties of the flexible wall have been changed from those in the Richter's case, in order to obtain visible, but yet not extreme deformations. The wall is too thick to obtain significant vibrations with any realistic material. Its density, ρ_s, has been set equal to 8000 kg/m^3, Poisson's ratio, ν_s, has been set to 0.3, while the elastic modulus has been chosen to be E = 4 MPa, which is very soft. The flexible wall is fixed ($\mathbf{d}_s = 0$) at the bottom, but is otherwise free to move.

4.2. Initial conditions

It is chosen to start the solution from a fully developed turbulent flow, and the structure at rest; a technique which has also been used in [18, 42]. The chosen initial velocity field is shown in Figure 4. The structure is allowed to move during the first computational step. This initial condition is preferred, because it requires little computational effort, compared to the alternatives.

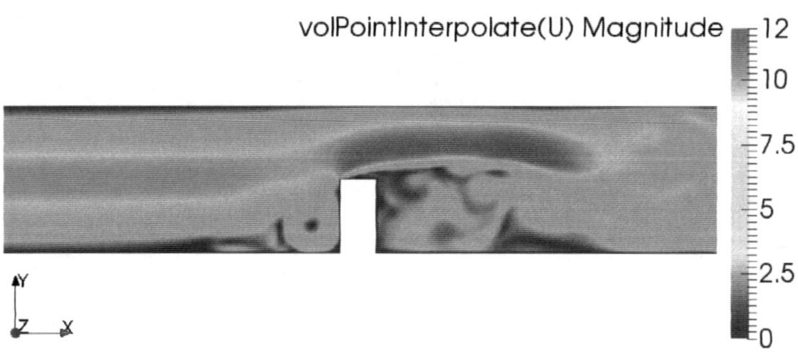

Figure 4. Initial velocity field [m/s] at $t = 0$ (volume point interpolated).

However, as illustrated in [18, 42], the sudden exposure of the flexible structure to a fully developed fluid flow induces an initial transient to its vibration. In the context of fatigue life estimation, we are mostly interested in the stabilised vibrations remaining when this transient effect has been damped out. In order to reduce the transient effect, the inlet velocity could have been slowly increased to the full flow. However, as very many FSI iterations are required even at low velocities, this procedure is computationally demanding. Furthermore, the PISO solver implemented in the FSI toolkit does not contain any ability to automatically adjust the time step to keep the Courant number within a given limit, meaning that the time step would have needed to be manually decreased as the velocity increased.

4.3. Computational meshes

While Richter [13] uses finite elements to discretise the computational domain, both Mehl, et al. [14] and Gillebaart, et al. [17] use hexahedral finite volume cells which are approximately 2.5 cm long in each direction, for the flow domain. In a fully three-dimensional model, this corresponds to 40 448 finite volume fluid cells. This is quite a coarse mesh for a large eddy simulation. In the current study, the cell size has therefore been decreased to 1.0 cm in each direction, resulting in a total of 632 000 cells in the fluid domain. In Figure 5, the vibration history for the flexible wall, computed by both meshes, is plotted for 1.5 seconds, and it is clear that the coarser mesh is not able to capture all the small eddies which significantly influence the vibration of the flexible wall. The fine mesh, which is used to obtain the remainder of the results presented, is shown in Figure 6.

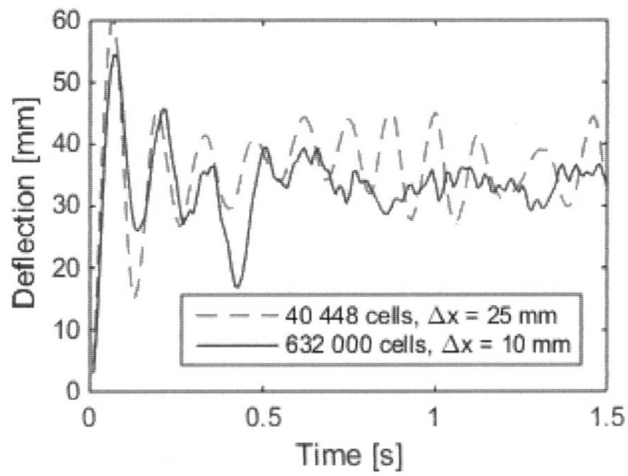

Figure 5. Deflection history for the wall, for two different meshes.

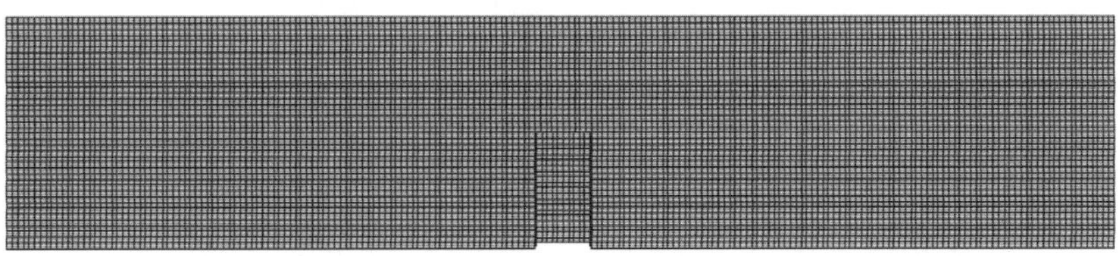

Figure 6. Section of the mesh used in the simulations.

The flexible wall has in all cases been discretised by 512 hexahedral finite volume cells, each 2.5 cm long in each direction. Both meshes have been decomposed for 20 processors using the metis algorithm. A computational time step Δt of 250 μs has been used to keep the maximum Courant number around 0.53. Simulation results have been saved every 0.01 s, i.e. at a sampling frequency of 100 Hz.

In order to assess the quality of the simulation, the dependence of these parameters should be further investigated. Such studies could not be conducted within the present work however, due to time and CPU constraints.

IOP Conf. Series: Materials Science and Engineering **276** (2017) 012031 doi:10.1088/1757-899X/276/1/012031

5. Simulation results

Results from the simulations are illustrated in this section, and further discussed in section 6. If otherwise is not mentioned, the figures show the computed fields at the plane z = 0.4 (ref. Figure 3), which cuts through the middle of the flexible wall. Finite volume results are saved in the nodes in the centre of each cell, and the field values are thereby constant throughout each cell. However, volume point interpolation has been used to smooth out the fields shown in the figures.

5.1. Computation time

Due to time constraints, the simulation was run for 2.32 simulated seconds only. For the first 1.5 simulated seconds, the computation time was 96 hours (clock time) using 20 of 80 available Intel Xeon CPU E7- 8870 @ 2.40 GHz processor cores. The remaining cores were idle. This corresponds to an average of 38 min 33 s computation time per 0.01 second simulated. On average, 17.7 FSI iterations had to be made for each time step of 250 μs, as shown in Figure 7. Simulation data was saved every 0.01 simulated second, each time step being composed of approximately 293 MB of data.

For the remainder of the simulation, the 80-core computer was subjected to an increased workload, and did also randomly crash. Continuous log files are therefore not available between t = 1.5 s and t = 1.88 s. Between t = 1.88 s and t = 2.32 s, the computation time was 96 hours (clock time) using 20 cores, i.e. 130 min 51 s per 0.01 second simulated, on average. As the average number of FSI iterations was more or less the same as for the first 1.5 s, the increased computation time is thought to result from the increased workload on the computer, which had around 120-150 concurrent jobs at this time.

For reference, similar simulations without FSI did only require 33 s clock time per 0.01 second simulated, using 20 processor cores.

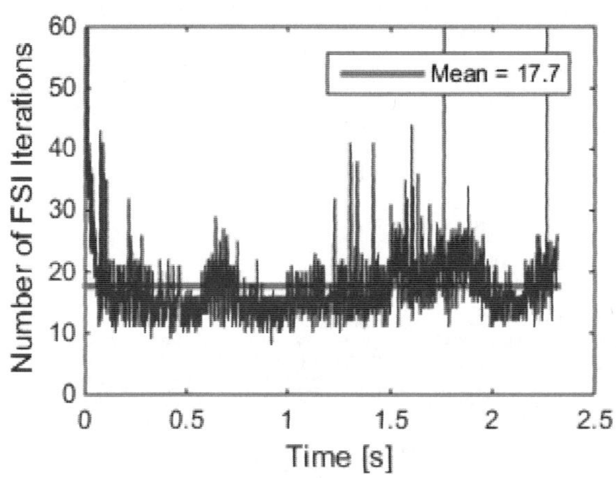

Figure 7. Number of FSI iterations required at each time step to reach a relative FSI tolerance of 10^{-5}.

5.2. The fluid flow

The local velocity reaches its maximum at t = 0.079 s. The nearest time step which is saved is t = 0.08 s, and the velocity field at this time step is shown in Figure 8(a). It is clear that the flexible wall has become significantly curved, and that the relatively smooth flow path (continuous red field in Figure 4) has been disturbed.

The velocity field for the last time step is shown in Figure 8(b). The wall is not as curved as at t = 0.08 s, but the amount of turbulence has increased. A lot of mixing is taking place in the wake of the wall, and the velocity profile is no longer smooth upstream the wall either.

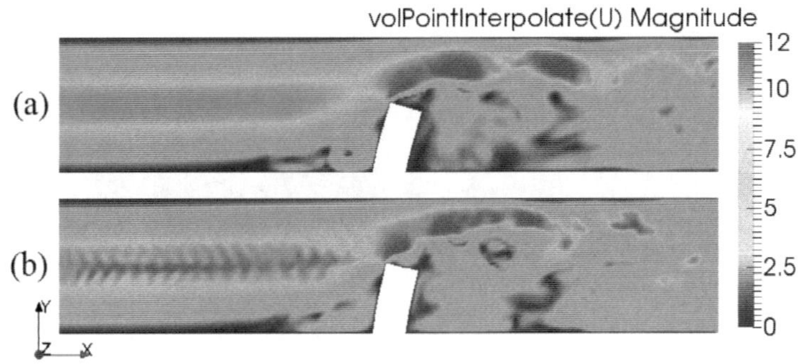

Figure 8. Velocity field [m/s] at (a) $t = 0.08$ s, and (b) $t = 2.32$ s.

The time average (mean) velocity fields are shown in Figure 9 for $t = 1.80$ s and for $t = 2.32$ s. It is observed that these two fields look relatively similar; indicating that a stabilised flow has been reached.

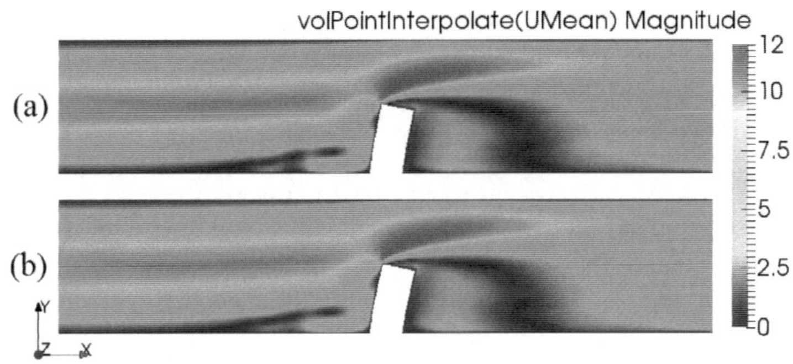

Figure 9. Time averaged velocity field [m/s] up to (a) $t = 1.80$ s, and (b) $t = 2.32$ s.

The increase in turbulence due to the vibration of the flexible wall is also evident from Figure 10, where the SGS kinematic viscosity fields are shown for $t = 0$ and $t = 2.32$ s. The flow downstream the flexible wall is clearly turbulent, and the upstream turbulence has also been increased. The turbulent field at $t = 2.32$ s is illustrated by isosurfaces of the second invariant of the velocity gradient tensor, Q, in Figure 11. In addition to the turbulence downstream the flexible wall, it indicates that a spiralling flow pattern has developed near the inlet.

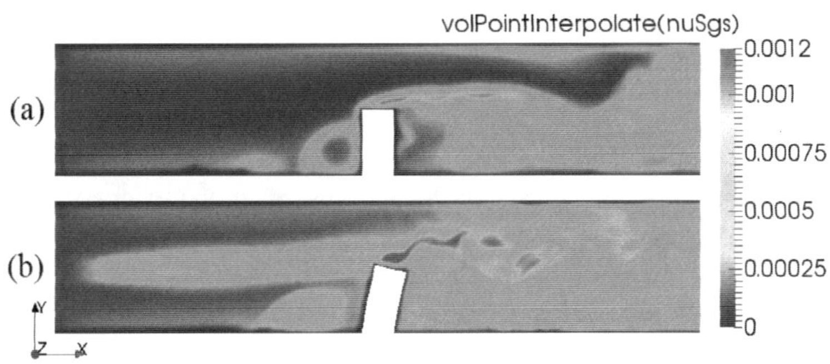

Figure 10. The SGS kinematic viscosity fields [m^2/s] at (a) $t = 0$, and (b) $t = 2.32$ s.

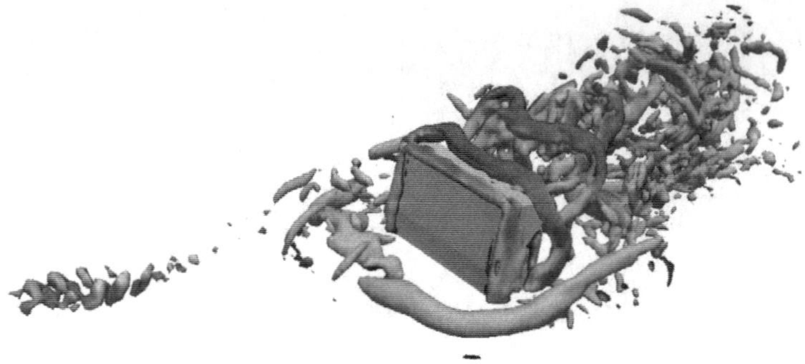

Figure 11. Isosurfaces for Q = 10^4, at t = 2.32 s, coloured by velocity magnitude.

5.3. The vibration of the flexible wall

The deflection history for the point (x, y, z) = (0.45, 0.2, 0.4), which is positioned at the middle of the upper front edge of the wall, is shown in Figure 12. The transient initial vibration due to the sudden freedom of the wall at t = 0 is clearly visible. The time history is too short to determine whether these transient effects have disappeared during the simulated interval, but the curves do indicate a possible stabilised vibration starting a bit before t = 2 s.

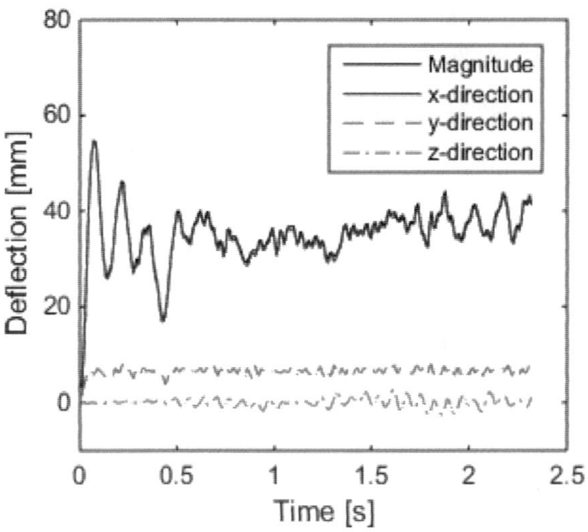

Figure 12. Deflection history for the wall, at (x, y, z) = (0.45, 0.2, 0.4). The difference between the histories of the component in x-direction, and the absolute value/magnitude of the displacement, is indistinguishable.

With a sampling frequency of 100 Hz, the deflection magnitude history has been found to be dominated by frequencies around 7.3 Hz, with lesser peaks at 2.2 Hz and 27 Hz.

For the purpose of estimating the fatigue life of a structure, it is essential to know its stress history. The stress state at a point is often represented by the von Mises equivalent stress, which may be calculated as:

$$\sigma_{\mathrm{VM}} = \left\{ \frac{1}{2}\left[\left(\sigma_{xx} - \sigma_{yy}\right)^2 + \left(\sigma_{yy} - \sigma_{zz}\right)^2 + \left(\sigma_{zz} - \sigma_{xx}\right)^2 \right] + 3\left(\sigma_{xy}^2 + \sigma_{yz}^2 + \sigma_{xz}^2\right) \right\}^{1/2} \quad (15)$$

The von Mises stress distribution in the wall at t = 2.32 s is shown in Figure 13. The asymmetry along the z-axis is assumed to be a result of the turbulence of the flow.

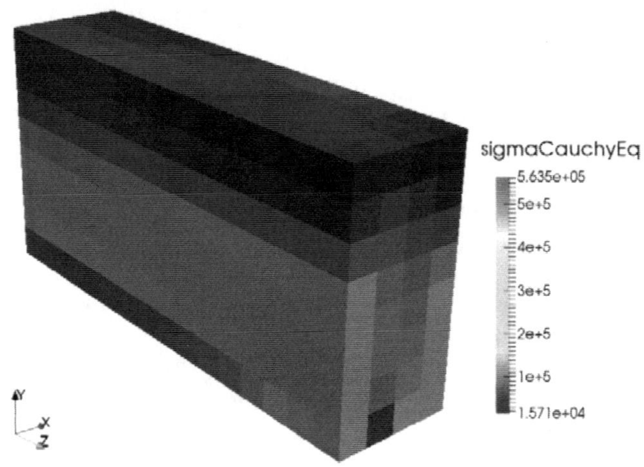

Figure 13. Von Mises stress distribution [Pa] at $t = 2.32$ s.

As illustrated in Figure 13, the most critically loaded part of the wall is the strip along the front, close to the bottom. It is likely that any fatigue crack will nucleate along this strip. The time histories of the different stress components at the middle of the strip, $(x, y, z) = (0.45, 0.0125, 0.4)$ are shown in Figure 14.

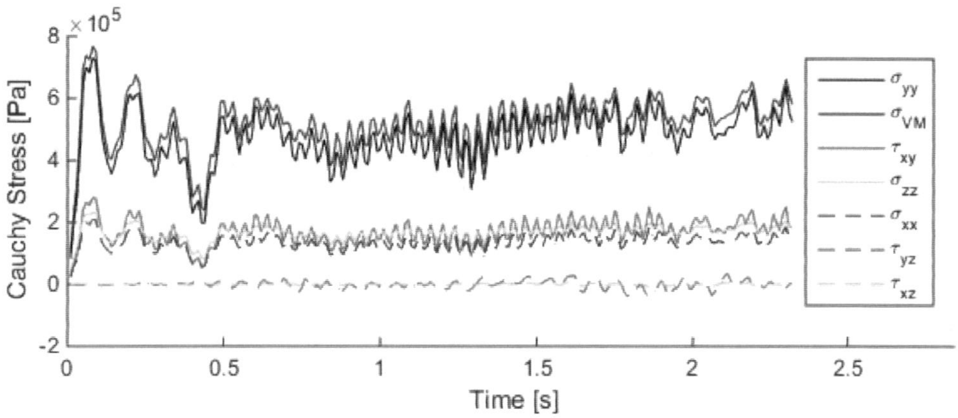

Figure 14. Stress histories at $(x, y, z) = (0.45, 0.2, 0.4)$.

6. Discussion

Figures 12 & 14 indicate that the FSI toolkit for foam-extend 4.0 is capable of producing a vibration history, which is appropriate for evaluating the fatigue life of the structure. However, some important aspects need to be taken into account, and will be addressed in this section.

6.1. The initial transient

As predicted in Section 4.2, the sudden application of fluid-structure interaction in the fully developed flow induces an initial transient to the vibration history, as observed in Figure 12. As momentum is dissipated from the structure to the fluid, the vibration of the wall, due to the sudden application, is damped, and will diminish over time. This means that a "steady-state" or stabilised vibration will be reached after a certain time. If the wall is subjected to the turbulent flow for most of its life, it is this stabilised vibration which is interesting for fatigue purposes. When the vibration is stabilised, a relatively short vibration response may be extrapolated to represent the full vibration history of the wall. Then, its fatigue life may be estimated by using a S-N curve for the material and applying a cycle counting technique and a cumulative damage law [1], like the well-known linear Palmgren-Miner rule [45] or more recent, nonlinear rules, e.g. [46].

However, as shown in Figures 12 & 14, it is not possible to indisputably assess whether the transient effect has diminished completely within the 2.32 seconds simulated in the current work. There is still a lot of variation. Still, the last three major oscillations of the deflection magnitude shown in Figure 12 seem to have a relatively constant mean value and frequency, with only small deviations in the amplitude. This may indicate that the stabilised vibration either has been reached, or that it is coming. This indication is further strengthened by the observation made of the time averaged velocity fields shown in Figure 9, where the flow seems to have stabilised at about $t = 1.80$ s.

This indicates that even though the simulated interval of 2.32 s is too short to make any sensible fatigue life estimations, we are not very far away from having sufficient data to make such estimations. It is absolutely possible that having results all the way up to $t = 3$ or 4 s would have made such estimations possible. The major problem is the time required to obtain these results. Based on the obtained results, somewhere between 44 and 148 hours of computation would be required to reach $t = 3$ s.

6.2. The need for simulation of the fluid-structure interaction

Throughout section 5, it is noted that the amount of turbulence has increased during the time interval from $t = 0$ to $t = 2.32$ s, whereas the amount of turbulence stabilised in similar simulations where FSI was not employed. This turbulence increase may be directly related to the fluid-structure interaction. While the turbulence induces vibrations to the flexible wall, the vibrations of the wall do also induce additional disturbances and mixing, i.e. turbulence, to the flow. This illustrates the importance of applying (strong) fluid-structure interaction when simulating the vibration of flexible structures being subjected to a turbulent flow.

6.3. Assessment of the simulation results

As no experimental data is available for the case investigated, the validity of the simulation results cannot be confirmed. Furthermore, the authors are not aware of other implementations of LES to this case. The flexible wall is relatively thick and short, and can therefore be analysed neither as a plate nor as a cantilever beam. The next step would therefore be to perform simulations of a more realistic problem, where an analytical solution of the natural frequencies of the structure, and preferably, experimental data, are available.

Several sources of errors exist for the computed results. First and foremost, a full convergence check with respect to mesh refinement and time step length has not been conducted within the present work, due to time and CPU constraints. The results may therefore be sensitive to these parameters. Secondly, while the influence of the inlet length has been investigated, it is possible that the length from the flexible wall to the outlet may be too short. Also, the sampling frequency of 100 Hz may be too low to resolve the highest vibration frequencies. Furthermore, the accuracy of the vibration history may have been increased by using the second-order backward temporal discretisation scheme for the structure.

In addition, as the flow is confined by the duct on four sides, the results rely on the quality of the ν_{SGS} wall function, which may be questionable. Further refinement of the mesh near the wall, so that the first cell centre would lie at a non-dimensional distance, y^+, of ~1 from the wall (instead of $y^+ \approx 12$, as used here), would make it possible to omit the use of the wall function [41]. The spiralling flow developing from a unidirectional inlet may indicate that some numerical problem exists.

6.4. Computational effort

With the resolution and settings used in the current work, between 64 and 218 hours of computation are required to compute one simulated second, depending on the workload of the computer. The computational cost is therefore high. However, as indicated above, even three to four seconds of data may be sufficient to estimate the full fatigue life of the structure. For critical structures, performing

FSI simulations in foam-extend may be a good alternative to the empirical and analytical tools often used today [7].

Within the current FSI toolkit for foam-extend, a couple of parameters could be tuned which could possibly further decrease the computational effort. In the current work, the computation has been performed in parallel using 20 processor cores, each core handling 31 600 fluid cells and 25–26 structure cells, decomposed by the metis algorithm. It is possible that this is a low amount of cells per core, meaning that excessive time is used to transfer information between the cells. However, computations with even fewer fluid cells per processor core are found in the literature, e.g. Tuković and Jasak's [29] simulation of a droplet using 8 parallel processors, each handling about 7777 cells.

Nevertheless, some computation time could probably be saved by using fewer processor cores to compute the structure. Here, the toolkit's standard setting of using all of the same cores for the computation of both regimes was chosen for simplicity. The fluid and structure are usually computed in serial, i.e. in a staggered pattern, meaning that all the processor cores cannot be fully utilised when running a parallel computation, no matter how the structure is decomposed. Mehl, et al. [14] have proposed two techniques for simultaneous parallel computation of both regimes at each time step, which could be implemented to maximise the utilisation of each processor core.

A relatively large amount of the computation time is spent moving and deforming the fluid mesh [31]. In the current work, the FVM-based "refVelocityLaplacian" solver with 2 non-orthogonal correctors has been used to deform the mesh. The FEM-based solver developed by Jasak and Tuković [25] is also available. It is stated to be more robust, and may influence the computation time.

The number of FSI iterations does also contribute significantly to the computation time. With the Aitken scheme, an initial under-relaxation factor of 0.4 and required relative tolerance of 10^{-5}, 17.7 FSI iterations were required at each time step, on average. This is a significant increase from the laminar case. Changing any of these parameters would influence the number of FSI iterations, and thereby the computation time. Limiting the number of FSI iterations will be of major importance for reducing the computational effort for simulating turbulent FSI. van Zuijlen and Bijl [42] have suggested a technique where the flow is solved on alternating coarse and fine meshes during the FSI iteration within each time step, in order to reduce the computational effort for RANS simulations. However, this technique has limited application to LES simulations, where the fine mesh always is required to give a sufficient resolution.

The last solution to reduce the computational effort we will mention is the use of a more sophisticated turbulence model. Implicit large eddy simulation [16], hybrid RANS-LES models [32] and detached eddy simulation techniques [7] have all been used for this purpose to solve turbulent FSI problems.

7. Conclusions

In this paper, a large-eddy simulation (LES) turbulence model has been implemented to Richter's fluid-structure interaction (FSI) test case, and 2.32 seconds of turbulent fluid-structure interaction has been simulated, using the FSI toolkit for foam-extend 4.0. It has been shown that this toolkit is capable of producing a vibration history, which is appropriate for evaluating the fatigue life of a flexible structure.

The computational effort required to perform such simulations is large, however. Even for the relatively simple case investigated here, 632 000 cells were required to discretise the fluid domain. An average number of 17.7 FSI iterations were required at each time step, due to the relatively slow FSI convergence rate. The FSI convergence rate is significantly lower for turbulent flows modelled by LES, than for laminar flows. Because of this, over 200 computation hours was required for the simulation of 2.32 seconds of turbulent FSI.

This indicates the need of applying more sophisticated turbulence models and/or faster FSI iteration schemes. The FSI toolkit for foam-extend could also require the possibility of automatically adjusting the time step according to the Courant number, and the implementation of additional solvers.

Especially a two-phase solver for the simulation of ocean waves and a compressible solver to simulate water hammer induced vibrations of pipelines would be appreciated.

In future work, the factors influencing the computational effort, as mentioned in section 6.4, should be further studied. The studies should also consider a more realistic problem, with experimental and/or analytical data available.

Acknowledgements
This work was supported by the Norwegian Ministry of Education and Research. Dr. K.E. Giljarhus is acknowledged for constructive suggestions.

References

[1] Suresh S 1998 *Fatigue of Materials 2nd ed* (Cambridge: Cambridge University Press)
[2] Saeed R A, Galybin A N and Popov V 2010 Modelling of flow-induced stresses in a Francis turbine runner *Adv. Eng. Softw.* **41** 1245-55
[3] Nebenführ B and Davidson L 2017 Prediction of wind-turbine fatigue loads in forest regions based on turbulent LES inflow fields *Wind Energy* **20** 1003-15
[4] Trim A D, Braaten H, Lie H and Tognarelli M A 2005 Experimental investigation of vortex-induced vibration of long marine risers *J. Fluids Struct.* **21** 335-61
[5] Holmes S, Oakley jr O H and Constantinides Y 2006 Simulation of riser VIV using fully three dimensional CFD simulations *Proc. of OMAE2006 25th Int. Conf. on Offshore Mech. and Arctic Eng. (Hamburg, Germany, 4-9 June)* vol 4 pp 563-70
[6] Kamble C and Chen H-C 2016 3D VIV fatigue analysis using CFD simulation for long marine risers *Proc. of the 26th (2016) Int. Ocean and Polar Eng. Conf. (Rhodes, Greece, 26 June – 1 July)* vol 3 pp 1094-100
[7] Nguyen V-T and Nguyen H H 2016 Detached eddy simulations of flow induced vibrations of circular cylinders at high Reynolds numbers *J. Fluids Struct.* **63** 103-19
[8] Shepherd J E and Inaba K 2010 Shock loading and failure of fluid-filled tubular structures *Dynamic Failure of Materials and Structures* ed A Shukla, et al. (New York: Springer) chapter 6 pp 153-90
[9] Tijsseling A S 1996 Fluid-structure interaction in liquid-filled pipe systems: A review *J. Fluids Struct.* **10** 109-46
[10] Benra F-K, Dohmen H J, Pei J, Schuster S and Wan B 2011 A comparison of one-way and two-way coupling methods for numerical analysis of fluid-structure interactions *J. Appl. Math.* **2011** 853560
[11] Woldemariam E T and Lemu H G 2016 Comparative analysis of computational methods in fluid-structure interaction: Temporal discretization and coupling techniques *Proc. of the 6th Int. Workshop of Adv. Manuf. and Automation (Manchester, UK 10-11 November)* pp 144-50
[12] Hübner B, Walhorn E and Dinkler D 2004 A monolithic approach to fluid-structure interaction using space-time finite elements *Comput. Methods Appl. Mech. Eng.* **193** 2087-104
[13] Richter Th 2012 Goal-oriented error estimation for fluid-structure interaction problems *Comput. Methods Appl. Mech. Eng.* **223-224** 28-42
[14] Mehl M, Uekermann B, Bijl H, Blom D, Gatzhammer B and van Zuijlen A 2016 Parallel coupling numerics for partitioned fluid-structure interaction simulations *Comput. Math. Appl.* **71** 869-91
[15] Campbell R L and Paterson E G 2011 Fluid-structure interaction analysis of flexible turbomachinery *J. Fluids Struct.* **27** 1376-91
[16] Cesur A, Carlsson C, Feymark A, Fuchs L and Revstedt J 2014 Analysis of the wake dynamics of stiff and flexible cantilever beams using POD and DMD *Comput. Fluids* **101** 27-41
[17] Gillebaart T, Blom D S, van Zuijlen A H and Bijl H 2016 Time consistent fluid structure interaction on collocated grids for incompressible flow *Comput. Methods Appl. Mech. Eng.* **298** 159-82

[18] Münsch M and Breuer M 2010 Numerical simulation of fluid-structure interaction using eddy-resolving schemes *Fluid Structure Interaction II, Lecture Notes in Computational Science and Engineering 73* ed H-J Bungartz, et al. (Berlin/Heidelberg: Springer) pp 221-53

[19] Menter F, Sharkey P, Yakubov S and Kuntz M 2006 Overview of fluid-structure coupling in ANSYS-CFX *Proc. of OMAE2006 25th Int. Conf. on Offshore Mech. and Arctic Eng. (Hamburg, Germany, 4-9 June)* vol 4 pp 579-85

[20] Marzban A, Lakshmiraju M, Richardson N, Henneke M, Wu G, Vargas P M and Oakley O 2012 Offshore platform fluid structure interaction simulation *Proc. of the ASME 2012 31st Int. Conf. on Ocean, Offshore and Arctic Eng. OMAE2012 (Rio de Janeiro, Brazil, 1-6 July)* vol 2 pp 235-42

[21] Paczkowski K W, Zhang P, Rogers R and Richardson N 2014 Fluid-structure interaction study on dynamic response of a capped drilling riser filled with mud *Proc. of the ASME 2014 33rd Int. Conf. on Ocean, Offshore and Arctic Eng. OMAE2014 (San Francisco, US, 8-13 June)* vol 2 p V002T08A018

[22] Nilsson P, Lillberg E and Wikström N 2012 LES with acoustics and FSI for deforming plates in gas flow *Nucl. Eng. Des.* **253** 387-95

[23] Weller H G, Tabor G, Jasak H and Fureby C 1998 A tensorial approach to computational continuum mechanics using object-oriented techniques *Comput. Phys.* **12** 620-31

[24] Tuković Ž, Cardiff P, Karač A, Jasak H and Ivanković A 2014 OpenFOAM library for fluid structure interaction *9th OpenFOAM Workshop (Zagreb, Croatia, 23-26 June)*

[25] Jasak H and Tuković Ž 2006 Automatic mesh motion for the unstructured finite volume method *Trans. FAMENA* **30**(2) 1-20

[26] Tuković Ž and Jasak H 2007 Updated Lagrangian finite volume solver for large deformation dynamic response of elastic body *Trans. FAMENA* **31**(1) 55-70

[27] Beaudoin M and Jasak H 2008 Development of a generalized grid interface for turbomachinery simulations with OpenFOAM *Open Source CFD Int. Conf. (Berlin, Germany, 4-5 December)*

[28] Jasak H and Tuković Ž 2010 Dynamic mesh handling in OpenFOAM applied to fluid-structure interaction simulations *V Eur. Conf. on Comput. Fluid Dyn., ECCOMAS CFD 2010 (Lisbon, Portugal, 14-17 June)*

[29] Tuković Ž and Jasak H 2012 A moving mesh finite volume interface tracking method for surface tension dominated interfacial fluid flow *Comp. Fluids* **55** 70-84

[30] Cardiff P 2012 Development of the Finite Volume Method for Hip Joint Stress Analysis, Ph.D. dissertation, School Mech. Mater. Eng., University College Dublin, Dublin, Ireland

[31] Ponweiser T, Stadelmeyer P and Karásek T 2014 Fluid-structure simulations with OpenFOAM for aircraft designs *Partnership for Adv. Comput. in Eur. (PRACE)* White Paper 172

[32] Šekutkovski B, Kostić I, Simonović A, Cardiff P and Jazarević V 2016 Three-dimensional fluid-structure interaction simulation with a hybrid RANS-LES turbulence model for applications in transonic flow domain *Aerosp. Sci. Technol.* **49** 1-16

[33] Turek S and Hron J 2006 Proposal for numerical benchmarking of fluid-structure interaction between an elastic object and laminar incompressible flow *Fluid Structure Interaction, Lecture Notes in Computational Science and Engineering 53* ed H-J Bungartz and M Schäfer (Berlin/Heidelberg: Springer) pp 371-85

[34] Pereira Gomes J and Lienhart H 2010 Experimental benchmark: Self-excited fluid-structure interaction test cases *Fluid Structure Interaction II, Lecture Notes in Computational Science and Engineering 73* ed H-J Bungartz, et al. (Berlin/Heidelberg: Springer) pp 383-411

[35] Versteeg H K and Malalasekera W 2007 *An Introduction to Computational Fluid Dynamics 2nd ed* (Essex: Pearson)

[36] Tian X, Ong M C, Yang J and Myrhaug D 2014 Large-eddy simulation of the flow normal to a flat plate including corner effects at a high Reynolds number *J. Fluids Struct.* **49** 149-69

[37] van Driest E R 1956 On turbulent flow near a wall *J. Aeronaut. Sci.* **23** 1007-11

[38] Belytschko T, Liu W K, Moran B and Elkhodary K I 2014 *Nonlinear Finite Elements for Continua and Structures 2nd ed* (Chichester: Wiley)

[39] Hjertager B H 2014 *Computational Analysis of Fluid Flow Processes* (Stavanger: University of Stavanger)

[40] Greenshields C J 2016 *OpenFOAM User Guide version 4.0* (OpenFOAM Foundation Ltd.)

[41] Hjertager B H 2009 *Lecture notes in OpenFOAM* (Stavanger: University of Stavanger)

[42] van Zuijlen A H and Bijl H 2010 Multi-level accelerated sub-iterations for fluid-structure interaction *Fluid Structure Interaction II, Lecture Notes in Computational Science and Engineering 73* ed H-J Bungartz, et al. (Berlin/Heidelberg: Springer) pp 1-25

[43] Mok D P 2001 Partitionierte Lösungsansätze in der Strukturdynamik und der Fluid-Struktur-Interaktion, Ph.D. dissertation, Institut für Baustatik, University of Stuttgart, Stuttgart, Germany

[44] Finnemore E J and Franzini J B 2002 *Fluid Mechanics with Engineering Applications 10th ed* (New York: McGraw-Hill)

[45] Miner M A 1945 Cumulative damage in fatigue *J. Appl. Mech.* **12** A159-64

[46] Rege K and Pavlou D G 2017 A one-parameter nonlinear fatigue damage accumulation model *Int. J. Fatigue* **98** 234-46

Evaluating stress analysis and failure criteria for offshore structures for Pechora Sea conditions

S Nesic[1]*, Y Donskoy[2] and A Zolotukhin[2,3,4]

[1]NIS GAZPROM Neft, Novi Sad, Serbia,
[2]Gubkin Russian State University (NRU) of Oil and Gas, Moscow, Russia
[3]Northern Arctic Federal University, Arkhangelsk, Russia
[4]University of Stavanger, Stavanger, Norway

* Contact Author: slavkonesic@yahoo.com

Abstract. Development of Arctic hydrocarbon resources has faced many challenges due to sensitive environmental conditions including low temperatures, ice cover and terrestrial permafrost and extreme seasonal variation in sunlight. Russian offshore field development in Arctic region is usually associated with annual ice cover, which can cause serious damage on the offshore platforms. The Pechora Sea has claimed as one of the most perspective oil and gas region of the Russian Arctic with seven discovered oil and gas fields and several dozens of structures. Our rough assessment, based on in-place hydrocarbon volumes and recovery factor evaluation concept, indicates that Pechora Sea alone has in-place volumes amounting to ca. 20 billion barrel oil equivalent (BOE). This quantity is enough to secure produced volumes by 2040 exceeding 3 billion BOE [1] that indicates huge resource potential of the region. The environmental conditions are primarily function of water dynamics and ice cover. The sea is covered by the ice for greatest part of the year. In this article, the ice load simulations were performed using explicit dynamic analysis system in ANSYS software to determine best shape and size of an offshore platform for the Pechora Sea ice conditions. Different gravity based structures (GBS) were analyzed: artificial island, hollow cylindrical and conical concrete structures and four-leg GBS. Relationships between the stress, deformations and time were analyzed and important observations from the simulation results were a basis for selecting the most preferable structures.

1. Introduction

In the Arctic conditions, the ice is presented in a form of ice floes, packs, ridges and icebergs. The ice forces can be static and dynamic, depending of the weather conditions. Moving of ice and collision with offshore structures can create damage or destroy a platform. Offshore structure must be dimensioned with high strength and stability to avoid issues with ice moving and collision. Hence, the offshore platforms are over dimensioned, and construction of platforms requires large quantities of steel and concrete as well as construction costs. To optimize platform construction, maximum ice loads must be evaluated and present a base parameter for design of the offshore structure. Analytical, numerical and simulation methods can be used for prediction of ice loads [2]. Field measurements are the best practice to investigate full scale ice loading and results present a reliable image of whole process [3]. Ice loading measurements has started in 1970's by installing special panels on the offshore structures and recording ice loads and measure interaction between ice and offshore structures [4-17]. Finite elements analysis (FEA) solvers are in commercial use like ANSYS and ABAQUS. Simulations from those software's have very reliable results [18]. In FEA, ice properties present a critical factor for evaluation of the ice interaction with offshore structures [19]. To simulate ice floe collision on the offshore structures, FE simulation is used in this article. The ice collision will create non-linear deformation of the concrete offshore structure. ANSYS FEA software is used with explicit dynamic solver of AUTODYN.

2. Pechora Sea conditions

Development of Arctic hydrocarbon resources has faced with many challenges due to sensitive environmental conditions [20]. Characteristic of Arctic climate are:

- low temperature,
- ice cover and terrestrial permafrost and
- extreme seasonal variation in sunlight

The forming of ice in the Pechora Sea is depended by the season. The most extensive ice period is from March to April when covers the whole sea. Ice cover duration is from the end of October/mid November until the end of July/early August [21-23]. Pechora Sea ice parameters are shown in Table 1 [24].

Table 1. Pechora sea ice parameters

Beginning of ice freeze-up		**Duration of ice-covered season**	
early	25.X	minimum	131 days
average	18.XI	average	213 days
late	23.XII	maximum	272 days
Fast ice freeze-up		Fast ice freeze-up	
early	23.XII	extent	3 to 15 km
average	22.II	average	110 cm
late	11.IV	thickness	
Beginning of fast ice break-up		Drift ice thickness	
early	5.IV	average	80 cm
average	23.V	maximum	145 cm
late	7.VII	continuity	10
Total disappearance of ice cover		hummocks	60-90 %
early	10.IV	mass of	47-130 x10^{3}t
average	19.V	hummocks	
late	30.VIII		

The storm season in the Pechora Sea is from October until the end of December. The highest waves are spreading from northwest to eastern direction. Extreme waves may be 11.5 meters high at the depth from 20-30 meters during October and November [25]. Waves with 2-3 m during the storm season are very dangerous for ships and vessels carrying out offshore operations in the sea.

3. Explicit dynamics solver in ANSYS

Degree of finite element modeling, realized with the help of similar systems, depends on many factors: the degree of discretization of the simulated bodies; type (shell, membrane, volumetric) and the formulations (Lagrangian, Euler, arbitrary) elements; type of contact algorithm, etc. At the same time, the main conditions for the reliability of the calculations are the correct formulation of the defining relations for materials of bodies (mathematical description of the behavior of materials under load), criteria destruction and the task of physical and mechanical characteristics of materials. The physical and mechanical characteristics of ice are determined by its temperature, salinity, availability impurities, formation conditions, age. Therefore, in its natural state, ice is a material, extremely heterogeneous in its properties. There are several methods of ice modeling. One of the first CAD methods is described by Zhang and Hibler [26]. Due to this method sea ice motion is governed by the following momentum balance:

$$m\frac{Du}{Dt} = -mfk \cdot u + \tau_a + \tau_w - mg\nabla_H p(0) + F \tag{1}$$

where $u = u_i + v_j$ is ice velocity vector, m is the ice mass per unit area, F is the Coriolis parameter, g is the gravity acceleration, $p(0)$ is the sea surface dynamic height, τ_a is the force due to air stress, τ_w is the nonlinear water drag, F is the ice interaction force, and i, j, and k are the unit vectors in the x, y, and z directions, respectively. The air stress and water stress terms are given by:

$$\tau_a = \rho_a C_g \left|U_g\right| (U_g \cos\phi + k \cdot U_g - \sin\phi \tag{2}$$

$$\tau_w = \rho_w C_w \left|U_w - u\right| \left[(U_w - u)\cos\theta + k \cdot (U_w - u)\sin\theta = \tau_w(u)\right] \tag{3}$$

where U_g is the geostrophic wind, U_w is the geostrophic ocean current, C_a and C_w are the air and water drag coefficients, ρ_a and ρ_w are the air and water densities, and θ and ϕ are the air and water turning angles, F is the force due to internal ice interaction is given by:

$$F = \nabla \cdot \sigma_{ij} \tag{4}$$

Here σ is the stress tensor σ_{ij} which for an isotropic system is related to the ice strain rate and strength via a nonlinear viscous-plastic constitutive law:

$$\sigma_{ij} = 2\eta(\varepsilon_{ij}, P)\varepsilon_{ij} + \left[\xi(\varepsilon_{ij}, P) - \eta(\varepsilon_{ij}, P)\right]\varepsilon_{kk}\delta_{ij} - \frac{P}{2}\delta_{ij} \tag{5}$$

In the above equation, ε_{ij} is the ice strain rate, given by:

$$\varepsilon_{ij} = \frac{1}{2}\left[\left(\frac{\partial u_i}{\partial x_i}\right) + \left(\frac{\partial u_i}{\partial x_i}\right)\right] \tag{6}$$

Here P is the ice strength depending on the ice compactness and thickness, and η and θ are nonlinear bulk and shear viscosities. These "viscosity" parameters are functions of the ice strain rate invariants and the ice strength and take on some maximum "creep" value when the deformation rate becomes very small. The idea here is to approximate a rigid behavior by a state of a very slow creep. For the elliptical yield curve used here, the nonlinear viscosities differ from each other by a constant factor.

For simplicity, the numerical method for solving is described in rectangular coordinate system. The method is however, applicable to the momentum equations in an arbitrary orthogonal curvilinear coordinate system. These curvilinear equations are briefly presented in the next section to demonstrate the structural similarity.

From the constitutive law, the ices for interaction force components in a Cartesian coordinate system are derived as:

$$F_x = \frac{\partial}{\partial x}\left\{[\eta + \xi]\frac{\partial u}{\partial x} + [\xi - \eta]\frac{\partial v}{\partial x} - \frac{P}{2}\right\} + \frac{\partial}{\partial y}\left[\eta\left(\frac{\partial u}{\partial y} + \frac{\partial v}{\partial x}\right)\right] \tag{7}$$

$$F_y = \frac{\partial}{\partial y}\left\{[\eta + \xi]\frac{\partial v}{\partial y} + [\xi - \eta]\frac{\partial u}{\partial x} - \frac{P}{2}\right\} + \frac{\partial}{\partial x}\left[\eta\left(\frac{\partial u}{\partial y} + \frac{\partial v}{\partial x}\right)\right] \tag{8}$$

The component equations can be organized and simply written as:

$$-\frac{\partial}{\partial x}[\eta + \xi]\frac{\partial u}{\partial x} - \frac{\partial u}{\partial y}\left(\eta\frac{\partial u}{\partial y}\right) + m\frac{\partial u}{\partial t} + C_d u$$
$$= \tau_x + C_s v + \frac{\partial}{\partial x}[\xi - \eta]\frac{\partial v}{\partial y} + \frac{\partial}{\partial y}\left(\eta\frac{\partial v}{\partial y}\right) - \frac{\partial P}{\partial x} \tag{9}$$

$$-\frac{\partial}{\partial y}[\eta+\xi]\frac{\partial v}{\partial y}-\frac{\partial}{\partial x}\left(\eta\frac{\partial v}{\partial y}\right)+m\frac{\partial v}{\partial t}+C_d u$$
$$=\tau_y-C_s v+\frac{\partial}{\partial y}[\xi-\eta]\frac{\partial u}{\partial x}+\frac{\partial}{\partial x}\left(\eta\frac{\partial u}{\partial y}\right)-\frac{\partial P}{\partial y} \quad (10)$$

where C_d and C_s are functions of the ice velocity and the viscosities, ζ and η are functions of the strain rate invariants.

The application of ANSYS makes it possible to set the real properties of ice in analytic and graphical form, which makes it possible to accurately simulate the behavior of ice under different conditions.

4. Model properties

Since the ice floe play critical role in this article, determination of ice properties is very important. Special emphasize is on mechanical properties due to importance on the final results of simulations. Ice is very complex material because its temperature, salinity, age etc. can have significant effect on the mechanical properties [27]. Ice properties are defined based on the current results of comprehensive researches. Ice properties used in this article are listed in Table 2. For this article, the general properties of concrete defined in ANSYS database are used. Concrete does not have any steel or other reinforcements. Concrete properties are listed in Table 3.

Table 2. Ice properties used in this study

Density (kg/m^3)	915
Young's modulus (GPa)	10
Poisson's ratio	0,3
Bulk modulus (GPa)	8,33
Shear modulus (GPa)	3,8462
Compressive strength (MPa)	7
Tensile strength (MPa)	2,5

Table 3. Concrete properties used in this study

Density (kg/m^3)	2300
Young's modulus (GPa)	30
Poisson's ratio	0,18
Bulk modulus (GPa)	15,625
Shear modulus (GPa)	12,712
Compressive ultimate strength (MPa)	41
Tensile ultimate strength (MPa)	5

5. Simulation study

In this article, ice floe collision on the different shapes of the concrete offshore structure. Offshore structures are set as static and the ice floe will have initial velocity. Ice layer is set to be homogeneous with the 1m thickness and ice velocity is set to 1m/s. There are three objects to model, water layer, ice layer and concrete offshore structure. Water layer is used to be 25 m depth and offshore structures have 50 m height. Concrete offshore structures with cylindrical, conical and rectangular shape were considered in this study. The ice layer is moving in one direction except four-leg structures and rectangular structures where two directions of the ice flow were simulated. The focus of the simulation is a collision between the ice layer and offshore structure.

In every simulation set (i.e., number of simulations performed for a specified type of structure) the first run is made for a solid concrete structure followed by runs for the same structure but with different wall thickness to determine the minimum wall thickness for the Pechora Sea conditions. The output of this simulation will be equivalent stresses vs time and potential deformations, visually detected in the graphical output of the software.

IOP Conf. Series: Materials Science and Engineering **276** (2017) 012032 doi:10.1088/1757-899X/276/1/012032

5.1. Cylindrical offshore concrete structure

Simulation setup for the cylindrical offshore concrete structure is shown in the Table 4. The wall thicknesses of the platform are 0,5 m, 1 m and 2 m. Simulation model of the cylindrical platform is shown in Figure 1.

Table 4. Simulation setup for the cylindrical platform used in this study

Objects	Material	Dimensions (m)
Cylindrical platform	Concrete	r = 100, h = 50
Ice layer	Ice	200 x 200, h = 1
Water layer	Water	350 x 200, h = 25

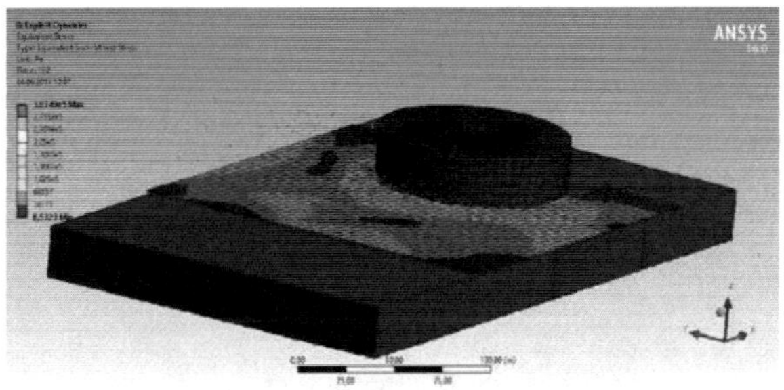

Figure 1. Simulation model of the cylindrical platform

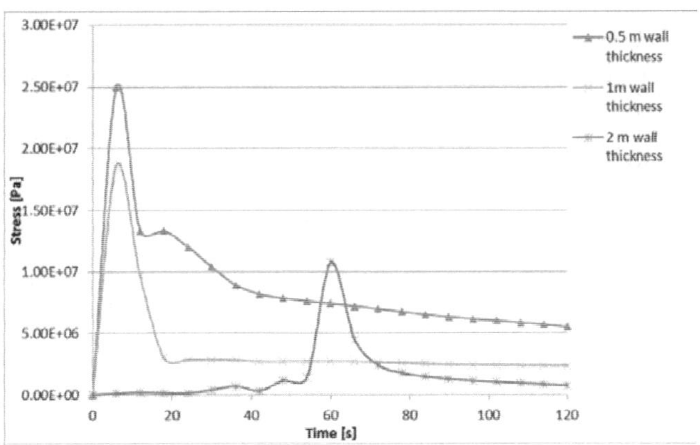

Figure 2. Equivalent stress vs time for different wall thicknesses of the cylindrical platform

Equivalent stress vs time for different wall thicknesses of the cylindrical platform is shown in Figure 2. For wall thicknesses of 0,5 m and 1 m, the platform is collapsed under the ice floe pressure. For wall thickness of 2 m there is no deformation of the platform for the whole 120 seconds of simulation. Different wall thicknesses vs stress for different wall thicknesses of the cylindrical platform is shown in Figure 3.

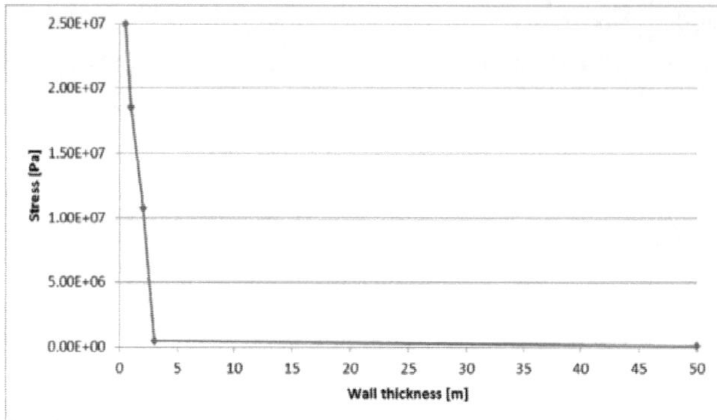

Figure 3. Different wall thicknesses vs stress for different wall thicknesses of the cylindrical platform

5.2. Rectangular offshore concrete structure – one side loading

For the rectangular offshore structure, the position of the ice layer will be guided directly to the one side of the platform and in second run, on the edge side of the model. The wall thicknesses of the platform are 0,5 m, 1 m, 2 m and 3 m. Simulation setup for the rectangular offshore concrete structure – one side loading is shown in Table 5.

Simulation model for the rectangular offshore concrete structure – one side loading is shown in the Figure 4. Equivalent stress vs time for different wall thicknesses the rectangular offshore concrete structure – one side loading is shown in Figure 5.

Unlike cylindrical offshore structure, the rectangular structures show very unfavorable behavior because the ice layer collides with the platform with an angle of 90 degrees and there is no space for ice moving. A lot of ice is accumulated on the one side of the platform contributing on the cumulative stress. For wall thicknesses of 0,5m and 1m, the platform is collapsed under the ice floe pressure. For wall thickness of 2m there is a minor deformation of the platform and with the wall thickness of 3m there is no deformation for the whole 120 seconds of simulation. Different wall thicknesses vs stress for rectangular offshore concrete structure – one side loading is shown in Figure 6.

Table 5. Simulation setup for for the rectangular offshore concrete structure – one side loading

Objects	Material	Dimensions (m)
Rectangular platform	Concrete	a =100, h = 50
Ice layer	Ice	200 x 200, h = 1
Water layer	Water	350 x 200, h = 25

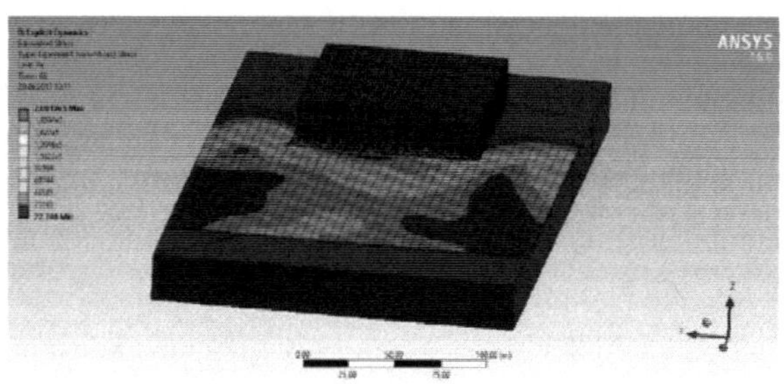

Figure 4. Simulation model for the rectangular offshore concrete structure – one side loading

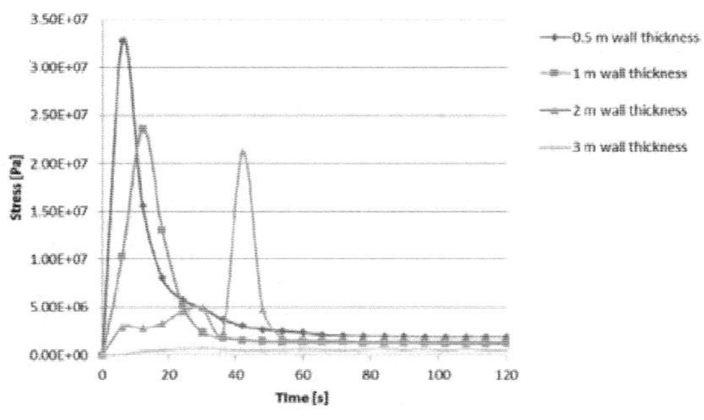

Figure 5. Equivalent stress vs time for different wall thicknesses the rectangular offshore concrete structure – one side loading

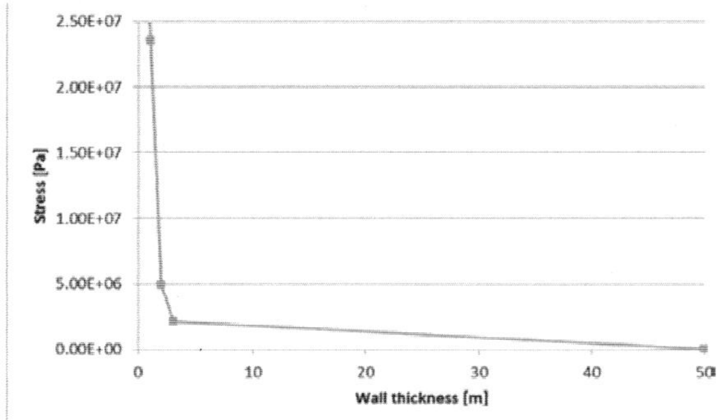

Figure 6. Different wall thicknesses vs stress for rectangular offshore concrete structure – one side loading

5.3. Rectangular offshore concrete structure – edge loading

In the second run, the ice floe is directed on the edge side of the structure. The wall thicknesses are 0,5 m, 1 m and 2 m. Simulation setup for the rectangular offshore concrete structure – edge loading is shown in Table 6. Simulation model for the rectangular offshore concrete structure – edge loading is shown in the Figure 7 and the equivalent stress vs time for the rectangular offshore concrete structure – edge loading is shown in Figure 8.

Table 6. Simulation setup for the rectangular offshore concrete structure – edge loading

Objects	Material	Dimensions (m)
Rectangular platform	Concrete	a = 100, h = 50
Ice layer	Ice	200 x 200, h = 1
Water layer	Water	350 x 200, h = 25

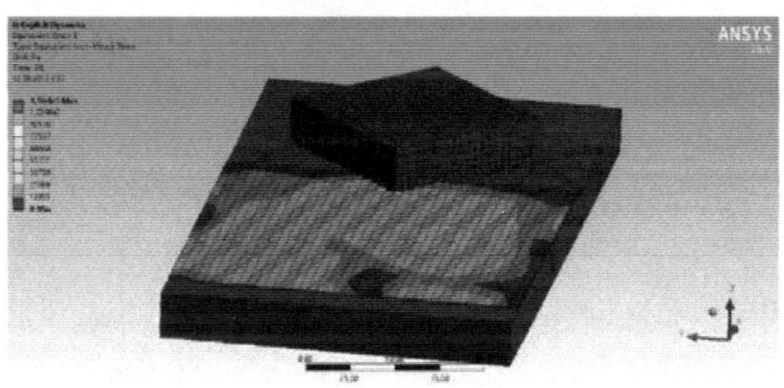

Figure 7. Simulation model for the rectangular offshore concrete structure -edge loading

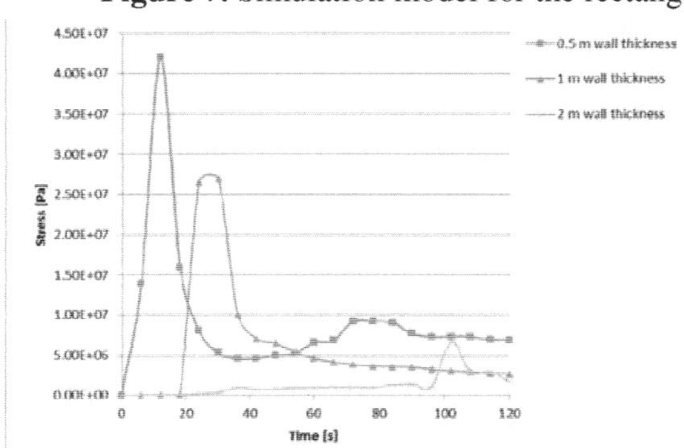

Figure 8. Equivalent stress vs time for the rectangular offshore concrete structure – edge loading

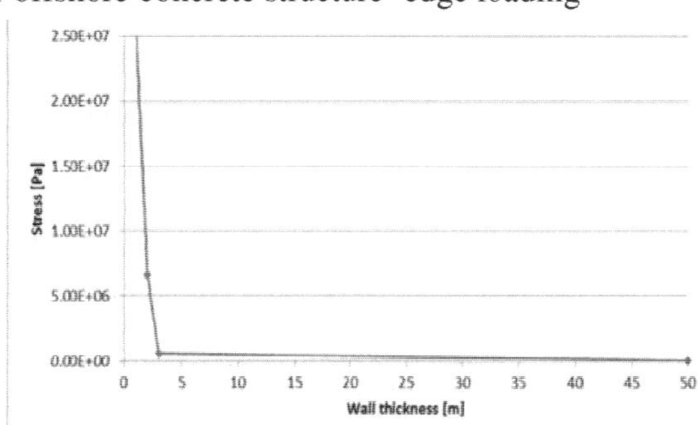

Figure 9. Different wall thicknesses vs stress for the rectangular offshore concrete structure – edge loading

This direction of ice moving creates the smallest stresses on the platform. For wall thicknesses of 0,5m and 1m, the platform is collapsed under the ice floe pressure. For wall thickness of 2 m there is no deformation of the platform for the whole 120 seconds of simulation. Different wall thicknesses vs stress for the rectangular offshore concrete structure – edge loading is shown in Figure 9.

5.4. Sloped offshore conical structures

Based on the assumption that the tensile strength of the ice is much lower from compressive strength sloped structures were developed. In this study, cylindrical platform is modelled with 100m diameter of the base and three angles, 45°, 60°, 70°. Simulation setup for the sloped conical offshore concrete structure is shown in Table 7. Simulation model for the sloped conical offshore concrete structure is shown in Figure 10 and the equivalent stress vs time for a conical offshore concrete structure is shown in Figure 11.

Table 7. Simulation setup for the sloped conical offshore concrete structure

Objects	Material	Dimensions
Conical platform	Concrete	r =100 m, h=50 m, α = 45°, 60 °, 70°
Ice layer	Ice	200 m x 200 m, h=1 m
Water layer	Water	350 m x 200 m, h=25 m

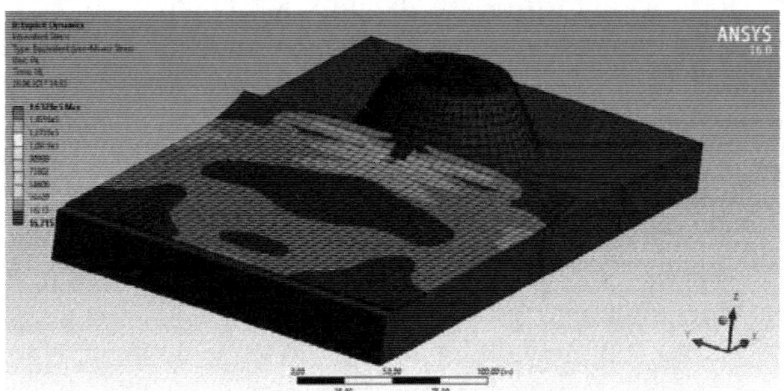

Figure 10. Simulation model the sloped conical offshore concrete structure

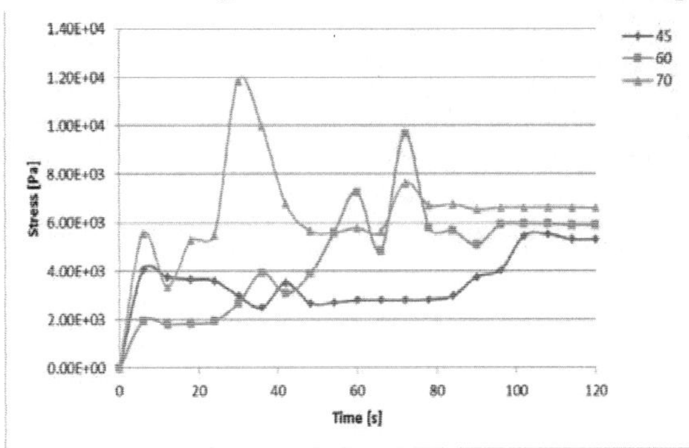

Figure 11. Equivalent stress vs time the sloped conical offshore concrete structure

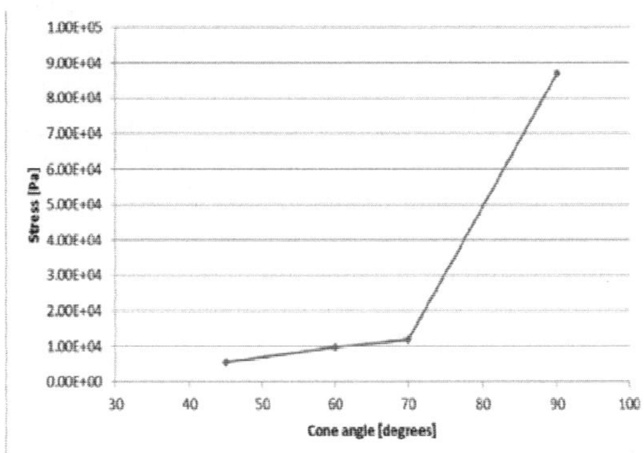

Figure 12. Equivalent stress vs different cone angles for the sloped conical offshore concrete structure

Simulation results show that with an increasing the angle of the structure, the equivalent stress is increasing. With the inclined structures, the mechanisms of ice loading are very complex and present a combination of compressive and tensile stress because the ice floe tend to move up on the conical part of the structure. A solid sloped structure undergoes much lower stresses than hollow cylindrical structures with different wall thicknesses due to mechanism of ice loading because compressive strength of the ice is much bigger than its tensile stress. Equivalent stress vs different cone angles for a conical offshore concrete structure is shown in Figure 12.

5.5. Four legs concrete structures

Four legs concrete offshore structure is modelled in rectangular pattern with 50 m distance between each leg. There are two directions of ice floe, one to first pair of legs and diagonally on the first leg. Simulation setup for the four leg offshore concrete structure is shown in Table 8.

Table 8. Simulation setup for the four legs offshore concrete structure

Objects	Material	Dimensions (m)
Conical platform	Concrete	r = 20, h = 50
Ice layer	Ice	200 x 200, h = 1
Water layer	Water	350 x 200, h = 25

Simulation model of the four legs offshore concrete structure where the ice floe is directed to the first pair of legs is shown in Figure 13. The equivalent stress vs time of the four legs offshore concrete structure where the ice floe is directed to the first pair of legs is shown in Figure 14.

From the stress analysis, it can be concluded that the first two legs receive most of the stress caused by ice floe loading. Simulation model of the four leg offshore concrete structure in case of the ice floe movement diagonally to the first leg is shown in Figure 15. Equivalent stress vs time of the four-leg offshore concrete structure for such a case is shown in Figure 16. In this configuration of the model, the first leg receives most of the stress caused by the ice floe loading.

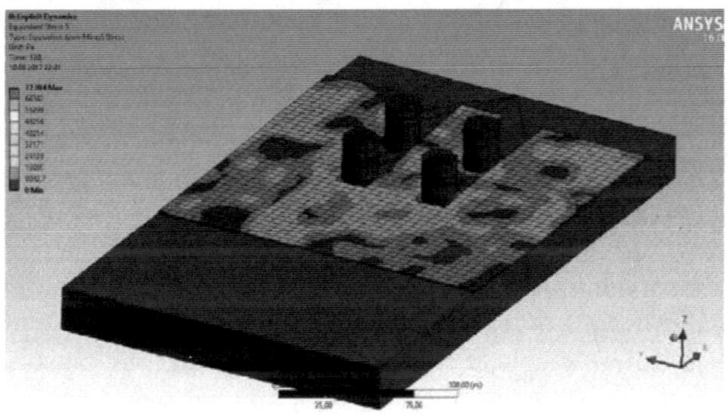

Figure 13. Simulation model of the four legs offshore concrete structure where the ice floe is directed to the first pair of legs

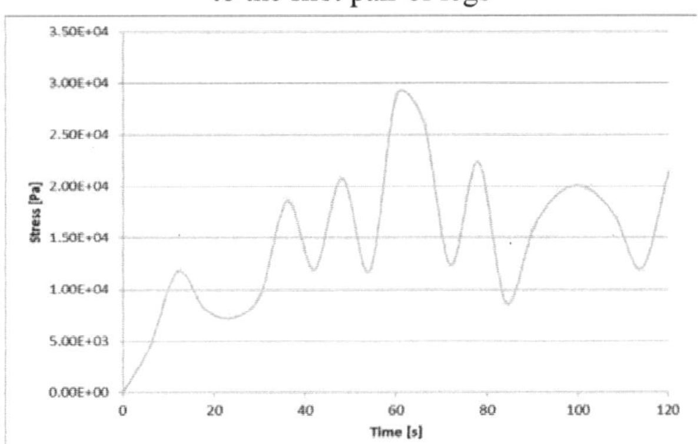

Figure 14. Equivalent stress vs time of the four legs offshore concrete structure where the ice floe is directed to the first pair of legs

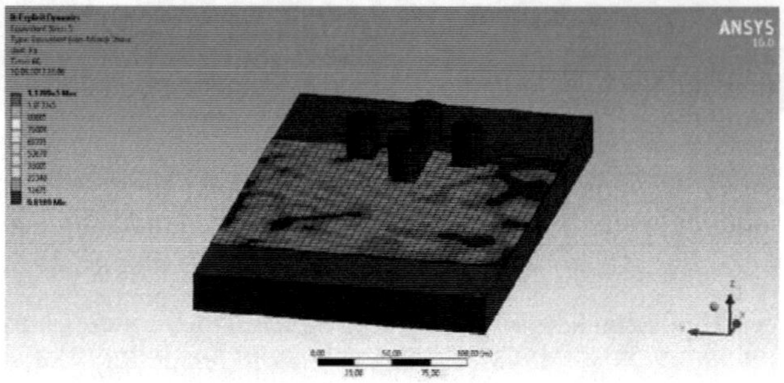

Figure 15. Equivalent stress vs time of the four legs offshore concrete structure where the ice floe is directed to the first legs

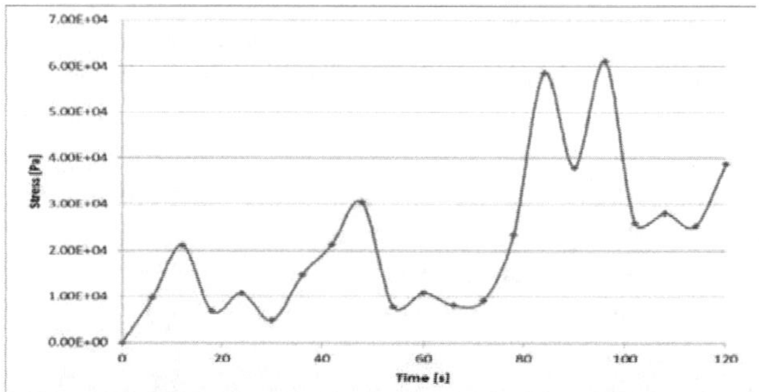

Figure 16. Equivalent stress vs time of the four leg offshore concrete structure in case of the ice floe movement diagonally to the first leg

6. Conclusions

In this article, simulation results of the ice floe loading on different offshore structure are presented. The main goal was to determine an optimum shape and dimension of the offshore structure for the Pechora Sea conditions. Multiple simulation runs were performed for different gravity-based structures including cylindrical, rectangular, sloped cylindrical and four leg structures. The grid for ice model in this article is slightly coarse and there is an additional space for improving the modelling. Due to the size of the model, simulation time will be very long, up to seven days for one run and require very strong computer. However, the results of simulation in this article are very usable for understanding of the ice impact on different shapes of the offshore structures, analyzing of stresses and determination of minimum wall thickness for safe operations. Based on the simulation results, the critical thickness of the offshore structures in the Pechora Sea condition is found to be 2 m for pure concrete without any steel reinforcement. The critical wall thickness will be lesser in a case of steel-reinforced concrete. Based on the simulation results and Pechora Sea ambient data it can be concluded that gravity based fixed cylindrical and sloped offshore structures are recommended.

Acknowledgement

The work presented here was initiated and sponsored by Gubkin Russian State University (National Research University) of Oil and Gas.

References

[1] Gubaidullin M G et al. 2016 *Simulation of Oil Spill in the Western Sector of the Russian Arctic* ed M G Gubaidullin (Arkhangelsk-Northern (Arctic) Federal University) p. 219

IOP Conf. Series: Materials Science and Engineering **276** (2017) 012032 doi:10.1088/1757-899X/276/1/012032

[2] Molyneuh D and Spencer D 2013 *Predicting ice loads on offshore structures* Available at: http://www.sname.org/HigherLogic/System/DownloadDocumentFile.ashx?DocumentFileKey=7ffd2c57-c147-4fe6-b1a2-79d4b479a791
[3] Xu N and Yue Q *J. Offshore Mech. Arct. Eng* **136(1)** p. 014501
[4] Blenkam K A 1970 *Proc. Offshore Technology Conf.* (Houston)
[5] Nordlund O P, Kama T, and Järvinen E 1998 *Proc. 9th IAHR Ice Symp.* (Sapporo) pp. 537-48.
[6] Brown T G 1997 *Proc. CSCE Annual Conf.* (Sherbrooke) **1** pp. 177-86.
[7] Yue Q J and Bi X J 1998 *Proc.14th Int. Association for Hydraulic Research (IAHR) Symp. on Ice* **2** pp. 939-45.
[8] Yue Q J, Bi X J, Sun B, Zhang T and Chen X 1996 *Proc. IAHR Ice Symp.* (Beijing) pp. 282-9.
[9] Yue Q J and Bi X J 2000 *J. Cold Reg. Eng.* **14(2)** pp. 81-92.
[10] Timco G and Johnston G M 2003 *Cold Reg. Sei. Technol.* **37** pp. 51-68.
[11] Sodhi D S 1987 *Proc. 6th Int. Conf. on Offshore Mechanics & Arctic Eng.* (Houston) **4** pp. 281-4.
[12] Izumiyama K, Kitagawa H, Koyama K and Uto S 1991 *Proc. 11 th Int. Conf. on Port and Ocean Eng. under Arctic Conditions (POAC)* (St. John's) pp. 155-66.
[13] Nevel D E 1992 *Proc. 11th IAHR Ice Symp.* (Alberta-Banff) **3** pp. 1391-1404.
[14] Maattanen M, Hoikkanen A N and Avis J 1996 *Proc. 13th Int. Symp on Ice (IAHR)* (Beijing) **1** pp. 8-17.
[15] Mayne D C and Brown T G 2000 *Proc. 19th Int. Conf. on Offshore Mechanics and Arctic Eng.*, (New Orleans).
[16] Shkhinek K and Uvarova E 2001 *Proc. 16th Int. Conf. on Port and Ocean Eng. Under Arctic Conditions* (Ottawa) **2** pp. 639-48.
[17] Qu Y and Yue Q J 2003 *J. Glaciol. Geocryol.* **25(2)** pp. 326-29.
[18] Lin M W and Abatan A O 1994 *Intelligent Material Systems and Structures* **5(6)** pp. 869-75.
[19] Derradji-Aouat A 2005*Proc.18th Int. Conf. on Port and Ocean Eng. Under Arctic Conditions* **1** pp. 225-38
[20] Belkina N 2013 *SPE Arctic and Extreme Environments Technical Conf. and Exhibition.* (Russian).
[21] Gudmestad O T, Zolotukhin A B, Ermakov A I, Jakobsen R A, Michtchenko I T, Vovk V S, Loeset S and Shkhinek K N 1999 *Basics of Offshore Petroleum Engineering and Development of Marine Facilities with Emphasis on the Arctic Offshore* (Stavanger/Moscow/St. Petersburg/Trondheim- Publishing house "Oil and Gas) pp.112-115, 124-126
[22] Gorshkov S G, and Faleev V I 1980 *Atlas of the Oceans. The Arctic Ocean* (Moscow-USSR Ministry of Defence)
[23] Spichkin V and Egorov A 1995 *Proc. of the Second Int. Conf. on Development of the Russian Arctic Offshore (RAO-95)* (St.Petersburg).
[24] Danilov A I and Efremkin I M 1998 *Natural-climatic conditions in the regions of oil-gas fields on the Arctic shelf* (St.Petersburg-Osvoenie shel'fa arkticheskikh morei Rossii (Exploration of the Russian Arctic shelf)) pp. 479-487 (in Russian).
[25] Mischenko S M 1996 *Wave Parameters in the Kara and Pechora Seas* (Report-St. Petersburg State Technical University) p. 21
[26] Zhang J and Hibler W D 1997 *J. Geophys. Res.* **102** pp. 8691-702.
[27] Liu Z 2013 *PhD thesis-Analytical and numerical analysis of iceberg collisions with ship structures* (Trondhiem-Norwegian Unviersity of Science and Technology).

IOP Conf. Series: Materials Science and Engineering **276** (2017) 012033 doi:10.1088/1757-899X/276/1/012033

Integrity management of offshore structures
and its implication on computation of structural action effects and resistance

T Moan
Centre for Autonomous Marine Operations and Systems and the Department of Marine Technology, Norwegian University of Science and Technology (NTNU), Trondheim, Norway

E-mail: torgeir.moan@ntnu.no

Abstract. An overview of integrity management of offshore structures, with emphasis on the oil and gas energy sector, is given. Based on relevant accident experiences and means to control the associated risks, accidents are categorized from a technical-physical as well as human and organizational point of view. Structural risk relates to extreme actions as well as structural degradation. Risk mitigation measures, including adequate design criteria, inspection, repair and maintenance as well as quality assurance and control of engineering processes, are briefly outlined. The current status of risk and reliability methodology to aid decisions in the integrity management is briefly reviewed. Finally, the need to balance the uncertainties in data, methods and computational efforts and the cautious use and quality assurance and control in applying high fidelity methods to avoid human errors, is emphasized, and with a plea to develop both high fidelity as well as efficient, simplified methods for design.

1. Introduction

The oceans contribute to energy [1], mineral resources [2] and food [3] and provide means for transport and other infrastructure. About 1/3 of hydrocarbons, 15% of the total energy supply in the world, are produced from subsea reservoirs. Moreover, the energy industry, like other man-made activities, is challenged by the demand for sustainability [4]. New solutions for attaining clean and renewable energy combined with cleaning of fossil energy are required. Sustainable energy, especially from wind, is expected to be a part of this. Subsea rare minerals also represents a potential for a new ocean industry. Due to the stagnation of wild fish catch at about 90 mill tones, the world has to rely on aquaculture which already produces about the same as wild catch and annually increases by 8%.

The exploitation of the oceans takes place in increasingly harsh environments, including oil and gas production and transport in arctic conditions. Moreover, the demanding industrial environment of, e.g., in the oil and gas industry, has an inherent risk of blowout, severe fires/explosions and other undesirable events.

Figure 1 shows some structures used for various functions. The variety of shapes and sizes clearly shows the many opportunities for innovation. The use of the oceans requires approaches that are socially, environmentally, economically and energetically sustainable. The success of marine technology in the future requires exploration and implementation of relevant enabling technologies such as information and communication technology, automatic control, materials, fabrication and logistics.

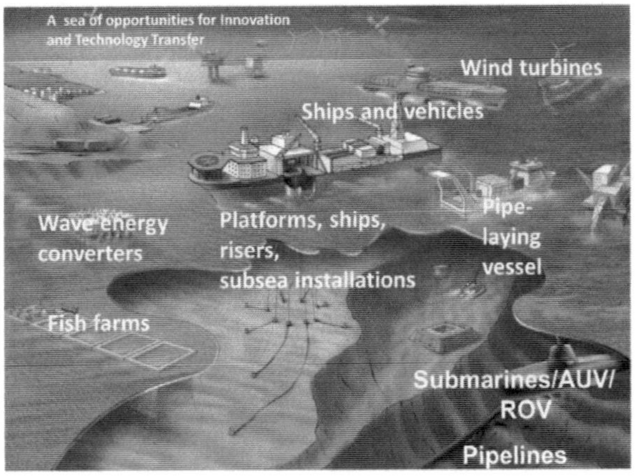

Figure 1. A variety of offshore structures.

The continuous innovation to deal with new serviceability requirements and demanding environments as well the inherent potential of risk of fires and explosions in the offshore oil and gas industry, have led to an industry which has been in forefront of development of design and analysis methodology for ocean systems, applicable to other ocean industries.

This paper deals with structural integrity management during the life cycle - design, fabrication, installation and operation (Figure 2), mainly referring to the oil and gas industry. Various types of offshore structures experience different failure modes which ultimately can lead to fatalities, pollution or property loss.

Current industry practice is implemented in standards for design and reassessment of structures, e.g. [6-14] for offshore oil and gas platforms and [15-18] offshore wind turbines and [19] for (floating) bridges. It is noted that the standards for offshore wind turbines and floating bridges are emerging, and refer to a large extent to the principles and methods specified in the standards for offshore petroleum facilities, however, with different target safety levels due to different consequences of failure. In addition, many classification societies have specified design codes and guideline. The most advanced standards and codes are characterized by:

- design criteria formulated in terms of serviceability and safety limit states, considering payloads, environmental and accidental actions
- semi-probabilistic methods for ultimate strength design which have been calibrated by reliability or risk analysis methodology
- fatigue design checks depending upon consequences of failure (damage-tolerance) and access for inspection
- explicit accidental collapse design criteria to achieve damage-tolerance for the system
- global and local structural analysis by finite element methods for ultimate strength and fatigue design checks
- nonlinear analyses to demonstrate damage tolerance in view of inspection planning and progressive failure due to accidental damage
- life cycle feature, with strong link between design, and inspection, monitoring, maintenance and repair (IMMR)

Moreover, reassessment of design is required during operation, for instance, because of planned change of platform function and, hence, the need to increase payload or over-load damage due to hurricanes in the Gulf of Mexico, subsidence of North Sea jackets, after accidental loads such as e.g. explosions, fires and ship impacts, as well as in connection with extension of service life.

Basically, the reassessment involves the same assessments as carried out during initial design. However, depending upon the inherent damage tolerance ensured by the initial design, the measures that have to be implemented to improve the strength of an existing structure may be much more expensive than for a new structure. This fact commonly justifies a more advanced mechanics and probabilistic assessment than in the initial design [1, 8, 13, 14, 20].

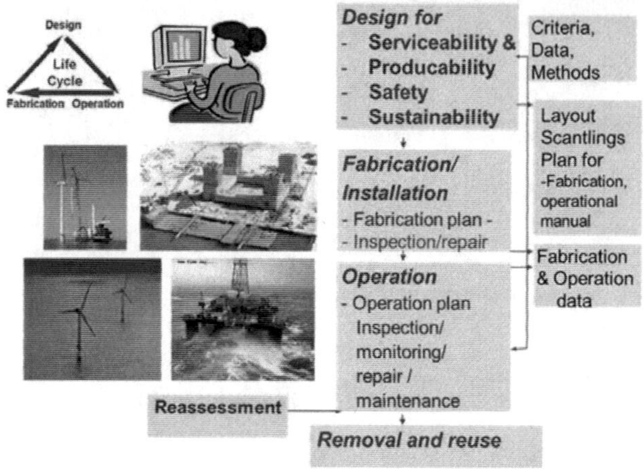

Figure 2. Life cycle phases of offshore structures

The offshore industry early recognized that design took place under uncertainty and adopted risk and reliability methods to make more rational decisions. In addition to the uncertainties affecting the predicted behaviour under extreme and fatigue conditions, inspection is subjected to uncertainty. Reliability methods are, hence, crucial to support decisions about safety and economy of degrading structures. Significant developments of structural reliability methodology, including Bayesian updating techniques, have taken place since the 1980s, as outlined e.g. in [21-25].

While current design and inspection procedures have been established on a reliability basis of the generic information available at the design stage, it is important that information obtained during operation of individual structures, e.g. by inspections during operation, is used to update the inspection plan. This fact implies dedicated IMMR efforts to each individual structure, in contrast to ultimate limit state (ULS) criteria that are based on a generic code calibration. However, service experiences from other structures with similar geometry, loading and site of operation could obviously also be utilized to update IMMR plans for a given structure, e.g. Moan [26-27]. In particular, fatigue depends on very local geometrical features, and gross errors might appear as relatively small geometrical deviations (Section 3.3). To justify a probabilistic approach addressing normal uncertainties it is important that such anomalities are corrected based on the results from inspection of the as-built structure.

Moreover, service experiences show that accidental actions and abnormal strength due to gross errors or omissions made during design, fabrication or operation, contribute significantly to the risk. Clearly, control of the risk associated with those kinds of events needs a broad safety management approach during the life cycle, including a proper measure of the risk; with considerations of the expenditure to achieve the desired safety level, e.g. [28].

In the following sections the basic requirements and in-service experiences of oil and gas structures are briefly described, followed by a formulation of the structural integrity management approach. This includes a quantitative limit state for robustness to limit the likelihood of progressive failure, as well as quality assurance and control (QA/QC) of the engineering process, fabrication and operation. Use of probabilistic methods to aid the design and decision process during operation is addressed, considering the fact that human errors play an important role on the safety.

2. The basic requirement to offshore structures

The design is the crucial life cycle phase (Figure 2) in which important decisions can be made regarding:

- serviceability during operation
- fabrication and installation
- safety during fabrication, installation and use

In the design phase, the optimal choice of layout, materials and equipment is to be made based on the fulfilment of requirements with regard to serviceability, ease of fabrication and installation as well as safety.

Serviceability of a petroleum production facility requires among other properties, a structure with proper payload capacity and deck area, limited motions and station-keeping to avoid damage to risers and ensure an efficient drilling or production process [29]. Moreover, platforms for exploratory drilling need to be mobile and thus have limited resistance in transit.

To achieve competitive fabrication costs, design for producibility is also crucial. This especially implies designing and outfitting the structure with a simple geometry, preferably as a system built up by repeatable modules to facilitate parallel production and cost savings through serial production. Another issue is the design for automated fabrication. This goal can only be achieved if powerful processes and tools are used for the design, production and planning/logistics based on computer tools to allow efficiency, speed and reliability.

Production yards are becoming increasingly "digital" by using numerical methods to predict imperfections and residual stresses, and logistic methods for integrating product and production information that reduce costs and error potential. [30].

A description of design and engineering analyses of different offshore oil and gas structures, such as jackets, semi-submersibles, tension-leg, spar platforms may be found in [29].

The metocean, seismic and hydrocarbon hazards make safety an important issue for oil and gas platforms. Safety requirements are specified to avoid fatalities, environmental damage and property, and are related to the following failure modes:

- overall, rigid body instability (capsizing) under permanent, variable and environmental loads
- failure of (parts of) the structure, foundation or mooring systems, considering permanent, variable, and environmental as well as accidental loads

The location far off shore makes evacuation and rescue difficult, but on the other hand accidents on offshore facilities do not affect the general public in the same manner as accidents on land do.

3. In-service failure experiences

3.1. General

Safety may be defined as the absence of accidents or failures. Hence, useful insight about the safety features can be gained from the detailed investigations of catastrophic accidents, as described e.g. in [26-28] for offshore oil and gas structures. In addition, statistics about offshore accidents, regularly compiled in WOAD(-),[31] which provides an overview of accident rates. Data for wind turbines are found e.g. in [32-33], essentially referring to the rotor-drivetrain assembly for land-based turbines.

Detailed investigations have been carried out for accidents with significant consequences in the oil and gas industry; e.g. namely the three-legged jack-up Ranger [34], semi-submersibles Alexander Kielland platform [35] and Ocean Ranger [36], P-36 [37], jacket. Piper Alpha [38] and the semisubmersible Deepwater Horizon [39].

Capsizing/sinking and global structural failures normally develop in a sequence of technical and physical events. Structural damage can cause progressive structural failure or flooding, which may result in capsizing of buoyant structures. However, to fully explain accident event sequences, it is necessary to interpret them in view of the human and organizational factors (HOF) of influence as advocated in [40] for structures in general and for offshore structures in [41-42].

Basically structural failure occurs when the resistance, R is less than the load effect, S. From an HOF point of view this can be due to too small safety factors to account for the normal uncertainty and variability in R and S relating to ultimate limit state (ULS) and fatigue limit state (FLS) criteria. But the main cause of actual structural failures is abnormal resistance or accidental or abnormal loads due to human errors and omissions.

Design errors materialise as a deficient (or excessive) resistance, which cannot be derived from the parameters affecting the "normal" variability of resistance. Fabrication imperfections (such as cracks, plate misalignment, etc.), which also affect the resistance, are influenced by human actions. The "normal" variability of welders performance, environmental conditions, etc. lead to a "normal" variability in the imperfection size, characterised by a smooth variation of the relevant imperfection parameter. Sometimes an abnormal deviation from this behaviour occurs, e.g., caused by using a wet electrode, etc. or another gross fabrication error. Thus, the initial fatigue failure of a brace in the Alexander L. Kielland platform was due to lack of fatigue design checks as well as inadequate inspection [43]. Even though the fatigue failures that had occurred in semi-submersibles in the period 1965-70 resulted in fatigue requirements, the requirements were not properly implemented even for platforms built in the 1970's. Many platforms built in the 1970s had joints with design fatigue lives as low as 2-5 years. Fatigue failure has been rediscovered many times, even in marine technology, since Wöhler's initial "scientific" discovery more than 150 years ago.

3.2. Overload damages and failures of hull structures

Man-made live loads have a "normal" and an abnormal component, while some loads, notably fires and explosions, ship collisions, etc. do not have a normal counterpart. They are simply caused by operational errors or technical faults. The mobile platform Ocean Ranger capsized offshore New-foundland in 1982. The accident was initiated by control room window breaking due to wave slamming. The water entering the control room lead to short circuit of the ballast valve system, thereby leading to spurious operation of ballast valves. The resulting accidental ballast condition could not be controlled partly because of lack of crew training and partly because of inadequate ballast pumps, and open chain lockers [36].

The significant damage to the jacket in Figure 3a was caused during the hurricane Lilli in the Gulf of Mexico by waves hitting platform deck due to apparently too small deck clearance.

The catastrophic explosion and fire on the Piper Alpha platform (Figure 3b) in 1988 was initiated by a gas leak from a blind flange of a condensate pump that was under maintenance and not adequately shut [37]. The gas ignited and the initial explosion lead to damage of an oil pipe and subsequent oil fire and explosion.

In 2001 the platform P-36 in Brazil experienced a burst collapse of the emergency drainage tank, accidental explosion and subsequent flooding, capsizing and sinking. A series of operational errors were identified as the main cause of the first event and also the sinking. [38]. Other accidents and structural failures are discussed e.g. by Vinnem [43].

3.3. Fatigue and other degradation of hull structures

Degradation due to corrosion and fatigue crack growth would normally develop slowly to failure. However, fatigue can cause catastrophic accidents if the fatigue life is not sufficient to make IMMR effective or if there is lack of robustness, as for the Ranger I [34] and Alexander Kielland [35] platforms.

Crack is the most critical degradation phenomenon, because it could result in rupture. The stages of crack growth depend upon the layout of the structure. For a frame or truss structure consisting of slender members, it is natural to consider crack growth in the following stages: visible crack, through-thickness crack, and failure (rupture) of member. In monocoque structures the situation is different in that cracks in the main hull girder can grow continuously until global rupture of the hull [45]. Fatigue failure, as inferred from SN curves, is normally taken to be a crack through the plate thickness. In this case it is desirable to have a criterion for rupture (for large cracks).

(a) (b)

Figure 3. (a) Jacket in the Gulf of Mexico after the Lilli hurricane and (b) PiperAlpha after the explosion/fire [37]

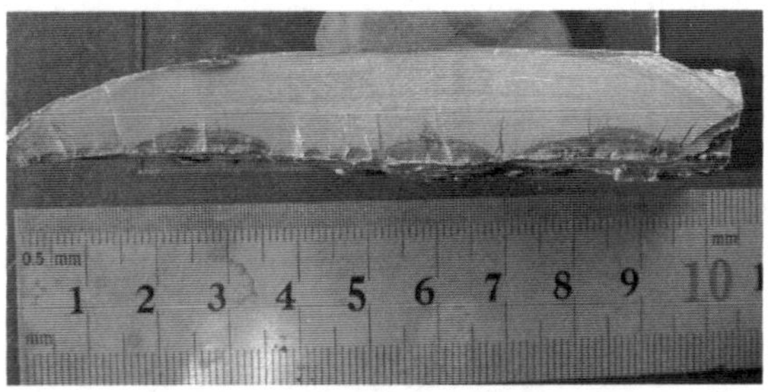

Figure 4. Crack surface of a steel plate, with a fillet weld on the lower side. The dark areas indicate a 2 mm deep hydrogen crack initiated during fabrication [47]

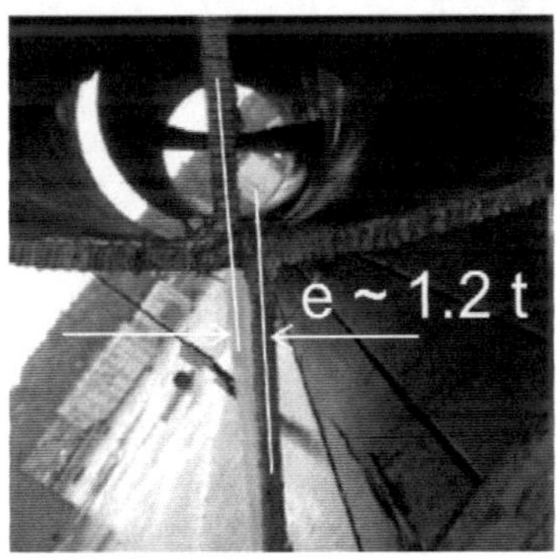

Figure 5. Illustration of misalignment in plates and measured misalignment in semi-submersible platforms [47].

Cracks are for instance always present at welds. The development of cracks depends on their initial size (normally a fraction of a mm), uncertainties in the stresses driving the crack, fatigue resistance and the inspection method used to detect the crack. Due to the very nature (small size) of initial (fabrication) defects, an abnormal size of such defects often have been the cause of fatigue cracks. Hence, cracks have been caused by e.g. [26-28].

- not carrying out proper fatigue design checks and inspections and possible repairs, e.g. [35]
- abnormal initial defect size, e.g. due to wet electrodes
- abnormal crack propagation e.g. due to corrosion fatigue effects
- fatigue caused by overprotection or loss of protection

- abnormal local geometry due to deviation between as-built structure deviating and design or bad design. Hence inspection engineering should always refer to the as-built condition
- error in load (stress) analysis (environmental conditions, method, stress concentration factor) and particular phenomena like VIV.

During fabrication the QA/QC addresses control of material and geometry, especially tolerances relating to misalignment etc., possible damage that could occur etc. It is noted that defects in welds are primarily controlled indirectly by the specification of welding procedure, environment etc. This is because the normal defect depth, say of the order of 0.1 mm are not likely to be detected by non-destructive examination (NDE) inspections with a mean detectable crack size which is significantly larger. However, gross errors, like cracks of a depth 2.0 mm or more might be detected by NDE. Larger defects than the 0.1 mm could be due to e.g. wet electrodes or other deviations from normal welding procedures. Figure 4 shows an example with a 2 mm deep initial defect. The implication is a 80-90% reduction in the fatigue life for this plate under membrane stresses. Based on underwater inspection results for tubular joints in jackets Moan et al. [46] back-calculated the initial crack depth to be 0.9 mm. The collected experience data [47] clearly show shortcomings in local design and fabrication quality than normally accounted for in fatigue management based on traditional fatigue analyses

Extensive experiences regarding cracks in North Sea jacket platforms and semi-submersible drilling platforms have for instance been reported in [26-27, 46-48].

The occurrence of cracks in jackets have been correlated to fatigue predictions. The most important lessons learnt are that the initial methods used in general were conservative, but 2-3 % of the cracks detected were not predicted. The latter fact indicates that gross fabrication defects occur. The average crack depth of propagating cracks detected was 4.8 mm, with a small percentage of through thickness cracks. Another lesson was the significant difference in relative crack occurrences in platforms installed before and after 1978.

All data for semisubmersibles have not yet been correlated with predictions. However, comparisons made show a generally good agreement, but that deviation between observations and predictions are especially due to discrepancies between design drawings and specification and as-built structural geometry as well as excessive weld defects.

Figure 5 shows an example of a misalignment between plates which is larger than the plate thickness, while the fabrication tolerances normally accounted for in the design and inspection planning normally is of the order of 30-50% of the plate thickness. This geometrical effect significantly reduces the fatigue life of the associated welded joint under transverse loading.

During fabrication detected cracks are repaired. Post welding treatment might also be applied. However, while the effect of weld toe grinding after fabrication corresponds to approximately a factor of 2 in terms of fatigue (design) life, the effect is more significant if it is done at a late stage during operation because the crack has grown to a larger size and is “removed”.

Even if the most fatigue-prone and critical areas of jackets and semi-submersibles are much more limited in extent than for ships, there might be several hundred hotspots to follow up.

It should be noted that limited experiences are available for novel concepts like Tension-leg and Spar platforms. The importance of updating the information of the geometry based on the as-built geometry is crucial as basis for the follow-up during operation.

3.4. Unknown phenomena

In some cases, accidents have been caused due to lack of knowledge in the engineering profession at large, i.e. unknown phenomena. Recently discovered new phenomena, such as ringing and springing of Tension-leg platforms, deterioration failure mechanism of flexible risers, and vortex induced motions of floating platforms, were observed in time before any catastrophic accident occurred, and the necessary precautions could be made. However, sometimes the accident cause can be attributed to design codes not being up to actual knowledge in the profession.

4. Safety management

4.1. General

Basically the integrity of structures is ensured by

- designing them to withstand environmental and intended operational loading throughout their life cycles according to limit state criteria listed in Table 1 – for a service life of 20 years or more.
- providing measures to detect through inspection or monitoring, hazards or damage and mitigate them at an early time to avoid accident escalation.

Relevant criteria for ULS, FLS and ALS (Table 1) are reviewed in [49], with emphasis on steel structures. It is emphasized that there is no criteria that account for the interaction between fatigue (crack growth) and ultimate failure (rupture). For this reason it is important that the limit (failure mode) considered in FLS and ULS is properly defined.

The ULS check is commonly made by determining action effects, S due to G (gravity actions), P (payloads) and E (environmental actions) by a linear global structural analysis while the component resistances R are determined by laboratory experiments and nonlinear structural analysis and given by parametric formula in design standards. The ULS component design check is the made typically by:

$$\begin{aligned} &S_d \le R_d ; where \\ &S_d = \gamma_G \, S(G_c) + \gamma_P \, S(P_c) + \gamma_E S(E_c) \\ &R_d = R\left(X_c, a_{nom}\right) / \gamma_R \end{aligned} \tag{1a-c}$$

where the γ's are action and material factors. The subindices *c* and *d* denote characteristic and design values, respectively. A system design check is also emerging [47]. (NORSOK N-003). The characteristic environmental action for ULS refers to an annual probability, of exceedance of 10^{-2}– considering the short- and long- term variability of the actions.

Table 1. Safety criteria for design of structures.

Limit states	Description	Remarks
Ultimate (ULS)	Overall "rigid body" instability Ultimate strength of structure, mooring or possible foundation	Different for bottom – supported, or buoyant structures. Component design check of strength
Fatigue (FLS)	Failure of welded joints	Component design check depending on residual system strength (see Table 4)
Accidental collapse (ALS)	Ultimate capacity[1)] of damaged structure with "credible" damage	System design check

1) Capacity to resist "rigid body" instability or total structural failure

A simple expression for the cumulative fatigue damage can be obtained by assuming that the SN-curve is defined by $NS^m = K$ and the number, n(s) of stress ranges is given by the Weibull distribution:

$$F_S(s) = 1 - \exp\left[-\left(\frac{s}{A}\right)^B\right] \quad \text{for s > 0} \tag{2}$$

where *A* and *B* are the scale and shape parameters of the distribution, respectively. The corresponding uncertainty is modelled by the parameters *A* and *B*, which then depend upon uncertainties associated with environmental conditions, wave load model and structural modelling. These parameters might be expressed as: $A = s_0/(\ln N_0)^{1/B}$; with s_0 corresponding to $P[\, S \ge s_0 \,] = 1/N_0$, where N_0 is the number of cycles in a reference period for s_0. The cumulative damage, *D* in a period, *t* with N_T cycles, is then

$$D=\sum_i \frac{n_i}{N_i}=\sum \frac{n(s_i)}{N(s_i)}=\frac{N_T}{K}\left[\frac{s_0}{(\ln N_0)^{1/B}}\right]^m \Gamma(m/B+1) \tag{3}$$

where $\Gamma(\ )$ is the Gamma function. The assumption of Weibull distribution of stress ranges is relevant for platforms operating in extratropical regions. The expression in Equation (3) can be refined by considering other models for the load effects (stress ranges) and SN curves, e.g. [50-51].

4.2. Human and Organizational Factors

From a human and organizational error point of view, however, the causes of failures and mitigation measures can be categorized as shown in Table 2. In general, the measures include quality assurance and control (QA/QC) relating to the engineering process as well as the hardware and operational procedures as well as the Accidental Collapse Limit State criterion.

Gross errors and their effects should be avoided by adequate competence, skills, attitude and self-checking of those who do the design, fabrication or operation in the first place; and by exercising "self-checking" of their work. In addition, quality assurance and control (QA/QC) should be implemented in all stages of design, fabrication and operation.

The quality assurance and control of the engineering process have to address two different situations, which require different type of attention, namely:

- detect, control and mitigate errors and omissions made in connection with technology that is known in the engineering community as such. With the increasing use of computers in the design, construction, and operation of oil and gas structures, software errors are of particular concern
- identify possible unknown phenomena, e.g. associated with load, response and resistance, and clarify the basis for accounting for such phenomena in design

In the future the aging of existing offshore structures, especially due to crack growth, needs attention, see Section 4.3.

Table 2. Causes of structural failures and risk reduction measures.

Cause	Risk Reduction Measure
Less than adequate safety margin to cover "normal" inherent uncertainties.	Increase safety factors or margins in ULS, FLS; Improve inspection of the structure(FLS)[1)]
Gross error or omission during: - design (d) - fabrication (f) - operation (o)	Improve skills, competence, self-checking (for d, f, o) QA/QC of engineering process (for d) Direct design for damage tolerance (ALS) – and provide adequate damage condition (for f, o) Inspection/repair of the structure (for f, o)[2)]
Unknown phenomena	Research & Development

1) Measure by Structural Reliability Analysis
2) Measure by Risk Assessment

As mentioned above, operational errors typically result in fires or explosions or other accidental actions. Such events may also be controlled by appropriate measures such as detecting the gas/oil leakage and activating valve shut in; extinguishing of a fire by a deluge system activated automatically etc. These actions are often denoted "Event Control".

In the treatment of the effect of the gross errors in design one might separate between identifiable/quantifiable versus unidentifiable/unquantifiable hazards. Thus errors resulting in accidental loads belong to the first category while design errors belong to the second category.

Depending upon the function (oil and gas platform, wind turbine, bridge) of the structure and regulatory regime, the acceptance criteria in terms of consequences, such as fatalities, environmental damage or property damage, are established. While fatalities caused by structural failures would be related to global failure, i.e. capsizing or total failure of deck support, smaller damages may result in pollution; or property damage which is expensive to repair, especially for the underwater part of a permanent structure.

4.3. Design for Robustness: Accidental Collapse Limit State

Accidental Collapse Limit State (ALS) criteria are introduced to limit the corresponding residual risk i.e. to prevent progressive failure [28]. This goal could be achieved by designing the structure locally to sustain accidental actions and other relevant actions. Alternatively, local damage may be accepted and the ALS requirement should focus on survival of the damaged structure to relevant actions, to avoid escalation.

Accidental Collapse Limit State (ALS), initially called the Progressive Collapse Limit State, requirements, are motivated by the design philosophy that "small damages, which inevitably occur, e.g. due to ship impacts, fires and explosions, and other accidental loads, as well as environmental hazards, should not cause disproportionate consequences". The ALS criterion was formally first introduced for all failure modes of offshore structures in Norway in 1984 [52]. The background and practicing of the ALS criterion is described in [28, 49].

The introduction of the quantitative robustness check in terms of ALS, was possible because FEMs had been developed for accomplishing the necessary nonlinear structural analysis to estimate the damage and the strength of the damaged structure. See current versions in [10-12].

ALS checks apply to all relevant failure modes, like, structural failure as well as instability/capsizing (in terms of intact and damage stability requirements for floating platforms).

The structural integrity criterion in [10] is a two-step procedure as illustrated in Figure 6. First, the initial damage due to accidental actions (fires, explosions, ship collisions, dropped objects as well as abnormal distribution of payload or ballast) with an annual exceedance probability of 10^{-4} is estimated by a nonlinear analysis. The initial damage in some cases need to be specified, based on consideration of the vulnerability of the structural components, difficulty of detecting the damage or other experiences, as failure of a slender member in the hull, a mooring line etc. Based on the experience that crack type defects which are larger than the normal initial defects could occur and escape detection at the fabrication stage, Moan [28] suggested to include a corresponding specified damage condition. Other barriers to deal with the occurrence of abnormal defects are presented in Table 5. Abnormal sea loads, at an annual probability level of 10^{-4} also need to be considered in the ALS check – in a single step.

The accidental loads are to be determined by risk analysis, see e.g. [44], accounting for relevant factors that affect the accidental loads. In particular, risk reduction can be assumed to take place by reducing the probability or consequences of hazards.

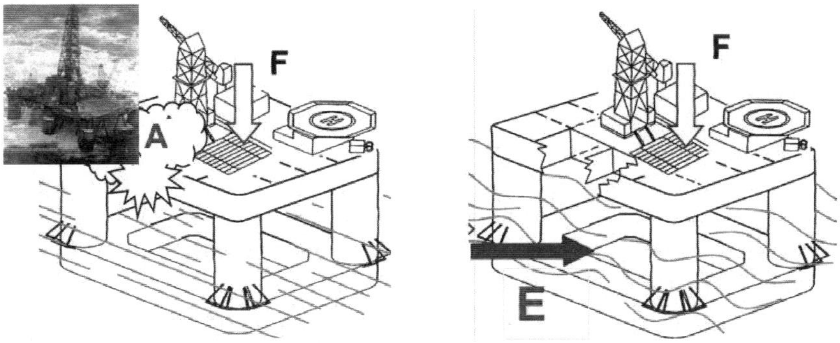

Figure 6. Schematic illustration of the two-step ALS check: (Left) assessment of damage caused by an accidental action (A) ship impact –considering relevant permanent and functional (F) loads and possibly environmental loads, (Right) a check of the survival of the damaged platform under environmental load, E as well as F loads.

The second step in the NORSOK ALS procedure is to demonstrate that the damaged structure resists relevant functional and environmental loads an annual exceedance probability of 10^{-2} or a lower probability depending on the correlation between the accidental event and the environmental condition – without global failure.

More details about accidental loads, risk analysis to estimate their magnitude, and their effect on structures may be found in e.g. [11-12, 44, 53-54].

The ALS criterion is supposed to include "abnormal" wave loading as well. Obviously, this check will involve a survival check based on a load event corresponding to an exceedance probability of 10^{-4} only. In this connection the focus is on possible "abnormal" waves, with high crest or other unusual shape – which is not a simple "extrapolation" of the 10^{-2} event [55-56].

4.4. Inspection, Monitoring, Maintenance and Repair

Inspection, Monitoring, Maintenance and Repair (IMMR) should address all kinds of damages as well as conditions that could lead to damages. The occurrence of accidental damage due to fires/explosions and ship impacts should on one hand be controlled by surveillance of hydrocarbon leaks and ship traffic but on the other hand the damage once it has happened, easy to identify.

Corrosion and crack growth need continuous surveillance. Corrosion damage is indirectly monitored by the quality of the coating or cathodic protection system; and eventually by thickness reduction. Moreover, there is normally ample time for repair in case of corrosion damage.

The challenge is cracks. It is important that the IMMR reflects the fact that fabrication errors occur and cause abnormal initial cracks. In view of the fact that normal welding induced cracks are of the order of 0.1 mm, an abnormal crack is not that large. In general, therefore, it becomes important to ensure that there is ample time between visible (detectable) crack and fracture. A general remedy to ensure this feature is to limit the stress level by designing for fatigue by a large FDF or to ensure ultimate strength after member failure.

For a given inspection (with normal quality) with no crack detection in a joint with an abnormally large defect, the updating of Pf will be conservative. However, the inspection itself can be subjected to gross errors. Thus, if abnormal defects are present in joints which are not inspected, the estimate of the failure probability will obviously be non-conservative. This fact should be accounted for in the estimate of the risk of system failure in view of the coverage of the inspection program.

5. Reliability and reliability based design and inspection approach

5.1. General

In principle risk based design could be carried out, by achieving a total system (structural layout, scantlings and equipment, procedures and personnel) that complies with a certain acceptable risk level. This is, however, not feasible in practice. In reality, different subsystems, like:

- loads-carrying structure & mooring system
- process equipment
- evacuation and escape system

are designed according to criteria given for the particular subsystems. Safety criteria for the structural design of offshore structures may be classified as shown in Table 1. While ULS and FLS criteria are generally applied, the introduction of a quantitative ALS criterion by NPD [52] was a significant step towards the generally agreed concept of making structures damage-tolerant (robust). Figure 7 illustrates the design check involving the different safety criteria.

The assessment of the structure during the initial design and operation are subjected to uncertainties, namely

- normal variability and uncertainty due to inherent variability or lack of data (also to establish analysis models)
- human errors

affecting both the actions and their effects as well as resistance. Design is based on generic measures about uncertainties while the inspection of the as-built structure and its behaviour during operation provide improved information about abnormal geometry, including defects, and, hence, the component strength.

In-service condition monitoring by e.g. acceleration and stress measurements may be used for validation of the structural analysis and design assumptions and the occurrence of damages. Measurements of the change of vibration properties can be used to detect damage [57].

Normally in-service measurements of stresses focuses on nominal stresses, while laboratory studies would be needed to validate local (hot spot) stresses as well as both ultimate and fatigue strength. Fatigue strength in design is normally based on SN curves, but crack initiation and propagation features would be required in a high fidelity analysis e.g. for inspection planning.

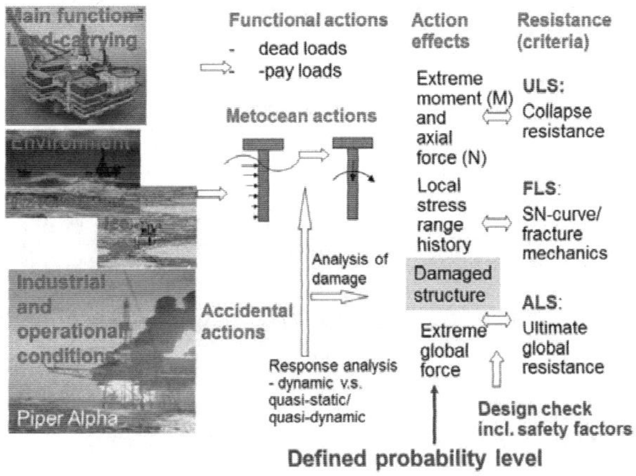

Figure 7 Analysis of action effects and resistances for design checks.

Structural reliability methods (SRMs) have been developed as a tool to assess failure probability. However, the SRM does not account for human errors. Hence, the probabilities should be considered notional and not "real" – relating to the generic uncertainties known at the design stage. However, the inspection of the as-built structure makes it possible to (partly) eliminate the effect of gross errors that caused deviations from the planned structure – relating to the occurrence of cracks. For use of SRM in inspection planning it is important to identify and "eliminate" as far as possible abnormal defects and modelling the geometry of the as-built structure. These features are further treated for jackets and semisubmersibles in [47]. They show how SRM analyses give measure of the capability of the analyses to describe the actual fatigue failure potential.

The fabrication quality is one of the main design assumptions. Deviations in the fabrication quality are one of the reasons for reduced correlation between observed failures and the probability for failure given by the analyses.

The planned use of information from the inspection, repair, operation and modification history will determine the requirement for the process securing the data for the inspection, repair and modification history. However, this report describes how the inspection and repair history is applied as part of the assessment of structural integrity and define requirement for the inspection and modifications scope. Bayesian updating plays and important role in the analysis of information and data from inspections carried out.

5.2. Ultimate Limit States

5.2.1 General: Ultimate Limit State (ULS) criteria for overall stability of bottom-supported and floating structures as well as strength formulations are given e.g. in [29] for relevant load conditions.

Extreme loads with an annual exceedance probability of 10^{-2} are normally required for ULS check, while for FLS check the local stress range history is needed for welded connections while the joint variation in stress range and mean stress is required e.g. for base material.

Load effects for ULS design checks are typically obtained by considering all or a carefully selected sea states and applying relevant wave kinematics, hydrodynamic load and structural models [11]. The challenges in predicting extreme wave load effects for ULS design of offshore structures relate to the modelling freak or rogue wave or crest and their occurrence rate as well as dealing with severe nonlinear hydrodynamic effects; e.g. involving phenomena indicated in Figure 8. The latter effects include wave-in-deck loading, slamming, column run-up, green water on deck and excitation by steep, high waves, e.g. causing dynamic ringing effects. Figure 3a illustrates a damaged platform due to wave loads on the deck structure. Methods for predicting the corresponding hydrodynamic loads are still quite uncertain, and experiments are necessary, see e.g. [11]. The structural analysis is normally hierarchical, starting with a global and obtaining the local stresses in principle by using substructures in the global analysis or zooming/sub-modelling techniques.

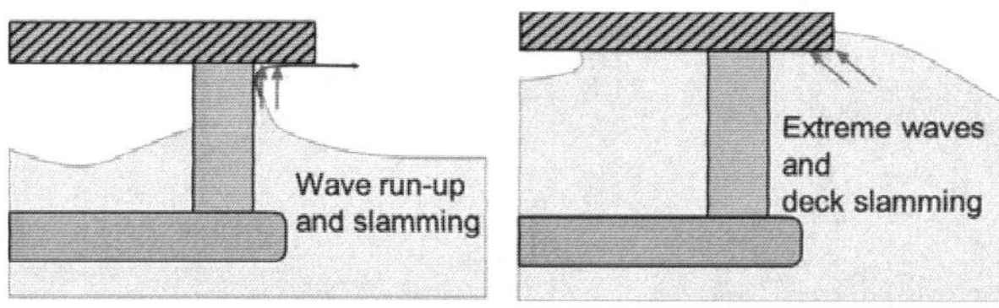

Figure 8. Wave action effects with significant uncertainty.

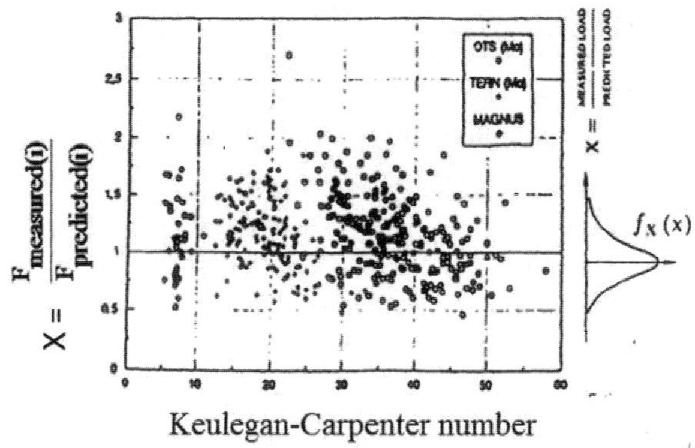

Figure 9. Wave by wave comparison of measured and predicted base shear of the Magnus platform in the North Sea [58].

Figure 9 shows the uncertainty in the base shear according to the API/ISO jacket wave load recipe based on a regular wave obtained by Heideman and Weaver [58]. The mean and CoV of the model uncertainty, X, are 1.06 and 0.25, respectively. It is noted that the uncertainty can be significantly reduced if a random wave model is applied [59]. On the other hand the uncertainty increases if the behaviour of the jacket is dynamic.

Ultimate strength formulations used in design are traditionally based on strength of material formulations and substantiated by extensive test results. However, design based on direct ultimate strength analysis, by using finite element methods and by accounting for nonlinear geometric and material effects, is emerging.

A simplified model for the total uncertainty in the base shear can by estimated by modelling the base shear, S as:

$$S = X\, h^{\alpha} \tag{4a}$$

where α is 1-3 depending on the relative magnitude of drag and inertia wave actions; h - wave height and X - model uncertainty of action. It is seen that the CoV of S is approximately:

$$V_s^2 \approx V_X^2 + \alpha^2 V_h^2 \tag{4b}$$

showing that the uncertainty in the wave height can contribute significantly the that in the base shear.

5.2.2 Reliability measure: Structural reliability methods (SRMs) have been used to ensure that ULS requirements are consistent with the desired target safety level, especially by the efforts by Fjeld, Moses, Lloyd and Karsan, Moan and Jordan and Maes [60-64], to calibrate the safety in design codes to a certain reliability level. An evaluation of previous efforts on calibration of offshore codes was provided in [65] in conjunction with the ISO effort to harmonize codes for offshore structures. Assessment of uncertainties in load effects and resistances was a crucial issue in these studies.

SRM is applied to determine the failure probability considering fundamental variability, as well as uncertainties due to lack of knowledge in loads, load effects and resistance. The state of art methods for calculating the failure probability are numerical First Order and Second Order Reliability as well as Monte Carlo simulation methods, e.g. [66]. However, analytical solutions exist for a few cases, for instance, when failure is expressed by $g(\,) = R - S \leq 0$ and both the resistance R and the load effect S are normal or lognormal random variables.

The failure probability is in general expressed by:

$$P_f = P(g() \leq 0) = \Phi(-\beta) \quad \text{or} \quad \beta = -\Phi^{-1}(P_f) \tag{5}$$

where $\Phi(-\beta)$ is the standard cumulative normal distribution and β is the reliability index. Numerical values of this relationship are shown in Table 3. The reliability index $\beta = \beta_{LN}$ can be exactly written as follows when R and S have lognormal distributions, see e.g. [66-67]:

$$\beta_{LN} = \frac{\ln\left[\frac{\mu_R}{\mu_S}\sqrt{\frac{1+V_S^2}{1+V_R^2}}\right]}{\sqrt{\ln\left[\left(1+V_S^2\right)\left(1+V_R^2\right)\right]}} \approx \frac{\ln\left[\frac{\mu_R}{\mu_S}\right]}{\sqrt{V_R^2+V_S^2}} = \beta_{LN}^{'} \tag{6}$$

This simple expression has turned out to be useful and was applied in the API reliability based code calibration [64]. However, using the lognormal format to estimate the failure probability for cases where e.g. the distribution of S is not lognormal, may result in significant discrepancy in the Pf. This fact shows that the target reliability level needs to be intimately associated with the reliability model applied.

The reliability estimate critically depends on the assessment of uncertainties in structural loads and strength, as indicated above and exemplified subsequently. It is important that authoritative values of the uncertainties are applied in reliability analyses.

Structural reliability methods provide notional estimates of probabilities. These probabilities are small (current ULS requirements imply the order of 10^{-3} –10^{-5}) and are lower than the experienced P_f for offshore structures, e.g. [65, 68]:

Another matter is that gross errors that often cause failures (Table 2), is not recognized in the SRM. For this reason the structural reliability method (SRM) does not provide a measure of the actuarial total risk level associated with offshore facilities. Yet, SRM is useful in ensuring that the ultimate strength and fatigue design criteria are consistent by calibrating safety factors for ULS and FLS criteria (Section 5.3). However, to achieve a true measure of safety, a risk assessment methodology is needed. See also overviews in [69-70].

Table 3. Relation between β and P_f.

β	1.0	1.4	1.8	2.2	2.6
P_f	0.16	0.081	0.036	0.014	$0.47 \cdot 10^{-2}$
β	3.0	3.4	3.8	4.2	4.6
P_f	$0.14 \cdot 10^{-2}$	$0.34 \cdot 10^{-3}$	$0.72 \cdot 10^{-4}$	$0.13 \cdot 10^{-4}$	$0.21 \cdot 10^{-5}$

Since offshore structures are subjected to time variant loads, the failure probability should refer to a time interval, e.g. a year or the service life. This can be achieved by considering S as a random variable, referring to an annual or service life time maximum value. However, this approach is not straight forward when the failure event is described by multiple load effect variables. Then more advanced methods are needed [67, 71]:

5.2.3 Relationship between failure probability and safety factors *given the uncertainty in R and S:* By assuming the simple ULS design format (Equations 1a-c) and the following relations between characteristic and design values:

$$R_d = {R_c}/{\gamma_R} = \gamma_S S_c = S_d$$

$$\begin{array}{ll} \text{Mean} & \text{CoV} \\ \mu_S = B_{S1} S_c, B_S \le 1; & V_S = 0.15 - 0.3 \\ \mu_R = B_{R1} R_c, B_R \ge 1; & V_R = 0.1 \\ B_S = B_{S1} B_{S2} & B_{S2} - \text{ load model bias} \end{array} \tag{7a-d}$$

the following relationship results:

$$\beta'_{LN} \approx \frac{\ln(\mu_R/\mu_S)}{\sqrt{V_R^2 + V_S^2}} = \frac{\ln(B_R \gamma_R \gamma_S / B_S)}{\sqrt{V_R^2 + V_S^2}} \tag{8}$$

With the relationship between P_f and β'_{LN} given above, the desired relationship between the uncertainty measures and the safety factors, γ are obtained. This expression also shows the relative importance of systematic (B) versus random (V) uncertainty.

5.3. Fatigue Limit State

5.3.1 General: Fatigue is an important consideration for structures in areas with more or less continuous storm loading (such as the North Sea) and especially for dynamically sensitive structures. Even if fatigue was known since Wöhler's "scientific" account of the fatigue phenomenon in the 1850s, it was not until the 1980s and 1990s the criterion was fully recognized for offshore (and ship) structures.

Fatigue crack growth is primarily a local phenomenon. Fatigue strength is commonly described by SN-curves, i.e. the number of cycles to failure as a function of the stress (range) level that have been obtained by laboratory experiments. Failure is often defined as through thickness crack, visible crack and total rupture of the test specimen. Fracture mechanics analysis of fatigue strength has been adopted to assess more accurately the different stages of crack growth including calculation of residual fatigue life beyond what is defined as fatigue failure in conjunction with SN-curves. Such detailed information about crack propagation is also required to plan inspections and repair, e.g. [50, 51]. However, it is important to ensure that the fracture mechanics model is consistent with the SN-approach with respect to fatigue failure. This is because the initial crack size needs to be modelled in the fracture mechanics while it is implicitly represented in the experimental data that serve as basis for the SN-approach , e.g. [72].

5.3.2 Reliability measure: The elementary reliability format (6) may also be used to obtain an estimate for the fatigue reliability based on the SN formulation. By expressing fatigue failure by

$$g(\ )=\Delta-D \tag{9}$$

Equation (6) can be used with $R = \Delta$ (cumulative damage at failure) and $S = D$. Moreover, if only the dominant variables s_0 and K are taken as random variables with lognormal distribution as advocated by Wirsching [73] a lognomal reliability format can be formulated (see further discussion in [68]). For a fatigue design with allowable damage Δ_{all} = 1.0 and 0.1; the implied P_f's are about 0.1 and $2{\cdot}10^{-3}$ – $3{\cdot}10^{-5}$, depending on uncertainty measures, respectively, in the service life.

With a notional probability of fatigue failure in the service life of the order of 0.1 when the fatigue design factor is taken to be 1.0, inspection and repair are clearly required to ensure adequate lifetime safety [26]. In addition, gross errors could occur.

5.3.3 Reliability estimates by account of inspection: The failure probability, P_f, of a specific structural component can be updated through additional information obtained by response measurement or observed damages and a fracture mechanics model of the crack growth, such as the Paris-Erdogan formulation, namely

$$\frac{da}{dN}=\begin{cases} C\,(\Delta K)^m & for\quad \Delta K > \Delta K_{th} \\ 0 & for\quad \Delta K \le \Delta K_{th} \end{cases} \tag{10}$$

where a is crack depth, N is number of cycles, C is crack growth parameter, m is the inverse slope of the SN curve, and ΔK_{th} is a threshold of the stress intensity factor range ΔK given as a function of a .

For instance when inspections are made, say with no detection of cracks in joint i after a time, t in-service, the failure probability of joint j may be updated as follows

$$\begin{gathered} P_{f_i UP j} = P\left[\left(g_i(\cdot)\le 0\right)\middle| IE_j \le 0\right] = \\ P\left[\left(g_i(\cdot)\le 0\right)\cap\left(IE_j \le 0\right)\right]/P\left[IE_j \le 0\right] \end{gathered} \tag{11a}$$

where the failure event is

$$g_j(\ )=a_{cj}-a_j(t) \tag{11b}$$

and the inspection event is

$$IE_j(\ )=a_j(t)-a_{dj} \tag{11c}$$

for the j'th joint. a_c and a_d are the critical and detectable crack size, respectively. See e.g. [21-22].The updated P_f of joint i based on the inspection of joint j depends on the correlation between the $g_i(a)$ and IE_j events. Equations (11a-c) can be recasted in a convenient form for analysis as shown by Madsen et al. [21-22].

5.3.4 Inspection quality: The quality of non-destructive examination (NDE) methods for detection of cracks in metal structures is expressed by a probability of detection (POD) curve, which corresponds to the distribution function of detectable crack size ad. Various sources of POD data are discussed in [26, 74]. In practice it is important to define the quality of NDE inspections in classes in the range of a mean POD from 0.4 to 1.6 mm [47], depending upon the competence of the inspection crew and environmental conditions; e.g. air versus underwater.

The reliability of assessing corrosion damage relates to the accuracy of estimating the change of thickness due to corrosion.

5.3.5 Uncertainties: The failure probability is crucially dependent upon the uncertainty measures of load and resistance parameters. It is important that authoritative values of the stochastic variables are used. Relevant uncertainty measures are given e.g. in [25, 74].

5.3.6 Calculation of reliability: It is noted that the FORM/SORM methodology is approximate and should be validated by using "converged" Monte Carlo simulation. This is particularly important for calculating the probability of intersections of events.

5.3.7 Guidelines on inspection planning with respect to fatigue cracks: Recently DNV GL [74] issued a comprehensive guidance for use of probabilistic methods for inspection planning with respect to fatigue cracks in jacket structures, semi-submersibles and floating production vessels. The approach is rather general may be used also for inspection planning of other structures subjected to significant dynamic loading such as jack-ups.

5.3.8 Validation – or rather estimate of uncertainty: The various models that constitute the reliability analysis method may be validated by: using model tests or full scale measurements to validate wave induced nominal stress; model tests to validate hot spot stress and fatigue strength and the reliability model at large by direct comparison of predicted and observed crack occurrences, see e.g. [26]. The comparison in [75] showed that the number of predicted propagating cracks identified in 3411 inspections of tubular joints in jackets, was typically 3 to 10 times larger than the number observed. This discrepancy can be traced back to uncertainties (conservativism) in the prediction of the (nominal) stress.

5.3.9 The effect of inspection on the fatigue reliability: The effect of inspection may be estimated in two different ways: a) at the design stage, before inspections are done or b) after the actual inspection has been made. If the effect of inspections is estimated at the design stage, the two possible outcomes: detection: D and no detection: ND need to be their probability of occurrence based on the reliability method. After conducting an inspection the outcome (D or ND) is known.

Moreover, during the service-life, there may be several inspections. Equations (11a-c) can be generalised to cover cases with several inspections, with two alternative out-comes. Moan et al. [76] show how the allowable fatigue damage, Δ_{all} at the design stage can be relaxed when inspections and necessary repairs are carried out; i.e. as a basis for Table 4.

Figure 10 shows the reliability index β as a function of time based on prediction of the effect of inspections every 5th year with no crack detection using an NDE inspection method with mean detectable crack size of 1.5 mm - at the design stage and based on the outcome of actual inspections. The latter obviously is more efficient in increasing the reliability index β.

The updating methodology is useful in connection with extension of service life for structures with joints governed by the fatigue criterion [77]. In such cases, the design fatigue life is in principle exhausted at the end of the planned service life. However, if no cracks have been detected during inspections, a remaining fatigue life can be demonstrated. But it is not possible to bring the structure back to its initial condition by inspection only. This is because the mean detectable crack depth typically is 1.0–2.0 mm, while the initial crack depth is 0.1–0.4 mm.

5.3.10 Inspection planning of wind turbines: While the support structure of wind turbines can be treated by the methods used for oil and gas platforms, the blades and drive train involve mechanical equipment and are difficult to access and condition monitoring is primarily used to follow up during operation. Damage is detected by using vibration analysis, Damage to the gears and bearings is identified by the metal particle content in the lubrication oil. See [57].

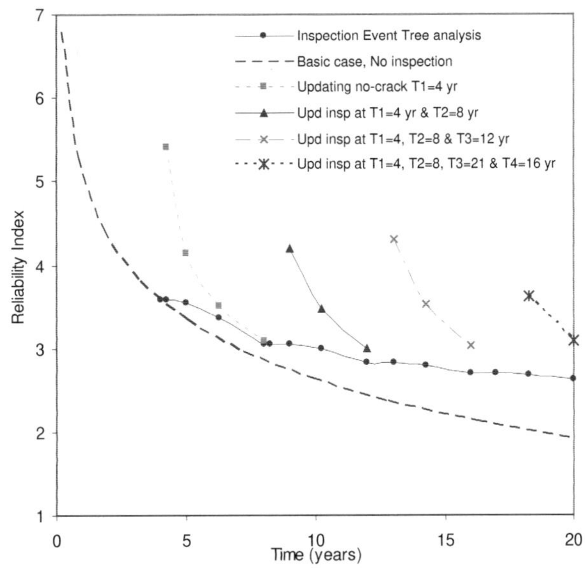

Figure 10. Reliability index as a function of time and inspection strategy. Inspection Event Tree analysis is based on prediction at the design stage. The other curves are based on inspections with known outcome during the service life.

5.4. Structural System Reliability

Systems reliability approaches are attractive because the most significant platform failure consequences, and especially fatalities, are associated with system failure and hence are useful for making decisions which affect safety and the significant costs of replacing or modifying the structure.

System failure can be expressed mathematically by load and resistance parameters relating to all failure modes for all structural components (members, joints, piles) and the probabilistic properties of these parameters. Broadly speaking, this may be achieved by a failure mode (or survival mode) analysis or direct-simulation methods, considering a sequence of fatigue and overload failures, see e.g. [78-80] and the overview in [23].

A significant challenge is to describe the failure modes, the probabilistic features of the random variables, and especially the correlation between them.

However, De [81] and Wu and Moan [82] demonstrated for jackets that accurate estimates of the systems failure probability under extreme sea loading can be achieved by considering a single system failure mode, i.e. by referring both the load (S) and resistance (R) to a given load pattern and using the (overall) base shear as variable.

A first approximation to the failure probability, P_{fSYS}, of a jacket system consisting of discrete elements, considering both overload as well as fatigue failure followed by over-load failure modes, may be expressed as :

$$P_{fSYS} = P\left[FSYS\right] \approx P\left[FSYS(U)\right] + \sum_{j=1}^{n} P\left[F_j\right] \cdot P\left[FSYS(U) \mid F_j\right] + ... \quad (12)$$

where $FSYS(U)$ and $FSYS(U)/F_j$ are treated as pure over-load system failures readily calculated by efficient methods for framed structures [83-84] and F_j the fatigue failure of component j.

The approximations made and the use of Equation (12) is discussed in [26]. Despite the approximation made in connection with Equation (12), use of this system reliability measure can give reasonable decisions if the target reliabilities are determined consistently for the same expressions.

The formulation expressed by Equation (12) is particularly applicable for structures with clearly defined components such as members and joints on jackets, jack-ups, drilling semi-submersibles (with braces) [68] while cracks in monocoque structures like ship hulls are more challenging to deal with.

There is an important link between fatigue and ALS design criteria and inspection, by the fact that the acceptable fatigue failure probability should depend on the consequences – i.e. potential progressive failure, see Table 4.

5.5. Target reliability or risk level for design codes
The target safety level should depend upon the following factors, e.g. [85-86]:

- initiating events (hazards) such as environmental loads versus accidental loads (which are caused by human errors or omissions)
- method of SRM or structural risk analysis, especially which uncertainties are included
- failure cause and mode
- the possible consequences of failure in terms of risk to life, injury, economic losses and the level of social inconvenience.
- the expense and effort required to reduce the risk of failure.

A main issue is that target levels for notional failure probabilities relating to SRM should be clearly distinguished from the true failure probabilities relating to risk assessment considering human factor effects also. Hence, Jordaan and Maes [64]: argued that the target failure probability in the context of SRM should be a fraction of the true failure rate.

Moreover, rather that setting an overall target level it is more practical to establish target levels for each hazard separately, see e.g. [41]. Moreover, target reliability levels for ULS and FLS criteria should be based on SRM accounting for normal variability and uncertainties in load effects and resistances, while ALS criteria needs to be based on a broader risk assessment.

In general it is recommended to establish target levels in a consistent manner based on inferring the target level by using the same reliability or risk analysis method to estimate the inherent reliability level implied by a reference design code that is considered acceptable and the method used to demonstrate compliance with the target level [85]. Herein the focus is on fatigue target levels.

Table 5 shows semi-probabilistic fatigue acceptance criteria established by NPD/NORSOK in Norway and API [87] (- essentially for jackets), respectively. A reference target failure probability level was first established based on that inherent in semi-probabilistic fatigue criteria for the cases of two consequence cases; namely where fatigue implied total loss and the structure surviving a 100 years storm after fatigue failure. By accounting for the effect of inspection [41] and using a simplified system approach by considering each term in the sum in Equation (12) for jacket structures [48, 76, 85] fatigue design criteria considering the effect of inspection, have been derived and compared with codified criteria in Table 4.

6. Inspection, monitoring, maintenance and repair during fabrication, installation and operation

6.1. General
As indicated in Section 4.3 IMMR needs to address structural damage but also other operational and hardware aspects, especially event control relating to fires/explosions, ship impacts that can result in damages or failures. Herein the focus is on crack control.

The main life cycle efforts to prevent that cracks develop into failures, include:

- adequate design with respect to
- fatigue design (Fatigue Design Factor)
- residual fatigue life
- ultimate reserve strength (to overload)
- adequate inspection, or
- adequate monitoring of leak, when relevant

Table 5 shows qualitatively which role different safety measures play regarding crack control for different types of structures. During design structural scantlings and local geometry to achieve a desired fatigue life, inspection, maintenance and repair plan are decided.

Fatigue design requirements depend upon inspectability and failure consequences are considered, e.g. as shown in Table 4. It is noted that the costs of the underwater inspections on jackets and the number of possible crack sites on FPSOs make it necessary to prioritize inspections. For semi-submersibles some welded joints will have a high priority because of high likelihood of crack occurrence and failure consequences.

Whether the inspection should be chosen to aim at detecting barely visible cracks, through-thickness cracks (e.g. by leak detection) or member failures would depend on how much resources are spent to make the structure damage tolerant. This choice again would have implication on the inspection method; the main inspection methods being non-destructive examination (NDE) methods, detection of through-thickness crack by e.g. leak detection, or failed members by visual inspection. The quality of visual or NDE methods depends very much upon the conditions during inspection. Large volume offshore structures are normally accessible from the inside, while, e.g., small diameter tubular tethers in TLPs and maybe some joints in semi-submersibles, are not.

Permanent repairs are made by grinding to remove small cracks, cut out and butt welding a new section, re-welding, adding or removing scantlings, brackets, stiffeners, lugs or collar plates.

Table 4. Fatigue design factor, FDF to multiply with the planned service life to obtain the design fatigue life: NPD-NORSOK/API and (reliability analysis).

Classification of structural components based on damage consequence[1)]	**Access for inspection and repair**		
	No access or in the splash zone	Accessible (inspection according to generic scheme is carried out)	
		Below splash zone	Above splash zone or internal
Substantial consequences	10/10 (10)	3/5 (4)	2/5 (3)
Without substantial consequences[2] (i.e. satisfaction of ALS req.)	3/5 (4)	2/2 (2)	1/2 (1)

1) NPD(1984)-NORSOK N-001(2002) – general requirements while; API[87]; and reliability estimates [48, 76, 85]
2) The consequences are substantial if the Accidental Collapse Limit State (ALS) criterion is not satisfied in case of a failure of the relevant welded joint considered in the fatigue check.

Table 5. Crack control measures.

Type of structure	**Type of joint**	**FDF[1)]**	**Residual fatigue life**	**Ultimate reserve strength**	**Inspection method**
Jacket	Tubular joint	2-10	Some- Sign.	Normally	NDE,U[2)]
Semi-Submersible	Plated brace	1-3	Some	By ALS[4)]	LBB[3)], NDE
	Plated column-pontoon	1-3	Some	Limited	LBB, NDE
TLP	Tether Plating	10	Small	By ALS	IM[5)]
	Column-pontoon plates.	1-3	Some	Limited	LBB NDE
Ship	Longitud.comp.	1-3	Sign.	None	Close Visual

1) FDF - Fatigue Design Factor – by which the service life is to be multiplied with to achieve the design fatigue life
2) NDE - Non Destructive Examination Method; U-underwater
3) LBB - Leak before break monitoring
4) ALS - Accidental Collapse Limit State
5) IM - Instrumental monitoring (by "an intelligent rat")

The design criteria (Table 4) are based on criteria that account for normal variability and uncertainties. However, these measures are generic values; i.e. with the same criteria for all kinds of offshore structures. The initial inspection plan is also based on generic information available at the design stage. Besides differences between the features of different structures, the fabrication process might induce additional uncertainties, especially on features regarding local geometry that affect fatigue performance as described below. Hence, it is crucial to update the model used to predict crack growth based on the particular features of each structure.

During fabrication the QA/QC addresses control of material and geometry, especially tolerances relating to misalignment etc., possible damage that could occur etc. It is noted that defects in welds are primarily controlled indirectly by the specification of welding procedure, environment etc. This is because the normal defect depth, say of the order of 0.1 mm are not likely to be detected by NDE inspections with a mean detectable crack size which is significantly larger. However, gross errors, like cracks of a depth 2.0 mm or more might be detected by NDE.

The importance of updating the information of the geometry based on the as-built geometry is crucial as a basis for the follow-up during operation.

Finally, achieving robust offshore structures by applying ALS criteria is an important issue. In the present context a damage condition for an ALS check associated with ab-normal cracks is envisaged, as commented upon in Section 4.3, see also [28].

During operation inspections are carried out to detect cracks and deterioration. An inspection plan involves:

- prioritizing which locations are to be inspected
- selecting inspection method (visual inspection, Magnet Particle Inspection, Eddy Current) depending upon the damage of concern
- scheduling inspections
- establishing a repair strategy (size of damage to be repaired, repair method and time aspects of repair)
- assessment methodology for crack detections

Typically major inspections of offshore structures (special surveys, renewal surveys) are carried out every 4 - 5th year, while intermediate and annual inspections normally are less extensive. Further refinement of the inspection planning has been made by introducing probabilistic methods as described below. As mentioned above it is important then to account for the findings regarding the geometry and cracks etc. during inspections after fabrication and during operation. Moreover, repairs might involve structural modifications that need to be considered. Besides making the necessary repairs of damages etc. it is important to make a record of the deviation of the as-built structure from the as-designed structure, cracks, corrosion, inspection history and repairs in order to obtain a more general assessment of the quality of the fabrication, for use in integrity management of other structures as well..

Cracks in Floating Production Storage and Offloading (FPSO) vessels may be repaired during operation, but operational procedures need to be changed to avoid hot work in tanks with combustible gases. However, more extensive repairs such as by introducing some thousands of additional brackets, grinding of cracks etc. need to be done in yards.

Semi-submersibles have to be dry-docked to carry out inspection and repair of the hull in North Sea operations. A major cost issue in connection with inspection of semi-submersibles, is the demobilization offshore, transport to yard, yard set-up, transport to offshore site and offshore mobilization. Production semis and TLPs, however, would normally have to be inspected and repaired on site.

Tethers are particularly critical components. Each tether is a series system with multiple potential crack sites. Inspection is carried by a remotely operated instrument and possibly detection of leak before break. Repair usually requires replacement of tether. For tether systems it is hence crucial to achieve safety by restrictive fatigue design criteria as well as an elaborate quality assurance at the fabrication stage.

The inspection strategy would normally be to inspect pre-selected potential crack sites. If the detected damage exceeds the acceptable level, the inspection would have to be extended to cover more possible crack sites, in the same or other "sister" platforms.

Inspection, monitoring and repair measures can contribute to the safety only when there is a certain damage tolerance. This implies that there is an interrelation between design criteria (fatigue life, damage tolerance) and the inspection and repair criteria, as shown in Table 4.

However, during operation the situation is different. Modification of the design and hence the scantlings is very expensive, and the most relevant measure to influence safety is the inspection and repair. The following section briefly describes how fatigue design and inspection plans (based on an assumed inspection method) can be established by reliability analysis to ensure an acceptable safety level.

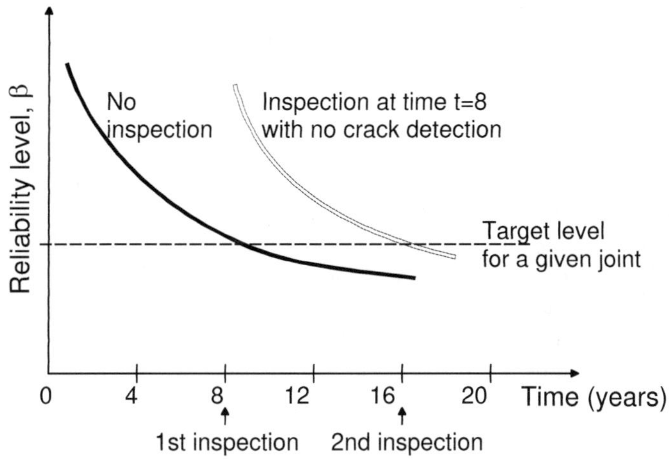

Figure 11. Scheduling of inspections to achieve a target safety level of $P_{FfT}(i)$.

6.2. Reliability (or risk) based inspection (RBI)

The first step from empirical to more rational procures to plan inspections, was to use fatigue analyses based on SN-data. Moreover, as mentioned above, design and inspection criteria (Table 6) relating to fatigue crack growth, are based on generic information. Inspections during operation of the specific structure yields more information – that can be used in planning future inspections. Reliability methods can serve as tools for the decision making under uncertainty relating to fatigue crack growth – to maintain an acceptable safety level. However, it is crucial that such analyses are based on information from inspections during fabrication and operation in order to refer to an as realistic model of the structure and its behaviour as possible. This includes in-formation about crack type defects with excessive size.

Moreover, any information about the environmental conditions and the global nominal stress level based on monitoring, will be useful.

By assuming that the fatigue failure probability in Equation (12) is updated based on inspection of the relevant joint (member), - before its (rupture) failure. Moreover, the target reliability is assumed by allocating a certain target probability level, $P_{fSYS}(T)$ to each term in the sum of Equation (12), i.e.

$$P_{fSYS(i)} = P_{FfT(i)} \cdot P\left[FSYS(U)| F_i\right] \leq P_{fSYS(T)} \tag{13}$$

where the system failure probability, $P_{fSYS}(i)$ is associated with a fatigue failure of member (i) followed by an ultimate system failure. $P_{fSYS}(T)$ is obtained by generalizing e.g. acceptance criteria implied by Table 4. $P_{FfT}(i)$ the represents the target safety level for the fatigue failure of the individual joint.

This approach has been implemented for template, space frame structures [88, 48]. Given the target level for a given joint, inspections and repairs by grinding or other modifications are scheduled to maintain the reliability level at the target level as shown in Figure 11.

IOP Conf. Series: Materials Science and Engineering **276** (2017) 012033 doi:10.1088/1757-899X/276/1/012033

7. Requirements to computational tools for structural integrity management

Managing of the integrity of offshore structures require tools to cover the following items:

- Geometry representation, visualization of the relation-ship between the parts and the whole – while accounting for life cycle changes by means of CAD/CAM/CAE tools..
- Data about environmental conditions,
- Methods to determine overturning and stabilizing moment, calculate gravity and payloads, hydrodynamic, aerodynamic- and accidental actions and their effects, structural resistance,
- Approaches for carrying out code checks according to different limit states
- Reliability and risk analysis with respect to structural integrity in the life cycle in a wide sense to manage uncertainties due to normal variability and human errors

in view optimal Capex and Opex

Such data and methods are applied in the initial design and during operation in connection with reassessment. In the operational phase the initial design analyses and data gathered of the as-built structure and during operation need to be available in a data base. While obviously data for each structure is needed, use of data (especially relating to failure and damages) of other (especially related) structures, is also of interest. The challenges in striking a balance between sharing data between the stakeholders involved in the lifecycle and protecting the "IPR" of the data, and standardization of data bases, are important issues in this connection.

The hierarchy of methods at different fidelity levels and efficiency is needed for the different phases: conceptual via engineering to detailed design. It is therefore important to highlight the need to carry out R&D to develop methods at different level of refinement and computational efforts.

To illustrate the hierarchy of refinements consider for instance the determination of extreme wave-induced action effects at an annual exceedance probability level of 10^{-2} for ULS design: a) the wave condition may be described by a full long-term model of metocean data, selected sea states or even selected regular waves, clearly with different inherent uncertainties in representing the probability of the resulting action effect; b) Different wave theories may be applied; c) Hydrodynamic actions may be estimated by using a semi-empirical Morison formula, linear or nonlinear potential theory or full CFD methods; d) The determination of action effects can be done typically using Finite Element methods with different refinements, based on a static or dynamic frequency or time domain approach, considering linear or nonlinear behaviour. e) Uncertainties in the methods applied, beyond the fundamental variability of the wave condition, might be dealt with in design by introducing conservative simplifications, or a more formal semi-probabilistic approach with safety factors or a reliability analysis. In general, high fidelity methods, validated against model tests or in-service measurement might be applied to calibrate simplified methods. An important aspect of choosing methods is the fact that high fidelity methods require "high fidelity data" to perform better than simplified methods and hence require expert users to avoid a false impression of accuracy or even human errors.

Similar features apply for analyses relating to other environmental actions, and for FLS and ALS design checks. Moreover, design is an iterative process – in the first place to satisfy all design criteria and also account for the balance between the design and inspection/monitoring, maintenance and repair plans. An additional challenging iterative approach is to satisfy all design targets in an optimal manner according to a certain measure, such as life cycle costs. To obtain a realistic optimum, it is important that all serviceability and safety constraints, objectives together with realistic cost data are explicitly accounted for. If the design is based on direct analysis of action effects, resistance and use of a goal-based setting with safety in terms of reliability or risk acceptance, the computational effort becomes excessive. Hence, there needs to be a balance of the refinement in the methods/tools applied in view of the uncertainties in the end solution. It might also be convenient to apply a sequential approach for the design, starting with a global design and gradually including detailed design relating to fatigue.

For the above reasons it is crucial that information about the uncertainty and efficiency of the hierarchy of mechanics and probabilistic methods for different purposes, is available, also as a basis for reliability based design. While apparently the same methods are used by different organizations, there are differences that need to be identified by benchmarking. The increasing complexity of the tools also increases the chance of errors in the software or by the users; i.e. use of simplified methods might also reduce the risk of gross errors. Besides providing information from research in terms of research papers, research results might be disseminated in a more easily accessible way for users, in the form of standards and guidelines for data bases and methods.

8. Concluding remarks

The main life cycle efforts to prevent that cracks develop into failures, include: adequate design with respect to fatigue life (Fatigue Design Factor), residual fatigue life and ultimate reserve strength (to overload); adequate inspection, or adequate monitoring of leak, when relevant. This structural integrity management approach especially depends a follow-up during operation based on continuously establishing an inspection, modification and repair history. This is important because the initial structural design and inspection and repair planning is based on generic information for a class of structures and don't reflect the particular features of a given structure with regard to design, fabrication and operational features. Some of the features are result of gross errors.

The importance of properly accounting for uncertainties in the integrity management is crucial. Structural reliability methods can provide an improved basis for ULS and FLS design and inspection and repair planning by accounting for the normal uncertainties in actions and resistances and inspection quality, to ensure that these criteria comply with an intended target failure probability. However, observations made during inspections might identify deviations between geometry, including fabrication defects that need to be considered in the structural integrity management during operation. The advantage of reliability approaches can only be realized if the modelling is updated based on data gathered in connection with fabrication and operation.

However, it is recognized that the main cause of damages and accidents is human and organizational errors and omissions. To achieve an acceptable safety level, therefore, requires quality in all phases of carrying out engineering, fabrication and operation, as well as QA and QC of the engineering process; inspection, monitoring and repair of the structure; and design for structural robustness; e.g. through.an ALS criterion. In connection with crack control for aging structures, robustness is required to rely on a favorable effect of inspection and repair on the safety.

The development of high fidelity tools to aid the integrity management is clearly welcome. On the other hand, design requires repetitive analyses, especially in connection with optimal and reliability-based design. Moreover, the increasing refinement and automatization of analyses, increases the chance of errors in the software and its users. Development of efficient, robust and simplified methods is therefore also important. Besides providing information from research in terms of research papers, research results might be disseminated in a more easily accessible way for users, in the form of standards and guidelines for data bases and methods.

Acknowledgement

I appreciate excellent cooperation with many people in carrying out the research as well as code development for the Norwegian Petroleum Directorate, ISO and other regulatory bodies, that are reported in this paper. They particularly include J. Amdahl, R.G. Bea, the late C.A. Cornell and I. Holand, D. Karsan, J. Lloyd, I. Lotsberg, H.O. Madsen, J.E. Vinnem and O. T. Vårdal as well as my many PhD and post-doctor candidates, who are referenced in this paper. The opinions expressed, however, are those of the author.

References

[1] International Energy Agency 2016 *Key World Energy Statistics* (Paris).

[2] https://en.wikipedia.org/wiki/Deep_sea_mining, accessed 15.09.2017.

[3] FAO 2016 *The State of World Fisheries and Aquaculture* (Rome).

[4] *The Paris Agreement by the 21th Conf. of the Parties of the UNFCCC in Paris and adopted by consensus on 12 December 2015*. United Nations Treaty Collection.

[5] Working Group III of IPCC 2011 IPCC *Special Report on Renewable Energy Sources and Climate Change Mitigation* (Cambridge-Cambridge University Press).

[6] ISO 2394 2015 *General Principles on Reliability for Structures* Int. Standards Organisation (London).

[7] ISO 19900 2013 *Petroleum and Natural Gas Industries – Offshore Structures – Part 1: General Requirements, Part 2: Fixed Steel Structures Int. Standardization Organization* (London).

[8] API 1993/1997 *Recommended Practice for Planning, Designing and Constructing Fixed Offshore Platforms. API RP2A-WSD July 1993 with Supplement 1 with Sect., 17.0, Assessment of Existing Platform* American Petroleum Institute (Dallas).

[9] HSE 1992 *Safety Case Regulations* Health and Safety Executive HMSO (London).

[10] NORSOK N-001 2012 *Structural Design* Norwegian Technology Standards (Oslo).

[11] NORSOK N-003 2017 *Actions and Action Effects* Norwegian Technology Standards (Oslo).

[12] NORSOK N-004 2013 *Steel Structures* Norwegian Technology Standards (Oslo).

[13] NORSOK N-005 2017 *Condition monitoring of load bearing structures* Norwegian Technology Standards (Oslo).

[14] NORSOK N-006 2015 *Assessment of structural integrity for existing offshore load-bearing structures Norwegian Technology Standards* (Oslo).

[15] IEC 61400-1 2010 *Wind turbine. part 1: Design requirements* Int. Electrotechnical Commission (Geneva).

[16] IEC 61400-3 2009 *Wind turbines. part 3: Design requirements for offshore wind turbines* Int. Electrotechnical Commission (Geneva).

[17] IEC 61400-4 2012 *Wind turbines. part 4: Design requirements for wind turbine gearboxes* Int. Electrotechnical Commission (Geneva).

[18] IEC Draft Technical Specification 88/649/DTS 2017 *Wind energy generation systems – Part 3-2: Design requirements for floating wind turbines* Int. Electrotechnical Commission (Geneva).

[19] Statens vegvesen 2015 *Design of bridges* Handbook N400 (in Norwegian) The Norwegian Public Road Administration.

[20] Moan T 2000 *J. Offshore Mech. Arct. Eng.* **122** pp. 20-32.

[21] Madsen H O, Krenk S and Lind N C 1986 *Methods of structural safety* (New Jersey-Englewood Cliffs) pp. 45-56.

[22] Madsen H O, Skjong R K and Tallin A G 1987 *Proc. Marine Struct Symp.* (Arlington-Virginia) pp. 45-55.

[23] Moan T 1994 *Proc. 6th ICOSSAR Conf.* **1** (AA Balkema-Rotterdam) pp.21-43.

[24] Yang Y N 1994 *Proc 6th ICOSSAR Conf.* **1** (AA Balkema, Rotterdam) pp.3-18.

[25] Lotsberg I and Sigurdsson G 2013 *Rev. 2. DNV Report No. 2011-0666* (Høvik-Oslo).

[26] Moan T 2005 *Struct. Infrastruct. Eng.* **1 (1)** pp. 33-62.

[27] Moan T 2007 *Int. J. of Offshore and Polar Engineering* **17(1)** pp.1-21.

[28] Moan T 2009 *Structural Safety* **31(2)** pp. 124-135.

[29] Chakrabarti S 2005 *Handbook of Offshore Engineering* **1 & 2** (Amsterdam-Elsevier)

[30] Collette M et al. 2015 *Proc. 19th ISSC Congress* **1** pp 460-518 (Cascais) Proc. published by CRC Press, Boca Raton.

[31] DNVGL WOAD: *Worldwide Offshore Accident Database* (Oslo) updated continuously.

[32] Tavner P 2012 *Multi-Science No.13 in Renewable Energy Series* The Institution of Engineering and Technology.

[33] Wilkinson M, Harmann K, Spinato F, Hendriks B and van Delft T 2011 *EWEA Conference* pp. 1-8 (Brussels).

[34] U.S. Coast Guard 1981 *Collapse and Sinking of Mobile Offshore Drilling Unit Ranger I in the Gulf of Mexico on 10 May 1979 with Loss of Life* (Washington D.C.)

[35] Utredninger NO 1981 *The" Alexander L. Kielland" Accident* **11** (Norwegian- English translation available) (Oslo).

[36] Canada Royal Commission *The Ocean Ranger Marine Disaster: Report One: The Loss of the Semi-submersible Drill Rig Ocean Ranger and its Crew* (St. Johns Newfoundland).

[37] Inquiry Commission HMSO *The Public Inquiry in the Pipe Alpha Disaster* (London).

[38] Petrobas and Coppe (eds) *Workshop on the Accident with the P-36 Platform* (Rio de Janeiro).

[39] *Macondo Well: Deepwater Horizon Blowout. Marine Board.* National Academy of Engineering and National Research Council 2012 (Washington D.C).

[40] Pugsley A G 1966 *The Safety of Structures* (London-Edward Arnold).

[41] Moan T 1983 *Proc.4th ICASP Conf*, (Firenze).

[42] Bea R G 2000 *Achieving step change in Risk Assessment & Management (RAM)* (WA-University of Western Australia)

[43] Moan T G 1985 *The Progressive Structural Failure of the Alexander L. Kielland Platform in Case Histories in Offshore Engineering* ed Maier (Berlin-Springer-Verlag)

[44] Vinnem J E 2014 *Offshore Risk Assessment* (London-Springer).

[45] Bach-Gansmo O, Carlsen C A and Moan T 1987 *Proc. Conf. on Mobile Offshore Units* (London-City University).

[46] Moan T, Vårdal O T, Hellevig N C and Skjoldli K 1997 *Proc. OMAE-97* (Yokohama)

[47] Vårdal O T and Moan T 2016 *Proc. 3rd OSRC* (Stavanger).

[48] Moan T, Vårdal O T and Johannesen J M 1999 *Proc.8th ICASP* pp. 191-200 (AA Balkema-Rotterdam).

[49] Moan T 2016 *Proc. 3rd OSRC* (Stavanger).

[50] Almar-Naess (ed.) 1985 *Fatigue Handbook for Offshore Steel Structures* (Trondheim-Tapir Publ.).

[51] Lotsberg I 2016 *Fatigue Design of Marine Structures* (London-Cambridge University Press).

[52] The Norwegian Petroleum Directorate 1984 *Regulations for load-carrying structures for extraction or exploitation of petroleum* (Stavanger).

[53] Storheim M and Amdahl J 2014 *Marine Structures* **37** pp. 135–72.

[54] Czjuko J et al. 2015 *Proc. 19th ISSC Congress* (Cascais) Proc. published by CRC Press, Boca Raton.

[55] Haver S 2016 *Proc. 3rd OSRC* (Stavanger).

[56] Tromans P S 2015 *Proc. Offshore Structural Reliability Conf.* (Houston).

[57] Coronado D and Fischer K 2015 *Report Fraunhofer IWES* (Bremerhafen).

[58] Heideman J C and Weaver T O 1992 *Proc. Civil Engng. in the Oceans V* pp. 496-517 (Texas-College station).

[59] Efthymiou M and Van De Graaf J 2011 *Proc. 30th OMAE Conf.* (New York).

[60] Fjeld S 1977 *Proc. 9th Offshore Technology Conf.* (Houston).

[61] Moses F 1987 *Report API PRAC 87-22* (Texas-Dallas).

[62] Lloyd J R and Karsan D 1988 *Proc. 20th. Offshore Technology Conf.* **4** pp. 593-600 (Houston).

[63] Jordaan I J and Maes M A 1991 *C.J. Civil Engng.* **18 (3)** pp. 454-64.

[64] Moan T 1988 *SINTEF Report No. F88043* (Trondheim).

[65] Moan T 1995 *SINTEF Report STF70 A95210* (Trondheim).

[66] Melchers R E 1999 *Structural Reliability Analysis and Prediction* (Chichester-John Wiley & Sons).

[67] Naess A and Moan T 2012 *Stochastic dynamics of marine structures* (Cambridge University Press).

[68] Moan T 2004 *9th PRADS Conference* (Travemünde-Lübeck).

[69] Moan T 2014 *Proc. of the Offshore Structural Reliability Conf.* 2014OSRC (Texas-Houston).

[70] Wisch D 2014 *Proc. of the Offshore Structural Reliability Conf.* 2014OSRC (Texas-Houston).

[71] Videiro P M and Moan T 1999 *Proc. OMAE-99* (St. John's Newfoundland).

[72] Ayala-Uraga E and Moan T 2007 *Int. J. of Fatigue* **29(3)** pp. 444–56.

[73] Wirsching P H 1983 *Report API-PRAC Project No. 81-15* (Tucson-University of Arizona).

[74] DNVGL-RP-0001 2015 *Probabilistic methods for planning of inspection for fatigue cracks in offshore structures.*

[75] Vårdal O T and Moan T 1997 *Proc. 16th OMAE Conf.* (Yokohama).

[76] Moan T, Hovde G O and Blanker A M 1993 *Proc. 25th Offshore Technology Conf.* **2** pp. 591-99 (Houston).

[77] Vårdal O T, Moan T and Bjørheim L G 2000 *Proc. 19th OMAE Conf.* (New Orleans).

[78] Dalane J I 1993 *PhD thesis-System Reliability in Design and Maintenance of Fixed Offshore Structures* (Trondheim-The Norwegian Institute of Technology).

[79] Karamchandani A, Dalane J I and Bjerager P 1993 *Marine Structures* **4** pp. 353-79.

[80] Shetty N K 1992 *Ph.D.Thesis-System Reliability of Fixed Offshore Structures under Fatigue* Deterioration (London: Imperial College, Univ. of London).

[81] De R S Karamchandani A and Cornell CA 1989 *Proc. 5th ICOSSAR* **2** pp. 975-82 (New York)

[82] Wu Y L and Moan T 1989 *Proc. 5th ICOSSAR*, San Francisco, pp. 887-94.

[83] Skallerud A and Amdahl J 2002 *Nonlinear Analysis of Offshore Structures* (Research Studies Press).

[84] DNVGL-RP-C208 2016 *Determination of Structural Capacity by Non-linear FE analysis Methods* (Oslo).

[85] Moan T 1998 *Target Levels for Structural Reliability and Risk Analysis of Offshore Structures* ed Guedes Soares C (A. A. Balkema-Rotterdam).

[86] Aker Offshore Partner A.S. 2002 *Target levels for reliability-based assessment of offshore structures during design and operation* (Sandsli).

[87] Karsan D I 2005 *Fixed Offshore Platform Design. Chap 6, Handbook of Offshore Eng* ed Chakrabarti S (Amsterdam-Elsevier).

[88] Kirkemo F 1988 *Appl. Mech. Rev* **41(2)** pp. 61-84.

Analytical and numerical investigation of bolted steel ring flange connection for offshore wind monopile foundations

C A Madsen[1]*, J-C Kragh-Poulsen[2], K J Thage[3] and M J Andreassen[4]

[1,2] COWI A/S, Department of Marine & Foundation Engineering, Parallelvej 2, DK-2800 Kgs. Lyngby, Denmark
[3]COWI A/S, Department of Major Bridges International, Parallelvej 2, DK-2800 Kgs. Lyngby, Denmark
[4]Technical University of Denmark, Department of Civil Engineering, Brovej, Building 118, DK-2800 Kgs. Lyngby, Denmark

*Contact author: chan@cowi.com

Abstract. The monopile foundation is the dominant solution for support of wind turbines in offshore wind farms. It is normally grouted to the transition piece which connects the foundation to the turbine. Currently, the bolted steel ring flange connection is investigated as an alternative. The monopile--transition piece connection has specific problems, such as out-of-verticality and installation damage from driving the MP into the seabed and it is not fully known how to design for these. This paper presents the status of the ongoing development work and an estimate of what still needs to be covered in order to use the connection in practice. This involves presentation of an analytical and non-linear FE analysis procedure for the monopile-transition piece connection composed of two L flanges connected with preloaded bolts. The connection is verified for ultimate and fatigue limit states based on an integrated load simulation carried out by the turbine manufacturer.

1. Introduction

Monopile foundations are the prevailing substructure for offshore wind turbine generators (WTG) supporting 81% of the WTGs in operation at the end of 2016 [1]. The monopile (MP) substructure consists of circular steel tubes welded together and driven to target penetration by hydraulic hammering. The bottom of the WTG tower section is bolted directly to the MP or to the top of the transition piece (TP) by the use of flanges pre-welded to the tubular shells. The majority of the installed MPs have a TP, as shown in Figure 1. The loads between the MP and TP are typically transferred by a grouted connection. Settlements of the TP on several offshore wind farms (OWF) in 2010 [2] with the traditional cylindrical grouted connection led to the development of today's most widely used conical grouted connection and alternatives such as the bolted ring flange connection or slip/double slip joints. Current tendencies are that an increased number of offshore wind farms are moving towards the bolted flange solution between the MP-TP [3].

Figure 1. Conceptual sketch of the MP foundation incl. a TP joined to the bottom of the WTG tower. The top of the monopile is joined to the transition piece and the bottom of the monopile is fixed in the seabed. Figure: © COWI A/S.

This paper presents the ongoing work of extending the established design procedures of the bolted steel ring flange connection for the WTG tower to the requirements of the MP-TP connection. The WTG tower flange is typically located as shown in 0, pos. A-D. The position of the MP-TP connection, pos. E, leads to specific considerations of the MP-TP flange connection:

- During installation the MP is driven into the seabed by direct hammering on the pile top. This may cause deformation of the flange exceeding limiting values not taken into account in established design procedures.
- Driving the MP into the seabed can result in the MP not being perfectly vertical and the WTG mounting flange being inclined. This is not covered in existing design procedures.

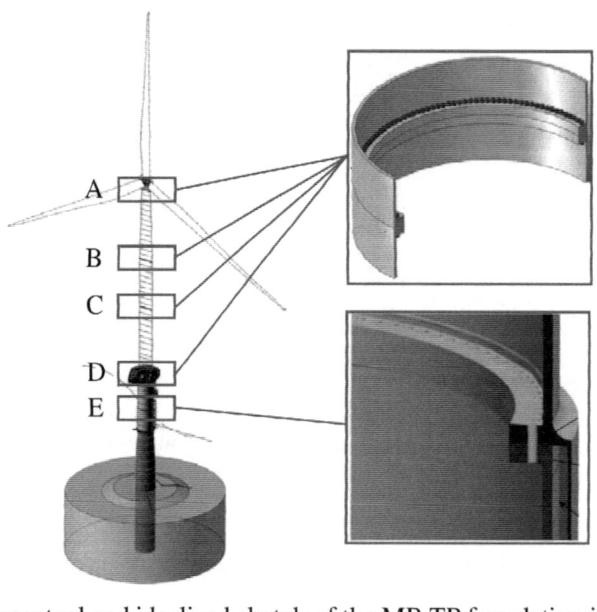

Figure 2. Conceptual and idealized sketch of the MP-TP foundation indicating potential positions for application of a bolted steel ring flange connection. Figure: © COWI A/S.

The work described in this paper is limited to investigation of the ultimate (ULS) and fatigue (FLS) limit states of the flange connection during operation of the WTG. ULS and FLS are investigated by well-known analytical methods considering the shells and the two flanges of the MP-TP. The analytical analysis is accompanied by definition of similar numerical models. This is done to compare the methods and to gain further knowledge of the safety levels included in the analytical approach. An ideal MP-TP flange geometry is assumed in the numerical study, and local and global tolerances are not considered. ULS and FLS loads applied for the investigation are obtained from an offshore wind farm designed by COWI A/S.

2. Analytical approach

The tensile ULS and FLS resistance of the circular flange connection used in present paper are approximated by the analytical methods of Petersen-Seidel and Schmidt-Neuper [4]-[6]. The methods have already been studied by several researchers in relation to the design of the WTG tower or similar tubular flange connections. The review presented in the following, highlights the existing methods and assumptions. Furthermore, it will form the basis for evaluating the key issues associated with the specific requirements of the MP-TP connection.

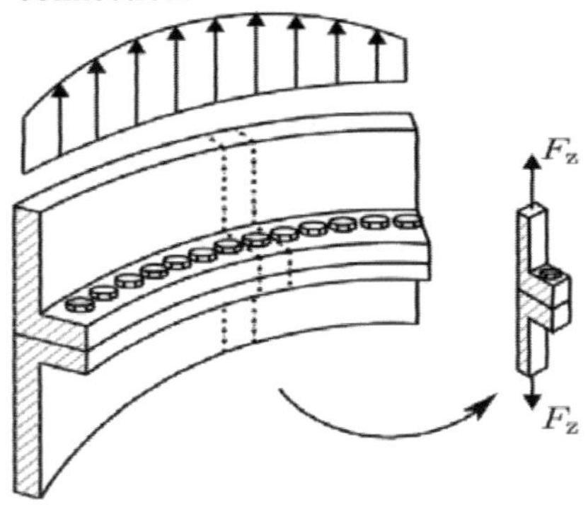

Figure 3. Bolted ring flange connection and equivalent single bolt segment. Figure from [7].

2.1. Ultimate limit state

The tensile ULS resistance of the flange connection by Petersen and Seidel consider a segment cut out of the entire flange connection (L segment) including a single bolt, see Figure 3. The approach assume linear bolt-load behaviour as preload is not considered. Furthermore, perfectly aligned flanges are assumed.

The L segment is investigated for the equivalent tensile normal force, F_z specified in Equation (1). The expression is derived from Navier's formula considering the foundations global bending moment, M_{Ed} and axial force, N_{Ed} at the position of the circular flange connection. Additional verification is needed for torsional and shear forces which are not considered in present paper. The loads are scaled to the L segment by the number of bolts, n_b in the flange connection and the mid-thickness diameter, $D_{sh,m}$ of the tubular shells.

$$F_z = \frac{4 \cdot M_{Ed}}{n_b \cdot D_{sh,m}} + \frac{N_{Ed}}{n_b} \tag{1}$$

The tensile resistance of the L segment, F_{ult}, Equation (2) is found from static force equilibrium considering the three failure modes defined in Figure 4.

$$F_z \leq F_{ult} = \min(F_{U,A} \; ; \; F_{U,B} \; ; \; F_{U,C1} \; ; \; F_{U,C2}) \tag{2}$$

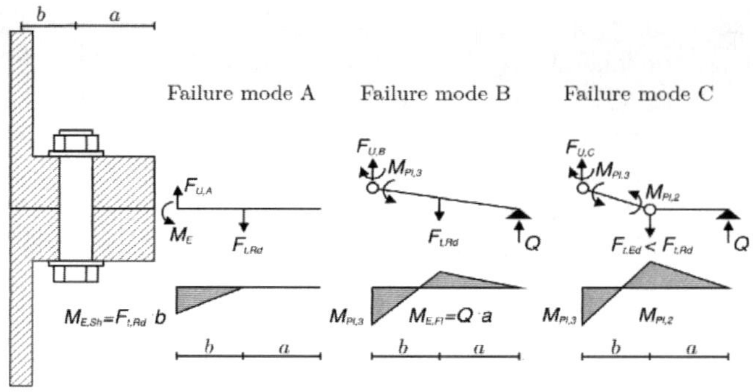

Figure 4. Approximated failure modes for L flange (segment) connection according to Petersen's method [5]. Figure modified from [7].

2.1.1 Failure of the bolt: The tensile resistance for failure mode A is linked to tensile failure of the bolt. The bending deformation of the flange caused by the eccentric tensile load is assumed small combined with the assumption of a sufficiently large flange thickness resulting in absence of prying forces, Q. The resistance of the segment flange for failure mode A, $F_{U,A}$ is therefore equal to the design tensile resistance of the bolt, $F_{t,RD}$ according to Equations (3) and (4). $f_{y,bolt}$ and $f_{u,bolt}$ is the design yield and ultimate tensile strength of the bolt. A_s is the tensile stress area of the bolt and γ_{M2} and γ_{M7} partial safety factors.

$$F_{U,A} = F_{t,Rd} \tag{3}$$

$$F_{t,Rd} = \min\left(\frac{f_{y,bolt} \cdot A_s}{\gamma_{M2}}; \frac{0.9 \cdot f_{u,bolt} \cdot A_s}{\gamma_{M7}}\right) \tag{4}$$

Failure mode A assumes that the shell is able to resist the moment, $M_{E,Sh}$ according to Equation (5),

$$\frac{M_{E,Sh}}{W_{E,Shell}} < f_{yd,Shell} \tag{5}$$

where $W_{E,Shell}$ is the elastic moment of resistance and $f_{yd,Shell}$ the design yield strength of the shell.

2.1.2 Failure of the bolt combined with plastic hinge (partial yielding) of the shell: Failure mode B considers combined failure due to plastic yielding of the shell and failure of the bolt in tension. The resistance is obtained from force equilibrium according to Equation (6) assuming the flange can resist the moment, $M_{E,Fl}$ Equation (7).

$$F_{U,B} = \frac{F_{t,Rd} \cdot a + M_{pl,3}}{a + b} \tag{6}$$

$$M_{E,Fl} \leq W_{E,Flange} \cdot f_{yd,Flange} \tag{7}$$

2.1.3 Plastic hinge in shell and flange: Failure mode C examines the tensile resistance, $F_{U,C}$ for the occurrence of failure of the shell and flange due to yielding. This supposes that the considered bolt has sufficient capacity to resist the axial tension load, $F_{t,Ed}$ including effects of prying forces. The resistance is determined from Equation (8). Petersen's method outlined in [5] is different from [4] as it further divides failure mode C into (C1) plastic hinge in the bolt hole center plane and (C2) plastic hinges next to the bolt hole.

$$F_{U,C} = \frac{M_{pl,2} + M_{Pl,3}}{b} \tag{8}$$

2.2. Fatigue limit state

The FLS verification of the circular flange connection is in this paper limited to investigating the resistance of the bolts. The Schmidt-Neuper method [4] links the cyclic load in the form of a max./min. equivalent normal force in the shell, $F_{z\,(max/min)}$ to the bolt force, F_b. The stress range in the bolt is determined from the bolt force from which the fatigue capacity of the bolt can be calculated. The method is selected as it considers the effect of preload and resilience of the joint. The method, however, does not consider bending stresses in the bolt which are taken into account during selection of the fatigue reference strength (FAT class) for the bolt. The Schmidt-Neuper method is devised to be valid for fabrication/installation imperfections of the flange within the specified range defined in Table 1 and illustrated in Figure 5. This implies that the bolt loads determined by the Schmidt-Neuper method include additional prying forces as a consequence of the gap, k. Furthermore, it is assumed that additional bending stresses in the bolts due to the inclination of the flange, α_s are sufficiently small.

The Schmidt-Neuper method approximates the non-linear relation between the shell and bolt forces due to preload by a piecewise trilinear function. The three straight lines are derived by Equations (10)-(12) and presented in Figure 6 as three regions:

Region 1: The bolt load for region 1, F_{b1}, Equation (10) is determined for the flange zone under compression due to the preload. During the increase of the shell load the compressive zone moves in the direction of the edge of the flanges. The bolt load is for region 1 a function of the tensile shell force multiplied by the load factor for the total stiffness of the bolt, p.

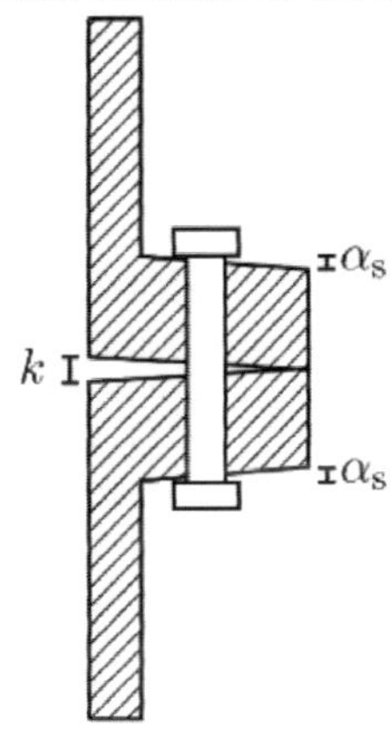

Figure 5. Specification of flange tolerances. Figure from [9].

Table 1. Allowable flange tolerances. According to [9].

Criteria	Limiting value
Flatness deviation, k per flange around the entire circumference	2.0 mm
Flatness deviation, k per flange over a segment of 30°	1.0 mm
Taper, α_s to the inside of the connecting surface of each flange	0.0° to 0.7°
Outer flange surfaces limit, α_s	2°

The preload, $F_{P,C*}$ is added, representing the initial condition before the external load is imposed on the joint. As a consequence, the bolt force relative to the shell force increases in region 1 according to the total stiffness of the joint.

$$F_{b1} = F_{P,C*} + p \cdot F_Z \tag{9}$$

Region 2: The tensile zone increases in region 2 and the flange connection begins to open, leading to the bolt load, F_{b2} given by Equation (10), to increase more rapidly than the shell force. The first part of Equation (10) represents the maximum bolt force for region 1, and the last part the increase in magnitude of the bolt force relative to the increase in shell force.

$$F_{b2} = F_{P,C*} + p \cdot F_{Z,I,\max} + \left[\lambda^* \cdot F_{Z,II,\max} - (F_{P,C*} + p \cdot F_{Z,I,\max})\right] \cdot \frac{Fz - F_{Z,I,\max}}{F_{Z,II,\max} - F_{Z,I,\max}} \tag{10}$$

Region 3: Region 3 is where the tensile shell load has exceeded the compressive preload, $F_{z,IImax}$ and the flanges are in contact only at their edges. The bolt force, F_{b3} given by Equation (11), increases as a function of the shell force multiplied by the auxiliary parameter λ^* (16). λ^* represents the additional contribution from prying forces due to eccentric load of the connection.

$$F_{b3} = \lambda^* \cdot F_Z \tag{11}$$

It is decided in the present paper to reduce the characteristic preload, $F_{p,C}$ to 90% according to [8], see Equation (12), to account for relaxation of the connection. Assuming a 90% reduction requires re-tightening of the bolts 6 months after commissioning. The torque applied to the bolt should minimum be 70% of the ultimate tensile strength of the bolt.

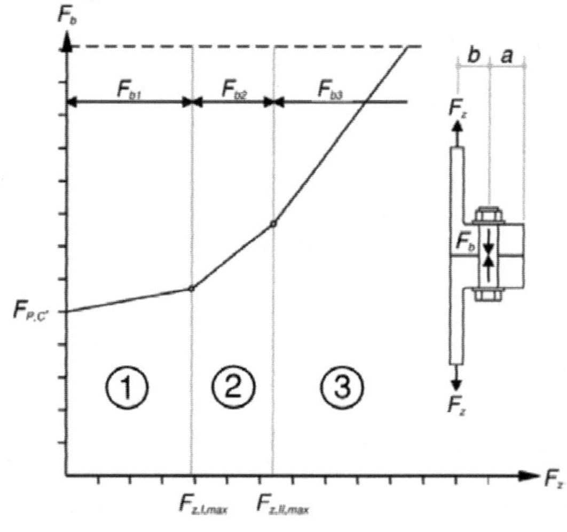

Figure 6. Trilinear approximation of the non-linear relation between the shell force and the bolt force. Figure modified from [7].

$$F_{P,C*} = 0.7 \cdot A_s \cdot f_{u,bolt} \cdot 0.9 \tag{12}$$

The maximum limit shell force for region 1 and 2 is given in Equations (13) and (14).

$$F_{Z,I,\max} = \frac{a - 0.5 \cdot b}{a + b} \cdot F_{P,C*} \tag{13}$$

$$F_{Z,II,\max} = \frac{1}{\lambda^* \cdot q} \cdot F_{P,C*} \tag{14}$$

$$\lambda^* = \frac{0.7 \cdot a + b}{0.7 \cdot a} \tag{15}$$

The load factor for the total stiffness of the bolt, p, and the flanges, q, are obtained from Equation (16) and (17) where C_s and C_D is the spring stiffness of the tension- and compression spring. δ_p and δ_s represents the elastic resilience of the clamped parts and the bolt. The spring characteristics of the members in the connection are derived according to methods outlined in [10].

$$p = \frac{C_s}{C_S + C_D} \qquad q = \frac{C_D}{C_S + C_D} \tag{16}$$

$$C_S = \frac{1}{\delta_s} \qquad C_D = \frac{1}{\delta_p} \tag{17}$$

3. Numerical approach

The ULS and FLS resistance of the one bolt segment is in the present paper investigated by using Abaqus CAE. A geometry equivalent to the analytical approaches presented in Section 2.1 and 2.2 is used and no tolerances are included. Non-linear material properties are defined for the shell, flange, and bolt assembly along with contact interaction specified between the parts according to Table 2. The bolt and nut geometry is simplified where an equivalent shank diameter is specified for the bolt. Linear brick elements (C3D8R) are assigned to the geometry. Different element sizes are used as part of the refinement study. Two types of loads are defined, see Figure 7; a bolt load representing the preload and a pressure load applied at the TP shell representing the equivalent normal force.

The mechanical boundary conditions specified for the L flange segment are shown in Figure 8. This includes cylindrical symmetry conditions applied on both sides of the section cut of the segment and a pinned support at the bottom of the MP shell.

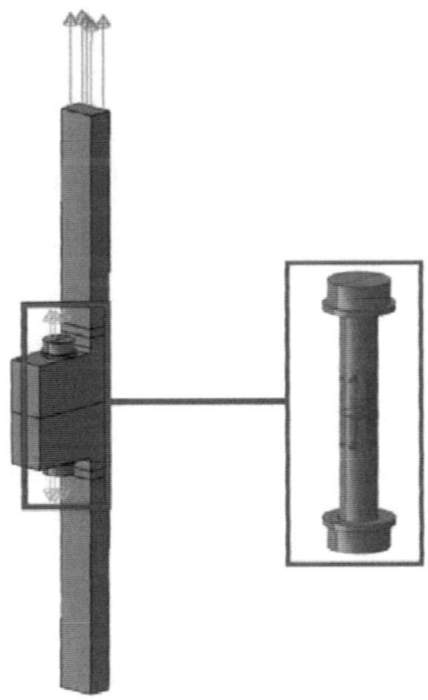

Figure 7. Definition of loads. Left: Bolt load. Right: Segment shell load. Figure from [7].

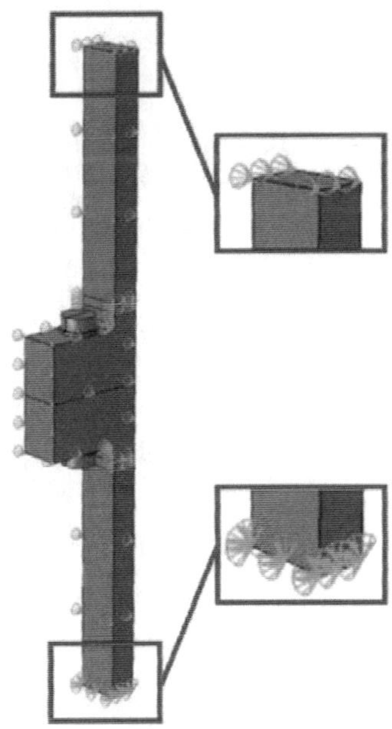

Figure 8. Mechanical boundary conditions specified for the segment model. Left: Pinned support. Right: Cylindrical symmetry condition. Figure from [7].

IOP Conf. Series: Materials Science and Engineering **276** (2017) 012034 doi:10.1088/1757-899X/276/1/012034

Table 2. Contact surface properties.

Definition	Flange to flange, Bolt to washer Flange to washer
Friction coefficient	0.2
Behaviour	Small sliding
Formulation	Augmented Lagrange
Interface treatment	Adjust to remove overclosure

4. Case study

The connection is designed as a one sided preloaded flange connection located on the inside of the foundation structure. The connection has 140 bolts specified as hot dip galvanized M64, 10.9 grade (HV M64-10.9 tZn) bolts. The bolt size exceeds M36, hence [11] is consulted for specification of the geometry of the bolt, nut and washer. The flange and shell are specified as S355ML. No tolerances are included in the case study. Non-linear material properties are estimated based on values defined in [12].

4.1. Joint geometry and material parameters

The main geometrical and material parameters of the connection are defined in Figure 9 and listed in Table 3. The material properties including strain hardening, considered in the numerical analysis, are given in Figure 10.

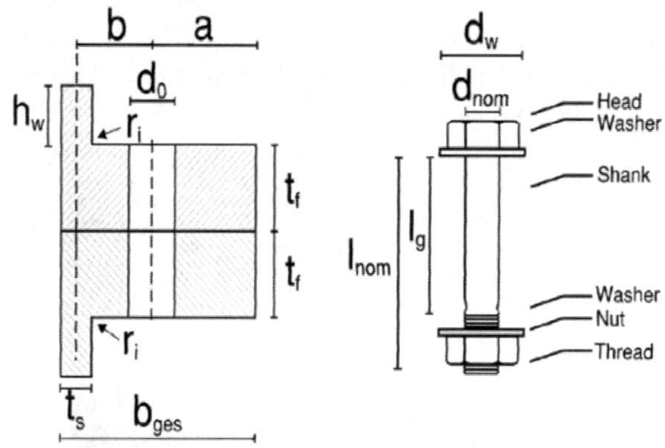

Figure 9. Definition of joint geometry.

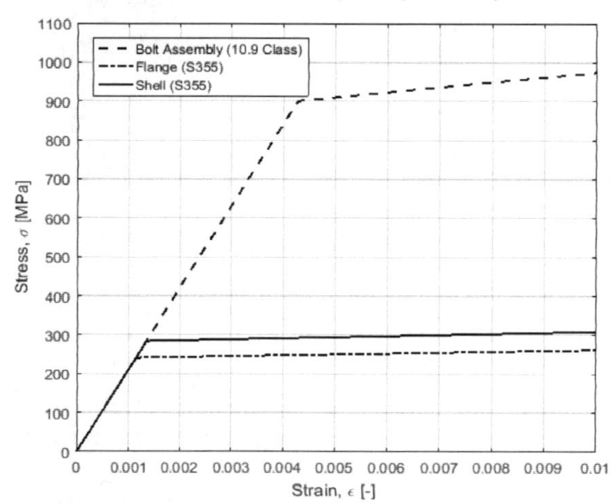

Figure 10. Properties of bolt, head, nut, washers (bolt assembly), shell and flange materials.

Table 3. Geometrical and material parameters.

Description	Symbol	Value	Unit
Outer diameter of shell	OD	6	[m]
Nominal shell thickness	t_s	0.08	[m]
Nominal flange thickness	t_f	0.18	[m]
Nominal flange width	b_{ges}	0.31	[m]
Dist. bolt center to flange edge	a	0.15	[m]
Dist. bolt center to shell center	b	0.12	[m]
Inner radius	r_i	15	[mm]
Number of bolts	n_b	140	[-]
Nominal bolt size	d_{nom}	64	[mm]
Nominal diameter of washer	d_w	115	[mm]
Nominal washer thickness	h	10	[mm]
Grip length	l_g	380	[mm]
Nominal bolt length	l_{nom}	450	[mm]
Shank length	l_{sch}	332	[mm]
Cross section area of shank	A_{sch}	3,217	[mm^2]
Cross section area of bolt thread	A_{bd3}	2,520	[mm^2]
Nominal diameter of washer	d_w	115	[mm]
Characteristic yield strength of shell (t=80mm)	$f_{yk,sh}$	325	[MPa]

4.2. Analytical approach

ULS and FLS resistance of the L segment are found analytically in this section.

4.2.1 Ultimate limit state: The analytical ULS determination of the L segment joint is carried out according to Section 2.1. The found tensile resistance of the L segment and bolt configuration using the corresponding equilibrium equations is summarized in Table 4. The L segment is seen to exhibit a load deformation characteristic leading to failure mode B. Indicating development of a plastic hinge in the flange at the transition to the shell and subsequently tensile failure of the bolt. Equivalent design ULS loads at the MP-TP joint from a COWI A/S OWF project are presented in Table 5 including the tensile equivalent normal force determined from Equation (1). The utilization ratio representing the ratio between the equivalent tensile normal force and the resistance of the L segment is presented in Equation (18).

$$UR_Z = \frac{F_z}{F_{ult}} = 0.88 \tag{18}$$

4.2.2 Fatigue limit state: The analytical FLS determination of the L segment according to Section 2.2 is limited to an investigation of the cyclic loads in the bolt during operation of the WTG. Parameters for establishment of the Schmidt-Neuper curve reflecting the bolt force as a function of the tensile force in the shell wall are presented in Table 6. The spring stiffness of the bolt assembly and L-segment under compression are specified according to procedures in [10].

Table 4. Tensile resistance for failure mode a-c for the l flange single bolt connection.

Description	Symbol	Value	Unit
Plastic limiting moment of flange without hole	M_{pl2}	334	[kNm]
Plastic limiting moment of flange with hole	M'_{pl2}	150	[kNm]
Plastic limiting moment of tubular section	M_{pl3}	56	[kNm]
Design value of limit tension force	$F_{t,Rd}$	1,927	[kN]
Limit tension force failure mode A	$F_{U,A}$	1,927	[kN]
Limit tension force failure mode B	$F_{U,B}$	1,279	[kN]
Limit tension force failure mode C1	$F_{U,C1}$	1,980	[kN]
Limit tension force failure mode C2	$F_{U,C2}$	2,448	[kN]

Table 5. Resulting ULS design loads for full flange and segment model

Global design bending moment	Global design Axial force	Equivalent normal force
M_{Ed} *[kNm]*	N_{Ed} *[kN]*	F_z *[kN]*
248,475	10,674	1,123

Table 6. Parameters for definition of bolt force as function of tensile force in shell wall.

Description	Symbol	Value	Unit
Spring stiffness, bolt	p	0.17	[-]
Spring stiffness, clamped flange	q	0.83	[-]
Design preload	$F_{p,C*}$	1,680	[kN]
Shell force max. limit (region 1)	$F_{z,I,max}$	568	[kN]
Shell force max. limit (region 2)	$F_{z,II,max}$	954	[kN]
Bolt force max. limit (region 1)	$F_{b,I,max}$	1,775	[kN]
Bolt force max. limit (region 2)	$F_{b,II,max}$	2,021	[kN]
Auxiliary parameter	λ^*	2.1	[-]

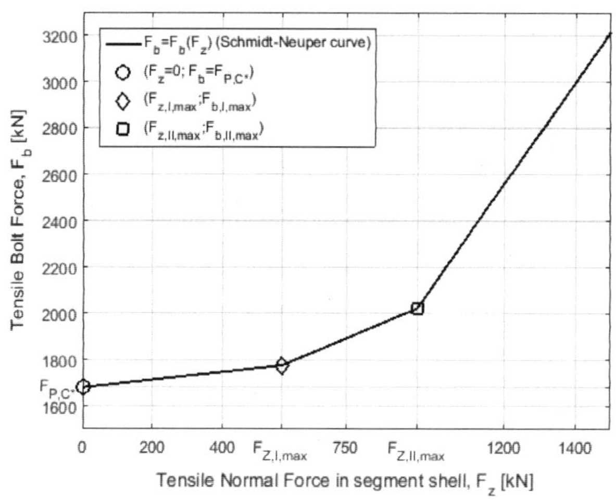

Figure 11. Characteristic tensile bolt force as a function of the shell force by the Schmidt-Neuper approach. Figure modified from [7].

The trilinear relation between shell- and bolt force for the segment flange is presented in Figure 11. The tensile shell force, F_z, is determined from Markov matrices containing information on the moment load cycles at the MP/TP joint due to the cyclic actions on the foundation. The Markov matrix containing the governing directional cyclic loads relative to north is used as basis for the verification. The moment loads in the Markov matrix are sorted into bins, j, based on the magnitude of the moment range. Each bin contains n_i number of load cycles. The extent of the Markov matrix used in this paper is given in the format presented in Table 7. The Markov matrix includes information on min. and max. moment ranges and min. and max. mean moment values for the load cycles included in each bin [13]. The Markov matrix contains $9.95 \cdot 10^8$ load cycles divided into j = 5,373 bins which in this study are assumed to reflect the expected load history for the foundation.

From the information in the Markov matrix the min. and max. values of the shell force are obtained by Equation (19) to (24). A_{shell} is the cross sectional area of the shell for the L segment, $W_{E,shell}$ the section modulus of shell, and A_{bd3} the minor cross section area of the bolt at the thread-root section.

Table 7. Markov matrix incl. bending moments and number of cycles sorted into bins. Example values given in top row.

Min. moment range $\Delta M_{min,j}$ ***[kNm]***	**Max. moment range** $\Delta M_{max,j}$ ***[kNm]***	**Min. mean moment** $M_{mean,min,j}$ ***[kNm]***	**Max. mean moment** $M_{mean,max,j}$ ***[kNm]***	**No. of cycles** $n_{j,1}$ ***[-]***
0	$1.398 \cdot 10^3$	$-1.053 \cdot 10^5$	$-1.036 \cdot 10^5$	9
$\Delta M_{min,2}$	$\Delta M_{max,2}$	$M_{mean,min,2}$	$M_{mean,max,2}$	n_2
$\Delta M_{min,3}$	$\Delta M_{max,3}$	$M_{mean,min,3}$	$M_{mean,max,3}$	n_3
$\Delta M_{min,4}$	$\Delta M_{max,4}$	$M_{mean,min,4}$	$M_{mean,max,4}$	n_3

$$\Delta M_{mean,j} = \frac{\Delta M_{\min,j} + \Delta M_{\max,j}}{2} \tag{19}$$

$$M_{m,j} = \frac{\Delta M_{mean,\min,j} + M_{mean,\max,j}}{2} \tag{20}$$

$$\Delta F_{z,j} = A_{shell} \cdot \frac{\Delta M_{mean,j}}{W_{E,Shell}} \tag{21}$$

$$F_{z,mean,j} = A_{shell} \cdot \frac{M_{m,j}}{W_{E,Shell}} \tag{22}$$

$$F_{z,\max,j} = F_{z,mean,j} + 0.5 \cdot \Delta F_z \tag{23}$$

$$F_{z,\min,j} = F_{z,mean,j} - 0.5 \cdot \Delta F_z \tag{24}$$

The non-linear dependency of the shell force and the bolt force illustrated in Figure 11 result in the maximum and minimum tensile bolt force being determined for each bin. The stress range in the bolt is found according to Equation (25) to (27) for the minor cross section area of bolt, A_{bd3} considering the reduction of the bolt diameter due to the thread.

$$\Delta F_{b,j} = F_{b,\max,j} - F_{b,\min,j} \tag{25}$$

$$\Delta\sigma_{\mathrm{b,max,j}} = \frac{\mathrm{F_{b,max,j}}}{\mathrm{A_{bd3}}} \qquad \Delta\sigma_{\mathrm{b,min,j}} = \frac{\mathrm{F_{b,min,j}}}{\mathrm{A_{bd3}}} \tag{26}$$

$$\Delta\sigma_{b,j} = \left|\Delta\sigma_{b,\max,j} - \Delta\sigma_{b,\min,j}\right| \tag{27}$$

The sum of the fatigue damage in the bolt is based on the S-N approach in [9] under assumption of linear cumulative damage (Palmgren-Miner rule):

$$D = \sum_{i=1}^{j} \left(\frac{n_i}{N_i}\right) \leq 1.0 \tag{28}$$

As bending moments in the bolt are not considered, the detail category reference strength of the bolt, 36* (40 MPa) is used for damage accumulation according to [5].

Available S-N curves are limited to nominal bolt sizes up to 30mm. The M64 bolts studied in the present paper exceeds this value, hence the additional size effect leading to a reduction of the fatigue strength, is to be considered. Multiplying the reduction factor, k_s defined in Equation (29) with the reference strength of the bolt account for this effect according to [5].

$$k_s = \left(\frac{30}{d_{nom}}\right)^{0.25} \tag{29}$$

The obtained fatigue damage for the bolt exposed to the loads presented in the Markov matrix result in a summarized fatigue damage in Equation (30).

$$\sum D_{bolt} = 0.83 \leq 1.0 \tag{30}$$

4.3. Numerical approach

The ULS and FLS capacity for the numerical L flange model are determined. The model includes the flange with connected shell, bolt and washer. The equivalent normal force is defined as a variable pressure load applied at the top of the TP shell.

4.3.1 Ultimate limit state: The maximum allowable tensile load, f_{ult}, according to the analytical approach in Section 2.1 is applied as reference in the numerical analysis to obtain the corresponding stresses and plastic strains. Using this approach result in the safety level inherent in the analytical solution are ensured in the numerical segment model. Positions A-E in Figure 12 are selected for the analysis as they represent the points of interest for failure mode B.

The calculated axial stresses in the bolt and the flange segment, for the segment force equal the analytically derived ultimate resistance, are presented as contour plots in 0 and 0. Stresses and plastic strains as a function of the segment shell force for positions A-E are given in Figure 15 and Figure 16. The obtained maximum allowable plastic strains are listed in Table 8.

The development of plastic strain shown in Figure 16 indicates that the elastic limit is exceeded for a tensile segment force equal to 1,100kN and consequently a overutilization according to Equation (31).

$$UR_{elastic} = \frac{F_{z,elastic,\max}}{F_{ult}} = 1.16 \tag{31}$$

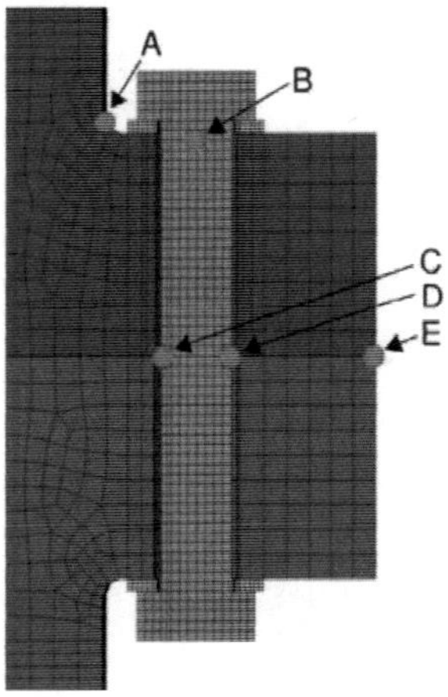

Figure 12. L segment model geometry incl. idealized MP/TP flanges, equivalent bolt diameter, identical bolt head and nut, and washers. Pos. A-E indicate investigated positions.

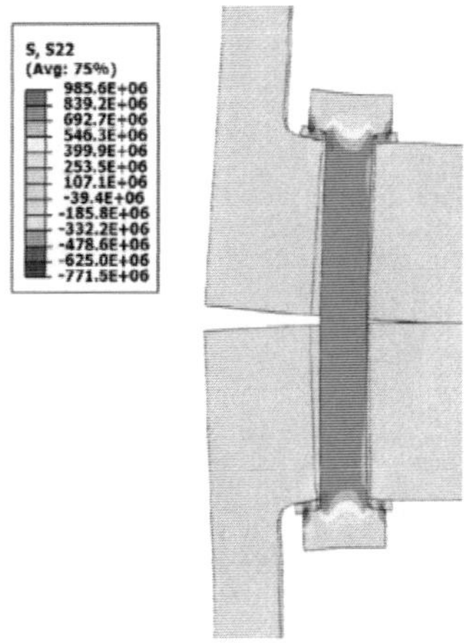

Figure 13. Axial stresses in the bolt for $F_z = F_{ult}$

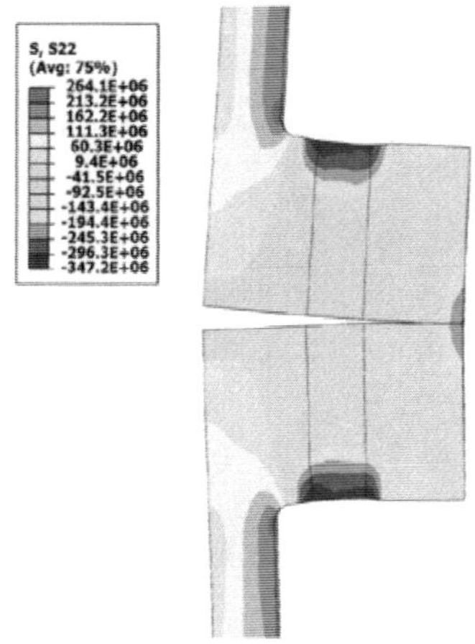

Figure 14. Axial stresses in the shell and flange for $F_z = F_{ult}$

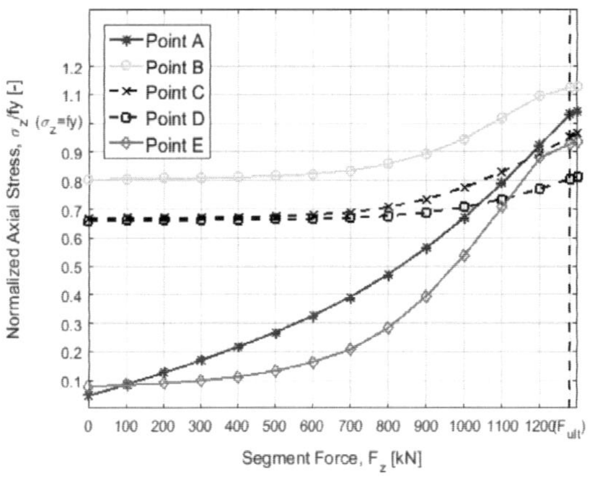

Figure 15. Normalized axial stresses at pos. A-E as function of tensile shell segment force.

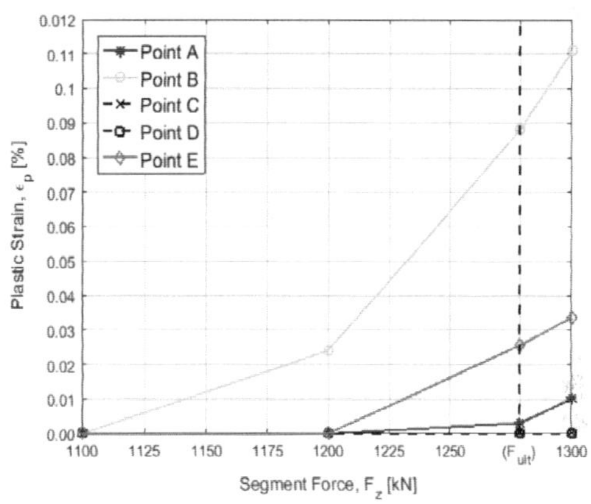

Figure 16. Plastic strain at pos. A-E as function of tensile shell segment force.

Table 8. Max. allowable plastic strains for positions A-E.

Position	Max. Plastic strain [%]
A	0.003
B	0.088
C-D	0
E	0.026

4.3.2 Fatigue Limit State: The fatigue capacity of the bolt is investigated by a finite element analysis (FEA) of the bolt center, positions C-D, and near the transition from bolt head to shank, position B shown in Figure 12. The segment shell force is defined in the numerical model for a range of -1,000kN to 1,500kN, covering the full load range of the Markov matrix.

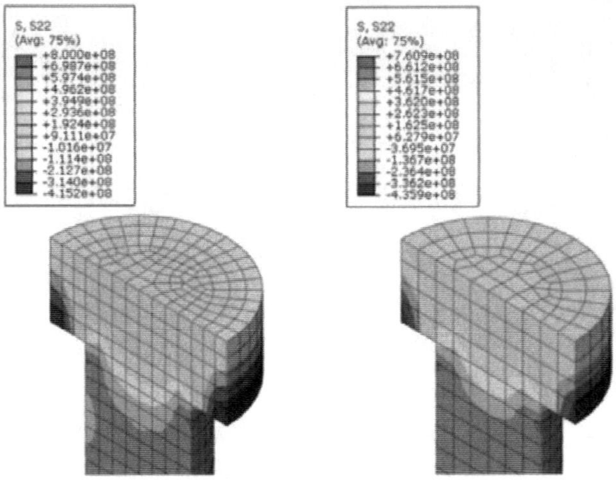

Figure 17. Output stresses for Fz = 600kN for a mesh size of 8mm and 10mm. Figure from [7].

The stresses obtained from the FEA at the bolt center and near the bolt head are extracted for the defined segment load range and for two mesh configurations as illustrated in Figure 17.

The position-specific bolt stresses as functions of the shell force are utilized to determine the stress range for each load bin in the Markov matrix according to (28). The stresses are hereby a function of the obtained relation between the segment and bolt force, shown in Figure 18. Damage accumulation is calculated equivalent to Section 2.2, except that a detail category reference strength of 50 MPa is chosen. This is a consequence of the numerical model considers both axial and bending stresses. The fatigue damage sum is given in Table 9.

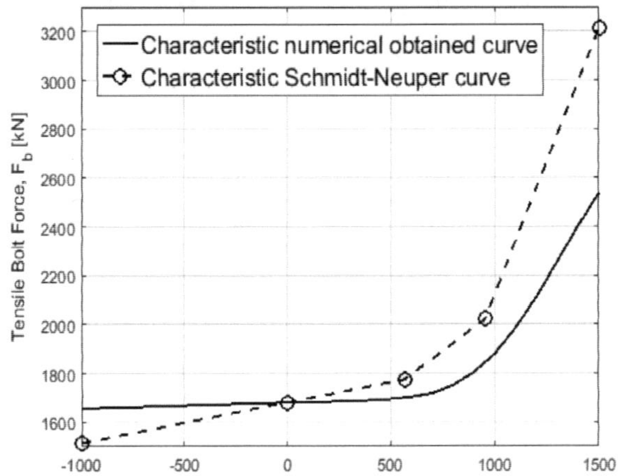

Figure 18. Characteristic tensile bolt force as a function of the shell force in the segment shell wall. The normal force range covers all load values derived from the Markov matrix.

Table 9. Accumulated fatigue damage in the bolt for two element sizes

	Fatigue damage (element = 8 mm)	**Fatigue damage (element = 10 mm)**
Bolt head	0.36	0.04
Bolt Shank	0.041	0.028

5. Comparison and discussion

The analytical and numerical results of the L flange segment are compared and discussed in the following.

5.1. Ultimate limit state

The minimum derived tensile resistance in Table 4 is obtained for failure mode B. The computed stresses shown in Figure 13 and Figure 14 confirm this behavior. As seen at the transition of the pile wall to the flange, local yielding develops and subsequently load transfer via the inner edge of the flanges at the limit tension force of the bolt. As a result positions A-E, defined in Figure 12, are specified as hot spots for establishing a link between the tensile shell resistance and the corresponding stresses and plastic strains. From the comparison of the analytically and numerically obtained results in positions A-E the following is highlighted:

- Maximum stresses and corresponding development of plastic strains are obtained at point B in Figure 15 and Figure 16. The presented result indicates that a more detailed study of the stresses in this area is needed along the entire circumference.
- As the tensile segment force shown in Figure 15 increases the slope of the compressive stresses, representing the prying force, at (pos. E) develop more rapidly. As result, the slope of the graphs for the tensile stresses in the bolt (pos. B-D) becomes steeper. As the analytical approach does not include any tolerances, the prying force presented in Figure 15 is seen as minimum value.
- The difference in stress found at the two sides of the bolt C-D in Figure 15 is seen to increase with increasing tensile segment load. The difference is explained by the fact that the secondary bending moment in the bolt is not considered in the analytical approach.
- The plastic strains obtained in Table 8 are the answer to an extended requirement that ensures the integrity of the connection according to the analytical approach. As the analytical L flange concept ignores the fact that opening of the connection and plastic deformation will lead to some redistribution of forces, this requirement is to be maintained for verification of a full flange model.

5.2. Fatigue limit state

The numerically determined accumulated fatigue damage sum in the bolt is found by the numerical approach to be approximately 50% lower than the analytically obtained value. Furthermore, the numerical model indicates that the maximum fatigue damage in the bolt occurs in the bolt shank near the bolt head. The difference between the analytically and numerically obtained fatigue damage is seen to be caused by:

- Figure 18 shows the numerically and analytically derived bolt forces as a function of the segment shell force. For a given tensile shell force it can be seen that the numerically predicted bolt force is significantly lower than the analytically predicted.
- The result shows the effect of the built-in fabrication tolerances in the Schmidt-Neuper method which are not included in the numerical model.
- An initial study of the fabrication tolerances identifies the estimated contact surface between two flanges as a crucial parameter for the slope of the curve.
  Using the numerical model as an alternative to the analytical approach includes consideration of following:
- The numerically obtained damage values are seen to be very sensitive to the mesh size. In order to directly apply the stresses obtained in the numerical model further investigations are required. In addition, the impact of simplifying the model bolt geometry should be clarified.
- The numerical procedure is as consequence found only to be applied for specification of the relation between segment shell force and the bolt force. Following this approach the integrity of the bolt can be verified for fabrication tolerances exceeding the values included in the Schmidt-Neuper method or sought optimized by consideration of the full flange model.

IOP Conf. Series: Materials Science and Engineering **276** (2017) 012034 doi:10.1088/1757-899X/276/1/012034

6. Conclusion

An analytical and numerical ULS and FLS examination of the L flange segment is presented as the basis for the ongoing development work on the design of the bolted MP/TP steel ring flange connection. The L flange segment is investigated due to the availability of analytical approaches making it possible to establish numerical criteria applicable to a full flange model. The analytically determined failure mode B in the analytical ULS determination is confirmed by the identical numerical model leading to the conclusion that the derived tensile resistance can be applied to specify plastic strain criteria. However, as the analytical approach does not consider tolerances, the tolerances influence on the plastic strain values is required before the criteria can be applied to a full flange model. Fabrication tolerances are included in the analytical FLS determination by the trilinear curve relating the segment force with the tensile bolt force. Comparison of the analytically and numerically determined accumulated fatigue damage shows the damage found by the numerical approach (which does not include tolerances) too low. Furthermore, the local obtained stress values in the numerical model are very sensitive to the choice of element size as well as the level of detailing of the bolt. The numerical FLS approach is therefore suggested to be preferred for determining the relation between the segment load and the tensile bolt force. Implementation of the numerical FLS model to a full flange model will require that sufficient fabrication tolerances can be included in the numerical model.

The presented results indicate that focus for future development work must include handling fabrication and installation tolerances in the numerical models.

References

[1] Wind Europe 2017 *The European Offshore Wind Industry – Key Trends and Statistics 2016* vol 1, ed Pineda I and Tardieu P (Belgium: Wind Europe) p 7

[2] Vries D 2010 *Offshore Monopile Failure – A solution may be in sight Windpower* [Online]. Available: http://www.windpowermonthly.com/article/1011507/offshore-monopile-failure---solution-may-sight

[3] Offshore Wind Industry 2016 *Waiting for new tenders* number 2, ed Buddensiek V (Germany: Offshore Wind Inustry) p 24

[4] Damiani R R 2016 *Offshore Wind Farms: Technologies, Design and Operation* vol 1, ed Chong Ng and Ran L (United Kingdom: Woodhead Publishing) chapter 10 pp 263-357

[5] IEC 61400-6 2016 *Wind Energy Generation Systems – Part 6: Tower and Foundation Design Requirements*

[6] Petersen C 2013 *Grundlagen der Berechnung und baulichen Ausbildung von Stahlbauten* vol 4, ed Harms R and Prenzer A (Munich: Springer Vieweg) chapter 16 pp 1165-1260

[7] Thage K J 2017 *Investigation of Flange Connection Capacity for Offshore Wind Turbine Foundations* M.Sc. Eng. Thesis (Lyngby: Technical University of Denmark)

[8] DiBt 2012 *Richtlinie für Windenergieanlagen Einwirknungen und Standsicherheitsnachweise für Turm und Gründung*

[9] GL COWT 2012 *Guideline for the Certification of Offshore Wind Turbines, Rules and Guidelines, IV-Industrial Services, Part 2, Guideline for the Certification of Offshore Wind Turbines*

[10] VDI 2230-1 2015 *Systematic Calculation of High Duty Bolted Joints – Joints with One Cylindrical Bolt*

[11] DASt 021 2013 *Schraubenverbindungen aus Feuerverzinkten Garnituren M39 bis M72 entsprechend DIN 14399-4, DIN EN 14399-6*

[12] DNVGL-RP-C208 2016 *Determination of Structural Capacity by Non-linear Finite Element Analysis Methods*

[13] Hobbacher A F 2016 IIW-2259-15 2016: *Recommendations for Fatigue Design of Welded Joints and Components* vol 2, ed Mayer C (Switzerland: Springer International Publishing) chapter 10 pp 113-138

Natural frequency and vibration analysis of jacket type foundation for offshore wind power

Y-C Hung*, Y-Y Chang and S-Y Chen

Marine Industrial Department, Ship and Ocean Industries R&D Center, New Taipei City, Taiwan

*Contact author: ych0407@hotmail.com

Abstract. There are various types of foundation structure for offshore wind power, engineers may assess the condition of ocean at wind farm, and arrange the transportation, installation of each structure members, furthermore, considering the ability of manufacture steel structure as well, then make an optimum design. To design jacket offshore structure, unlike onshore cases, offshore structure also need to estimate the wave excitation effect. The aim of this paper is to study the difference of natural frequency between different kinds of structural stiffness and discuss the effect of different setting of boundary condition during analysis, besides, compare this value with the natural frequency of sea wave, in order to avoid the resonance effect. In this paper, the finite element analysis software ABAQUS is used to model and analyze the natural vibration behavior of the jacket structure.

1. Introduction

The offshore wind energy plays an important role in green power industries, therefore lots of countries engage in this field. Building offshore wind farm is not only the task of power industry but also includes civil engineering, material engineering and marine engineering. Due to the large scope, each detail and risk should be concerned.

To construct offshore wind farm is much more difficult than onshore condition, so the structural design should concern the strength and the feasibility of construction. But most of offshore structures are huge, it is very difficult to carry out the real scale test. Computer-aided engineering is a way to help engineer to simulate and arrange their design. Natural vibration frequency is an important index among the structural strength and integrity. The boundary conditions, including the effect of soil structure interaction may affect the behavior of vibration a lot. Most of the models may simplify the boundary condition between soil and structure to fixed end, because the stiffness of soil is very complicated to describe, but the fixed end assumption may be unconservative in soft soil conditions. This study uses finite element software ABAQUS to analyze natural frequency, and adopt spring elements to simulate soil condition and simplify the problem.

2. Parameters

The piles under each legs of jacket foundation are installed by pre-piling method, the connecting method between leg and pile may be different from post-piling type. The diameter and length of the pile may considered by the bearing capacity of soil, in addition, the distance between the bottom of legs also need to considered by the stability of overturning.

The size of the model in this paper is refer to the design for the offshore wind field in Taiwan Strait, and some parameters are reasonably assumed if they are not available. The site is set at the sea area of 30 m water depth. In order to avoid the action force of waves to disturb the transition piece and tower, the height of the jacket needs to exceed the depth of sea and also reserve the air gap.

According to the products and previous cases [1], the tower height of 6 MW wind turbine is about 90 m to 140 m, this paper assumes the tower height is 110 m. The simulation of seabed is using the parameters from Taiwan Power Company which is obtained by drilling test at the wind field in Taiwan Strait. Integrated above items and refer to some real design, the size, material and soil condition (Table 1) are as follows:

Structure Scale

Jacket Height : 50.0 m
Jacket Upper Width : 8.0 m
Jacket Footprint Width : 20.0 m
Pile Diameter : 2.0 m
Pile Length : 40.0 m
Tower Height : 110.0 m
Rotor & Nacelle Mass : 400.0 t

Material

Steel Density : 7.85 t/m^3
Steel Elastic Modulus : 210.0 GPa
Poisson ratio : 0.3
Concrete Density : 2.4 t/m^3

Table 1. *Soil Parameters*

Depth (m)	**Description**	γ_t **(t/m^3)**	**c (t/m^2)**	ϕ' **(deg.)**
0.00 ~ 21.5	SM	1.93	4.98	27.6
21.5 ~ 24.5	SC	1.95	14.9	--
24.5 ~ 31.25	SM	1.91	--	38.3
31.25 ~ 33.45	CL, SC	1.97	12.1	--
33.45 ~ 35.75	SM	2.01	--	40.5
35.75 ~ 39.75	ML	1.89	13.1	--
39.75 ~ 44.00	SP-SM	1.98	--	43.5

3. Model establishment

3.1. Comparison against a reference study

In this paper, the model is established by finite element software ABAQUS. Because the mechanical behavior of each members in jacket structure is close to beam element, besides, due to jacket structure includes a lot of connecting part, and the setting for those connection may concern with the contact behavior between each members, these setting may spend much more operation times, and also increase the difficulties to get convergent result. In order to simplify the problem, beam element is used to build piles, tower and the members in jacket structures.

In the first part of this analysis, a comparative model is built as a reference, the natural frequency of this comparative model was already analyzed by other researchers by using the finite element software like ANSYS and SubDyn [2], and the model is shown in Figure 1. By comparing the value of natural frequency computed by ABAQUS with the reference article [2], the accuracy of this analysis and all the modeling steps are assured.

In this study, only the foundation part is considered, including jacket and part of piles (5 meters length), the boundary condition is to fix all degree of freedom at the bottom of piles as Figure 2 shows.

The result from reference article [2] is shown in Table 2 and Figure 3. After analysis, the natural frequency of first eight modes extracted from ABAQUS. By comparing the result between corresponding modes, the value and tendency are very similar, that means this modeling method and analytic step is workable.

Table 2. Natural frequency (Cycles/sec.)

Mode	1st Mode	2nd Mode	3rd Mode	4th Mode
ABAQUS	2.0831	2.0831	5.1435	5.3890
References	2.5500	2.5500	5.0500	7.5000
Mode	5th Mode	6th Mode	7th Mode	8th Mode
ABAQUS	7.4018	7.4018	7.6830	7.9759
References	7.5500	7.5500	8.5000	9.0500

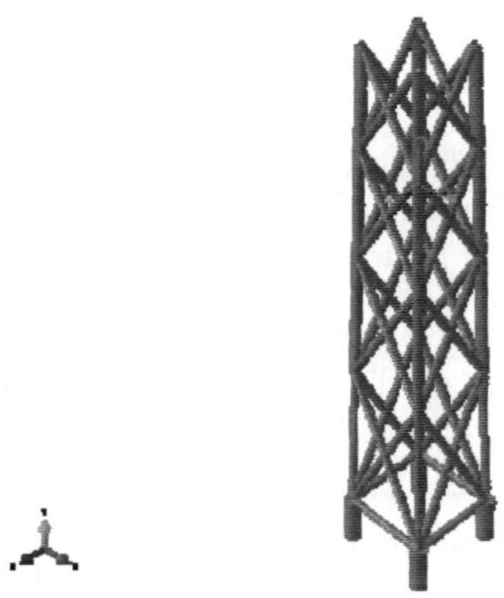

Figure 1. Jacket structure model of control group

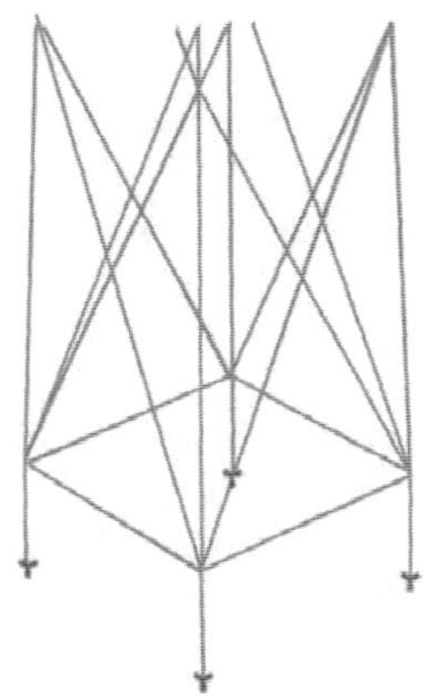

Figure 2. Fixed boundary condition of control group

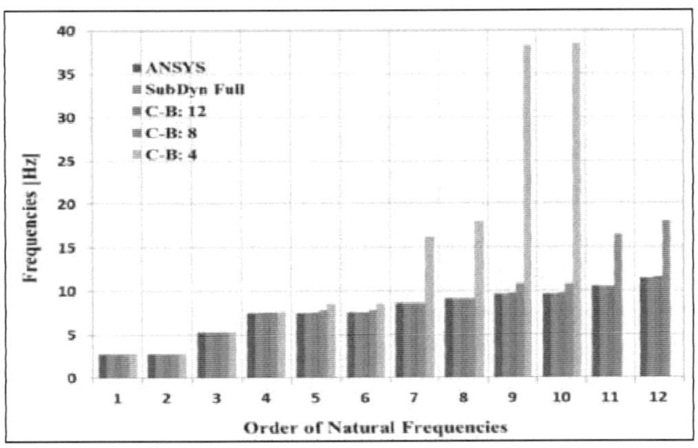

Figure 3. Natural frequency [2]

3.2. Boundary Condition and Pile-Soil Interaction Study

The following study analyzes the natural frequency in two conditions, one is the boundary surface around pile fixed, and the other case includes soil effect.

With the parameters of model described previously, besides, the mass of rotor nacelle assembly is assumed as a concentrated mass point on the top of tower whose arrangement of the model is shown in Figure 4.

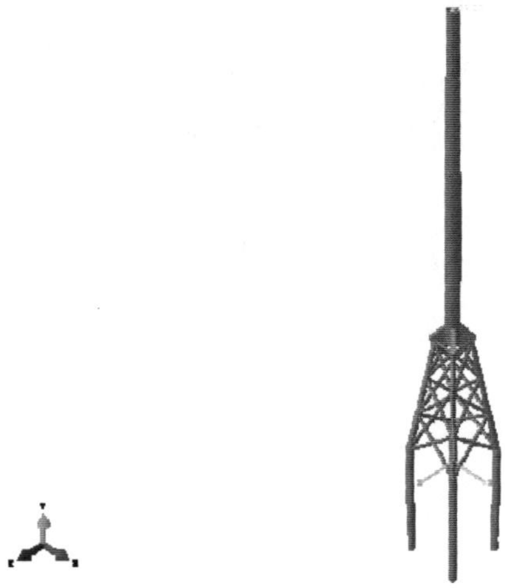

Figure 4. Jacket structure model

For the purpose of simulating the interaction between soil and pile structure, the empirical formulas for p-y curve provided by API (American Petroleum Institute) [3] are used to display the stiffness behavior of soil around piles. With the empirical formulas and soil parameters of offshore wind field show in Table 1, the soil spring with non-linear stiffness, which is the resistance-deflection relationship of the soil, can be simulated. The relationship between lateral bearing capacity and lateral deflection of soil may different along with the depth, for this reason the p-y curves corresponding to each depth need to be calculated one by one, Figure 5 shows the p-y curve of sand in 10 m depth and p-y curve of clay in 22 m depth.

The pile length penetrating into seabed is 40 m, and the spacing between soil springs is 2 m, so there will have 21 different stiffness soil springs per pile in one direction. For the soil characteristic that only taking compression, the nonlinear elastic property of each spring is only input the compression part, so it will need four springs on x-z plane to provide lateral support, the arrangement of soil springs around pile is shown in Figure 6. However, the friction force on the lateral surface of pile is ignored, so the degree of freedom in vertical direction (y direction) is set to free, but at the bottom of pile is set to fixed, in order to support the structure and simplify the model.

During analysis, first eight modes from fixed pile model and soil spring model are extracted, and compared with each other. As the result shown in Table 3 indicates, the deformation tendency of both models are similar, but the values from soil spring model are a little bit smaller than the values from fix-pile model, the reason of this phenomenon may be the stiffness of soil-spring must smaller than fixed boundary, so the structural behavior of soil-spring model may be more flexible, and the corresponding natural vibration period will be longer.

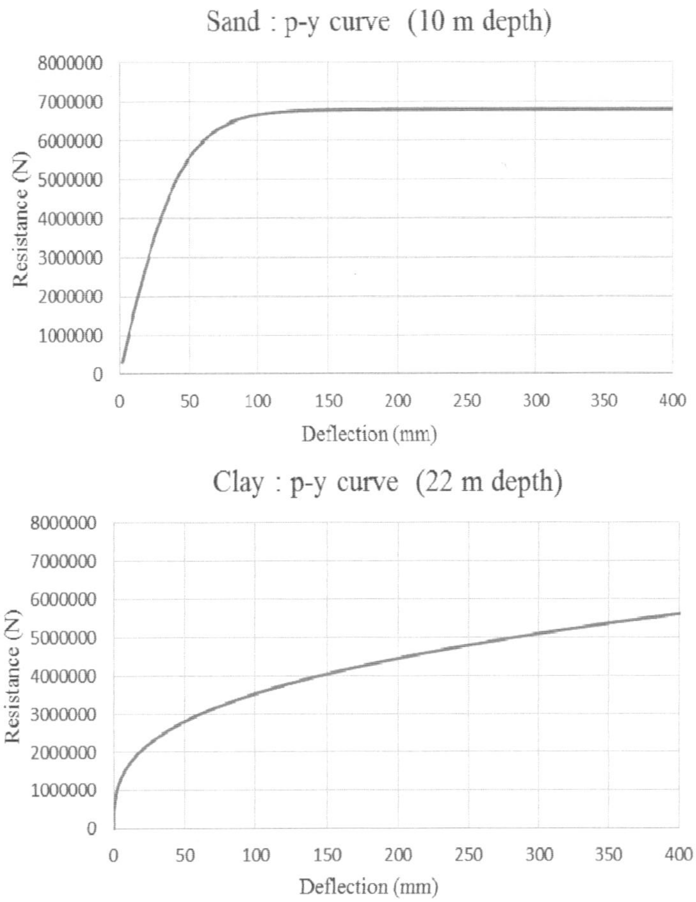

Figure 5. P-y curve

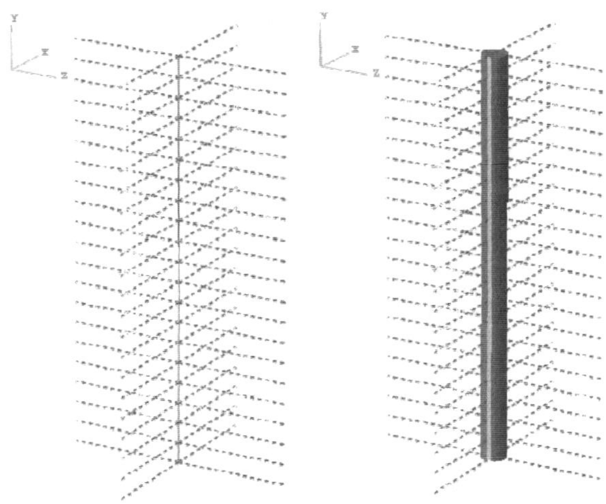

Figure 6. Soil-spring around pile

Furthermore, applying a cyclic load which the amplitude is 100 kN on the top of jacket for 50 sec with the load frequency equals to the value of first mode of structure, in order to check the response of structure, and also test others load frequency to compare, including 0.1, 0.15, 0.25, 0.3, 0.4 (Hz).

Figure 7 shows the displacement of tower tip over time when the load frequency is equal to the natural frequency, the tendency is getting divergent, means resonance phenomenon occurs. The results of maximum displacement at tower tip in each above cases are shown on the graph in Figure 8. The outcome clearly shows that the structural response will increase obviously if the load frequency is close to the structural natural frequency.

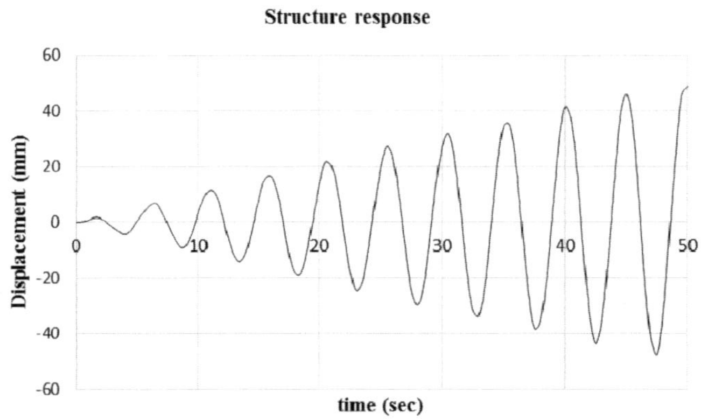

Figure 7. Resonance response

Table 3. Natural frequency (cycles/sec)

Mode	1st Mode	2nd Mode	3rd Mode	4th Mode
Fix-pile model	0.2038	0.2038	1.4459	1.4459
Mode	1st Mode	2nd Mode	3rd Mode	4th Mode
Soil-spring model	0.2036	0.2036	1.4109	1.4110
Mode	5th Mode	6th Mode	7th Mode	8th Mode
Fix-pile model	3.4334	3.4334	4.3930	5.5029
Mode	5th Mode	6th Mode	7th Mode	8th Mode
Soil-spring model	2.8968	2.8968	4.2362	4.4512

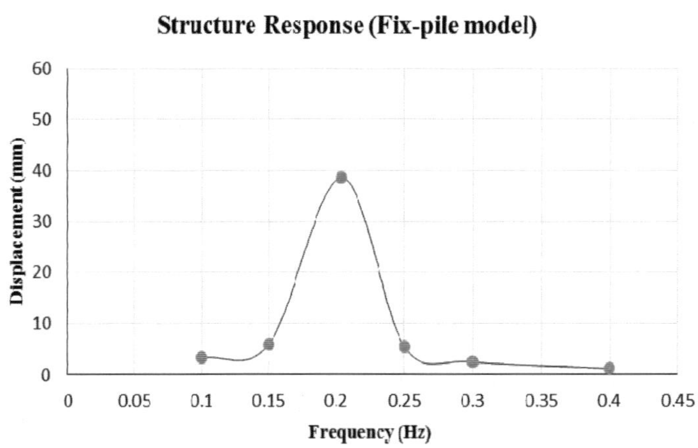

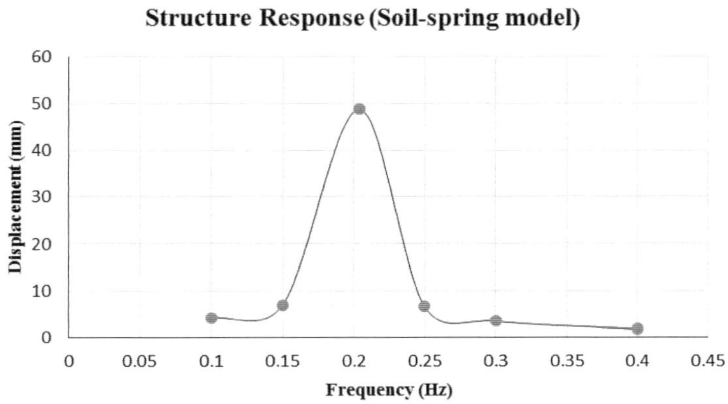

Figure 8. Structure response

3.3. Advanced Apply

Comparing the structural frequency from previous analysis to the probability about the significant wave height and the corresponding mean period at one of wind farms in Taiwan Strait (in Appendix A), the natural frequency is in the region of high occurrence rate of wave frequency. In order to avoid resonance occur, the mass or stiffness of structure need to be modified

First, after comparing to other jacket structures, a mud-brace is added at the bottom of jacket, and the difference of natural frequency is observed. From the result shown in Table 4, the value has increased a small amount, but it is still not enough to improve the whole structure.

The other approach is to decrease the tower height to 90 m, from the result in Table 5, the natural frequency has changed obviously. These two approaches and the simulation in last section are in purpose to figure out the effect of different members in structure. In reality, engineer may consider the requirement and condition, and design the structure cautiously.

Table 4. Natural frequency (Structure with mud brace)

Mode	1st Mode	2nd Mode	3rd Mode	4th Mode
Freq. (cycles/sec)	0.2040	0.2040	1.4328	1.4328

Table 5. Natural frequency (Structure with 90 meters tower height)

Mode	1st Mode	2nd Mode	3rd Mode	4th Mode
Freq. (cycles/sec)	0.3048	0.3048	2.2564	2.2564

4. Conclusions

For the result of this study, the method of soil spring can present the deformation behavior of soil without build the soil element. In addition, the mode shape of high frequency showed existence of different behavior between these two models. In real environment condition, the seabed cannot provide perfect fixed restraint, especially in shallow region, the capability of lateral support from soil is lower at the shallow part than deeper part, so in this situation the fixed boundary setting will be unreal.

Once the foundation design requires pile-soil interaction calculation, soil spring method can save a lot of computational resource and easy to get convergent result. In order to simplify the problem, the friction force on the surface between pile and soil are ignored. Future studies should consider the soil spring in vertical direction, and the behavior of structure will be more accurate.

Acknowledgement
Thanks to all my coworkers and partners providing me a lot of information and help me to overcome all the difficulties, and arouse me many ideas. Thanks to the section leader for giving us this chance to conduct this research and attend the conference abroad. Thank you all again.

Appendix A Joint probability of wave heights and periods

	$H_{1/3}$ (m)											
T_m (sec)	0.0~0.5	0.5~1.0	1.0~1.5	1.5~2.0	2.0~2.5	2.5~3.0	3.0~3.5	3.5~4.0	4.0~4.5	4.5~5.0	>=5	Total
0~1	0.00	0.00	0.00	0.00	0.00	0.00	0.00	0.00	0.00	0.00	0.00	0.00
1~2	0.55	0.00	0.00	0.00	0.00	0.00	0.00	0.00	0.00	0.00	0.00	0.55
2~3	8.28	0.08	0.00	0.00	0.00	0.00	0.00	0.00	0.00	0.00	0.00	8.37
3~4	6.03	11.23	0.00	0.00	0.00	0.00	0.00	0.00	0.00	0.00	0.00	17.26
4~5	0.37	13.87	8.53	0.00	0.00	0.00	0.00	0.00	0.00	0.00	0.00	22.77
5~6	0.01	2.20	8.96	8.07	0.07	0.00	0.00	0.00	0.00	0.00	0.00	19.30
6~7	0.00	0.21	0.92	4.05	10.31	1.35	0.00	0.00	0.00	0.00	0.00	16.84
7~8	0.00	0.05	0.05	0.11	0.77	7.62	2.82	0.00	0.00	0.00	0.00	11.43
8~9	0.00	0.00	0.03	0.01	0.01	0.05	1.88	1.17	0.02	0.00	0.00	3.18
9~10	0.00	0.00	0.00	0.00	0.00	0.00	0.00	0.07	0.19	0.01	0.00	0.29
>=10	0.00	0.00	0.00	0.00	0.00	0.00	0.00	0.00	0.00	0.01	0.01	0.02
Total	15.24	27.64	18.50	12.24	11.17	9.02	4.70	1.24	0.21	0.02	0.01	100.00

References

[1] Twidell J and Gaudiosi G 2014 Offshore wind power *Shy Mau publishing Co.*
[2] Song H, Damiani R, Robertson A and Jokman J 2013 A new structural-dynamics module for offshore multimember substructures within the wind turbine computer-aided engineering tool FAST *Offshore and Polar Engineering Conference-ISOPE*
[3] API 2014 Planning, designing, and constructing fixed offshore platforms-working stress design *American Petroleum Institute*
[4] Hu H T, Ye D S and Yang C 2015 Finite element frequency analysis of offshore wind turbine structure under soil and structure interaction *SIMULIA Regional User Meeting*

Fracture mechanics analysis for a mooring system subjected to tension and out-of-plane bending

X Xue and N-Z Chen*

School of Engineering, Newcastle University, Newcastle upon Tyne, UK

*Contact author: nianzhong.chen@ncl.ac.uk

Abstract. A fracture mechanics analysis for the mooring system of a semi-submersible accounting for out-of-plane bending (OPB) is presented in this paper. Stress ranges acting on the mooring chain links are calculated based on tension and OPB of mooring chain links induced by motions of wave frequency (WF) and low frequency (LF). The narrow-banded method is used for predicting the combined mooring loading process. Initial cracks are assumed to propagate from surfaces of chain links and stress intensity factors are then calculated in terms of stress ranges determined by a finite element analysis. The influence of the OPB on the remaining service life of mooring chain links is investigated and the results show that the remaining service life of mooring chain links connecting to fairleads is significantly reduced due to the OPB effects.

1. Introduction

Fatigue failure is one of critical failure modes of offshore permanent mooring systems. Fontaine et al. [1] reviewed 107 offshore mooring accidents and they pointed out that the 29 mooring accidents are primarily due to fatigue. In particular, the occurrence of 3 of 29 mooring fatigue accidents was induced by the out-of-plane bending (OPB).

Traditionally, only mooring tensions are considered in mooring chain fatigue analysis. However, for fatigue analysis of mooring chain links connecting to chain wheels at fairleads, chain links passing over bending shoes, or chain linkers provided by chain hawse or chain stoppers, the effects of the OPB on the mooring fatigue should be considered [2, 3].

Vargas et al. [4] modelled mooring chains and a mooring chain wheel with seven pockets, and preformed a finite element (FE) analysis to investigate the influence of the OPB on mooring chain stresses. Ter Brake et al. [5] presented an S-N curve based fatigue analysis for mooring chains, in which the effects of the OPB on mooring chain stresses were determined by a FE analysis of chain links. Melis et al. [6] presented experimental results of mooring chain stresses induced by OPB. Vargas and Jean [7] compared mooring chain stresses calculated by FE method with experimental results of Melis et al. [6]. Lassen et al. [8] and Lassen et al. [9] performed a mooring chain fatigue life using an S-N curves based approach, and the behaviour of chain links subjected to tension and OPB was investigated by both experimental and FE methods.

In this paper, a fracture mechanics analysis for the mooring system of a semi-submersible accounting for OPB is presented. Stress ranges acting on the mooring chain links are calculated based on tension and OPB of mooring lines induced by motions of wave frequency (WF) and low frequency (LF). Initial cracks are assumed to propagate from surfaces of chain links and stress intensity factors are then calculated in terms of stress ranges determined by a finite element analysis. The influence of the OPB on the remaining service life of mooring chain links is investigated.

2. Mooring system

A semi-submersible with 16 mooring lines operated at OWA (Offshore Western Africa) is utilized for the mooring analysis in this paper [10].

The semi-submersible hull comprises a ring pontoon and four columns and it is symmetric in both east-west and south-north directions. The draft of the semi-submersible is 26 m. The cross-sections of the pontoon and columns are both rectangular. The cross-section of the pontoon is 22 m (Width) × 10 m (Height). The cross-sections of the columns are 22 m × 22 m with 4m corner radius.

The semi-submersible is operated at the water depth of 1829 m. A taut mooring system is designed for the semi-submersible. The semi-submersible is spread moored with 16 mooring lines in 4 clusters. Mooring lines are numbered counter-clockwise and the Line 1 is laying 37.5 degrees from north. Each adjacent mooring line in the same cluster is in 5 - degree separation. The layout of mooring system is shown in Figure 1.

Each mooring line consists a 6-inch R4 class studless bottom chain, a 267 mm polyester rope and a 6-inch R4 class studless top chain. The pretension of each line is set as 3152 kN (18% of minimum breaking strength of mooring chain link). The length of the top chain is 150 m, the bottom chain 210 m. The polyester rope is divided into four segments and the length of each segment is 650 m. In the fairlead of the mooring system, seven-pocket chain wheels are installed to guide the top chains operating around the semi-submersible. The sketch of a mooring chain wheel at the fairlead is shown in Figure 2 [11].

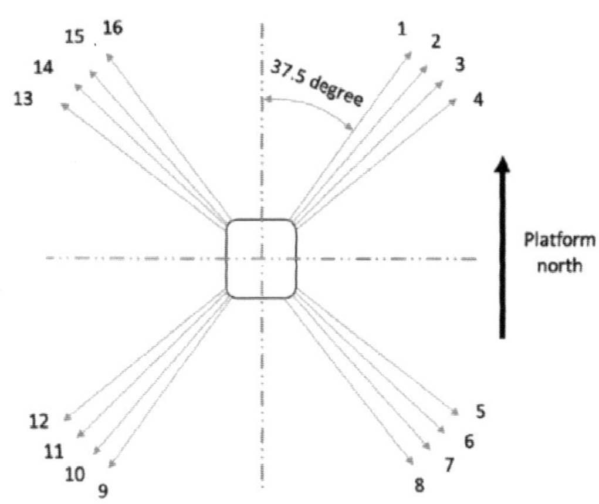

Figure 1. Layout of mooring system

Figure 2. Sketch of mooring chain wheel at fairlead [11]

The semi-submersible is operated at offshore West Africa (OWA). OWA environment is characterized by long period swells. In this paper, the wave, wind and swell effects are taken into account to examine the remaining service life of mooring chains as shown in Table 1. The environmental conditions can be referred to Wang et al. [12] and Wu et al. [10]. Ochi-Hubble spectrum [13] is used herein for the simulation of the semi-submersible motion induced by wave and swell, and Harris spectrum [14] is used for the simulation of the vessel motion induced by wind.

Table 1. Environmental conditions at OWA

Bin	Swell		Wave		Wind speed (m/s)	Direction (degree)	Probability (100%)
	Hs(m)	Tp(s)	Hs(m)	Tp(s)			
1	1.75	14	0.3	2.7	8	0	30.4
2	1.75	22	0.7	3.8	14	0	7.6
3	2.75	14	0.3	2.7	8	0	9.2
4	2.75	22	0.7	3.8	14	0	2.3
5	3.75	14	0.3	2.7	8	0	0.4
6	3.75	22	0.7	3.8	14	0	0.1
7	1.75	14	0.3	2.7	8	45	30.4
8	1.75	22	0.7	3.8	14	45	7.6
9	2.75	14	0.3	2.7	8	45	9.2
10	2.75	22	0.7	3.8	14	45	2.3
11	3.75	14	0.3	2.7	8	45	0.4
12	3.75	22	0.7	3.8	14	45	0.1

3. Mooring chain tension

The function of mooring chain tensions can be written as:

$$T = T_M + T_P + T_D \tag{1}$$

where T_M is the mooring tension due to the mean environmental loads including mean wind, swell and wave forces. T_P is the pretension of mooring chain. T_D is the dynamic mooring tension due to environmental loads, which is usually considered as the combined tension induced by low frequency (LF) and wave frequency (WF) motions.

Both LF and WF mooring chain tension processes are assumed to be Guassian processes. LF motions are estimated by a quasi-static method and WF motions are calculated by FE method through a numerical simulation.

In this paper, the narrow-band method is applied to predict the combined tension process, T_D. The probability density function of the amplitude of combined tension process, T_A, can be given as:

$$f(t_a) = \frac{t_a}{\sigma_D^2} e^{\left(-\frac{t_a^2}{2\sigma_D^2}\right)} \tag{2}$$

where σ_D is the standard deviation of T_D, and it can be written as

$$\sigma_D = \sqrt{\sigma_W^2 + \sigma_L^2} \tag{3}$$

where σ_W is the standard deviation of WF tension process, and σ_L is the standard deviation of LF tension process.

The mean zero up-crossing frequency of T_D is then formulated as

$$v = \sqrt{\frac{\sigma_W^2 v_W^2 + \sigma_L^2 v_L^2}{\sigma_W^2 + \sigma_L^2}} \tag{4}$$

where v_W is the mean zero up-crossing frequency of WF tension process, and v_L is the mean zero up-crossing frequency of LF tension process.

4. Out-of-plane bending (OPB) of mooring chain

It is indicated by industry experience that chain links connecting to fairleads, chain links passing over bending shoes, or chain linkers provided by chain hawses or chain stoppers, may be subjected to additional OPB, which may induce fatigue failures out of expectation [2, 3].

In traditional mooring fatigue analysis, the bending stiffness of the chain segments is neglected because of rolling and sliding between chain links. However, when the chain links are installed with high pretension, the high frictions between chain links would arise. Then the interlink angle between chain links would be locked, which leads to bending moments between the contact chain links. The definitions of OPB and interlink angle, applied load angle are shown in Figure 3.

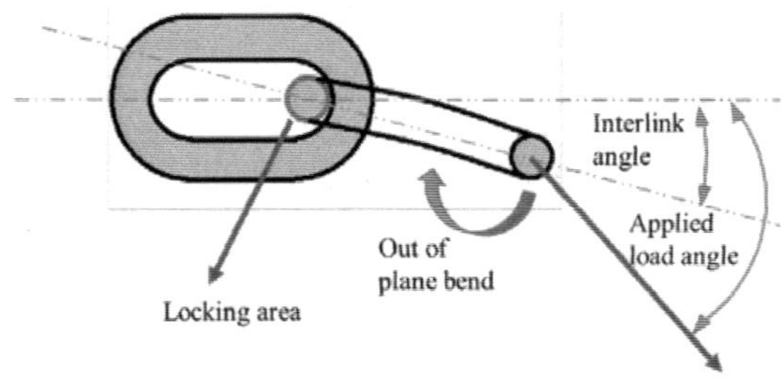

Figure 3. Sketch of out-of-plane bending (OPB) of mooring chain with deviation angles

5. Fracture mechanics analysis

In fracture mechanics analysis, the chain link is treated as a round bar and an initial semi-elliptical surface crack is assumed to propagate at the surface of the chain link.

The crack growth is predicted based on the Paris-Erdogan equation [15]:

$$\frac{da}{dn} = C(\Delta K)^m \tag{5}$$

where a is crack depth, n is the number of stress cycles. $C = 2.3\times10^{-12}$ and $m = 3$, which can be referred to BS7910 [16]. ΔK is the stress intensity factor range and stress intensity factor K can be calculated as

$$K = Y\sigma\sqrt{\pi a} \tag{6}$$

where Y is the stress intensity correction factor. In this paper, both membrane stress σ_m and bend stress σ_b acting on the cross-sections of chain links are considered in FM analysis, so $Y\sigma$ is defined as [16]:

$$Y\sigma = Mf_w(M_m k_{tm} M_{km}\sigma_m + M_b k_{tb} M_{kb}\sigma_b) \tag{7}$$

where k_{tm} and k_{tb} are stress concentration factors. For a semi-elliptical surface crack in a round bar, $M = f_w = M_{km} = M_{kb} = 1$, M_m and M_b can be referred to BS7910 [16].

The service life N_S based on fracture mechanics approach can be estimated by an integration procedure:

$$N_S = \int_{a_0}^{a_c} \frac{da}{C(\Delta K)^m} \tag{8}$$

where a_c is the critical crack depth and a_0 is the initial crack depth.

6. Finite element (FE) analysis

A FE analysis is performed to predict the stresses acting on the cross-sections of a studless chain link. R4 class common studless mooring chain links are main components suffered to fatigue damage in the designed mooring system. The chain links are modelled with standard dimensions where the length L and breadth B are given by:

$$L = 6D, B = 3.35D \tag{9}$$

where D is the nominal diameter of the chain link. The corrosion and wear allowance for 25 years is also considered for each link.

Three scenarios are considered in the FE analysis. In the Scenario 1, only tension loads are considered. In the Scenarios 2 and 3, the effects of OPB are taken into account.

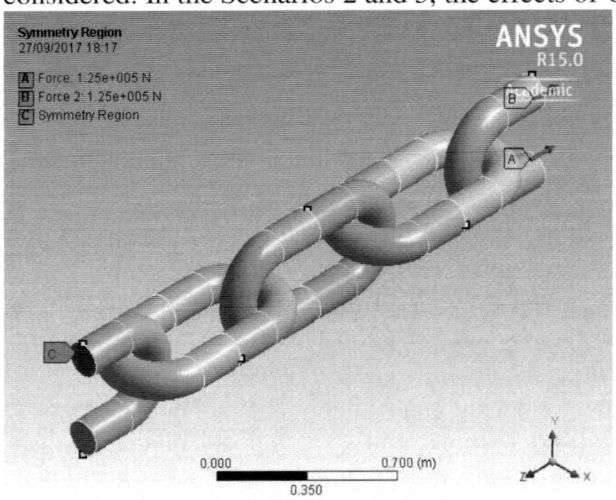

Figure 4. Finite element model of mooring chain links of Scenario 1

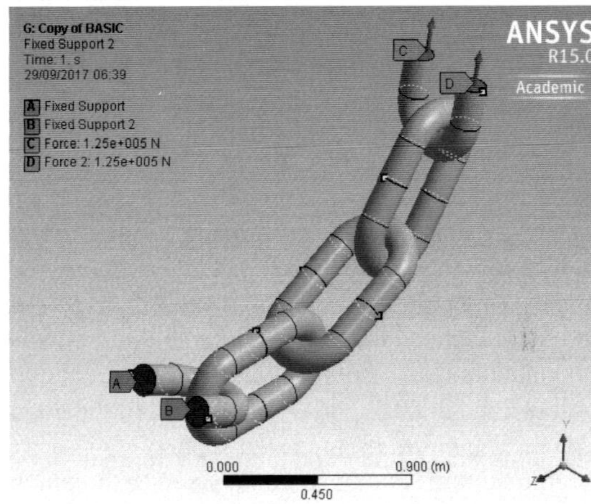

Figure 5. Finite element model of mooring chain links for Scenario 2

For Scenario 1, the interlink link angles of mooring chains are considered as 0 degrees, as the model shown in Figure 4. End face C of the half chain shown in Figure 4 is defined as a symmetry plane. The tension forces are applied at the end faces A and B, as shown in Figure 4.

For Scenarios 2 and 3, the chain links are assumed to lay on the seven-pocket chain wheel in the fairlead, and interlink link angles of mooring chains are considered as 25.71 degrees, as shown in Figure 5. The contact type between mooring chain links is set as bonded, which means neither sliding nor separation is available between contact surfaces and the angle between chain links is locked. In Scenario 2, end faces A and B are fixed. The tensions are applied at the end faces C and D. The surface contract effects between two mooring chains are considered and interlink angles between chain links are locked. Compared with Scenario 2, end faces A and B in Scenario 3 are simple supported. The details of each scenario are listed in the Table 2.

Table 2. Details of each scenario

	Boundary condition	Whether consider OPB?
Scenario 1	Symmetry	No
Scenario 2	Fixed support	Yes
Scenario 3	Simple support	Yes

7. Results and discussion

Due to the space limitation, only the results of frequency-domain mooring line analysis of mooring line 1 is presented in the Table 3, where T_W and T_L are mean time period of WF and LF motions, respectively.

Table 3. Mooring line tension at fairleads of Line 1

BIN	T_W(s)	σ_W(kN)	T_L(s)	σ_L(kN)
1	13.0	15.2	94.9	7.6
2	22.0	29.8	90.1	27.0
3	13.4	25.4	103.1	9.9
4	22.0	48.0	90.1	27.0
5	13.6	36.6	111.4	14.6
6	22.0	67.5	90.1	27.0
7	13.5	16.9	57.9	3.1
8	22.1	22.3	55.9	11.2
9	13.6	27.5	58.7	3.1
10	22.0	36.4	55.9	11.2
11	13.7	39.0	61.0	3.3
12	22.0	51.7	55.9	11.2

For Scenario 1, the outside of crown cross-section (C-C cross-section), the inner surface of the beginning of the bend area (B-B cross-section) and the inner surface of weld section (A-A cross-section), as shown in Figure 6, are treated herein as three potential locations that are prone to fatigue failure [6].

After the FE analysis, it is found that the stresses at the crown sections are larger than stresses at the bend sections or weld sections without considering stress concentration factors due to weld quality. A typical normal stress distribution of the crown section of a mooring chain is shown in Figure 7.

For Scenarios 2 and 3, the most critical location of mooring chain links prone to fatigue damage moves to the middle of crown section (D-D cross-section) as shown in Figure 8 [6-8]. Figure 9 shows a typical normal stress distribution over the cross-section described in Figure 8.

Then, a linearization of stress distribution over the cross-section is performed herein to divide the normal stresses into membrane and bending stresses and the detailed method can be referred to BS7910 [16]. These stresses are then used for the FM analysis for predicting the remaining service life of mooring chains at the fairleads.

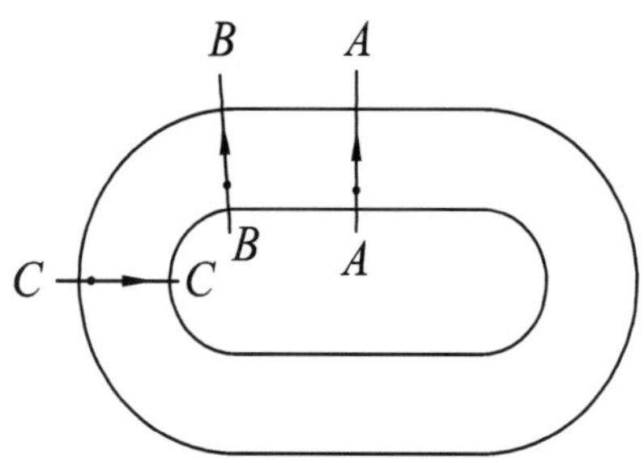

Figure 6. Scenario 1: locations of the chain link prone to fatigue damage in pure tension

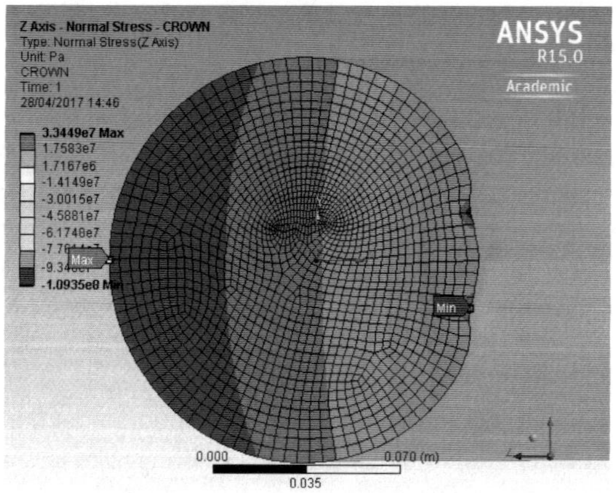

Figure 7. Normal stresses at the crown section of the chain link

IOP Conf. Series: Materials Science and Engineering **276** (2017) 012036 doi:10.1088/1757-899X/276/1/012036

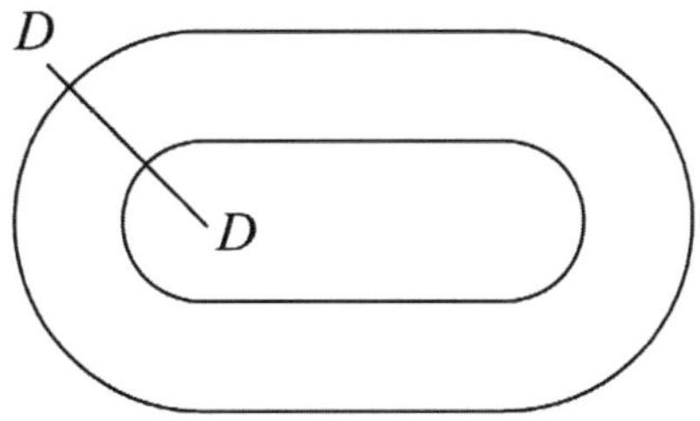

Figure 8. Scenarios 2 and 3: the critical location of the chain link prone to fatigue damage in coupled tension and out-of-plane bending (OPB)

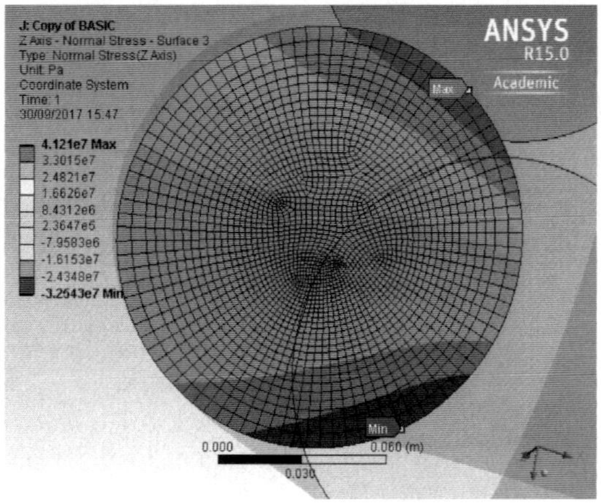

Figure 9. Normal stresses at the crown section of the chain link

As recommended by Mathisen and Larsen [17], the critical crack depths are set as 12% of the chain diameter at weld sections, 30% of the chain diameter at bend section, and 15% of the chain diameter at crown section. The initial crack depth is assumed to be 0.5 mm. The ratios of remaining service life of mooring chain links connecting to fairleads of the semi-submersible in Scenarios 2 and 3 to Scenario 1 are plotted in Figure 10.

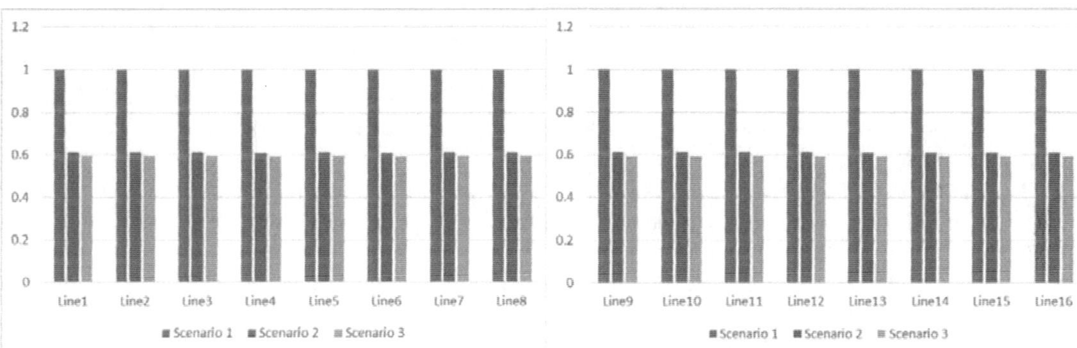

(a) (b)

Figure 10. Ratios of remaining service life of mooring chain links in Scenarios 2 and 3 to Scenario 1: (a) Line 1 to Line 8; (2) Line 9 to Line 16

The result shows that the OPB effects on the remaining service life of mooring chain links at the fairleads is obvious. The service lives of mooring chains in Scenarios 2 and 3 are decreased as 39% and 41%, respectively, compared to those of Scenario 1. The effect of boundary condition setting in the FE analysis of Scenarios 2 and 3 is also evident, the remaining service life with the fixed support in Scenario 2 is 5% longer than those with the simple support.

8. Conclusion

This paper was to perform a fracture mechanics based analysis for mooring chains of a semi-submersible taking account into the OPB effects. The WF and LF load processes were regarded as two random processes. The load combination of these two processes was considered and the narrow-banded method were used for calculating the combined mooring chain loading process. The effects of the OPB on mooring chain pricipal stresses were determined by a FE analysis. The remaining service life of mooring chain links was calcuated based on the FM analysis.The influence of the OPB on the remaining service life of mooring chain is investigated and the results show that the remaining service life of mooring chain was significantly decreased due to the OPB effects.

References

[1] Fontaine E, Kilner A, Carra C, Washington D, Ma K T, Phadke A and Kusinski G 2014 Industry survey of past failures, pre-emptive replacements and reported degradations for mooring systems of floating production units *Proc. Offshore Technology Conf. 2014 (Houston)*

[2] API RP 2SK 2008 *Design and analysis of stationkeeping systems for floating structure* (American Petroleum Institute)

[3] DNVGL OS E301 2015 *Offshore standard DNVGL-OS-E301: Position mooring* (Det Norske Veritas and Germanischer Lloyd SE)

[4] Vargas P M, Hsu T M and Lee W K 2004 Stress Concentration Factors for Stud-Less Mooring Chain Links in Fairleads *Proc. 23rd Int. Conf. on Ocean, Offshore and Arctic Eng. (Vancouver)*

[5] ter Brake E, van der Cammen J and Uittenbogaard R 2007 Calculation Methodology of out of Plane Bending of Mooring Chains *Proc. 26th Int. Conf. on Ocean, Offshore and Arctic Eng. (San Diego)*

[6] Melis C, Jean P and Vargas P 2005 Out-of-plane bending testing of chain links *Proc. of 24th Int. Conf. on Ocean, Offshore and Arctic Eng. (Halkidiki)*

[7] Vargas P M and Jean P 2005 FEA of Out-of-Plane Fatigue Mechanism of Chain Links," *Proc. of 24th Int. Conf. on Ocean, Offshore and Arctic Eng. (Halkidiki)*

[8] Lassen T, Storvoll E and Bech A 2009 Fatigue Life Prediction of Mooring Chains subjected to Tension and Out of Plane Bending *Proc. of 28th Int. Conf. on Ocean, Offshore and Arctic Eng. (Honolulu)*

[9] Lassen T, Aarsnes J and Glomnes E 2014 Fatigue Design Methodology for Large Mooring Chains Subjected to Out-of-Plane Bending *Proc. of 33rd Int. Conf. on Ocean, Offshore and Arctic Eng. (San Diego)*

[10] Wu Y, Wang T, Eide Ø and Haverty K 2015 Governing factors and locations of fatigue damage on mooring lines of floating structures *Ocean Eng.* **96** pp 109-124

[11] Ihc website: https://www.royalihc.com.

[12] Wang J, Luo Y H and Lu R 2002 Truss Spar structural design for west Africa environment *Proc. of 21st Int. Conf. on Ocean, Offshore and Arctic Eng. (Oslo)*

[13] Ochi M K and Hubble E N 1976 Six-Parameter Wave Spectra *Proc. of 5st Coastal Eng. Conf. (Honolulu)*

[14] Deaves D M and Harris R I 1978 *A mathematical model of the structure of strong wind* (CIRIA Report no.76)

[15] Paris P C and Erdogan F 1963 A critical analysis of crack propagation laws *J. Basic Eng.* **85(4)** pp 528–534

[16] BS 7910 2013 *Guide on methods for assessing the acceptability of flaws in fusion welded structures* (British Standard Institute)

[17] Mathisen J and Larsen K 2004 Risk-based inspection planning for mooring chain *J. Offshore Mechanics and Arctic Eng.* **126** pp 250-257

IOP Conf. Series: Materials Science and Engineering **276** (2017) 012037 doi:10.1088/1757-899X/276/1/012037

Effect of corrosion on the buckling capacity of tubular members

F H Øyasæter*, A Aeran, S C Siriwardane and O Mikkelsen

Department of Mechanical and Structural Engineering and Materials Science, University of Stavanger, Norway

*Contact author: fridtjof@bergoya.no

Abstract. Offshore installations are subjected to harsh marine environment and often have damages from corrosion. Several experimental and numerical studies were performed in the past to estimate buckling capacity of corroded tubular members. However, these studies were either based on limited experimental tests or numerical analyses of few cases resulting in semi-empirical relations. Also, there are no guidelines and recommendations in the currently available design standards. To fulfil this research gap, a new formula is proposed to estimate the residual strength of tubular members considering corrosion and initial geometrical imperfections. The proposed formula is verified with results from finite element analyses performed on several members and for varying corrosion patch parameters. The members are selected to represent the most relevant Eurocode buckling curve for tubular members. It is concluded that corrosion reduces the buckling capacity significantly and the proposed formula can be easily applied by practicing engineers without performing detailed numerical analyses.

1. Introduction

Offshore structures are subjected to a very harsh and corrosive environment during their lifetime [1]. Corrosion is a major hazard for offshore structures, especially for structures in the splash zone [2]. Corrosion, as a time-dependent degradation process is highly relevant to aging offshore structures [3]. Ageing is becoming an increasing concern in the offshore industry and the majority of fixed offshore platform on the Norwegian Continental Shelf (NCS) and United Kingdom Continental Shelf (UKCS) are currently being operated beyond the design life [4]. It is therefore important to investigate the effect of structural degradation on the buckling capacity of aging offshore installations.

Structural degradation can occur in many forms but is mainly characterized by corrosion and reduction of member thickness with time. Corrosion is one of the major structural degradation phenomenon especially in the marine environment and can reduce the buckling capacity of members [5]. Corrosion can broadly be classified into localized and uniform corrosion. Localized corrosion such as crevice and pitting may not have a significant effect on the overall strength but can cause local stress concentrations and fatigue cracking [5]. Uniform is the most common form of corrosion [6] and causes uniform reduction of member thickness. Corrosion of offshore structures is normally counteracted using protective coating or a cathodic protection system (CPS) consisting of several sacrificial anodes. Also, a corrosion allowance is generally taken for the members in the splash zone where the rate of corrosion is the highest [3]. However, CPS has a typical life of only 5 to 15 years [5] and is ineffective in the splash zone due to intermittent action of waves and tides [4]. Moreover, the maintenance and repair of CPS for fixed offshore structures are generally very costly and sometimes impractical [7]. In addition, the corrosion allowance is normally taken for the initial design life and not for the extended life, as is the case with ageing offshore structures [8].

Tubular members are often used to construct the primary load bearing system of offshore structures, especially for bottom fixed jacket structures and wind turbines [9]. The advantage of tubular members is low drag resistance and higher utilization of bearing capacity against the self-weight. The effect of corrosion on the ultimate load capacities of tubular members has become a growing concern [1]. Over the past decades, a significant amount of research has been carried out on the buckling capacity of corroded tubular members. Ostapenko et al. [10] studied the residual strength of corroded and dented tubular members by conducting test on several salvaged, fabricated, and small-scale manufactured column specimens. A procedure to estimate corroded cross-sections for local buckling was also proposed. Ostapenko and Gulec [11] analyzed the effect of multiple corrosion patches on the buckling capacity of tubular members and proposed a simplified procedure for estimating the capacity of such members. Lutes et al. [12] assessed the residual strength of corroded tubular using approximate formulas given by the American Petroleum Institute (API) [13]. These formulas consider global buckling accounting for inelasticity. Hebor and Ricles [1] carried out an analytical and experimental study of the strength of patch-corroded short tubular steel bracing members. The result of this study was used to derive a semi-empirical formula for estimating the strength of such members.

Okada et al. [14] investigated the deterioration of strength of tubular members with perforation due to corrosion and proposed a simplified method to estimate the residual strength of such members. Yamane et al. [15] examined the ultimate strength of steel tubular members with uniform corrosion and patch corrosion under coating paint, both numerically and experimentally. The applicability of existing design formulas for finding residual strength was also examined. Nazari et al. [9] investigated the strength of patch corroded tubular members and performed a parametric study for the different parameters describing a corrosion patch. A semi-empirical formula to predict the capacity was derived. Most of the proposed formulas in previously mentioned studies are semi-empirical. These semi-empirical relations were either proposed based on very limited experimental data or using results from numerical analyses on a few cases. Moreover, most of the earlier proposed formulas does not considered the effect of initial imperfections for assessing the buckling capacity of corroded tubular members.

In addition to research articles, several codes are available for assessment of existing structures [16] [17]. These assessment codes recommend uniform thickness wastage of the member surface. However, corrosion located in some parts of the member can be more critical for buckling capacity due to the load eccentricity induced by such patch corrosion. There are no generalized guidelines available for finding the buckling capacity of tubular members subjected to patch corrosion. Also, it is computationally demanding and time consuming for practicing engineers to model patch corrosion in detail to assess the buckling capacity of these tubular members. It is therefore necessary to have an accurate analytical formula to determine the buckling capacity of corroded tubular members which can be easily applied by practicing engineers, without performing advanced numerical analyses.

This paper proposes a new analytical formula for estimating the buckling capacity of corroded tubular members. This formula also includes the effect of initial geometrical imperfections and can be applied to members subjected to uniform corrosion. The proposed formula is verified using result from finite element analysis (FEA) performed on several corroded members having different combinations of patch corrosion. One of the most relevant buckling curve from Eurocode 3 (EC 3-1-1) [18] for hollow cross-sections is used as reference for comparing results from corroded and uncorroded tubular members. This buckling curve is reproduced using finite element (FE) simulation results from a set of uncorroded tubular members. The paper starts by discussing the relevant Eurocode buckling curve for uncorroded tubular members and deriving it using results from FEA. Then, the procedure for performing nonlinear buckling analyses using FEA on corroded tubular members is explained and the result from a study of corrosion patch parameters is presented. Next, the proposed formula is derived and presented in detail. Finally, the verification study of the proposed formula is presented.

2. Eurocode buckling curves for uncorroded tubular members

The Eurocode buckling curves consider flexural buckling of columns with initial imperfections and is used to find buckling capacity resistance of columns according to EC 3-1-1. The various initial imperfections are represented using a single equivalent imperfection with sinusoidal shape [18]. The different buckling curves represent different values for the amplitude of this sinusoidal shaped imperfection.

Buckling curve "a" given in EC 3-1-1 [18] represents hollow cross-sections with yield stress (f_y) less than 460 MPa that has been produced by hot-rolling [19]. This is one of the most relevant curves for tubular members used in offshore structures and is therefore used as a reference for comparing the buckling capacity for uncorroded and corroded tubular members.

2.1. Considered cases of tubular members

Curve "a" is derived using results from FEA of a set of uncorroded members. The set consist of eight members with different values for relative slenderness ($\bar{\lambda}$) and is selected to represent buckling curve "a". Table 1 shows the dimensions such as dimeter (D), thickness (t) and length (L) of these members. The cross-sectional class and relative slenderness ($\bar{\lambda}$) of the selected tubular members are also shown. Each value of relative slenderness has a corresponding value for the reduction factor (χ). The reduction factor (χ) is multiplied with the axial yield capacity to find the flexural buckling capacity.

Table 1. Dimensions of the selected tubular members

No.	D (mm)	t (mm)	L (m)	Class	$\bar{\lambda}$
1	1800	35	14.307	3	0.3
2	1800	35	19.076	3	0.4
3	1200	30	18.971	2	0.6
4	1200	30	25.294	2	0.8
5	1000	40	25.957	1	1.0
6	1000	40	31.148	1	1.2
7	500	35	20.156	1	1.6
8	500	35	25.195	1	2.0

The buckling curve "a" given in EC 3-1-1 [18] and the corresponding slenderness ratios/points of eight selected member geometries on the curve is shown in Figure 1. The selected members have a relative slenderness ranging from 0.3 to 2.0. Values between 2.0 and 3.0 are excluded to only represent the "usable" part of the buckling curve. The curve "a" is only valid for members with cross-sectional classes 1, 2 and 3 that are subjected to global/flexural buckling.

2.2. Finite element analysis of uncorroded members

Initial imperfections are included as a single geometrical imperfection with sinusoidal shape. This imperfection is introduced by first performing an eigenvalue buckling analysis of each of the uncorroded members, which result in a large sinusoidal shaped deformation of the members and finally scaling the deformation to achieve the same amplitude of the initial imperfections as in EC 3-1-1 [18]. Figure 2 shows the deformed shape after performing an eigenvalue buckling analysis.

The selected material is steel grade of S355 and material properties are given in Table 2. The material behavior is assumed to be elastic-perfectly plastic. The members are modelled as pinned-ended columns and loaded by controlling displacement. The members are meshed as a single body using the sweep-method in ANSYS version 17.2. This method creates a hexahedral mesh, where the number of nodes and elements is much less than if a free meshing method is applied, which makes it very efficient [20]. The element type is SOLID186 (i.e. 20-node brick element) [20] and element size is set based on a mesh convergence study.

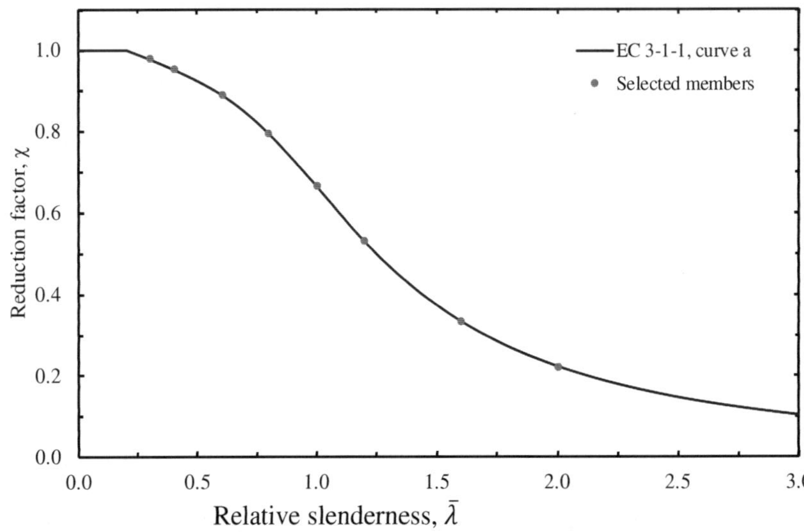

Figure 1. Eurocode buckling curve "a" with selected members indicated along the curve

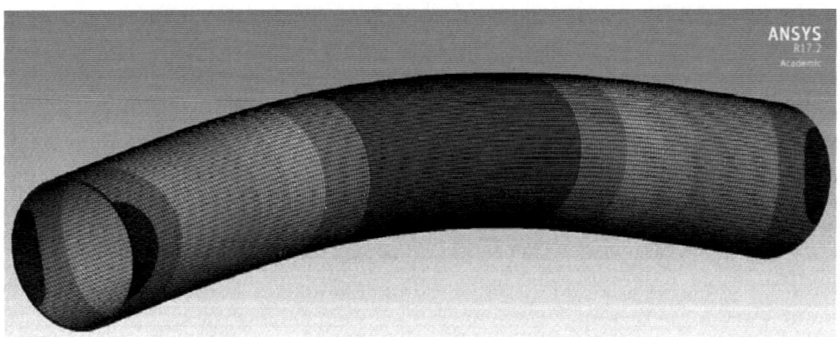

Figure 2. Deformation of a tubular member subjected to Euler's critical buckling load in an eigenvalue buckling analysis

Table 2. Material properties of reference material

Steel type	Yield stress (MPa)	Young' modulus (MPa)	Poison's ratio
S355	355	210000	0.3

2.3. Comparison of EC buckling curves with FEA results

The nonlinear buckling analysis is performed for each of the considered members shown in Table 1. The results are used to reproduce buckling curve using ANSYS employed FE simulations. Figure 3 shows both the reproduced buckling curve and the buckling curve from EC 3-1-1 [18]. The maximum deviation between the two curves is around 2 %, where buckling curve from EC 3-1-1 [18] gives slightly lower buckling capacity than from FEA.

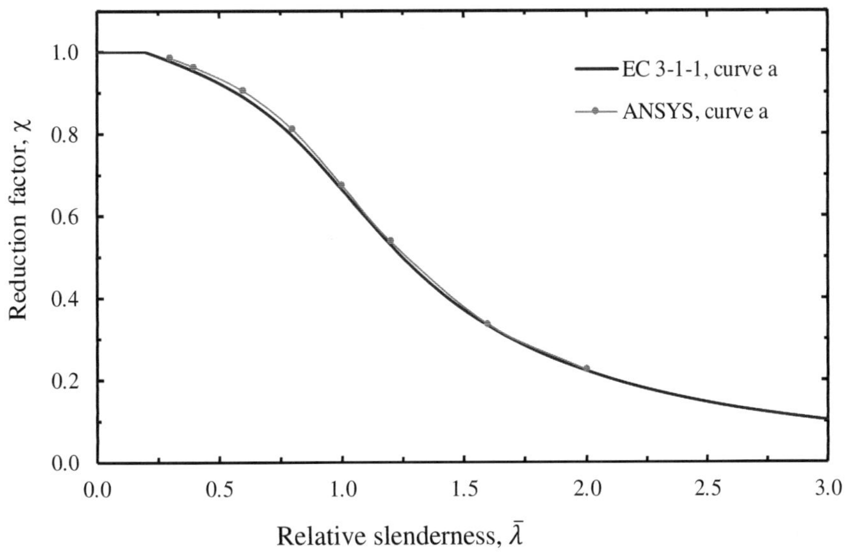

Figure 3. Buckling curve "a" using analytical formulas from EC 3-1-1 [18] and FE-analysis in ANSYS

3. Finite element analysis of corroded tubular members and parametric study of corrosion patch parameters

This section describes the procedure for modelling corroded tubular members and performing nonlinear FEA on these members using ANSYS. The result from a study of the corrosion patch parameters is also presented. The aim this study is to investigate the effect of patch parameters on buckling capacity of corroded tubular members. The corrosion patch is conservatively assumed to be rectangular shaped by combining several irregular shaped patches.

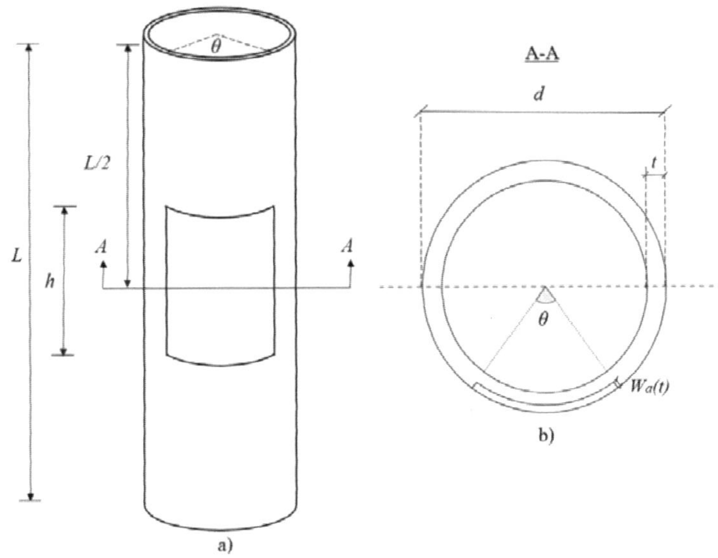

Figure 4. Schematic representation of corroded tubular members: a) Tubular member with a corrosion patch. b) Cross-section at mid-span

The patch can be described by three different parameters. These are height (*h*), angular extent (*θ*) and the thickness wastage (W_a). Figure 4 show a tubular member with outer diameter (*d*), wall thickness (*t*) and a corrosion patch centered in the mid-span.

3.1. Corrosion wastage model

The thickness wastage from corrosion is found using a recently proposed corrosion wastage model [5]. The model is given in equation (1):

$$W_a(t) = R_{corr}\alpha(t - t_{pt}) \tag{1}$$

where, $W_a(t)$ is the thickness wastage as a function of time, R_{corr} is the rate of corrosion (mm/year), α is parameter that is equal to 1 or 2 depending if the corrosion is one-sided or two-sided, *t* is the time in years and t_p is the corrosion protection time in years. The corrosion rate (R_{corr}) in the splash zone is taken as 0.3 mm/year according to design standards [21]. Thickness wastage can also be expressed as an equivalent number of years (or age).

3.2. Modelling of corroded members in ANSYS

The models of the tubular members to be used in the nonlinear buckling analyses include both the sinusoidal shaped initial imperfection and the corrosion patch. The corrosion patch is assumed to be positioned at the outer surface (single-sided corrosion) at the mid-span of the tubular member. Circumferentially, the patch is placed in the opposite direction of the lateral deflection from the initial imperfection.

3.2.1. Modelling of corrosion patch

The corrosion patch is modelled using the CAD-software SolidEdge. The patch is first projected onto the surface of the member and then created using a cut-out tool in SolidEdge, as shown in Figure 5.

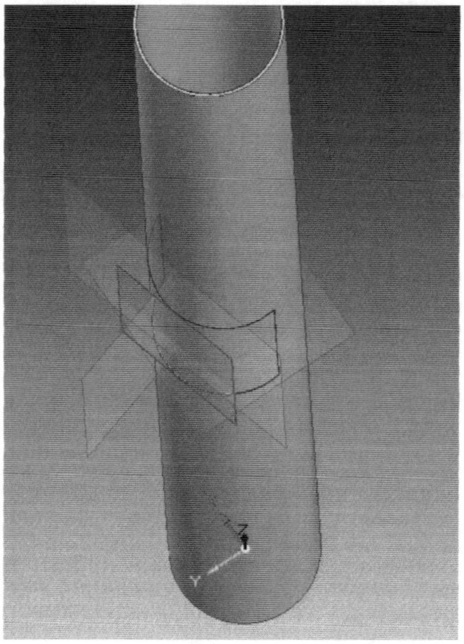

Figure 5. Modelling corrosion patch on a tubular member

3.2.2. Meshing procedure

The initial imperfection and corrosion patch, results in a very irregular shape that it is not possible to mesh as a single body using the "sweep-method". Instead, the geometry of the tubular members is split into "sweepable" bodies, as shown in Figure 6. The "sweepable" bodies are mostly meshed with SOLID186 and SOLID187 (i.e. 10-node tetrahedral element). The two mentioned element types support both geometric and material non-linearities [20].

The optimal element size is determined by performing a convergence study. The convergence study is limited to varying element size in the "sweepable" bodies and using a fixed element size in the small intermediate body. The geometry of a corroded tubular member consists of four different "sweepable" bodies. Two cylindrical bodies in each end of the tubular member and two half cylinders in the corroded part of tubular member. The two bodies in the corroded parts is meshed with a finer mesh than the two bodies in each end. Table 3 shows the different combination of mesh sizes and results from the mesh convergence study where the critical load is taken as the axial load that initiates yielding. The mesh convergence study is performed on the first member from Table 1. The corrosion patch had a height $h = L/2$, angle $\theta = 180^0$ and wasted thickness $W_a = 7.5$ mm. Case 3 from Table 3 is selected as a combination of mesh sizes.

Table 3. Critical buckling load for different combinations of mesh sizes

Case	Mesh size (mm), end bodies	Mesh size (mm), bodies in corroded part	Critical load (MN)
1	300	150	54.028
2	200	100	53.950
3	100	50	53.689
4	80	40	53.523
5	60	30	53.534

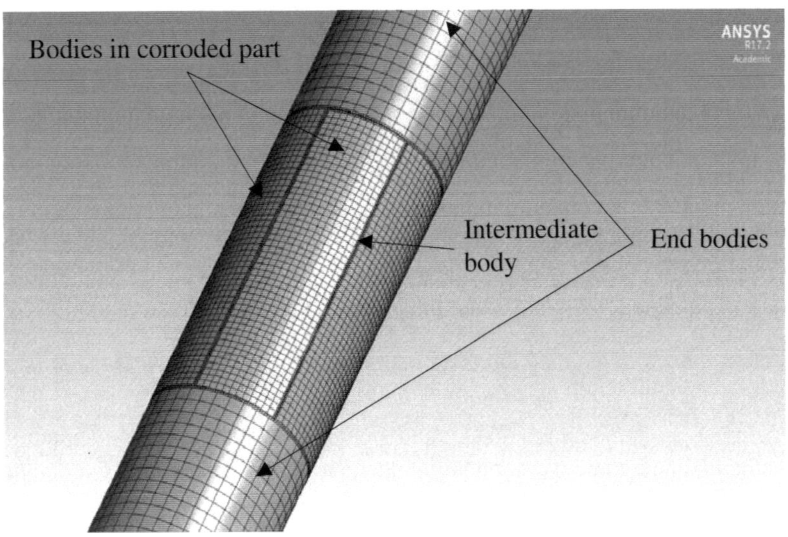

Figure 6. FE mesh of a corroded member (member split into multiple bodies)

3.3. Considered cases for parametric study

The aim of this parametric study is to investigate the effect of corrosion patch parameters (W_a, h and θ) on the buckling capacity of a tubular member. It is also an aim to find the most critical position of the patch on the tubular member. The study is carried out by varying the value of each patch parameter, keeping the two other parameters constant and performing a FEA for each case. All testing of patch parameters is done on the same member, which is the first of the selected member in Table 1. Each of the test is given a test number and all the results are listed in Table 4.

Table 4. Result from corrosion patch parameter testing

Parameter	Test No.	*h*/L	θ (^{0})	W_a (mm)	N_{cr} (MN)
	1	1	180	7.5	54.069
	2	1/2	180	7.5	53.693
	3	1/4	180	7.5	53.717
h	4	1/8	180	7.5	53.262
	5	1/16	180	7.5	53.145
	6	1/32	180	7.5	53.653
	7	1/64	180	7.5	57.066
	8	1/4	30	7.5	64.738
	9	1/4	60	7.5	61.650
θ	10	1/4	120	7.5	56.072
	11	1/4	180	7.5	53.717
	12	1/4	360	7.5	49.899
	13	1/4	180	1.5	65.069
	14	1/4	180	4.5	59.937
W_a	15	1/4	180	7.5	53.693
	16	1/4	180	10.5	47.073
	17	1/4	180	13.5	41.014

3.3.1. Effect of corrosion patch height on the critical load. The result from testing the effect of height on the critical load is plotted and shown in Figure 7. It can be seen that the critical load is almost independent of the corrosion patch height.

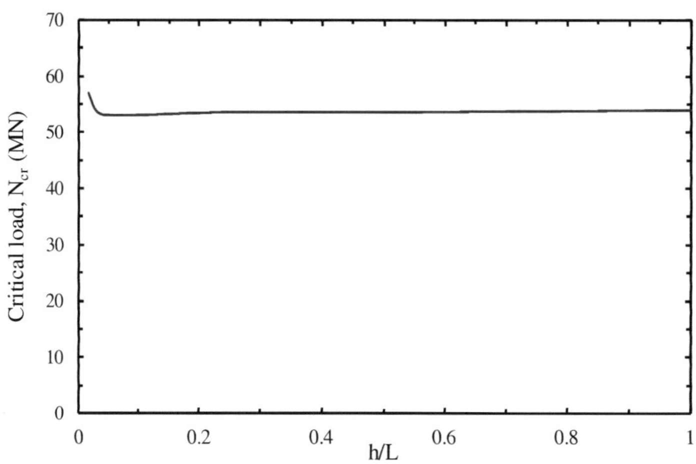

Figure 7. Variation of critical load with height of corrosion patch

3.3.2. Effect of corrosion patch angle on the critical load. Figure 8 shows the way critical load varies with angular extent of the corrosion patch. It can be seen that the load is very dependent on the angular extent of the corrosion patch. This shows that increasing angle results in a lower buckling capacity.

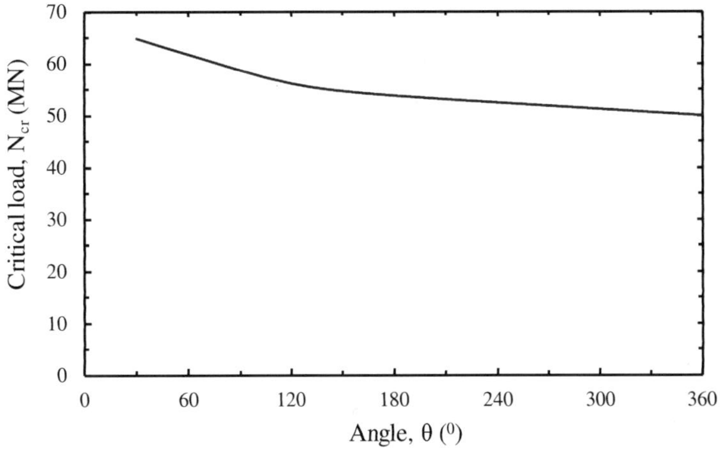

Figure 8. Variation of critical load with the corrosion patch angle

3.3.3. Effect of corrosion thickness wastage on the critical load. The variation of the critical load with different values for thickness wastage is plotted in Figure 9. It can be seen that the critical load is significantly dependent on the wasted thickness of the corrosion patch and indicates a linear relationship between these two parameters.

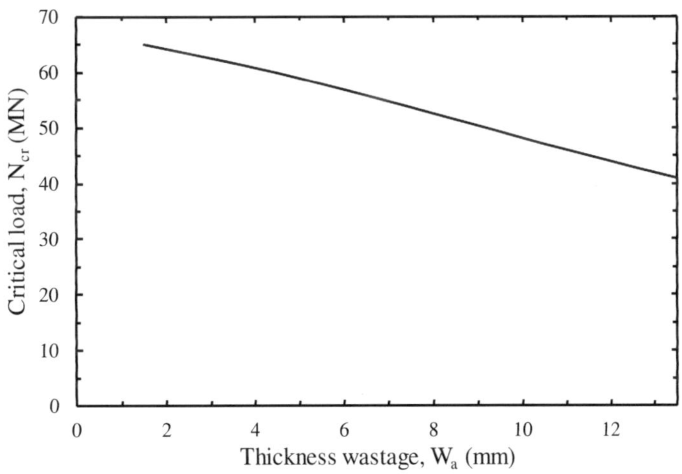

Figure 9. Variation of critical load with thickness wastage of corrosion patch

3.3.4. Critical position of the corrosion patch. As mentioned earlier, the initial imperfections are represented by a single equivalent imperfection with a maximum value of load eccentricity in the mid-span. The most critical position of the corrosion patch is theoretically when the patch is centered at the mid-span of the member. This has been confirmed in a study by Nazari et al. [9] and therefore not tested any further here. The corrosion patch can be placed anywhere around the circumference of the cross-section. The maximum eccentricity of corroded cross-section can be achieved when the corrosion patch is located most far away from initial longitudinal axis of the member.

The maximum eccentricity governs the buckling capacity of the corroded tubular member. Two different positions are tested using results from FEA. The position where the corrosion patch is located most away from initial longitudinal axis is called "back" and the position where it is placed nearest is called "front". The two back and front side positions have been tested for three different cases and the results are shown in Table 5. The results show that the most critical position of the patch is located circumferentially opposite side of the initial geometric imperfection ("back") and thus most for away from initial longitudinal axis. This position is used for modelling corrosion for the verification study in the Section 5. Figure 10 shows a tubular member with a corrosion patch located at his position.

Table 5. Result from testing of patch position

d (m)	*t* (mm)	*L* (m)	θ (^{0})	W_a (mm)	*h/L*	Critical load, N_{cr} (MN)	
						Back side	Front side
1.8	35	14.307	180	7.5	0.5	53.693	55.074
1.8	35	14.307	180	7.5	0.25	53.717	55.582
0.5	35	20.156	180	7.5	0.5	3.4190	3.7709

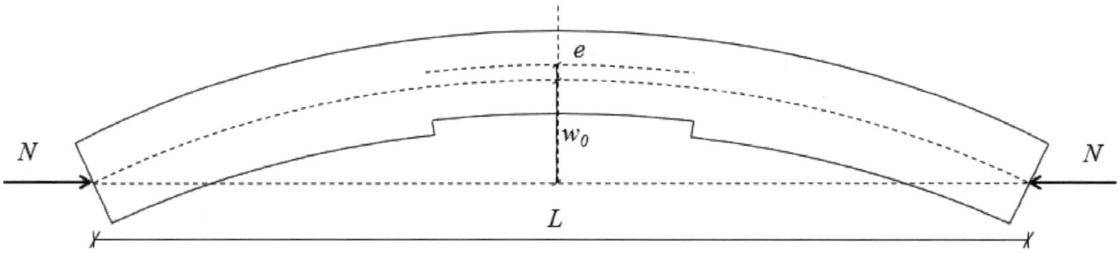

Figure 10. Tubular member with an equivalent initial imperfection and a corrosion patch centred at the mid-span and on the "back" position.

4. Derivation of formula for estimating buckling capacity of corroded tubular members

This paper derives a new formula to include the effect of corrosion on the buckling capacity of tubular steel members. The proposed formula considers a tubular member with a single rectangular shaped patch of uniform corrosion centered at the mid-span, as shown in Figure 4. The patch is described by the height (h), angular extent (θ) and the thickness wastage (W_a) as described in the previous section. Initial imperfections are included by assuming that the member have an equivalent geometrical imperfection with sinusoidal shape. The corrosion patch is centered at the mid-span of the member because the initial bending stresses from the initial imperfection has a maximum value at the mid-span and this will be the critical cross-section (i.e. most stresses cross-section). The corrosion patch is positioned in the opposite direction of the lateral deflection to have maximum eccentricity of the axial load, resulting in maximum stresses. The formula is derived on the assumptions of linear elastic material behavior and corresponding failure occurs when the most heavily loaded cross-section reaches yield strength of the material.

4.1. Cross-sectional properties of corroded tubular members

A corroded cross section where W_a is the thickness wastage, θ is the angular extent and e is the shifted distance of the neutral axis is shown in Figure 11. The neutral axis is shifted because the corrosion patch leads to non-uniform reduction of cross-sectional area. The surface of the corrosion patch is assumed to be uniformly corroded.

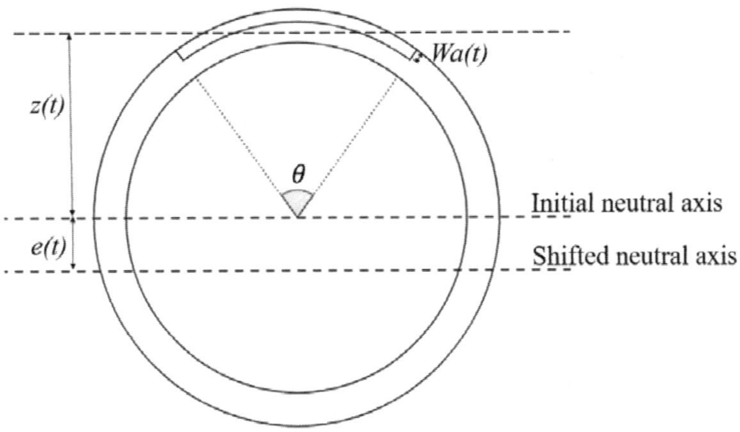

Figure 11. Cross-section of the corroded part of a tubular member

The reduced cross-sectional area as a function of time t is given in equation (2):

$$A_{red}(t) = L_{red} W_a(t) \tag{2}$$

where L_{red} is the arc-length of the corrosion patch and $W_a(t)$ is the thickness wastage. The distance from the neutral axis of the corrosion patch to the neutral axis of the original cross-section (z) is found using equation (3).

$$z(t) = \frac{d \sin\left(\frac{\theta}{2}\right)}{\theta_{rad}} \tag{3}$$

where A_{eff} is the effective area of the corroded cross-section. The effective moment of inertia can be found using equation (4).

$$I_{eff}(t) = I_0 + A_0 e(t)^2 - \left[\Delta I + A_{red}\left(z(t) + e(t)\right)^2\right] \tag{4}$$

where I_0 is the moment of inertia of the uncorroded cross-section, A_0 is the cross-sectional area of the uncorroded cross-section and ΔI is the moment of inertia of the corrosion patch. The section modulus of the corroded cross-section is found using equation (5):

$$W_{eff}(t) = \frac{I_{eff}(t)}{y_{max}(t)} \tag{5}$$

where y_{max} is the maximum distance from the neutral axis to the outmost fiber of the cross-section.

4.2. Derivation of proposed formula

The corrosion patch is positioned so that the initial imperfection (w_0) and the shifting of neutral axis (e) causes maximum stresses at the mid-span (as shown in Figure 9). The first order bending moment (M) in the mid-span is found using equation (6).

$$M = N\left(w_0 + e\right) \tag{6}$$

The second order bending moment in the mid-span is found by using the amplification factor (*AF*) according to EC 3-1-1 [18]. The amplification factor can be found using equation (7):

$$AF = \left(1 - \frac{N}{N_E}\right)^{-1} \tag{7}$$

where N_E is Euler's critical load found using corroded cross-sectional properties. *AF* is exact for the equivalent initial imperfection, but not for the shifting of neutral axis. As a simplification, the amplification factor in equation (7) is used for both the initial imperfection and the shifting of neutral axis. The total bending moment in the mid-span is found using equation (8):

$$M = N\left(w_0 + e\right)\left(1 - \frac{N}{N_E}\right)^{-1} \tag{8}$$

The following failure criteria is used for a tubular member with elastic material behavior subjected to combined loading:

$$\frac{N}{N_{Rk}} + \frac{M}{M_{Rk}} \leq 1.0 \tag{9}$$

where N_{Rk} is the characteristic axial load capacity and M_{Rk} is the characteristic bending moment capacity of the corroded cross-section. The failure criteria can be written as shown in equation (10):

$$\frac{N}{A_{eff} f_y} + \frac{N\left(w_0 + e\right)\left(1 - \frac{N}{N_E}\right)^{-1}}{W_{eff} f_y} \leq 1.0 \tag{10}$$

where f_y is the yield stress for the material used in the tubular member.

The critical axial load is found by solving equation (11) with respect to the unknown axial load (N). This results in a second order equation with the following possible solution:

$$N_{1,2} = \frac{1}{2W_{eff}}\left(A_{eff} N_E e + A_{eff} N_E w_0 + A_{eff} W_{eff} f_y + N_E W_{eff} \pm \sqrt{Z}\right) \tag{11}$$

$$Z = \begin{array}{c} A_{eff}{}^2 N_E{}^2 e^2 + 2A_{eff}{}^2 N_E{}^2 e w_0 + A_{eff}{}^2 N_E{}^2 w_0{}^2 + 2A_{eff}{}^2 N_E W_{eff} e f_y \\ +2A_{eff}{}^2 N_E W_{eff} f_y w_0 + A_{eff}{}^2 W_{eff}{}^2 f_y + 2A_{eff} N_E{}^2 W_{eff} e \\ +2A_{eff} N_E{}^2 W_{eff} w_0 - 2A_{eff} N_E W_{eff}{}^2 f_y + N_E{}^2 W_{eff}{}^2 \end{array} \tag{12}$$

The critical load (N_{cr}) is the lowest load that satisfies the failure criteria given in Equation (11) and is the second solution of Equation (11). To simplify this equation, the following imperfection parameter is introduced:

$$\eta_{corr} = \frac{A_{eff}(e+w_0)}{W_{eff}} \tag{13}$$

The imperfections parameter is a modified version of the imperfection parameter from Perry-Roberson's equation [22]. Introducing this parameter into second solution in equation (11) result in formula with similar format as the Perry-Robertson's equation that takes corrosion into account:

$$N_{cr} = \frac{N_{Rk} + N_E(1+\eta_{corr}) - \sqrt{\left(N_{Rk} + N_E(1+\eta_{corr})\right)^2 - 4N_{Rk}N_E}}{2} \tag{14}$$

The proposed formula in Equation (14) can easily be used by practicing engineers to find the critical buckling capacity of corroded tubular members without the need for performing advanced numerical analyses.

5. Verification of proposed formula

The proposed formula is verified using result from FE-analyses of tubular members with different combinations of patch corrosion. The combinations are created by varying the values of the different corrosion patch parameters.

5.1. Selection of cases for verification study

The set of tubular members given in Table 1 is also used in the verification study. Curve "a" is therefore used for quantifying the effect of corrosion on the buckling capacity and to quantify the difference between result obtained from the proposed formula and the corresponding result from FEA. The material data from Table 2 is also used for the verification study. The effect of corrosion is studied by performing FE analyses for different cases of patch corrosion. As mentioned in Section III, corrosion case is a combination of values for the three different patch parameters (h, θ and Wa). The patch parameter study indicated that the critical load is independent of patch height. Therefore, the patch height is kept fixed to a value of 1/2 of the length of the tubular member. For the other two parameters, it is selected four different values for patch angle and four values for thickness wastage. Combining all these, 128 different corrosion cases of the tubular members are considered for nonlinear FEA. The values in Table 6 are selected based on the patch parameter study conducted in the Section 3.

Table 6. Overview of the number of FE analyses in the verification study

No. of members	θ	h	W_a (mm)	Age (years)	Number of analyses
8	60, 180, 120, 380	L/2	1.5, 4.5, 7.5, 10.5	10, 20, 30, 40	128

5.2. Comparison of results obtained from proposed formula and from FEA

The results of the verification study are presented as four different plots for the different values of the patch angle, as shown in Figure 12 to Figure 15. The plots show different buckling curves, where each curve represents a specific value of thickness wastage. Each of the plots presents the buckling curves from FEA results and corresponding curves obtained using the proposed formula. The buckling curves from the proposed formula are shown as a solid thick line and the buckling curves from FEA results is shown as dotted thin lines.

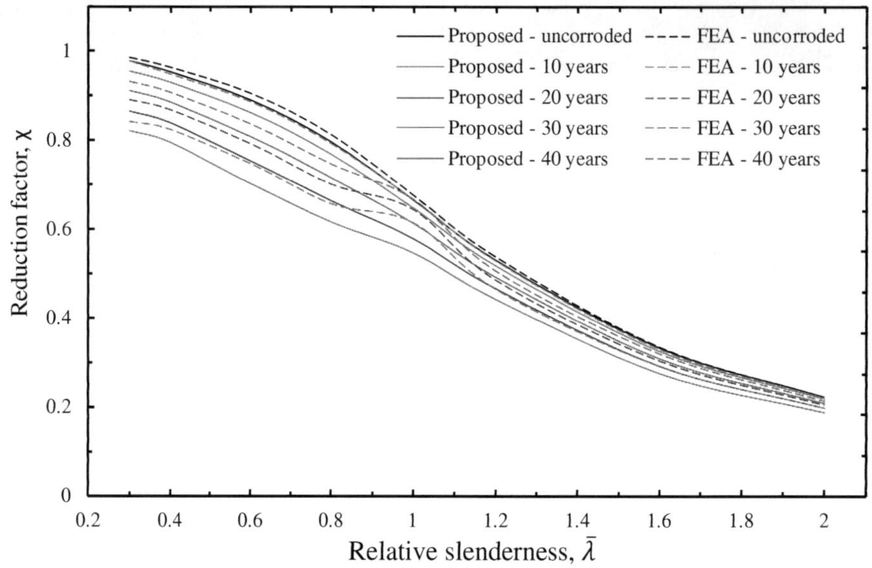

Figure 12. Comparison of results from FEA and using the proposed formula for h = L/2 and $\theta = 60^0$

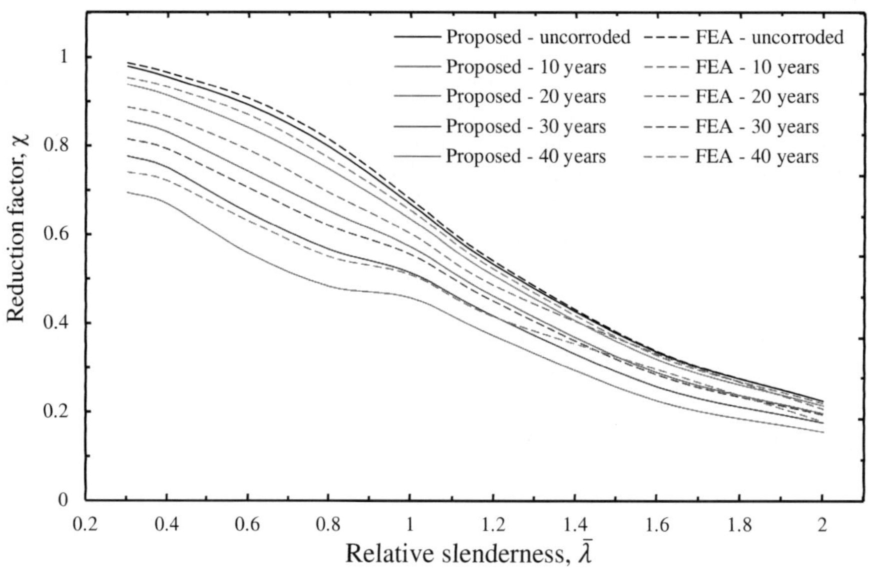

Figure 13. Comparison of results from FEA and using the proposed formula for h = L/2 and $\theta = 120^0$

5.3. Discussion of the results.
This section discusses the validity of the proposed formula for determining buckling capacities of corroded tubular members after comparing the results obtained by FEA. The effects of corrosion patch angle and thickness wastage on critical load are also discussed in the later part of the section.

5.3.1. Comparison of proposed formula and results from FEA. The critical load obtained from the proposed formula is always lower than the load from FEA, as shown in Figure 16. The deviation between result from FEA and the proposed formula seems to be at maximum at a patch angle of 180^0 and minimum at an angle of 360^0. The maximum, minimum and average deviation is 22,7%, 0,8% and 6,1%

respectively. The deviation is possibly due to some degree of inelastic buckling in the FEA results, which is not considered in the proposed formula.

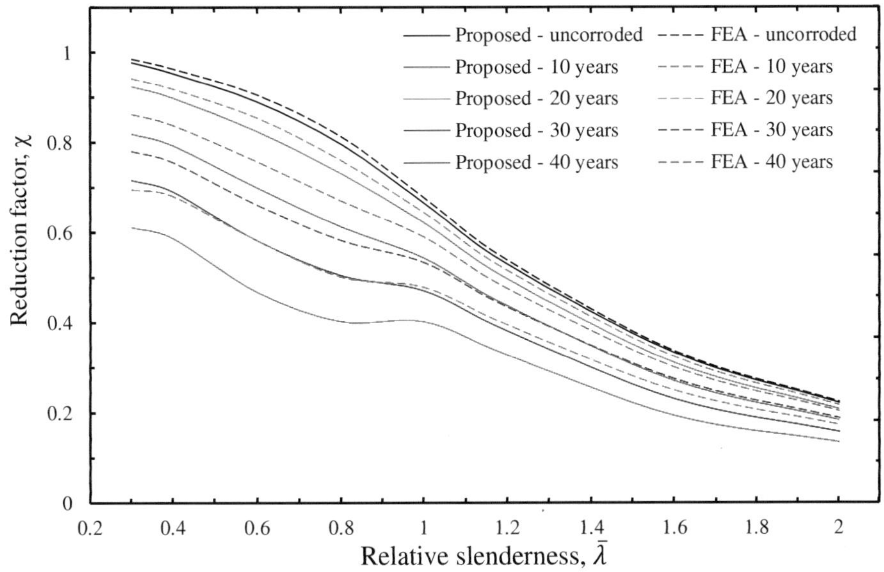

Figure 14. Comparison of results from FEA and using the proposed formula for h = L/2 and $\theta = 180^0$

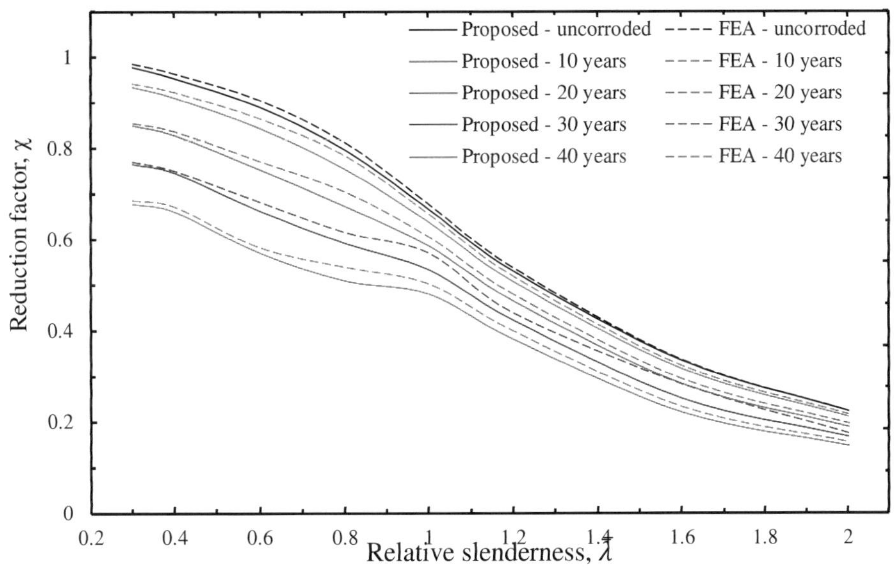

Figure 15. Comparison of results from FEA and using the proposed formula for h = L/2 and $\theta = 360^0$

5.3.2. Effect of corrosion patch angle on the critical load. The patch parameter study indicated that the critical load decreases with increasing angular extent (θ). The effect of patch angle is further investigated by comparing the buckling curves from different values for patch angle. This is done by keeping thickness wastage constant. The largest value for thickness wastage is selected for this comparison to maximize the differences between the curves. This resulted in a plot shown in Figure 17. An angular extent of 120^0 is more critical than 60^0. Also, an angle of 180^0 or 360^0 is more critical than 120^0. Increasing the angular extent (θ) from 0^0 to 180^0 leads to both reduction of cross-sectional area and shifting of the neutral axis. The shift is reduced when increasing the angle 180^0 to 360^0 and is zero for angle equal to 360^0. It can be concluded that increasing angle results a lower critical load for angles up to 180^0.

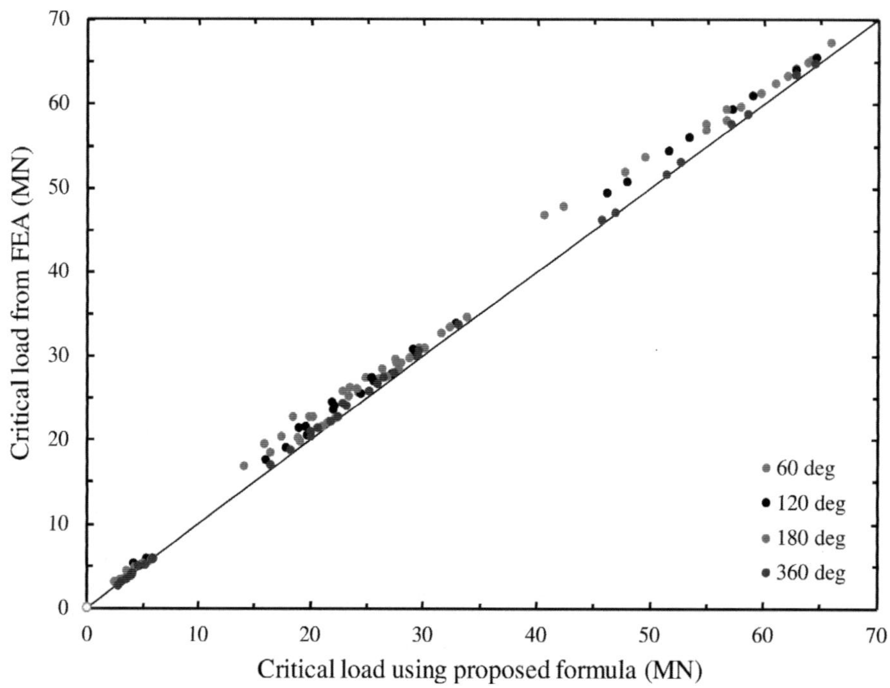

Figure 16. Comparison of critical load from FEA and using proposed formula for patch height h = L/2

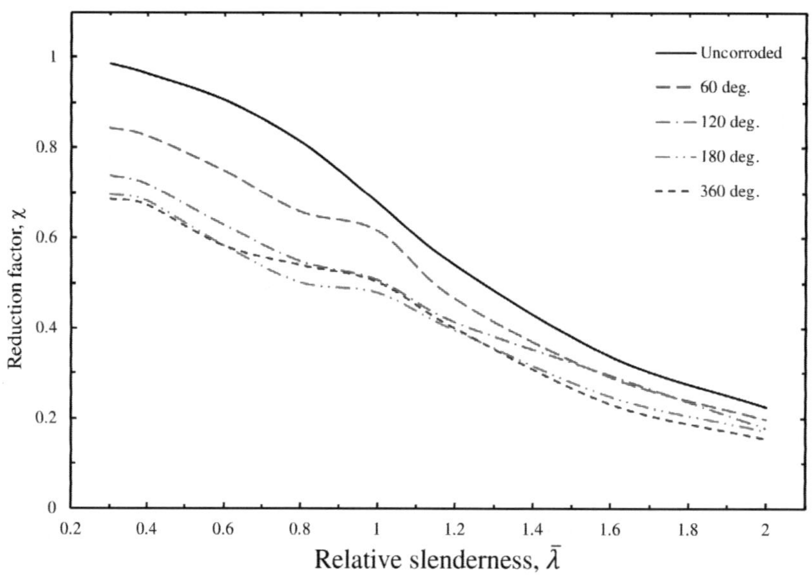

Figure 17. Effect of corrosion patch angle on critical load for h = L/2 and W_a = 10.5 mm (40 years)

5.3.3. Effect of corrosion thickness wastage on critical load. The maximum reduction of the critical load can be found using the FEA result from uncorroded members and members with corrosion thickness wastage after an age of 40 years (W_a = 10.5 mm). This has been found for all the different values for angular extent (θ) of patch, for all the eight members. The results are presented in Table 7 and both the maximum, average reduction and maximum reduction factor (χ_{max}) for the eight different members is presented. The results from Table 7 shows that the maximum reduction of axial load capacity after 40 years of patch corrosion is 38.3 %, which is a quite significant reduction.

Table 7. Reduction of buckling capacity after 40 years of age

Angle, θ (^{0})	Maximum reduction of critical load (%)	Average reduction of critical load (%)	Maximum reduction factor, χ_{max}
60	19.1	14.2	0.657
120	32.5	24.3	0.548
180	38.3	29.8	0.501
360	35.7	30.5	0.582

6. Conclusions

In this paper, a formula is proposed to estimate the buckling capacity of corroded tubular members. This can be used as a tool to evaluate the residual life time of aged offshore structures built with tubular members. The proposed formula is verified with result from FE analyses. It is concluded that patch corrosion significantly reduced the buckling capacity with ageing and that the proposed formula can provide both an accurate and conservative prediction for the buckling capacity of corroded tubular member. The height of the corrosion patch has minor impact on the buckling capacity as buckling occurs in the weakest cross-section. However, both the patch angle and thickness wastage reduce the buckling capacity significantly. Finally, it is concluded that the proposed formula can be easily applied by practicing engineers for a reasonable accurate conservative estimation of residual buckling capacities of corroded tubular members, without the need of any advanced numerical analyses. It is recommended that the proposed formula should be further verified with experimental tests and more numerical analyses.

Acknowledgement

The funding for writing this article is supported by the Program Area for research "COTech - Computational Methods in Offshore Technology" which is founded by Faculty of Science and Technology, University of Stavanger.

References

[1] Hebor MF and Ricles JM 2002 Local buckling strength of patch corrosion damaged steel tubular bracing *J. Steel Struct.* **2** *pp 59-70*

[2] Smith M, Bowley C and Williams L 2002 In situ protection of splash zones - 30 years on *Mat. Perf.* **41** pp 30-33

[3] Stacey A, Birkinshaw M and Sharp JV 2008 Life extension for ageing offshore installations *Proc. of the 27th Int. Conf. on Off. Mech. and Art. Eng. (Portugal)*

[4] Aeran A and Gudmestad OT 2017 Guidelines for estimating remaining fatigue life of ageing offshore jacket structures *Proc. of the 36th Int. Conf. Ocean, Off. Artic Eng.* (*Trondheim*)

[5] Dong W, Moan T and Gao Z 2012 Fatigue reliability analysis of the jacket support structure for offshore wind turbine considering the effect of corrosion and inspection *Reliability Engineering & System Safety* **106** pp 11-27

[6] Melchers RE 1999 Corrosion uncertainty modelling for steel structures *J. of Construct. Steel Res.* **52** pp 3-19

[7] DNV-RP-B401 2010 *Offshore standard DNV-RP-B401: Cathodic protection design* (Det Norske

IOP Conf. Series: Materials Science and Engineering **276** (2017) 012037 doi:10.1088/1757-899X/276/1/012037

Veritas and Germanischer Lloyd SE)

[8] Faulkner D, Cowling JM and Incecik A 1991 *Integrity of offshore structures* (Glasgow: *CRC Press)*

[9] Nazari M, Khedmati MR and Khalaj FA 2014 A numerical investigation into ultimate strength and buckling behavior of locally corroded steel tubular members *Lat. Am. J. Sol. Struct.* **11**, pp 1063-76

[10] Ostapenko A, Wood BA et al 1993 *Residual strength of damaged and deteriorated offshore structures* (Betlehem: ATLSS Reports)

[11] Ostapenko A and Gulec O 1999 *Tubular columns with multiple corrosion patches* (Bethlehem: ATLSS Reports)

[12] Lutes LD, Kohutek TL, Ellison BK and Konen KF 2001 Assessing the compressive strength of corroded tubular members *Appl. Ocean Res.* **23** pp 263-68

[13] RP 2A-LRFD 1993 *International standard RP 2A-LRFD: Recommended practice for planning, designing and constructing fixed offshore platforms - load and resistance factor design* (American Petroleum Institute)

[14] Okada H, Masaoka K, Katsura S and Matsui T 2004 A study of deterioration of strength and reliability of aged jacket structures *Proc. of the OCEANS '04. MTTS/IEEE TECHNO-OCEAN '04 (Kobe*)

[15] Yamane M, Tanaka K, Matsuda N, Fujikubo M, Yanagihara D and Iwao N 2006 Residual strength evaluation of corroded steel members in marine environments *Proc. of the 16th Int. Off. Polar Eng. Conf. (San Francisco)*

[16] N-006 2015 *Offshore standard N-006: Assessment of structural integrity for existing offshore load-bearing structures* (NORSOK Standard)

[17] API RP 2SIM 2014 *API RP 2SIM Recommended Practice for Structural Integrity Management of Fixed Offshore Platforms* (American Petroleum Institute)

[18] Boissonnade N, Greiner R, Jaspart JP and Lindner J 2006 *Rules for Member Stability in EN 1993-1-1: Background documentation and design guidelines* (European Convention for Constructional Steelwork)

[19] Eurocode 3 2005 *Design of Steel Structures - Part 1-1: General rules and rules for buildings, NS-EN 1993-1-1* (European Standard)

[20] ANSYS Meshing User's Guide 2013 (ANSYS Inc.)

[21] ISO 19902 2007 *International standard ISO 19902: Fixed steel offshore structures* (European Committee for Standardization)

[22] Robertson A 1925 The strength of struts *Selected Engineering Papers* **1** (28)

IOP Conf. Series: Materials Science and Engineering **276** (2017) 012038 doi:10.1088/1757-899X/276/1/012038

An accurate fatigue damage model for welded joints subjected to variable amplitude loading

A Aeran*, S C Siriwardane, O Mikkelsen and I Langen

Department of Mechanical and Structural Engineering and Materials Science, University of Stavanger, Norway

*Contact author: ashish.aeran@uis.no

Abstract. Researchers in the past have proposed several fatigue damage models to overcome the shortcomings of the commonly used Miner's rule. However, requirements of material parameters or S-N curve modifications restricts their practical applications. Also, application of most of these models under variable amplitude loading conditions have not been found. To overcome these restrictions, a new fatigue damage model is proposed in this paper. The proposed model can be applied by practicing engineers using only the S-N curve given in the standard codes of practice. The model is verified with experimentally derived damage evolution curves for C 45 and 16 Mn and gives better agreement compared to previous models. The model predicted fatigue lives are also in better correlation with experimental results compared to previous models as shown in earlier published work by the authors. The proposed model is applied to welded joints subjected to variable amplitude loadings in this paper. The model given around 8% shorter fatigue lives compared to Eurocode given Miner's rule. This shows the importance of applying accurate fatigue damage models for welded joints.

1. Introduction

Fatigue is one of the main causes of failures in both onshore and offshore steel structures subjected to variable amplitude loading (VAL) [1]. The fatigue damage is generally determined using the Miner's rule due to its simplicity and ease of application [2]. Moreover, the current design codes and standards such as Eurocode [3], Det Norske Veritas [4] etc. also recommend its use in the design practice. However, Miner's rule may lead to inaccurate life predictions as it does not consider the damage due to loading sequence accurately [5][6]. Several damage theories have been proposed since then to overcome these shortcomings.

Among the first improvements are the nonlinear Palmgren-Miner rule [7] and the Marco-Starkey model in 1954 [8]. The models have a constant parameter C that depends on the physical variables of the material and requires fatigue tests [9]. Several other models were proposed subsequently towards the end of the 1950s, but they required the determination of similar material parameters. Manson proposed a double linear damage rule (DLDR) in 1966 by replacing the linear Miner's rule with a set of two lines converging at a knee point [10]. The synergetic effect of the application of this rule in conjunction with a probabilistic approach based on P-S-N curves is recently shown by Correia et al.

[11]. The DLDR was further improved and the double damage curve approach (DDCA) was proposed by Manson and Halford in 1981 [12]. Many other damage models were proposed in the 1980s and 1990s such as models by Lemaitre and Plumtree [13] and by Chaboche and Lesne [14]. These models were again based on material parameters p, α, β which can be determined only through extensive material testing. Although some of the above-mentioned models have shown good agreement with the experimental data for specific materials, determination of the knee point location and additional material parameters restricted their use in applied engineering problems [15]. A more detailed review of the damage models developed before 1998 can be found in an article by Fatemi and Yang [16]. These damage models are based on crack growth concepts, damage curve modifications and energy based theories, as well as on continuum damage mechanics. However, the application of these models is not found in any of the design standards as they require testing for the determination of material parameters.

Material testing was performed by some researchers to establish fatigue damage behaviour in 1999. These tests were based on the exhaustion of material ductility and estimated the instantaneous damage in material for a given stress amplitude or range [17]. As a result, the damage evolution curves for several materials were established, and these represent the variation in experimental observed damage with number of cycles to failure. Testing was also performed to develop damage models based on hardness increase during the fatigue of material [18]. Subsequently, damage models were proposed to fit such experimental data but were again dependent on material parameters with application to specific materials. In 2005, Mesmacque et al. proposed a sequential law, which does not require any material parameters other than the full-range S-N curve [6]. The application of this model was demonstrated in steel bridges in 2008 [19]. Though the model can capture the loading sequence and predicts the fatigue life accurately, its requirement for a full-range S-N curve restricts industrial application by practicing engineers. Also, the design codes given S-N curves are based on detail categories, and the physical meaning of the intercept used for such details is not clear. Moreover, this law cannot be used with the design codes and standards having bilinear and trilinear S-N curves. Some other proposed models do not require material testing and overcome the shortcomings of Miner's rule using load interaction factors [20-22]. However, both the damage evolution curves and the fatigue life predictions differ significantly from experimental results, as will be shown later.

Recently, some probabilistic approaches have been proposed by considering a probabilistic S-N field and providing a statistical distribution of the Miner's damage based on log-normal distribution [23] [24]. The Miner's damage is related to a normalized variable V, which represents percentile curves in the S-N field unequivocally associated to probability of failure. These models can be applied for fatigue design of structural components subjected to variable amplitude loading and a recent application has been shown on riveted connection made of puddle iron from a bridge [25]. Also, a one parameter fatigue damage model has been proposed based on the concept of iso-damage curves [26]. As mentioned in the discussion of this paper, the model is based on S-N curves with a constant slope and is difficult to apply with bilinear or trilinear S-N curves in the current codes and standards. Also, the model parameter b is verified only for four types of steel and not for any other material. The model has only been compared to two step cyclic loading. Practical applications of the model to structural details subjected to variable amplitude loading have not been presented. Another non-linear fatigue damage model based on strain life curve has been proposed by Huffman and Beckman [27]. However, the applications of this model are not found as the design standards are based on stress life curves (i.e. S-N curves). Many other damage models exist in the literature, based on the concepts of continuum damage mechanics, energy conservation and entropy change. More details about these models can be found in several of the recently reviewed articles regarding fatigue damage theories such as by Santecchia et al. [28] and by Silitonga et al. [29].

Although many fatigue damage models have been proposed in the literature, the existing codes and standards still use Miner's rule because of its simplicity and ease of application. This indicates the need for an equally simple, easy to apply and accurate fatigue damage model for engineering applications/structural engineering problems. This becomes even more important for estimating the remaining life of ageing steel structures, where loading histories might be known as well. Such a model will not only predict the remaining life but will also help maintain the existing infrastructure more efficiently. The risk of failure and subsequent consequences will also be reduced using a more accurate model.

To overcome above mentioned problems, an accurate and easy-to-apply fatigue damage model is proposed in this paper. The proposed damage model does not require any material testing and depends only on the commonly available S-N curve of the material or its corresponding detail category. Moreover, it does not require a full-range S-N curve and can be easily applied by practicing engineers using the partially known S-N curves given in design standards. The proposed model can be applied to structural details subjected to variable amplitude loading conditions unlike most of the earlier models applied only to block loading cases. As a result, the proposed model can be easily implemented by practicing engineers for fatigue analysis of several practical problems involving design detail categories subjected to variable amplitude loadings. The proposed model considers the loading sequences, along with the interactions between them, and provides better agreement with experimental data compared to previously proposed models. Due to its unique features and better accuracy, the proposed model can provide a platform to design and maintenance communities for an efficient use of existing ageing steel infrastructure by predicting their remaining safe life more accurately. Initially, the paper presents the proposed damage model in detail. The model is verified with experimentally derived damage evolution curves for C 45 and 16 Mn and gives better agreement compared to previous models. The practical application of the proposed model is shown on welded joints used in several engineering applications. The model is applied to butt and fillet welded joints subjected to block loading and predicted lives are compared with experimental results. The application of the model under variable amplitude loading is also shown on these joints and the results are compared with Eurocode predicted lives. Hence, the applicability, validity and significance of the proposed model is confirmed.

2. Proposed fatigue damage model

A new and easy-to-apply fatigue damage model is proposed as shown in equation (1). The damage for a given stress level is determined using the damage evolution curve of the considered material. This damage is then transferred to the next stress level by determining the effective number of cycles using the proposed load interaction factor. The physical meaning of the concept is to assume the same damage state of the material, while transferring the loading state from one stress level to the next and determining the effective number of cycles required to cause this damage.

$$D_i = 1 - \left[1 - \frac{n_i}{N_i}\right]^{\delta_i} = 1 - \left[1 - \frac{n_{(i+1),eff}}{N_{i+1}}\right]^{\frac{\delta_{i+1}}{\mu_{i+1}}} \tag{1}$$

where D_i is the damage at load level i when member is subjected to a certain stress amplitude (or range) σ_i for n_i number of cycles and $n_{(i+1)\,eff}$ is the effective number of cycles corresponding to the stress range σ_{i+1} at level $i+1$. N_i and N_{i+1} are the corresponding number of cycles to failure and can be determined from the S-N curve given in design standards and codes. δ_i and μ_i are the model parameters and depends only on N and given stress levels as shown in equation (2) and equation (3). The proposed model also represents some of the earlier proposed damage models for specific values of δ_i as shown in Table 1.

$$\delta_i = \frac{-1.25}{\ln N_i} \tag{2}$$

$$\mu_i = \left(\frac{\sigma_{i-1}}{\sigma_i}\right)^2 \tag{3}$$

The damage at load level $i+1$ can be determined using equation (4) and equation (5).

$$D_{i+1} = 1 - \left[1 - \frac{n_{(i+1),total}}{N_{i+1}}\right]^{\delta_{i+1}} \tag{4}$$

where

$$n_{(i+1),total} = n_{(i+1),eff} + n_{(i+1)} \tag{5}$$

where n_{i+1} is the number of cycles for stress state σ_{i+1}. Also, $n_{(i+1)}$, total is the total number of cycles for stress state σ_{i+1} obtained using the proposed concept.

Table 1. Parameters of the proposed model and its special cases [2] [3] [17].

δ_i	Damage model	Remarks
$-1.25/\ln N_i$	Proposed model	only based on commonly available S-N curves
1	Miner's rule, 1945 [2]	most commonly used, but is inaccurate under VAL
$1/p+1$	Lemaitre and Plumtree, 1979 [13]	p is a material constant determined by damage evolution curves based on cyclic stress ranges
$1/1-\alpha$	Shang and Yao, 1999 [17]	α is a material constant determined by damage evolution curves based on static ductility change

This damage transfer technique is continued until the fatigue damage D becomes one, denoting fatigue failure. The proposed damage transfer concept is explained graphically, as shown in Figure 1. More details of the proposed model and associated parameters can be found in recently published work by the authors [30].

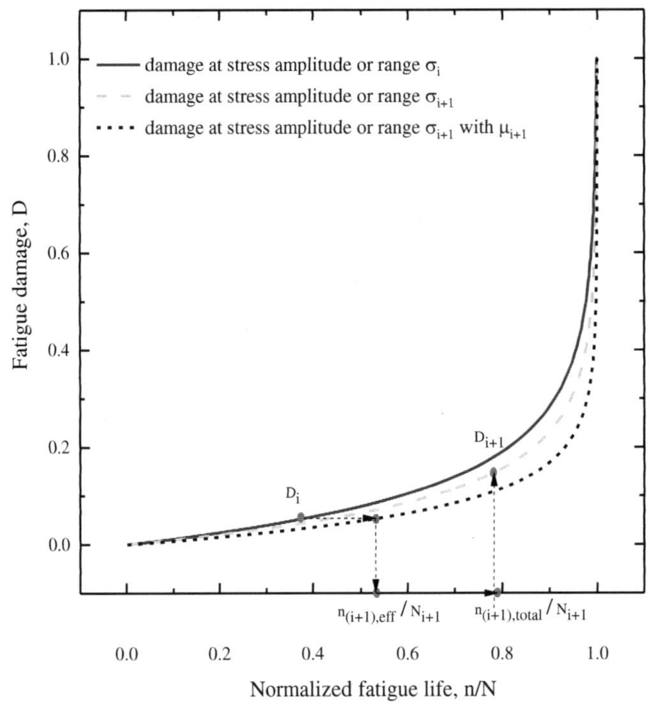

Figure 1. Graphical representation of the proposed damage transfer concept

2.1. Verification of proposed model with experimental damage evolution

The proposed fatigue damage model is verified by comparing the experimental results for damage evolution curves. C 45 and 16 Mn steels are used for this verification [17].

2.1.1. C 45 steel. This material is of interest to the authors due to its currently increasing use in offshore structures [31]. The yield strength and ultimate tensile strength for this material are 371.7 MPa and 598.2 MPa, respectively [17]. In the past, experiments were conducted on this material for constant amplitude stresses, and the fatigue damage was determined by measuring the static relative ductility change in the material [17]. The experiments were conducted at 330.9 MPa and 405.8 MPa stress amplitudes (with zero mean stress); the results are shown in Figure 2 (a) and Figure 2 (b), respectively. Fatigue damage curves are also plotted for each of the stress amplitudes using the proposed model. The proposed model is also compared with some of the recently developed models which do not require additional material testing, other than their S-N curve.

2.1.2. 16 Mn steel. The damage evolution curves were also determined experimentally for 16 Mn steel by measuring the static relative ductility change in material [17]. For this material, the yield strength is 382.5 MPa, and the ultimate tensile strength is 570.7 MPa. The experiments were conducted at 337.1 MPa and 373.5 MPa stress amplitudes (with zero mean stress); the damage evolution curves are shown in Figure 3 (a) and Figure 3 (b), respectively. Fatigue damage curves are also plotted for each of the stress amplitudes using the proposed model. The proposed model is also compared with some of the recently developed models which do not require additional material testing, other than their S-N curve.

2.1.3. Discussion of verification study. From Figure 2 (a) and Figure 2 (b), it is seen that the proposed model gives good prediction of the damage curves for material C 45 and is in good agreement with the experimental results. The damage evolution curves for 16 Mn steel are shown in Figure 3 (a) and Figure 3 (b) for the two stress amplitudes. Again, it is seen that the proposed model is in good agreement with the experimental results and can predict the damage behaviour more accurately. The proposed model is also compared with some of the other recently developed models which do not require additional material testing, other than their S-N curve. It is seen that the model developed by Kwofie and Rahbar [20] is not in good agreement with the experimental data and is equivalent to Miner's rule while predicting the damage behaviour. The models proposed by Liakat and Khonsari [21] and Lv et al. [22] seem to give the same damage evolution curves for both considered materials. The sequential law proposed by Mesmacque et al. [6] is also compared with the experimental data. It is seen that the model gives good agreement with the experimental results. However, the model requires the determination of the full-range S-N curve, thereby restricting its use by practicing engineers. After comparing all the curves shown in Figure 2 and Figure 3, it is concluded that the proposed model predicts the real damage evolution behaviour quite accurately for the two materials considered, when compared to previous models. The damage curves can be plotted using the proposed model without determining additional material parameters and using only code-given S-N curves.

3. Application of proposed model to welded joints

The proposed model is applied on welded joints subjected to both block loading as well as variable amplitude loading. These welded joints are used in several engineering applications such as vehicles, electrical mobile units (EMUs), etc., and an accurate prediction of life is necessary for the safety of both passengers and vehicles. There has been a significant advancement in technology over the past couple of decades. As a result, conventional diesel locomotives are increasingly being replaced by faster electrical mobile units (EMUs). These EMUs can reach speeds of 250 km/h and above, as compared to the diesel locomotives which used to have typical speeds of 160 km/hr [32]. While the advancement of the vehicle technology is good for the fast commuting of passengers, it also puts vehicle safety, as well as the safety of people, at huge risk. Such risks arise from the increasingly daily passenger loads, inefficient maintenance of vehicles, as well as ageing of the car bodies, which, in these high-speed trains are generally made from aluminium alloy due to its light weight, security and comfort [33].

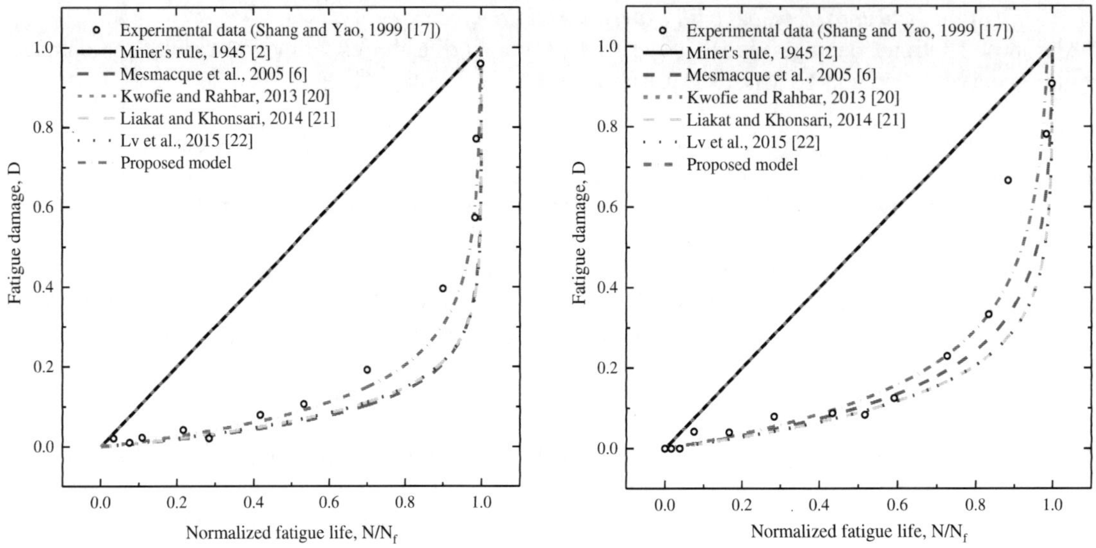

(a) (b)

Figure 2. Comparison of theoretically predicted fatigue damage evolution with experimental damage of C 45 steel for stress (a) $\sigma_a = 330.9$ MPa (b) $\sigma_a = 405.8$ MPa

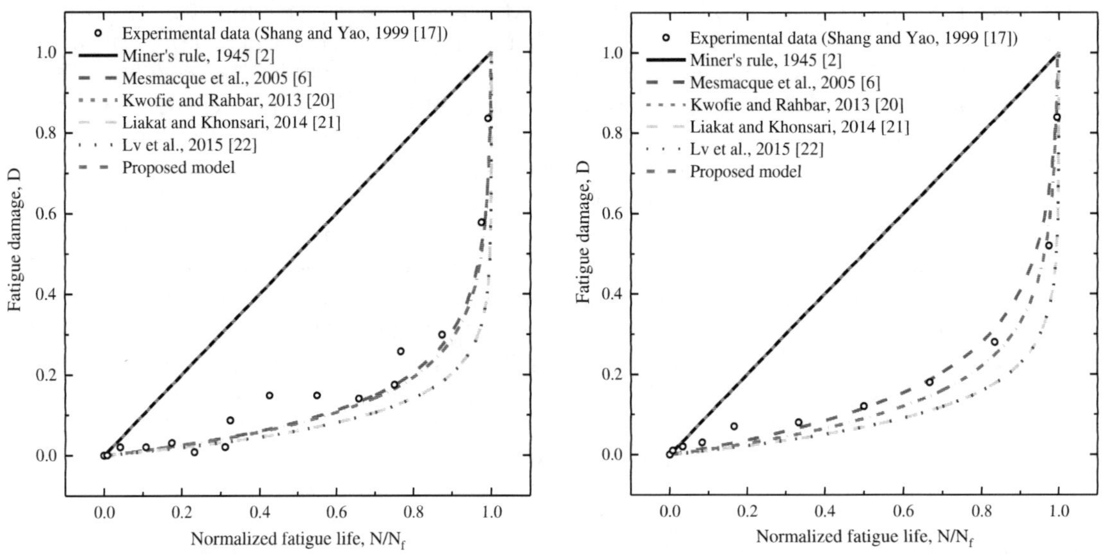

(a) (b)

Figure 3. Comparison of theoretically predicted fatigue damage evolution with experimental damage of 16 Mn steel for stress (a) $\sigma_a = 337.1$ MPa (b) $\sigma_a = 373.5$ MPa

However, as a result of the high thermal conductivity coefficient and high linear expansion coefficient, aluminum alloys can have complex welding deformations and stresses at the welded joints. The softening and embrittlement of these welded joints can have a serious effect on the structural strength of the car body. It is reported that the failure of the car body usually occurs in the welded joint, and the major cause is fatigue loading [33] [34], which is most critical for the welded connections; it is therefore important to estimate the fatigue life accurately to minimize the risks.

Much research has been carried out in the past on the fatigue properties of welded aluminum joints. Recently, a group of researchers performed fatigue tests on such aluminum joints to thoroughly understand the fatigue properties and predict the life more accurately under the block loading conditions [33] [34]. Firstly, the experimental setup and considered welded joints are explained in detail in this section. The proposed model is then applied to these joints under the given block loading conditions and predicted lives are compared with experimental results. The applicability and accuracy of proposed model under block conditions is henceforth confirmed. The application of the proposed model is also shown on these welded joints subjected to random variable amplitude loadings (VAL). An in-situ measured VAL is considered for this study. The results are compared with Eurocode recommended Miner's rule. Hence, the applicability of the proposed model under VAL is also confirmed in this study.

3.1. Considered welded joint details

The fatigue tests were performed for both butt and fillet welded joints by using the PLG-200 fatigue testing machine. The test specimens are shown in Figure 4 (a) and Figure 4 (b) respectively. The base material of the specimen is aluminum alloy ENAW6005 having the main chemical composition in mass percentage as: 0.35Fe, 0.6-0.9Si, 0.1Mn, 0.1Ti, 0.4-0.6Mg, 0.1Cu, 0.1Zn and 0.1 Cr [33] [34]. The tensile strength and yielding strength of the base material are 270 and 225 MPa, respectively. The specimens were tested under a four-point bending load cycle with cyclic stress ratio R = -1. The load-time history follows the sine function. The loading frequency is 110 Hz, and room temperature is 20~30°C [33] [34].

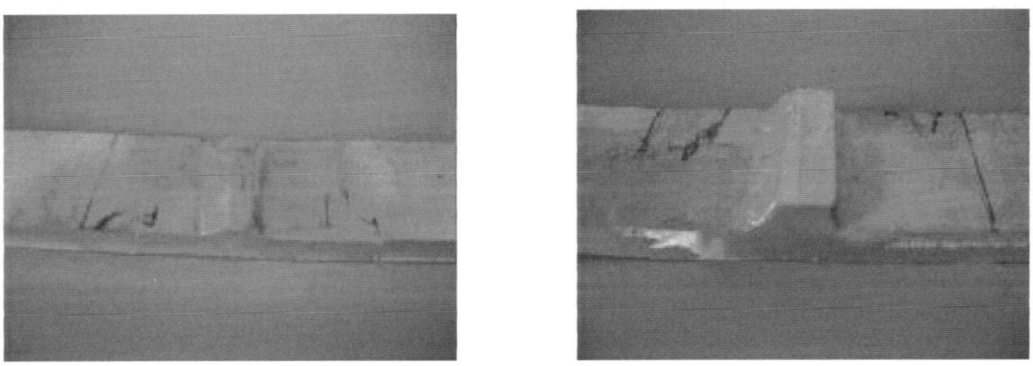

(a) (b)

Figure 4. Welded joint test specimens (a) Butt welded joint (b) Fillet welded joint [33] [34]

3.2. Application of proposed model under block loading

The proposed model is applied to welded joints under block loading conditions and the predicted lives are compared with experimental lives. The experiments were conducted by previous researchers under two-stage block loading for each of the considered welded joints [33] [35]. The load sequence considered for the butt-welded joint are 104–74 MPa and 89–74 MPa for high-low loading sequence, and 74–89 MPa and 74–104 MPa for low-high loading sequences. The load sequences considered for the fillet welded joint are 93–73 MPa and 83–73Mpa under high-low loading, and 73–83 MPa and 73–93 MPa under low-high loading, respectively. The experimentally obtained fatigue lives for the butt and fillet welded joint are shown in Table 2 and Table 3, respectively [33] [35].

The proposed damage model is compared with experimentally obtained fatigue lives for the considered welded joints. The loading sequences are given in Table 2 and Table 3. Some of the recently proposed models have also been included in the comparison. The comparison of fatigue life is shown in Figure 5 (a) and Figure 5 (b) for the butt and fillet welded joints, respectively. The maximum and average deviations from the experimental results are 20% and 9% respectively. The solid line

IOP Conf. Series: Materials Science and Engineering **276** (2017) 012038 doi:10.1088/1757-899X/276/1/012038

refers to ideal conformity of the results. The dotted lines represent a scatter band with a coefficient of two.

Table 2. Experimental fatigue lives for fatigue testing on butt welded joint [33] [35].

Load Set	σ_1 (MPa)	σ_2 (MPa)	n_1 (*1000)	n_2 (*1000)	N_{f1}	N_{f2}
Set 1	104	74	109.9	797.6	549300	1540100
Set 2	89	74	176.1	1029.2	880500	1540100
Set 3	74	89	770.1	545.6	1540100	880500
Set 4	74	104	770.1	418.9	1540100	549300

Table 3. Experimental fatigue lives for fatigue testing on fillet welded joint [33] [35].

Load Set	σ_1 (MPa)	σ_2 (MPa)	n_1 (*1000)	n_2 (*1000)	N_{f1}	N_{f2}
Set 1	93	73	309.9	587.5	619800	1546100
Set 2	83	73	476.1	681.1	952300	1546100
Set 3	73	83	509.2	708.2	1546100	952300
Set 4	73	93	773.0	426.4	1546100	619800

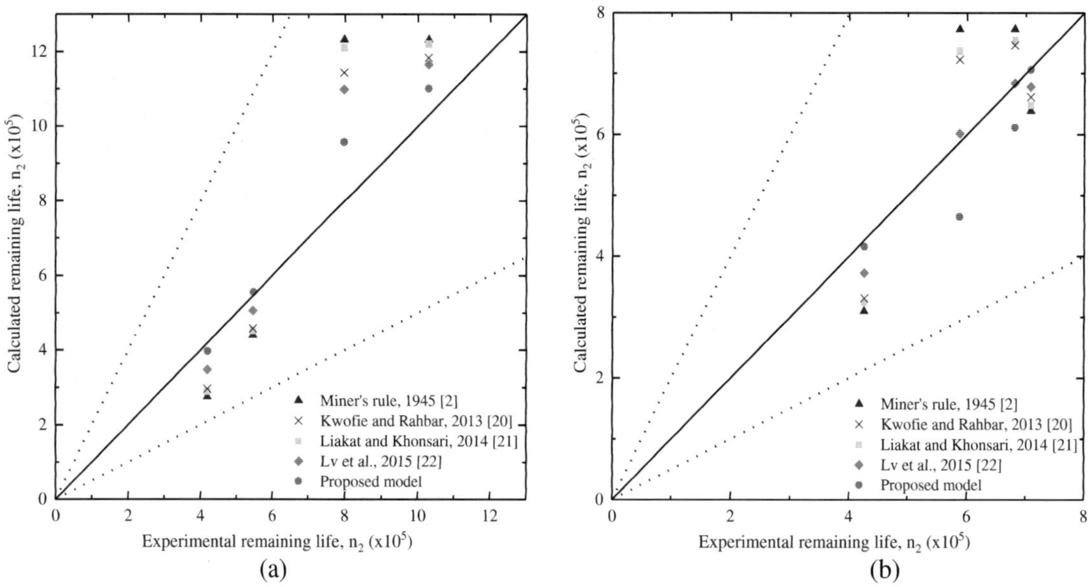

Figure 5. Comparison of calculated fatigue life with experimental life for (a) butt welded joint (b) fillet welded joint

3.3. Application of proposed model under variable amplitude loading (VAL)

The proposed model is also applied to considered welded joints under variable amplitude loading conditions. The fatigue life of both butt and fillet welded joints is predicted using the proposed model. The lives are also predicted using the Miner's rule recommended by standard codes and standards such as Eurocode. The detail category S-N curves for the joints are taken from the Eurocode [36]. The cycle counting is performed using the rainflow counting method [37]. Finally, the damage accumulation curves are obtained and fatigue lives from proposed model are compared with those obtained using Miner's rule.

3.3.1. Considered variable amplitude loading (VAL). The variable amplitude loading considered for this study is shown in Figure 6. The stress values are shown for every 0.13 seconds and for a period of one hour. The same stress history is assumed to be repeated until the end of the fatigue life (i.e. the same loading block is assumed to be repeated until the failure point).

3.3.2. Rainflow counting and mean stress correction. The cycle counting is performed on the considered VAL using the rainflow counting method [37]. This is followed by the mean stress correction as recommended by Goodman rule [5]. The resulting stress amplitude history is shown in Figure 7.

3.3.3. Considered detail category S-N curves for welded joints. The detail category S-N curves of the welded joint details are taken from Eurocode [36]. The parameters of these curves including the fatigue endurance stress limit are shown in Table 4. The stress history after the removal of endurance limit are shown in Figure 8 and Figure 9 for the butt welded and fillet welded joint respectively.

Table 4. S-N curve parameters for considered joints given in Eurocode [36]

Joint	S-N curve parameters				
	$\Delta\sigma_L$	$\Delta\sigma_D$	$\Delta\sigma_C$	m_1	m_2
Butt	20.1	32.3	40	4.3	3.4
Fillet	12.3	21.4	28	6.3	5.4

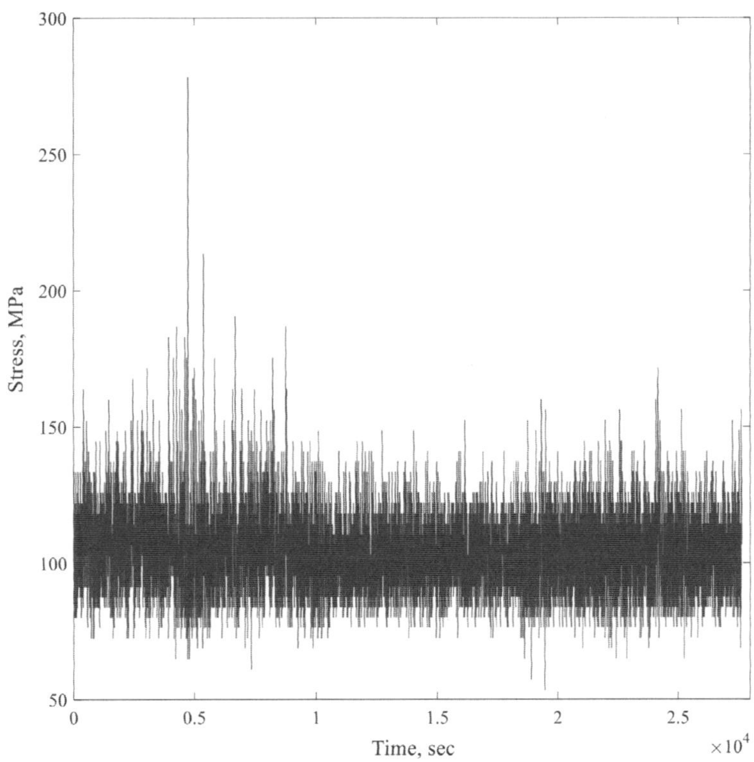

Figure 6. Considered variable amplitude loading for welded joints

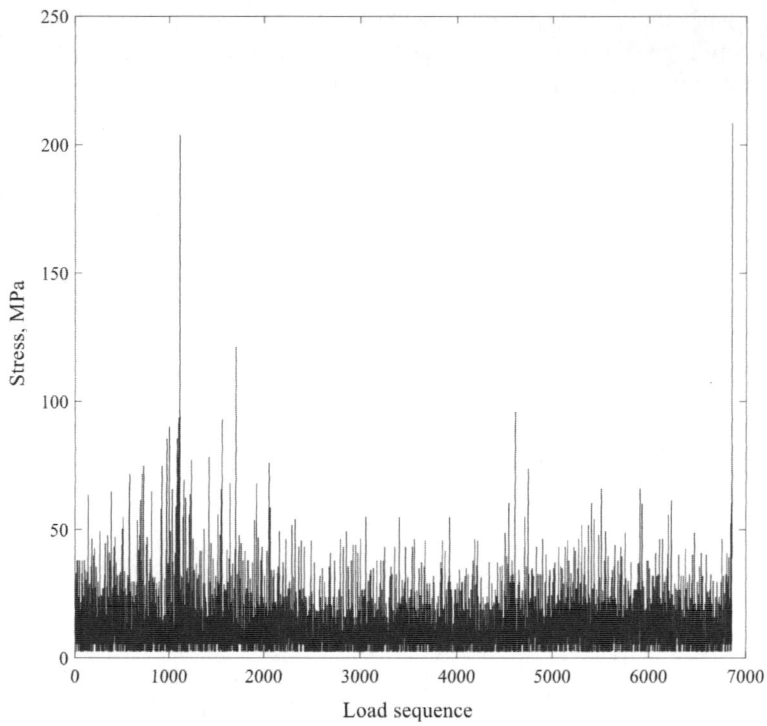

Figure 7. Stress history after mean stress correction

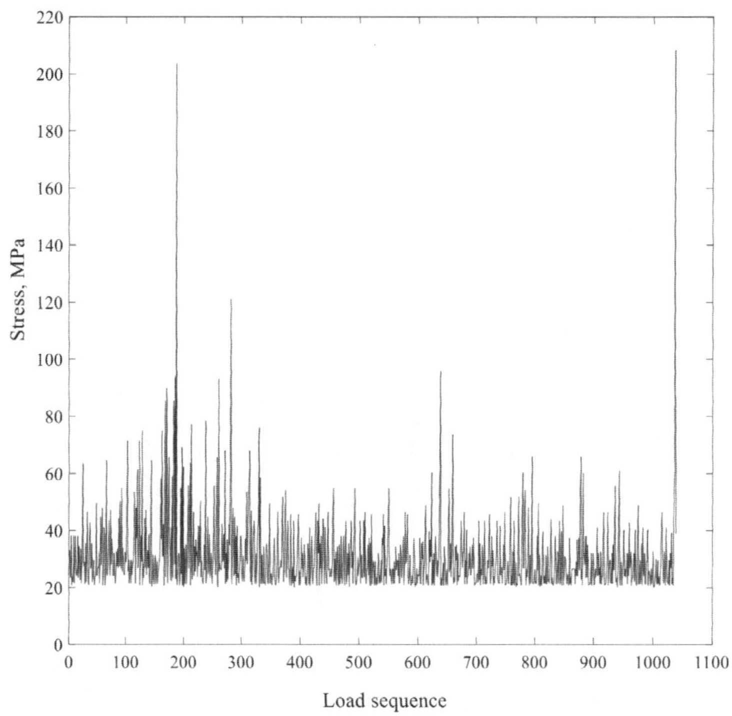

Figure 8. Stress history above endurance for butt welded joint

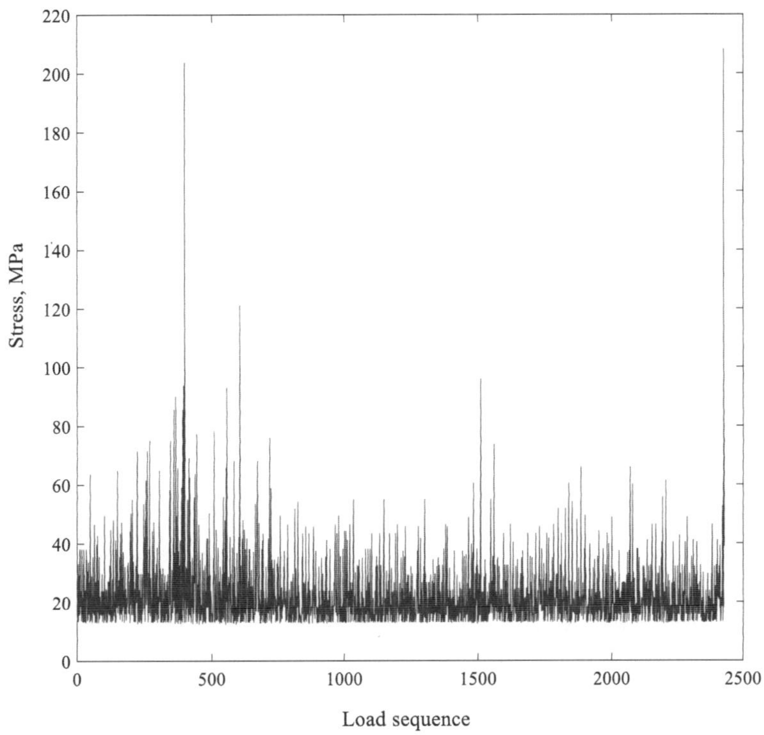

Figure 9. Stress history above endurance for fillet welded joint

3.3.4. Comparison of damage curve and fatigue life. The proposed model is applied to calculate fatigue life of the considered welded joints. The stress amplitude histories shown in Figure 8 and Figure 9 are used for these calculations. The interaction factor in the proposed model is taken as one for this study and the accumulated fatigue damages until failure is shown in Figure 10 and Figure 11 for butt and fillet welds respectively. The damage accumulations using the Eurocode recommended Miner's rule are also shown in Figure 10 and Figure 11.

The fatigue life for the considered joints can be determined from Figure 10 and Figure 11. The fatigue lives are calculated in terms of number of blocks of the VAL considered (i.e. number of repeated blocks) and are shown in Table 5.

Table 5. Fatigue life comparison using proposed model and Eurocode (in blocks)

Joint	Number of blocks of considered VAL		
	Eurocode	Proposed	% difference
Butt	899	829	7.80
Fillet	544	514	5.3

3.4. Discussion and comparison of the results

The proposed model is applied to predict the fatigue lives of welded joints used in several engineering applications. Butt and fillet welded joints were considered in the case study. The application of the fatigue damage model to these joints is shown under both block loadings and variable amplitude loading conditions. The comparisons of the fatigue lives under block loading conditions are shown in Figure 5 (a) and Figure 5 (b) for butt welded and fillet welded joint respectively.

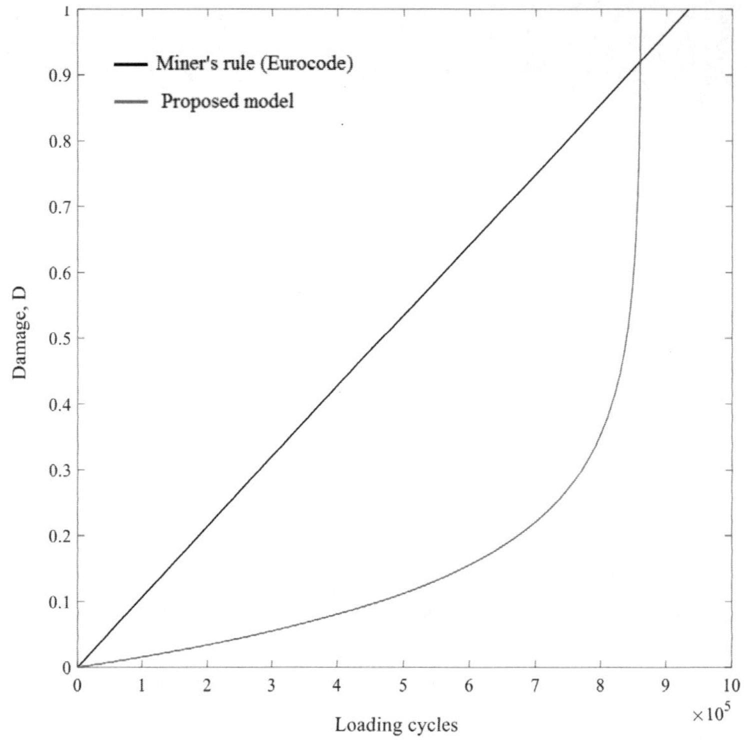

Figure 10. Fatigue damage accumulation using proposed model and Eurocode for butt welded joint

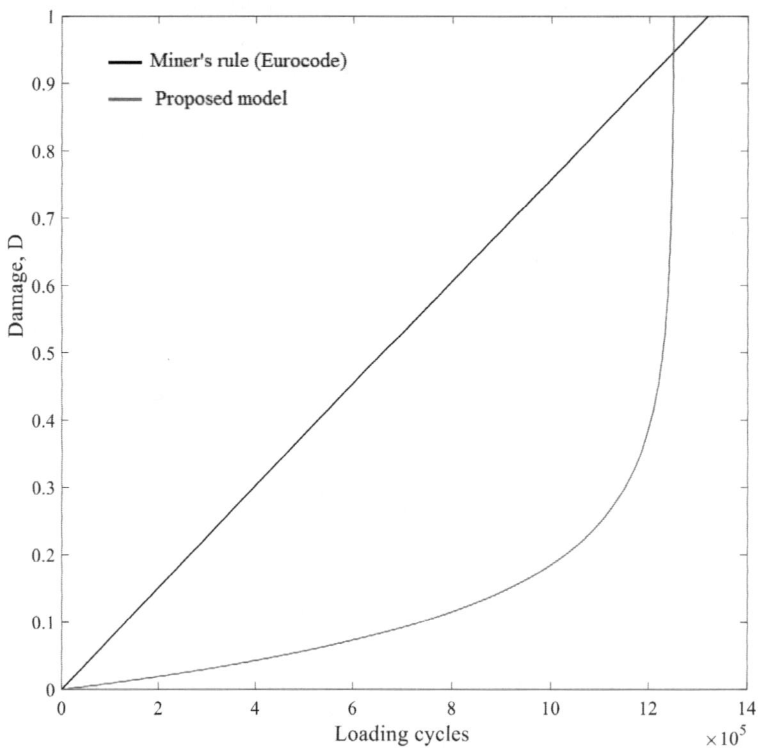

Figure 11. Fatigue damage accumulation using proposed model and Eurocode for fillet welded joint

The predicted lives using the proposed model are in good agreement with the experimental lives compared to Eurocode given Miner's rule. The proposed model predictions are also better compared to earlier models. The sequential law model could not be applied to these joints since full range S-N curve is not available for these detail categories. Using the proposed model, the maximum and average deviations from the experimental results are 20% and 9% respectively for these joints. The scatter bands with a coefficient two is also plotted in all the figures where experimental fatigue life is compared with predicted life. The predicted fatigue lives are all lying within the scatter band.

The application of proposed model is also shown under variable amplitude loading conditions. The variable amplitude loading considered is taken from in-situ measurements for a period of one hour and is assumed to be repeated until the failure. The mean stress zero equivalent stress amplitude history is then derived using the rainflow counting algorithm followed by the mean stress correction. The proposed model requires only the S-N curve parameters of the detailed categories and these were taken as given in the Eurocode. The fatigue damage accumulation curves are shown in Figure 10 and Figure 11 for the butt welded and fillet welded joint respectively. These are compared for both proposed model and the Eurocode (i.e. using the Miner's rule). The predicted fatigue lives are shown in Table 5.

From Figure 10 and Figure 11, it is seen that the damage accumulation curves using the proposed model are nonlinear and are very similar to the damage evolution curves discussed earlier in this paper. This further confirms the real physical damage accumulation behaviour in the material instead of the linear accumulation as assumed by the commonly used Miner's rule given in Eurocode and other design standards. From Table 5, it is seen that the predicted lives using the proposed model are around 8% shorter compared to the Eurocode. This difference may not seem too significant for the design of new structures. However, the proposed model can be used for a more accurate prediction of remaining lives of existing infrastructures such as old bridges, ageing oil/gas platforms and industrial buildings. The remaining life calculations based on Eurocode might result in longer lives than the structures can stand. This shows the importance of applying an accurate fatigue damage model.

4. Conclusions

The paper presents a new fatigue damage model. The proposed model does not require any material parameters, other than the commonly available S-N curve parameters, which are generally used with Miner's rule. The major advantage of proposed model is that it does not require detail material testing or modifications to the S-N curve. Also, unlike earlier models, the proposed model can be applied to design detail categories using the corresponding partially known S-N curve in the design standards. Therefore, the proposed model can be easily implemented by practicing engineers for fatigue analysis of several engineering problems. The model is verified with both the damage evolution curves and fatigue life estimations. It is concluded that the damage curves plotted using the proposed model are in good agreement with the available experimental data for two considered materials. It is also concluded that the fatigue life predictions are more accurate using the proposed model compared to Eurocode given Miner's rule. The application of the proposed model is shown to welded joints used in several engineering applications. The both butt and fillet welded joints are commonly subjected to both block loading and variable amplitude loading conditions. The obtained damage accumulation curves using proposed model confirms the real physical damage accumulation behaviour in the material instead of the linear accumulation as assumed by the commonly used Miner's rule given in Eurocode. It is therefore concluded that the proposed model can be used for better fatigue life predictions and can be easily applied by practicing engineers using only the code-given S-N curves. The applicability, significance and validity of the proposed fatigue damage model should be further verified in the future by applying the model to more case studies.

References

[1] Naess AA 1985 *Fatigue Handbook* (Trondheim: Tapir)

[2] Miner MA Cumulative damage in fatigue 1945 *J. Appl. Mech.* **A** pp 159–64

[3] Eurocode 3 2010 *Design of steel structures – Part 1-9: Fatigue, NS-EN 1993-1-9:2005+NA:2010* (European Standard)

[4] DNVGL RP C203 2016 *Offshore standard DNVGL-RP-C203: Fatigue Design of Offshore Steel Structures* (Det Norske Veritas and Germanischer Lloyd SE)

[5] Suresh S 1998 *Fatigue of Materials* (Cambridge: Cambridge University Press)

[6] Mesmacque G, Garcia S, Amrouche A and Rubio-Gonzalez C 2005 Sequential law in multiaxial fatigue, a new damage indicator *Int. J. Fatigue* **27** pp 461-7

[7] Leve HL 1969 *Cumulative damage theories. In Metal Fatigue: Theory and Design* (New York: Wiley) p 170

[8] Marco SM and Starkey WL 1954 A concept of fatigue damage *Trans. of the ASME* **76** 627-32

[9] Hwang W and Han K S 1986 Cumulative damage models and multi-stress fatigue life prediction *J. Compos. Mater.* **20** pp 125-53

[10] Manson S, Freche J and Ensign C 1967 Application of a double linear damage rule to cumulative fatigue *Fatigue Crack Propagation, ASTM STP 415* pp 384-412

[11] Correia O, Jesus A, Blasón S, Calvente M and Fernández-Canteli A. Probabilistic non-linear cumulative fatigue damage of the P355NL1 pressure vessel steel *ASME 2016 pressure vessels and piping conference* **6** pp 17-21

[12] Manson S and Halford G 1981 Practical implementation of the double linear damage rule and damage curve approach for treating cumulative fatigue damage *Int. J. Fract.* **7** 169-92

[13] Lemaitre J and Plumtree A 1979 Application of damage concepts to predict creep-fatigue failures *J. Eng. Mater. Technol.* **101** pp 284-92

[14] Chaboche JL and Lesne PM A non-linear continuous fatigue damage model *Fatig. Fract. Eng. Mater. Struct.* **11** pp 1-7

[15] Goodin E, Kallmeyer A and Kurath P 2004 Evaluation of nonlinear cumulative damage models for assessing HCF/LCF interactions in multiaxial loadings *Proc. of 9th Nat. Turbine Engine HCF Conf. (North Carolina)*

[16] Fatemi A and Yang L 1998 Cumulative fatigue damage and life prediction theories: a survey of the state of the art for homogeneous materials *Int. J. Fatigue* **20** pp 9-34

[17] Shang DG and Yao WX 1999 A nonlinear damage cumulative model for uniaxial fatigue *Int. J. Fatigue* **21** pp 187-94

[18] Pavlou DG 2002 A phenomenological fatigue damage accumulation rule based on hardness increasing, for the 2024-T42 aluminum *Eng. Struct.* **24** pp 1363-68

[19] Siriwardane S, Ohga M, Dissanayake R and Taniwaki K 2008 Application of new damage indicator-based sequential law for remaining fatigue life estimation of railway bridges *J. Construct. Steel. Res.* **64** pp 228-37

[20] Kwofie S and Rahbar N 2013 A fatigue driving stress approach to damage and life prediction under variable amplitude loading *Int. J. Damage Mech.* **22** pp 393-404

[21] Liakat M and Khonsari MM 2014 An experimental approach to estimate damage and remaining life of metals under uniaxial fatigue loading *Mater. Des.* **57** pp 289-97

[22] Lv Z, Huang HZ, Zhu SP, Gao H and Zuo F 2015 A modified nonlinear fatigue damage accumulation model *Int. J. Damage. Mech.* **24** pp 168-81

[23] Fernández-Canteli A, Blasón S, Correia JA and De Jesus AM 2014 A probabilistic interpretation of the Miner number for fatigue life prediction *Frattura ed Integrita Strutturale* **30** pp 327-39

[24] Blasón S, Correia JA, De Jesus AM, Calçada RA and Fernández-Canteli A 2016 A probabilistic analysis of Miner's law for different loading conditions *Struc. Eng. Mech.* **60** pp 71-90

[25] Correia JA, De Jesus AM, Fernández-Canteli A and Calçada RA 2015 Fatigue damage

assessment of a riveted connection made of puddle iron from the Fão Bridge using the modified probabilistic interpretation technique *Procedia Eng.* **114** pp 760-7

[26] Rege K and Pavlou DG 2017 A one-parameter nonlinear fatigue damage accumulation model *Int. J. Fatigue* **98** pp. 234-46

[27] Huffman PJ and Beckman SP 2013 A non-linear damage accumulation fatigue model for predicting strain life at variable amplitude loadings based on constant amplitude fatigue data *Int. J. Fatigue* **48** pp. 165-169

[28] Santecchia E, Hamouda AM, Musharavati F, Zalnezhad E, Cabibbo M, El Mehtedi M and Spigarelli S 2016 A review on fatigue life prediction methods for metals *Adv. Mat. Sci. Eng.* **9** pp 1-26

[29] Silitonga S, Maljaars J, Soetens F and Snijder HH 2013 Survey on damage mechanics models for fatigue life prediction *Heron* **58** pp 25-60

[30] Aeran A, Siriwardane SC, Mikkelsen O and Langen I 2017 A new nonlinear fatigue damage model based only on SN curve parameters *Int. J. Fatigue* **103** pp 327-341

[31] Abed FH, Saffarini MH, Abdul-Latif A and Voyiadjis GZ 2017 Flow stress and damage behavior of C45 Steel Over a Range of Temperatures and Loading Rates *J. Eng. Mater. Techno.* **139** pp 1-8

[32] Fang J, Zhao WZ and Zhang J 2012 Fatigue life prediction of CRH3 carbody based on rigid-flexible coupling model and the master SN curve *Proc. of 1st Int. Work. on High-Speed and Intercity Railways (Berlin)*

[33] Wang W, Li Q, Liu Z and Wang B 2010 Experimental study on fatigue performance and damage model of aluminum alloy welding joints for high-speed train car body *Proc. of 4th Int. Conf. on Experimental Mechanics (Singapore)*

[34] Zou H, Li W, Li Q and Wang P 2010 Probabilistic fatigue property of aluminium alloy welded joint for high-speed train *Adv. Mater. Res.* **118** pp 527-31

[35] Tian J, Liu ZM and He R 2012 Nonlinear fatigue-cumulative damage model for welded aluminum alloy joint of EMU *J. Chin. Railw. Soc.* **34** pp 40-3

[36] Eurocode 3 2011 *Design of aluminium structures – Part 1-3: Structures susceptible to fatigue, NS-EN 1999-1-3.2007+NA1:2011* (European Standards)

[37] Downing S and Socie D 1982 Simple rainflow counting algorithms *Int. J. Fatigue* **4** pp 31-40

Fatigue strength degradation of metals in corrosive environments

N D Adasooriya*, T Hemmingsen and D Pavlou

Department of Mechanical and Structural Engineering and Materials Science, University of Stavanger, Norway

*Contact author: mudiyan.n.adasooriya@uis.no

Abstract. Structures exposed to aggressive environmental conditions are often subjected to time-dependent loss of coating and loss of material due to corrosion; this causes reduction in the cross-sectional properties of the members, increased surface roughness, surface irregularities and corrosion pits, and degradation of material strengths. These effects have been identified and simulated in different research studies. However, time and corrosive media dependent fatigue strength curves for materials have not been discussed in the design or assessment guidelines for structures. This paper attempts to review the corrosion degradation process and available approaches/models used to determine the fatigue strength of corroded materials and to interpolate corrosion deterioration data. High cycle fatigue and full range fatigue life formulae for fatigue strength of corroded materials are proposed. The above formulae depend on the endurance limit of corroded material, in addition to the stress-life fatigue curve parameters of the uncorroded material. The endurance limit of corroded material can either be determined by a limited number of tests in the very high-cycle fatigue region or predicted by an analytical approach. Comparison with experimentally measured corrosion fatigue behavior of several materials is provided and discussed.

1. Introduction

Corrosion is one of the principal deterioration processes that affect the integrity of structures. Major effects can be listed as thickness reduction, irregularities of the surfaces, surface roughness, and stress concentration due to the pitting and degradation of material due to the electrochemical process. The cracking of members especially starts at highly stressed locations. Thickness reduction, surface roughness and cyclic loading accelerate the crack initiation under different types of corrosion [1]. Cracks caused by corrosive media are mainly governed by environmental assisted cracking (EAC) such as stress corrosion cracking (SCC), corrosion fatigue (CF), sulfide stress cracking, etc. CF and SCC play a significant role in the degradation of metal structures. Fractures due to cyclic stress in corrosive media are designated as CF. SCC is generally defined as failure caused by continuously applied stress in corrosive media. Researchers are still highlighting the significance of further study on CF [2-4], due to the inherent nature of corrosion and its effects on random fatigue behavior.

Recent failures of key structures such as Silver Bridge in 1967, Mianus River Bridge in 1990, Minnesota Bridge in 2007, etc., emphasize the importance of more accurate simulation of CF in different corrosive environments in respect to different structural materials. Therefore, the consideration of the corrosion effects on structural integrity is important [3,5].

This paper reviews the effects of CF in structural degradation. Initially, the paper describes the corrosion degradation process of metallic materials. Then methods for modeling CF are comprehensively reviewed. New formulae for determining the fatigue strength of corroded material are proposed as an original contribution of this paper. The proposed formulae are compared with published results of the experimentally measured CF behavior (i.e. fatigue test of corroded specimens) of several steel types.

2. Corrosion degradation process of metalic materials

Corrosion of steel is an electrochemical process, which starts with the presence of oxygen and water. Common agents such as carbon dioxide, sodium chloride and hydrogen sulfide affect this process. The corrosion mechanism of iron can be described as follows: oxidation of iron takes place when dissolution of iron produces ferrous iron (Fe^{2+}) at the anode with the presence of various impurities. At the cathode, oxygen produces hydroxide irons(OH^-).Then Fe^{2+} and OH^- iron combine to form ferrous hydroxide ($Fe(OH)_2$).Further oxidation of ferrous hydroxide produces rust ($Fe_2O_3.H_2O$). Corrosion pits, corrosion products on surfaces, uneven surfaces, loss of material and degradation of material strength can be seen as outcomes of the above electrochemical process [3,6]. Generally, structures are subjected to five of the most important forms of corrosion [6,7], the most common of which is general (uniform) corrosion, which is uniformly distributed on the surface. General corrosion often occurs in locations where water accumulates [6,8]. Pitting corrosion is a localized form of corrosion which is restricted to a small area and usually begins with an irregularity on the surface. This type is dangerous because it may cause local stress concentrations. Crevice corrosion is also a form of localized corrosion, occurring where different components of the structure are close together, leading to narrow spaces. Galvanic corrosion takes place when two different metals are placed in an electrolyte and are electrically connected, as is possible at bolted or welded connections. The last dominant form of corrosion is CF, which is the mechanical degradation of a material under the combined action of localized corrosion and cyclic loading [8]. Fractures due to corrosion fatigue are often transgranular. The effect of CF on structural integrity is reported as significant for most situations [2,3,6-8,10].

The effects of corrosion deterioration to metal structures yields; (i) reductions of the effective cross-sectional area, moment of inertia, torsional and warping constants [6,11-14]. This may change the overall stiffness of the structures and hence change the stress state in the remaining part of the material/member; (ii) Interaction of corrosion (i.e. mostly pitting and crevice) and cyclic loading affects the mechanical properties of metal and, hence, cracks initiate from the corrosion-induced surface (i.e. CF); thereby a significant decrease in the fatigue strength (i.e. degrade the S-N curve in air) can be observed [7, 15-17] as shown in Figure 1. Finally, these may cause a reduction of the fatigue life of metal structures which are subjected to loss of coating, rust or corrosion wastage.

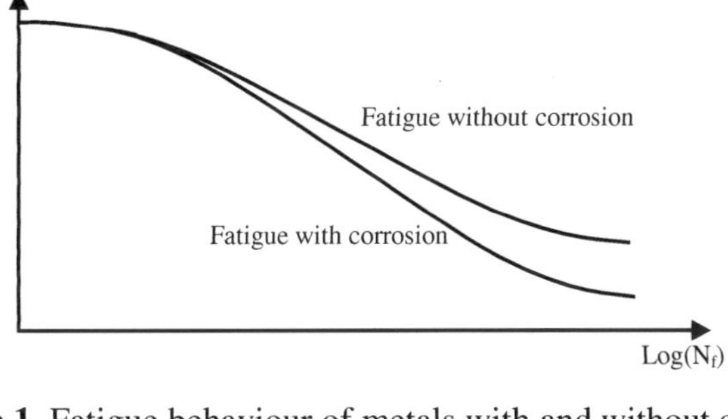

Figure 1. Fatigue behaviour of metals with and without corrosion

The behavior of the fatigue strength of corroded metals depends on various parameters such as types of corrosion, type and state of corrosive environment, rate of corrosion, chemical composition of the metal, etc. [18]. Generally, laboratory fatigue testing of corroded samples is popular for simulating the fatigue strength behavior of corroded materials [3]. However, it is obvious that testing corroded specimens is a time-consuming process, and it is still complicated to simulate a real corrosion situation in the environment. For this reason, a limited number of experimental studies are found, especially in the field of civil engineering, related to structural steels [2,3]. Therefore, there is a need for simple models or methodologies to model the fatigue strength of corroded materials. To fulfill this need, the major objective of this review paper is to illustrate/check the possibilities of proposing an interpolation formula for the fatigue strength of corroded material.

3. Modeling of fatigue damage due to corrosion

Approaches for modeling the fatigue damage of materials/structures due to corrosion can be mainly divided into two parts. The first is geometrical simulation of thickness loss and pitting. The other part is to model the degradation of material properties due to the corrosive environment (CF) by a stress-life fatigue model that simulates the fatigue strength of corroded material. The following subsections will review the aforementioned two parts of the modeling. At the latter part of this section, modeling of corrosion rate (i.e. intensity of corrosivity) is discussed, as it has a relation to the corresponding stress-life fatigue behavior of the corroded material.

3.1. Modeling of fatigue damage due to degradation of material properties

Fatigue damage due to corrosion pits and cracks is analytically modeled by fracture mechanics theories. Pitting corrosion damage models are widely used, based on stress intensity factor [19,20]. Moreover, fatigue crack propagation is determined by using stress intensity factor based fracture mechanics approaches [21, 22]. Using the modified corrosion-fatigue stress intensity threshold, ΔKth, for the material to determine critical length L for corroded materials is reported as one method of modeling CF. Many researchers [2, 3] report that it is very difficult to get a clear value for ΔKth in corrosive environments, in which the fatigue limit is not usually clear. The determination of critical length has not been well defined, as it is not clear in corrosive environments. Therefore, detailed research should be performed to fill this gap. A new strain-life model, based on the Smith–Watson–Topper model, was proposed in 2012 [2] to simulate the CF behavior of metals. The model takes into account the state of corrosive environment, the stress level, and the corrosive behavior of the material [2]. The parameters of the above CF model, which are modified Basquin's exponent and Coffin-Manson exponent, should be determined by a series of fatigue tests of materials in 3.5 wt/wt% NaCl which is a highly corrosive environment. As these model parameters are material dependent, and due to the complexity of fatigue tests, fewer applications of this model are found. This again emphasizes the importance of having a simple formula or methodologies to model the fatigue strength of corroded materials

3.2. Geometrical simulation of thickness loss and pitting

In order to estimate fatigue damage, it is essential to determine the stress histories generated by random loading of the structure. Therefore, it is necessary to know the stress cycles of all the corroded members in the past, present and future. The stress histories of the corroded members are evaluated by considering (i) past, present and future loadings and (ii) time-dependent change of cross-sectional shapes due to loss of material caused by the corrosion deterioration and related change in rigidities (i.e. change of axial, bending, torsional and warping rigidities), which finally cause change to the overall stiffness of the structure. These simulations associated fatigue damage/life assessment methods were discussed in recent literature [6, 11-14]. The proposed procedure with related theoretical formulae for structural elements has been comprehensively presented in recent literature [23]. For precise prediction of time-dependent stress histories, it is necessary to accurately model the time-dependent growth of corrosion wastage (i.e. rate of corrosion).

3.3. Modeling of corrosion rate

The environment makes a huge contribution to corrosion and changes the rate of corrosion. Corrosion rate is the speed of deterioration of material (loss of metal). It is the thickness reduction per year and can be calculated in general by metal loss divided by time. As corrosion rate is one of the key parameters, it is necessary to study the way corrosion effects in the integrity of steel structures and then the CF. Corrosion rate is one of the main factors affecting the service life of the structure. The intensity of the corrosivity depends on the corrosive environment. The corrosion rate is also presented for five categories of steel [23].

The experimental studies on corrosion rates include the following: (i) corrosion rates of low carbon steels in potable water conducted in this study, and (ii) effects of environmental factors on corrosion rates of previous research [24-29]. El Aghoury [2] proposed a new relationship between the logarithm of the penetration versus the proposed environmental corrosivity intensity factor (γcorr). The corrosive environments of the materials have been divided into corrosivity categories in ISO 9224, based on average corrosion rate [2].

4. Proposed formulae for fatigue strength of corroded material

Two S-N curve formulae for corroded metals are proposed, based on Basquin's law [21] and the full-range curve proposed by Kohout and Vechet [30]. Conventional S-N curves for uncorroded steel have been used for assessing structures and components in a non-corrosive environment. Vacuum is the perfect non-corrosive environment. The difference in the fatigue strength of metals such as mild steel in the vacuum and in dry air is less than 5% [1]. Therefore, dry air can also be considered as non-corrosive and, hence, Basquin's law and Kohout and Vechet's proposed full-range S-N curve can be used in dry air without any modifications. It is necessary to use degraded S-N curves, when fatigue life is estimated for structures and components in a corrosive atmosphere (rain, various gases, water, marine, etc.).

4.1. Concept used for the proposed formulae

Gliding may occur in some of the grains when materials are subjected to alternating stresses. Gliding is halted when dislocation reaches a grain boundary. In the presence of corrosion, disorganized atoms are in motion along a gliding plane with less activation energy than in the absence of corrosion. This phenomenon may reasonably be expected, even below the fatigue limit. This means that there is no safe stress level at which the fatigue life is infinite, as shown in Figure 1 [18]. However, it is convenient to determine an endurance limit for corroded steel corresponding to a specified number of cycles by assuming the material will endure some specified number of cycles which must be stated. Previously experimentally obtained fatigue lives show that the difference between fatigue strengths in corrosive and non-corrosive environments in the low-cycle fatigue (LCF) region is very low, compared to the high-cycle fatigue (HCF) and very-high-cycle fatigue (VHCF) regions [1,9,22]. The difference is significantly large in the HCF and VHCF regions.

4.2. Proposed formula for high-cycle fatigue region

The relation between the applied stress amplitude, σ_a, and the corresponding number of cycles to fatigue failure, N_f, follows Basquin's law, as presented in (1), for uncorroded material when the stress is below the yield strength, σ_y [21],

$$\sigma_a = \sigma'_f N_f^b \tag{1}$$

where σ'_f is the fatigue strength coefficient and b is Basquin's exponent.

The linear variation in the difference between the fatigue strengths in the uncorroded and corroded materials has been observed in previous fatigue test results [18], and the relative difference in log scale is linearly deducted, as shown in Figure 2, from the corrosion-free material fatigue strength to obtain the corroded material fatigue strength. The difference between corroded and uncorroded S-N curves at the stress amplitude just above yield strength (σ_y) can be neglected, when compared to the difference at the very-high-cycle fatigue region. Hence, it is assumed that corroded and uncorroded S-N curves are intercepted when stress amplitude is equal to yield strength, as shown in Figure 2.

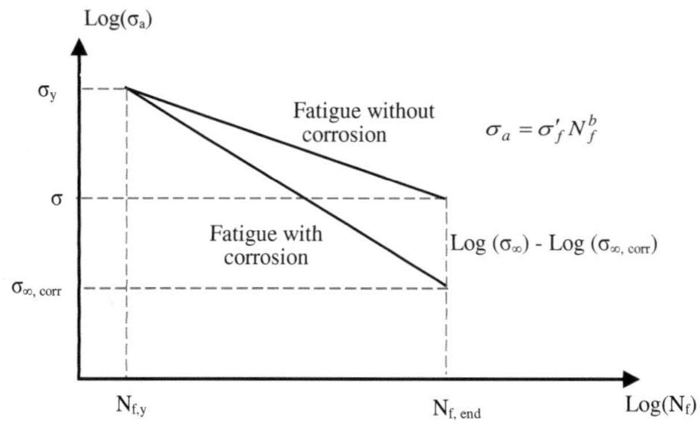

Figure 2. Schematic representation of fatigue strength degradation of corroded materials in HCF region

Hence, the fatigue strength of corroded material, $\sigma_{a,corr}$, which corresponds to number of cycles to fatigue failure, Nf, can be derived as,

$$\log(\sigma_{a,corr}) = \log(\sigma_a) - \frac{(\log\sigma_\infty - \log\sigma_{\infty,corr})}{(\log N_{f,end} - \log N_{f,y})}(\log N_f - \log N_{f,y}) \tag{2}$$

where σ_∞ is endurance limit (i.e. fatigue limit for high-cycle fatigue) and $\sigma_{\infty,corr}$ is endurance limit for corroded material, which corresponds to a specified number of cycles, $N_{f,end}$. The $N_{f,y}$ is the number of cycles to fatigue failure of uncorroded materials when stress amplitude is σ_y. The σ_a is fatigue strength of uncorroded material, which corresponds to the number of cycles to fatigue failure, N_f,, as defined in Eq. (1). The proposed formula for fatigue strength of corroded materials can be simplified as,

$$\log(\sigma_{a,corr}) = \log(\sigma_a) - \frac{\log\left(\sigma_\infty / \sigma_{\infty,corr}\right)}{\log\left(N_{f,end} / N_{f,y}\right)}(\log N_f - \log N_{f,y}) \tag{3}$$

or

$$\log(\sigma_{a,corr}) = \log(\sigma_a) - c(\log N_f - \log N_{f,y}) \tag{4}$$

where c is constant for a given material corresponding to the current state of corrosion and is given as,

$$c = \frac{\log\left(\sigma_\infty / \sigma_{\infty,corr}\right)}{\log\left(N_{f,end} / N_{f,y}\right)} \tag{5}$$

The proposed formula for fatigue strength can be obtained as,

$$\sigma_{a,corr} = (\sigma'_f N_{f,y}^{c}) N_f^{(b-c)} \tag{6}$$

The only parameter required for the above formulae is the endurance limit for the corroded material, $\sigma_{\infty,corr}$, which can be either experimentally obtained by a limited number of tests in the VHCF region or can be predicted if there is a relationship between fatigue endurance degradation and rate of corrosion of the material.

4.3. Proposed formula for full range

The relationship between the applied stress amplitude, σ_a and number of cycles to fatigue failure, N_f, in full range (i.e. S-N curve for ultra-low cycle fatigue, LCF, HCF and VHCF regions) for uncorroded material is given as [31],

$$\sigma_a = \sigma_\infty \left(\frac{N+B}{N+C} \right)^b \tag{7}$$

where B and C are the number of cycles corresponding to the points of intersection of the tangent for the region of finite life described by Basquin's law, with the horizontal line across the ultimate strength and horizontal line across the endurance limit, respectively.

The linear variation in the difference between the fatigue strengths of uncorroded and corroded materials has been observed in the previous fatigue test results as described in the previous section. The relative difference is then linearly deducted to obtain corroded material fatigue strength for full range as follows:

$$\sigma_{a,corr} = \sigma_a - \frac{(\sigma_\infty - \sigma_{\infty,corr})}{\log C} \log N_f \tag{8}$$

The only parameter which is required for this formula is also the endurance limit for corroded material, $\sigma_{\infty,corr}$, which can be determined in the same way as described in the previous sub-section.

5. Experimental Verification of Proposed formulae

The proposed fatigue strength formulae for corroded materials are verified by comparing the experimental fatigue lives of corroded specimens of different materials, as shown below. The corrosion fatigue test results of three different steel types [18] and one type of aluminum alloy [31] are selected and compared with the predicted fatigue lives by the proposed formulae, as shown in the following sections. The fatigue tests have been conducted for constant corrosion rate.

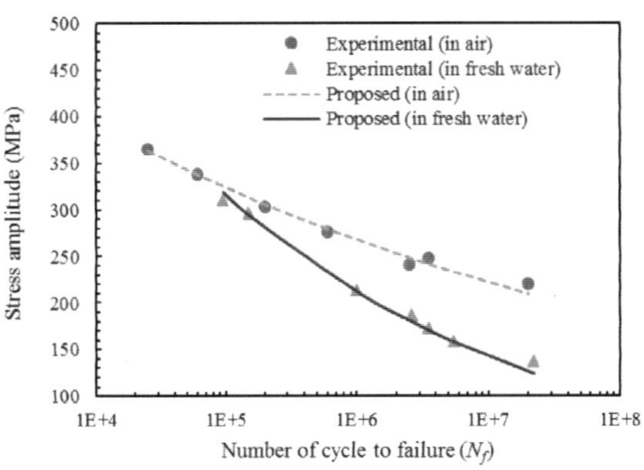

Figure 3. Comparison of proposed S-N curve with experimental test made in air and fresh water for annealed copper steel.

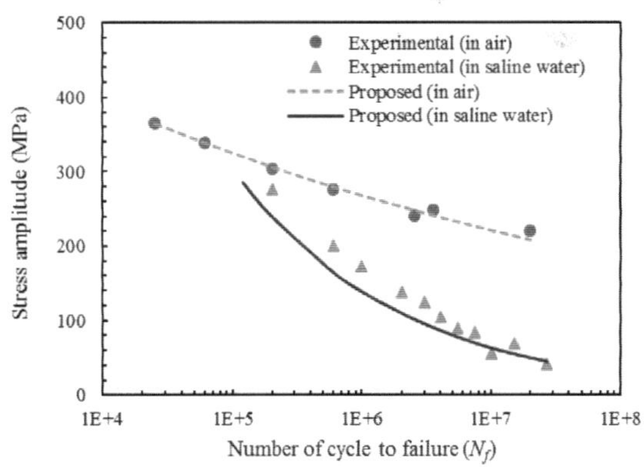

Figure 4. Comparison of proposed S-N curve with experimental test made in air and saline water for annealed copper steel.

5.1. Experimental verification of proposed formula for HCF

Figures 3 to 8 show the comparison of experimental fatigue lives of annealed copper steel (0.14% C, 0.98% Cu), hardened and tempered 0.16% carbon steel, hardened and tempered chrome-vanadium steel (0.88% Cr, 0.14% Va, 0.46% C) [18] and 6061-T6 aluminum alloy [31] specimens in different corrosive environments with interpolated fatigue lives by the proposed interpolation formula for the high-cycle fatigue region shown in Eq (6).

Figures 3 to 7 show a good agreement between interpolated fatigue lives and experimentally observed lives for all three types of steel. This shows that the proposed formula in HCF provides a conservative rule regarding the fatigue strength of corroded steels. However, there are not conservative results for the aluminum alloy, as shown in Figure 8. One reason is the lack of an experimental data point to determine the endurance limit for corroded specimens. However, the proposed formula gives a better calculation of the mean fatigue strength of corroded aluminum alloy specimens.

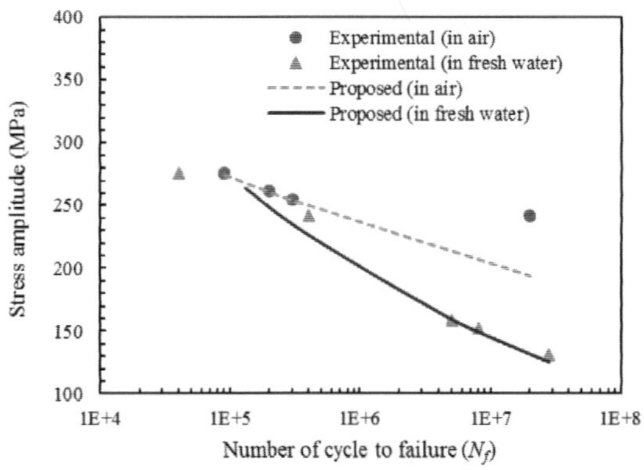

Figure 5. Comparison of proposed S-N curve with experimental test made in air and fresh water for 0.16% carbon steel (hardened and tempered).

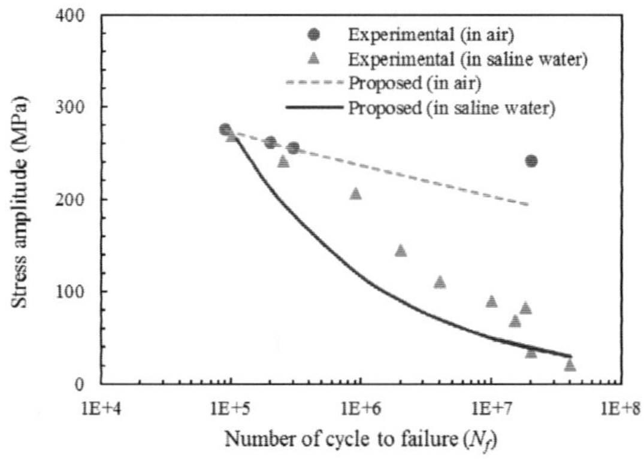

Figure 6. Comparison of proposed S-N curve with experimental test made in air and saline water for 0.16% carbon steel (hardened and tempered).

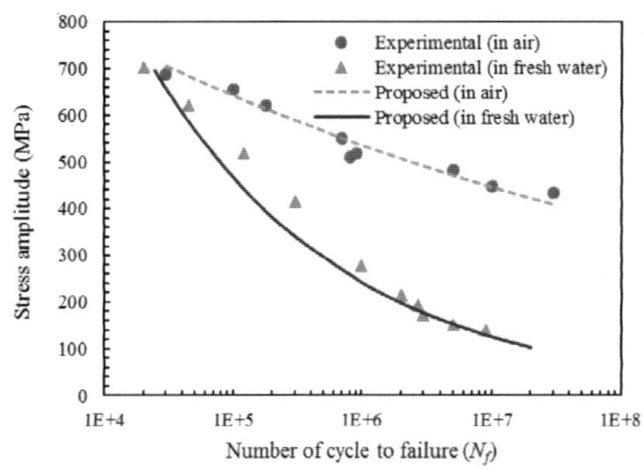

Figure 7. Comparison of proposed S-N curve with experimental test made in air and fresh water for chrome-vanadium steel (hardened and tempered).

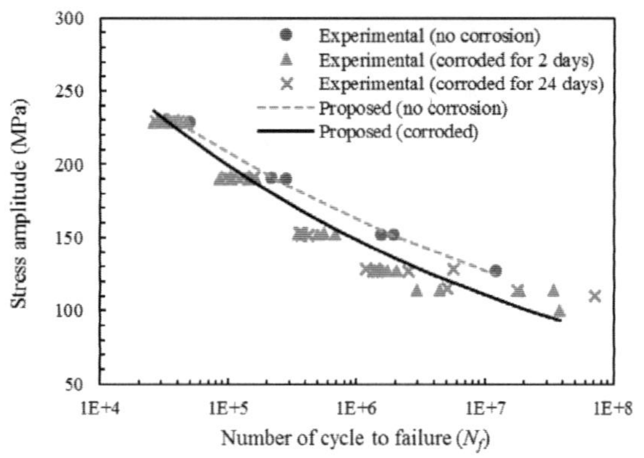

Figure 8. Comparison of proposed S-N curve with experimental test made with corrosive and non-corrosive situations of 6061-T6 aluminum alloy.

5.2. Experimental verification of proposed formula for full range

Figures 9 to 14 show the comparison of experimental fatigue lives of annealed copper steel (0.14% C, 0.98% Cu), hardened and tempered 0.16% carbon steel, hardened and tempered chrome-vanadium steel (0.88% Cr, 0.14% Va, 0.46% C) [18] and 6061-T6 aluminum alloy [31] specimens in different corrosive environments with interpolated fatigue lives by the proposed formula for full range shown in Eq. (8). Figures 9 to 14 show that the proposed formula for full range provides a conservative calculation of the fatigue strength of corroded materials. The proposed formula does not give better results when material is subjected to severe corrosion. This concludes that improvement of the proposed formula is required.

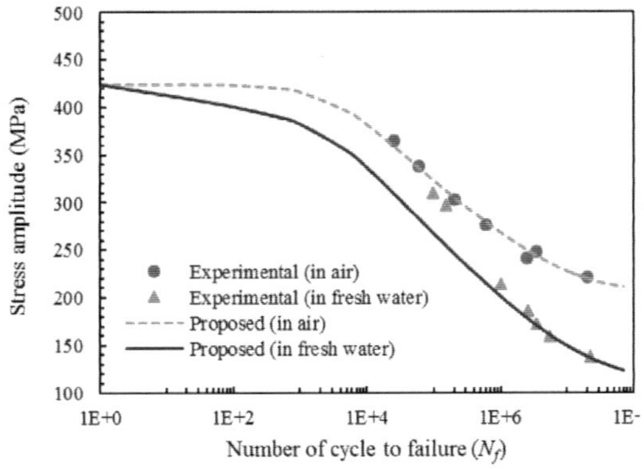

Figure 9. Comparison of proposed S-N curve with experimental test made in air and fresh water for annealed copper steel.

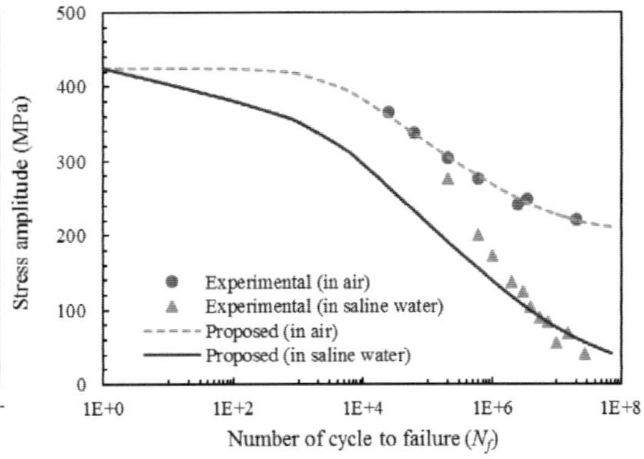

Figure 10. Comparison of proposed S-N curve with experimental test made in air and saline water annealed copper steel

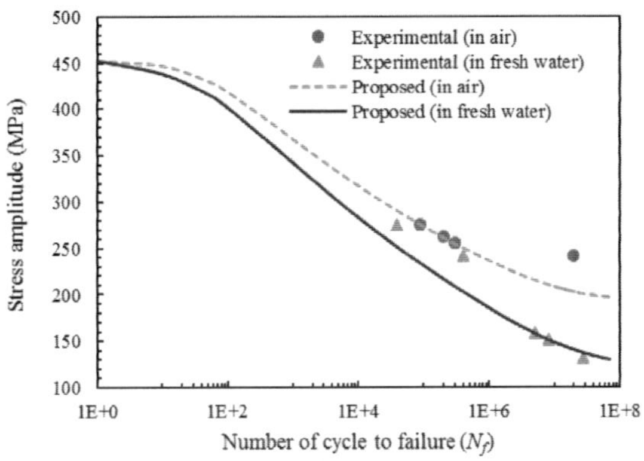

Figure 11. Comparison of proposed S-N curve with experimental test made in air and fresh water for 0.16% carbon steel (hardened and tempered).

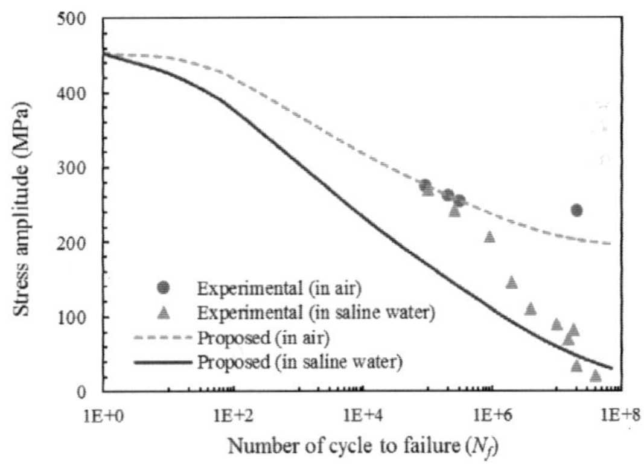

Figure 12. Comparison of proposed S-N curve with experimental test made in air and saline water for 0.16% carbon steel (hardened and tempered).

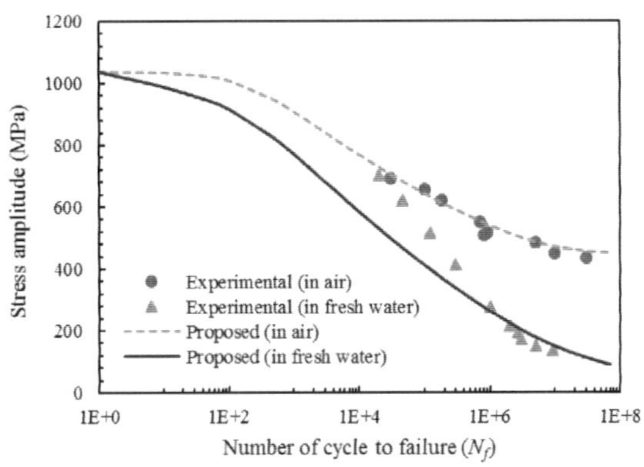

Figure 13. Comparison of proposed S-N curve with experimental test made in air and fresh water for chrome-vanadium steel (hardened and tempered).

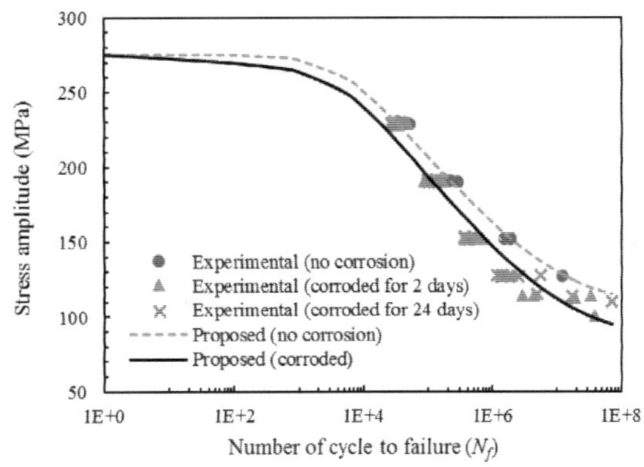

Figure 14. Comparison of proposed S-N curve with experimental test made with corrosive and non-corrosive situations of 6061-T6 aluminum alloy.

6. Discussion and conclusions

The corrosion degradation process of metallic materials has been reviewed by highlighting the mechanism of corrosion fatigue. Time-dependent geometrical changes in structural components and degradation of material strength are identified as key effects of corrosion. Corrosion fatigue, which is the degradation of materials under the combined action of localized corrosion and cyclic loading, significantly affects the integrity of structures. The fatigue damage due to corrosion pits and cracks can be modeled by fracture mechanics theories. Many researchers have reported the shortcomings of this approach in corrosive environments, as fatigue limit is not usually clear in corrosive environments. Generally, laboratory fatigue testing of corroded samples is popular for simulating CF. Testing corroded specimens is costly, and experimentally simulating the real corrosion situation in the environment is also complicated. A new strain-life model was proposed in 2012. Some of the model parameters should be determined by a series of complicated fatigue testing and, therefore, very few applications of this model have been reported.

Finally, new formulae for the fatigue strength of corroded materials were proposed in the paper. The proposed formulae only depend on the endurance limit of corroded material ($\sigma_{\infty,corr}$), in addition to the stress-life fatigue curve parameters of the uncorroded material. The $\sigma_{\infty,corr}$ can be either determined by a limited number of tests in the VHCF region or predicted by analytical approach. The interpolation formulae give fatigue lives were compared with experimentally obtained fatigue lives of a few materials in different corrosive environments. The comparisons show that the proposed formula for the HCF region gives a conservative calculation to the fatigue strength of corroded materials. It also reveals that the new formula for full range provides a better agreement with the fatigue strength of corroded materials. Knowledge of corrosion rates in the region where the structure is located is extremely important for damage assessments. Further verification of the proposed formulae is required for variable corrosion rates, as verification of the proposed formulae has been performed by comparing the results of fatigue tests conducted at a constant corrosion rate.

References

[1] Revie R W and Uhlig H H 2008 *Corrosion and Corrosion Control, An Introduction to Corrosion Science and Engineering* (Hoboken New Jersey: 4th edn., Wiley and Sons)

[2] Aghoury I El 2012 *Numerical tool for fatigue life prediction of corroded steel riveted connections using various damage models* (PhD thesis, Concordia University, Montreal)

[3] Aghoury I M El and Galal K 2014 Corrosion-fatigue strain-life model for steel bridge girders under various weathering conditions *J. Struct. Eng.* **140(6)** pp

[4] Roberge P R 2000 *Handbook of Corrosion Engineering* (New York: 1st ed., McGraw Hill)

[5] Mikkelsen O, Rege K, Hemmingsen T H and Pavlou D 2007 Numerical estimation of the stop holes-induced fatigue crack growth retardation in offshore structures taking into account the corrosion effect, *Proc. of 27th International Ocean and Polar Engineering Conference* (San Francisco, California)

[6] Yasser Y and Jeom K.M 2011 Ultimate strength reliability analysis of corroded steel-box girder bridges *Thin Wall. Struct* **49** pp 157-166

[7] Kayser J R 1988 *The effects of corrosion on the reliability of steel girder bridges* (PhD thesis, University of Michigan, Ann Arbor, Michigan, USA)

[8] Paik J K, Lee J M and Ko M J 2004 Ultimate shear strength of plate elements with pit corrosion wastage *Thin Wall. Struct* **42** pp 1161-1176

[9] Glaser W and Wright L G 1992 *Mechanically assisted degradation* (Ohio: ASM Handbooks, ASM Internationals) pp 137-144

[10] Wang Y Z 2011 *Corrosion fatigue, Corrosion Handbook* 3rd edn, ed R W Revie and H H Uhligs (New York: John Wiley and Sons)

[11] Li Z X and Chan T H T 2006 Fatigue criteria for integrity assessment of long span steel bridge health monitoring *Theo. & App. Fract. Mech.* **46** pp 114-127

[12] Lima K, Robson N, Oosterhof S, Kanji S, DiBattista J and Montgomery C J 2008 Rehabilitation

of a 100-year-old steel truss bridge *Proc. of Canadian society for civil engineering. annual conference* (Quebec: Canada)

[13] Nakamura S and Suzumurab K 2009 Hydrogen embrittlement and corrosion fatigue of corroded bridge wires *J. Constr. Steel. Res.* **65** pp 269-277

[14] Yukikazu Y and Makoto K 2002 Maintenance of steel bridges on Honshu- Shikoku crossing *J. Constr. Steel. Res.* **58** pp 131-150

[15] Mohamed E L M, Thierry P L, Nicolas S and Olivier D 2013 Effect of corrosion on the high cycle fatigue strength of martensitic stainless steel X12CrNiMoV12-3 *Int. J. Fatigue* **47** pp. 330-339

[16] Wahab M A and Sakano M 2003 Corrosion and biaxial fatigue of welded structures *J. Mat. Proc. Tech.* **3** pp 410-415

[17] Nguyen K T, Garbatov Y and Soares C G 2013 Spectral fatigue damage assessment of tanker deck structural detail subjected to time-dependent corrosion *Int. J Fatigue* **48** pp 147-155

[18] Boyer H E 1986 *Atlas of Fatigue Curves* (Ohio: USA, ASM Internationals)

[19] Hoeppner D W, Chandrasekaran V and Taylor A M H 1999 Review of pitting corrosion fatigue models *Proc. of ICAF International Conf.* pp 549

[20] Acuña N, González-Sánchez J, Ku-Basulto G and Domınguez L 2006 Analysis of the stress intensity factor around corrosion pits developed on structures subjected to mixed loading *Scripta Materialia* **55** pp 363-366

[21] Suresh S 1998 *Fatigue of Materials* (Cambridge: 2nd ed. Cambridge University Press, Cambridge, UK)

[22] Gangloff R P 2005 *Environmental cracking - corrosion fatigue* eds. R Baboian, S W Dean Jr., H P Hack, E L Hibner and J R Scully (Pennsylvania: USA ASTM Internationals, Corrosion Tests and Standards Manual)

[23] Adasooriya N D and Siriwardane S C 2014 Remaining fatigue life estimation of corroside bridge members *Fatigue Fract. Eng. Mater. Struct.* **37** pp 603-622

[24] Roberge P R 2011 *Atmospheric corrosion, Uhligs Corrosion Handbook* ed. R W Revie (New York: USA 3rd edn. John Wiley and Sons)

[25] Vashi R T, Kadiya H K and Champaneri V A 2008 Corrosion study of metals in rural atmosphere *J. Env. Research and Dev.* **3** pp 1

[26] Natesan M, Selvaraj S and Manickam T 2008 Corrosion behaviour of metals and alloys in marine – industrial environment *Science and Technology of Advance Materials* **9**, Page no. 045002 (7pp)

[27] Matsushima I 2011 *Carbon steel corrosion by sea water, Uhligs Corrosion Handbook* ed. R W Revie, (New York: 3rd edn. John Wiley and Sons) pp 601-608

[28] Parekh S P, Pandya A V and Kadiya H K 2012 Progressive atmospheric corrosion study of metals like mild steel, Zn and Al in urban station of Ahmedabad district *Int. J. Chemtech. Res.* **4(4)** pp 1770-1774

[29] Marref A, Basalamah S and Al Ghamdi R 2013 Evolutionary computation technique for predicting atmospheric corrosion *Int. J Corrosion* **2013** Article ID 805167

[30] Kohout J and Vechet S 2001 A new function for fatigue curve characterization and its multiple merits *Int. J. Fatigue* **23** pp175-83

[31] Matthew W, Paul D E, Huseyn O and Murat T 2017 The effect of surface corrosion damage on the fatigue life of 6061-T6 aluminum alloy extrusions *Mater. Sci. Eng.: A-Struct.* **690** pp 427-432